Use of Humic Substances to Remediate Polluted Environments:
From Theory to Practice

NATO Science Series

A Series presenting the results of scientific meetings supported under the NATO Science Programme.

The Series is published by IOS Press, Amsterdam, and Springer (formerly Kluwer Academic Publishers) in conjunction with the NATO Public Diplomacy Division.

Sub-Series

I. Life and Behavioural Sciences	IOS Press
II. Mathematics, Physics and Chemistry	Springer (formerly Kluwer Academic Publishers)
III. Computer and Systems Science	IOS Press
IV. Earth and Environmental Sciences	Springer (formerly Kluwer Academic Publishers)

The NATO Science Series continues the series of books published formerly as the NATO ASI Series.

The NATO Science Programme offers support for collaboration in civil science between scientists of countries of the Euro-Atlantic Partnership Council. The types of scientific meeting generally supported are "Advanced Study Institutes" and "Advanced Research Workshops", and the NATO Science Series collects together the results of these meetings. The meetings are co-organized by scientists from NATO countries and scientists from NATO's Partner countries — countries of the CIS and Central and Eastern Europe.

Advanced Study Institutes are high-level tutorial courses offering in-depth study of latest advances in a field.
Advanced Research Workshops are expert meetings aimed at critical assessment of a field, and identification of directions for future action.

As a consequence of the restructuring of the NATO Science Programme in 1999, the NATO Science Series was re-organized to the four sub-series noted above. Please consult the following web sites for information on previous volumes published in the Series.

http://www.nato.int/science
http://www.springeronline.com
http://www.iospress.nl

Series IV: Earth and Environmental Series – Vol. 52

Use of Humic Substances to Remediate Polluted Environments: From Theory to Practice

edited by

Irina V. Perminova
Lomonosov Moscow State University,
Moscow, Russia

Kirk Hatfield
University of Florida,
Gainesville, FL, U.S.A.

and

Norbert Hertkorn
Institute of Ecological Chemistry,
GSF-Research Center for Environmental Health,
Neuherberg, Germany

Springer

Published in cooperation with NATO Public Diplomacy Division

Proceedings of the NATO Advanced Research Workshop on
Use of Humates to Remediate Polluted Environments: From Theory to Practice
Zvenigorod, Russia
23–29 September 2002

A C.I.P. Catalogue record for this book is available from the Library of Congress.

ISBN 1-4020-3250-1 (HB)
ISBN 1-4020-3252-8 (e-book)

Published by Springer,
P.O. Box 17, 3300 AA Dordrecht, The Netherlands.

Printed on acid-free paper

This volume is dedicated to **Galina Moiseevna Varshal** – the prominent Soviet-Russian scientist, Professor of the Institute of Geochemistry and Analytical Chemistry (GEOKHI) of the Russian Academy of Sciences.

She passed away on July 16, 2001; her lecture on the geochemical role of humics in metal migration was to open the Workshop.

Galina M. Varshal was an outstanding scientist and a bright and generous person. This volume is dedicated to Galina Moiseevna in appreciation for her devoted service to the science of humic substances.

CONTENTS

vii

PREFACE

Effective remediation of polluted environments is a priority in both Eastern and Western countries. In the U.S. and Europe, remediation costs generally exceed the net economic value of the land. As a result, scientists and engineers on both sides of the Atlantic have aggressively tried to develop novel technologies to meet regulatory standards at a fraction of the costs. *In situ* remediation shows considerable promise from both technical and economic perspectives. *In situ* technologies that deploy natural attenuating agents such as humic substances (HS) may be even more cost effective. Numerous studies have shown humics capable of altering both the chemical and the physical speciation of the ecotoxicants and in turn attenuate potential adverse environmental repercussions. Furthermore, the reserves of inexpensive humic materials are immense. Which suggests HS portend great promise as inexpensive amendments to mitigate the environmental impacts of ecotoxicants and as active agents in remediation.

To elucidate emerging concepts of humics-based remediation technologies, we organized the NATO Advanced Research Workshop (ARW), entitled "Use of humates to remediate polluted environments: from theory to practice", held on September 23-29, 2002 in Zvenigorod, Russia (see the web-site http://www.mgumus.chem.msu.ru/arw). The purpose of this ARW was to bring for the first time a league of experienced scientists who have studied humics structures, properties and functions in the environment, together with an association of environmental engineers who have developed novel remediation technologies. The workshop created among the participants an awareness of the current status of research in remediation chemistry and in humics technology. At the meeting, 20 oral and 29 poster presentations were given followed up by multiple discussions that were both engaging and constructive.

This book summarizes the proceedings of the Workshop, and is dedicated to Professor G.M. Varshal (deceased). Four chapters appearing in this volume are from authors who could not attend the workshop, but kindly agreed to prepare written contributions (J.F. De Kreuk, J.A. Field, D.S. Gamble, and V. Moulin). Three reports from Working Group discussions prepared by rapporteurs designated at the ARW are included in the Introduction. All papers have been subject to peer review by at least two referees. We thank all the authors for their helpful collaboration.

We are grateful for the financial support from NATO Science Affairs Division, which made it feasible to organize this Workshop. We thank Muefit Tarhan (Humintech Ltd.) for the monetary awards to the best poster presenters. The hard work of Joy Drohan (Eco-Write) on editing the translations from Russian is deeply acknowledged. We also wish to express our sincere appreciation to Dr. Natalya A. Kulikova and Alexey V. Kudryavtsev for their invaluable assistance in organizing the Workshop and in preparing this volume for publishing.

Irina V. Perminova   Kirk Hatfield   Norbert Hertkorn

INTRODUCTION

1. Objectives of the book

The aim of this book is to establish a linkage between scientists who have studied humics structures, properties and functions in the environment, and environmental engineers who have developed novel remediation technologies. In pursuing this goal, the engineer interested in remediation is provided with sufficient information on the basics and state-of-art of humics research pertinent to remediation. On the other hand, for the interested scientist familiar with the basic research on humics, this text provides sufficient information on the methods, needs, and limitations of existing remediation technologies, and on the latest developments in applied humic research. It is the objective of the authors that this text be sufficiently thorough that it serves to facilitate the development of a common language and stimulate discussions among scientists and engineers. The diverse expertise brought together to create this book promises desirable synergetic effects of solving remediation problems by revealing new fields of applied humic research and extending current technical application of humics.

The principal objective of the book is to elaborate a systematic characterization of the mediating effects of HS by categorizing these effects according to their impact on the fate of ecotoxicants and on the physiological functions of living organisms. Given the subtitle: 'from theory to practice', another substantial objective is to assess the status of knowledge pertaining to the function or role humics may serve in the remediation of polluted environments, and to show the limits of current knowledge on the properties and functions of HS. Concerning the practice of remediation, principal objectives are to assess the scope of current applications, to define promising directions of technological developments, and to formulate the research needs. Hence, the volume pays particular attention to *in situ* remediation technologies as the most viable option for the application of humics as active agents in remediation.

2. Organization of the book

This book contains 24 chapters that are structured into five parts. Part 1 gives a general overview of the remedial properties of humic substances and identifies various challenges pertinent to their application in remediation. The first part also introduces the topics of subsequent sections. For example, the focus of Part 2 is the interaction that occurs between HS and heavy metal and radionuclide via complexation. Part 3 addresses sorption/partitioning as a primary interaction between humics and organic ecotoxicants. Then there is Part 4, where the topic is humics interactions with the physiological functions of living organisms and their effects on microbially mediated ecotoxicant transformations. Finally, in Part 5 analytical approaches for quantifying HS structure and properties are discussed, as are studies where humic materials have been modified to acquire desired properties.

2.1. REMEDIAL PROPERTIES OF HUMIC SUBSTANCES: GENERAL CONSIDERATIONS AND PROBLEMS IN ADDRESSING NEEDS OF ENVIRONMENTAL REMEDIATION

The first chapter gives a general introduction to the field of remediation chemistry in context of humic substances potential implications. It provides an overview of the interactions encountered between HS, ecotoxicants, and living organisms in a polluted environment. The most important interactions identified include: binding interactions affecting chemical speciation and bioavailability of contaminants; interfacial interactions altering physical speciation or interphase partitioning of ecotoxicants; abiotic-biotic redox interactions that influence metabolic pathways coupled to pollutants; and finally direct and indirect interactions coupled to various physiological functions of living organisms. All of these interactions possess significant utility for *in situ* remediation; consequently, several humic-based reactions are examined in detailed that are pertinent to permeable reactive barriers, *in situ* flushing, bioremediation, and phytoremediation. Finally, this chapter introduces the novel concept of "designer humics" which are a special class of customized humics obtained by chemically modifying and cross-linking the humic backbone such that this new humic-based material acquires specified reactive properties. Designer humics, as described herein possess the potential for achieving enhanced remediation and for quantifying remediation performance. The latter is described in the context of the passive flux meter technology.

Chapter 2 presents the general theory of humification, describes the biogeochemical rules of humus formation in soil, and defines the functions of HS in the biosphere. The kinetic nature of humification process is given particular consideration. The following features inherent to the process of forming humus are identified: first, the process is biomineral in nature; second, transformations of organic matter are biothermodynamically driven; and third, the quality of humus generated is determined by the prevailing environmental condition at each stage of humification. Chapter 2 defines five principal functions displayed by HS in the biosphere; these are the accumulative, transport, regulatory, physiological and protective functions. The protective function is considered in detail. Experimental evidence is provided for the example of interactions occurring between soil humics and heavy metals.

Chapter 3 sets the stage for improvements the quantitative prediction of reaction equilibrium between HS and different reagents. The reviewed literature clearly shows that the principles of classical chemistry, as understood for monomeric reagents, are to some extent adaptable to humic complex mixtures. Demonstrations of predictive chemical calculations have been published and the practical implications are discussed herein. However, this chapter emphasizes that there is an urgent need for a new generation of models (perhaps stochastic in formulation) capable of predicting the reactive properties of humics. These new models will play a crucial role in the design of novel humics-based remediation strategies.

2.2. COMPLEXING INTERACTIONS OF HUMIC SUBSTANCES WITH HEAVY METALS AND RADIONUCLIDES AND THEIR REMEDIAL IMPLEMENTATION

The topic of humics/heavy metals interactions is developed in Chapter 4. Authors provide the requisite theoretical background and review data on the complexing capacity of humics for various heavy metals. The role of HS interactions in metal uptake by higher plants is given particular attention. Impacts of HS on heavy metal speciation, migration, and toxicity in aquatic environments are discussed in Chapters 5 and 6. The former focuses on changes in the complexing properties of UV-oxidized natural organic matter (NOM). It is shown that UV oxidation reduces the molecular size of the NOM and in turn decreases its inherent Al-complexing capacity. Of interest is that UV-oxidation does not effect the same change in the complexing properties of NOM towards Pb and Zn. Relationships between the structure and the complexing properties of NOM are further addressed in Chapter 5. Studies of seasonal variations evident in the properties of NOM in the natural waters reveal significant decreases in complexing capacity in late spring and early summer, when NOM is enriched with the newly produced organic matter. In this context of particular importance is the demonstrated coupling between NOM complexation with ecotoxicants and the resultant detoxification. Model systems with copper show that a decrease in water toxicity is correlated with a reduction in the concentration of free metal produced when copper binds to NOM. The given results elucidate a coupled relationship between the structure of humics and their ability to complex heavy metals.

Chapter 7 reviews a vast pool of data on HS interactions with the most dangerous metal contaminants, radionuclides. The data originates from research conducted at the Centre of Atomic Energy (CEA) in France. The chapter provides a solid analytical background of investigations focused on the complexing properties of HS with radionuclides. Data on the stoichiometry of the complexes and the respective stability constants are given for actinides, iodine and lanthanides. Results presented are obtained for Aldrich humic acid, and aquatic fulvic acid. Discussed is the influence HS have on the retention of trace element on mineral surfaces. The latter topic is the focus of Chapter 8 where HS are considered as components of geochemical barriers for actinide migration. Original results on sorptive and redox interactions in a mineral-HS-actinide system are given. A conclusion is drawn regarding the suitability of using humics as agents to immobilize actinides in contaminated environments. This conclusion is confirmed by results of a case study presented in Chapter 9. Both laboratory and field studies demonstrate that treating polluted soils with brown coal composed of up to 80% humic acids induces a drastic reduction in the mobility and toxicity of heavy metals. This shows that humic materials can be used to ameliorate heavy metals toxicity in soils.

In summary, the above chapters clearly demonstrate HS can affect metal speciation and migration in soil and aquatic environments. The structure features found in the humic macromolecule are of particular importance when assessing the binding affinity for metals. The more oxidized, low molecular weight humics (i.e., fulvic acids) produce less stable and more mobile complexes; whereas, the less oxidized and higher molecular weight humics (i.e., humic acids) produce more stable and less mobile complexes. Hence, humic acids function as both detoxifying and binding agents; whereas fulvic acids may be more effective as flushing agents useful for remediation metal contaminated soils and aquifers.

2.3. SORPTIVE-PARTITIONING INTERACTIONS OF HUMIC SUBSTANCES WITH ORGANIC ECOTOXICANTS AND THEIR REMEDIAL IMPLEMENTATION

Chapter 10 is devoted to examining the sorptive interactions of HS with minerals and with organic contaminants and to discussing the prospects of their use in permeable reactive barrier (PRB) technologies. Basic concepts are addressed relevant to the *in situ* remediation of hydrophobic organic contaminants (HOC) contaminated aquifers using sorptive PRBs. Particular attention is given to an original concept of creating an *in situ* sorptive PRB through a sequential process of first coating aquifer materials with an iron precipitate that is then covered with humics. The obtained humic coating induces sorptive interactions with dissolved HOC, and thus, retards their migration. The results of significant laboratory experiments are given. A design of a corresponding field experiment is discussed. The considerations given thus far on the sorptive interactions of humic coatings with HOC and on the use of humics in sorptive PRBs are complemented by the contents of Chapter 11. This chapter examines the partitioning interactions between humics and HOC in the framework of using concentrated humic solutions as flushing agents for the remediation of polluted aquifers. Colloidal dispersions of HS are shown to enhance the removal of HOC from aquifers. It is clearly demonstrated that a concentrated solution of Aldrich® humic acid can be used to flush diesel fuel from a pilot-scale model sand aquifer. Experiment shows elevated organic constituent solubility producing an accelerated rate of plume remediation. Practical recommendations on a use of HS as flushing agents are given.

   Chapter 12 complements considerations given to the use of HS as flushing agents by identifying problems of soil bioremediation caused by the mass transfer limitations. Several factors are identified as constraining the performance of bioremediation in the field including: the groundwater flow rate; the soil structure heterogeneity; the binding of contaminants to organic matter; the binding of essential nutrients to soil; and the presence of contaminants in an immiscible phase (i.e., DNAPL). The chapter concludes with a proposed strategy of using concentrated humic solutions to facilitate groundwater flow through hydrophobic zones of contaminated aquifer. With nutrients and electron donors/acceptors supplied, local indigenous microorganisms are stimulated to degrade contaminants, and bioremediation resumes. Chapter 13 introduces another feasible strategy of using the interfacial activity of HS to meet the needs of *in situ* remediation. This chapter describes an original approach to improving the biodegradability of organic contaminants in wastewaters using cationic surfactant modified zeolites. Zeolite particles are good carriers of bacteria, but the formation of the bacteria layer on the zeolite surface is slow. The approach requires the attachment of cationic surfactants to the surfaces of zeolite particles. The sorbed polyelectrolytes alter the surface charge on the zeolite particles which in turn accelerates the surface sorption of bacteria. The resultant increase in sludge density produces a desirable increase in sludge activity. The efficiency of modified zeolites has been proven in the laboratory and in full-scale experiments. Use of HS for modification of zeolites as an alternative to artificial surfactants is discussed.

In summary, it can be concluded that when fixed on mineral surfaces, HS can retard the migration of organic contaminants, and when dissolved in water, humics can facilitate the transport of the contaminants in the subsurface. The former process heralds opportunities to use humics as sorbents in PRB, while the latter process is of particular value where humics can function as reactive agents in flushing technologies and in bioremediation.

2.4. IMPACT ON PHYSIOLOGICAL FUNCTIONS OF LIVING ORGANISMS AND ON MICROBIAL TRANSFORMATIONS OF ECOTOXICANTS

Chapter 14 presents a broad overview of the mitigating effects exerted by HS over living organisms in the polluted environments. Several major pathways of mitigation are identified including, impacts on organism development (as an organic carbon source and as hormone-like compounds), impacts on nutrient transport across cell membranes, interactions with enzymes including impacts on biochemical reactions, and finally, antioxidant activity. The original data on enhanced transport of nutrients and on hormone-like and antioxidant activity are presented. Chapter 15 introduces original concepts of how humics impact cell physiology. It is based on the assumption that the physiological interactions of interest occur after HS have crossed the cell membrane and are engaged in particular metabolic activities, such as protein synthesis. This precludes any damage to the system responsible for protein synthesis that might otherwise results from negative environmental factors. Given the critical role of protein synthesis when cells adapt to unfavourable conditions, HS are considered natural adaptogens. Experimental results are given on the adaptogenic activity of HS.

In Chapter 16 the focus is on humic interactions that provide stability to extracellular enzymes in the soils Presented is a vast pool of experimental data on HS induced protease inhibition in the presence of different metals. It is shown that humics are non-specific inhibitors of proteases, and that metals, possessing a higher binding affinity for HS, exhibit a pronounced effect on the stability of enzyme-HS complexes.

The multiple roles of HS in redox reactions catalysed by various microorganisms are considered in Chapter 17. HS are ascribed to have three distinct roles as electron carriers that support biotransformation of priority pollutants, nominally, as electron acceptors for respiration; as redox mediators for reduction processes; and as electron donors to microorganisms. The evidence provided in this review indicates that humics can stimulate the anaerobic (bio)transformation of a wide variety of organic and inorganic compounds, including priority pollutants. Chapter 18 presents a practical example of a bioremediation technology based on the ability of HS to participate in microbial redox reactions. The corresponding technology was developed for the remediation of soil and water sites polluted with explosives (2,4,6 trinitrotoluene, TNT). In reducing environments TNT was reduced to amino-metabolites that produced bound residues with soil organic matter. The experiments were performed on a pilot-scale and in the field. Best results were obtained using molasses as the source of organic carbon for biological reduction. The technical feasibility using this bioremediation strategy at sites contaminated with explosives was clearly demonstrated.

Chapters 19 and 20 present studies of the impacts of commercial humates on soil properties and higher plant development under conditions where heavy metals, radionuclides, and herbicides are the primary pollutants. Chapter 19 reports on the mitigating actions of humates extracted from coal and peat obtained both in the lab and the field. Conclusions are formulated on the prospects of using humics to detoxifying polluted soils and particularly, for the case of combined pollution. Chapter 20 describes application of three commercial humates produced by the same company as adjuvants of the polluted soil. The structure and soil amending properties of all three humate samples studied were quantified. Conclusions were drawn on substantial structural differences between the humate samples studied. Of particular interest was that two structurally similar humates exerted different stimulating effects on plant growth. The results obtained demonstrate that only under standardized production schemes it is possible to produce humic materials of predictable quality.

2.5. QUANTIFYING STRUCTURE AND PROPERTIES OF HUMIC SUBSTANCES AND EXAMPLE STUDIES ON THE DESIGN OF HUMIC MATERIALS OF THE DESIRED PROPERTIES.

Chapter 21 and 22 are devoted to describing the use of high-resolution analytical techniques such as nuclear magnetic resonance (NMR) spectroscopy and capillary zone electrophoresis (CZE) for the investigation of humic structures and of humic interactions with contaminants. These Chapters are complementary with respect to the topics and data discussed. The NMR-spectroscopy approach presented in Chapter 21 gives detailed chemical information on structural components; whereas, the CZE provides information on the molecular behaviour of HS in solution. By defining the relative amounts and structural details of fundamental building blocks, multinuclear quantitative one-dimensional NMR spectroscopy provides the key margin of any structural model of NOM/HS. A large array of higher dimensional NMR spectra permits improved definition beyond the extended substructures of NOM/HS. CZE primarily offers information on the charge, conformation, and charge density of HS; all of which are important parameters when investigating their interactions with pollutants. A detailed description of CZE separation processes is given to interpret the behaviour of humics in aqueous media. Based on the numerous measurements of different HS samples, the conclusion is drawn that humics in aqueous solutions behave as molecular associates showing a distribution of the electrophoretic mobilities in the anionic range. CZE offers a quantitative description of the charge density distribution among the molecular associates. The aspects of structural dynamics are essential for modelling humic-pollutant interactions.

Chapters 23 and 24 address the problem of producing humic materials of desired quality using a novel approach of directed chemical modification. The presented direction can be considered a significant shift is HS-related research, because previous investigations involving humic modifications were focused on developing an understanding of humics structure, not to enhance or change reactive properties.

Chapters describe ozonation and sulfonation as two methods of introducing functional groups into humics derived from coal and peat. Sulfonation increases the solubility of humic-metal complexes, whereas ozonation enriches HS with hydroxyl groups of enhanced chelating ability. An examination of the metal-complexing properties of sulfonated derivatives comply well with desired changes in the properties of the parent humics. It was concluded that chemical modification holds considerable promise as a means of producing humics of the desired properties.

Concluding on the papers overview, two fundamental reasons can be cited as to why HS have not been widely used in remediation technologies. First, few natural HS possess the specific reactive properties required to treat selected environmental contaminants. Second, humics are by definition polydisperse and heterogeneous; consequently, their reactive properties vary between natural sources and between industrial suppliers. Lacking the specificity of action, HS can be referred to as low efficient adjuvants. In addition, because humics possess an ill-defined structure, their properties are difficult to predict. It is believed that the nature and the quality of humics need to be assessed quantitatively, and that such an assessment could facilitate the prediction of humic properties. Improving tools to predict humic properties, and in addition, developing the novel concept of the producing humic materials of the desired properties, portend new opportunities for the broader application of humic-based products in the field of environmental remediation.

3. Outcomes of the round-table discussions: From practice to theory?

Outcomes of the round-table discussions reflected the findings and conclusions of three working groups (WG). Discussion topics were chosen in support of the main task of the workshop and each focused on the application of humics in remediation. The topic of WG 1 was "Humics as binding agents and detoxicants", for WG 2 "Humics as sorbents and flushing agents", and for WG 3 "Humics as biologically active substances". The outcomes of each WG were summarized below.

3.1. FINDINGS OF WORKING GROUP ONE: "HUMICS AS BINDING AGENTS AND DETOXICANTS" (RAPPORTEURS: F.H. FRIMMEL, I.V. PERMINOVA)

This working group examined first the question of how best to use the binding properties of HS as a tool or function in remediation. The binding functions considered included:
a) Detoxification by complexation (reversible binding, mostly metals);
b) Detoxification by incorporation into humics structure (irreversible covalent binding of organic contaminants);
c) Intelligent mobilization (binding/release): to have humics function at the correct place and at the opportune moment as scavengers for metals and other contaminants (recycling technology), and as carriers for nutrients (bioremediation technologies).
The working group next sought to determine what is not known, or what is needed to characterize the binding properties of HS in such a fashion that their function in the polluted environment becomes predictable. The issues and concerns raised included:

a) Can we define the reactive properties of humics that are consistent with the needs of various remediation technologies?
b) Can we predict these remedial properties of humics?
c) Can we quantify the key remedial properties of humics?
d) Do we have the tools to control the remedial properties for producing humics with predictable reactive properties suitable for a certain remediation technology?
   The answers to the above issues and concerns were:
a) A specific knowledge is lacking on the remedial properties of HS; while, the requisite systematic studies are missing.
b) Quantitative structure-property relationship (QSPR) approach is a promising tool for predicting the remedial properties of HS. To make it work, we need to develop consistent descriptors (parameters) of the remedial properties.
c) We need appropriate analytical techniques and data treatment tools to measure these parameters and predict the properties.
d) Directed chemical modification could be a promising tool for producing humics with the desired remedial properties.
   Conclusions: If the above issues and concerns were addressed in a way that key humic properties pertinent to remediation could be defined, and that the descriptors/parameters of these properties and the techniques to assess them could be developed, then humics of appropriate properties could be match to a given remediation problem. Moreover, new humic materials could be tailored or designed to possess properties suitable for remediation.

3.2. FINDINGS OF WORKING GROUP TWO: "HUMICS AS SORBENTS AND FLUSHING AGENTS" (RAPPORTEUR: G.U. BALCKE)

This working group sought to identify how the sorptive properties of HS could be used to immobilize or mobilize contaminants from an engineering perspective.
   The group proposed a general list of actors involved in mobilization/immobilization technologies. This list included both contaminants: metals; organometallics; organics; and humics: humic preparations; humics-containing and humics-derived materials.
   Where HS could be used for immobilization, the contaminants of interest included:
a) Heavy metals; radionuclides; metal oxoforms.
b) Polychlorinated and polycyclic aromatic compounds.
c) Pesticides, dyes, anilines, etc.
   Where humics could be used for mobilization, the identified contaminants were:
a) Moderately hydrophobic compounds such as naphthalene.
The working group proposed potential areas where humics could be applied to address environmental problems including:
a) Pollution prevention: for example, a landfill liner;
b) Environmental remediation (flushing/immobilization applications): contaminant retardation (immobilization); contaminant mobilization (flushing); reactant retardation (*in situ* PRB construction).
Finally, the working group formulated a list of the advantages and limitations of using humics for mobilization/immobilization based remediation strategies.

The perceived advantages were:
a) Humics are cost effective because 1) different production processes produce a large available supply of humics-containing waste, 2) there exist enormous reserves of raw humics, and 3) humics can be recycled.
b) Humics are generally soluble and thus easily introduced into the subsurface.
c) Humics are environmentally friendly (non-toxic).
d) Humics are environmentally recalcitrant (non-biodegradable).
e) Some reactions with contaminants are irreversible.
    The identified disadvantages/limitations included:
a) Unknown stability of humics-contaminant complexes (feasible release of the contaminant due to dissociation of the formed complexes).
b) Plugging of the porous media due to uncontrolled humic precipitation.
    Conclusions: The working group surmised that humics could be used to facilitate both contaminant mobilization and immobilization. Responsible use, however, requires that sorption be treated as a reversible process. This means that contaminant release or remobilization is feasible and must be considered in an assessment of risks associated with the remediation. Finally, long-term studies of the performance of humic sorbents and flushing agents are needed under a broad range of environmental conditions.

3.3. FINDINGS OF WORKING GROUP THREE: "HUMICS AS BIOLOGICALLY ACTIVE SUBSTANCES" (RAPPORTEUR: O.V. KOROLEVA)

The third working group examined first, humic functions pertinent to *in situ* bioremediation. Particular attention was given to functions which facilitate long-term stability within the subsurface microbial consortia. The functions examined were:
a) Protective (detoxifying).
b) Transporting (carrier of metals: sorption/desorption and distribution among the microorganisms of a consortium).
c) Stabilizing (immobilization of enzymes or microbes on the minerals resulting in better growth, and enhancing metabolic reactions of methanogenic, nitrificators, cellulolytic, lignolytic, and other microorganisms).
d) Nutritional (bacteria are able to utilize the low molecular weight compounds formed as a result of irradiation of humics; the similar compounds are formed during the growth of some fungi, e.g. basidiomycetes).
Next, the working group sought to determine what is not known or what is needed to assess and to predict humic functions coupled to biological activity. The issues examined were:
a) Can we define the biochemical functions of HS if their synthesis has no genetic code?
b) What is an appropriate approach to assess biological activity of humics?
c) Can we define the biological effects pertinent to bioremediation?
d) How can we characterize the desired biological effects of humics?
e) How can we quantify biological activity of HS?

The answers obtained to the above issues and concerns were:
a) General biological terms can be applied to characterize the biological activity of humic substances.
b) The most promising approach leading to a comprehensive assessment of biological activity is a hierarchical evaluation (from molecule to organism, from organism to ecosystem) of biological effects on living organisms.
c) A broad array of the bioassays should be applied for defining the biological effects of HS, e.g. in case of HS interaction with plant cells, the following processes should be monitored: sorption onto a cell membrane, changes in the Na/K balance, and changes in the Ca/Mg ratio of cell membrane transport.
d) The bioassays should be performed under standardized conditions ensuring observation of the direct impact of HS on the organisms, and not of their indirect impact due to interactions with components of cultivation media; the studies for determining and standardizing such conditions for bioassays are needed.
e) Ranking could be a desirable quantitative tool for assessing parameters of biological activity from results of different bioassays.

Conclusions: The working group surmised that humics could be used for bioremediation to ensure the long-term stability of microbial consortia. Efficient use, however, requires humic materials exhibiting effects that are substantially beneficial to the microbial consortia. Comprehensive studies of the biological activity coupled to HS are needed that would examine the various biological effects at each organizational level (molecules, organelles, cells, etc.) using a broad array of bioassays. Clearly, standardization of these bioassays is needed.

3.4. SUMMARY OF THE ROUND TABLE DISCUSSIONS

From both theoretical and practical perspectives, the above findings of the working groups clearly disclosed gaps in existing knowledge and simultaneously defined future research needs. The summary of shortcomings evident in current humics research was perfectly complemented with the substantial comments from remedial engineers (Thomas and Dias). They gave an overview of practical demands and needs in the context of the knowledge gained on the properties and potential uses of humics. This contribution inspired vivid discussions; furthermore, it motivated participants to consider a follow-up workshop with a working theme of "use of humics to remediate polluted environments: from practice to theory." The proposed workshop would focus on the limitations of remedial practices and would reveal novel opportunities for the use of humics in solving those problems.

The Editors

CONTRIBUTORS

Balcke, G.U. Division of Hydrogeology, UFZ Center for Environmental Research, Leipzig-Halle, Theodor-Lieser-Str. 4, 06120 Halle/S. Germany.

Bezuglova, O. S. Soil Science and Agrochemistry Department, Rostov State University, B. Sadovaya St., 105, 344006 Rostov-on-Don, Russia.

Bruccoleri, A. G. Department of Chemistry, University of Calgary, 2500 University Drive N. W., Calgary Alberta T2N 1N4, Canada.

Bulgakova, M. P. State Agrarian University, Voroshilov St. 25, 49600 Dnepropetrovsk, Ukraine.

Cervantes, F. J. Departamento de Ciencias del Agua y del Medio Ambiente, Instituto Tecnologico de Sonora, 5 de Febrero 818 Sur, 85000 Cd. Obregon, Sonora, Mexico.

Chen, Y. Faculty for Agricultural, Food and Environmental Quality Sciences, Department of Soil and Water, Hebrew University of Jerusalem, P.O. Box 12, 76100 Rehovot, Israel.

De Kreuk, J. F. BioSoil R&D. Nijverheidsweg 27, 3341 LJ Hendrik Ido Ambacht, the Netherlands.

Field, J. A. Department of Chemical and Environmental Engineering, University of Arizona, P.O. Box 210011, Tucson, Arizona 85721-0011, USA.

Frimmel, F. H. Engler-Bunte Institute, Department of Water Chemistry, University of Karlsruhe, Postfach 6980, 76128 Karlsruhe, Germany.

Gamble, D. S. Department of Chemistry, Saint Mary's University, Halifax Nova Scotia B3H 3C3, Canada.

Georgi, A. UFZ Center for Environmental Research Leipzig-Halle, Permoserstr. 15, 04318 Leipzig, Germany.

Gerth, A. WASAG DECON GmbH, Deutscher Platz 5, 04103 Leipzig, Germany.

xxiii

Gorova, A. Ecology Department, National Mining University, Karl Marx Ave 19, 49027Dnepropetrovsk, Ukraine.

Hatfield, K. Department of Civil and Coastal Engineering, University of Florida, P.O. Box 116580, Gainesville FL 32611-6580, USA.

Hertkorn, N. Institute of Ecological Chemistry, GSF, Ingolstaedter Landstrasse 1, 85758 Neuherberg, Germany.

Hudaybergenova, E. Institute of Chemistry and Chemical Technology, National Academy of Sciences of Kyrgyz Republic, Chui prospect 267, 720071 Bishkek, Kyrgyzstan.

Iakimenko, O. S. Soil Science Department, Lomonosov Moscow State University, 119992 Moscow, Russia.

Jorobekova, S. Institute of Chemistry and Chemical Technology, National Academy of Sciences of Kyrgyz Republic, Chui prospect 267, 720071 Bishkek, Kyrgyzstan.

Kalmykov, S. N. Department of Chemistry, Lomonosov Moscow State University, 119992 Moscow, Russia.

Karbonyuk, R. A. State Agrarian University, Voroshilov St. 25, 49600 Dnepropetrovsk, Ukraine.

Kaschl, A. Interdisciplinary Department of Industrial and Mining Landscapes, UFZ Centre for Environmental Research, Leipzig-Halle, Permoserstr. 15, 04318 Leipzig, Germany.

Kettrup, A. Institute of Ecological Chemistry, GSF, Ingolstaedter Landstrasse 1, 85758 Neuherberg, Germany.

Kharytonov, M. M. State Agrarian University, Voroshilov St. 25, 49600 Dnepropetrovsk, Ukraine.

Khasanova, A. Department of Chemistry, Lomonosov Moscow State University, 119992 Moscow, Russia.

Klimkina, I. Ecology Department, National Mining University, Karl Marx Ave 19, 49027 Dnepropetrovsk, Ukraine.

Kopinke, F.-D. UFZ Center for Environmental Research Leipzig-Halle, Permoserstr. 15, 04318 Leipzig, Germany.

Koroleva, O. V. Bach Institute of Biochemistry of the Russian Academy of Sciences, Leninsky prospect 33, 119071 Moscow, Russia.

Ksenofontova, M. M. Department of Chemistry, Lomonosov Moscow State University, Leninskie Gory, 119992 Moscow, Russia.

Kudryavtsev, A. V. Department of Chemistry, Lomonosov Moscow State University, Leninskie Gory, 119992 Moscow, Russia.

Kulikova, N. A. Department of Soil Science, Lomonosov Moscow State University, Leninskie Gory, 119992 Moscow, Russia.

Kydralieva, K. Institute of Chemistry and Chemical Technology, National Academy of Sciences of Kyrgyz Republic, Chui prospect, 267, 720071 Bishkek, Kyrgyzstan.

Langford, C. H. Department of Chemistry, University of Calgary, 2500 University Drive N. W., Calgary Alberta T2N 1N4, Canada.

Lesage, S. National Water Research Institute, P.O. Box 5050, Burlington, Ontario L7R4A6, Canada.

Linnik, P. N. Department of Hydrochemistry, Institute of Hydrobiology, National Academy of Sciences, 04210 Kiev, Ukraine.

Litrico, M. E. Department of Civil and Coastal Engineering, University of Florida, P.O. Box 116580, Gainesville FL 32611-6580, USA.

Lunin, V. V. Department of Chemistry, Lomonosov Moscow State University, Leninskie Gory, 119992 Moscow, Russia.

Matorin, D. N. Department of Biology, Lomonosov Moscow State University, Moscow 119992, Russia.

Mitrofanova, A. N. Department of Chemistry, Lomonosov Moscow State University, Leninskie Gory, 119992 Moscow, Russia.

Molson, J. Département des génies civil, géologique et des mines, École Polytechnique de Montréal, C.P. 6079, Succ. Centre-ville, Montréal, Québec, H3C 3A7 Canada.

Moulin, V. Commissariat à l'Energie Atomique, CEA, Nuclear Energy Division & UMR CEA-CNRS-UEVE, Laboratory of Analysis and Environment, 91191 Gif–sur-Yvette, France.

Novikov, A. P. Vernadsky Institute of Geochemistry and Analytical Chemistry, 119991 Moscow, Russia.

Olah, J. Living Planet Environmental Research Ltd., H-2040 Budaors, Szivarvany u. 10, Hungary.

Orlov, D. S.	Department of Soil Science, Lomonosov Moscow State University, Moscow 119992, Russia.
Pankova, A. P.	Department of Chemistry, Lomonosov Moscow State University, Moscow 119992, Russia.
Pavlichenko, A.	National Mining University, Ecology Department, Karl Marx Ave 19, 49027 Dnepropetrovsk, Ukraine.
Perminova, I. V.	Department of Chemistry, Lomonosov Moscow State University, Leninskie Gory, 119992 Moscow, Russia
Petrosyan, V. S.	Department of Chemistry, Lomonosov Moscow State University, Moscow 119992, Russia.
Poerschmann, J.	UFZ Center for Environmental Research Leipzig-Halle, Permoserstr. 15, 04318 Leipzig, Germany.
Princz, P.	Living Planet Environmental Research Ltd., H-2040 Budaors, Szivarvany u. 10, Hungary.
Pryakhin, A. N.	Department of Chemistry, Lomonosov Moscow State University, Leninskie Gory, 119992 Moscow, Russia.
Rusanov, A. G.	Department of Biology, Lomonosov Moscow State University, Moscow 119992, Russia.
Sadovnikova, L. K.	Department of Soil Science, Lomonosov Moscow State University, Moscow 119992, Russia.
Sapozhnikov, Yu. A.	Department of Chemistry, Lomonosov Moscow State University, 119992 Moscow, Russia.
Scherbina, N.	Department of Chemistry, Lomonosov Moscow State University, 119992 Moscow, Russia.
Schmitt, D.	Engler-Bunte Institute, Department of Water Chemistry, University of Karlsruhe, Postfach 6980, 76128 Karlsruhe, Germany.
Schmitt-Kopplin, Ph.	Institute of Ecological Chemistry, GSF, Ingolstädter Landstraße 1, 85764 Neuherberg, Germany.
Shestopalov, A. V.	Soil Science and Agrochemistry Department, Rostov State University, B. Sadovaya St., 105, 344006 Rostov-on-Don, Russia.

Skvortsova, T. Ecology Department, National Mining University, Karl Marx Ave 19, 49027 Dnepropetrovsk, Ukraine.

Smith, S. E. Department of Civil and Coastal Engineering, University of Florida, P.O. Box 116580, Gainesville FL 32611-6580, USA.

Stepanova, E. V. Bach Institute of Biochemistry of the Russian Academy of Sciences, Leninsky prospect 33, 119071 Moscow, Russia.

Thomas, H. WASAG DECON GmbH, Deutscher Platz 5, 04103 Leipzig, Germany.

Van Stempvoort, D. R. National Water Research Institute, P.O. Box 5050, Burlington, L7R 4A6 Ontario, Canada.

Vasilchuk, T. A. Department of Hydrochemistry, Institute of Hydrobiology, National Academy of Sciences, 04210 Kiev, Ukraine.

Vercammen, K. Engler-Bunte Institute, Department of Water Chemistry, University of Karlsruhe, Postfach 6980, 76128 Karlsruhe, Germany.

Woszidlo, S. UFZ Center for Environmental Research Leipzig-Halle, Permoserstr. 15, 04318 Leipzig, Germany.

Yudov, M. V. Department of Chemistry, Lomonosov Moscow State University, Moscow 119992, Russia.

Zhilin, D. M. Department of Chemistry, Lomonosov Moscow State University, Moscow 119992, Russia.

Part 1

Remedial properties of humic substances:
general considerations and problems
in addressing needs of environmental remediation

REMEDIATION CHEMISTRY OF HUMIC SUBSTANCES: THEORY AND IMPLICATIONS FOR TECHNOLOGY

I.V. PERMINOVA[1], K. HATFIELD[2]

[1]*Department of Chemistry, Lomonosov Moscow State University, Leninskie Gory, Moscow 119992, Russia <iperm@org.chem.msu.ru>*
[2]*Department of Civil and Coastal Engineering, University of Florida, P.O. Box 116580, Gainesville FL 32611-6580, U.S.A. <khatf@ce.ufl.edu>*

Abstract

An overview is given of the interactions encountered between humic substances (HS), ecotoxicants, and living organisms in the context of environmental remediation. The most important interactions identified include: binding interactions affecting chemical speciation and bioavailability of contaminants; interfacial interactions altering physical speciation or interphase partitioning of ecotoxicants; abiotic-biotic redox interactions that influence metabolic pathways coupled to pollutants; and finally direct and indirect interactions coupled to various physiological functions of living organisms. Because humics are polyfunctional, they can operate as binding agents and detoxicants, sorbents and flushing agents, redox mediators of abiotic and biotic reactions, nutrient carriers, bioadaptogens, and growth-stimulators. It is shown that these functions possess significant utility in the remediation of contaminated environments and as such humic-based reactions pertinent to permeable reactive barriers, *in situ* flushing, bioremediation, and phytoremediation are examined in detail. Finally, this chapter introduces the novel concept of "designer humics" which are a special class of customized humics of the reduced structural heterogeneity and of the controlled size. They are developed and deployed to carry out one or more of the above *in situ* functions in an optimum manner and for the purpose of enhancing the efficacy of one or more remediation technologies. Designer humics possess specified reactive properties obtained by chemical modification and cross-linking of the humic backbone. This new class of reactive agents portend new opportunities for achieving enhanced remediation and for quantifying remediation performance. The latter is described in the context of the passive flux meter technology developed for direct measuring fluxes of contaminants and biomass.

1. Introduction

Effective remediation of polluted environments is one of the crucial issues on Agenda 21, which lists priorities for achieving sustainable development [1]. Eastern and Western

I. V. Perminova et al. (eds.),
Use of Humic Substances to Remediate Polluted Environments: From Theory to Practice, 3–36.
© 2005 *Springer. Printed in the Netherlands.*

countries alike are currently facing environmental reclamation costs that are increasing exponentially. In the U.S. and Europe, remediation costs generally exceed the net economic value of the land, and often threaten responsible companies with bankruptcy. Given this perspective it is not surprising that scientists and engineers on both sides of the Atlantic have aggressively tried to develop novel technologies to meet regulatory standards at a fraction of the costs associated with traditional approaches (incineration, pump-and-treat, etc.) [2-7].

New remediation technologies are often discovered in process of overcoming limitations of current technologies, and *in situ* remediation is one novel class of technologies that shows considerable promise from both technical and economic perspectives [8-10]. *In situ* remediation relies upon natural and enhanced processes that govern the fate and transport of chemicals released in environment. To a large extent, the reliance on natural processes is predicated on a desire to control costs [11]. Thus, *in situ* technologies that deploy natural attenuating agents such as humic substances (HS) may be even more cost effective.

HS are ubiquitous in the environment and comprise the most abundant pool of non-living organic matter [12]. Their peculiar feature is polyfunctionality, which enables them to interact with both metal ions and organic chemicals. The palette of potential interactions includes ion exchange, complexation, redox transformations, hydrophobic bonding, etc. As a result, numerous studies have shown humics capable of altering both the chemical and the physical speciation of the ecotoxicants (ET) and in turn affecting their bioavailability and toxicity [13]. Hence, HS hold great promise functioning as amendments to mitigate the adverse impacts of ET and as active agents in remediation.

The goal of this chapter is to elucidate emerging concepts of HS-based remediation technologies. Thus, the objectives are: (1) to categorize the interactions encountered between humics, ecotoxicants and living organisms in a polluted environment in the context of remediation chemsitry; (2) to assess the scope of current remedial applications of humics, and (3) to define promising directions of technological developments for remedial implementation of humics.

2. Basic definitions and main features of humic substances

2.1. GENESIS, SOURCES, AND RESERVES OF HUMIC SUBSTANCES IN THE ENVIRONMENT

Humification is the chemical-microbiological process of transforming debris from living organisms into a general class of refractory organic compounds otherwise known as humic substances. It is the second largest process after photosynthesis and involves 20 Gton C/a [12]. Humic substances account for 50 to 80% of the organic carbon of soil, natural water, and bottom sediments [14-16].

Humic materials are typically derived on an industrial scale from peat, sapropel, and coal. *Peat* is a heterogeneous mixture of more or less decomposed plant material (humus) that accumulated in a water-saturated environment in the absence of oxygen [17]. *Coalification* of plant debris preserved in peat mires leads to the formation of *humic coals*. Terms like *peat, lignite, subbituminous, bituminous* and *anthracite* indicate different stages of the coalification process, and they also denote the rank of various coals. The term *"brown coal"* is often used for lignite and subbituminous coals, while *"hard coal"* indicates coals of higher rank. The net result of coalification is an extension of the humification process to include a continuous enrichment of fixed carbon with increasing rank. The relevant increments of carbon content, or % of the total mass, range from: 10-30 (peat), 30-40 (lignites), 40-65 (subbituminous), 65-80 (bituminous), and over 80 (anthracite) [18]. *Sapropel* is an unconsolidated sedimentary deposit rich in *bituminous* substances [19]. It is distinguished from peat in being rich in fatty and waxy substances and poor in cellulosic material. When consolidated into rock, sapropel becomes oil shale, bituminous shale, or *sapropelic* (boghead) *coal*.

The richest source of HS is *leonardite*, a soft brown coal-like deposit usually found in conjunction with deposits of lignite. Leonardite is the most widely used raw material for production of commercial humic preparations [20] followed by other low-rank coals, peat, and sapropel. Table 1 shows the reserves of inexpensive humics-rich materials are immense; however, these reserves are not currently being tapped for environmental remediation.

Table 1. Reserves of humic materials of industrial value.

Source	Amount, Gton C	Ref.
Lignite and Subbituminous coal (Total/Recovered)	1120/512	[21]
Anthracite and Bituminous coal (Total/Recovered)	3880/571	[21]
Peat	400-500	[22]
Sapropel	800	[23]

2.2. CLASSIFICATION, STRUCTURE AND REACTIVITY OF HUMIC SUBSTANCES

Being the products of stochastic synthesis, HS have an elemental composition that is non-stoichiometric, and structure which is irregular and heterogeneous [24]. Aiken et al. [14] defined HS as "a general category of naturally occurring, biogenic, heterogeneous organic substances generally characterized as yellow to black in colour, of high molecular weight, and refractory". MacCarthy and Rice [25] hypothesized that the structural heterogeneity of humics may explain their resistance to biodegradation as longevity of HS in soils is typically on the order of thousands of years. The recalcitrant nature of humics is of practical relevance particularly when the objective is to develop soil/aquifer remediation technologies predicated on a reactive matrix that is not consumed by microorganisms during remediation.

The best illustration of the stochastic nature of HS is provided by the structural model of Kleinhempel (1970) [26] depicted in Figure 1. Clearly, as shown in this figure, a single structural formula cannot be ascribed to any humic sample; consequently, current definitions and classifications of HS are based on isolation procedures rather than on specific molecular features. Thus, the most commonly applied classification is based on humic constituent solubilitiy in dilute acids and bases [27]: *humic acids* (*HA*) represent the fraction that is insoluble at pH<2, *fulvic acids* (*FA*) constitute the fraction soluble under all pH conditions, and *humin* is the fraction insoluble under all pH conditions. Alkali extraction is the most common industrial technique of preparing humics from brown coal or peat and the resultant salts of humic acids are called *humates*. Humates of sodium, potassium, and ammonium comprise the major fraction of the commercially available humic products. The term "humates" is often used to designate any commercially available humic-based product; however, in this chapter, the term is used only to designate the alkali/alkali-earth metals or ammonium salts of HA.

Despite its stochastic nature, HS from different sources share common elements of structural organization. The average humic macromolecule consists of a hydrophobic aromatic core that is highly substituted with functional groups (mostly carboxyl and hydroxyl), and with side aliphatic chains. The core is ensconced in a periphery of hydrolysable carbohydrate-protein fragments [15, 28]. The mass fraction of peripheral fragments decreases with humification; hence, the contribution of labile fragments is greatest among humics derived from composts followed by peat, soil, and finally coal. Coal-derived humics are enriched in condensed aromatic structures and depleted in aliphatic carbohydrate moieties; thus, these humics are much more hydrophobic and less biodegradable than their peat-based counterparts.

The structural complexity inherent in HS creates opportunities for a broad range of chemical interactions as indicated in Figure 2. Humics can be oxidized by strong oxidants; act as reducing agents; take part in protolytic, ion exchange, and complexation reactions; participate in donor-acceptor interactions; engage in hydrogen bonding; and take part in van-der-Waals interactions [29 and citations in it]. Hence, HS can interact practically with all chemicals released in the environment. More pertinent, however, is that humics interact with all classes of ecotoxicants including: heavy metals, petroleum and chlorinated hydrocarbons, pesticides, nitroaromatic explosives, azo dyes, actinides, etc. as shown in Figure 3. Indeed, humics are known to form stable complexes with heavy metals [30-34] and adducts with hydrophobic organic compounds [35-38]; produce charge-transfer complexes [39, 40]; act as electron shuttles [41, 42] and mediate redox reactions of transition metals [43], of chlorinated and nitrated hydrocarbons [44, 45]; adsorb onto mineral surfaces [46, 47]; and influence the interphase distribution of the contaminants [48, 49]. Finally, humics can strengthen the resistance of living organisms against non-specific stress factors [50, 51].

This unique constellation of reactive features strongly suggests HS have the potential to address a broad spectrum of needs within the focus area of environmental remediation [53]. This theoretical statement is confirmed by multiple examples of actual applications in remediation [54-58]. However, to ensure optimum and systematic application, an expanded knowledge base is needed concerning interactions between humics, ecotoxicants, and living organisms.

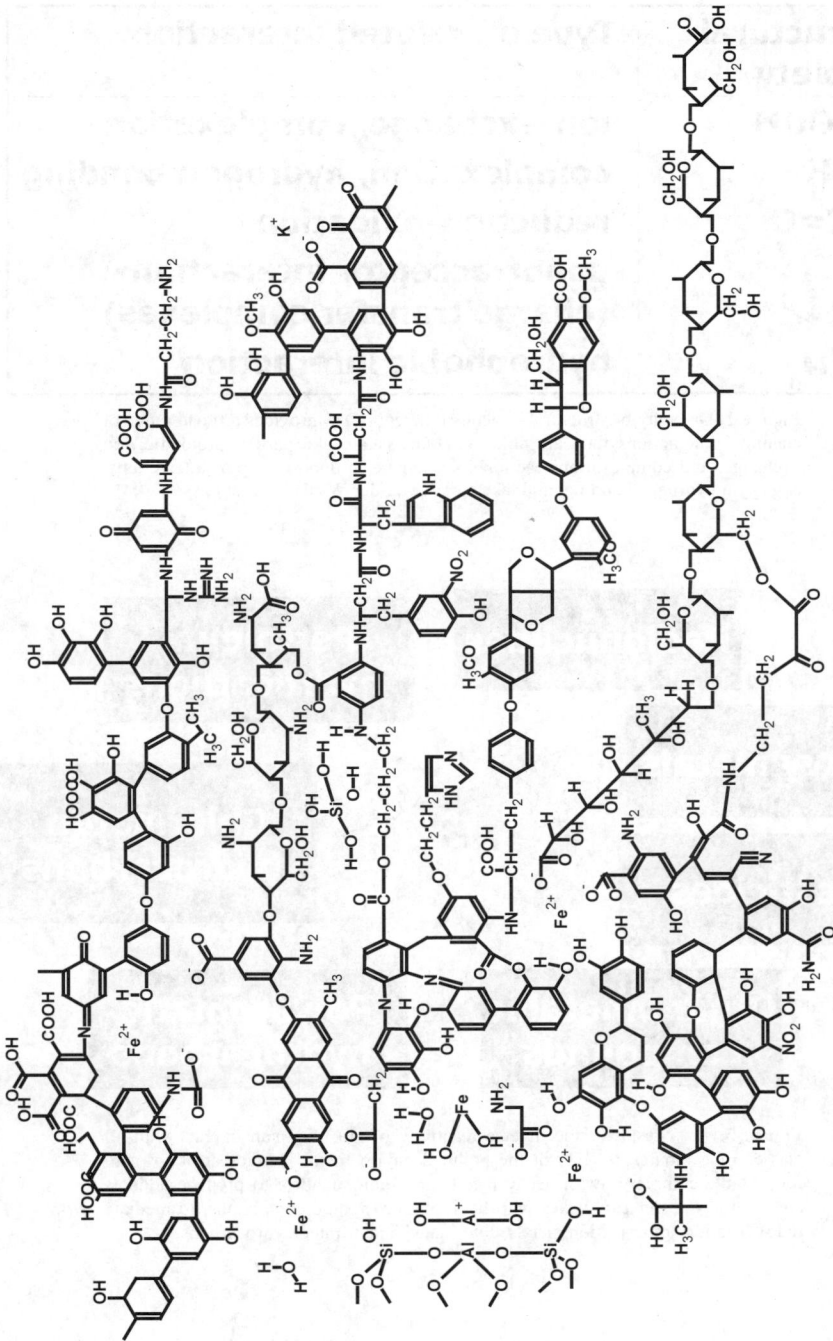

Figure 1. Hypothetical structural fragment of soil humic substances by Kleinhempel [26] that illustrates stochastic nature of humics.

8

Structural moiety	Type of related interaction
- COOH	ion-exchange, complexation
- OH	complexation, hydrogen bonding
>C=O	reduction-oxidation
⬡	donor-acceptor interaction (charge transfer complexes)
- CH_n	hydrophobic interaction

Figure 2. Diversity of structural moieties inherent in HS provides a broad range of chemical interactions they are able of. Humics can take part in protolytic, ion exchange, and complexation reactions; participate in donor-acceptor interactions; engage in hydrogen bonding; and take part in van-der-Waals interactions

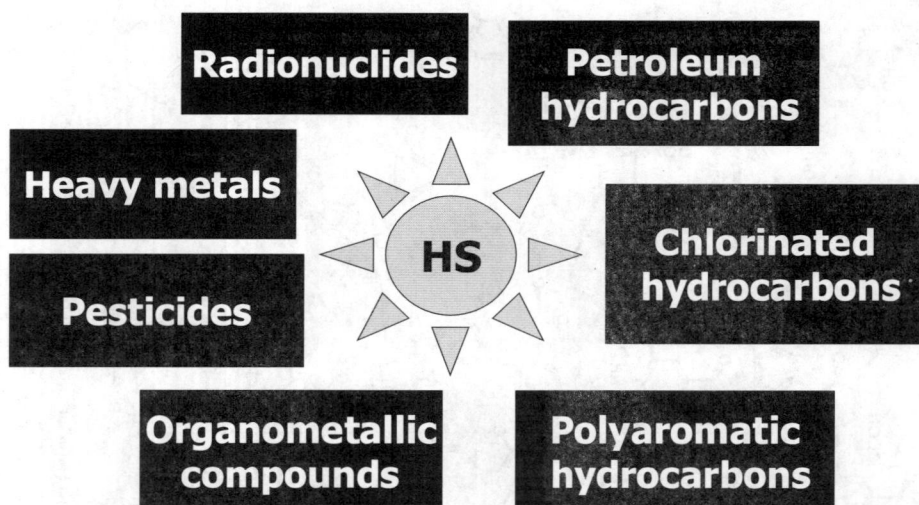

Radionuclides   **Petroleum hydrocarbons**

Heavy metals

HS   **Chlorinated hydrocarbons**

Pesticides

Organometallic compounds   **Polyaromatic hydrocarbons**

Figure 3. As a result of the diverse reactivity of HS, they can interact with all classes of ecotoxicants (ET) in the polluted environment. Humics are known to form stable complexes with heavy metals and radionuclides; to produce adducts and charge transfer complexes with hydrophobic organic compounds; to mediate redox reactions of transition metals, chlorinated and nitrated hydrocarbons.

3. Remedial properties of HS in polluted environments

Multiple interactions between HS, ecotoxicants, and living organisms may be organized to include:
- binding interactions that impact chemical speciation and bioavailability of ecotoxicants;
- sorptive interactions affecting physical speciation or interphase partitioning of ecotoxicants;
- abiotic-biotic redox interactions that impact metabolic pathways coupled to ecotoxicants; and,
- direct and indirect interactions with various physiological functions of living organisms.

To assess the extent to which the above interactions translate into properties pertinent to environmental remediation, each will be considered in the context of the needs and the limitations existing among *in situ* remediation technologies.

3.1. BASIC CONCEPTS AND NEEDS OF *IN SITU* REMEDIATION TECHNOLOGIES

In situ remediation relies upon natural or enhanced processes and does not imply the removal of contaminated soil or the extraction of polluted groundwater [59]. The various *in situ* remediation technologies can be organized to include [60]:

physical treatment: air sparging; directional wells; electrokinetics; fracturing (blast-enhanced, hydraulic, pneumatic); thermal enhancements; vacuum extraction, etc.;

chemical treatment: flushing; permeable reactive barrier (PRB) and treatment walls; immobilization/solidification; etc.; and

biological treatment: intrinsic bioremediation, enhanced bioremediation, phytoremediation, etc.

Most remedial technologies can be applied as a combination of physical-chemical, chemical-biological, or physical-chemical-biological treatments. The details of the above technologies will not be reviewed here but can be found elsewhere [59, 60], while corresponding reviews and case studies are available from relevant web-sites [4-7]. The most promising opportunities for the application of humics-based products and for the development of new humics-based remediation technologies are those predicated on strategies of *in situ* chemical and biological treatment. Examples of these technologies are given in Figure 4 and briefly described in the following paragraphs.

PRBs are replaceable or permanent units installed across the flow path of a contaminant plume. The plume is allowed to migrate passively through the PRB and in the process contaminants are precipitated, sorbed, or degraded [61, 62]. PRBs are filled with different reactive materials such as metals or metal-based catalysts for degrading volatile organics, chelators or ion exchangers for immobilizing metal ions, nutrients and oxygen for microorganisms to enhance bioremediation, or other agents. The reactions that take place in barriers are dependent upon parameters such as pH, oxidation/reduction potential, concentrations, and kinetics. Reactive materials used must demonstrate sufficiently rapid kinetics to remove target contaminants from ground water under natural gradient conditions. In addition, these reactive materials must be inexpensive and functional over an extended time horizon. Finally, pertinent chemical reactions must not produce and release toxic by-products. To date, a limited number of reactive materials satisfy these restrictions including zero valent iron (ZVI) – the most frequently utilized medium, zeolites, peat, lime and ferric oxyhydroxide [62, 63]. Humic-based materials show considerable promise as refractory and inexpensive reactive PRB components. This is particularly true wherever remediation involves a complex array of contaminants, and the reactive material must treat both soluble heavy metals and hydrophobic organics [54]. To evaluate the potential for using HS in PRBs, it is important to understand the sorptive and the redox properties of humics, and both are considered later in this chapter. For a case study on the application of humics in a sorptive PRB see Balcke et al. [64].

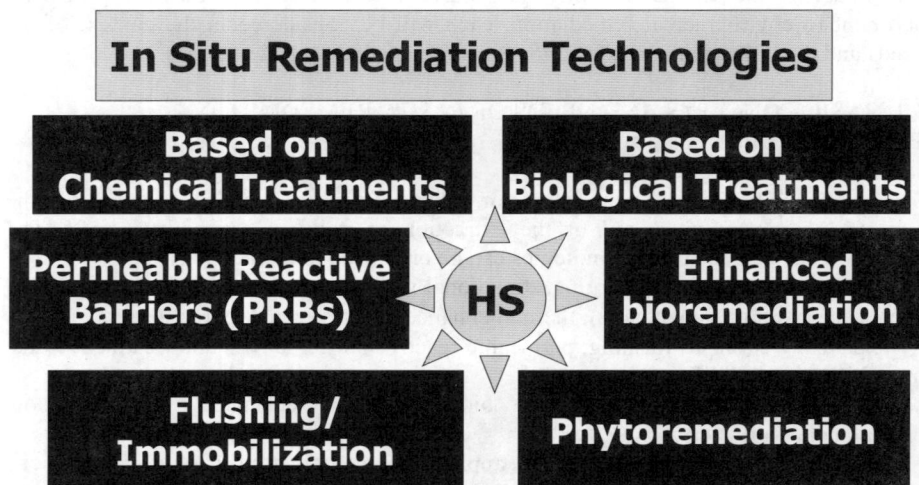

In Situ Remediation Technologies

Based on Chemical Treatments	Based on Biological Treatments
Permeable Reactive Barriers (PRBs)	**Enhanced bioremediation**
Flushing/ Immobilization	**Phytoremediation**

HS

Figure 4. The technologies predicated on strategies of *in situ* chemical and biological treatment can be considered as target remediation technologies for application of HS-based products. The diagram shows the most promising examples of those.

In situ flushing involves the injection or infiltration of an aqueous solution into a zone of contaminated soil or aquifer [60]. The injected fluid functions to increase the mobility and/or solubility of immobilized contaminants. Co-solvents and surfactants are most often the active agents used in flushing solutions [65]. *In situ* flushing has been used to treat soils and aquifers contaminated with halogenated volatiles, nonhalogenated semivolatiles, and nonvolatile metals [66]. The technology can encounter various problems stemming from the flushing agent. Flushing solutions can adhere to the soil or the aquifer matrix, accelerate microbial growth, and cause dissolved constituent precipitation within the porous matrix and thereby reduce system permeability. Furthermore, difficulties can occur with separating co-solvents and surfactants from the elutriate. Left unaddressed, the above problems will manifest themselves in the form of greater demands for reactive agents and additional costs to treat and dispose waste fluids [66 and citations within]. However, such is not the case for flushing solutions comprised of HS: first because the cost of materials is low, and second because humics are biologically recalcitrant and not expected to support microbial growth. However, the use of humics does present its own challenges. For example, under certain conditions dissolved humics can adhere to soils and aquifer materials, and as a consequence, undermine efforts to flush by intercepting and immobilizing soluble contaminants. For a case study on the use of concentrated HS solutions for flushing technologies see review of Van Stempvoort et al. [67] and other related publications [68, 69].

Enhanced bioremediation is an active strategy whereby microbial processes are used to degrade or transform contaminants to less toxic or non-toxic forms. The technology specifically promotes microbial growth for the purpose of harnessing natural processes that effect direct and indirect contaminant degradation or transformation [70, 71]. Soils and aquifers contaminated with organic compounds such as petroleum hydrocarbons, volatile organic compounds, pesticides, wood preservatives, etc. have been treated successfully with bioremediation [72]. The technology has also been used to change the valence state of inorganics such as metal oxoforms for the purpose of inducing adsorption or uptake by microorganisms [73]. However, bioremediation can fail if the supply of nutrients, oxygen, or other electron acceptors is insufficient to support microbial growth. Furthermore, this type of remediation can be inhibited by high concentrations of contaminants or by the presence of other soluble constituents toxic to microorganisms. Other conditions such as pH or the presence of dissolved constituents more amenable to biodegradation can also affect the rate and efficiency of remediation. If used to enhance bioremediation, humics function as reactive agents that are not susceptible to degradation or expected to undergo co-metabolism with target contaminants. HS can serve as extracellular electron shuttles and accelerate microbial redox reactions [41, 42]. Humic-based products can in some cases ameliorate contaminant toxicity by transforming pollutants into less-toxic forms or by sequestering them in a separate phase and reducing bioavailability. With toxicants sequestered, microbial growth is stimulated, and from this the formation of bound residues may be intensified resulting in contaminants covalently bonded to humics [74-77]. Additional discussion on the detoxifying properties of humics is given later in this chapter. For a case study on intensified humification of TNT, see Thomas & Gerth [78].

Phytoremediation uses plants to intercept, accumulate, and/or degrade contaminants in soil and groundwater [79, 80]. The technology is applicable to a broad range of contaminants including numerous metals and radionuclides, various organic compounds (such as chlorinated solvents, petroleum hydrocarbons and their monoaromatic components benzene, toluene, ethylbenzene and xylene (BTEX), polychlorinated biphenyls (PCB), PAH, pesticides, explosives, nutrients, and surfactants). The technology requires plants which have specific characteristics including; tolerance to elevated contaminant concentrations; the tendency to produce significant root biomass; the capability of immobilizing contaminants through uptake, precipitation, or reduction; and the characteristic of retaining target contaminants within the roots such that special handling and disposal of shoots may be avoided [81, 82]. In this context, the application of humics-based products poses several advantages, because as indicated above, HS are known to reduce the toxicity of contaminants and increase a tolerance of the plants to chemical stress [83-85]. In addition, they stimulate the development of roots [86, 87] and as a result, bring about a desirable increase in root biomass. For more details on the direct effects of HS on plants, see reviews by Kaschl & Chen [88] and Kulikova et al. [89].

Clearly, a case can be made for the use of humics and humic-based products to enhance chemical and biological *in situ* remediation. However, in order to develop systematic humics-based applications, the properties of HS must be studied in the context of each remedial technology. In the following sections the fundamental interactions between humics, ecotoxicants and living organisms in the polluted environments are considered.

3.2. IMPACTS OF HS ON SPECIATION AND BIOAVAILABILITY OF ECOTOXICANTS

As indicated above, the broad-spectrum reactivity of HS exists because the humic structure contains numerous functional groups and hydrophobic moieties. This enables humics to bind with both metal ions and organic chemicals. The general binding of HS to ecotoxicants in homogeneous system can be described by the following formalized equation:

$$HS + ET \leftrightarrow HS \cdot ET \tag{1}$$

The equilibrium constant K is commonly used to characterize this interaction:

$$K = \frac{[HS\text{-}ET]}{[ET] \times [HS]} \tag{2}$$

where [HS], [ET], and [HS-ET] represent equilibrium concentrations of the reagents and the reaction product.

Due to the stochastic nature of humics, the stoichiometry of interaction (1) is unknown; hence, specific assumptions are introduced to facilitate the use of eq. (2). In the case of organic compounds, the most common approach is to treat humic substances as "dissolved sorbents" and the HS-ET interaction as a phase partitioning [90-92]. This approach assumes the equilibrium constant K equates to a partition coefficient characterizing a sorbate-sorbent interaction in a heterogeneous system. To account for the mass:volume ratio, the partition coefficient is normalized to the mass concentration of humics in solution, thus:

$$K_{OC} = \frac{1-\alpha}{\alpha} \times \frac{1}{C_{HS}} \tag{3}$$

where K_{OC} is a partition coefficient of ET normalized to mass concentration of soluble HS; α is the portion of the freely dissolved ET in the presence of HS, $\alpha = [ET]/C_{ET}$; and C_{HS} is a mass concentration of HS expressed on an organic carbon basis (kg C/L).

From eq. (3), K_{OC} can be found by determining the fraction of the freely dissolved ET in the presence of HS. This can be done using common analytical techniques with or without preliminary separation of the freely dissolved and the HS-bound species of ET [93-97]. The reported K_{OC} values for polycyclic aromatic hydrocarbons (PAH) vary from 10^2 to 10^6 L/kg C [35, 36, 98-101]. The largest values of 10^5-10^6 L/kg C are observed for partitioning of PAH having four and more rings in their structure (pyrene, benz(a)pyrene, fluoranthene, and others) [101-103]. Much lower binding affinity is observed for small polar molecules like triazines, anilines, phenols, etc. [104-107]. In addition, a very strong dependence of binding affinity on the structural properties of humics is worth noting. When compared to aquatic humic or fulvic acids, considerably higher K_{OC} values are observed for humic acids derived from soil, peat and coal or for commercial humates enriched with aromatic moieties [100, 101, 108]. These findings are indicative of hydrophobic binding that is therefore a governing mechanism of interactions between humics and organic ET. The aromatics enriched humics from coal and mollisol are generally the most hydrophobic; consequently, they are more likely to bind organic contaminants than HS derived from other sources. This association between enhanced binding affinity and enriched levels of aromatic moieties within the humic structure was confirmed by the quantitative structure-activity relationship (QSAR) studies [101, 104, 109, 110].

The binding interactions with ET are of particular importance in remediation, as such interactions reduce concentrations of freely dissolved ET; and as a result leave the offending contaminant less available and perhaps less toxic to living organisms. This is shown in studies of the bioaccumulation [109, 111, 112] and the toxicokinetics [113-115] of hydrophobic organic contaminants in the presence of dissolved humics. For example, the bioconcentration factor (BCF) of PAH in the presence of HS is directly proportional to the fraction of freely dissolved PAH [109, 116]. Furthermore, the partition coefficients determined from bioaccumulation matches those measured by equilibrium dialysis. Similar results are also obtained in acute toxicity studies of three PAHs (pyrene, fluoranthene and anthracene) and a wide range of HS samples from water, soil, peat and coal [117]. Hence, the partition coefficients determined by analytical methods can be used as reliable predictors of the capacity of humics to bind and detoxify (or sequester) organic ET in aquatic environments. This is further confirmed by QSAR-studies revealing a direct correlation between soluble humics aromaticity and the ability to detoxify and/or sequester target hydrophobic contaminants e.g., PAH [109, 117]. In contrast, similar studies show the detoxification effects of HS are less consistent with polar organic compounds [118-120].

The discussed impact of binding interactions on bioavailability of organic ET is even more pronounced where metals are concerned. As in case of organic ET, metal toxicity is related to the free aqua metal ion concentration rather than the total metal concentration. The binding of heavy metals to HS causes a change in metal speciation followed by a change in toxicity and bioaccumulation [121]. This is confirmed by a number of publications (see review [122] and the related recent publications [85, 123-128]).

Figure 5 conveys the concept that HS can bind with ET and as a result, reduce the bioavailability and the toxicity of ET. As indicated in Fig. 5, contaminant detoxification is most relevant in biologically based technologies such as phytoremediation and with *in situ* bioremediation. Depending on the contaminant, elevated concentrations can be toxic to microorganisms and plants and as such undermine remediation efforts unless local concentrations are reduced. The function of humics in this case is simply to reduce concentrations of freely dissolved ET while bioremediation or phytoremediation resumes. Thus, in the case of hydrophobic organic ET, hydrophobic humics from coal are likely to be more effective than HS extracted from other sources. This is an important practical outcome of the above noted QSAR studies. In case of metals, the structure-activity relationship is much more complex, and desired results may depend on the availability of "designer" humics or humics customized to enhance metal complexing properties; the concept of designer humics is discussed further in the final section of this chapter.

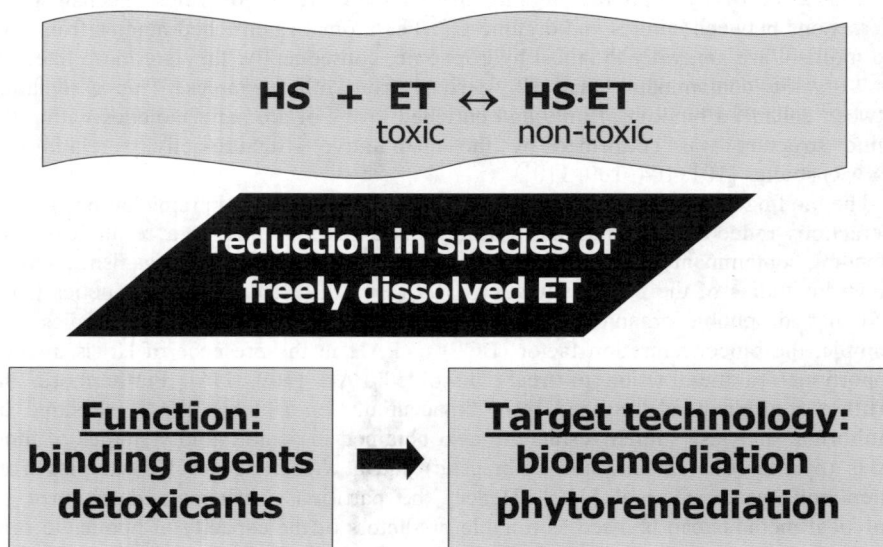

$$HS + ET \leftrightarrow HS \cdot ET$$
$$\text{toxic} \quad \text{non-toxic}$$

reduction in species of freely dissolved ET

Function: binding agents detoxicants	⟹	**Target technology:** bioremediation phytoremediation

Figure 5. HS can bind ecotoxicant (ET) and as a result reduce the bioavailability and the toxicity of ET. This concept has immediate relevance in biologically based remediation technologies such as *in situ* bioremediation and phytoremediation.

An extremely important issue constraining the application of humic substances in remediation is the unknown stability and longevity of humic complexes and/or adducts with ecotoxicants under environmental conditions. Clearly, quantitative studies on the dissociation kinetics of HS complexes with organic chemicals and heavy metals are particularly important for the proper evaluation of humics as reactive materials for PRB and other technologies. Similarly, the same is true of detoxification, long-term experiments are needed to evaluate humics as detoxicants. All investigations regardless of focus (i.e., humic/ET complexation, partitioning, or detoxification), need to be conducted under a range of environmental conditions and using a broad variety of humic samples.

3.3. IMPACTS OF HS ON THE INTERFACIAL INTERACTIONS OF ECOTOXICANTS WITH MINERALS

The previously discussed binding of ET to HS in an aqueous system is often treated as a homogeneous reaction between a dissolved contaminant and a dissolved humic macromolecule. In the soil and the subsurface, however, the leading interaction is simply the heterogeneous sorptive partitioning of ET between the water and the solid phase. The water phase contains soluble inorganic ions and dissolved humic components, whereas the solid phase is represented by minerals and organo-mineral complexes. The latter are formed due to the sorption of humics onto mineral surfaces; the resultant humic coating functions as a natural sorbent with regard to contaminants. Hence, when fixed on mineral surfaces, HS can retard migration of trace metals and organic contaminants; but when dissolved in water, humics can facilitate the transport of the contaminants in the subsurface. Both processes are intensively discussed in the literature. For example, immobilization on organo-mineral particles has gathered considerable attention among researchers investigating the migration of hydrophobic organic contaminants (HOC) in soils and sediments [129], while facilitated transport with organo-mineral colloids has been the focus of the studies concerned with the subsurface migration of heavy metals and radionuclides [130, 131].

The immobilization of HOC by humic coatings recently captured the interest of scientists and engineers when it was shown the binding affinities for HOC were several orders of magnitude higher for humics immobilized on sediments compared to those dissolved in water [98, 132-134], and that the sorption capacity for HOC was proportional to the soil/sediment organic carbon mass fraction [135]. Typically, linear, equilibrium partitioning models are used to quantifying organic ET sorption [135]. These models employ a distribution coefficient K_D to describe the partitioning between aqueous–phase and solid-phase concentrations of ET at equilibrium. As in the homogeneous system, the coefficient K_D is normalized to the soil/sediment organic carbon mass fraction, (f_{OC}), to yield a relatively constant partition coefficients $K_{OC} = K_D/f_{OC}$ for a given ET. Furthermore, as previously indicated, the sorption or partitioning is assumed to be linear, instantaneous, and reversible, that is not subject to competition among different HOC solutes. However, reported findings on HOC sorption by organo-mineral complexes reveal: sorption is indeed characterized by substantial non-linearity and hysteresis [136-140]; and that the sorption affinity of bound humics is more complicated than the simple structure-property relationships revealed by QSARs for homogeneous systems [141-143].

To explain the non-ideal sorption phenomena and the complicated character of the QSPRs, a dual reactive domain model was developed and introduced almost simultaneously by Weber and his group [144] and Xing and Pignatello [145]. The formulation of this model assumed the organic (humic) material bound to mineral surfaces was comprised of two principal organic domains: one a highly amorphous domain (rubbery domain) and the other a relatively condensed domain (glassy domain). Sorption of HOC coupled to the amorphous domain was linear, fast and completely reversible; thus, it could be described using the linear equilibrium partitioning model. However, in the glassy domain, sorption was slow, non-linear and hysteretic [146].

Numerous studies have since validated the two-domain model. However, disparate opinions exist in the literature as to the structural moieties (aliphatics or aromatics) responsible for the hysteretic sorption observed in the glassy domain. A number of recent publications [147-150, and citations in them] claim previous studies have overestimated the importance of the aromatic moieties present in humics and most specifically with regards to explaining the sorption of hydrophobic ET. Simultaneously, these contemporary investigators have espoused the concept, that humics possess an aliphatic component which plays a major role in controlling the sorption of organic contaminants. This assessment is predicated on results of comparative sorption studies involving non-polar probes (phenanthrene, pyrene, etc.) and an array of sorptive matrices including polymers enriched with aliphatics (as opposed to aromatic structures) and different humic fractions. The aliphatics thesis has acquired additional support from solid-state CPMAS ^{13}C NMR data gathered on the presence of poly(methylene)-rich aliphatic domains in the different humic fractions and from correlations developed between aliphatic moieties and non-polar organic sorption [151, 152]. However, it must be noted that in these studies substantial differences exist between the organic matter found in humic coatings and the solid humics used in the NMR studies. The former represents heterogeneous surface complexes of minerals and organic macromolecules, whereas the latter represents a condensed polymeric phase. Hence, the discussed poly(methylene) aliphatic domains were detected in the condensed polymeric phase of humics, and as such caution should be exercised with regards to extrapolating their existence in humic coatings on mineral surfaces. Of course, insight into this question could be provided by conducting analogous NMR studies of model humic-clay complexes obtained using well-characterized humics.

From the above considerations, the following strategies can be formulated for the use of HS in remediation. Humics enriched with glassy rigid domains in their structure (supposedly, rich in aliphatics) are preferential in applications as reactive materials in sorptive barriers designed to intercept and retain non-polar organic ET. The corresponding concept is shown in Fig. 6. These humics have the highest sorption affinity for organic ET and provide the slowest desorption kinetics, or the highest retardation of the organic ET. In terms of the "designer" humics, the best candidates for a use as HOC-sorbents would be cross-linked humics rich in rigid or glassy domains. A potentially cost effective method of creating humic-based sorptive PRBs is to construct them without excavation, in other words, use an *in situ* process of attaching humics to the aquifer matrix. G. Balcke et al. [64] are among the first to investigate an *in situ* approach of coating mineral surfaces with injected humics. QSAR studies provide considerable insight into the adsorption mechanism responsible for the adherence of humics onto mineral surfaces [153]. These studies show the highest affinity for mineral surfaces is seen for humics enriched with aromatics, and that sorption reversibility is inversely proportional to the molecular weight of the humics used. Hence, aromatic-rich humics of high molecular weight are likely to be the best candidates for producing reactive coatings on mineral surfaces.

Illustrated in Fig. 7 is yet another process whereby an ecotoxicant bound to organic-mineral complex is mobilized due to the formation of adducts with dissolved humics. The surfactant properties of HS can be used to develop humic-based flushing agents suitable for the *in situ* remediation of soils and aquifers contaminated with hydrophobic organic chemicals. Aromatics-rich dissolved humics could be the best candidates for this purpose. The reduction of ET concentrations in groundwater will cause a shift in the partition equilibrium towards groundwater and ultimately result in total contaminant removal. Van Stempvoort et al. [67] present a case study on organic ET mobilization using concentrated Aldrich humic acid solutions.

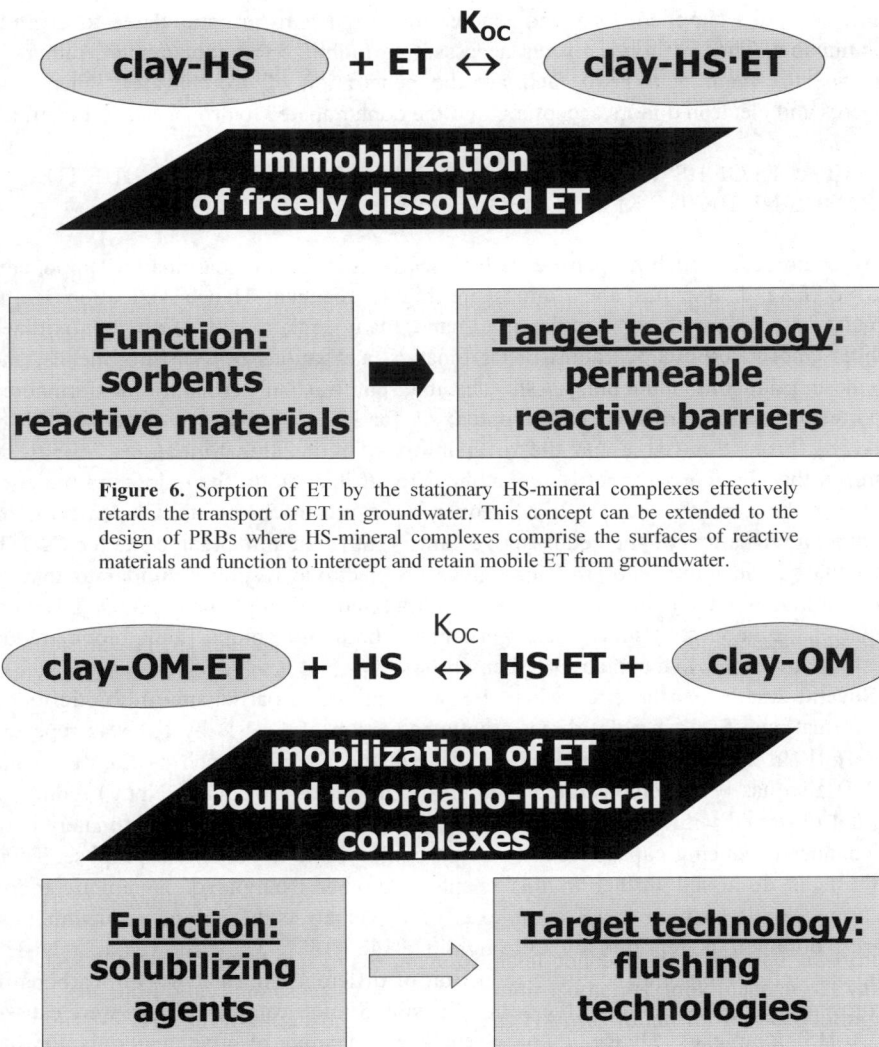

$$\text{clay-HS} \quad + \text{ET} \quad \overset{K_{OC}}{\leftrightarrow} \quad \text{clay-HS·ET}$$

immobilization of freely dissolved ET

| **Function:** sorbents reactive materials | ➡ | **Target technology:** permeable reactive barriers |

Figure 6. Sorption of ET by the stationary HS-mineral complexes effectively retards the transport of ET in groundwater. This concept can be extended to the design of PRBs where HS-mineral complexes comprise the surfaces of reactive materials and function to intercept and retain mobile ET from groundwater.

$$\text{clay-OM-ET} \quad + \text{HS} \quad \overset{K_{OC}}{\leftrightarrow} \quad \text{HS·ET} + \quad \text{clay-OM}$$

mobilization of ET bound to organo-mineral complexes

| **Function:** solubilizing agents | ⇨ | **Target technology:** flushing technologies |

Figure 7. Dissolved humics form HS-ET complexes and enhance ecotoxicant desorption from the mineral surfaces. Under this scenario, HS solutions may be used as flushing agents to facilitate transport of contaminants through an aquifer.

Another application of the surface active properties of concentrated solutions of HS is to facilitate the flow of water through hydrophobic zones of contaminated aquifer. For example, in areas contaminated with residual immiscible hydrocarbons and dense non-aqueous phase liquids the aquifer matrix is sufficiently hydrophobic that the flow of water through the medium is inhibited. Humics introduced into the mobile water phase will reduce the interfacial tension at the surface of porous matrix and permit the aqueous phase to penetrate the hydrophobic medium. With nutrients and electron donors/acceptors supplied, local indigenous microorganisms are stimulated to degrade contaminants. Thus, it is not always necessary to mobilize the contaminant; rather, in this case the main goal is to facilitate the penetration of groundwater, laden with nutrients and electron donors/acceptors, into the contaminated hydrophobic soil matrix.

3.4. IMPACTS OF HS ON THE METABOLISM OF ECOTOXICANTS DUE TO ABIOTIC AND BIOTIC REDOX MEDIATION

Many ecotoxicants such as petroleum hydrocarbons, their monoaromatic components (e.g., BTEX), hydrazines and amines are highly reduced. Hence, oxidation is the primary path of degradation. On the other hand, many contaminants are highly oxidized such as chlorinated hydrocarbons, nitroaromatics, and anions of transition metals, and for these pollutants reduction is the feasible pathway of terminal transformation. Reported values of formal electrode potentials for HS vary from +0.15 to +0.79 [28, 154-158]. From this range and the reversibility of redox transformations, it may be surmised that the redox properties of humics are attributable to the quinonoid moieties present in the aromatic core [159]. Moreover, direct electrochemical evidence exists on the quinonoid nature of the redox-active units [160]. Natural organic matter (NOM) (particularly, the polyphenol fraction) gives an electrode response similar to that of model quinones such as juglone, lowsone, anthraquinone disulphonate (AQDS). Hence, similar to quinones, humics can participate both in abiotic and biotic redox transformations of ET in contaminated environments.

Several studies can be cited where HS were shown to participate in abiotic redox transformations. For example, direct abiotic reduction of Cr(VI) by HS was reported [161-164]. In addition, reduction of highly oxidized actinides such as Pu(VI,V) and Np(VI) also has been demonstrated [165, 166]. However, U(VI) and Np(V) reduction was not observed with HS of natural origin [167]. The customized humic materials of the enhanced reducing capacity can be of particular value to serve that purpose. Their synthesis is discussed further in this chapter. Another example of an abiotic redox reaction where transition metal complexes of HS catalyse the abiotic reduction of a priority pollutant is reported by O'Loughlin et al. [168]. They showed that Ni-HS complexes effectively enhanced the reduction of different chlorinated hydrocarbons in the presence of Ti(III) citrate as the bulk reductant. Similar catalytic effects were caused by Cu-HS complexes. Hence, humics represent potential reactive materials for the immobilization of highly oxidized species of radionuclides and heavy metals, and for the reduction of highly oxidized organics (Fig. 8).

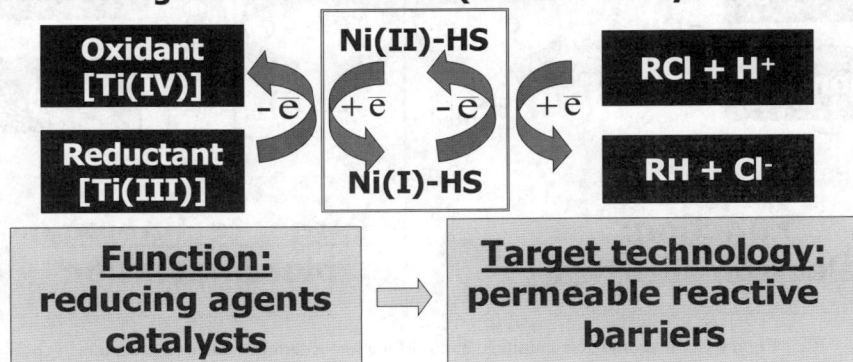

Figure 8. Reported values of formal electrode potential for HS vary from +0.15 to +0.79 [157, 158]. Over this range HS can facilitate both direct and indirect abiotic reduction of highly oxidized contaminants. The given reactions demonstrate direct abiotic reduction of oxoforms of high valence metals on the example of Cr(VI) [161-164] and show mechanism of catalytic impact of transition metal complexes of HS on kinetically slow abiotic reduction of organic contaminants by the bulk reductant as reported by O'Loughlin et al. [168].

Recently, considerable attention has been focused on the ability of humics to mediate the microbial degradation of various contaminants (see review [169]). Humics possess the unique capability of functioning as both an electron acceptor and a donor depending on environmental conditions [41, 42]. This ability permits humics to facilitate both oxidative and reductive biodegradation as is shown in Figs. 9 and 10. Under anoxic conditions, humics operate as terminal electron acceptors supporting the mineralization of various organic pollutants to CO_2 by anaerobic microbial communities [44, 170, 171]. Fig. 9 illustrates the feasibility of HS functioning as redox mediators within a technology designed to bring about *in situ* oxidative bioremediation [56].

It has also been widely reported that HS facilitate reductive biodegradation by shuttling electrons from microorganisms to various highly oxidized organic contaminants (e.g., chlorinated hydrocarbons and azo dyes [172-174] as well as to high valence metals (e.g., Cr(VI), U(VI), and Tc(VII)) [43, 175, 176]. Fig. 10 reveals that HS can mediate reductive biodegradation both directly via shuttling electrons from microorganisms to high valence metals or oxidized organics, and indirectly via interactions with different Fe(III) oxide minerals [177]. Hence, it is plausible for humics to function as redox mediators within technologies designed to bring about *in situ* reductive bioremediation [177].

Reductive biodegradation

Highly oxidized contaminants (e.g., high valence metals):

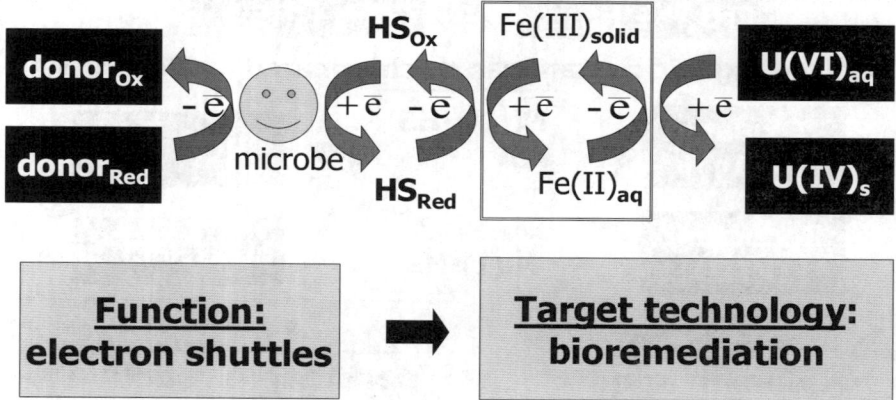

| Function: electron shuttles | ⟹ | Target technology: bioremediation |

Figure 9. Under anoxic conditions, humics function as terminal electron acceptors or redox mediators supporting the oxidative biodegradation of the reduced organic pollutants to CO_2 by anaerobic microbial communities. An example of the corresponding technological development is published on the site of the USGS [56].

Oxidative biodegradation under anoxic conditions

Highly reduced contaminants (petroleum hydrocarbons):

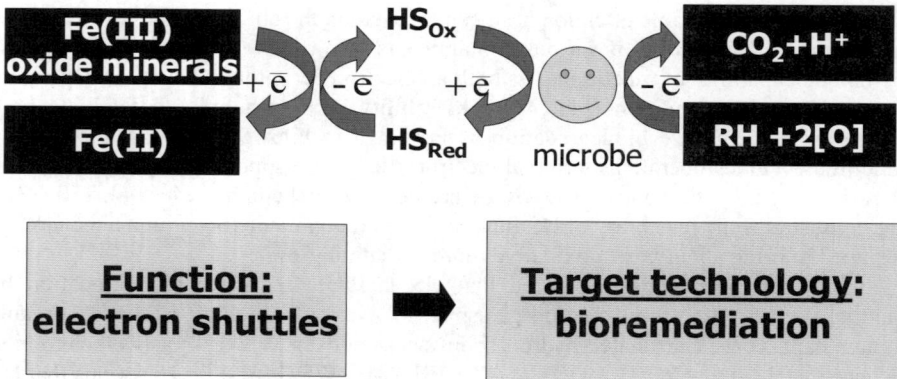

| Function: electron shuttles | ⟹ | Target technology: bioremediation |

Figure 10. HS can participate in the reductive biodegradation of various highly oxidized contaminants (i.e., metal oxoforms or chlorinated hydrocarbons) either by direct shuttling electrons from microorganisms or via interaction with Fe-oxides. Enhanced bioreduction of U(VI) was described in [43]. Operating as redox mediators, it is plausible for HS to function in technologies designed to bring about *in situ* reductive bioremediation [177].

Given the above described electron shuttling properties, it is conceivable humics can be used to facilitate or stimulate a suite of microbially mediated redox reactions pertinent to *in situ* bioremediation. However, the primary factor limiting their effectiveness is the structural polydispersity and heterogeneity which translates into reactive properties that are highly variable between natural sources and amongst different humic fractions. Gu and Chen [43] revealed a wide disparity in redox mediating properties among humic samples of different origin and fractional composition. For example, the best performance in the abiotic reduction of Cr(VI) and Fe(III) are observed with a polyphenols-rich fraction of natural organic matter (NOM). However, soil humics enriched with polycondensed aromatic moieties are more effective in mediating the microbial reduction of Cr(VI) and U(VI). To overcome the problem of structural heterogeneity and polydispersity, directed modification of HS may be advantageous, e.g. the incorporation of additional quinonoid moieties to bring about a desired enhancement in redox-capacity [178].

Finally, it is necessary to point out that HS exhibit yet another kind of mediating effect on the biotic transformation of organic ET; this is the covalent bonding to humics also known as oxidative coupling [179]. This process includes the oxidation, generation, and rearrangement of free radicals, and the incorporation of the ecotoxicant into the HS structure. To introduce oxidative coupling into the practice of remediation, biostimulators of this process must be designed. Early efforts to initiate oxidative coupling with low molecular weight initiators (substrates of oxidoreductive enzymes) proved inefficient [180]. An alternative approach, using high-molecular weight humics-based promoters deserves further consideration. These promoters can be prepared by directed modification of humic materials, given the properties of such materials are clearly defined by environmental microbiologists.

3.5. IMPACTS OF HUMICS-ECOTOXICANTS INTERACTIONS ON LIVING ORGANISMS

The pool of data describing the direct biological effects of HS is vast, miscellaneous, and controversial. The array of biotargets studied includes pure cell cultures, bacteria, algae, fungi, higher plants, animals, and humans. Some authors show humics stimulating the growth of higher plants [51, 181-183] and microbial communities [184-186]. Other investigators find HS strengthen the resistance of higher plants under stress [50, 52, 87] and as a consequence, define humics as natural adaptogens. Mazhul et al. [187] and Vigneault et al. [188] believe humics have a direct impact on cells by changing the permeability of the cell membrane; whereas others claim HS increase the bioavailability of nutrients via the formation of metal-HS complexes [181, 189].

Given the growth stimulating and adaptogenic (anti-stressor) activity of humics, these substances may be useful as agents to enhance bioremediation and, in particular, phytoremediation. Fig. 11 conveys a concept of humics use in phytoremediation technologies. The beneficial effects of HS on plant growth in the polluted environment can be related to an increase in nutrient supply, to an improvement of the overall plant development, and to an increased resistance to chemical stress [190, 191]. In the context of phytoremediation, it is well documented that humics cause an increase in root growth superior over that of the shoot [192]. Taken together, the combined effects humics suggest they are engaged facilitating nutrient transport, promoting plant growth, and serving as bioadaptogens.

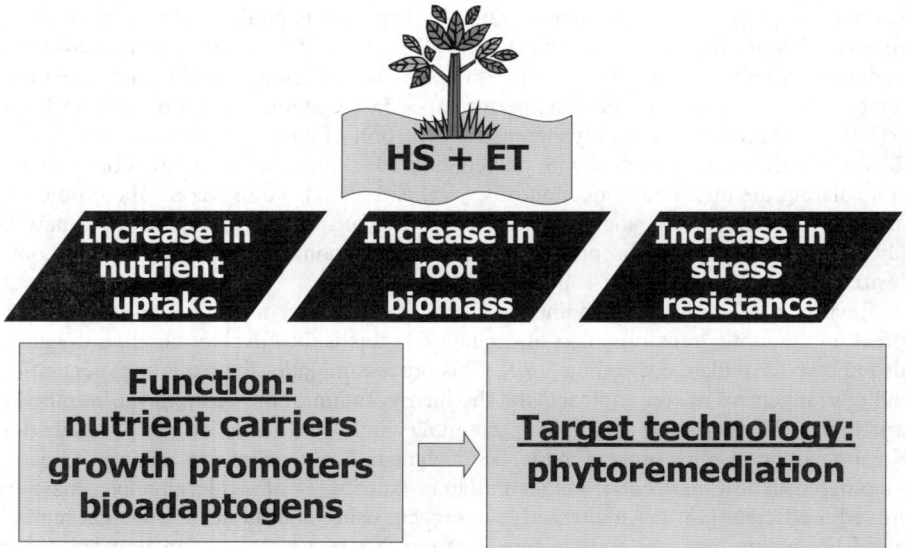

Figure 11. Potential functions performed by HS for higher plants in the polluted environments and their associated biological responses.

Existing studies on quantitative-structure activity relationships for beneficial effects of humics on higher plants deliver contradicting results [51]. The most consistent are the studies on the effects exerted by high and low molecular weight humic fractions. It is shown that high molecular weight fractions promoted the plant growth, but decrease enzyme activity [86, 193]; accelerate root differentiation [194]; and readily adsorb onto the cell wall, but do not enter the cell [195]. At the same time, low molecular weight fractions were shown to reach the plasmalemma of root cells and, in part, were translocated into the shoots [196]. Such observations brought Nardi with co-workers to review the physiological functions of HS on higher plants [51] and to conclude that it was the lower molecular weight humic fractions that acted at the symplast and directly influenced plant metabolism, whereas the higher molecular weight fractions operated mainly on the cell surface where they influenced differentiation and growth at the apoplast. Hence, the reported positive effects on plant growth were induced by the lighter fractions of humic matter.

The conditions of polluted environment require meticulous consideration, when rationalizing the use of humics to enhance phytoremediation. Another recent review on molecular size dependent impacts of humics-ecotoxicants interactions on biota by Perminova et al. [122] should be particularly mentioned here. This review demonstrates that the adverse effects of the ecotoxicants are always lower if those contaminants possessed a strong affinity for humics representing the higher molecular weight fraction. Whereas for ecotoxicants having a greater affinity for lower molecular weight fraction, toxicity can decrease, remain the same or increase. It is hypothesized that for humics of high molecular weight, the complex with the ET forms that cannot penetrate the cell membrane and, hence, reduces bioavailability of the ecotoxicant. In case of humics of low molecular weight, the corresponding complex (or associate) is translocated to the interior of the cell where transformations of the humic-ET complex occur under the prevailing conditions inside the living cell interior. If the complex (associate) does not dissociate under cellular conditions, no additional toxicity occurs. However, if the complex (associate) breaks down, the toxic effects of the ET will be enhanced. Typical contaminants likely to share a high affinity for the heavier humics are: HOC like polycyclic aromatic and polychlorinated hydrocarbons, and metals like Cu and Pb. ET among the group most likely to bind with lighter humics are: trace metals and organic chemicals which bind to humic molecules via ion exchange or hydrogen binding, or the other mechanisms involving not the core but functional moieties of the humic compound. Examples of this group of ET include cadmium, Cr(VI), substituted phenols/anilines, and others.

Hence, to ensure positive biological effects of HS on higher plants in the polluted environments the preference should be given to a use of humics of higher molecular weight. More research on the biological effects of well characterized HS is needed to develop sound scientific base for humics applications in phytoremediation.

4. Design of humic materials of the desired properties, or how to make humics work for remediation technologies

Despite the diverse protective functions transferable to humics in a polluted environment, application of humics-based products for remediation remains limited. Two fundamental reasons can be formulated as to why HS are not widely used. First, few natural HS possess the specific reactive properties required to treat selected contaminants. Second, humics by definition are polydisperse and heterogeneous, which translates into properties that vary between natural sources and between industrial suppliers. Hence, the structural heterogeneity needs to be reduced or controlled to the extent that reactive properties become predictable; this will facilitate the use of humic materials in remediation. From this perspective, Perminova and co-workers are developing "designer humics" in other words, they are chemically modifying humic materials to acquire desired reactive properties [178, 197].

Fig. 12 conveys a conceptual model for designing reactive humic materials based on idea of reducing structural heterogeneity and polydispersity. This model introduces the concept of incorporating specific reactive moieties into the humic backbone for purposes of acquiring desired reactive properties, and of cross-linking humic materials for producing a desired reactive form (soluble, colloidal, solid).

24

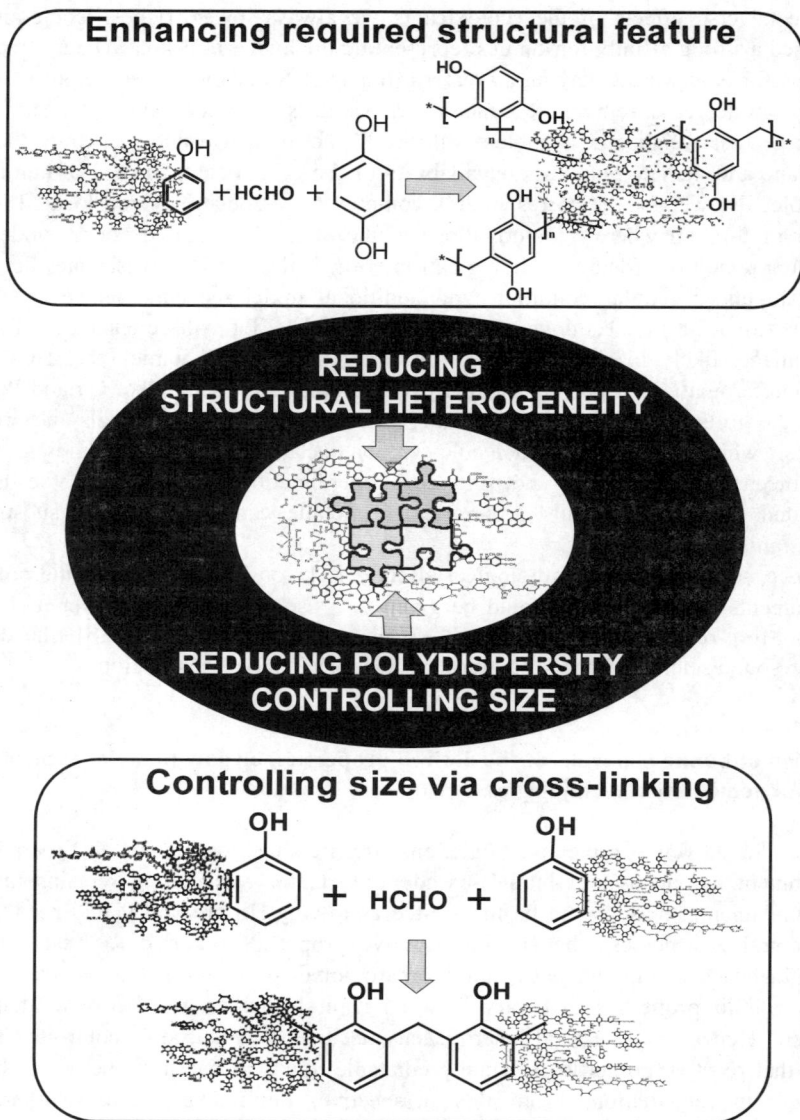

Figure 12. Conceptual model of the design of humic reactive materials. The examples are shown for tailoring redox-active and cross-linked structures.

using phenolformaldehyde-like condensation reactions with coal-based HS, and hydroquinone and catechol monomers (see Fig. 12). The modified humics possess reducing capacities with respect to Fe(III) that are greater than parent humic materials [178]. Cathecholic derivatives are more effective at detoxifying Cu(II) [198] than humics modified with hydroquinone monomers; and this confirms the formation of non-bioavailable Cu(II) chelates as the main mechanism of detoxification. The demonstrated consistency of the changes in structure and properties of the humic derivatives lay bare the feasibility of employing chemical modification for preparing humic materials of the desired properties. For example, an available supply of humics possessing a variety of quinonoid reactive centres and covering a wide range of electrode potentials presents a unique opportunity to probe a selection of electron shuttling interactions mediated by humics between ecotoxicants and diverse microorganisms. As a result, a strategy can be formulated for a use of humics as biostimulators to accelerate the microbial redox transformation of the contaminants.

The different physical forms of the designed humics - soluble, colloidal, and solid, - can ensure control over interfacial interactions of HS with contaminants. The soluble forms can be used either for retarding heavy metals and radionuclides migration in soil or as biostimulators of metal reducing microbial communities. The colloidal and solid forms may be embedded into permeable polymer matrixes (e.g., polyurethane foams) for purposes of producing easily deployable sorbents suitable for vehicle and building decontamination. Solid forms can also be used as sorbents for PRBs or they can be applied to other forms of engineered flow-through systems. It is feasible that humic reactive materials be developed to suit dual needs of both site remediation and remediation monitoring. In particular, humics can be used as sorbents and reactive materials in novel strategies to quantify remediation performance and to better characterize remediation potential as described below.

Hatfield with co-workers [199, 200] present a passive flux meter (PFM) method for measuring in situ groundwater flux and contaminant mass flux. The PFM is a self-contained permeable unit that is inserted into a well. The interior composition of the flux meter is a matrix of hydrophobic and hydrophilic permeable sorbents. As ground water flows passively through the device, the internal sorptive matrix retains dissolved contaminants present in the volume of water intercepted. The total mass of contaminant intercepted and retained on the sorbent permits direct determination of the local in situ contaminant mass flux. The latter yields a quantitative assessment of the source loading or strength, – a change of which is the primary indicator of the effectiveness of remediation. Fig. 13 illustrates the PFM as simply a permeable cartridge that is inserted into a well located down gradient from a contaminant source. The sorptive matrix is also impregnated with known amounts of one or more water-soluble resident tracers. These tracers are displaced from the sorbent at rates proportional to the groundwater flux; hence, the resident tracers are used to quantify groundwater flux.

To enhance characterization of biological subsurface remediation, humics can be used to create new sorbents for a new class of PFM's designed to measure microbial mass fluxes. Novel humic-based sorbents serve to intercept and retain microbial biomass. Used in conjunction with the traditional PFM, the passive biomass flux meter

(PBFM) offers an opportunity to better characterize spatial variations in microbial biomass and ecology - indicators of bioremediation.

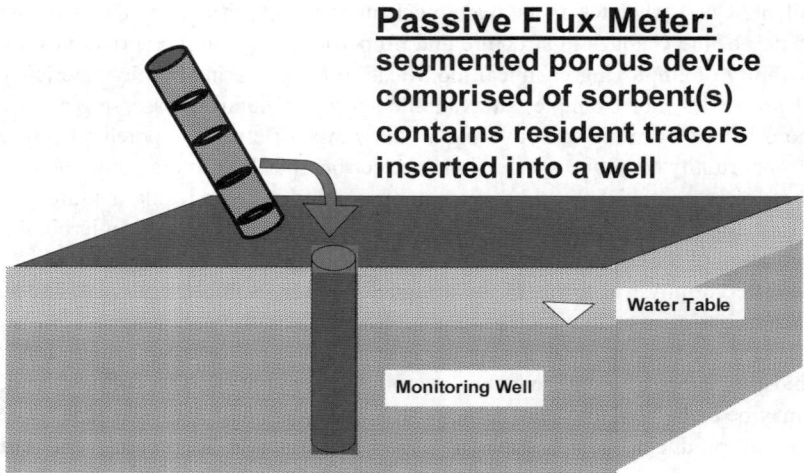

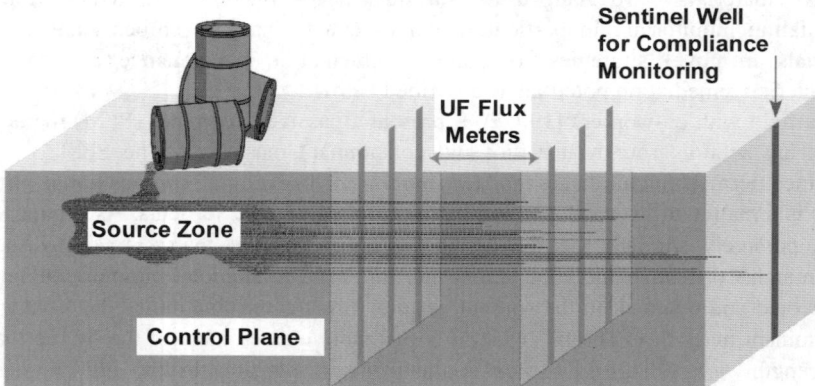

Figure 13. Conceptual model and application of flux meter technology.

Given the biocompatibility of humics together with their resistance to biodegradation, solidified humic matrixes represent ideal sorbents for bacteria and can be inserted into a monitoring well with little concern for possible negative repercussions. Hence, modified humics combined with passive monitoring methods presents new opportunities for the combined chemical and biological characterization of subsurface remediation.

5. Conclusions

An overview of the interactions of HS with ecotoxicants and living organisms in the polluted environments in the context of remediation chemistry allows us to conclude that HS can perform perform multiple functions within a remedial strategy. For example, it is shown that HS can function as: binding agents and detoxicants; sorbents and flushing agents; redox mediators of abiotic and biotic degradation; and nutrient carriers, bioadaptogens and growth-stimulators. With the above functions characterized, it is then proposed that humic-based products hold great promise as reactive agents for in situ remediation. The most promising technologies include: enhanced bioremediation; permeable reactive barriers; *in situ* flushing; and phytoremediation. However, to develop successful humic-based applications, it is stressed that the properties of HS must be studied in the context of each remedial technology. For example, research is needed to define the kinetics, stability, and longevity of humic complexes and/or adducts with organic chemicals and heavy metals. Results from such research are salient for the proper evaluation of humics as reactive matrices for PRB's and other technologies.

In the final section of this chapter, a novel concept is introduced for the development and application of designer humics. For the most part, natural humics possess an elemental composition that is non-stoichiometric and a structure that is irregular and heterogeneous; consequently, reactive properties vary widely between humic samples from various sources. Using example methods of chemical modification described herein, it is proposed that the properties of humics be tailored to satisfy the needs of a given remediation technology. Pursuing this approach, it is expected that the utility and the value of humics will expand.

6. Acknowledgements

The authors wish to acknowledge the support of NATO CLG-programme (grant # 980508) and of ISTC (project KR-964).

7. References

1. The United Nations Programme of Action from Rio (1993) *Agenda 21, Earth Summit*, United Nations Publisher.
2. *United Nations Industrial Development Organization* (UNIDO), <http://www.unido.org>.
3. *United Nations Environment Programme* <http://www.unep.org>.
4. *U.S. Federal Remediation Technologies Roundtable* (FRTR), <http://www.frtr.gov>.
5. *U.S. Remediation Technologies Development Forum*, <http://www.rtdf.org>.
6. *U.S. EPA's Office of Superfund Remediation and Technology Innovation* (OSRTI), <http://www.clu-in.org>.
7. *NETC-EPA Ground-Water Remediation Technologies Analysis Center* (GWRTAC), <http://www.gwrtac.org>.
8. U.S. Environmental Protection Agency (EPA) (1999) *Groundwater Cleanup: Overview of Operating Experience at 28 Sites*, EPA-542-R-99-006, <http://www.clu-in.org>.
9. U.S. Environmental Protection Agency (EPA) (1999) *Field Applications of In Situ Remediation Technologies: Permeable Reactive Barriers*, EPA-542-R-99-002, <http://www.clu-in.org>.

28

10. U.S. Environmental Protection Agency (EPA) (1998) *Remediation Case Studies*, Volume 8, *In Situ Soil Treatment Technologies* (Soil Vapor Extraction, Thermal Processes), EPA-542-R-98-012.

11. U.S. Environmental Protection Agency (EPA) (2001) *Cost Analyses for Selected Groundwater Cleanup Projects: Pump and Treat Systems and Permeable Reactive Barriers*, EPA 542-R-00-013, <http://www.clu-in.org>.

12. Hedges, I.J. and Oades, J.M. (1997) Comparative organic geochemistries of soils and marine sediments, *Org. Geochem.* **27**, 319-361.

13. Schnitzer, M. and Khan, S.U. (1972) *Humic substances in the environment*, Marcel Dekker, New York.

14. Aiken, G.R., McKnight, D.M., and Wershaw, R.L. (eds.) (1985) *Humic substances in soil, sediment and water*, Wiley, New York.

15. Orlov, D.S. (1990) *Soil humic acids and general theory of humification*, Moscow, State University Publisher, (In Russian).

16. Thurman, E.M. (1985) *Organic geochemistry of natural waters*, Martinus Nijhof/Dr. W. Junk Publishers, Dordrecht, The Netherlands.

17. *Peat*, retrieved October, 2003 from Encyclopædia Britannica Premium Service, <http://www.britannica.com/eb/article?eu=67401>.

18. Edbrooke, S. (1999) Coal, in *Mineral Commodity Report 18*, Institute of Geological and Nuclear Sciences Ltd, New Zealand, p. 15.

19. *Sapropel*, retrieved October, 2003 from Encyclopædia Britannica Premium Service, <http://www.britannica.com/eb/article?eu=60369>.

20. Ozdoba, D.M., Blyth, J.C., Engler, R.F., Dinel, H. and Schnitzer, M. (2001) Leonardite and humified organic matter, in *Proc Humic Substances Seminar V*, Boston, MA, March 21-23, 2001.

21. Lou, J. (2003) *World Estimated Recoverable Coal. United States*, Energy Information Administration, <http://www.eia.doe.gov/emeu/iea/table82.html>.

22. Markov, V.D., Olunin, A.S., Ospennikova, L.A., Skobeeva, E.I., Khoroshev, P.I. (1988) *World Peat Resources*, Moscow, Nedra (in Russian).

23. Belyakov, A.S. and Kosov, V.I. (2002) *Rational use of peat and sapropel in Russia*, Russian State Duma Committee on rational use of nature and resources, Moscow, Russia, <http://www.mineral.ru/Chapters/Production/Issues/18/Issue_Files.html>.

24. Hayes, M.H.B., MacCarthy, P., Malcolm, R.L., Swift, R.S. (eds) (1989) *Humic Substances II: In Search of Structure*, Wiley, Chichester.

25. MacCarthy, P. and Rice, J.A. (1991) An ecological rationale for the heterogeneity of humic substances, in S. H. Schneider, P. J. Boston (eds.), *Proceedings of Chapman Conference on the Gaia Hypothesis*, San Diego, CA, March 7-11, 1988, MIT Press, Cambridge, pp. 339-345.

26. Kleinhempel, D. (1970) *Albrecht-Thaer-Archiv* **14**, 3-14.

27. Stevenson, F.J. (1994) *Humus Chemistry: Genesis, Composition, Reactions*, John Wiley and Sons, Inc., NY.

28. Ziechmann, W. (1980) *Huminstoffe*, Verlag Chemie, Weinheim, Deerfield Beach, Basel.

29. Clapp, C.E., Hayes, M.H.B., Senesi, N. and Griffith, S.M. (eds.) (1996) *Humic Substances and Organic Matter in Soil and Water Environments: Characterization, Transformations and Interactions*, IHSS Inc., St. Paul, MN, USA.

30. Weber, J.H. (1988) Binding and transport of metals by humic materials, in F.H. Frimmel and R.F. Christman (eds.), *Humic Substances and Their Role in the Environment*, J. Wiley & Sons Ltd, pp. 165-178.

31. Varshal, G.M., Velyukhanova, T.K. and Koshcheeva, I.Ya. (1993) Geochemical role of humic acids in elements migration, in *Humic Substances in Biosphere*, Moscow, Nauka, pp. 97-117 (in Russian).

32. Benedetti, M.F., Van Riemsdijk, W.H., Koopal, L.K., Kinniburgh, D.G., Gooddy, D.C. and Milne, C.J. (1996) Metal ion binding by natural organic matter: from the model to the field, *Geochim. Cosmochim. Acta* **60**(14), 2503-2513.

33. Linnik, P.I. and Nabivanets, B.I. (1986) *Migration forms of metals in fresh surface waters*, Leningrad, Gidrometeoisdat (in Russian).

34. Croué, J.-P., Benedetti, M.F., Violleau, D. and Leenheer, J.A. (2003) Characterization and copper binding of humic and non-humic organic matter isolated from the South Platte River: Evidence for the presence of nitrogenous binding site, *Environ. Sci. Technol.* **37**(2), 328-336.

35. Gauthier, T.D., Seitz, W.R., Grant, C.L. (1987) Effects of structural and compositional variations of dissolved humic materials on pyrene KOC values, *Environ. Sci. Technol.* **21**, 243-248.

29

36. McCarthy, J.F. and Jimenez, B.D. (1985) Interactions between polycyclic aromatic hydrocarbons and dissolved humic material: binding and dissociation, *Environ. Sci. Technol.* **19**, 1072-1075.
37. Gevao, B., Semple, K.T. and Jones, K.C. (2000) Bound residues in soils: A review, *Environmental Pollution* **180**, 3-14.
38. Kristoffer, E.N., Jonassen, K.E.N., Nielsen, T. and Hansen, P.E. (2002) The application of high-performance liquid chromatography humic acid columns in determination of Koc of polycyclic aromatic compounds, *Environ. Toxicol. Chem.* **22**(4), 741-745.
39. Muller-Wegener, U., and Ziechmann, W. (1980) Elektronen-Donator-Akzeptor-Komplexe zwischen aromatischen Stickstoffheterocyclen und Huminsaure, *Z. Pflanz. Bodenk.* **143**, 247-249.
40. Senesi, N. and Testini, C. (1982) Theoretical aspects and experimental evidence of the capacity of humic substances to bind herbicides by charge-transfer mechanism, *Chemosphere* **13**, 461-468.
41. Lovley, D.R., Coates, J.D., Blunt-Harris, E.L., Phillips, E.J.P. and Woodward, J.C. (1996) Humic substances as electron acceptors for microbial respiration, *Nature* **382**, 445-448.
42. Lovley, D.R., Fraga, J.L., Coates, J.D., Blunt-Harris, E.L. (1999) Humics as an electron donor for anaerobic respiration, *Environ. Microbiol.* **1**, 89-98.
43. Gu, B.H. and Chen, J. (2003) Enhanced microbial reduction of Cr(VI) and U(VI) by different natural organic matter fractions, *Geochim. Cosmochim. Acta* **67**, 3575-3582.
44. Bradley, P.M., Chapelle, F.H., and Lovley, D.R. (1998) Humic acids as electron acceptors for anaerobic microbial oxidation of vinyl chloride and dichloroethene, *Appl. Environ. Microbiol.* **64**, 3102-3105.
45. Tratnyek, P.G., and. Macalady, D.L. (1989) Abiotic reduction of nitroaromatic pesticides in anaerobic laboratory systems, *J. Agri. Food Chem.* **37**, 248-254.
46. Murphy, E.M., Zachara, J.M., Smith, S.C., and Phillips, J.L. (1992) The sorption of humic acids to mineral surfaces and their role in contaminant binding, *Sci. Total Environ.* **117/118**, 413-423.
47. Vermeer, A.W.P and Koopal, L.K. (1998) Adsorption of humic acids to mineral particles. 2. Polydispersity effects with polyelectrolyte adsorption, *Langmuir* **14**, 4210-4216.
48. Laird, D.A., Yen, P.Y., Koskinen, W.C., Steinheimer, T.R., and Dowdy, R.H. (1994) Sorption of atrazine on soil clay components, *Eviron Sci. Technol.* **28**, 1054-1061.
49. Murphy, E. M. and Zachara, J.M. (1995) The role of sorbed humic substances on the distribution of organic and inorganic contaminants in groundwater, *Geoderma* **67**, 103-124.
50. Khristeva, L.A. (1970) Theory of humic fertilizers and their practical use in the Ukraine, in Robertson R.A. (ed.), *2-nd International Peat Congress, Leningrad, HMSO,* Edinburgh, pp. 543-558.
51. Nardi, S., Pizzeghello, D., Muscolo, A., and Vianello, A. (2002) Physiological effects of humic substances on higher plants, *Soil Biol. Biochem.* **34**, 1527-1536.
52. YuLing, C., Min, C., YunYin, Li, and Xie, Z. (2000) Effect of fulvic acid on ABA, IAA and activities of superoxide dismutase and peridoxase in winter wheat seedling under drought conditions, *Plant Physiol. Communications* **36**, 311-314.
53. Fukushima, M. and Tatsumi, K. (2001) Functionalities of humic acid for the remedial processes of organic pollutants, *Analytical Sci.* **17**, i821-i823.
54. Sanjay, H.-G., Fataftah, A.K., Walia, D., Srivastava, K. (1999) Humasorb CS: A humic acid-based adsorbent to remove organic and inorganic contaminants, in G. Davies, and E.A. Ghabbour (eds.), *Understanding humic substances: advanced methods, properties and applications.* Royal Society of Chemistry, Cambridge, pp. 241-254.
55. Pennington, J.C., Inouye, L.S., McFarland, V.A., Jarvis, A.S., Lutz, C.H., Thorn, K.A., Hayes, C.A., and Porter, B.E. (1999) *Explosives conjugation products in remediation matrices: final report prepared for U.S. Army Corps of Engineers,* Strategic Environmental Research and Development Program. Technical Report SERDP-99-4, 50 pp.
56. U.S. Geological Survey (2002) *Using humic acids to enhance oxidative bioremediation of chlorinated solvents,* <http://toxics.usgs.gov/topics/rem_act/remediation_testing.html>.
57. U.S. Department of Energy (2002) *Factors controlling in situ uranium and technetium bio-reduction and reoxidation at the NABIR Field Research Center,* NABIR-2002 award to Istok, J, <http://www.esd.ornl.gov/nabirfrc/>
58. Sawada, A., Tanaka, S., Fukushima, M., Tatsumi, K. (2003) Electrokinetic remediation of clayey soils containing copper(II)-oxinate using humic acid as a surfactant. *J. Hazard. Mater,* **B96**, 145-154.
59. Sara, M.N. (2003) *Site assessment and remediation handbook,* 2nd Ed. Boca Raton, FL: CRC Press, 1160 pp.

30

60. Nyer, E.K (2000) *In situ treatment technology*, 2nd Ed. Boca Raton, FL: CRC Press, 552 pp.
61. Vidic, R.D. (2001) Permeable reactive barriers: case study review, *Technology evaluation report, GWRTAC*, < http://www.gwrtac.org>.
62. Naftz, D., Morrison, S.J., Fuller, C.C., and Davis, J.A., eds. (2002) *Handbook of groundwater remediation using permeable reactive barriers. Applications to radionuclides, trace metals, and nutrients*, San Diego, Calif., Academic Press, 539 pp.
63. Scherer, M.M., Richter, S., Valentine, R.L., Alvarez, P.J.J. (2000) Chemistry and microbiology of permeable reactive barriers for *in situ* groundwater clean up. *Crit. Rev. Environ. Sci. Technol.* **30**, 363-411.
64. Balcke, G.U., Georgi, A., Woszidlo, S., Kopinke, F.-D., and Poerschmann, J. (2005) Utilization of immobilized humic organic matter for *in situ* subsurface remediation, in I.V. Perminova, N. Hertkorn, K. Hatfield, *Use of humic substances to remediate polluted environments: from theory to practice*, Chapter 10, pp. 203-232 (this volume).
65. Jalvert, C.T. (1996) *Surfactants/cosolvents, Technology Evaluation Report TE-96-02*, Ground-Water Remediation Technologies Analysis Centre, Pittsburg, PA. <http://www.gwrtac.org>.
66. Roote, D.S. (1997) *In situ flushing. Technology status report. TS-98-01*, Ground-Water Remediation Technologies Analysis Centre, Pittsburg, PA, <http://www.gwrtac.org>.
67. Van Stempvoort, D., Lesage, S., Molson, J. (2005) The use of aqueous humic substances for in-situ remediation of contaminated aquifers, in I.V. Perminova, N. Hertkorn, K. Hatfield, *Use of humic substances to remediate polluted environments: from theory to practice*, Chapter 11, pp. 233- 265 (this volume).
68. Lesage, S., Brown, S., Millar, K. and Novakowski, K.S. (2001) Humic acids enhanced removal of aromatic hydrocarbons from contaminated aquifers: Developing a sustainable technology, *J. Environ. Sci. Health,* **A36**(8), 1515-1533.
69. Van Stempvoort, D.R., Lesage, S., Novakowski, K S., Millar, K., Brown, S. and Lawrence, J.R. (2002) Humic acid enhanced remediation of an emplaced diesel source in groundwater: 1. Laboratory-based pilot scale test, *J. Contam. Hydrol.* **54**, 249-276.
70. Alexander, M. (1994) *Biodegradation and Bioremediation*, Academic Press, San Diego.
71. Cookson, J.T. (1995) *Bioremediation Engineering; Design and Application*. McGraw Hill, New York.
72. Van Cauwenberghe, L., and Roote, D.S. (1998) *In situ bioremediation. Technology overview report*, GWRTAC, <http://www.gwrtac.org>.
73. Battelle Memorial Inst. (1994) *Emerging Technology for Bioremediation of Metals*, Boca Raton, FL, CRC Press, 160 pp.
74. Achtnich, C., Fernandes, E., Bollag, J.-M., Knackmuss, H.-J., Lenke, H. (1999) Covalent binding of reduced metabolites of [15N]TNT to soil organic matter during a bioremediation process analyzed by 15N NMR spectroscopy, *Environ. Sci. Technol.* **33**, 4448-4456.
75. Rüttimann-Johnson, C., and Lamar, R.T. (1997) Binding of pentachlorophenol to humic substances in soil by the action of white rot fungi, *Soil Biol. Biochem.* **29**(7), 1143-1148.
76. Berry, D.F., Boyd S.A. (1985) Decontamination of soil through enhanced formation of bound residues, *Environ. Sci. Technol.* **19**, 1132-1133.
77. Gevao, B., Semple, K.T., Jones, K.C. (2000) Bound residues in soils: A review, *Environ. Pollution* **180**, 3-14.
78. Thomas, H., and Gerth, A. (2005) Enhanced humification of TNT, in Perminova, I.V., N. Hertkorn, K. Hatfield, *Use of humic substances to remediate polluted environments: from theory to practice*, Chapter 18, pp. 353-364 (this volume).
79. Miller, R.R. (1996) *Phytoremediation. Technology overview report*, GWRTAC, <http://www.gwrtac.org>.
80. U.S. EPA. (2000) *Introduction to phytoremediation. Report EPA/600/R-99/107*, National Risk Management Research Laboratory, Office of Research and Development, U.S. EPA, Cincinnati, OH.
81. Nanda Kumar, P.B.A., Dushenkov, V., Motto, H., Raskin, I. (1995) Phytoextraction: the use of plants to remove heavy metals from soils, *Environ. Sci. Technol.* **29**, 1232-38.
82. Schnoor, J.L., Licht, L.A., McCutcheon, S.C., Wolfe, N.L., and Carreira, L.H. (1995) Phytoremediation of organic and nutrient contaminants, *Environ. Sci. Technol.* **29**, 318A-323A.
83. Genevini, P.L., Saxxhi, G.A. and Borio, D. (1994) Herbicide effect of atrazine, diuron, linuron and prometon after interaction with humic acids from coal, in N. Senesi and T.M. Miano (eds.), *Humic*

substances in the global environment and implications on human health, Elsevier Science B.V., pp. 1291-1296.

84. Perminova, I.V., Kovalevsky, D.V., Yashchenko, N.Yu., Danchenko, N.N., Kudryavtsev, A.V., Zhilin, D.M., Petrosyan, V.S., Kulikova, N.A., Philippova, O.I., and Lebedeva, G.F. (1996) Humic substances as natural detoxicants, in C.E. Clapp, M.H.B. Hayes, N. Senesi and S.M. Griffith (eds.), *Humic substances and organic matter in soil and water environments: characterization, transformations and interactions,* IHSS Inc., St. Paul, MN, USA, pp. 399-406.

85. Haynes, R.J., and Mokolobate, M.S. (2001) Amelioration of Al toxicity and P deficiency in acid soils by additions of organic residues: a critical review of the phenomenon and the mechanisms involved, *Nutrient Cycling in Agroecosystems* **59**, 47–63.

86. Visser, S.A. (1986) Effects of humic substances on plant growth, in *Humic substances, effects on soil and plants,* REDA, Rome, pp. 89–135.

87. Cooper, R.J., Liu, C., and Fisher, D.C. (1998) Influence of humic substances on rooting and nutrient content of creeping bentgrass, *Crop Sci.* **38**, 1639-1644.

88. Kaschl, A., Chen, Y. (2005) Interactions of humic substances with trace metals and their stimulatory effects on plant growth, in I.V. Perminova, N. Hertkorn, K. Hatfield, *Use of humic substances to remediate polluted environments: from theory to practice,* Chapter 4, pp. 83-114 (this volume).

89. Kulikova, N.A., Stepanova, E.V., Koroleva, O.V. (2005) Mitigating activity of humic substances: direct influence on biota, in I.V. Perminova, N. Hertkorn, K. Hatfield, *Use of humic substances to remediate polluted environments: from theory to practice,* Chapter 14, pp. 285-309 (this volume).

90. Schwarzenbach, R.P., Gschwend, P.M., Imboden, D.M. (1993) *Environmental Organic Chemistry,* John Wiley & Sons, New York.

91. Kopinke, F-D., Georgi, A., Mackenzie, K., Kumke, M. (2002) Sorption and chemical reactions of PAHs with dissolved humic substances and related model polymers, in F.H. Frimmel (ed.), *Refractory organic substances in the environment,* John Wiley, Heidelberg, Germany, pp. 475-515.

92. Chiou, C.T., Porter, P.E., Schmedding, D.W. (1983) Partition equilibria of nonionic organic compounds between soil organic matter and water, *Environ. Sci. Technol.* **17**, 227-231.

93. Chiou, C.T., Malcolm, R.L., Brinton, T.I. and Kile, D.E. (1986) Water solubility enhancement of some organic pollutants and pesticides by dissolved humic and fulvic acids, *Envir. Sci. Technol.* **20**, 502-508.

94. Doll, T.E., Frimmel, F.H., Kumke, M.U., Ohlenbusch, G. (1999) Interaction between natural organic matter (NOM) and polycyclic aromatic compounds (PAC) – comparison of fluorescence quenching and solid phase microextraction (SPME), *Fres. J. Anal. Chem.* **364**, 313-319.

95. Burkhard, L. P., (2000) Estimating dissolved organic carbon partition coefficients for non-ionic organic chemicals, *Environ. Sci. Technol.* **22**, 4663-4668.

96. Landrum, P.F., Nihart, S.R., Eadie, B.J. and Gardner, W.S. (1984) Reverse-phase separation method for determining pollutant binding to Aldrich humic acid and dissolved organic carbon of natural waters, *Environ. Sci. Technol.* **18**, 187-192.

97. Kopinke, F.-D., Poerschmann, J., Georgi, A. (1999). Application of SPME to study sorption phenomena on dissolved humic organic matter, in J. Pawliszyn (ed.), *Applications of SPME, RSC Chromatographic Monographs,* Cambridge, UK, pp 111-128.

98. Jones, K.D. and Tiller, C.L. (1999) Effect of solution chemistry on the extent of binding of phenanthrene by a soil humic acid: A comparison of dissolved and clay bound humic, *Environ. Sci. Technol.* **33**, 580-587.

99. Schlautman, M.A., and Morgan, J.J. (1993) Effects of aqueous chemistry on the binding of polycyclic aromatic hydrocarbons by dissolved humic materials, *Environ. Sci. Technol.* **27**, 961-969.

100. Chin, Y.P., Aiken, G.R., Danielsen, K.M. (1997) Binding of pyrene to aquatic and commercial humic substances: the role of molecular weight and aromaticity, *Environ. Sci. Technol.* **31**, 1630-1635.

101. Perminova, I.V., Grechishcheva, N.Yu., Petrosyan, V.S. (1999) Relationships between structure and binding affinity of humic substances for polycyclic aromatic hydrocarbons: relevance of molecular descriptors, *Environ. Sci. Technol.* **33**, 3781-3787.

102. McCarthy, J.F., Roberson, L.E., and Burris, L.W. (1989) Association of benzo(a)pyrene with dissolved organic matter: prediction of K_{DOM} from structural and chemical properties of the organic matter. *Chemosphere* **19**(12), 1911-1920.

103. Morehead, N.R., Eadie, B.J., Lake, B., Landrum, P.F. and Berner, D. (1986) The sorption of PAH onto dissolved organic matter in lake Michigan waters, *Chemosphere* **15**, 403-412.

32

104. Kulikova, N.A. and Perminova, I.V. (2002) Binding of atrazine to humic substances from soil, peat, and coal related to their structure, *Environ. Sci. Technol.* **36**, 3720-3724.
105. Wang, Z., Gamble, D.S., Landford, C.H. (1991) Interaction of atrazine with Laurentian humic acid, *Anal. Chim. Acta* **244**, 135-143.
106. Celis, R., Cornejo, J., Hermosin, M.C., Koskinen, W.C. (1998) Sorption of atrazine and simazine by model associations of soil colloids, *Soil Sci. Soc. Am. J.* **62**, 165-171.
107. Weber, E.J., Spidle, D.L., Thorn, K.A. (1996) Covalent binding of aniline to humic substances. 1. Kinetic studies, *Environ. Sci. Technol.* **30**, 2755-2763.
108. Chiou, C.T., Kile, D.E., Brinton, T.I., Malcolm, R.L., and Leenheer, J.A. (1987) A comparison of the water solubility enhancements of organic solutes by aquatic humic materials and commercial humic acids, *Environ. Sci. Technol.* **21**, 1231-1234.
109. McCarthy, J.F., Jimenez, B.D., and Barbee, Th. (1985) Effect of dissolved humic material on accumulation of polycyclic aromatic hydrocarbons: structure-activity relationships, *Aquat. Toxicol.* **7**, 15–24.
110. Steinberg, C.E.W., Haitzer, M.; Brueggemann, R., Perminova, I.V., and Yashchenko, N.Yu. (2000) Towards a quantitative structure activity relationship (QSAR) of dissolved humic substances as detoxifying agents in freshwaters, *Int. Rev. Hydrobiol.* **85**(2–3), 253–266.
111. Leversee, G.J., Landrum, P.F., Giesy, J.P, and Fannin, T. (1983) Humic acids reduce bioaccumulation of some polycyclic aromatic hydrocarbons, *Can. J. Fish. Aquat. Sci.* **40**, 63-69.
112. Kukkonen, J. (1991) Effect of pH and natural humic substances on the accumulation of organic pollutants in two freshwater invertebrates, in B. Allard (ed.), *Humic Substances in the Aquatic and Terrestrial Environment*, pp. 413–422.
113. Landrum, P.F., Reinhold, M.D., Nihart, S.R., and Eadie, B.J. (1985) Predicting the bioavailability of organic xenobiotics to *Pontoporeia hoyi* in the presence of humic and fulvic materials and natural dissolved organic matter, *Environ. Toxicol. Chem.* **4**, 459–467.
114. Gensemer, R.W., Dixon, D.G., and Greenberg, B.M. (1998) Amelioration of the photo-induced toxicity of polycyclic aromatic hydrocarbons by a commercial humic acid, *Ecotoxicol. Environ. Safety* **39**, 57-64.
115. Day, K.E. (1991) Effects of dissolved organic carbon on accumulation and acute toxicity of fenvalerate, deltamethrin and cyhalothrin to *Daphnia magna* (Straus), *Environ. Toxicol. Chem.* **10**, 91–101.
116. Kukkonen, J., and Oikari, A. (1987) Effects of aquatic humus on accumulation and toxicity of some organic micropollutants, *Sci. Total Environ.* **62**, 399–402.
117. Perminova, I.V., Grechishcheva, N.Yu., Kovalevskii, D.V., Kudryavtsev, A.V., Petrosyan, V.S., and Matorin, D.N. (2001) Quantification and prediction of detoxifying properties of humic substances to polycyclic aromatic hydrocarbons related to chemical binding, *Environ. Sci. Technol.* **35**, 3841-3848.
118. Stewart, A.J. (1984) Interactions between dissolved humic materials and organic toxicants, in K.E. Cowser (ed.), *Synthetic Fossil Fuel Technologies*, Boston, Butterworth Publisher, pp. 505-521.
119. Oikari, A., Kukkonen, J., and Virtanen, V. (1992) Acute toxicity of chemicals to *Daphnia magna* in humic waters, *Sci. Total Environ.* **117/118**, 367–377.
120. Mézin, L.C., Hale, R.C. (2003) Effect of humic acids on toxicity of DDT and chlorpyrifos to freshwater and estuarine invertebrates, *Environ. Toxicol. Chem.* **23**(3), 583–590.
121. Buffle, J. (1988) *Complexation reactions in aquatic systems*, Ellis Horwood Ltd.
122. Perminova, I.V., Kulikova, N.A. Zhilin, D.M., Gretschishcheva, N.Yu., Kholodov, V.A. Lebedeva, G.F., Matorin, D.N., P.S. Venediktov, V.S. Petrosyan. (2004) Mediating effect of humic substances in aquatic and soil environments, in *Environmentally-Acceptable Reclamation and Pollution Endpoints: Scientific Issues and Policy Development*, NATO Science Series, Kluwer Academic Publisher, Dordrecht, (in press).
123. Buchwalter, D.B., Linder, G., and Curtis, L.R. (1995) Modulation of cupric ion activity by pH and fulvic acid as determinants of toxicity in *Xenopus laevis* embryos and larvae, *Environ. Toxicol. Chem.* **15**(4), 568–573.
124. Lorenzo, J.I., Nieto, O., Beiras, R. (2002) Effect of humic acids on speciation and toxicity of copper to *Paracentrotus liidus* larvae in seawater, *Aquatic Toxicol.* **58**, 27–41.
125. Ma, H., Kim, S.D., Cha, D.K., and Allen, H.E. (1998) Effect of kinetics of complexation by humic acid on toxicity of copper to *Ceriodaphnia dubia*, *Environ. Toxicol. Chem.* **18**(5), 828–837.

126. Mandal, R., Hassan, N.M., Murimboh, J., Chakrabarti, C.L., and Back, M. (2002) Chemical speciation and toxicity of nickel species in natural waters from the Sudbury area (Canada), *Environ. Sci. Technol.* **36**, 1477-1484.

127. Voets, J., Bervoets, L., and Blust, R. (2004) Cadmium bioavailability and accumulation in the presence of humic acid to *Zebra mussel, Dreissena polymorpha, Environ. Sci. Technol.* **38**, 1003-1008.

128. Weng, L.P., Wolthoorn, A., Lexmon, T., Temminghoff, E.J.M., and Van Riemsdijk, W.H. (2004) Understanding the effects of soil characteristics on phytotoxicity and bioavailability of nickel using speciation models, *Environ. Sci. Technol.* **38**, 156-162.

129. Weber, W.J. Jr., Huang, W., Le Boeuf, E.J. (1999) Geosorbent organic matter and its relationship to the binding and sequestration of organic contaminants, *Colloids & Surfaces A.* **151**, 167-179.

130. Moulin, C., Moulin, V. (2004) Fate of actinides in the presence of humic substances under conditions relevant to nuclear waste disposal, *Appl. Geochem.* **10**, 573-580.

131. Schuessler, W., Artinger, R., Kienzler, B., Kim, J.I. (2000) Conceptual modeling of the humic colloid-borne americium(III) migration by a kinetic approach. *Environ. Sci. Technol.* **34**, 2608-2611.

132. Terashima, M., Tanaka, S., Fukushima, M. (2003) Distribution behavior of pyrene to adsorbed humic acids on kaolin, *J. Environ. Qual.* **32**, 591-598.

133. Hura, J., Schlautman, M.A. (2004) Effects of mineral surfaces on pyrene partitioning to well-characterized humic substances, *J. Environ. Qual.* **33**, 1733-1742.

134. Laor, Y., Farmer, W.J., Aochi, Y., Strom, P.F., (1998) Phenanthrene binding and sorption to dissolved and mineral-associated humic acid, *Water Res.* **32**, 1923-1931.

135. Schwarzenbach, R.P., Gschwend, P.M., Imboden, D.M. (1993) *Environmental organic chemistry*, John Wiley & Sons, New York, 680 pp.

136. Karickhoff, S.W., Brown, D.S., Scott, T.A. (1979) Sorption of hydrophobic pollutants on natural sediments, *Wat. Res.* **13**, 241-248.

137. Chiou, G., Kile, D., Rutherford, D., Sheng, G., Boyd, S. (2000) Sorption of selected organic compounds from water to a peat soil and its humic-acid and humin fractions: potential sources of the sorption nonlinearity, *Environ. Sci. Technol.* **34**, 1254-1258.

138. Celis, R., Cornejo, J., Hermosin, M.C. and Koskinen, W.C. (1998) Sorption of atrazine and simazine by model associations of soil colloids, *Soil Sci. Soc. Am. J.*, **62**, 165-171.

139. Weber, W.J.Jr., Huang, W., Yu, H. (1998) Hysteresis in the sorption and desorption of hydrophobic organic contaminants by soils and sediments. 2. Effects of soil organic matter heterogeneity, *J. Contam Hydrol.* **31**, 149-165.

140. Spurlock, F.C., Biggar, J.W. (1994) Thermodynamics of organic chemical partitioning in soils. 2. Nonlinear partition of substituted phenylureas from aqueous solution, *Environ. Sci. Technol.* **28**, 996-1002.

141. Von Oepen, B., Kordel, W., Klein, W., Schuurmann, G. (1991) Predictive QSPR models for estimating soil sorption coefficients: potential and limitations based on dominating processes, *Sci.Total Environ.* **109/110**, 343-354.

142. Huang, W., Weber, W.J. Jr. (1997) A distributed reactivity model for sorption by soil and sediments. 10. Relationships between desorption, hysteresis, and the chemical characteristics of organic domains, *Environ. Sci. Technol.* **31**, 2562-69.

143. Anmad, R., Kookana, R., Alston, A., Skjemstad, J. (2001) The nature of soil organic matter affects sorption of pesticides. 1. Relationships with carbon chemistry as determined by 13C CPMAS NMR spectroscopy, *Environ. Sci. Technol.* **35**, 878-884.

144. LeBoeuf, E.J., Weber, W.J. Jr. (1997) A distributed reactivity model for sorption by soils and sediments. 8. Sorbent organic domains: discovery of a humic acid glass transition and an argument for a polymer-based model, *Environ. Sci. Technol.* **31**, 1697-1702.

145. Xing, B., Pignatello, J.J. (1997) Dual-mode sorption of low polarity compounds in glassy poly(vinylchloride) and soil organic matter, *Environ. Sci. Technol.* **31**, 792-799.

146. Leboeuf, E., Weber, W. Jr. (2000) Macromolecular characteristics of natural organic matter. 2. Sorption and desorption behavior, *Environ. Sci. Technol.* **34**, 3632-3640.

147. Gunesakara, A.S., Xing, B. (2003) Sorption and desorption of naphthalene by soil organic matter: importance of aromatic and aliphatic components, *J. Environ. Qual.* **32**, 240-246.

148. Simpson, A., Chefetz, B., Hatcher, P. (2003) Phenanthrene sorption to structurally modified humic acids, *J. Environ. Qual.* **32**, 1750-1758.

34

149. Khalaf, M., Kohl, S.D., Klumpp, E., Rice, J., Tombacz, E. (2003) Comparison of sorption domains in molecular weight fractions of a soil humic acid using solid-state 19F NMR, *Environ. Sci. Technol.* **37**, 2855-60.

150. Chefetz, B., Deshmukh, A.P., Hatcher, P.G., Guthrie, E.A. (2000) Pyrene sorption by natural organic matter, *Environ. Sci. Technol.* **34**, 2925-2930.

151. Mao, J.D., Hundal, L., Thompson, M., Schmidt-Rohr, K. (2002) Correlation of poly(methylene)-rich amorphous aliphatic domains in humic substances with sorption of a nonpolar organic contaminant phenanthrene, *Environ. Sci. Technol.* **36**, 929-936.

152. Salloum, M.J., Chefetz, B., Hatcher, P. (2002) Phenanthrene sorption by aliphatic-rich natural organic matter, *Environ. Sci. Technol.* **36**, 1953-1958.

153. Balcke, G.U., Kulikova, N.A., Kopinke, F.-D., Perminova, I.V., Hesse, S., Frimmel, F. H. (2002) Adsorption of humic substances onto kaolin clay related to their structural features, *Soil Sci. Soc. Am. J.* **66**, 1805-1812.

154. Visser S.A. (1964) Oxidation-reduction potentials and capillary activities of humic acids, *Nature*, **204**, 581.

155. Helburn, R.S., MacCarthy, P. (1994) Determination of some redox properties of humic acid byalkaline ferricyanide titration, *Anal. Chim. Acta* **295**, 263–272.

156. Skogerboe, R.K., Wilson, S.A. (1981) Reduction of ionic species by fulvic acid, *Anal. Chem.* **53**,228–232.

157. Oesterberg, R., and Shirshova, L. (1997) Non-equilibrium oscillating redox properties of humic acids, *Geochim. Cosmochim. Acta* **61**, 4599-4604.

158. Struyk, Z., Sposito, G. (2001) Redox properties of standard humic acids, *Geoderma* **102**, 329–346.

159. Scott, D.T., McKnight, D.M., Blunt-Harris, E.L., Kolesar, S.E., and Lovley, D.R. (1998) Quinone moieties act as electron acceptors in the reduction of humic substances by humics-reducing microorganisms, *Environ. Sci. Technol.* **32**, 2984-2989.

160. Nurmi, J.T. and Tratnyek, P.G. (2002) Electrochemical properties of natural organic matter (NOM), fractions of NOM, and model biogeochemical electron shuttles, *Environ. Sci. Technol.* **36**, 617-624.

161. Fukusima, M., Nakayasu, K., Tanaka, Sh., Nakamara, H. (1997) Speciation analysis of chromium after reduction of chromium (VI) by humic acid, *Toxicol. Environ. Chem.* **62**, 207-215.

162. Wittbrodt, P.R. and Palmer, C.D. (1996a) Reduction of Cr(VI) by soil humic acids, *Eur. J. Soil Sci.* **47**, 151-162.

163. Wittbrodt, P.R. and Palmer, C.D. (1996b) Effect of temperature, ionic strength, background electrolytes and Fe(III) on the reduction of hexavalent chromium by soil humic substances, *Environ. Sci. Technol.* **30**, 2470-2477.

164. Zhilin, D.M., Schmitt-Kopplin, P., Perminova, I.V. (2004) Reduction of Cr(VI) by peat and coal humic substances: implication for remediation of contaminated sites, *Env. Chem. Let.* [Published on-line September 7, 2004].

165. Bondietti, E.A., Reynolds, S.A., Shanks, M.N. (1976) *Transuranic nuclides in the environment*, IAEA, Vienna.

166. Andre, C., Choppin, G.R. (2000) Reduction of Pu(V) by humic acid, *Radiochim. Acta* **88**, 613-616.

167. Rao, L., and Choppin, G.R. (1995) Thermodynamic study of the complexation of neptunium(V) with humic acids, *Radiochim. Acta* **69**, 87-95.

168. O'Loughlin, E., Ma, H., Burris, D. (2002) Catalytic effects of Ni-humic complexes on the reductive dehalogenation of chlorinated alanes and alkenes, *Proceedings of the 11th Int. Meeting of IHSS "Humic substances: nature's most versatile materials"*, July 21-26, 2002, Northeastern University, Boston, MS, USA, pp. 415-417.

169. Field, J.A., and Cervantes, F.J. (2005) Microbial redox reactions mediated by humus and structurally related quinones, in I.V. Perminova, N. Hertkorn, K. Hatfield, *Use of humic substances to remediate polluted environments: from theory to practice*, Chapter 17, pp. 343-364 (this volume).

170. Cervantes, F.J., Dijksma, W., Duong-Dac, T., Ivanova, A., Lettinga, G. and Field, J.A. (2001) Anaerobic mineralization of toluene by enriched sediments with quinones and humus as terminal electron acceptors, *Appl. Environ. Microbiol.* **67**, 4471-4478.

171. Finneran, K.T., and Lovley, D.R. (2001) Anaerobic degradation of methyl tert-butyl ether (MTBE) and tert-butyl alcohol (TBA), *Environ. Sci. Technol.* **35**, 1785-1790.

172. Curtis, G.P., and Reinhard, M. (1994) Reductive dehalogenation of hexachlorethane, carbon-tetrachloride, and bromoform by anthrahydroquinonedisulfonate and humic acid, *Environ. Sci. Technol.* **28**, 2393-2401.

173. Keck, A., Klein, J., Kudlich, M., Stolz, A., Knackmuss, H.J. and Mattes, R. (1997) Reduction of azo dyes by redox mediators originating in the naphthalenesulfonic acid degradation pathway of *Sphingomonas* sp. strain BN6, *Appl. Environ. Microbiol.* **63**, 3684-3690.

174. Fu, Q.S., Barkovskii, A.L. and Adriaens, P. (1999) Reductive transformation of dioxins: An assessment of the contribution of dissolved organic matter to dechlorination reactions, *Environ. Sci. Technol.* **33**, 3837-3842.

175. Fredrickson, J.K., Kostandarithes, H.M., Li, S.W., Plymale, A.E. and Daly, M.J. (2000) Reduction of Fe(III), Cr(VI), U(VI), and Tc(VII) by *Deinococcus radiodurans* R1, *Appl. Environ. Microbiol.* **66**, 2006-2011.

176. Finneran, K.T., Anderson, R.T., Nevin, K.P. and Lovley, D.R. (2002) Potential for bioremediation of uranium-contaminated aquifers with microbial U(VI) reduction, *Soil Sedim. Contam.* **11**, 339-357.

177. Lloyd, J.R., and Macaskie, E. (2000) Bioremediation of radionuclide-containing wastewaters, in D.R. Lovley (ed.), *Environmental Microbe-Metal Interactions*, ASM press, Washington, DC. p. 277-327.

178. Perminova, I.V., Kovalenko, A.N., Kholodov, V.A., Youdov, M.V., Zhilin, D.M. (2004) Design of humic materials of a desired remedial action, in L. Martin-Neto, D. Milori, W. Silva (eds), *Humic substances and soil and water environment*, Proceedings of the 12th International Meeting of IHSS, Sao Pedro, Sao Paulo, Brazil, July 25-30, 2004, Sao Pedro, Sao Paulo, Embrapa Instrumentacao Agropecuaria, pp. 506-508.

179. Bollag, J.-M. (1999) Effect of humic constituents on the transformation of chlorinated phenols and anilines in the presence of oxidoreductive enzymes or birnessite, *Environ. Sci. Technol.* **33**, 2028-2034.

180. Bollag J.-M., and Mayers C. (1992) Detoxification of aquatic and terrestrial sites through binding of pollutants to humic substances, *Sci. Total Environ.* **117/118**, 357-366.

181. Chen, Y., and Avaid, T. (1990) Effect of humic substances on plant growth, in P. MacCarthy, C.E. Clapp, R.L. Malcom, and P.R. Bloom (eds.), *Humic substances in soils and crop science: selected readings*, Soil Sci. Soc. Am., Madison, pp.161-186.

182. Mackowiak, C.L., Grossl, P.R. and Bugbee, B.G. (2001) Beneficial effects of humic acid on micronutrient availability to wheat, *Soil Sci. Soc. Am. J.* **65**, 1744-1750.

183. Visser, S.A. (1986) Effects of humic substances on plant growth, in *Humic substances, Effects on Soil and Plants*, REDA, Rome, pp. 89–135.

184. Dehorter, B. and Blondeau, R. (1992) Extracellular enzyme activities during humic acid degradation by the white rot fungi *Phanerochaete chrysosporium* and *Trametes versicolor*, *FEMS Microbiol. Let.* **94**, 209–216.

185. Gramss, G., Ziegenhagen, D., and Sorge, S. (1999) Degradation of soil humic extract by wood- and soil-associated fungi, bacteria, and commercial enzymes, *Microbiol. Ecol.* **37**, 140–151.

186. Kirschner, R.A. Jr., Parker, B.C., and Falkinham, J.O. III. (1999) Humic and fulvic acids stimulate the growth of *Mycobacterium avium*, *FEMS Microbiol. Ecol.* **30**, 327-332.

187. Mazhul, V.M., Prokopova, Zh.V., and Ivashkevich, L.S. (1993) Mechanism of peat humic acids action on membrane structural status and functional activity of the yeast cells, in *Humic Substances in Biosphere*, Moscow, Nauka, pp. 151–157 (in Russian).

188. Vigneault, B., Percot, A., Lafleur, M., and Campbell, P.G.C. (2000) Permeability changes in model and phytoplankton membranes in the presence of aquatic humic substances, *Environ. Science Technol.* **34**, 3907-3913.

189. Clapp, C.E., Chen, Y., Hayes, M.H.B., Cheng, H.H. (2001) Plant growth promoting activity of humic substances, in R.S. Swift, K.M. Sparks (eds.), *Understanding and Managing Organic Matter in Soils, Sediments, and Waters*, International Humic Science Society, Madison, pp. 243–255.

190. Dell'Agnola, G., Ferrari, G., Nardi, S. (1981) Antidote action of humic substances on atrazine inhibition of sulphate uptake in barley roots, *Pestic. Biochem. Physiol.* **15**, 101–104.

191. Varanini, Z., and Pinton, R., (2001) Direct versus indirect effects of soil humic substances on plant growth and nutrition, in R. Pinton, Z. Varanini, P. Nannipieri (eds.), *The Rizosphere*, Marcel Dekker, Basel, pp. 141–158.

192. Vaughan, D., Malcom, R.E., (1985) Influence of humic substances on growth and physiological processes, in D. Vaughan, R.E. Malcom (eds.), *Soil Organic Matter and Biological Activity*, Martinus Nijhoff/ Junk W, Dordrecht, The Netherlands, pp. 37–76.

193. Nardi, S., Arnoldi, G., Dell'Agnola, G. (1988) Release of the hormone-like activities from *Allolobophora rosea* and *A. caliginosa* faeces, *Can. J. Soil Sci.* **68**, 563–567.
194. Nardi, S., Pizzeghello, D., Reniero, F., Rascio, N. (2000) Chemical and biochemical properties of humic substances isolated from forest soils and plant growth, *Soil Sci. Soc. Am. J.* **64**, 639–645.
195. Vaughan, D., Ord, B.G., (1981) Uptake and incorporation of 14 C-labelled soil organic matter by roots of *Pisum sativum* L., *J. Exp. Bot.* **32**, 679–687.
196. Prat, S. (1963) Permeability of plant tissues to humic acids, *Biologia Plantarum* **5**, 279–283.
197. Kovalenko, A., Youdov, M., Perminova I., Petrosyan, V. (2004). Synthesis and characterization of humic derivatives enriched with hydroquinoic and catecholic moieties. In: *Humic substances and soil and water environment. Proceedings of the 12th International Meeting of IHSS*. Sao Pedro, Sao Paulo, Brazil, July 25-30, 2004. Martin-Neto, L., Milori, D., Silva, W. (Eds). Sao Pedro, Sao Paulo. Embrapa Instrumentacao Agropecuaria, pp. 472-474.
198. Kholodov, V.A., Kovalenko, A.N., Kulikova, N.A., Lebedeva, G.F., and Perminova, I.V. (2004) Enhanced detoxifying ability of hydroquinones-enriched humic derivatives with respect to copper, in L. Martin-Neto, D. Milori, D., W. Silva (eds), *Humic substances and soil and water environment*, Proceedings of the 12th International Meeting of IHSS, Sao Pedro, Sao Paulo, Brazil, July 25-30, 2004, Sao Pedro, Sao Paulo, Embrapa Instrumentacao Agropecuaria, pp. 189-191.
199. Hatfield, K., Annable, M., Cho, J., Rao, P.S.C., Klammler, H. (2004) A direct method for measuring water and contaminant fluxes in porous media, *J. Contam. Hydrol.* (in press).
200. Hatfield, K., Annable, M. (2003) New approach to quantify remediation, in *Biotechnology: state of the art and prospects of development*, Proceedings of the 2-nd Moscow Int. Congress, Nov. 10-14, 2003, Moscow, Russia, P&I JSC "Maxima" Part II, p. 13.

Chapter 2

SOIL ORGANIC MATTER AND PROTECTIVE FUNCTIONS OF HUMIC SUBSTANCES IN THE BIOSHERE

A Critical Review and Prospects

D.S. ORLOV, L.K. SADOVNIKOVA
*Department of Soil Science, Lomonosov Moscow State University,
Moscow 119992, Russia <sadov@soil.msu.ru>*

Abstract

Soil organic matter (SOM) is the most important source of nutrients and regulates the main physical, chemical, and biological properties of different soils. A fundamental set of issues related to the humus formation conditions are the zone-genetic approach, the biogeochemical nature of the humification process, the effects of the thermodynamics and kinetics directing the synthesis and degradation of humic substances (HS), the variability of the properties of SOM and humic substances. In this respect, important features are the physico-chemical diagnostics and identification of humic acids, the special role of mineral components in the humification process, the fundamental questions regarding the authenticity humic acids, and the knowledge derived from the study of altered and denatured products (e.g. industrially manufactured humic-like substances. Five important functions of humic substances provide the opportunity to evaluate the ecological role of humus: 1) accumulative, 2) transport, 3) regulatory, 4) physiological, and 5) protective. The latter involves the ability of humic substances to bind different pollutants: heavy metals, pesticides, radionuclides, and the surplus of fertilizers. Various detoxification mechanisms operate in the soil-plant system in highly diverse natural environments. The relative stability of soil as a whole mainly depends on the stability of humic substances and its fractions, of which humic acids and humin seem to be the most significant.

1. Introduction

In order to fulfil life-supporting functions against variable external conditions, the biosphere and the hydrosphere have to maintain a certain stability over a broad range of space and timescales. Undoubtedly, during evolution on geochemical time spans, dynamic equilibrium has been established via processes of natural selection under the constraints of thermodynamics and kinetics. As an epigraph to his influential book "*The Humus*", Waksman [1] quoted A. Thaer's words "Humus is the product of live matter as

37

I. V. Perminova et al. (eds.),
Use of Humic Substances to Remediate Polluted Environments: From Theory to Practice, 37–52.
© 2005 *Springer. Printed in the Netherlands.*

well as its source". The role of humus in the biosphere is so multifarious that it attracts attention not only of soil scientists, but also of geologists, geochemists, biologists, chemists, and engineers.

2. Soil Organic Matter (SOM): Sources, Reserves, Properties and Functions

2.1. SOURCES AND RESERVES OF SOIL ORGANIC MATTER

Carbon reserves stored in humus substantially exceed those in the biomass of living organisms. According to Kononova [2], the humus reserves in the top layer of soil extending to one hundred 100 cm depth account for 282-426 t/ha. Because 100 cm depth is not the lowest boundary of humus accumulation, the actual humus reserves are expected to be even larger. For the biosphere as a whole, the carbon content of the humic acids alone has been estimated by some authors to be $6x10^{11}$ t, whereas the carbon content in living organisms accounts for $7x10^{11}$ t [3].

A large fraction of plant and animal residues decomposes fairly quickly and becomes eventually mineralised. The chemical composition of plant and animal residues show remarkable similarity in bulk features, but displays considerable variation in the content of the components (Table 1). Carbohydrates (e.g. cellulose, hemicellulose, pectin, etc.), lignin and proteins are predominant constituents of SOM. The other ones are present in minor quantities. When living organisms die, all the complex biomolecules and polymers reach the soil surface or enter the soil. Subsequently, they are decomposed or transformed into humic substances whose composition reflect the respective ecosystem. A part of the soluble SOM fraction is transported by surface or subsurface flow.

Table 1. Chemical composition of organic residues, % of dry, ash-free material.

Organisms	Ash, %	Proteins	Hemicellulose, pectin	Cellu-lose	Lignin	Lipids, tannins
Bacteria	2-10	40-70	negligible	0	0	1-40
Algae	20-30	10-15	50-60	5-10	0	1-3
Lichens	2-6	3-5	60-80	5-10	8-10	1-3
Moss	3-10	5-10	30-60	15-25	0	5-10
Ferns	6-7	4-5	20-30	20-30	20-30	2-10
Conifers:						
Stem	0.1-1	0.5-1	20-30	40-50	20-25	5-15
Needles	2-5	4-10	10-20	15-25	20-30	5-15
Deciduous:						
Wood	0.1-1	0.5-1	20-30	40-50	20-25	5-15
Leaves	3-8	4-10	10-20	15-25	20-30	5-15
Perennial grasses:						
Cereals	5-10	5-12	25-35	25-40	15-20	2-10
Beans	5-10	10-20	15-25	25-30	15-20	2-10

Despite the long history of research, not all functions of humus in the biosphere are yet clearly understood. For instance, it is unclear how humification relates to the ecological conditions necessary for the survival of the species and associations of organisms characteristic of terrestrial ecosystems that were formed on land in the early and recent earth history. The intriguing aspect of current interest, regarding enhanced UV radiation and temperature on the earth's surface is, whether and how will the present life forms adapt to the conditions of accelerated disintegration of remains into the ultimate products, i.e., CO_2, H_2O, and mineral salts?

Soil humus is one of the essential links in a continuous chain of trophic relations between different life forms, and plays simultaneously two roles in a self-regulating process. The equilibrium of humus disintegration and accumulation defines the conditions of local, regional and global ecosystems. One of the essentials of a dynamic equilibrium resides in the fact that input and output are balanced; in the humic synthesis/degradation this balance is always maintained when considering sufficiently large space and time scales. The thermodynamically stable end products are CO_2, H_2O and mineral salts, as stated above; humic substances are transient intermediates in this process, whose fate is governed by the combined restraints of thermodynamics (driving force) and kinetics (time dependence).

2.2. FACTORS AND MECHANISMS GOVERNING HUMUS FORMATION

2.2.1. Basic Stipulations for Understanding Conditions Of Humus Formation
It is well known that SOM regulates the main physical, chemical, and biological properties of different soils, and is the most important source of nutrients. There are some principal stipulations for understanding, *at first stage*, the conditions of humus formation and, *at last stage*, for adequate evaluation of the experimental results.

These are:

A. **The zone-genetic approach**

This methodological approach is a prerequisite both for providing a practical use of the data obtained, and for understanding the particular features of humus chemistry in the context of the conditions of its formation.

B. **Biochemical nature of the humification process**

The predominantly biochemical nature of humification is widely accepted; however, data are also available on the purely abiotic stages of this process associated with, for instance, the catalytic action of the mineral soil components. We assume that the principal indices to evaluate the trend and the rate of humification relate to the biochemical (biological) soil activity, which is a function of both microorganisms and higher living organisms population [4, 5]. The relative levels of soil biochemical activity can be estimated from various indirect features such as the population structure and biomass of microorganisms, soil respiration, and enzymatic activity. These indices have distinct zonal nature and are closely related to the content, composition, and properties of humus.

C. **Thermodynamic stability**

The refractory nature of humics has been emphasized by many authors. It has been postulated that in the complex soil environment, the compounds accumulate according to local thermodynamic and kinetic control. Alexander [6] has discussed various aspects of the stability of organic substances. Thermodynamically and kinetically controlled processes lead to formation of humic acids, representing the most stable forms of organic substances under the conditions of a dynamic equilibrium [7]. From this point of view, humus formation should be regarded mostly as a process of synthesis involving a unique "natural selection": unstable and high-yield substrates quickly decompose, then they are assimilated by short-living organisms, and the continuous sequence of transformation is delayed at the stage represented by the most stable compounds. According to Kleinhempel [8], the recalcitrance of humic acids may be associated rather with steric hindrance of the enzymatic activity than with the high stability of chemical bonds. These considerations led him to the idea of structural entropy.

D. **Statistical variations within SOM properties**

The description of SOM and humus, in particular, could be performed only on a statistical basis. The variability of the properties of humics is mainly caused by two factors: 1) random nature of humus formation that involves a variety of reactions and precursors and is governed exclusively by the recalcitrance of the formed products. This is a main difference between the synthesis of humics and the (enzymatic) synthesis of organic compounds in living organisms; 2) high spatial and temporal variability of the soil properties.

The content and composition of humus in soil as well as the properties of individual soil-humic-complexes vary greatly. The humus content ranges from 5-10 to 40-50%; the elemental and functional group composition of humic acids ranges: C: from 3 to 30, N: from 8 to 18, the methoxy group: from 11 to 21 percent. For example, the analysis of 40 soil samples showed a direct correlation between total organic carbon and humic acid content ($r = 0.88$, $P = 0.99$) [3]. Similar variability of the humus indices was observed across the whole bandwidth of soil types (Table 2).

Table 2. Variability of the various humus indices.

Index	Mean	Variance, %
C_{total}, %	1.8	57.7
C_{HA}, % of C_{total}	24.8	10.6
C_{FA}, % of C_{total}	32.3	27.6
$C_{HA}:C_{FA}$	0.8	45.6

The relative content of humic acids appeared to be most conservative; at the same time, the $C_{HA}:C_{FA}$ ratio should be used with caution due to much higher variability. The distribution of the values of many soil indices (e.g., humus content) was found to be close to normal.

E. **Physico-chemical diagnostics and identification of humics**

Substantial progress in the chemistry of soil humics depends closely on the development of the appropriate analytical techniques. Standardized extraction and purification protocols, like those proposed by the International Humic Substances Society (IHSS) are a great step forward to facilitate comparison of research performed by different research groups and disciplines. A minimum requirement is the proper and detailed description of the experimental protocol, when humics have been isolated from natural source materials. As an example, it is not correct to include different brown products in the category of "soil humus" only because they are found in soils or cultural media, or they are soluble in alkaline media and can be precipitated by acids.

F. **The special role of mineral components in the humification process**

The special role of the soil mineral components in humification is their **catalytic effect** – the initiation of oxidative coupling whose importance was discovered by the Russian scientist Troitskii in the 50-ies of the last century (cit. in [3]). A distinct catalytic effect on the oxidation of various phenols has been shown during the action of Mn, Fe and Si oxides, clays, and natural allophanes. It has been found that addition of montmorillonite (bentonite) increases the population and the biomass of microorganisms, their activity, the amount of humics formed, the capacity to form complex structures, and the incorporation of nitrogen into humus-like products. The presence of montmorillonite also influenced the dynamics of glucose decomposition [9]. In soil, the amount of minerals typically exceeds that of humics, and humification in soil always occurs in the presence of highly dispersed minerals, providing very large surface and, as a result, enhanced reactivity.

G. **The existence of humics as a class of chemical compounds and the possibility to study denatured humic materials**

The main question is whether humics are present in natural soils or whether they are formed in the course of soil treatment with alkaline solutions. The existence of humics as a distinct class of materials has been thoroughly established, but nevertheless, the above question is being discussed again and again for a variety of reasons. In any terrestrial ecosystem, humic substances are intimately bound to a mineral matrix, which typically exceeds the weight of organic material by one to several orders of magnitude. Breaking these bonds necessarily implies the use of rather harsh conditions. The options for obtaining humic substances include a use of rather soft extraction methods, which yield authentic, but not representative materials; another option is to use much harsher extraction methods, which provide higher yield, but are susceptible to alteration of original chemical structures of humics.

Without questioning the existence of soil humics, it is still necessary to emphasize that every method of their extraction inevitably modifies them to some extent. Therefore, it is only possible to study somewhat denaturated humic materials. We consider using "soft" methods to extract humics as undesirable protocol, especially when humic materials will be contaminated with different plant biomolecules, e.g., peptides, carbohydrates, etc.

2.2.2. Biogeochemical Rules of Humus Formation

Soil humus is a very well organized and continuously functioning system under biospheric conditions that ensures a supply of nutrients, nearly optimum acid-base and oxidation-reduction conditions with a high buffer capacity, and improves hydrophysical properties of soil for living organisms. The existence of such a system in the Earth's crust is the result of coevolution between the biotic habitat and the abiotic environment.

The ecological and evolutionary role of the humic substances is their capacity to buffer against any extreme conditions, thereby ensuring predominance of rather stable environmental conditions for the coevolution of life and the humic materials supporting it. This implies that on small size scale, any soil compartment is represented by specific humic substances; however, a quasi-continuum of substrates will be available for enzymatic activity or as nutrients. This fact has imparted stability and continuous functioning of this system. The well-balanced interaction of soil biota and humus could have been established only as the system evolved as a whole.

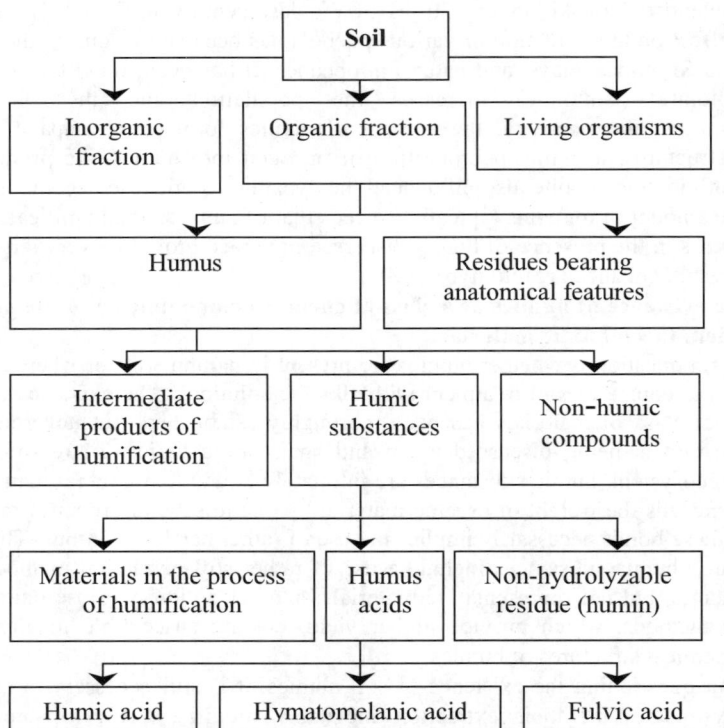

Figure 1. Scheme for subdivision of humic substances of soil (after [11]).

The generalized scheme of SOM and humic substances is shown in Figure 1. It is intimately connected with the methods of fractionation and with the classification of humic substances and the concepts used in soil organic matter studies. According to this scheme, the organic fraction of soil covers all organic substances present in the soil profile in free or organo-mineral complexed form, excluding biopolymeres and metabolites that are present in living organisms. Based on genesis, character, and functions, all non living organic materials are divided into two major groups: organic residues and humus. Organic residues include those dead tissues of living organisms that have not yet lost their anatomical structure; these are mainly the root residues in horizons H to C. These components are subjected to humification and, depending on local conditions, specific humic substances are formed in the soil compartments.

In order to assess humification and humic substances, the organic fraction of the soil should be examined separately from the inorganic compartment and living organisms. However, a major fraction of the soil humic substances is intimately associated with various inorganic materials, like metal cations, oxides, hydroxides, and silicates, thus forming various compounds such as simple and complex salts, or adsorbed complexes. Humic substances together with non-specific compounds, present in free form or as organo-mineral complexes, constitute soil humus, making it a combination of all the organic compounds, excluding those of living organisms.

Based on fundamental biogeochemical concepts we can formulate basic laws and "rules" governing the processes of humification, accumulation and composition of organic matter in soil:

- the biomineral nature of humus as a distinct natural material,
- the biothermodynamic direction of organic matter transformation,
- the leading role of the dominant factor (condition) during individual stages of humus formation.

The "rules" are:

(1) The humus formation in a given biogeodynamic environment occurs by selection of the most viable pathways under the fundamental constraints determined by kinetics and thermodynamics. The uniqueness of any humic substance is, for instance, reflected in the pronounced polydispersity and heterogeneity of this class of compounds. This rule also stipulates that the decomposition of organic residues of different structure (lignin, polysaccharides, proteins, pigments, etc.) always proceeds unidirectional, independent of the nature of the humic precursor material.

(2) The depth of transformation of organic residues to humic substances (humification degree) depends on the level of biological activity and is well described by an equation of humification kinetic theory. In a general form, the humification degree could be expressed as

$$H = f(Q, I, t) \tag{1}$$

where Q is the total volume of living organisms residues entered the soil and subjected to humification, I is the intensity of their transformation dependant on the rates of individual stages of the process (it is proportional to the soil biochemical activity), and t is the time of interaction of soil with the entered organic remains [10].

The intensity of transformation of organic remains is determined by the number of individual acts of the reaction (n) per init time t and can be expressed as:

$$I = n/t \tag{2}$$

(3) The dependable variable "humification degree (H)" can be approximated by the experimentally determined index of "humification level" (HL). The latter reflects contribution of humic acid (highly humified matter) versus fulvic acid (low humified matter) into the composition of soil humus, and can be defined as:

$$HL = C_{HA}:C_{FA} \tag{3}$$

where C_{HA} and C_{FA} is the content of humic and fulvic acids in the soil, respectively.

The HL index correlates strongly with the period of biological activity (PBA) and could be used to predict the humification degree. Its applicability and practical use is well illustrated by the relationship of the $C_{HA}:C_{FA}$ ratio versus PBA, established in our previous works for the principal soils of Russia (Figure 2) [11]. When HA substantially prevail in the total content of humus, the corresponding humus state can be defined as the humic-type (H); when fulvic acids strongly prevail in the humus composition, the humus state can be characterized as a fulvic type (F); the intermediate stages can be identified as humic-fulvic (HF), when FA slightly prevail over HA; and fulvic-humic (FH), when HA slightly prevail over FA. Hence, the humus can be ordered into the following sequence according to the humification degree: H > FH > HF > F.

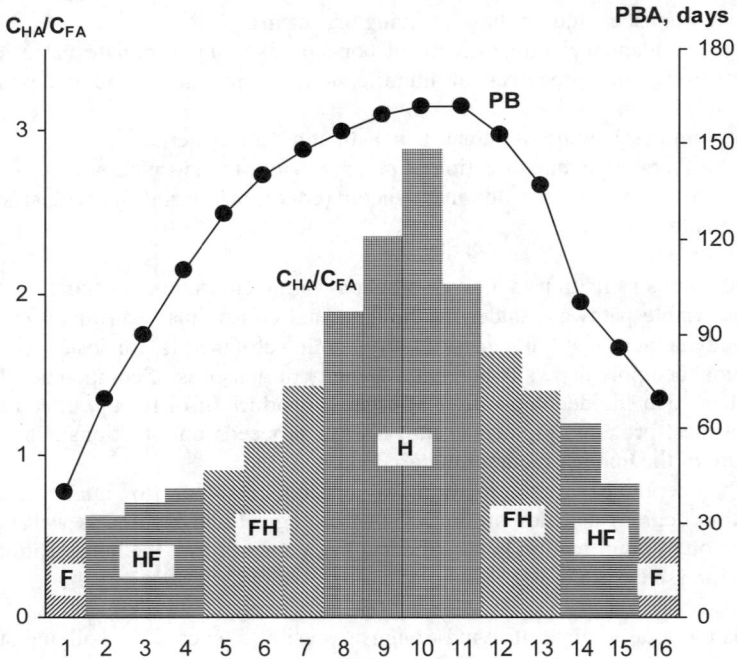

Figure 2. Relationship of HL index (C_{HA}/C_{FA} ratio) versus the period of biological activity (PBA) for different soils in Russia. The categories of axis X refer to the following types of soils: 1 – tundra; 2 – slightly-podzolic; 3 – podzol; 4 – sod-podzolic; 5 – light grey forest; 6 – grey forest; 7 – dark grey forest; 8 – leached chernozem; 9 – typical chernozem; 10 – common chernozem; 11 – southern chernozem; 12 – dark chestnut; 13 – chestnut; 14 – light chestnut; 15 – brown semi-desert; 16 – grey-brown. The capital letters on the graph designate the different humus types: F – fulvic; HF – humic-fulvic; FH – fulvic-humic, H – humic (see explanations in the text).

As it can be seen, for the Northern soils, the main role is played by the thermal regime or heat deficiency restricts the process. On the contrary, for the Southern soils, the humification level is governed by moisture deficiency. This rule operates within the certain limitations, which become most evident when the regions with equal duration of PBA are compared. The method of PBA calculation and relationship of PBA versus HL index are shown in Table 3 and Figure 3.

Table 3. *HL* and PBA of the principal soil types.

Soils	C_{total}	$C_{HA}:C_{FA}$	Number of days with T > 10°C (a)	Number of days with PMR<1-2% (b)	PBA duration, days (a - b)
Tundra	1.7	0.48	50	0	50
Gley and bog-podzolic	1.9	0.54	70	0	70
Podzols	0.4	0.70	92	0	92
Sod-podzolic	1.7	0.75	110	0	110
Grey forest	3.1	1.10	130	0	130
Chernozem:					
Leached	4.2	2.29	144	0	144
Typical	4.9	2.40	154	0	154
Southern	2.7	2.20	175	5	170
Chestnut	1.5	1.63	190	50	140
Brown semidesert	0.7	0.59	215	125	90
Grey-brown	0.3	0.44	210	137	73
Grey northern	0.4	0.53	210	137	73

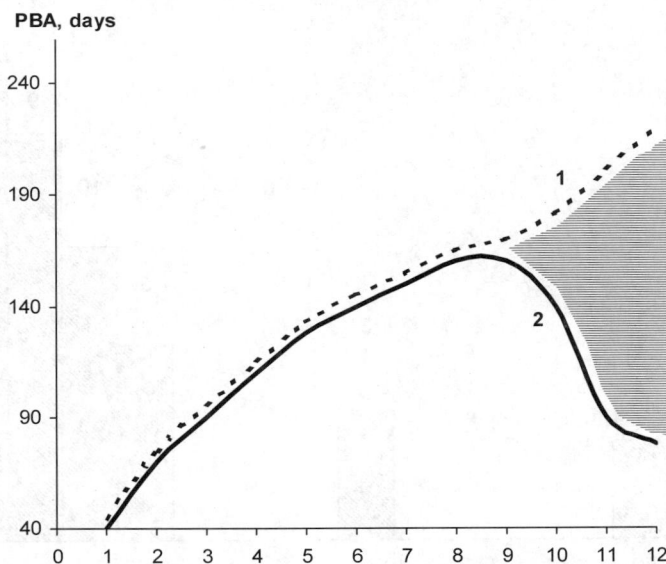

Figure 3. Schematic estimation of duration of PBA (after Biryukova, cit.[11]). 1 – number of days with temperature consistently above +10°C; 2 – duration of PBA. Horizontal lines define with the period with available water reserve less than 1-2%. Soils: 1 – tundra; 2 – gley-podzolic; 3 – podzol; 4 – sod-podzolic; 5 – grey forest; 6 – leached chernozem; 7 – typical chernozem; 8 – common chernozem; 9 – southern chernozem; 10 – chestnut; 11 – brown steppe; 12 – grey-brown; 13 – greyzem.

The degree of humification is directly (though not linearly) dependent on the duration of the PBA (Figure 4).

(4) The index "humification level" HL is a function of soil chemical and mineralogical composition and the rate of hydromorphism. For example, calcareous parent materials or hard, mineral-containing ground waters cause the formation of humic type humus. However, the chemical properties and mineralogical composition of soils acquire the significance of the dominant factor only under specific conditions. In agricultural practice, chemical treatments could become a dominant factor.

(5) Soil pH level and the rate of soil solution mineralization also control the humification degree. It means that the group and fractional humus composition could be independent. For instance, the group humus composition of sod-podzolic, gray-brown soils and serozems expressed as $C_{HA}:C_{FA}$ is the same; meanwhile, there is a total or partial absence of humic acids bound to calcium in sod-podzolic soils. At the same time, this fraction predominates in arid soils.

The five described rules are simple and the dominant conditions or reactions are easy to identify. Hence, on their basis it is possible to construct models of humus formation and develop measures for regulating the content of humus in cultivated and natural soils.

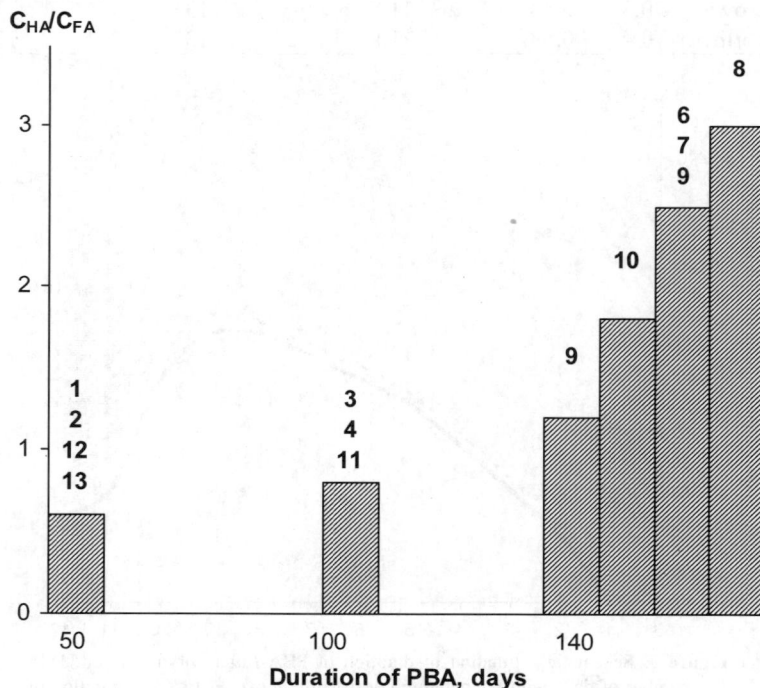

Figure 4. Dependence of depth of humification on duration of PBA. Soils: 1 – tundra; 2 – gley-podzolic; 3 – podzol; 4 – sod-podzolic; 5 – grey forest; 6 – leached chernozem; 7 – typical chernozem; 8 – common chernozem; 9 – southern chernozem; 10 – chestnut; 11 – brown steppe; 12 – grey-brown; 13 – greyzem.

2.3. PRINCIPAL FUNCTIONS OF SOIL ORGANIC MATTER

The functions of organic compounds in soils are diverse, and at times even contradictory. The low-molecular weight substances are usually readily available to microorganisms. They participate in the mobilization of mineral constituents of soils by extracting many elements from practically insoluble forms. Humics play a conservative role, to some extent imparting long-term stability of important properties such as humus reserve, cation exchange capacity, and buffer properties to soils. We distinguish five very important functions that allow us to evaluate the real ecological role of humic substances fairly well:

Accumulative function: involves accumulation of elements such as carbon, nitrogen, phosphorus, and others, including trace elements in soils, in the form of organic molecules and coordination compounds. This accumulative function is essential for vital activity. It should not be regarded as a passive stocking of nutrients because accumulation can occur even in soil solutions.

Transport function: enables geochemical fluxes of mineral and organic substances mostly in aqueous media due to the formation of stable but rather readily soluble complexes of humic substances with metal cations, metal oxohydrates and hydroxides, bioorganic molecules, or adsorption complexes of humic substances with partially cleaved aluminosilicates. Apparently, these are the forms in which organic and inorganic compounds migrate in the soil profile and in the landscape.

Regulatory function: This complex and multifaceted function may include: 1) regulation of soil structure and some hydrophysical properties [9]; 2) regulation of the equilibrium in ion exchange, acid-base, and oxidation-reduction processes [12]; 3) regulation of mineral plant nutrition; 4) regulation of the thermal regime of soil by the influence of the spectral reflectance and also by heat capacity and conductivity; 5) regulation of the process of chemical composition in intersoil differentiation.

Physiological function: involves both the direct specific effect on plants and the indirect protective action of humates. Humic substances can favourably influence the functioning of mitochondria and chloroplasts, which facilitate respiration and photosynthesis processes. The adverse effect of the environment on plants can be reduced by the use of humics. According to Gorovaya [13], the physiological effect of humics becomes more clearly manifested under unfavourable conditions. The physiological function of humic substances is important and is of great interest not only from the agronomic or medicinal point of view; this function compels us to think of the physiological activity and, consequently, the structure or molecular formulas of humics. The considerable and wide-ranging effects of humics are primarily due to the presence of many functional groups, not only such common ones as carboxyl, phenols, and alcohol, but also quinones, amines, and amides, which are capable of forming electrostatic and covalent bonds and intracomplex compounds. These groups ensure regulation of the ratio of free ions to bound ions in the soil solution and in the intracellular media.

Protective function: involves the ability of humic substances to bind different pollutants: heavy metals, pesticides, radionuclides, and the surplus of fertilizers. The protective function is widely spread in both the soil-plant system and the various landscape components. The relative stability of soil as a whole mainly depends on the stability of humic substances, presumably, humic acids and humin. It has been demonstrated that in soils with large reserves of humic acids and humin, the permissible limiting concentrations of heavy metals are much higher. Both heavy metals and radionuclides (e.g. strontium) are converted by humics into forms that are weakly available to plants. The adverse effects of high doses of fertilizers are also eliminated. Humic substances are capable of removing the adverse effect of pesticides on crop plants. The protective function involves not only the soil-plant system, but also the other components of the landscape and of the biocenosis. It has been shown that humus-rich soils act as a geochemical barrier and prevent the entry of many pollutants into ground water. The soil cover can hold a considerable amount of cations and anions and, thereby, maintain the quality of potable water at the desired level for a long time despite industrial pollution.

3. Protective Functions of Humus: Case Study for Heavy Metal Pollution

3.1. HUMICS-MINERAL INTERACTIONS IN SOIL AND THEIR REMEDIAL SIGNIFICANCE

The diversity and complexity of functions of humic substances are provided by their polychemical nature. Humic molecules differ in size and functional groups; they form a notable range of compounds differing in binding affinity and complexing capacity to metal ions. Polydispersity and multifunctionality ensure a high buffer capacity of humic substances in relation to acid-base, oxidation-reduction, and many other reactions.

A group of organomineral compounds (or so called adsorption complexes) occurring in soils is formed due to interaction of humic substances with the crystalline and amorphous soil minerals. The importance of organic-mineral interactions can be described as follows:

1) Organic matter facilitates dissolution of many low-soluble mineral components, thereby converting chemical elements into mobile forms readily available to plants.

2) Organic matter forms coatings on the surface of soil particles.

3) Organic matter directly or indirectly influences the oxidation state of mineral components.

The reactivity of humic substances toward a large variety of mineral compounds present in soil is provided by a wide range of their functional groups. Humic substances contain about 15 different structural units which are considered most important in realizing the protective functions of soil humus: amino groups, amide groups, alcoholic hydroxyls, aldehydes, carboxyls, keto groups, methoxyls, phenolic hydroxyls, quinone/hydroquinone fragments, peptidic units.

The largest fraction of soil humic substances is insoluble in aqueous phase [14]. In general, humic substances are macromolecular aggregates hold together by means of di- and trivalent metal bridges, provided usually by Ca^{2+}, Fe^{3+}, or Al^{3+}, or bound to clay minerals through electrostatic linkages (metal bridges of clay-metal-humus type), hydrogen bonds, or Van der Waals' forces. Humics are found to bind from 200 to 600 mmol/g of exchangeable metal ions in the form of coordination complexes [15]. Carboxylic and phenolic groups are the most important complexing groups. Lower molecular weight fractions of humics enriched with carboxyl and phenol groups, exhibit higher affinity for metal ions. It was reported by Weber (cit. in [16]) that humics extract

> 1 ppm of metal ions from various minerals. Therefore, the soil humics are very important metal complexing agents and can be considered as major contributors to the geochemical fluxes of metals in the bio- and geosphere.

The ability of humic acids to complex heavy metals is utilized in the remediation of polluted industrial landscapes. For this purpose, a number of humic preparations derived from natural sources, sewage and compost is produced by the industry. However, the physical and chemical properties of those preparations have been only scarcely investigated.

In this work, we have performed a comparative study of sorption properties of humic acids isolated from different sources and of commercially produced humic preparations. The objective of this study was to demonstrate the protective functions of humics in relation to heavy metals (Cu, Zn, and Pb) and establish their relationship with physico-chemical properties.

3.2. EXPERIMENTS ON ESTIMATING SORPTIVE PROPERTIES OF HUMICS TO HEAVY METALS

Humic materials. Various humic materials were isolated by standard methods [16] from natural sources: sod-podzolic soil – HA_S; deep peat – HA_P; organic rock or so called "mumie" (Mountain Altai) – HA_M. In addition, the industrially produced humic preparations were used: "humate" (brown coal) – HA_C, and a chemical reagent "humic acids" (commercial humate) – HA_H. The humics studied were purified by electrodialysis. The ash content ranged from 1.5 (HA_M) to 9% (HA_P).

Sorption experiments. Sorptive properties of humics were studied under steady state conditions in a stationary system. Aqueous solutions of metal salts were reacted with solid humic substances suspended in the solution. Foe this purpose, 25 mL of standard solutions of Cu^{2+}, Zn^{2+}, and Pb^{2+} in deionised water (20 to 80 μg ml^{-1}) were added to each sample of HA. 0.2 M $CaCl_2$ was used to determine the specific adsorption value. Tightly closed beakers were left for 4 days with periodic shaking for reaching equilibrium. Then, the samples were filtered, and the obtained supernatants were analyzed for the metal concentration using atomic absorption spectroscopy.

Data treatment. The obtained data were treated according to the adsorption Langmuir model. For this purpose, the difference between initial and equilibrium metal concentration was calculated to estimate Q value of adsorption, expressed in mg·g^{-1} HA. The maximum adsorption value (adsorption capacity) and the equilibrium constant of adsorption were determined by the linearized Langmuir equation.

3.3. SORPTIVE PROPERTIES OF HUMIC MATERIALS STUDIED IN RELATION TO HEAVY METALS

The data obtained showed that in the range of metal concentrations studied, adsorption of Cu, Zn, and Pb on humics according to Langmuir isotherms, took place at pH values of about 3-5. The strongest adsorption on various HA was found for Pb accounting for 21-38 mg g^{-1}. The sorption affinity of humics for Cu and Zn was lower and accounted for 11-14 and 6-12 mg g^{-1} HA, respectively. The maximum adsorption depended on the nature of the humic material and the type of metal (Table 4).

Table 4. Adsorption capacity Q_{Max} (mg/g HA) of humic materials used in this study in relation to Cu, Zn, and Pb.

HA sample	Q_{max}/Cu	Q_{max}/Zn	Q_{max}/Pb
HA_C	14.0	12.4	38.4
HA_P	13.9	10.3	33.8
HA_H	13.3	9.8	22.9
HA_S	13.4	6.3	21.2
HA_M	10.9	7.4	19.8

The ranking of HA samples used in this study according to their adsorption capacity (maximum adsorption) was similar for all metals studied:

Cu: $HA_C \geq HA_P > HA_S \geq HA_H > HA_M$;
Zn: $HA_P > HA_H > HA_M > HA_C > HA_S$;
Pb: $HA_P > HA_C > HA_H > HA_S > HA_M$;

The above orders were almost identical to those obtained for the relative specific adsorption value expressed as a percentage of the total adsorption (Table 5).

Table 5. Specific adsorption capacity – Q_{max-s} (mg/g HA) of humics materials used in this study in relation to Cu, Zn, and Pb.

Humics	Q_{max-s}/Cu		Q_{max-s}/Zn		Q_{max-s}/Pb	
	mg/g HA	% to Q_{max}	mg/g HA	% to Q_{max}	mg/g HA	% to Q_{max}
HA_C	9.0	64	5.4	64	16.4	43
HA_P	9.7	70	6.7	65	17.5	52
HA_H	10.0	75	6.6	67	13.4	59
HA_S	10.4	78	5.2	83	13.1	59
HA_M	8.6	79	6.0	80	11.6	62

The highest share of specific adsorption was observed for HA_S and HA_M; the lowest one was found for HA_C. Specific adsorption prevailed in all experiments, except for Pb and Zn adsorption on HA_C. The share of specific adsorption was lower for Pb than for Cu and Zn. The rather high adsorption capacity of HA_C can be explained by alterations in structure and properties, which have taken place during processing of the humate from the original raw material (brown coal). In addition to cleavage of peripheral units, conformation and molecular weight were changed, and the content of active oxygen functional groups increased. The adsorption capacity of peat derived HA was similar to that of HA_C being in line with a close resemblance of the infrared spectra of HA_P and HA_C. The values of general and specific adsorption capacities for HA_H, HA_S and HA_M were similar.

The experimental data obtained show only minor differences in the ability of humic materials specifically adsorb metals. This finding in consistent with the range of stability constants of Me-HA complexes obtained by Stevenson [14], who studied the complex formation of humic substances from soil, peat, and lignite with copper, lead, and cadmium.

It can be concluded that the value of general metal adsorption by humic substances is more affected by the nature of HA than the specific adsorption. Metal adsorption increased with aromaticity and with the content of complexing oxygen-containing functional groups. The differences in adsorption capacity regarding the metal nature were connected to specific features of the metal chemistry, i.e. their ability to form hydroxo complexes and their affinity for ion exchange or complexation.

4. Conclusion

The soils are shown to be among the most well organized and complex natural systems. They surpass all the known natural compartments with respect to variety of their constituents. Deep transformation of plant and animal residues and human-made substances take place during the soil formation process. Such intra-soil transformations lead to the accumulation of a specific group of organic substances named humic substances, which are composed of similar components such as humic acids, fulvic acids, hymatomelanic acids, and humin. Recent research attributes a leading role to humic acids and humin. The significance of fulvic acids and hymathomelanic acids is questioned today while they are considered as breakdown products produced during isolation from soil. Nevertheless, they are important objects for the study of the nature of humus as a whole. Particular feature of the chemical composition of modern soils is a large amount and diversity of industrial wastes and toxic compounds that greatly influence the soil chemistry. Among man-made compounds entering soil are various fertilizers and manures, surface-active compounds, and biologically active organic and inorganic substances. This ensemble of components could alter the soil from a natural body into an entity rather described as "natural-technogenic body." A more complete analysis of such a natural-technogenic body need to be fulfilled in the future.

5. Acknowledgements

This study was supported by the grants of the Russian Foundation for Basic Research 02-04-48022 and 02-04-48016.

6. References

1. Waksman, S.A. (1937) *Humus*, Sel'khozgiz-Publisher, Moscow (in Rusian).
2. Kononova, M.M. (1972) The present state of the problem of soil organic matter, in *Organic Matter of Virgin and Arable Soils*, Nauka-Publisher, Moscow (in Russian).
3. Orlov, D.S. (1992) *Soil Chemistry*, Russian Trans. Series, A. Balkema, Rotterdam.

52

4. Kononova, M.M. (1963) *Soil Organic Matter*, Academy of Sciences of the USSR (AN SSSR)-Publisher, Moscow (in Russian).
5. Kurcheva, G.F. (1971) *Role of Soil Fauna in Decomposition and Humification of Plant Remains*, Nauka-Publisher, Moscow (in Russian).
6. Alexander, M. (1965) Biodegradation: problems of molecular recalcitrance and microbial fallibility, *Adv. Appl. Microbiol.* **7**, 35.
7. Scheffer, F. and Ulrich, B. (1960) *Humus und Humusdüngung*, Stuttgart, Enke Verlag.
8. Kleinhempel, D. (1971) Theoretical aspects of the persistence of organic matter in soils, *Pedobiologia* **11(5)**, 65.
9. Flaig, W. (1971) Organic compounds in soil, *Soil Sci.* **111(1)**, 19-33.
10. Orlov, D.S. (1995) *Humic substances of soils and general theory of humification*, Oxford & IBH Publishing, New Delhi.
11. Orlov, D.S., Biryukova, O.N. and Sukhanova, N.I. (1996) *Organic matter of soils of Russian Federation*, Nauka-Publisher, Moscow (In Russian).
12. Franzmeier, D.P., Steinhardt, G.C. and Brasner, B.R. (1990) Relation of cation exchange capacity to clay and organic contents of Indian soils, *Proc. Ind. Acad. Sci.* **99**, 107-112.
13. Gorovaya, A.I., Orlov, D.S. and Scherbenko, O.V. (1995) *Humic Substances. Structure, Functions, Mode of Action, Protective Properties, Role in the Environment*, Naukova Dumka, Kiev (in Russian).
14. Stevenson, F.J. (1982) *Humus Chemistry, Genesis, Composition, reactions*, John Wiley & Sons, N.Y., Toronto, Singapore.
15. Rashid, M.A. (1971) Role of humic acids of marine origin and their different molecular weight fractions in complexing di- and tri-valent metal, *Soil Sci.* **111(5)**, 25-35
16. Orlov, D.S. (1985) *Humus Acids of Soils*, Oxonian Press Pvt. Ltd., N. Delhi-Calcutta.

Chapter 3

CHEMICAL STOICHIOMETRY AND MOLECULAR LEVEL MECHANISMS AS SUPPORT FOR FUTURE PREDICTIVE ENGINEERING CALCULATIONS

D.S. GAMBLE[1], C.H. LANGFORD[2], A.G. BRUCCOLERI[2]
[1] Department of Chemistry, Saint Mary's University, Halifax Nova Scotia B3H 3C3, Residence: 7412, Route 366 Northport Cumberland County Nova Scotia B0L 1E0, Canada <dgamble@ns.sympatico.ca>
[2] Department of Chemistry, University of Calgary, 2500 University Drive N. W., Calgary Alberta T2N 1N4, Canada

Abstract

The chemistry of humic materials has been under world wide investigation for about a 100 years. The pace of the research continues to increase. During at least the last 30 years, evidence has been accumulating in the literature that the principles of classical chemistry as they are understood for monomeric reagents can to some extent be adapted to humic polyelectrolyte mixtures. Published demonstrations of predictive chemical calculations for humic materials have resulted from this. There is a need for the predictive capability to be improved, and applied to environmental remediation and regulatory practice.

1. Introduction

The practical importance and scientific challenge of humic materials have attracted enough attention so that during the past century a quite large scientific and technical literature has accumulated. The accumulation has accelerated so that the number of important reviews is now also large. A consequence of this is that literature citations must be selective, while somehow avoiding the loss of important information that is relevant to particular research and developments objectives. A related consequence is that many publications not cited can be just as important as those selected as representative examples of the available science. When this large scientific literature is selectively consulted for a particular purpose, it is important that recent publications with new concepts and experimental methods be integrated with the relevant older literature, instead of simply substituted for it. There are at least three reasons for this. First, it is important that current investigators avoid "reinventing the wheel". Secondly, current publications sometimes comment on or even react to perceived knowledge gaps when in fact the information is already in the previous literature. Finally, humic materials research should move forward instead of going in circles. The research strategy that is required for this is based on a concept that has persisted in the literature

53

I. V. Perminova et al. (eds.),
Use of Humic Substances to Remediate Polluted Environments: From Theory to Practice, 53–79.
© 2005 *Springer. Printed in the Netherlands.*

54

for at least 30 years. A humic material is regarded as a chemical system, whose chemical properties are described by state functions as defined in classical physical chemistry. That is, dependent variables are defined in terms of all of the independent variables needed for complete descriptions. Inherent in this is the use of mathematical descriptions that link effects to the molecular level mechanism causes that drive them. In this way, classical chemistry is adapted to complicated mixtures.

Among the opportunities now available for moving forward is the development of new technology based on predictive engineering calculations and supporting test methods, for the protection of clean soil and water. The engineering calculations can include the use of computer models, and the test methods can include bench scale laboratory tests. Work time, costs, and environmental risks could all be reduced. The reason for this is that during at least the last 30 years, a strategy has emerged from the work of various investigators for adapting the concepts of classical chemistry to complicated humic polyelectrolyte mixtures. The trend continues, with the application of modern instrumentation and molecular modelling to humic mixtures having already begun. This is expected to further improve the predictive calculations that can be done for complicated humic mixtures.

Humic materials chemistry might be used for the protection of clean soil and water both by prevention, and by remediation. Two separate but related types of technology can be expected. There is a kind of cross connection between the research done on humic materials and that done on soils and sediments that contain them. This is especially true of peat soils. It might therefore be useful to exploit for humic materials, some of the work that has been done on soils and sediments. A few possibilities will be considered here.

2. Research Strategy: Description With Experimental Results

The research strategy consists of a combination of methods and supporting concepts, with which classical chemistry as understood for monomeric pure reagents, may be adapted to humic weak acid polyelectrolytes [1]. The motive for this is the desire for an ability to make quantitative predictions in the same sense that this is practiced for the chemistry of monomeric pure reagents. The chemical engineering of humic materials should then become more effective. Several characteristics of humic materials are taken into account:

(a) wide ranges of molecular weights so that molecular weight is not a practical basis of reaction stoichiometries;

(b) mixtures of similar but non identical reactive functional groups, resulting in distribution functions for equilibrium and kinetics;

(c) complicated distributions of reactive functional groups over the polymer molecules, causing fractionation effects during phase changes;

(d) hydrogen bonding conformations which can undergo changes that are kinetically controlled:

(e) because of hydrophilic and hydrophobic regions on the humic molecules, water influences the sorption or binding of monomeric reagents, including cations and organic chemicals.

(f) a macroscopic measurement made on a mixture gives a numerical value that is some kind of average of values for components.

When feasible, a number of means are employed for managing these humic properties:

(a) equilibriums are regarded as limiting conditions, allowing for kinetically controlled reactions or processes to go to completion;

(b) reversible and irreversible reactions or processes are experimentally distinguished;

(c) molecular level mechanisms are described by kinetics and equilibrium in terms of numbers of moles or equivalents of reactive sites and monomeric reagents, and not in terms of weights of humic sample or number average polymer molecules;

(d) a collection of reactive sites is described as a chemical system, mathematically by distribution functions for equilibrium and kinetics, and experimentally by scans over the ranges of numerical values for the inner variables, which are explained below;

(e) total numbers of reactive sites are measured, and related as much as possible to reactive functional groups, as outlined in the "Stoichiometry" section below.

(f) state functions, as the concept is understood in classical chemical thermodynamics, are used for describing kinetics and equilibrium.

(g) quantitative predictions are tested against independent experiments;

(h) a distinction is made between experimentally based weighted average functions that represent distributions of numerical values for a whole mixture, and differential functions that attempt to reflect the corresponding values for the components;

(i) weighted average distribution functions should be used for practical engineering, while differential functions are more correctly used for molecular level interpretations. As an example, thermodynamic formalism defines chemical potentials for monomeric pure reagents and the concept can be adapted to "location bound" reactive sites. But attempts to use chemical potential calculations with numerical values for a complicated mixture could give answers that are wrong by orders of magnitude.

2.1. CATEGORIES OF VARIABLES FOR THEORY AND EXPERIMENTS

When one goes from experiments through theory to engineering practice, three categories must be recognized and used differently [1].

A *Inner Variables.*

These variables are identified by theory as being required for totally specifying the state and/or behaviour of a humic mixture as a chemical system. For equilibrium functions and kinetic rate coefficients at constant temperature and pressure, the independent variables would describe the chemical compositions of a humic mixture. They directly link effects to the mechanistic causes that drive them. Inner variables should be used for correlations and predictions.

B *Outer Variables.*

Outer variables are the type that are manipulated operationally for the conduct of an experiment. Typical examples include the ratio of solid to solution, pH in some cases, and the total of a reagent in a mixture. Although often more useful than inner variables for conducting experiments, they are not properly predictive.

C *Background Descriptive Variables.*

This type of variable can provide useful qualitative insight, but are not used for direct calculations.

The acid catalysed hydrolysis of atrazine in aqueous solutions of fulvic acid provides a graphic example of a cause and effect link that should be used for predictive calculations [1, 2]. For the example in Figure 1, titration of the fulvic acid used had revealed a strongly acidic group of carboxyl groups, Type A, and a weakly acidic Type B of functional groups [3, 4]. The acid catalysis of atrazine hydrolysis by fulvic acid had been documented, and it was therefore important to predict the effect under practical conditions [5]. The anticipated Broensted acid catalysts were H^+ and protonated carboxyl groups. The kinetic rate coefficient for H^+ was already known from experiments with HCl [2]. Subtraction gave a rate coefficient expression that represented only the functional groups. The conventional plot with pH in Figure 1(a) gave a family of non-linear curves. With this kind of plot, an experiment having one concentration of fulvic acid cannot be used to predict the atrazine hydrolysis with a different concentration of fulvic acid. The reason is that pH reflects the activity of free H^+, which is not the direct cause of the catalytic effect being monitored. This plot also incorrectly implies that there are two independent variables, pH and fulvic acid concentration. Curve 1(a) is in fact simply a mathematical artefact that does not represent the chemistry. The linear family of curves in 1(b) simply shows that some of the functional groups were catalytic, and some of them were not. The non zero intercepts are mathematical artefacts caused by the inclusion of the non catalytic Type B functional groups. The real chemistry is again not properly represented. The single

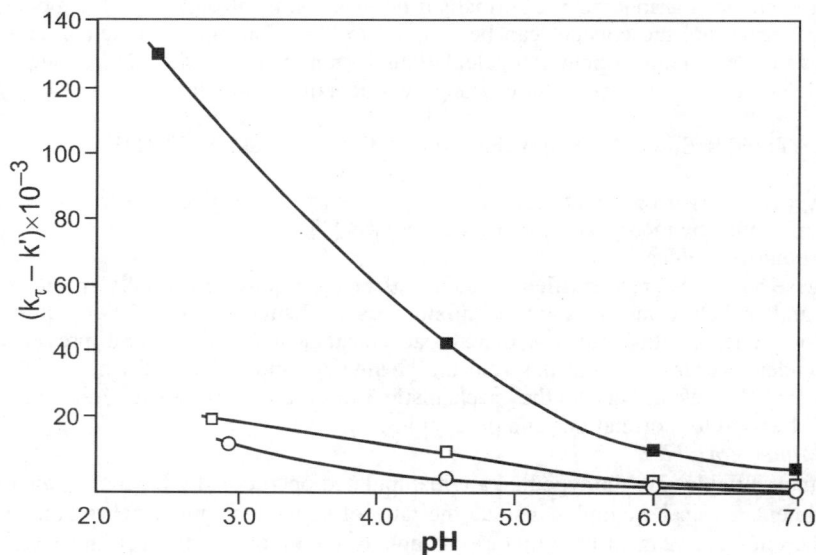

Figure 1 (a). Atrazine hydrolysis with fulvic acid catalyst at 25°C. (g fulvic acid/L): ○, 0.5; □, 1.0; ■, 5.0. A demonstration of the categories of variables. A conventional plot demonstrating the outer variables pH and (g fulvic acid/L) [1, 2, 3].

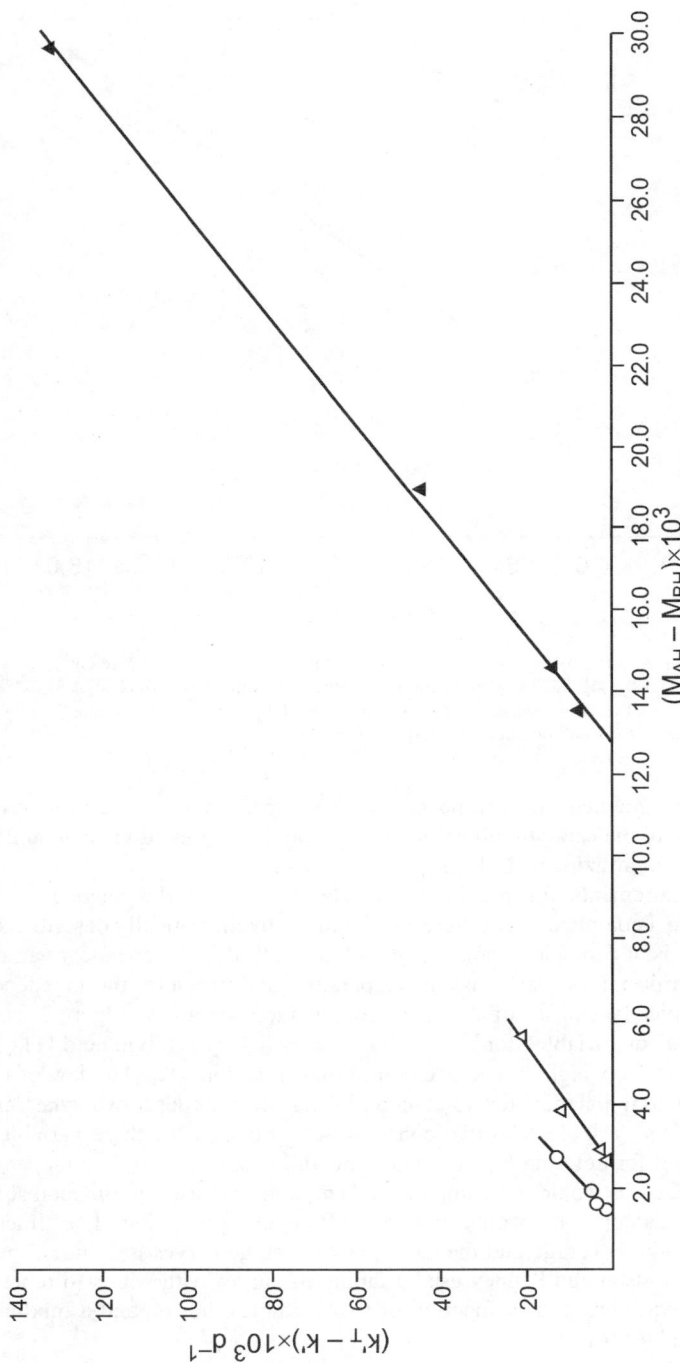

Figure 1 (b). Atrazine hydrolysis with fulvic acid catalyst at 25°C (g fulvic acid/L): $\bigcirc$, 0.5; $\triangle$, 1.0; $\blacktriangle$, 5.0. A demonstration of the categories of variables. A demonstration of catalytic and non catalytic functional groups. The total $(M_{AH} + M_{BH})$ is an incorrect choice of a variable [1, 2, 3].

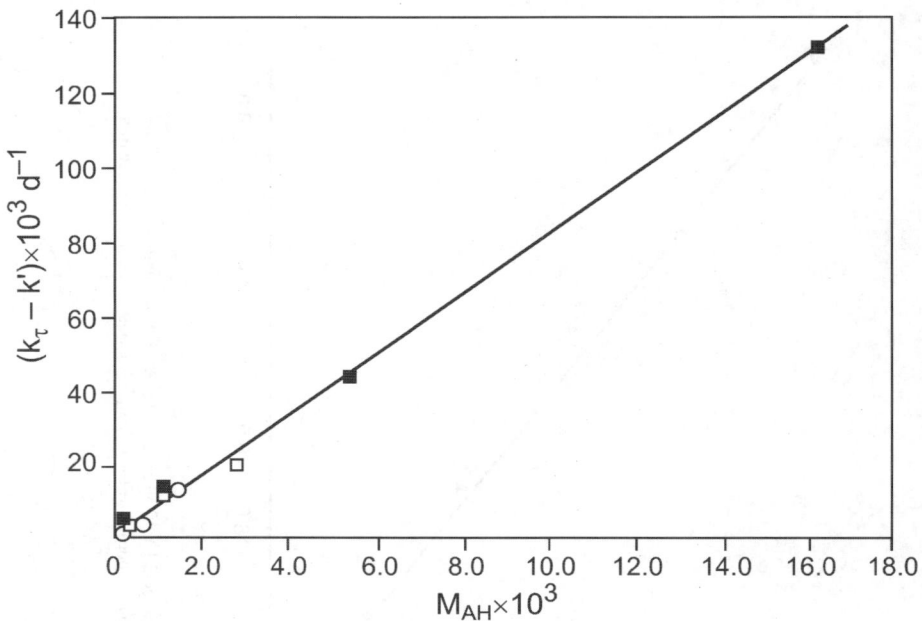

Figure 1 (c). Atrazine hydrolysis with fulvic acid catalyst at 25°C (g fulvic acid/L): ○, 0.5; □, 1.0; ■, 5.0. A demonstration of the categories of variables. A demonstration of the inner variable M_{AH}. The protonated Type A carboxyl groups are the cause of the catalytic effect [1, 2, 3].

linear plot with zero intercept in 1(c) means that the Type A acidic functional groups were the cause driving the catalytic effect being monitored. For this fulvic acid at 25°C, practical predictions of atrazine hydrolysis could be made.

One of the requirements for predictive calculations is that the phenomenon of interest, for example a complexing equilibrium, should be mathematically described by a state function. That is, a dependent variable should have all of its independent variables specified. For a humic mixture at constant temperature and pressure, the independent variables would typically consist of chemical composition variables. Figure 2 has an example of independent variables for the Cu(II) complexing by a fulvic acid [1] . The dependent variable is the weighted average equilibrium function , K_1. The law of mass action and the material balance for total complexing sites impose two constraints. Because there are four variables and two constraints, any two of the three variables in Figure 2 may be used for defining K_1 as a state function. If only one of the independent variables were used, cases would not compare well and unnecessary experimental scatter would result. The calculations would not be sufficiently predictive. For practical engineering, the H^+ site coverage and the Cu(II) site coverage curves in Figure 2 might be a useful choice. Wilson and Kinney used a family of curves with which to represent the Zn(II) - humic equilibrium as a function of two variables, but plotted a mixture of inner and outer variables [6].

59

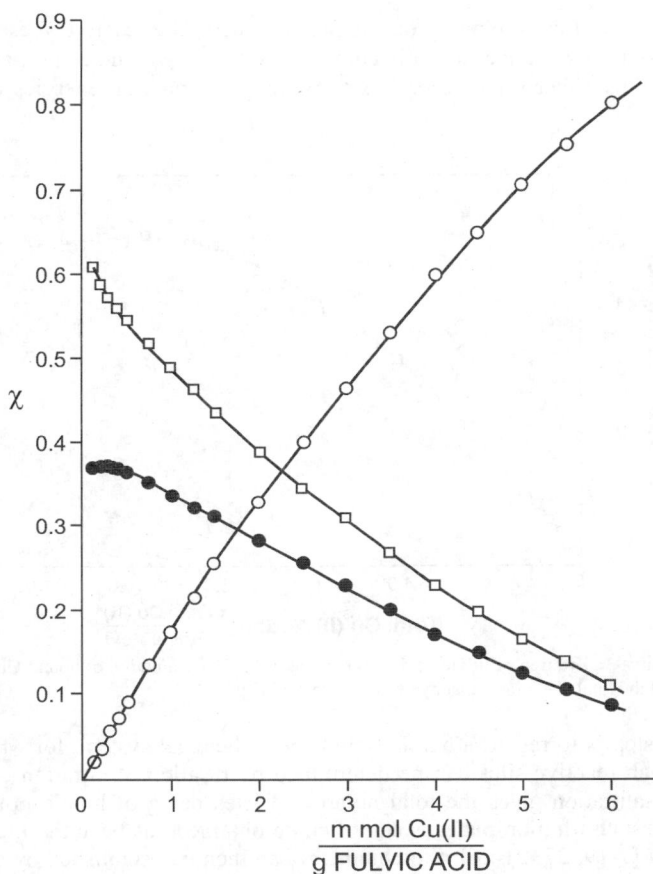

Figure 2. Mole fractions of unionised, ionised, and complexed chelation sites. ●, χ_{SH2} = mole fraction of unionised bidentate chelating sites. □, χ_{SH} = mole fraction of bidentate sites free from the 1st proton and from Cu^{++}.○, χ_C = mole fraction of bidentate sites occupied by chelated Cu^{++} [1].

2.2. CHEMICAL STOICHIOMETRY

Classical chemistry bases the calculations of equilibrium ands kinetics on balanced chemical reactions. Material balances among the appropriate collections of atoms, ions, molecules or free radicals must be maintained for the reactants and products. But when one moves from monomeric pure reagents to a polydisperse humic polyelectrolyte, stoichiometry is the first challenge. Valid stoichiometry cannot be based on either number average molecular weights or on nonchemical units such as g of humic material. Because they do not represent the real chemistry, these common practices do not support predictive calculations. But valid alternatives have been in the literature for at least three decades, and one example goes back 70 years [7-26]. The first step is to recognize that

60

cations react with discrete binding or complexing sites. The reactive sites are scattered in some irregular fashion over the collection of polyelectrolyte molecules, and consist of functional groups either separate or in sets. A range of binding energies can typically result.

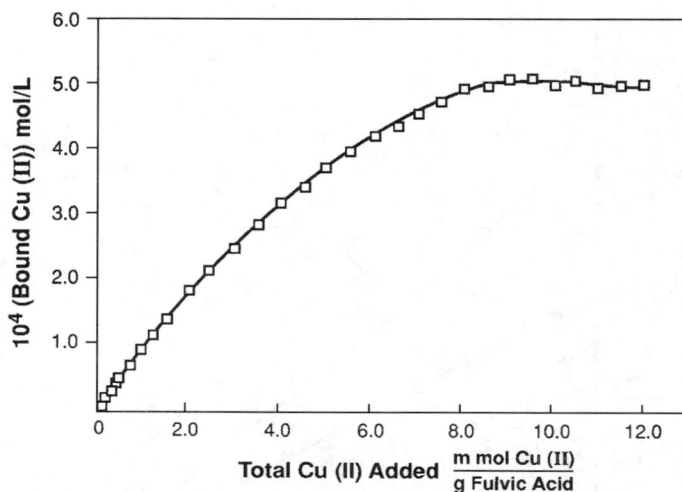

Figure 3. Titration of fulvic acid with Cu^{++} at 25°C. Determination of bound Cu^{++} and the total chelation capacity of the fulvic acid [7].

The next step is to regard a humic sample as a chemical system, for which the total number of such reactive sites can be determined by titration. A titration curve plateau representing saturation gives the total number of sites per g of humic polyelectrolyte. Evidence that such titration plateaus can often be obtained has been the literature for at least 45 years [7-19, 27, 21-26]. Stoichiometry can then be established by experimental values for empty and occupied reactive sites. Gamble and co-workers for example, have exploited this opportunity for equilibrium calculations [11, 15-17]. Cation cases were done first. Figure 3 is one of the examples [11]. For organic chemicals, well defined sorption capacities in (moles/g of humic material) have been reported for both dissolved and undissolved humic material. Figure 4 shows the titration of a fulvic acid with atrazine, with the complexing capacity defined by a plateau [17]. Figure 5 gives the complexing capacity as a function of the inner variable, which is the number of protonated carboxyl groups /g of fulvic acid. A plot of the same experiments using the outer variable pH gives a non predictive family of curves. Subtle combinations of hydrogen bonding, hydrophobic, hydrophilic, solvation, and conformation effects seem to be the causes. The net result of this is, as is seen in Figure 5, that only about 1% of the protonated carboxyl groups can bind the atrazine. Complexing capacities of dissolved fulvic acid for lindane, which is more hydrophobic, have also been reported [28]. In that example, the effect of carboxyl group protonation should be investigated further. Comparable complexing capacity trends have been observed for undissolved humic acid, although the complexing capacity for atrazine was about 3 to 4 times larger than that found with fulvic acid [29, 30].

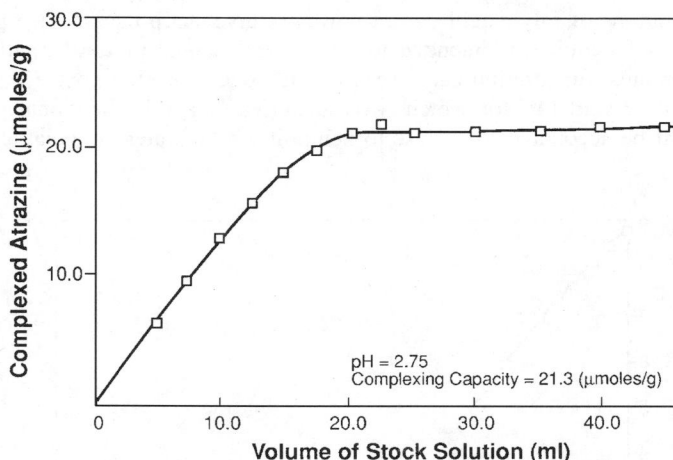

Figure 4. Atrazine complexing capacity of dissolved fulvic acid at pH = 2.70. Uncomplexed atrazine monitored by ultrafiltration and gas chromatography (1 g fulvic acid/L); atrazine stock solution, 1.1×10^{-4} M [17].

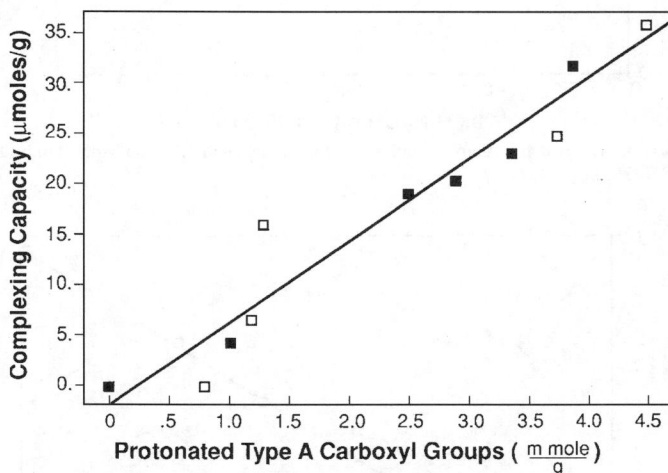

Figure 5. The effect of the internal variable (mmoles/g) of protonated carboxyl groups, on the atrazine complexing capacity of Armadale Bh horizon fulvic acid [12]. □, no KCl. ■, 0.1 M KCl. Each data point represents a titration like Figure 4 with about 2 dozen measurements. More than 130 total measurements [28]. $Y = 1.7563 + 8.1466X$ (μmoles/g), $\sigma = 3.55$ (μmoles/g), $r = 0.9621$.

Equivalence point plot methods with titration data have also been reported. A Gran's plot for the NaOH titration of a humic acid gave the sharp equivalence point in Figure 6, because dissolution of neutralized humic fractions removed the buffering effect [20, 29]. The sharp equivalence point plot in Figure 7 for Cu^{++} complexing by fulvic acid was produced by an ultrafiltration method that had also been used with atrazine [20, 29-31].

62

Acid base titrations of fulvic acid do not however, give sharp equivalence points. The original Gran's function was intended to give quick manual plots of relatively sharp titration end points, for titration curves having only one inflection point [32]. This was adapted to fulvic acid [4], for which two categories of acidic functional groups and buffering must be accounted for. Close to neutrality, even water dissociation has to be allowed for.

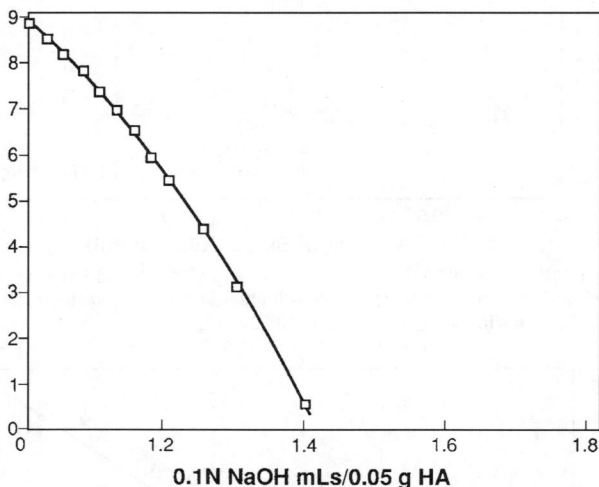

Figure 6. Equivalence point calculations for humic acid titration with standard NaOH [20, 29].

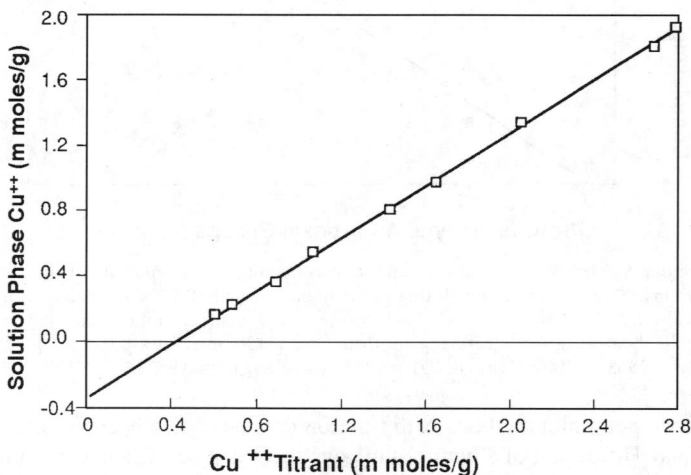

Figure 7. Cu^{++} complexing capacity of fulvic acid measured by compleximetric titration. Uncomplexed Cu^{++} after site saturation, was monitored by ultrafiltration and atomic absorption spectrophotometry [20, 29 - 31].

$$Y_1 = (Vm_H - Q_1) = K_A(V_{e1} - V)$$

$$Q_1 = [(W_0 + R_wR_vV)/N_B][m_B - m_H + (K_w/m_H)](K_A + m_H) \tag{1}$$

Y_1 is the Gran's function corrected for pH buffering. It is defined for the first of two equivalence points. The calculations did not impose the assumption of two categories of acidic functional groups on the data. Instead, the m_H vs. V titration curves imposed this condition on the calculations. V is the volume of standard base and V_{e1} is the volume of standard base at the equivalence point. K_A is the weighted average equilibrium function for acid dissociation. Q_1 is the buffering correction term that has been deduced from equilibrium and material balance conditions. It was produced by not introducing the approximations used for obtaining the original quick plot version of Y_1. In this case it has been derived for molal concentrations. W_0 is the initial k_g of water in the sample, and the term R_wR_vV accounts for the weight of water added by titration with the standard base. N_B is the normality of the standard base. m_H is the molality of H^+. m_B is the molality of the neutralized weaker type of acidic functional groups titrated after the equivalence point. K_w is the water constant.

The proper use of this equivalence point equation requires special measures for both the experimental work and for the calculations. Otherwise, reliable results might not be obtained. First, there are several problems inherent in conventional pH measurements [4]. To avoid them, titration curves were constructed by alternating replicate mvolt vs. V curves for sample aliquots with mvolt vs. V curves for standard solutions of HCl and NaOH. They were done with temperature control and CO_2 exclusion. By destroying hydrogen bonding in the fulvic weak acid polyelectrolyte, neutralization by standard base causes conformational changes. Because the conformational changes can be slow, stable voltage readings require finite amounts of time after additions of standard base. The preliminary data processing consisted of separate least squares fits of sample and calibration data to polynomials. Standard error estimated were found to be reliable indicators that the whole process could be repeated to give the same m_H vs. V titration curves to within the estimated error limits.

Because of the buffering correction term Q_1, the equivalence point determination required iterative calculations. For the calculations to converge within a reasonable number of iterations without mathematical artefacts, a section of the titration curve must be used that does not allow the ratio (Q_1/Y_1) to become too large. Ratios of 20% to 30% are a practical compromise. Because some doubts have been expressed about equivalence point methods for fulvic acid, an independent experimental test of this case has been published. The Broensted acid catalysis of atrazine by this number of strongly acidic functional groups is demonstrated in Figure 1 [2]. If the number of these functional groups were not correctly measured, Figure 1(c) would not have a linear plot with a zero intercept. This analytical chemical method is however, quite labour intensive. The practical implication is that instead of it being used frequently for numerous small batches of standard fulvic acid samples, it should instead be used as part of the characterization of a single very large standard fulvic acids sample. Such a sample could then be made universally available for research and for routine remediation, regulatory, and quality control purposes.

2.3. REACTION EQUILIBRIA OF REACTIVE SITES WITH CATIONS OR MOLECULES

A *Common Equilibrium Functions*

It is emphasized again here that in humic materials equilibrium is a limiting condition so that the kinetic approach to it needs to be monitored or otherwise allowed for. Chemical equilibrium are of course governed by the law of mass action as deduced from chemical potentials, and the importance of using it for describing equilibrium in humic materials has been understood for at least the last 70 years [33-35]. This is not always clearly reflected however in the various description found in the literature. Distribution coefficients, K_D, are widely used for tabulating experimental data and for attempts to do predictive calculations of equilibrium. In some cases other names and symbols are used for them. K_D data are also frequently tabulated and even used with non chemical units for mass and concentrations. There is a serious question about the practical usefulness of such data. Can they actually be used for anything, and if so, how correctly? Consider the simple hypothetical example of a monomeric reagent "X" being bound to a single type of reactive site "A" under conditions for which the activity coefficients are nearly 1. The arbitrary K_D definition in Equation (2) is different from the freshman textbook statement of the law of mass action in Equation (3).

$$K_D = M_{AX}/M_X \tag{2}$$
$$K = M_{AX}/(M_X M_A) \tag{3}$$

M_X, M_A, and M_{AX} are the molarities of X, of the unoccupied active sites A, and of the reaction product AX. K is i.e. usual equilibrium constant. The material balance Equation (5) for total

$$K_D \equiv K/M_A = M_{AX}/M_X \tag{4}$$

active sites C shows the relationship of the unoccupied and occupied active sites.

$$C = M_A + M_{AX} \tag{5}$$

Now the difficulties of applying K_D to real humic mixtures become apparent from these familiar simple relationships. The first is that the total number C of reactive sited can be different for different humic samples. Also the number of unoccupied reactive sites M_A can be orders of magnitude different under different conditions. The next difficulty is that K_D is frequently used as if it were a constant. This would happen under special conditions, such as very low loading of the reactive sites. For a humic mixture, the general case is that K_D would be a decreasing function of the fraction (M_{AX}/C), of total sites occupied. Predictions and comparisons might consequently have serious errors. When it has been scanned over all of the relevant independent variables, the law of mass action equilibrium function is a state function. The distribution coefficient K_D is not. In addition, the K for a humic mixture is also a weighted average function. In some cases experiments are restricted to fixed conditions instead of the relevant independent variables being scanned. The resulting K values are reported as "conditional stability constants". Attempts to use them for predictive calculations might produce doubtful extrapolations.

At least three data plot methods use rearranged forms of the law of mass action equation. The specialized forms for two of them are known as the Langmuir Isotherm [36-38] and the Henderson – Hasselbach Equation [39-42]. The graphical technique using the third form is called Scatchard Plot method. The observations made above for law of mass action functions apply to all of them, and they have sometimes been misunderstood or misused when applied to humic materials.

B *H⁺ Dissociation Of Fulvic Acid*

A quantitative description of the acidic properties of fulvic acid functional groups is a prerequisite for the investigations of its metal ion reactions. For this purpose the collection of acidic functional groups is mathematically defined as a chemical system, by a weighted average function. For the physical chemical state of the system to be totally specified, the weighted average function has to be described in terms of all of the relevant independent variables. The chemistry that governs the system of acidic functional groups determines what particular inner variables those independent variables must be. The mixture of acidic functional groups and irregular chemical structures give the chemical system ranges of numerical values for its physical chemical properties. In particular, the predictive inner variables can have ranges of numerical values. Now comes the experimental crunch. Analytical chemical methods are needed for scanning the ranges of those properties from which the inner variables can be calculated. Various titration techniques have been used for the double purpose of producing such scans and yielding equivalence points. Four general types of equilibrium functions for interpreting the experimental scans have been extensively reported and reviewed in the literature [43-45]. Two of them are related weighted average functions that represent alternate descriptions of the same equilibrium. They may be thought of as integral functions derived from macroscopic measurements on whole humic mixtures. In 1970, Gamble showed that the H^+ dissociation equilibrium of a fulvic acid can be rigorously described [3]. This was explained by Equations (6) and (7).

$$AH \rightleftarrows A^- + H^+$$

$$K_A = (m_A m_H / m_{AH})(\gamma_A \gamma_H / \gamma_{AH}) \tag{6}$$

$$K_A = [1/(1-\alpha)] \int_0^{(1-\alpha)} K^0 \Gamma e^{-\Delta G/RT} d(1-\alpha) \tag{7}$$

In these equations A represents a mixture of chemically similar but no identical functional groups, such as carboxyls. The concentrations are molalities and the "γ" terms are the corresponding activity coefficients. α is the degree of ionisation and K^0 is a differential equilibrium function not corrected for the electrostatic effect of a charged polyelectrolyte. The exponential term makes that correction. Γ is the activity coefficient ratio in Equation (6). From these equations K is seen to be a weighted average, with "$d(1 - \alpha)$" being the statistical weighting factor. An obviously parallel argument shows that α is also a weighted average of component degrees of ionisation. Macroscopic titration measurements of the whole mixture produce the two experimental average functions K and α. An experimental K distribution described by Equation (7) has been reported. It was obtained by a titration scan over the mixture of acidic functional groups.

Predictive calculations can be done with Equation (7). A $K_A = f(\alpha)$ calibration obtained for one set of conditions has been used for quantitatively predicting the acidic properties under other conditions. Manning and Ramamoorthy first reported that this distribution function successfully predicted acid dissociation for a different case [46]. Burch, Langford and Gamble tested this again with the successful results in Figures 8 and 9 [4, 43, 47]. The reason for the successful predictions is that K_A was used as a state function with predictive inner variables. Note that pH was not a predictive variable, and was instead the indirectly related experimental variable.

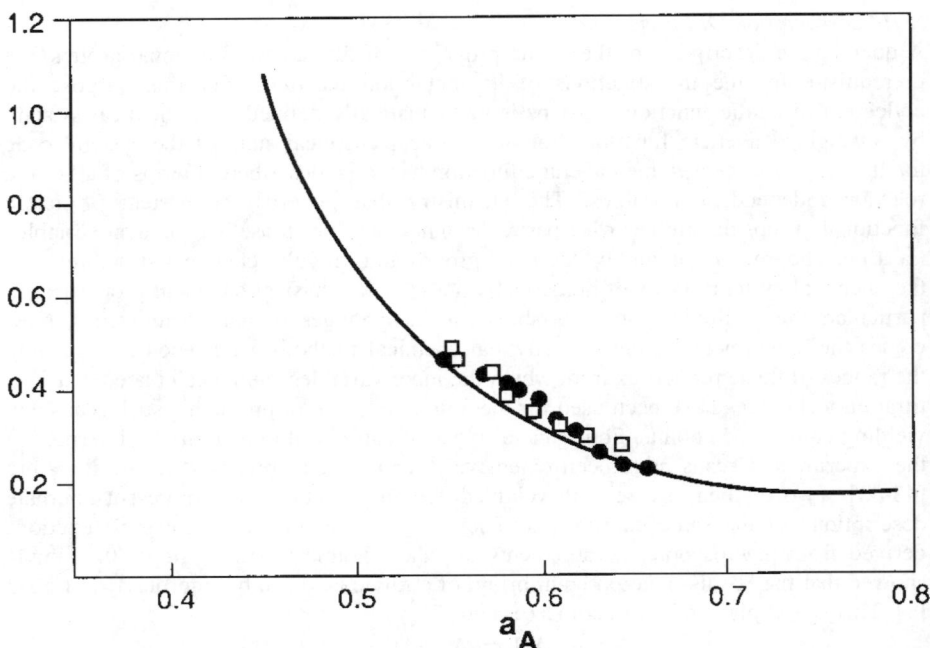

Figure 8. $K_A = f(\alpha_A)$ at different concentrations. Batch FA1 fulvic acid in 0.1 M KCl at 25°C. ●, 0.01073 (g fulvic acid/100 g H_2O). □, 0.01077 (g fulvic acid/100 g H_2O). – Predicted curves from experiments with 0.1000 (g fulvic acid/100 g H_2O) [4, 43, 47].

C Cu^{2+} Chelation By Fulvic Acid

Another 1970 publication by Gamble et al. Adapted the concepts to the Cu^{2+} chelation by fulvic acid bidentate sites [48]. In Equations (8) and (9), the symbols for concentrations are defined by the chemical reaction. "AH^-" represents a bidentate chelating site with one of the two functional groups ionised. Activity coefficients and electrostatic work terms can be written into the Cu^{2+} chelation equations, just as is shown for H^+ dissociation. This type of problem is quite general for any mixture of reactants, and is not peculiar to just humic materials.

$$AH^- + Cu^{2+} \rightleftarrows ACu + H^+$$

$$K_4 = m_c m_H / m_{AH} m_M \tag{8}$$

$$K_4 = (1/\alpha) \int_0^\alpha K_4 \, d\alpha \tag{9}$$

About a year later Klotz and Hunston independently published the same type of description of metal ion complexing by biological materials [49]. Their initiative deserves to be followed up in biochemistry. It should be noted that Equations (8) and (9) take H^+ competition for the metal ion explicitly into account, because Benedetti et al. found it necessary to point this out again 25 years later [50].

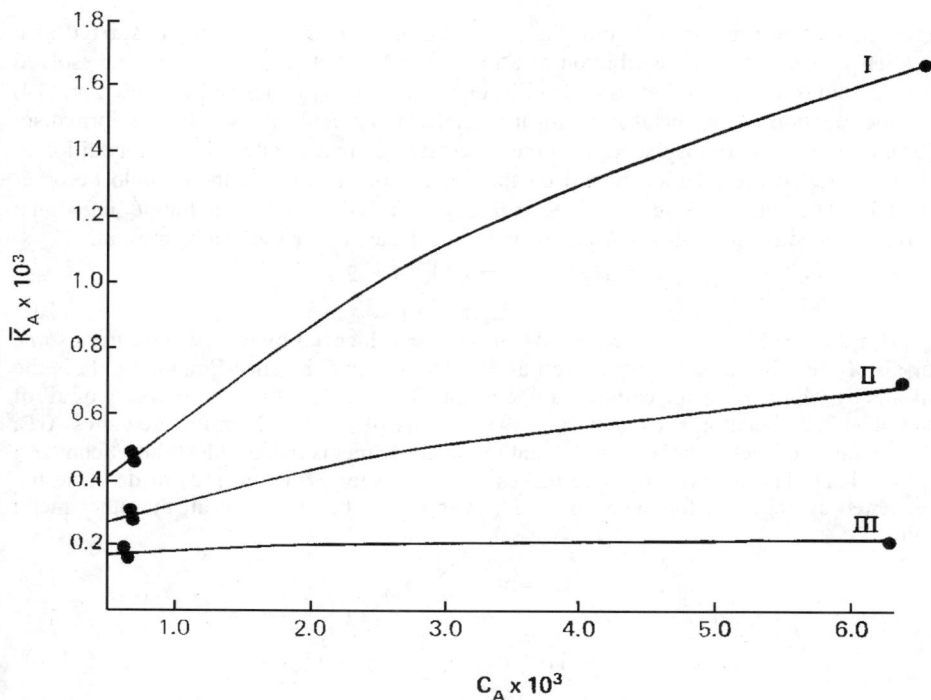

Figure 9. $K_A = f(C_A, Y_A)$. The predicted effects of carboxyl concentration and neutralization on the weighted average equilibrium function. Batch FA1 fulvic acid in 0.1 M KCl at 25°C. $Y = (C_{OH}/C_A)$: Curve I, 0.0000 (mole NaOH/mole A carboxyls); Curve II, 0.3946 (mole NaOH/mole A carboxyls); Curve III, 0.6533 (mole NaOH/mole A carboxyls). – Predicted curves. ●, Experimental test data [4, 43, 47].

The models of multiple equilibrium have been extensively developed in recent years. The limits on the differential equilibrium function given above have been mathematically explored. Precise models dealing with multiplicity of binding sites and electrostatic effects have been developed [64-68]. However, these still do not account for conformation and aggregation effects. The differential equilibrium function developed here is a simplification at the molecular level if there is a discrete distribution, but it is flexible enough to avoid omission of any of the contributions that are recognized to play a role.

D *Multiple Metal Ion Exchange Equilibrium By Humic Acid*

Under actual field conditions, humic materials such as those in peat bogs could already have a whole collection of metal ions even before some are added. More might be added either by pollution or by the use of peat for soil and water remediation. This raises at least two practical issues. One is the effect of acid rain on the elution of the mixture of metal ions. Another is the uptake and removal of metal ions from humic materials used for remediation processes. Equilibrium as a limiting condition can be considered as a reference state. For the distribution of a metal ion M^{n+} between water and undissolved humic acid, the weighted average equilibrium function $\overline{K}_M$ is defined by Equation (10) for the reaction of the metal ion with the whole humic acid mixture. In this formalism [51], two protons are displaced from the bidentate humic acid site "AH_2". In addition to these sites that are location bound on the humic polyelectrolyte, there could be other ligands. The other possible ligands include ^-OH, H_2O, and extra humic functional groups. An equation of this type can be written for each of the metal ions present.

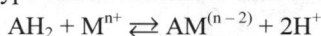

$$AH_2 + M^{n+} \rightleftarrows AM^{(n-2)} + 2H^+$$

$$\overline{K}_M = m_c m_H^2 / m_L m_M \qquad (10)$$

But the equilibrium for each metal ion will be influenced by competition from some or all of the other metal ions, as well as H^+. The material balance Equation (11) is the means by which all of the cations can be accounted for. C_S is the total molality of all of the bidentate chelation sites, and m_L is the molality of the doubly protonated sites AH_2. The number of metal chelates is "k" and in the experiments by Kerndorff and Schnitzer, $k = 11$ [51]. The material balance makes it possible for Equation (12) to describe the weighted average $\overline{K}_M$ for metal ion "M", in terms of the effects of all the other metal ions.

$$C_S = m_L + \sum_{j=1}^{k} m_{cj} \qquad (11)$$

$$K_M = -(1/\chi_L) \int_0^a \overline{K}_M d\chi_{C1} - \sum_{j=2}^{k} (1/\chi_L) \int_0^b \overline{K}_M d\chi_{C_j} \qquad (12)$$

$$a \equiv \chi_{C1}, \, b \equiv \chi_{Cj}$$

The first term on the right hand side of Equation (12) represents the metal ion "M", for which $\overline{K}_M$ has been defined. The summation accounts for the effects on this equilibrium, of all of the other metal ions. The χ_{Ci} symbols represent the mole fractions of chelating sites occupied by the various cations, including "L" for pairs of protons. For the Kerndorff - Schnitzer experiments, there are 11 of these equations. At constant temperature, the Equation (12) for each of the metal ions is a state function. That is, it fully defines the state of the humic acid - metal ion system in terms of inner variables. These inner variables are chemical composition variables in the form of mole fractions of chelating sites. Although Equation (12) is only a mathematical formalism, it can be used to deduce what kinds of calculations are needed to extract predictive chemical information from appropriate experiments. At least 3 kinds of information are of interest. The first is the distribution of K_M vs. χ_{Cj} values for each metal ion. Figure 10 shows the example for Cu^{++}. A plot of K_M vs. χ_L, the mole fraction of bidentate sites fully protonated simply gives a mirror image curve. These kinds of graphs could be useful for environmental protection and/or remediation. But the averaging that happens automatically when macroscopic measurements are made on a whole mixture obscures the high and low values of K_M, that exist for a metal ion reacting with a particular small portion of the mixture of chelating sites. There can sometimes be a difference between K_M and $\overline{K}_M$ of one or two orders of magnitude. The next kind of information reflects the complexing chemistry of location bound metal complexes at the molecular level. Another stage in the calculations is necessary to obtain this information.

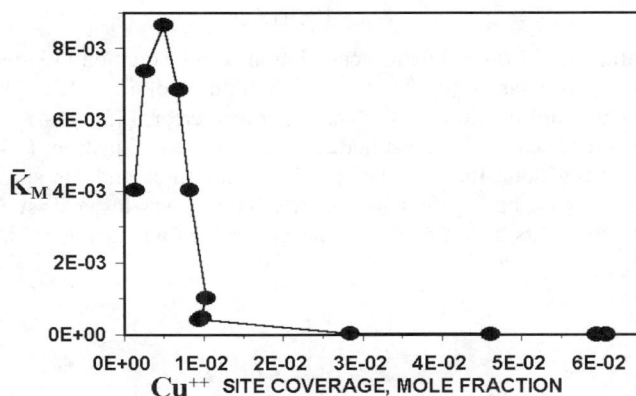

Figure 10. $Cu^{++} - 2H^+$ ion exchange reaction in the humic acid – multiple metal ion system at 25°C. The weighted average equilibrium function vs. the mole fraction of protonated exchange sites [51].

E *Binding Of Organic Chemicals*

In 1968 Walker and Crawford concluded from the evidence in the literature that triazine herbicides were bound or sorbed by soil organic matter [52]. Since then, a considerable number of authors have provided experimental evidence for the sorption or binding of organic chemicals by humic materials [53-60]. Hydrogen bonding, charge transfer, and hydrophobic interactions have all been reported as contributing causes. Haniff et al. Reported numerical values for the law of mass action equilibrium function for the solution phase complexing of atrazine by fulvic acid [58]. In that case, the equilibrium function was defined in terms of g of fulvic acid. But because the atrazine complexing capacity of the fulvic acid was also measured as a function of the degree of ionisation of the carboxyl groups, it would be possible to recalculate the equilibrium functions by using the complexing site stoichiometry. The feasibility of this has since been demonstrated for whole soils. The upgraded K values would then support predictive calculations much more reliably.

F *Molecular Level Descriptions As Chemical Systems*

In contrast to the practical engineering calculations for which the weighted average equilibrium functions should be used, any investigation or prediction of molecular level properties that reflect the components of a humic mixture requires that the averaging be undone. For equilibrium, two types of differential functions are used for describing humic mixtures as chemical systems at the molecular level. They are both intended to "undo the averaging" that happens automatically when macroscopic measurements are made on whole humic mixtures. These two types of differential functions are really only alternate descriptions of the same phenomena. They are derived from weighted average functions, one of which is the inverse function of the other.

The first type includes the distribution functions, Equation (13), which have been described above.

$$K_M = f(\chi_{C1}) \tag{13}$$

The differential equilibrium function calculation using Equation (13) was originally introduced for the particular limited purpose of obtaining individual K_M values, with the least mathematical amplification of the measurement errors. [3, 4, 48]. To test the validity of K_M numerical values estimated according to Equation (13), computer simulations have been done for hypothetical ligand mixtures and the single metal ion Cu^{++} [64]. A simple case having four monomeric ligands was tested first. Table 1 lists the Cu^{++} stability constants from Sillen and Martell for the four ligands [61].

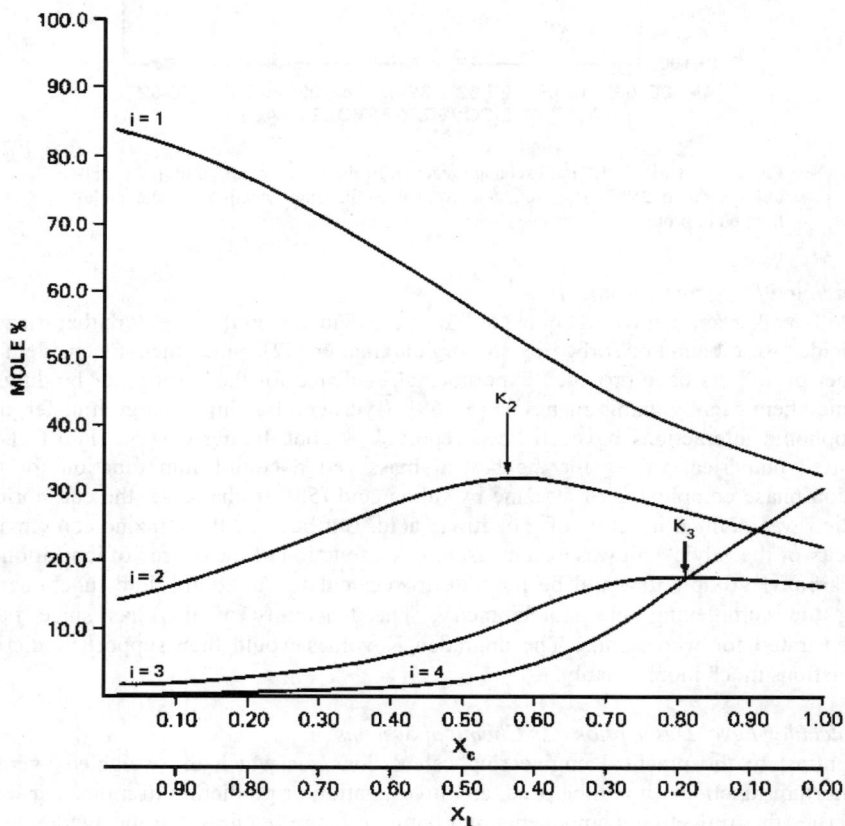

Figure 11. Mole percent contributions of components to the weighted average K_M. The effect of site coverage is shown [61, 65].

Good recoveries of the stability constants were obtained for three of the four ligands, with small systematic calculation errors. A reason for the fourth stability constant not being recovered is found in Figure 11. A ligand can only be detected if it responds to the probe cation. The percent contribution of the fourth complex was too weak to be effectively observed. A second simulation test with 100 hypothetical ligands was a more realistic model for real humic mixtures. In this example, one value missing from the end of the series was a rather small problem. It could be argued that in a real humic sample, if the weakest ligand at the end of a titration does not respond effectively to a probe metal ion, then it is not very important to the properties of the sample anyway.

Table 1. Computer simulation of Cu^{++} complexing in a mixed ligand system: The recovery from the K_M, of input values for stability constants [64, 65].

	Component Ligand	C_{Li}, Ligand Concentration	K_i		
i	Reagent	Molarity	Input Value	Recovered Value	Calculation Error, %
1	malonate	1.60×10^{-3}	3.548×10^5	3.521×10^5	−0.77
2	carbohydrazine	1.10×10^{-3}	8.318×10^4	8.053×10^4	−3.2
3	glycine ethyl ester	0.80×10^{-3}	1.380×10^4	1.367×10^4	−0.94
4	tartrate	1.50×10^{-3}	1.585×10^3	No Recovery	

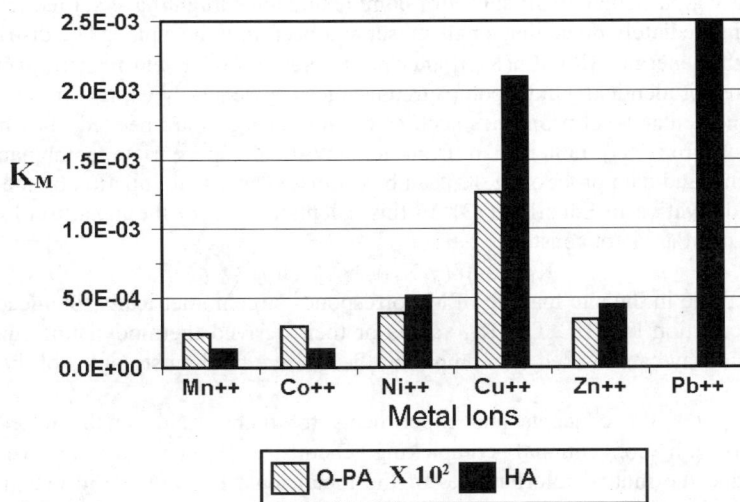

Figure 12. The Irving Williams series of metal ion complexing. O-PA; orthophthalic acid data from Martell and Smith. HA; the humic acid – multiple metal ion experiment [51, 61].

In the absence of interferences, a graph of Equation (13) will show K_M decreasing as χ_{cl} increases. The reason is that when a reagent is titrated onto a mixture of binding sites, the most strongly binding sites are occupied first. As the titration progresses, weaker sites are encountered. In fact for a true equilibrium, a Boltzmann distribution of site occupation is expected. That means that the humic mixture should have two kinds of distributions. Consider first that each location- bound site is characterized by its own K_M value. If it could be experimentally observed without a Boltzmann distribution, a histogram of K_M vs. n, the number of sites scanned up to the total N, would describe the collection of available complexing or sorption sites. The shape of such a histogram would depend on the sequence in which the sites were encountered as n, the number of them scanned, increased. This in turn would be determined by the hypothetical scan method. An imaginary titration without a Boltzmann distribution would produce a histogram in which the K_M bars become shorter as n, the number of sites scanned, increased. When the scan had reached the total, N, the histogram would represent the whole collection of location-bound sites. If the widths of the histogram bars approached the limit of zero, then the graph would approximate a smooth curve. Perdue [45] has suggested that for H^+ dissociation, the distribution of sites might be gaussian. But until this can be tested experimentally, other possibilities should be kept in mind. Natural humic mixtures might have truncated gaussian or non gaussian distributions. If a hypothetical titration curve could be observed without a Boltzmann distribution of site occupation, then a perfectly sharp titration end point would be revealed that would correspond exactly to the total number N of sites. In contrast to such a hypothetical case, a real equilibrium case would have a Boltzmann distribution of bound molecules or ions over the set of complexing sites characterized by a range of $[RT \times Ln(K_M)]$ values. Consequently a particular small subset of complexing sites might have started to fill up before the immediately preceding small subset had been fully occupied. The distribution of complexing energies $[RT \times Ln(K_M)]$ and the distribution of bound reagent over them are therefore not identical. Under some circumstances however, they might be similar.

When molecular level properties such as binding energies are needed, they have to be estimated from K_M rather than from K_M. With an appropriate combination of measurements and data processing, K_M can be estimated as a state function by the use of the partial derivative in Equation (13). In this calculation, all of the metal ion loadings except that of M are kept constant.

$$K_M = - [\partial(\chi_L K_M)/\partial\chi_{cl}] = f(\chi_{cl}) \qquad (13a)$$

Any increase in the site loading of M corresponds stoichiometrically to a decease of the site occupation by $2H^+$. The K_M values or their derived thermodynamic functions obtained this way can be used for comparing the relative binding strengths of the metal ions.

At this point, the concepts and experiments for humic acid can be tested by a comparison with conventional complexing chemistry. Because of distortion by averaging and the mutual interferences of the metal ions, K_M values did not properly follow the Irving Williams sequence. Assuming however that location bound chelating sites resemble typical monomeric ligands, then the differential K_M values estimated with Equation (13a) for the same site loadings should be consistent with this series. A comparison of the Kerndorff Schnitzer experiments with o-phthalic acid data from Sillen and Martell [61] is shown in Figure 12. It proves that classical metal complexing chemistry can be observed with location bound humic acid chelation sites, when the averaging effect is undone by using Equation (13). This implies that predictive calculations are possible, just as they are for the complexes of monomeric ligands. It is important to note however, that data processing and chemical calculations make heavy demands on the analytical chemistry.

Competition among metal ions for chelation sites is the third type of information that is obtainable from the multiple metal ion experiment. One way of doing this is to use Equation (14), which is a direct consequence of the material balance for total chelation sites. Each term in the summation has been evaluated by independent measurements for the mole fraction plot of site loading. The measurements and calculations have been tested in Table 2.

$$\sum_{j=1}^{k}\left(d\chi_j/d\chi_L\right)=-1 \tag{14}$$

Table 2. Acid Elution Equilibrium Of Metal Ions From Humic Acid. Tests for H^+ elution of metal ions and for metal ion competitions. $\chi_L \leq 0.94$ [44].

Metal Ion	$(d\chi c_j/d\chi_L)$	Comment
Cd^{++}	0.0118	Cations eluted by H^+ being replaced
Mn^{++}	0.0058	Cations eluted by H^+ being replaced
Co^{++}	0	No competitions
Zn^{++}	0	No competitions
Ni^{++}	−0.031	Eluted by H^+
Pb^{++}	−0.046	Eluted by H^+
Hg^{++}	−0.556	Eluted by H^+
Al^{+++}	0.0452	Cations eluted by H^+ being replaced
Cr^{+++}	0.0113	Cations eluted by H^+ being replaced
Cu^{++}	0.0089	Cations eluted by H^+ being replaced
Fe^{+++}	−0.443	Eluted by H^+
$\sum_{j=1}^{k}\left(d\chi c_j/d\chi_L\right)$	−0.982	Total: see Equation (14)
Error	0.02	2.0% Error

Equation (14) is supported to well within experimental error. Negative values indicate direct elution from chelation sites by H^+. A positive value for a metal ion indicates that it is replacing one or more other metal ions that are being eluted by H^+. Zero values represent a lack of net changes. Table 1 is an example of the kind of information that could be used for at least two types of remediation cases. Contaminated organic soils are the most obvious case. But if humic acids or peat soils containing them are used for scavenging metal ions from contaminated sites, then the subsequent treatment or disposal of the cleanup materials have to be considered. In all cases, processing methods and costs would depend on the behaviour of the metal ions. Table 1 data from inexpensive bench scale laboratory tests might allow engineering cost studies to determine budget estimates and feasibility on the industrial scale. One option is ultrafiltration stirred cells with atomic absorption spectrophotometry. Attempting to do this with field trials could cost an order of magnitude more.

When differential equilibrium functions had first been obtained from titrations, it was obvious that scans of them over the components of humic mixtures would be interesting. Although some real K values could be estimated from experimental titrations, there could be some distortion of the distribution curves. The attempts by several investigators to avoid the distortions have produced a considerable amount of published work using extensive mathematical transformations of the simple reaction mole fraction equations [62-66]. This other category of differential functions is the affinity spectrum. They are derived from the macroscopically measured mole fraction, χ_c, of binding sites that are occupied. χ_c is a weighted average function mole fraction that is explained by Equation (15).

$$\chi_c = K_M(m_M/m_H^2)/[1 + K_M(m_M/m_H^2] \tag{15}$$

Affinity spectrum calculation methods were first developed for acid dissociation equilibrium, defined for the whole mixture in the usual way by Equation (16).

$$K = [\alpha/(1 - \alpha)]M_H \tag{16}$$

In this case α, the degree of ionisation, is the mole fraction of the reaction that has been completed. By 1970 it was already evident from known relationships that α is a weighted average function for polyelectrolyte mixtures such as humic materials and proteins [3, 48]. A few reports gave descriptions of this similar to Equation (17) [43, 46, 49].

$$\alpha = 1 \bigg/ C_A \sum_{i=0}^{N} \alpha_i c_i \tag{17}$$

Continuing previous work by Klotz and Hunston for proteins, Hunston used a numerical approximation method for estimating an N vs. Log(K) distribution. The variable N was directly related α. This pioneered a large amount of subsequent work on the type of distribution functions that became known as affinity distribution spectra [44, 45, 63, 64, 66-68]. All of the calculation methods were based on weighted average mole fraction functions like Equation (17). The α_i values are summed up, with the c_i values being their statistical weights for the weighted average. In the literature for affinity distribution spectra, an alternate description interprets c_i as being related to the probability of the corresponding α_i value being encountered at a particular point on the scan of the affinity spectrum. The replacement of the summation in Equation (17) with an integration again models the mixture as a continuum, with the spectrum of α_i values being replaced by the differential mole fraction function $\beta = f(c)$. The statistical weight then becomes dc. But because it is the distribution of equilibrium constants that is of direct chemical interest, the integrals are mathematically transformed to give equations of the generic type represented by Equation (18).

$$\left. \begin{aligned} \alpha &= 1/C_A \int_{-\infty}^{\infty} \beta c(K) dLn(K) \\ \beta &= (K/M_H)/[1 + (K/M_H)] \end{aligned} \right\} \tag{18}$$

Equation (18) has been further elaborated with more mathematical transformations, for calculations that incorporated the Scatchard Plot strategy into the methods and applied inverse Laplace transform techniques. Important critical reviews of this research include especially those of Altman and Buffle [67], Buffle and Altman [44], and Perdue [45]. The reviews include comparisons of the differential equilibrium function and affinity spectrum types of calculation methods. At least four conclusions may be drawn from the reviews.

First, the differential equilibrium function is the quickest method for undoing the experimental averaging of titrations. There is however, some discussion about whether it is rather oversimplified. As a follow-up of these discussions, the mathematical conditions and limits have been explored with direct trial calculations [65]. Two key points emerged from the use of simulated mixtures having known or defined K_i values.

1. During the titration of a reactive mixture, the % contributions of its components to the weighted average K of the whole mixture go through maxima in sequence. At the maximum of each component peak, the K_i of that component can be estimated to within a calculation error of a few percent. Distortions at the extreme ends of the titration curve can be prevented. This use of the maxima with the weighted average K formalism and the use of local isotherms for Scatchard Plot calculations using the weighted average α formalism both give K_i values to within a few %. The reason for this is that they both give estimates at 50% reaction of individual components.

2. For experimental titration curves, the oversimplification is that the assignments of K_i values to components are not known. The experimental distribution functions do not effectively represent the assignments.

Secondly, an affinity spectrum calculation that integrates the Scatchard Plot technique into the procedure at the molecular level gives K distribution curves with the least distortion. Descriptions of the molecular level use of the Scatchard Plot technique refer to the use of "local Langmuir isotherms", which simply represent the reaction mole fractions of hypothetical small components of the humic mixture. This use of molecular level Scatchard Plot methodology should not of course, be confused with incorrect macroscopic Scatchard plots for the whole mixture.

The third important conclusion is that mathematical expressions and calculation procedures that become too complex can produce excessive errors in the answers, for two reasons. Not only are the measurement errors of real experiments amplified, but also calculation and/or computer truncation errors can be introduced. If care is not taken, significant mathematical artefacts can result. Balanced judgements must therefore be made about the practical trade-off between mathematical completeness and the avoidance of mathematical artefacts. This suggests the use of sensitivity analyses that are comparable to those used for computer models.

Finally, there is a fourth important observation that is not made in any of the reviews. A real remediation project will commonly encounter a combination of contaminants and naturally occurring chemicals. A mixture of metal ions is an example. This means that the predictive engineering calculations should account for relatively large numbers of chemical composition variables. The multiple metal ion complexing example described above shows that this needs a combination of chemical stoichiometry, mass balances, and partial differential calculus. As demonstrated by the example above, this can only be done with the differential equilibrium function calculation method. None of the affinity spectrum calculation methods can be adapted to a mixture of contaminants.

3. Conclusions: Progress Toward Quantitative Predictability

During the last 30 years, there has been steady progress toward the quantitative predictability of equilibrium for the chemical reactions and physical chemical interactions of humic materials with monomeric reagents. Ionisable and polar functional groups are the key to the weak acid polyelectrolyte, metal complexing, and organic chemical binding properties of humic materials. The weak acid polyelectrolyte properties that influence the equilibrium include hydrophobic interactions and hydrogen bonding aggregation. Literature reports accumulating for decades demonstrate that the calibration of chemical mechanism parameters under one set of conditions can be used for predictive calculations under other conditions, in the same sense that this is routinely done for monomeric pure reagents. This has been achieved by developing concepts and experimental methods that allow the principles of classical chemistry to be applied to complicated mixtures. There are some practical implications. One is that bench scale laboratory calibrations permit extrapolations to field conditions. Another is that the information is less site specific than the "chemically blind" raw data obtained from field trials, for example in a peat bog. The laboratory test costs can also be an order of magnitude less than those of the field trials. There is also an important point about environmental protection. Monitoring projects at the field level can only document at considerable continuing cost, the nature and scope of a problem after something has already gone wrong. On the other hand, predictive engineering calculations done at less cost should help to prevent or control such problems. Attempts should now be made to expand, adapt, and apply the known chemistry to regulatory and engineering practice.

The chemical kinetics of humic materials and their reactions with monomeric reagents deserve to be reviewed because of the important progress that had been made. But there is not enough time available for kinetics to be included in this review. The available evidence indicates that the redox properties of humic materials are also important, and deserve some more systematic investigations.

4. References

1. Gamble, D.S., Langford, C.H. and Webster, G.R.B. (1994) Interactions of pesticides and metal ions with soils: Unifying concepts, *Reviews of Environmental Contamination and Toxicology* **135**, 63-91.
2. Gamble, D.S., and Khan, S.U. (1985) Atrazine hydrolysis in soils: Catalysis by the acidic functional groups of fulvic acid, *Can. J. Soil Sci.* **65**, 435-443.
3. Gamble, D.S. (1970) Titration curves of fulvic acid: the analytical chemistry of a weak acid polyelectrolyte, *Can. J. Chem.* **48**, 2662-2669.
4. Gamble, D.S. (1972) Potentiometric titration of fulvic acid: Equivalence point calculations and acidic functional groups, *Can. J. Chem.* **50**, 2680-2690.
5. Khan, S.U. (1978) Kinetics of hydrolysis of atrazine in aqueous fulvic acid solution, *Pestic. Sci.* **9**, 39-43.
6. Wilson, D.E. and Kinney, P. (1977) Effects of polymeric charge variations on the proton - metal ion equilibria of humic materials, *Limnology & Oceanography* **22**, 281-289.
7. Chowdhury, A.N. and Bose, B.B. (1971) Role of "humus matter" in the formation of geochemical anomalies, in, R.W. Boyle and J.I. McGerrigle (eds.), *Geochemical Exploration,* Proceedings, 3rd International geochemical Exploration symposium, Toronto, April 16-18, 1970, The Canadian Institute Of Mining And Metallurgy, Special Volume 11.

8. Matsuda, K. and Schnitzer, M. (1971) Reactions between fulvic acid, a soil humic material, and dialkyl phthalates, *Bull. Environ. Contam. & Toxicology* **6**, 200-204.
9. Burns, G., Hayes, M.H.B. and Stacey, M. (1973) Some Physico-chemical interactions of paraquat with soil organic materials and model compounds. I Effects of temperature, time and adsorbate degradation on paraquat adsorption, *Weed Res.* **13**, 67-78.
10. Burns, G., Hayes, M.H.B. and Stacey, M. (1973) Some Physico-chemical interactions of paraquat with soil organic materials and model compounds. II Adsorption and desorption equilibria in aqueous suspensions, *Weed Res.* **13**, 79-90.
11. Gamble, D.S. Alan, W. (1980) Underdown and Cooper H. Langford. Copper(II) titration of fulvic acid ligand sites with theoretical, potentiometric and spectrophotometric analysis, *Anal. Chem.* **52**, 1901-1908.
12. Tummavuori, J. and Aho, M. (1980) On the ion exchanger properties of peat. Part I: On the adsorption of some divalent metal ions (Mn^{2+}, Co^{2+}, Ni^{2+}, Zn^{2+}, Cd^{2+} and Pb^{2+}) on the peat, *Suo* **31**, 45-51.
13. Rainville, D.P. and Weber, J.H. (1982) Complexing capacity of soil fulvic acid for Cu^{2+}, Cd^{2+}, Mn^{2+}, Ni^{2+}, and Zn^{2+} measured by dialysis titration: a model based on soil fulvic acid aggregation, *Can. J. Chem.* **60**, 1-5.
14. Ryan, D. K., Thompson, C. P., and Weber, J. H. (1983) Comparison of Mn^{2+}, Co^{2+}, and Cu^{2+} binding to fulvic acid as measured by fluorescence quenching, *Can. J. Chem.* **61**, 1505-1509.
15. Langford, C.H., Gamble, D.S. and Alan, W. (1983) Underdown and Stephen Lee. Interacion of metal iopns with a well characterized fulvic acid, Chapter 11, in R.F. Christman and E.T. Gjessing (eds.), *Aquatic and Terrestrial Humic Materials*, Ann Arbor Science Publishers, Ann Arbor MI 48106.
16. Haniff, M.I., Zienius, R.H., Langford, C.H. and Gamble, D.S. (1985) The solution phase complexing of atrazine by fulvic acid: Equilibria at 25°C, *J. Environ. Sci. Health* **B20(2)**, 215-262.
17. Sojo, L.E., Gamble, D.S., Langford, C.H. and Zienius, R.H. (1989) The reactions of paraquat and divalent metal ions with humic acid: Factors influencing stoichiometry, *J. Environ. Sci. Health* **B24(6)**, 619-646.
18. Lee, J.-F., Crum, J. R. and Boyd, S. A. (1989) Enhanced retention of organic contaminants by soils exchanged with organic cationes, *Environ. Sci. Technol.* **23**, 1365-1372.
19. Maes, A. and Cremers, A. (1990) Assessment of he capacity for complexation in natural organic matter, in, R. Merckx, H. Vereecken and K. Vlassak (eds.), *Fertilization and the environment*, Leuven University Press., pp. 101-115.
20. Mitchell, J. (1932) The origin, nature, and importance of soil organic constituents having base exchange properties, *J. Amer. Soc. Agron.* **24**, 256-275.
21. Lees, H. (1950) A Note on the copper - retaining power of a humic acid from a peat soil, *The Biochem. J.* **46**, 450-451.
22. Szalay, S. (1954) Studies on the adsorption of cations of large atomic weight on humus colloids, *Comm. Third Math. - Phys. Class Hung. Acad. Sci.* **4**.
23. Schlichting, E. (1955) Kupferbindung und -fixerung durch humusstoffe, *Acta Agr. Scand.* **5**, 313-356.
24. Szalay, A. (1957) The role of humus in the geochemical enrichment of U in coal and other bioliths, *Acta Phys. (Acad. Sci. Hung.)* **8**, 25-35.
25. Szabo, I. (1958) Adsorption of cations on humus preparations, *Hungarian academy of scienc, Budapest, Department of Mathematics & Physics* **8**, 393-402.
26. Basu, A. N., Mukherjee, D. C. and Mukherjee, S. K. (1964) Interaction between humic acid fraction of soil and trace element cations, *J. Indian Soc. Soil Sci.* **12**, 311-318.
27. Himes, F. L. and Barber, S. A. (1957) Chelating ability of soil organic matter, *Soil Sci. Soc. Amer. Proc.* **21**, 368-373.
28. Li, J., Gamble, D. S., Pany, B. C. and Langford, C. H. (1992) Interaction of lindane with a Laurentian fulvic acid: A complexation model, *Environ. Technol.* **13**, 739-749.
29. Wang, Z., Gamble, D. S., and Langford, C. H. (1991) Interaction of atrazine with Laurentian humic acid, *Anal. Chim. Acta* **244**, 135-143.
30. Gamble, D. S., Haniff, M. I. and Zienius, R. H. (1986) Solution phase complexing by fulvic acid: A Batch Ultrafiltration Technique, *Anal. Chem.* **58**, 727-731.
31. Gamble, D. S., Haniff, M. I. and Zienius, R. H. (1986) Solution phase complexing by fulvic acid: A theoretical comparison of ultrafiltration methods, *Anal. Chem.* **58**, 732-734.
32. Gran, G. (1952) *Analyst.* **77**, 661.

78

33. Mitchell, J. (1932) The origin, nature, and importance of soil organic constituents having base exchange properties, *J. Amer. Soc. Agron.* **24**, 256-275.
34. Buffle, J. and Staub, C. (1984) Measurement of complexation properties of metal ions in natural conditions by ultrafiltration: Measurement of equilibrium constants for complexation of zinc by synthetic and natural ligands, *Anal. Chem.* **56**, 2837-2842.
35. Fu, G., Allen, H. E. and Cao, Y. (1992) The importance of humkic acids to proton and cadmium binding in sediments, *Environ. Toxicol. Chem.* **11**, 1363-1372.
36. Szalay, A. and Szilagyi, M. (1961) Investigations concerning the retention of fission products on humic acids, *Acta Phys. Hung.* **13**, 421-436.
37. Szalay, A. (1964) Cation exchange properties of humic acids and their importance in the geochemical enrichment of UO_2^{++} and other cations, *Geochim. et Cosmochim. Acta.* **28**, 1605-1614.
38. Carter, M. C., Kilduff, J. E. and Weber Jr., W. J. (1995) Site energy distribution analysis of preloaded adsorbents, *Environ. Sci. Technol.* **29**, 1773-1780.
39. Khanna, S. S. and Stevenson, F. J. (1962) Metallo-organic complexes in soils:I. Potentiometric titration of some soil organic matter isolates in the presence of transition metals, *Soil Sci.* **93**, 298-305.
40. Chuveleva, E. A., Chmutov, K. V. and Nazarov, P. P. (1962). The ion-exchange sorption of radio-elements by soils. III. The dissociation constants of the carboxyl groups in humic acid, *Russian J. Phys. Chem.* **36**, 432-433.
41. Posner, A. M. (1964) Titration Curves of humic acid, in *8th International Congress of Soil Science*, Transactions, Vol. III, Publishing House of the academy of the Socialist Republic of Romania, Bucharest, Romania, pp. 161-174.
42. Mantoura, R. F. C. and Riley, J. P. (1975) The use of gel filtration in the study of metal binding by humic acids and related compounds, *Anal. Chim. Acta* **78**, 193-200.
43. Gamble, D. S. and Schnitzer, M. (1973) The chemistry of fulvic acid and its reactions with metal ions, Chapter 9, in P. C. Singer (ed.), *Trace metals and metal-organic interactions in natural waters*, Ann Arbor Science Publishers Inc., Ann Arbor, Michigan.
44. Buffle, J. and Altmann, R. S. (1987) Interpretation of metal complexation by heteeneous complexants, Chapter 13, in W. Stumm (ed.), *Aquatic Surface Chemistry*, John Wiley & Sons, New York.
45. Perdue, E. M. (1985) Acidic functional groups of humic substances, Chapter 20, in G. R. Aken, D. M. McKnight, R. L. Wershaw, and P. MacCarthy (eds.), *Humic Substances in soil, sediment, and water*, John Wiley & Sons, New York.
46. Manning, P. G. and Ramamoorthy, S. (1973) Equilibrium studies of metal - ion complexes of interest to natural waters - VII. Mixed-ligand complexes of Cu(II) involving fulvic acid as primary ligand, *J. Inorg. Nucl. Chem.* **35**, 1577-1581.
47. Burch, R. D., Langford, C. H. and Gamble, D. S. (1978) Methods for the compaison of fulvic acid samples: the effects of origin and concentration on the acidic properties, *Can. J. Chem.* **56**, 1196-1201.
48. Gamble, D. S., Schnitzer, M. and Hoffman, I. (1970) Cu^{2+} - fulvic acid chelation equilibium in 0.1 m KCl at 25.0°C, *Can. J. Chem.* **48**, 3197-3204.
49. Klotz, I. M. and Hunston, D. L. (1971) Properties of graphical representation of multipe classes of binding sites, *Biochem.* **10**, 3065-3069.
50. Benedetti, M. F., Milne, C. J., Kinniburgh, D. G., Riemsdijk, W. H. and Koopal, L. K. (1995) Metal ion binding to humic substances: Application of the non-ideal competitive adsorption model, *Environ. Sci. Technol.* **29**, 446-457.
51. Gamble, D. S., Schnitzer, M., Kerndorff, H. and Langford, C. H. (1983) Multiple metal ion exchange equilibria with humic acid, *Geochim. Et Cosmochim. Acta* **47**, 1311-1323.
52. Walker, A. and Crawford, D. V. (1968) The role of organic matter in adsorption of the triazine herbicides by soils, *Proceedings of the symposium on the use of isotopes and radiation in soil organic-matter studies*, International Atomic Energy Agency, Vienna, July 15-19, pp. 91-105.
53. Nearpass, D. C. (1972) Hydrolysis of propazine by the surface acidity of organic matter, *Soil Sci. Soc. Amer. Proc.* **36**, 606-610.
54. Khan, S. U. and Schnitzer, M. (1972) The retention of hydrophobic organic compounds by humic acid, *Geochim. et Cosmochim. Acta* **36**, 745-754.
55. Grice, R. E., Hayes, M. H. B., Lundie, P. R. and Cardew, M. H. (1973) Continous flow method for studying adsorption of organic chemicals by a humic acid preparation, *Chem. And Ind.*, 233-234.
56. Guy, R. D., Narine, D. R. and DeSilva, S. (1980) Organocation speciation. I. A comparison of the interaction of methylene blue and paraquat with bentonite and humic acid, *Can. J. Chem.* **58**, 547-554.

57.   Lindqvist, I. (1982) Charge-transfer interaction of humic acids with donor molecules in aqueous solutions, *Swedish J. Agric. Res.* **12**, 105-109.
58.   Haniff, M. J., Zienius, R. H., Langford, C. H. and Gamble, D. S. (1985) The solution phase complexing of atrazine by fulvic acid: Equilibria at 25°C, *J. Environ. Sci. Health.* **B20(2)**, 215-262.
59.   Gerstl, Z. and Kliger, L. (1990) Fractionation of the organic matter in soils and sediments, *J. Environ. Sci.* **B25(6)**, 729-741.
60.   Sojo, L. E. (1992) Ultrafiltration as speciation tool for paraquat in humic acids: effects of solution composition on membrane properties, *Anal. Chim. Acta* **258**, 219-227.
61.   Sillen, L. G. and Martell, A. E. (1971) *Stability constants of metal-ion complexes*, Special Publiction No. 25, The chemical Society, London.
62.   Hunston, D. L. (1975) Two techniques for evaluating small molecule-macromolecule binding in complex system, *Anal. Biochem.* **63**, 99-109.
63.   Shuman, M. S., Collins, B. J., Fityzgerald, P. J. and Olson, D. L. (1983) Distribution of stability constants and dissociation rate constants among binding sites on estuarine copper-organic complexes: rotated disk electrode studies and an affinity spectrum analysis of ion-selective electrode and potentiometric data, Chapter 17, in R. F. Christman and E. T. Gjessing (eds.), *Aquatic and Terrestrial Humic Materials*, Ann Arbor Science, Ann Arbor Michigan 48106.
64.   De Wit, J. C. M., Van Riemsdijk, W. H., Nederlof, M. M., Kinniburgh, D. G. and Koopal, L. K. (1990) Analysis of ion binding on humic substances and the determination of intrinsic affinity distributions, *Analytica Chimica Acta* **232**, 189-207.
65.   Gamble, D. S. and Langford, C. H. (1988) Complexing equilibria in mixed ligand systems: Tests of theory with computer simulations, *Environ. Sci. Technol.* **22**, 1325-1336.
66.   Černík, M., Borkovic, M. and Westall, J. C. (1995) Regularized least squares methods for the calculation of discrete and continuous affinity distributions for heterogeneous sorbants, *Environ. Sci. Technol.* **29**, 413-425.
67.   Altmann, R. S. and Buffle, J. (1988) The use of differential equilibrium functions for interprtation of metal binding in complex ligand systems: Its relation to site occupation and site affinity distributions, *Geochim. Cosmochim. Acta* **52**, 1505-1519.
68.   Nederlof, M. M., Van Riemsdijk, W. H. and Koopal, L. K. (1990) Determination of adsorption affinity distributions: A general framework for methods related to local isotherm approximations, *J. Colloid Interface Sci.* **135**, 410-426.

Part 2

Complexing interactions of humic substances with heavy
metals and radionuclides and their remedial implementation

INTERACTIONS OF HUMIC SUBSTANCES WITH TRACE METALS AND THEIR STIMULATORY EFFECTS ON PLANT GROWTH

A. KASCHL[1], Y. CHEN[2]

[1]*Centre for Environmental Research Leipzig-Halle (UFZ), Department of Groundwater Remediation, Permoserstr. 15, 04318 Leipzig, Germany*
[2]*Hebrew University of Jerusalem, Faculty for Agricultural, Food and Environmental Quality Sciences, Department of Soil and Water, P.O. Box 12, 76100 Rehovot, Israel <yonachen@agri.huji.ac.il.>*

Abstract

Humic substances (HS) are a quantitatively and qualitatively important component of soil organic matter (SOM). Due to their diverse associations with trace elements, they have a special relevance for many geochemical processes, the mobility and bioavailability of pollutants and micronutrients in the soil environment and can find special application as immobilisers of contaminants for remediation purposes.

In the current paper the role of HS in the soil environment is shortly addressed, followed by a review of the binding types and strengths of humic-trace metal complexes found in the literature. The most commonly employed methods for examining these interactions are described. Finally, evidence for stimulatory effects of HS on plant growth is presented, as the result of a better provision with the essential micronutrients Fe and Zn.

1. Introduction

As organic materials in the soil decay, macromolecules of a mixed aliphatic and aromatic nature are formed. The term *humus* is widely accepted as a synonym for soil organic matter (SOM) or natural organic matter (NOM, while dissolved organic matter (DOM) refers to water-borne organic substances). It is defined as the total of the organic compounds in soil and water, exclusive of non-decayed plant and animal tissues, their partial decomposition products, the biomass and defined biochemicals of small molecular weight.

The chemical and colloidal properties of NOM can be properly studied only in the free state, that is, when free of inorganic components. Thus, the first task of the researcher is to separate NOM (and SOM, in particular) from the inorganic matrix. Alkali, usually 0.1 to 0.5M NaOH, has been a popular extractant of SOM. A detailed discussion of different extraction procedures can be found elsewhere [1, 2] .

I. V. Perminova et al. (eds.),
Use of Humic Substances to Remediate Polluted Environments: From Theory to Practice, 83–113.
© 2005 *Springer. Printed in the Netherlands.*

Humic substances (HS) are ubiquitous in nature: they are found both in aqueous as well as terrestrial environments in almost all climate zones. They represent a significant proportion of total organic carbon in the global carbon cycle, while in soils they constitute the major organic fraction (70-80%) [3, 4]. More importantly, they constitute the most active fraction of the SOM due to their effects on soil ecology, soil structure, plant growth and complexation of metals [5].

HS are best described as a series of acidic, yellow-to-black coloured, moderately high-molecular weight polyelectrolytes that are formed by secondary synthesis reactions and have characteristics dissimilar to any of the compounds occurring in living organisms [1, 4, 6, 7].Chemically they have an aromatic core structure with attached or trapped aliphatic, carbohydrate and peptide moieties and a considerable degree of carboxyl functionalities [1, 8].

Researchers have traditionally divided HS into the operationally defined fractions: humic acid (HA), fulvic acid (FA) and humin. FA remains soluble even in acidic solutions, while HA is soluble only at pH>4 and humin is insoluble [1]. The higher solubility of FA is due to lower molecular weight and a higher content of acidic functional groups on the molecules [1, 7, 9]. In soil systems, HAs are predominantly insoluble and associated with the mineral phase [10]. Humin are HA-type molecules that are so strongly complexed by clays and hydrous oxides that they no longer can be extracted by dilute base or acid [11].

2. The Role of Trace Metal Binding by HS in the Soil Environment

One of the most striking characteristics of HS is their ability to interact with metal ions to form water-soluble, colloidal and water-insoluble complexes of varying properties and widely differing chemical and biological stabilities [1, 6]. Considering the binding of trace metals, HS may best be described as ligands with a variety of different functional groups that can bind trace metals in a number of ways. HS are quantitatively the most important organic fraction in soils and have equal and/or affinity towards trace metals compared to other organic ligands [12, 13]. The formation of metal-organic complexes has the following effects on the micronutrient cycle in soils [1, 7]:

A. Micronutrient cations that would ordinarily precipitate at the pH values found in most soils are maintained in solution through complexation with soluble organics. Complexes of trace elements with FA are water-soluble. Complexes of HA with trace metals such as Fe are likely to form colloids that will remain in solution or form suspended particles. Either of these conformations will contribute to the mobilization of the metal to the plant root.

B. Under certain conditions, metal ion concentrations may be reduced to a non-toxic level through complexation with HS. This is particularly true when the metal-organic complex has low solubility, such as in the case of complexes with HA and other high-molecular-weight components of organic matter (OM).

C. Various organic complexing agents mediate transport of trace elements to plant roots and, in some cases, to other ecosystems, such as lakes and streams.

D. Organic substances can enhance the availability of insoluble phosphates through the complexation of Fe and Al in acid soils, and Ca in calcareous soils.

E. Organic chelation plays a major role in the weathering of rocks and minerals.

In addition, it has been demonstrated recently, that HS actually interact in solution (in addition to metal ions) with free ligands and complexes of high stability without causing any ligand exchange[14-16]. Sorption of ligands either in their free or metal complexed form may occur, thus reducing the free transport of the chelate or ligand in the solution via sorption followed by aggregation and possibly precipitation. The contribution of HS to the migration of micronutrients and toxic elements might therefore be much more complex than previously believed.

The micronutrient cation pools in soils can be divided into [1]: (i) water soluble (free plus complexed); (ii) suspended organo-metal complexes; (iii) exchangeable; (iv) specifically adsorbed; (v) organically complexed, but water insoluble; (vi) insoluble inorganic precipitates; and (vii) held in primary minerals. The importance of the organically complexed pool (and HS as its main component) arises from findings indicating that organically bound forms of the micronutrient cations are more available to plants than those included in insoluble inorganic precipitates or held in primary minerals [17-19]. Organically bound micronutrient cations in soils are commonly determined using extraction with a complexing agent (e.g., pyrophosphate at 0.1M concentration), or by release to exchangeable forms via OM oxidation [20, 21].

In a study conducted by McLaren and Crawford [22], from 20 to 50% of the Cu in 24 diverse soils occurred as organic complexes. They concluded that the amount of Cu available to plants (exchangeable and soluble Cu) was controlled by equilibrium involving specifically adsorbed forms (Cu extracted with 2.5% CH_3COOH) and the organically bound fraction. The suggested relationship between the three forms was as follows:

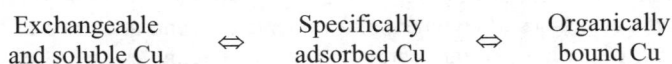

$$\text{Exchangeable and soluble Cu} \Leftrightarrow \text{Specifically adsorbed Cu} \Leftrightarrow \text{Organically bound Cu}$$

Shuman [23] obtained the following percentages of three micronutrients that occurred in organically bound species in 10 representative soils of the south-eastern U.S.A.: Cu, 1.0 to 68.6%; Mn, 9.5 to 82%; and Zn, 0.2 to 14.3%. A somewhat similar range for Zn (0.1-7.4%) was obtained by Ivengar [24] for 19 soils from the Appalachian, Coastal Plain and Piedmont regions of Virginia. Sedberry [25] found that an average of 2.6% of the Zn in 10 Louisiana soils was organically complexed. As one might expect, considerable variation exists in the distribution of organically bound forms of micronutrients among the various size fractions of soil and among soils differing in pH and other chemical properties [23].

Boron (B), Mo, Cr, As and a few other elements are unique among the micronutrient elements in that they normally occur in anionic forms (e.g. $H_2BO_3^-$ and MoO_4^{2-}) and thus they are subject to losses through leaching. However, the main form of B (the only non-metal in the group) may be combined with OM, primarily as borate complexes with compounds that contain the cis-hydroxyl group, such as carbohydrates [26, 27]. As OM is mineralised by microorganisms, B is released as readily available species. Temporary B deficiency in plants during periods of drought has been attributed to reduced mineralization of B in organically bound forms. Some studies [26, 27] have shown that the sorption capacity of composted OM for B (on a weight basis) was at least four times greater than for clay and soils. This was attributed to chemical association between B and organic molecules, such as HS. These binding reactions were shown to play a major role in the protection from excessive concentrations of B, which often occur in soils irrigated with reclaimed wastewater. This is one of the prominent and significant detoxifying beneficial effects of HS on plant growth.

Conversion of soluble forms of the micronutrient cations into insoluble organic forms can occur through solid-phase complexation by humates present as coatings on clay surfaces, as well as by the formation of soluble complexes that subsequently become associated with mineral surfaces through adsorption. Some polyvalent cations link humic complexes to clay surfaces; others occupy peripheral sites and are available for exchange with ligands of the soil solution.

3. Characterizing the Complexation of HS with trace metals: Binding Types, Complexing Capacity and Stability Constants

3.1. NATURE OF THE METAL BINDING SITES ON HS

Copper is the most commonly studied trace metal regarding its association with HS, probably due to its toxicity to biota, its release into the environment as a result of natural and anthropogenic sources and its known affinity for binding to organic matter [1, 28-47]. Other trace metals, whose binding properties regarding HS have commonly been addressed in recent research studies, are Cd, Zn and Pb, with fewer studies available for Hg, Ni, Cr and Co [1, 13, 28, 29, 32, 36, 38-40, 42-46, 48-59]. Some studies examined the binding characteristics of commercially valuable elements such as Ag, Au, Pt, Pd, U and V [30, 60, 61], while the rare earth elements and actinides La^{3+}, Eu^{3+}, Ce^{3+}, Sm^{3+}, Gd^{3+}, Am^{3+} and Y^{3+} have also been investigated [62-65].

A complex of a metal ion and an organic molecule arises when water molecules surrounding the metal ion are replaced by other molecules or ions, with the formation of a coordination compound. The organic molecule that combines with the metal ion is commonly referred to as the ligand. Functional groups having unshared pairs of electrons and able of forming coordination linkages with metal ions are shown below in the order of decreasing affinity for metal ions [1]:

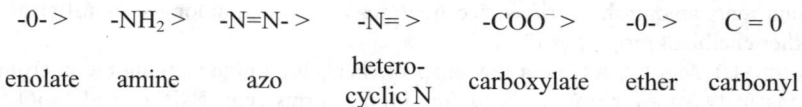

-0- >	$-NH_2$ >	-N=N- >	-N= >	$-COO^-$ >	-0- >	C = 0
enolate	amine	azo	hetero-cyclic N	carboxylate	ether	carbonyl

The most important metal binding groups of HS are generally believed to be O-containing functionalities like COOH and phenolic OH [34, 41, 66-72]. Depending upon source, location, parent material, climate and other factors, HS may have varying compositions. However, COOH and OH groups generally dominate in the humic molecules. A comprehensive statistical analysis of data showed that the content of different forms of C (alkyls, O-alkyls, aromatics, carbonyls) in HS from cultivated soils were similar, with carboxyl and carbonyl groups together constituting 15.7(±2.7)% of total C for HA and 26.9(±4.7)% of total C for FA [73]. The so-called 'model HA' for soils was listed as containing 3.6 mol COOH kg^{-1} dry matter and 3.9 mol OH kg^{-1} dry matter, and 'model FA' was given as 8.2 mol kg^{-1} COOH and 3.0 mol kg^{-1} OH [6]. The FAs are generally more oxidized than HAs, resulting in a higher content of carboxylic and phenolic groups [1].

HS from different sources usually show characteristic features. For instance, it was reported that HS derived from composted municipal solid waste (MSW) differed from soil HS in their structural and chemical composition [74, 75]. Particular features of compost-derived humic acids (HA) included lower aromaticity and higher aliphaticity, higher contents of H and N, lower levels of O-containing functional groups (carboxyl, carbonyl, phenolic C), and smaller macromolecular sizes [5, 46, 74, 75]. Similarly, NMR results for coal HA showed that it was more aromatic and contained less carboxylic groups than common soil HA [76]. These differences may noticeably affect the metal binding properties of HS.

In general, FA binds Zn and Cu in higher quantities than HA [1]. This can largely be explained by the higher content of carboxyl and phenolic groups in the FA molecules. In a study of metal binding to compost HS, the ratio of metal associated with the humic ligand and the acid-titrable groups present on the ligand was practically identical between HA and FA in the case of Zn and Cu, but not for Cd [46]. Stevenson [1] upholds that the maximum binding capacity for any given metal ion is approximately equal to the content of acidic functional groups, primarily COOH. However, in the above experimental study, the complexing capacity (CC) of HS constituted 65-75% (Cu) and 35-50% (Zn, Cd) of all acid-titrable groups, when assuming that both charges of the divalent metals were linked to acidic groups on the ligand [46].

The COOH and phenolic OH functional groups represent hard donors, whose interactions with trace metals are mostly driven by electrostatic forces and the entropy gains are derived from changes in orientation of H_2O molecules in the hydration sphere (outer-sphere complexes) [4]. In theory, groups containing the intermediate donors N or S should also be important for the complexation of borderline metals (which bind to hard and soft donors) such as Fe, Ni, Cu, Zn and Pb. Similarly, soft donors like Cd and Hg may be expected to preferably bind to N or S-containing functional groups [1, 4, 77, 78]. Therefore it is not surprising that N and S-containing groups have been shown to actively contribute to trace metal binding in several complexation studies [30, 61, 77, 79-83]. The N content in soil HS ranges on average from 3.2% (mass) dry matter for HA to 2.1% for FA. As has been pointed out for the content of acidic groups, HS from other sources may have a higher, or lower, content of N. For example, HS from sewage sludge or compost generally have a higher N content [46, 83]. This may have important consequences for the complexation of trace metals, when these organic fertilizers are added to the soil [13, 46, 84]. The content of S in HS on average lies between 0.8 to 1.9% (mass) dry matter for soil HA and FA, respectively [1]. While S-containing functional groups are almost absent in compost HS, sewage sludge HS may contain considerably higher amounts [85-87].

It is known that divalent or multivalent trace metals can form chelate rings with two or more adjacent sites on the same macromolecule [1, 7, 9]. If the chelating agent forms two bonds with the metal ion, it is called a bidentate ligand; similarly, there are tridentate, tetradentate, pentadentate and hexadentate complexes. Due to the complex structure and composition of HS, studies of trace metal binding have yielded differing results concerning the existence of inner and outer-sphere complexes as well as uni- and multidentate associations, even for the same trace element [29, 31, 34, 35, 71, 88-91]. Recent spectroscopic approaches such as X-ray spectroscopy indicate that tetra-, six- or octo-dentate coordination structures may be most relevant for Cu and Cd [31, 57, 92]. As the methodology is constantly improving, more clarity on the chemical binding type is likely to be achieved in the near future.

3.2. PARAMETERS INFLUENCING HUMIC METAL BINDING

Both pH and ionic strength have a strong influence on spatial conformation and the association of trace metals with HS. An increase in the solution pH leads to an increase in negative charge on the molecules, followed by electrostatic repulsion between ionised groups and molecular swelling, while an increase in the ionic strength improves the screening of charges and leads to molecular shrinkage [93, 94]. The pH dependence decreases with increasing electrolyte concentration and at high electrolyte concentrations the intrinsic viscosity is almost pH independent [93, 95]. The opening up of the molecule at high pH leads to a better exposition of the functional groups and therefore favours the complexation of trace metals, as has been frequently observed for many different trace elements [39, 42, 44, 62, 68, 69, 96-98]. The reverse is true for high ionic strength, in addition to competition effects from other cations in solution [39, 44, 53]. Therefore, the reported thermodynamic data on binding strength and complexing capacity of HS are only comparable under the same background conditions, namely ionic strength and pH of the solution in addition to temperature and ligand concentration. The latter is another important factor influencing the strength and type of association between trace metals and HS [1, 99, 100].

At high ionic strength and low pH, aggregation of HS occurs [101, 102]. Similarly, at high loadings of trace metals, aggregation and precipitation for several trace metals have been frequently observed [103-108]. It was found that there is a threshold concentration, below which no aggregation could be observed and whose magnitude depended upon nature of the metals studied (La, Cu, Cd) [107]. For Cu, this threshold was determined as 5×10^{-5} M [108]. The observed aggregation is believed to be a result of the neutralization of negatively charged groups on the HS molecule, thus reducing the inter-molecular repulsion, which in turn favours flocculation processes at specific threshold conditions [107].

3.3. QUANTIFYING TRACE METAL BINDING BY HS

In addition to characterizing the binding processes underlying the complexation of trace metals by HS, many efforts have been undertaken to predict the fate of trace elements in natural environments as influenced by HS. Hence, a primary objective of the research on metal-HS complexation has been to determine thermodynamic parameters [109]. Two parameters are generally used to characterize metal binding by a ligand [1, 7, 9]: the complexing capacity (CC), which refers to the amount of trace metal that maximally may be retained by the ligand, and the stability constant, characterizing the strength of association.

Table 1. Some reported values for the co

Ligand	Conc. [mg L^{-1}]	Method	Metal	Background electrolyte	pH	CC [μmol g^{-1} L$_C$]	Source
SFA	10	Dialysis	Cd^{2+}	0.001M KNO$_3$	7	3328*	[111]
SFA	10	Dialysis	Cd^{2+}	0.001M KNO$_3$	7	1672*	[104]
CFA	5 mgC L^{-1}	Dialysis	Cd^{2+}	0.001M KClO$_4$	7	2468	[46]
CHA	5 mgC L^{-1}	Dialysis	Cd^{2+}	0.001M KClO$_4$	7	2386	[46]
CFA	10 mgC L^{-1}	Cd-ISE	Cd^{2+}	0.01M KClO$_4$	7	2553	[13]
CHA	10 mgC L^{-1}	Cd-ISE	Cd^{2+}	0.01M KClO$_4$	7	2321	[13]
SFA	10	Dialysis	Cu^{2+}	0.001M KNO$_3$	5	2793*	[111]
SFA	10	Dialysis	Cu^{2+}	0.001M KNO$_3$	5	1862*	[104]
SFA	1-150	Cu-ISE	Cu^{2+}	0.5M NaClO$_4$	6.5	5200	[112]
SFA	100	Cu-ISE	Cu^{2+}		6	10345*	[105]
CFA	5 mgC L^{-1}	Dialysis	Cu^{2+}	0.001M KClO$_4$	5	5221	[46]
CHA	5 mgC L^{-1}	Dialysis	Cu^{2+}	0.001M KClO$_4$	5	3357	[46]
CFA	10 mgC L^{-1}	Cu-ISE	Cu^{2+}	0.01M KClO$_4$	5	5572	[12]
CHA	10 mgC L^{-1}	Cu-ISE	Cu^{2+}	0.01M KClO$_4$	5	2611	[12]
SHA (29$^+$)		Ion-exchange equilibrium	Zn^{2+}		7	655*-9155*	[110]
SFA (28$^+$)		Ion-exchange equilibrium	Zn^{2+}		7	1448*-30190*	[110]
CFA	5 mgC L^{-1}	Dialysis	Zn^{2+}	0.001M KClO$_4$	7	2809	[46]
CHA	5 mgC L^{-1}	Dialysis	Zn^{2+}	0.001M KClO$_4$	7	2167	[46]

(*) calculated value from literature data, assuming 58% carbon in the sample.
(+) number of samples investigated.
S/CFA: soil/compost fulvic acid, S/CHA: soil/compost humic acid; L$_C$: amount of ligand expressed as DOC; ISE: ion-selective ele

Since CCs are conditional values dependent on ionic strength, pH, and ligand concentration, comparisons of the different studies are only valid if the above mentioned parameters are similar. Table 1 shows some CC values for HS published in the literature, which were obtained under similar experimental conditions. In this case, CCs of most soil HS and compost HS appear to be in the same range of magnitude for a given element. However, as indicated by Matsuda and Ito's study of HAs and FAs extracted from 40 different soils [110], the complexing capacity for metals (in their case Zn) can vary from soil to soil. The CC was also shown to be element-specific: Hg, Pb, Al, Cu and Cr are characterized by higher CCs compared to Cd, Zn, Ni or Co [41, 42, 46, 88]. Since carboxylic and phenolic groups are considered to be the most relevant metal binding groups in terms of quantity, the total number of acid-titrable groups has been suggested to constitute the theoretical maximum CC of HS [1]. In a study on HA from compost, the measured CC for Cu (at pH = 5) constituted one third of the total content of acid-titrable groups, and a fifth of all acid-titrable groups for Cd and Zn (at pH = 7) [46]. In comparison to less humified (such as soluble FA molecules) and non-humic organic ligands present in the soil solution, HS demonstrated a larger holding capacity (=CC) for Cu and Cd [12, 13].

The second important parameter regarding the quantification of trace metal binding describes the strength of association between ligand and metal: the binding constant, usually given in logarithmic form as logK. Similar to the CC, binding affinity of HS varies greatly between sources. These differences have been traced to HS structure, aromaticity, or degree of humification in some studies [110, 113]. Mostly, however, they are a result of the different number and type of functional groups attached to the humic backbone in addition to sterical issues. The formation of more than one bond between the metal and the organic molecule usually results in high stability of the complex. In general, the stability of a metal-chelate complex is determined by such factors as the number of atoms that form a bond with the metal ion, the number of rings that are formed, the nature and concentrations of the metal ions, and pH. The Irving-Williams stability series has been confirmed for trace metal-humic associations by many studies: Generally, the experimental results agree on the following order of binding strengths for divalent ions: Hg > Pb > Cu > Cd $\geq$ Ni $\geq$ Zn > Fe (II) > Ca > Mn [32, 42, 46, 58,114–116]. More rarely, deviations from this series have also been reported [40]. Even though general agreement appears to exists on the stability series, the actual numerical values of stability constants found in the literature for the same element may vary several orders of magnitude (e.g. literature logK values for Cu vary between 3 and 10.8 [1]), even at similar experimental conditions of pH and salinity. Many early studies on trace metal binding of HS used the two-component Scatchard plot model, which arbitrarily fits two lines to a curvilinear plot, to derive stability constants [1, 68, 117]. In addition, the values obtained in these studies were also conditional on the range of concentrations within which the assays are carried out. It has been recognized that the association constants for metal-organic complexes derived in this way are often unrealistically low, since they are based on measurements at high metal loadings on the complexing sites of OM [118]. The metal binding at the strongest binding sites at low metal / ligand ratios is not described well by the discrete models such as the Scatchard plot due to a limitation of the sensitivity of the method. Trace metals, in fact, occupy different and more selective bonding sites at low metal loading levels [118].

In order to address this problem, some studies have used extrapolation to determine the binding constants of the strongest sites on the HS molecule [13, 97, 113] (see also Fig. 1). The methodology of this approach is discussed in greater detail below. Using this technique, the determined binding constants (log K_{int}) for the strongest binding groups for Cu ranged from 7.2 to 9.3 (pH = 5) for soil and compost HS [46, 97, 113, 119]. These values are at least 3 orders of magnitude higher than those reported by earlier studies on Cu binding by HS [1]. Correspondingly, recent studies have emphasized that organic complexation in the environment for such trace metals as Cd, Ni, and Zn may be more important for metal mobility than previously believed [10, 118, 120-123]. Researchers have also come to realize that binding sites on the HS molecule are difficult to describe using binding site strengths attributable to specific sites. Rather, the binding sites on the HS molecule represent a continuum of binding strengths, which is most appropriately described via continuous models. The latter are shortly discussed in the next section below.

4. Experimental Methods for the Study of Humic-Metal Interactions

4.1. SPECTROSCOPIC APPROACHES

The technological advances in spectrometry have led to an elaborate set of tools enabling the detailed study of large and complex organic molecules. Consequently, these tools have extensive application in the study of HS and their interaction with trace metals. Spectrometric methods are usually employed to observe and describe the interactions and type of association of a metal cation and a specific binding site on the macromolecule. They are not used to quantify binding parameters such as complexing capacities or binding strengths.

Fourier-transformed infrared spectrometry (FTIR) and diffuse-reflectance infrared Fourier-transformed spectrometry (DRIFT) have been useful in identifying carboxylic, phenolic and even N-containing groups as metal binding sites [1, 71, 127, 128]. The limitations of these methods are due to the complex absorption curve characteristically found for HS, making an interpretation of these spectra rather speculative.

The use of nuclear magnetic resonance (NMR) is a powerful tool in elucidating the structure of HS [5, 86, 129-132]. On the other hand, its application to the study of trace metals binding by HS has been limited to a few studies [4, 67, 133].

Electron spin resonance (ESR) spectrometry provides high sensitivity at low metal concentrations and allows to measure the samples directly with little pre-treatment [4]. With the ESR technique, paramagnetic signals due to unpaired electrons in transition metal ions are detected, allowing the study of the corresponding metal-HS complexes [1]. However, this technique is limited to the study of only some paramagnetic metals giving a detectable ESR signal such as Fe, Cu, Mn, V, or Mo, and by a limited signal resolution [4]. For the latter metals it has been employed to study the chemical and geometrical properties of their complexes [1, 4, 77, 90, 134, 135].

Paramagnetic metal ions are able to quench the fluorescence of organic ligands by enhancing the rate of energy dissipating processes that compete with fluorescence. This allows to distinguish between free and bound reactive sites on the ligand molecule [1, 4]. This technique is widely used in the field of humic-metal interactions, especially at low ligand-metal concentrations [1, 4, 136-142]. For this purpose both the continuous and synchronous fluorescence have been used in addition to a time-resolved laser-induced fluorescence (TRLIF) [140]. Fluorescence spectrometry allows the direct measurement of the CC of a ligand via the determination of the free ligand concentration [4]. However, its efficiency is usually limited to strongly binding, paramagnetic ions like Cu [2, 4].

Recently, with the help of X-ray absorption near edge structure (XANES) and extended X-ray absorption fine structure (EXAFS) spectroscopy, researchers have been able to explore the local geometry and structure of trace metal complexes with HS in great detail [29, 31, 57, 92]. EXAFS reveals the local atomic environment of the exited atom by analysing the measured oscillatory structure which appears 50-1000 eV above the absorption edge. It gives information on interatomic distances as well as on the number and chemical identity of the atoms within a 5 Å radius of the atom absorbing the X-ray photon [29]. XANES yields data on oxidation state of the excited atom, the coordination geometry and the bonding environment through a comparison with model compounds [29].

4.2. NON-SPECTROSCOPIC METHODS

In order to measure the strength of association between trace metals and HS and determine CC and binding constants, a variety of methods and models have been used. The quantification of metal binding data has been under scientific study for nearly three decades, and still creates controversy over the most suitable approach today. This is a result of several problems involved for quantifying metal complexation for this type of ligand as outlined in the section above: the heterogeneity of HS in shape, size, and composition; the strong influence of pH and salinity on binding strength; the heterogeneity of binding sites and electrostatic interferences between binding groups [1, 7, 9, 143, 144].

The ligand exchange method is based on using a competing ligand such as a cation exchange resin in a solution with the trace metal and the humic compound. The concentration of the free ion is determined by ICP or AAS and the dissociation kinetics of the metal complexes with HS may be calculated using a kinetic model [40, 53, 99, 145, 146]. Alternatively, ligand exchange may be combined with cathodic stripping voltammetry (CSV) or anodic stripping voltammetry (ASV) to determine the concentration of the free metal ion [36]. Another approach employs a gel chromatography column to separate the competing ligand (such as an organic acid) and the HS after their exposure to the trace metal ion. The amount of trace metal bound to both ligands is then determined subsequently [147].

Potentiometric titration have been used to characterize the strength of binding of trace metals by measuring the pH decrease in solution as a result of proton displacement on acidic groups with increasing concentrations of cations [50, 148-150]. However, Santons et al. found in a study interpreting literature data that hysteresis, varying sample moisture content, different calibration procedures or different equations used for the calculation of activity coefficients lead to large variations in the experimental results obtained with this technique [151].

A similar titration procedure is used in combination with the ion-selective electrode (ISE). It has been shown to be a reliable method for measuring low Ca, Cd, Cu, Zn or Pb concentrations in a variety of studies comprising HA and FA from many different various sources and environments [13, 41, 46, 68, 83, 113, 115, 152-155]. In a comparative study, the Cd-specific ISE has been shown to work well at concentrations ranging from 10^{-10} to $10^{-6.5}$ M in a complex organic matrix solution, well correlated with activities calculated using the GEOCHEM program [155]. The ISE allows the collection of many data points, which improves the quality of the binding data derived in the study [13]). It has been suggested that a use of the ligand (HS) rather than the trace metal as titrant improves data quality at low metal concentrations [113].

(Differential pulse) anodic stripping voltammetry (ASV) has been successfully applied to the study of metal complexation by HS by several researchers [36, 58, 114, 156-158]. The advantages of ASV are its sensitivity, broad variation of metal-to-ligand ratios, and suitability for determination of both thermodynamic and kinetic parameters [159]. The drawbacks are: precipitation of organic material at the electrode [48], and processes occurring during the stripping step that may produce artefacts. Similar to ASV, cathodic stripping voltammetry (CSV) may be used in conjunction with ligand exchange to obtain the free metal ion concentration and thereupon derive the complexation parameters [36, 53, 160].

Equilibrium dialysis (ED) has also been shown to be an effective tool to examine metal binding by organic ligands [44, 46, 104, 111, 161, 162]. A semi-permeable membrane is placed between the solution containing both trace metal and ligand (HS) and the solution of a background electrolyte, permitting the passage of the ionic metal and excluding the larger humic molecules. After equilibration the free metal concentration in solution is determined by ICP or similar methods. A direct comparison of the results obtained with ED and ISE for metal binding of compost HA and FA showed an excellent agreement for both Cu and Cd [12, 13].

The above methods enable the determination of a concentration of the 'free' trace metal in solution (M_F) and the amount of organically bound metal (M_B). The latter is often calculated as the difference between total (M_T) and free metal concentration. The complexing behaviour of a ligand towards a certain trace metal can then be quantified from the experimental data by calculation of the total CC and the determination of stability constants indicating the strength of binding at a given pH and ionic strength.

4.3. DERIVING BINDING SITE STRENGTH AND CAPACITY FROM EXPERIMENTAL DATA

A straightforward approach to determine the CC is to plot M_F/L_C (L_C: concentration of ligand) against M_T/L_C (Fig. 1). Due to metal complexation by the ligand, the line of data points will assume a slope = 1 when $M_T/L_C \gg$ CC. M_F will accumulate in the solution only after the CC of the ligand has been approached. Hence, a linear regression of all data points with $M_F > 0$ using the least-squares method approximates the CC as the x-axis intercept with the corresponding standard error.

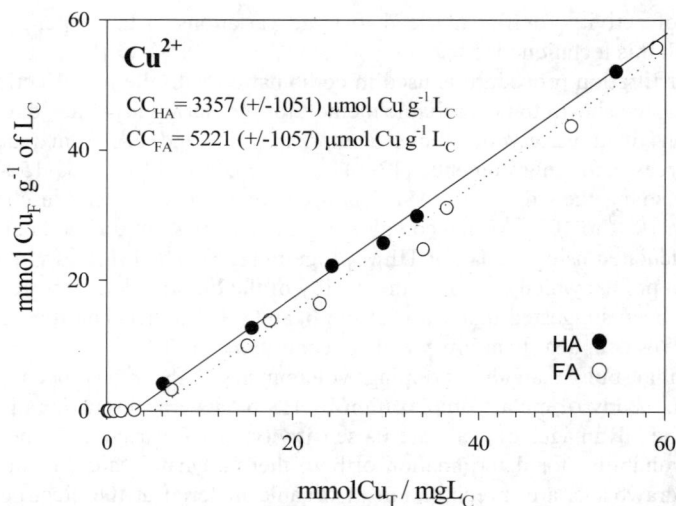

Figure 1. Deriving a complexing capacity from titration data: M_F/L_C is plotted against M_T/L_C followed by linear regression of all data with a free ion concentration > 0. In the example, humic acid (HA) and fulvic acid (FA) from compost were titrated with a Cu-perchlorate solution[46]. CC: complexing capacity.

To characterize the strength of bonding, stability constants are calculated. For a given ligand with discrete homogeneous binding sites for a metal ion, the titration data can be plotted in Hill or Scatchard plots, which yield the desired thermodynamic data as the slope and y-intercept of a linear fit [1]. For HS, however, the Scatchard plot is always curvilinear, as a result of the range of binding sites (BS) with slightly different binding energies rather than a discrete number of sites of defined binding strength. ML_2 (metal binding to functional groups on two ligands) complex formation has also been suggested as contributor to the curvilinear nature [1].

Since the two-component Scatchard plot yielded unsatisfying results for an interpretation of the curvilinear nature and the subsequent calculation of stability constants (see section above), different models have been developed to better quantify the binding strength and interpret the experimental data [1, 117, 163]. In general, researchers have moved away from using a discrete site approach to using models providing for a wide range of binding site strengths on the molecule, the so-called continuous models. Continuous models are employed to address the fact that metal binding by humic substances apparently cannot be explained by a limited number of binding sites with discrete binding strengths. One such continuous distribution model is based on a normal distribution of binding site strengths [1, 70, 117, 143, 163, 164]. To account for the fact that chemically different groups on the macromolecule are relevant for binding, Manunza et al. have proposed the use of a bimodal approach for the distribution model [165]. Such a model hypothesizes that two main types of binding groups, namely carboxyl and phenol groups, account for most of the complexation.

A successful approach to derive meaningful binding strength data calculates incremental stability constants from successive slope values in the Scatchard plot, which are subsequently plotted according to their frequency on the molecule using the Gaussian normal distribution [13, 68, 97, 113]. In this way, the strongest binding groups, the full range of binding site strengths on the molecule and the most abundant binding site strengths on the molecule may all be quantified. This methodology is described below.

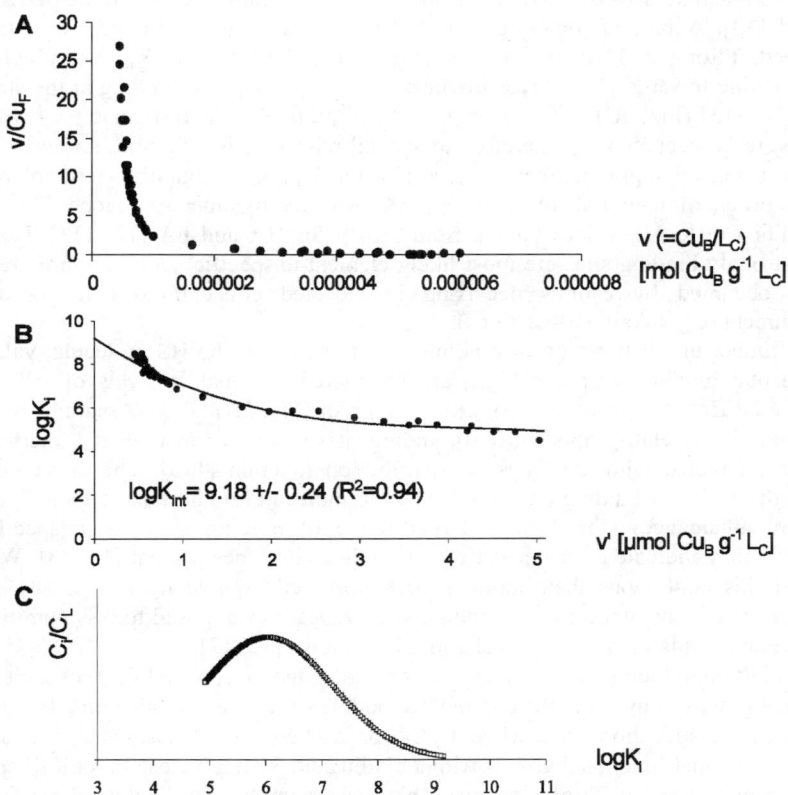

Figure 2. Derivation of $logK_{int}$ (pK_{int}) and a continuous distribution of binding sites from the Scatchard plot. A: Scatchard plot for titration data of compost fulvic acid and Cu (two replicates), B: Incremental stability constants $logKi$ calculated for each interval v' and polynomial fit, C: relative abundance of binding sites vs. $logK_i$ obtained with a continuous model based on a normal distribution. $logK_i$: incremental binding site strengths, C_i/C_L: mole fraction of binding sites at $logK_i$, L_c: ligand concentration as DOC.

It is assumed that the HS molecule (L) is the central group to which each metal ion (M) is bound to a single reactive site (formation of LM, LM_2, LM_n complexes) [1]. The slope of the Scatchard plot (v/M_F plotted against v with $v = M_B/L_C$; M_F = free 'ionic' metal, M_B = bound metal, L_C = ligand concentration as dissolved organic carbon (DOC)) is interpreted as the stability constant K_0 between ligand and M. For HS it forms a curve, for which successive (incremental) slope values may be calculated from neighbouring data points in the Scatchard plot (slope = $(v/M_F) / v$); Fig. 2A). These slope values are conditional stability constants K_i at a certain value of v', which in turn lies equidistant between the original neighbouring data points (used to calculate the incremental slope) on the v-axis of the Scatchard plot. This approach is superior to fitting a polynomial equation to the Scatchard plot data and calculating the derivative, since the inherent error of polynomial fitting would be multiplied when the derivative is formed [13]. A plot of $logK_i$ against v' displays the range of binding site strengths observed. Fitting a 3rd-degree polynomial to the data of this K_{int} vs. v' plot and extrapolating towards $v' = 0$ gives the intrinsic constant K_{int} for binding at the strongest sites [97, 113] (Fig. 2B). These sites may be of particular interest since they would be the first to be occupied and therefore of special relevance in a natural environment [1]. Studies using this approach have shown that the $logK_{int}$ for humic macromolecules is several orders of magnitude higher than logK measured by other approaches [12, 13, 97, 113]. For Cu, $logK_{int}$ values ranged from 7 to 9 for HA and FA [12, 113]. However, these strongly binding sites are most likely element-unspecific, and therefore are likely to be occupied by cations occurring in elevated concentrations in the natural environment (e.g. alkali-earth cations).

To model the distribution of binding site strengths on the HS molecule, values for incremental binding strengths $logK_i$ are calculated at equal intervals of y from the polynomial fit to the data in the ($logK_i$ vs. v')-plot (Fig. 2B) [113]. Assuming a normal distribution, the relative abundance of binding sites within the incremental interval δpK_i can be calculated with the Gaussian distribution function [113]. The curve obtained shows the range of binding sites on the humic molecule in addition to their frequency (relative abundance) (Fig. 2C). A larger range of binding sites for a specific HS molecule may indicate a greater variety of binding site types present [12, 13]. With the help of this continuous distribution model, humified organic ligands in DOM were shown to have a greater range of binding site strengths as opposed to less humified and non-humic ligands such as polysaccharides or proteins [12, 13].

In addition to the range of binding site strengths, the $logK_{int}$ (stability constant for the strongest groups) appears in the continuous model as the extreme value on the right hand side of each distribution curve where C_i/C_L approaches zero. Consequently it is obvious from Fig. 2C that functional groups with a binding site strength close to $logK_{int}$ represent only a minute percentage of total metal-binding groups on the ligands. Therefore, the quantification of the $logK_f$ may be more relevant for a characterization of HS binding strengths and for use in fate modelling. This parameter refers to the most abundant binding site strength on the molecule and constitutes the $logK_i$ value at the peak of the distribution curve. In a multi-species environment such as the soil solution, this parameter could be helpful to characterize the most likely binding strength of metal-humic associations. Its value ranged from 5 to 6 for Cu at pH = 5 and about 5 for Cd at pH = 7 in the studies using this modelling approach [12, 13, 113].

The distribution curves obtained for HS were skewed, indicating a deviation from the Gaussian distribution [12, 13, 113]. In addition, they were generally flatter than those obtained for other organic ligands in the soil solution, presumably indicating a higher number of differing reactive sites [12, 13]. Other recent modelling approaches for trace metal binding by HS have focused on thermodynamic considerations [166-168], a random structural approach to identify the likely metal binding sites [169], kinetic aspects [170], discrete binding sites [171, 172] and a combination of electrostatic considerations and site binding models [173, 173-177].

Modelling the binding process is the first vital step towards understanding the sorption phenomena in the field. Theoretically, the complexation of trace metals in the soil environment may result in two antipodal phenomena: an immobilizing of the trace metal (e.g. due to an association of the HS with the soil matrix or due to low solubility of the HS) or an increase of bioavailability and mobility (if the HS molecule remains in solution). The latter process is working against an effective containment of pollutants, but is important for plant nutrition as discussed in the following.

5. The Stimulatory Effects of HS-Metal Interactions on Plant Growth

5.1. INFLUENCE OF HS ON PLANT GROWTH: VIEWS AND OPINIONS

Man realized thousands of years ago that dark-coloured soils are more productive than light-coloured ones, and that soil fertility is closely associated with decaying plant and animal residues. This observation later evolved into the belief that humus is the only or the major soil constituent supplying nutrients to plants [178, 179]. This concept was the prevailing notion in the 19th century and referred to as the "humus theory". It was Liebig, who opposed the "humus theory" in a number of publications and provided fundamental information on the role of minerals in plant nutrition [180, 181].

The mineral theory was further supported by Lawes and Gilbert working at Rothamsted, England, who demonstrated in a long-term field study in the early 20th century that soil fertility may be maintained, for at least several years, by applying mineral fertilizers only [182, 183]. However, the controversy between the humus and mineral theories was not ended by these experiments. Scientists have realized that more exact experimentation is required to determine the benefit of humus to plant growth and to determine possible synergistic effects of HS and minerals.

In the early 20th century, Bottomley published a series of papers in which he showed that HS enhanced the growth of various plant species in mineral nutrient solutions (NS) [184-187]. Bottomley believed that HS acted as plant growth hormones and called them "auximones". Other investigators attributed the beneficial effects of HS on plant growth to the increased solubilization of some mineral ions such as Fe in either soluble or colloidal forms [188, 189].

The beneficial effects of HS on both root and shoot growth under conditions in which a complete ("optimal") NS was supplied is shown in Table 2. The data clearly show growth enhancement compared to pure water by additions of HA (50 mg L^{-1}), and further stimulation of growth in Hoagland's solution. The stimulation in the presence of HA exceeded that of Hoagland's solution alone by about 25%, which provides evidence for a synergistic effect of combined applications of mineral nutrition and HS.

Many studies show that HS can stimulate the growth of plant tissues and the total quantity of absorbed nutrients [190, 191]. Rauthan and Schnitzer [192] reported that the yield of cucumber plants grown for 6 weeks in a nutrient solution (NS) was significantly enhanced as a response to the addition of increasing concentrations of FA (Fig. 3). Concomitantly, an increase in the uptake of N, P, K, Ca, Mg, Cu, Fe, and Zn was observed in the shoots. Likewise, Tan and Nopamornbodi showed that shoots of maize plants grown for 16 days in NS to which 640 mg L^{-1} of HA was added, tripled their dry weight and the uptake of N, P, K, Ca, Mg, Cu, Fe, Zn, and Mn [193]. Other authors reported on a much smaller effect or none at all [190, 194]. Differences in the plant organ considered were shown by Dormaar in *Festuca scabrella* [195] and by Ernst et al. in *Scrophularia nodosa* [196]. The variability of results was often attributed to the variable sources of HS. Additional important factors, which have often been overlooked, are the extraction, purification and fractionation procedures of the HS preparation [197].

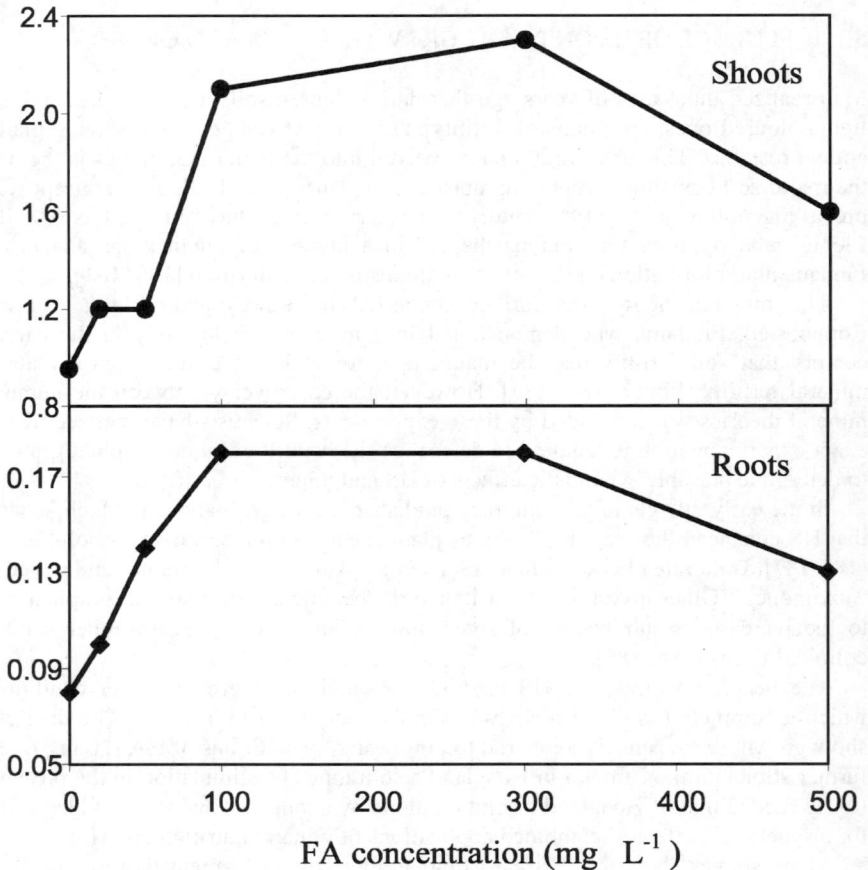

Figure 3. Influence of fulvic acid concentration on dry weight of cucumber (*Cucumis sativus L.*) shoots and roots [192].

Table 2. The effect of 50 mg L^{-1} of humic acid (HA) on growth of wheat in water and Hoagland's nutrient solution [190].

Culture medium	Plant organ	Fresh weight (mg/plant)	Stimulation (%)
Water	Root	93	0
	Shoot	185	0
Water + HA	Root	146	57.5
	Shoot	252	36.2
Hoagland	Root	182	96.3
	Shoot	342	84.9
Hoagland + HA	Root	203	119.0
	Shoot	390	110.8

The effects caused by HS on growth and nutrition were observed not only in hydroponics, but also when plants were grown on sand [198] or soil, especially those with low OM contents [197, 199], and in callus *in vitro* cultures [200]. Effects on roots and shoots differ, namely root growth is enhanced to a greater extent than shoot growth. Various plant organs were found to respond differently to the presence of HS in the NS [191]. A summary of the effects of HS on seeds, roots and shoots of higher plants and relevant concentrations of the HS is presented in Table 3.

Table 3. Effects of humic substances (HS) on seeds, roots and shoots of plants.

Plant organ response	HS Material	Concentration range (mg L^{-1})	Effect
Seed germination	HA	0-100	Enhanced rate; Accelerated water uptake
Root initiation and elongation	HA, FA	50-300	Stimulated root initiation; Lateral roots development
Excised root elongation	HA	5-25	Enhanced growth; Cell elongation
Intact plant growth	HA, FA	0-500 optimum 50-300	Enhanced growth of shoots and roots

A distinctive deficiency of many published reports on the effects of HS on plant growth stems from the fact that researchers paid little or no attention to the concentration of the HS in the solution. Most of them employed a single concentration in their studies. A close examination of the data presented by Rauthan and Schnitzer [192] showed, however, an optimum curve reaching a maximum at a range of concentrations of 100-300 mg L^{-1} HS in the NS, for both the roots and shoots (Fig. 3) [191]. Concentration ranges as well as optimum levels vary between reports and also in relation to the plant organ studied (Table 3). Two possible hypotheses may explain the contrasting effects displayed by HS at different concentrations. The first is that by increasing HS concentration, the solution equilibrium is shifted towards the formation of

HS-metal complexes of higher stability, thus reducing the availability of the micronutrient. This hypothesis, however, is not confirmed by data on stability constants of HS complexes with Fe or Cu, which are much lower than the stability constants of the EDTA or EDDHA, the most commonly used chelates in NS cultures. HS can, however interact with chelates without actually causing a ligand exchange of the metal ion between the chelate and HS molecules. This interaction is suggested here as a second hypothesis providing an explanation for the decreasing effect on high concentrations.

It was shown by UV-VIS absorbance studies [15] that for a number of Fe chelates, at concentrations above 100 mg L^{-1} (Fig. 4), the absorbance of the second charge transfer band (λ = 257 nm for EDDHA) starts to decrease in response to further additions of HS while the absorbance is fully suppressed at about 300 mg L^{-1}. The differential spectrophotometric titration of EDDHA and other Fe complexes with HS is therefore in accordance with the concentration effects found by Rauthan and Schnitzer [192], Chen et al. [201] and Chen [202]. This suggests that HS-chelates interactions are the cause of the lower availability of chelated nutrients to plants at high HS concentrations. Negative effects of this kind can hardly be expected to be observed for plants grown in soil as the solubility of HS in the soil solution is generally in the range where positive effects can be observed. Additions of composts rich in DOM could however act in this direction, as shown for Cd uptake and other heavy metals. In conclusion, FAs and/or HAs may stimulate shoot growth of various plants when applied either as foliar spray at concentrations of 50 to 300 mg L^{-1}, or when applied in NS at similar concentrations. The stimulatory effect on shoot growth usually correlates to root response regardless of the mode of application.

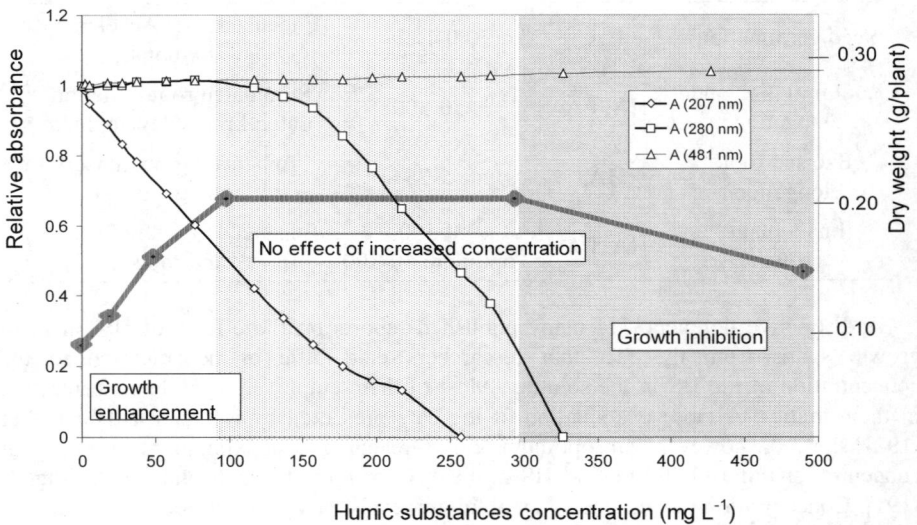

Figure 4. Relative absorbance and influence of humic substances concentration on dry weight of cucumber (*Cucumis sativus L.*) plants.

5.2. INFLUENCE OF HS ON Fe UPTAKE BY PLANTS: CASE STUDY

Since Fe is the main ion of concern and due to the complexity of its uptake mechanisms, a study was conducted by Chen et al. Two species were studied and their Fe uptake recorded: one representing monocotyledonous plants and the other the dicotyledonous plants, which differ in their Fe uptake strategies as follows [203-205]:
  A. Strategy I – mostly present in dicotyledons, represented in this study by melon plants (*Cucumis melo* L., *cv*. Ein Dor) – Experiment 1;
  B. Strategy II – mostly present in monocotyledons, especially in grasses, represented in this study by ryegrass (*Lolium perenne* L., *cv*. Omega) – Experiment 2.
In both experiments, no Zn deficiency symptoms were observed in any of the potentially deficient treatments, as opposed to Fe deficiency symptoms which were profoundly exhibited. Organically bound residual Zn that might have been added with the HS could explain this observation.

5.2.1. Melon plant growth (strategy I plants) – experiment 1

Plants were grown in a 0.5L NS which contained all essential effluents excluding Fe. The plants were grown in a growth chamber at $25°\pm0.5C$ during the light hours (14 h) and $15°\pm0.5C$ during the dark hours (10 h). Specific treatments (see Table 4) were given in the last 6 days of plant growth. During the final 6 days of treatment, plants which received Fe in their NS underwent a recovery and a re-greening process. Differences between the treatments became distinct. Chlorophyll concentration in the leaves (young yet mature leaves) differed in accordance with the treatments (Table 4). The measured chlorophyll levels were 0.2, 1.3, 1.7 and 2.1 $mg \cdot g^{-1}$ (fresh weight) for the NS, NS + $Fe(NO_3)_3$, NS + $Fe(NO_3)_3$, + HA, and NS + FeEDDHA, respectively. Remedy of the Fe-deficiency induced chlorosis was only partially achieved by a mineral form of Fe, whereas the presence of either FeHA or FeEDDHA greatly improved chlorophyll synthesis. FeEDDHA seemed to be more effective than the FeHA in enhancing chlorophyll formation (Table 4). This experiment clearly indicates that under neutral to slightly basic pH, HA stimulates plant growth and improves its health by enhancing Fe availability and chlorophyll synthesis. This is in accordance with other reports suggesting that HS are important to the Fe nutrition of dicotyledonous plants [191, 206, 207].

Table 4. Chlorophyll concentration in leaves of melon plants grown in nutrient solution after 6 days of treatment.

Nutrient solution (NS) treatment	Chlorophyll concentration ($mg \cdot g^{-1}$)
NS	0.2 d[†]
NS + $Fe(NO_3)_3$	1.3 c
NS + $Fe(NO_3)_3$ + HA	1.7 b
NS + FeEDDHA	2.1 a

[†] Statistical analysis – Tukey-Krammer test ($p = 0.05$): a, b, c, d indicate that the treatments are significantly different from each other.

5.2.2. Ryegrass plant growth (strategy II plants) – experiment 2

Ryegrass is a very common plant on golf courses and is therefore of importance to the industry. It is a monocotyledonous *Graminacea* and like other members of its family, takes up Fe from the soil or from NS via a ligand exchange mechanism between organo-Fe complexes and/or mineral Fe compounds and phytosiderophores [203, 204, 208]. As for the dicotyledonous plants, we hypothesized that the formation of organo-Fe complexes in the NS and the resulting maintenance of the Fe in solution will greatly improve the effectiveness of the Fe uptake mechanism via phytosiderophores. Chen et al. [205] therefore grew ryegrass plants in NS which did not contain Fe nor Zn (control), adding these elements in different forms according to the experimental design listed below (text and Table 5).

Chlorophyll concentrations for the ryegrass NS growth study are summarized in Table 5. Plants grown in a NS devoid of Fe and Zn exhibited very low chlorophyll levels and retarded growth resulting from a lack of proper photosynthesis. Additions of FA or HA to the NS (treatments NS + FA and NS + HA) did not result in a significant improvement in chlorophyll concentration, although a slight trend of improvement seemed to have taken place. This proves that the HS do not act as growth hormones when added as purified substances.

Table 5. Chlorophyll concentration of leaves of ryegrass plants grown in nutrient solution as specified.

Nutrient solution (NS) treatment	Chlorophyll concentration $(mg \cdot g^{-1})$
NS	0.66 c[†]
NS + FA	0.91 c
NS + HA	1.04 c
NS + $FeSO_4$ + $ZnSO_4$	1.76 b
NS + FeEDTA + ZnEDTA	3.28 a
NS + FA + $FeSO_4$ + $ZnSO_4$	3.39 a
NS + HA + $FeSO_4$ + $ZnSO_4$	3.47 a

[†] Statistical analysis – Tukey-Krammer test (p = 0.05): a, b, c indicate that treatments with different letters are significantly different from each other.

This observation contradicts the "hormone-like activity" or "plant growth regulators" concept suggested by a number of researchers [209] and supports the results reported by Chen et al. [201], who tried to extract plant hormones from HS and through their findings could exclude their presence.

Addition of Fe and Zn salts in the treatments NS + $FeSO_4$ + $ZnSO_4$ resulted in a significant, yet insufficient, improvement in plant chlorophyll synthesis. The plants were moderately green, but this was insufficient to induce improved plant growth. However, when the Fe and Zn were added to the solutions as organo-metal complexes, either as an EDTA chelated form or a HS complex (treatments: NS + FeEDTA + ZnEDTA; NS + FA + $FeSO_4$; NS + HA + $FeSO_4$ + $ZnSO_4$), the plants synthesized high levels of chlorophyll, reflecting those of healthy plants (almost twice as much biomass as those of the chlorophyll deficient plants – data not shown). Since we did not observe symptoms of Zn deficiency, we believe that the response of the plants can be attributed to improved Fe nutrition in solutions containing organo-Fe complexes. However, some contribution of the organo-Zn complexes cannot completely be ruled out at this stage and for this reason further research is being conducted in our laboratory to elaborate this issue. Since experiment 2 provides more comprehensive answers (for monocots) than experiment 1 (dicots), a similar design is employed at present in a study on dicots.

The conclusions from this study [205] were that growth enhancement of plants in NS and soil by HS should mostly be attributed to the maintenance of Fe and Zn in solution at sufficient levels. This effect is pH dependent and it becomes more prominent at high pH levels.

6. Conclusions

In the soil environment, HS are the most important organic ligands for trace metals, forming a myriad of both soluble and insoluble complexes. Micronutrients such as Zn and Fe may be held in solution bound to HS and thus remain available to biota, just as much as pollutants may be sorbed and immobilized by HS that bind tightly to the soil matrix. In order to make better use of HS as immobilizing agents for remediation purposes, it is absolutely necessary to develop a good understanding of the underlying processes of HS-trace elements interactions. Even more so since, as demonstrated by their role in plant nutrition, binding to HS does not necessarily result in a containment of trace metals, but may also serve the purpose of increasing mobility and bioavailability of trace elements.

Different experimental and modelling approaches have been used to describe these processes and an extensive amount of literature is now available on the topic. The functional groups responsible for the binding of metals have been mostly identified, even though the exact local geometry and structure of the ligand and metal complex is still under investigation. The use of X-ray spectroscopy promises to provide new insight on this aspect in the near future. The binding process itself is strongly affected by abiotic parameters such as pH and ionic composition of the soil solution, with important repercussions for immobilization processes.

To predict and model the effect of HS on metal solubility in the field, element-specific thermodynamic parameters describing binding strength and the total binding capacity of HS molecules have been obtained in numerous laboratory studies using several valid methodological approaches. These studies have to a large degree substantiated the Irving-Williams stability series, with the strongest complexation found in the case of Hg, Pb and Cu. The quantification of the binding process, however, is still being improved and researched today. The early approaches, used in the 1970s, to determine stability constants from experimental data such as the two-point Scatchard plot have seriously underestimated the strength of binding. In these studies, the highly soluble fraction of HS, composing soil or water DOM, have been overlooked. They were shown in recent studies conducted by our group (partially summarized in this article) to strongly bind various metals while maintaining high solubility and leachability. These studies have shown that HS binding sites on the same molecule vary in binding strength over a range comprising several orders of magnitude. Hence continuous distribution models appear to be most promising to more accurately describe the ability of HS to retain trace metals. A thorough understanding of the complexation process will help to model the occurring HS-trace metal interactions in the field and increase the chances of HS sorption processes to be used as immobilizing agents within enhanced natural attenuation approaches to contaminated site cleanup.

7. Acknowledgement

The authors wish to thank NATO for inviting Prof. Yona Chen to the ARW 'Use of Humates to Remediate Polluted Environments: From Theory to Practice' held in Zvenigorod, Sept. 22-29, 2002; and the Alexander von Humboldt Foundation for the Research Award granted to Prof. Yona Chen.

8. References

1. Stevenson, F.J. (1994) *Humus Chemistry: Genesis, Composition, Reactions,* John Wiley & Sons, Inc., New York.
2. Swift, R.S. (1996) Organic matter characterization, in D.L. Sparks (ed.), *Methods of Soil Analysis. Part 3. Chemical Methods,* Soil Sci. Soc. Am., Madison, WI., Madison, WI, USA, pp. 1011-1069.
3. Schnitzer, M. (1991) Soil organic matter - The next 75 years, *Soil Science* **151**, 41-58.
4. Senesi, N. (1992) Metal-humic substance complexes in the environment. Molecular and mechanistic aspects by multiple spectroscopic approach, in D.C. Adriano (ed.), *Biogeochemistry of trace metals,* Lewis Publishers, Boca Raton, pp. 429-494.
5. Chen, Y., Chefetz, B. and Hadar, Y. (1996) Formation and properties of humic substance originating from composts, in M. de Bertoldi, P. Sequi, B. Lemmes and T. Papi (eds.), *The Science of Composting,* Blackie Academic & Professional, Glasgow, pp. 382-393.
6. Schnitzer, M. (1978) Humic substances: chemistry and reactions, in M. Schnitzer and S. U. Khan (eds.), *Soil Organic Matter,* Elsevier, Amsterdam, pp. 1-64.
7. Chen, Y. and Stevenson, F.J. (1986) Soil organic matter interactions with trace elements, in Y. Chen and Y. Avnimelech (eds.), *The role of organic matter in modern agriculture,* Martinus Nijhof, Dordrecht, The Netherlands, pp. 73-116.
8. Chefetz, B. (1998) *Transformation of organic matter during composting of municipal solid waste,* Ph.D. Thesis, Hebrew University of Jerusalem, Rehovot, Israel.
9. Saar and Weber (1982) Fulvic acid: modifier of metal-ion chemistry, *Environmental Science & Technology* **16**, 510-518.
10. Temminghoff, E.J.M. (1998) *Chemical speciation of heavy metals in sandy soils in relation to availability and mobility,* Ph.D. thesis, Wageningen Agricultural University, The Netherlands.
11. Schulten, H.R. and Schnitzer, M. (1997) Chemical model structures for soil organic matter and soils, *Soil Science* 162, 115-130.
12. Kaschl, A., Römheld, R., Hadar, Y. and Chen, Y. (2000) Binding of copper by organic matter fractions from municipal solid waste compost, in *Proceedings 10th International Meeting of the International Humic Substances Society (IHSS 10),* July 24th-28th, Toulouse, France, pp. 511-514.
13. Kaschl, A., Römheld, R. and Chen, Y. (2002) Cadmium binding by fractions of dissolved organic matter and humic substances from municipal solid waste compost, *Journal Of Environmental Quality* **31**, 1885-1892.
14. Leita, L., De Nobili, M., Catalano, L., Moria, A., Fonda, E. and Vlaic, G. (2001) Complexation of iron-cyanide by humic substances, in R.S. Swift and K.M. Spark (eds.), *IHSS 9: Understanding and*

Managing Organic Matter in Soils, Sediments, and Waters, September 21st-25th, Adelaide, Australia, pp. 477.

15. De Nobili, M.D., Catalano, L., Siebner-Freibach, H. and Chen, Y. (2002) Sorption of microbial and synthetic ligands and chelates on organic matter and clays: a physico-chemical and iron supply study, in *XI ISINIP Book of Abstracts,* pp. 32.

16. Catalano, L., De Nobili, M., Siebner-Freibach, H. and Chen, Y. (2003) Effect of humic substances on the behaviour of iron siderophores in soil and water, in *Proceedings of the International Humic Substances Society twentieth anniversary conference,* Northeastern University, Boston, USA, pp. 268.

17. McLaren, R.G. and Crawford, D.V. (1973) The fractionation of copper in soils, *Journal of Soil Science* **24**, 172-181.

18. Murthy, A.S.P. (1982) Zinc fractions in wetland rice soils and their availability to rice, *Soil Science* **133**, 150-154.

19. Mandal, L.N. and Mandal, B. (1986) Zinc fractions in soil in relation to zinc nutrition of lowland rice, *Soil Science* **132**, 141-148.

20. Shuman, L.M. (1983) Sodium hypochlorite methods for extracting microelements associated with soil organic matter, *Soil Science Society of America Journal* **47**, 656-660.

21. Shuman, L.M. (1985) Fractionation method for soil microelements, *Soil Science* **140**, 11-22.

22. McLaren, R.G. and Crawford, D.V. (1973) Studies on soil copper. I. The fractionation of copper is soils, *Soil Science* **24**, 172-181.

23. Shuman, L.M. (1979) Zinc, manganese and copper in soil fractions, *Soil Science* **127**, 10-17.

24. Iyengar, S.S., Martens, D.C. and Miller, W.P. (1981) Distribution and plant availability of soil zinc fractions, *Soil Science Society of America Journal* **45**, 735-739.

25. Sedberry, J.E. and Reddy, C.N. (1976) The distribution of Zn in selected soils in Louisiana, *Communications in Soil Science & Plant Analysis* **7**, 10-17.

26. Yermiyahu, U., Keren, R. and Chen, Y. (1988) Boron sorption on composted organic matter, *Soil Science Society of America Journal* **52**, 1309-1313.

27. Yermiyahu, Keren and Chen (1995) Boron sorption by soil in the presence of composted organic matter, *Soil Science Society of America Journal* **59**, 405-409.

28. Spark, K.M., Wells, J.D. and Johnson, B.B. (1997) The interaction of a humic acid with heavy metals, *Australian Journal of Soil Research* **35**, 89-101.

29. Xia, K., Bleam, W. and Helmke, P.A. (1997) Studies of the nature of Cu^{2+} and Pb^{2+} binding sites in soil humic substances using X-ray absorption spectroscopy, *Geochimica et Cosmochimica Acta* **61**, 2211-2221.

30. Alberts, J.J. and Filip, Z. (1998) Metal binding in estuarine humic and fulvic acids: FTIR analysis of humic acid-metal complexes, *Environmental Technology* **19**, 923-931.

31. Korshin, G.V., Frenkel, A.I. and Stern, E.A. (1998) EXAFS study of the inner shell structure in copper(II) complexes with humic substances, *Environmental Science & Technology* **32**, 2699-2705.

32. Brown, G.K., MacCarthy, P. and Leenheer, J.A. (1999) Simultaneous determination of Ca, Cu, Ni, Zn and Cd binding strengths with fulvic acid fractions by Schubert's method, *Analytica Chimica Acta* **402**, 169-181.

33. Du, Q., Sun, Z.X., Forsling, W. and Tang, H.X. (1999) Complexations in illite-fulvic acid-Cu^{2+} systems, *Water Research* **33**, 693-706.

34. Osterberg, R., Wei, S.Q. and Shirshova, L. (1999) Inert copper ion complexes formed by humic acids, *Acta Chemica Scandinavica* **53**, 172-180.

35. Robertson, A.P. and Leckie, J.O. (1999) Acid/base, copper binding, and Cu^{2+}/H^+ exchange properties of a soil humic acid, an experimental and modeling study, *Environmental Science & Technology* **33**, 786-795.

36. Xue, H.B. and Sigg, L. (1999) Comparison of the complexation of Cu and Cd by humic or fulvic acids and by ligands observed in lake waters, *Aquatic Geochemistry* **5**, 313-335.

37. Carballeira, J.L., Antelo, J.M. and Arce, F. (2000) Analysis of the Cu^{2+}-soil fulvic acid complexation by anodic stripping voltammetry using an electrostatic model, *Environmental Science & Technology* **34**, 4969-4973.

38. Exner, A., Theisen, M., Panne, U. and Niessner, R. (2000) Combination of asymmetric flow field-flow fractionation (AF(4)) and total-reflexion X-ray fluorescence analysis (TXRF) for determination of

heavy metals associated with colloidal humic substances, *Fresenius Journal of Analytical Chemistry* **366**, 254-259.

39. Liu, A.G. and Gonzalez, R.D. (2000) Modeling adsorption of copper(II), cadmium(II) and lead(II) on purified humic acid, *Langmuir* **16**, 3902-3909.

40. Pandey, A.K., Pandey, S.D. and Misra, V. (2000) Stability constants of metal-humic acid complexes and its role in environmental detoxification, *Ecotoxicology and Environmental Safety* **47**, 195-200.

41. Abate, G. and Masini, J.C. (2001) Acid-basic and complexation properties of a sedimentary humic acid. A study on the Barra Bonita reservoir of Tiet(e)over-cap river, S(a)over-tildeo Paulo State, Brazil, *Journal of the Brazilian Chemical Society* **12**, 109-116.

42. Cezikova, J., Kozler, J., Madronova, L., Novak, J. and Janos, P. (2001) Humic acids from coals of the North-Bohemian coal field II. Metal-binding capacity under static conditions, *Reactive & Functional Polymers* **47**, 111-118.

43. Gomes, P.C., Fontes, M.P.F., da Silva, A.G., Mendonca, E.D. and Netto, A.R. (2001) Selectivity sequence and competitive adsorption of heavy metals by Brazilian soils, *Soil Science Society of America Journal* **65**, 1115-1121.

44. Hamilton-Taylor, J., Postill, A.S., Tipping, E. and Harper, M.P. (2002) Laboratory measurements and modeling of metal-humic interactions under estuarine conditions, *Geochimica et Cosmochimica Acta* **66**, 403-415.

45. Arias, M., Barral, M.T. and Mejuto, J.C. (2002) Enhancement of copper and cadmium adsorption on kaolin by the presence of humic acids, *Chemosphere* **48**, 1081-1088.

46. Kaschl, A., Römheld, R. and Chen, Y. (2002) Binding of cadmium, copper and zinc to humic substances originating from municipal solid waste compost, *Israel Journal of Chemistry* **42**, 89-98.

47. Ramos, M.A., Fiol, S., Lopez, R., Antelo, J.M. and Arce, F. (2002) Analysis of the effect of pH on Cu^{2+} – Fulvic acid complexation using a simple electrostatic model, *Environmental Science & Technology* **36**, 3109-3113.

48. Pinheiro, J.P., Mota, A.M., Goncalves, M.S. and van Leeuwen, H.P. (1994) Kinetics of Adsorption of Humic Matter on Mercury, *Environmental Science & Technology* **28**, 2112-2119.

49. Mathuthu, A.S. and Ephraim, J.H. (1995) Binding of cadmium to Laurentide fulvic acid. Justification of the functionalities assigned to the predominant acidic moieties in the fulvic acid molecule, *Talanta* **42**, 1803-1810.

50. Bolton (1996) Proton binding and cadmium complexation constants for a soil humic acid using a quasi-particle model, *Soil Science Society of America Journal* **60**, 1064-1072.

51. Christensen, J.B., Botma, J.J. and Christensen, T.H. (1999) Complexation of Cu and Pb by DOC in polluted groundwater: A comparison of experimental data and predictions by computer speciation models (WHAM and MINTEQA2), *Water Research* **33**, 3231-3238.

52. Masset, S., Monteil-Rivera, F., Dupont, L., Dumonceau, J. and Aplincourt, M. (2000) Influence of humic acid on sorption of Co(II), Sr(II), and Se(IV) on goethite, *Agronomie* **20**, 525-535.

53. Mandal, R., Salam, M.S.A., Murimboh, J., Hassan, N.M., Chakrabarti, C.L., Back, M.H. and Gregoire, D.C. (2000) Competition of Ca(II) and Mg(II) with Ni(II) for binding by a well-characterized fulvic acid in model solutions, *Environmental Science & Technology* 34, 2201-2208.

54. Melamed, R., Trigueiro, F.E. and Boas, R.C.V. (2000) The effect of humic acid on mercury solubility and complexation, *Applied Organometallic Chemistry* **14**, 473-476.

55. Pinheiro, J.P., Mota, A.M. and Benedetti, M.F. (2000) Effect of aluminum competition on lead and cadmium binding to humic acids at variable ionic strength, *Environmental Science & Technology* **34**, 5137-5143.

56. Datta, A., Sanyal, S.K. and Saha, S. (2001) A study on natural and synthetic humic acids and their complexing ability towards cadmium, *Plant and Soil* **235**, 115-125.

57. Liu, C., Frenkel, A.I., Vairavamurthy, A. and Huang, P.M. (2001) Sorption of cadmium on humic acid: Mechanistic and kinetic studies with atomic force microscopy and X-ray absorption fine structure spectroscopy, *Canadian Journal of Soil Science* **81**, 337-348.

58. Abate, G. and Masini, J.C. (2002) Complexation of Cd(II) and Pb(II) with humic acids studied by anodic stripping voltammetry using differential equilibrium functions and discrete site models, *Organic Geochemistry* **33**, 1171-1182.

59. Oste, L.A., Temminghoff, E.J.M., Lexmond, T.M. and van Riemsdijk, W.H. (2002) Measuring and Modeling zinc and cadmium binding by humic acid, *Analytical Chemistry* **74**, 856-862.

60. Sikora, F.J. and Stevenson, F.J. (1988) Silver complexation by humic substances; conditional stability constants and nature of reactive sites, *Geoderma* **42**, 353-363.

61. Wood, S.A. (1996) The role of humic substances in the transport and fixation of metals of economic interest (Au, Pt, Pd, U, V), *Ore Geology Reviews* **11**, 1-31.

62. Glaus, M.A., Hummel, W. and Van Loon, L.R. (2000) Trace metal-humate interactions. I. Experimental determination of conditional stability constants, *Applied Geochemistry* **15**, 953-973.

63. Gu, Z.M., Wang, X.R., Gu, X.Y., Cheng, J., Wang, L.S., Dai, L.M. and Cao, M. (2001) Determination of stability constants for rare earth elements and fulvic acids extracted from different soils, *Talanta* **53**, 1163-1170.

64. Peters, A.J., Hamilton-Taylor, J. and Tipping, E. (2001) Americium binding to humic acid, *Environmental Science & Technology* **35**, 3495-3500.

65. Alfassi, Z.B. (2002) On the complex of humic acid - Am^{3+}, *Journal of Radioanalytical and Nuclear Chemistry* **251**, 307-309.

66. Gamble, D. S., Underdown, A. W. and Langford, C. H. (1980) Copper titration of fulvic acid ligand sites with theoretical, potentiometric, and spectrometric analysis, *Analytical Chemistry* **52**, 1901-1908.

67. Larive, C.K., Rogers, A., Morton, M. and Carper, W.R. (1996) Cd-113 NMR binding studies of Cd - Fulvic acid complexes: Evidence of fast exchange, *Environmental Science & Technology* **30**, 2828-2831.

68. Logan, E.M., Pulford, I.D., Cook, G.T. and MacKenzie, A.B. (1997) Complexation of Cu^{2+} and Pb^{2+} by peat and humic acid, *European Journal of Soil Science* **48**, 685-696.

69. Spark, K.M., Wells, J.D. and Johnson, B.B. (1997) The interaction of a humic acid with heavy metals, *Australian Journal of Soil Research* **35**, 89-101.

70. Manunza, B., Deiana, S., Maddau, V., Gessa, C. and Seeber, R. (1995) Stability constants of metal-humate complexes: titration data analyzed by bimodal Gaussian distribution, *Soil Science Society of America Journal* **59**, 1570-1574.

71. Leenheer, J.A., Brown, G.K., MacCarthy, P. and Cabaniss, S.E. (1998) Models of metal binding structures in fulvic acid from the Suwannee River, Georgia, *Environmental Science & Technology* **32**, 2410-2416.

72. Evangelou, V.P. and Marsi, M. (2001) Composition and metal ion complexation behavour of humic fractions derived from corn tissue, *Plant and Soil* **229**, 13-24.

73. Mahieu, N., Powlson, D.S. and Randall, E.W. (1999) Statistical analyses of published carbon-13 NMR spectra of soil organic matter, *Soil Science Society of America Journal* **63**, 307-319.

74. Gonzalez (1985) Chemical structural characteristics of humic acids extracted from composted municipal refuse, *Agric. Ecosyst. Environ* **14**, 267-278.

75. Garcia, C., Harnandez, T. and Costa, F. (1992) Characterization of Humic Acids from Uncomposted and Composted Sewage Sludge by Degradative and Non-degradative Techniques, *Bioresource Technology* **41**, 53-57.

76. Dick, D.P., Mangrich, A.S., Menezes, S.M.C. and Pereira, B.F. (2002) Chemical and spectroscopical characterization of humic acids from two south Brazilian coals of different ranks, *Journal of the Brazilian Chemical Society* **13**, 177-182.

77. Goodman, B.A. and Cheshire, M.V. (1976) The occurence of copper-porphyrin complexes in soil humic acids, *Journal of Soil Science* **27**, 337-347.

78. Loux, N.T. (1998) An assessment of mercury-species-dependent binding with natural organic carbon, *Chemical Speciation and Bioavailability* **10**, 127-136.

79. Zunino, Aguilera, Caiozzi, Peirano, Borie and Martin (1979) Metal-binding organic macromolecules in soil: 3. Competition of Mg(II) and Zn(II) for binding sites in humic and fulvic-type model polymers, *Soil Science* **128**, 257-266.

80. Senesi, N. and Sposito, G. (1984) Residual Copper(II) complexes in purified soil and sewage sludge fulvic acids: electron spin resonance study, *Soil Science Society of America Journal* **48**, 1247-1253.

81. Gregor, J.E., Powell, H.K.J. and Town, R.M. (1989) Evidence for aliphatic mixed mode coordination in copper(II)- fulvic acid complexes, *Journal of Soil Science* **40**, 661-674.

108

82. Xiao, C., Ma, L.Q. and Sarigumba, T. (1999) Effects of soil on trace metal leachability from papermill ashes and sludge, *Journal Of Environmental Quality* **28**, 321-333.

83. da Silva, J.C.G.E. and Oliveira, C.J.S. (2002) Metal ion complexation properties of fulvic acids extracted from composted sewage sludge as compared to a soil fulvic acid, *Water Research* **36**, 3404-3409.

84. Gigliotti, G., Giusquiani, P.L., Businelli, D. and Macchioni, A. (1997) Composition changes of dissolved organic matter in a soil amended with municipal waste compost, *Soil Science* **162**, 919-926.

85. Baham, J., Ball, N. B. and Sposito, G. (1978) Gel filtration studies of trace metal-fulvic acid solutions extracted from sewage sludge, *Journal of Environmental Quality* **7**, 181-188.

86. Inbar, Y., Chen, Y. and Hadar, Y. (1989) Solid-state carbon-13 nuclear magnetic resonance and infrared spectroscopy of composted organic matter, *Soil Science Society of America Journal* **53**, 1695-1701.

87. Martinez, G.A., Traina, S.J. and Logan, T.J. (1998) Evaluation of proton and europium (III) binding reactions in a sewage sludge humic acid, *Journal of Agriculture of the University of Puerto Rico* **82**, 121-140.

88. Kerndorff, H. and Schnitzer, M. (1980) Sorption of metals on humic acid, *Geochimica et Cosmochimica Acta* **44**, 1701-1708.

89. Boyd, S.A., Sommers, L.E. and Nelson, D.W. (1981) Copper(II) and Iron(III) complexation by the carboxylate group of humic acid, *Soil Science Society of America Journal* **45**, 1241-1242.

90. Boyd, S.A., Sommers, L.E., Nelson, D.W. and West, D.X. (1983) Copper(II) binding by humic acid extracted from sewage sludge: An electron spin resonance study, *Soil Science Society of America Journal* **47**, 43-46.

91. Denecke, M.A., Pompe, S., Reich, T., Moll, H., Bubner, M., Heise, K.H., Nicolai, R. and Nitsche, H. (1997) Measurements of the structural parameters for the interaction of uranium(VI) with natural and synthetic humic acids using EXAFS, *Radiochimica Acta* **79**, 151-159.

92. Merdy, P., Guillon, E., Dumonceau, J. and Aplincurt, M. (2002) Spectroscopic study of copper(II) - Wheat straw cell wall residue surface complexes, *Environmental Science & Technology* **36**, 1728-1733.

93. Avena, M.J., Vermeer, A.W.P. and Koopal, L.K. (1999) Volume and structure of humic acids studied by viscometry pH and electrolyte concentration effects, *Colloids and Surfaces A-Physicochemical and Engineering Aspects* **151**, 213-224.

94. Wang, Y.G., Combe, C. and Clark, M.M. (2001) The effects of pH and calcium on the diffusion coefficient of humic acid, *Journal of Membrane Science* **183**, 49-60.

95. Barak and Chen, Y. (1992) Equivalent radii of humic macromolecules from acid-base titration, *Soil Science* **154**, 184-195.

96. Saar, R. S. and Weber, J. H. (1979) Complexation of cadmium (II) with water- and soil-derived fulvic acids: effect of pH and fulvic acid concentration, *Canadian Journal of Chemistry* **57**, 1263-1268.

97. Stevenson, F.J., Fitch, A. and Brar, M.S. (1993) Stability constants of Cu (II)-humate complexes: comparison of select models, *Soil Science* **155**, 77-91.

98. Christl, I. and Kretzschmar, R. (2001) Relating ion binding by fulvic and humic acids to chemical composition and molecular size. 1. Proton binding, *Environmental Science & Technology* **35**, 2505-2511.

99. Sekaly, A.L.R., Mandal, R., Hassan, N.M., Murimboh, J., Chakrabarti, C.L., Back, M.H., Gregoire, D.C. and Schroeder, W.H. (1999) Effect of metal/fulvic acid mole ratios on the binding of Ni(II), Pb(II), Cu(II), Cd(II), and Al(III) by two well- characterized fulvic acids in aqueous model solutions, *Analytica Chimica Acta* **402**, 211-221.

100. Filella, M. and Town, R.M. (2001) Heterogeneity and lability of Pb(II) complexation by humic substances: practical interpretation tools, *Fresenius Journal of Analytical Chemistry* **370**, 413-418.

101. Balnois, E., Wilkinson, K.J., Lead, J.R. and Buffle, J. (1999) Atomic force microscopy of humic substances: Effects of pH and ionic strength, *Environmental Science & Technology* **33**, 3911-3917.

102. Hosse, M. and Wilkinson, K.J. (2001) Determination of electrophoretic mobilities and hydrodynamic radii of three humic substances as a function of pH and ionic strength, *Environmental Science & Technology* **35**, 4301-4306.

103. Chen, Y. and Schnitzer, M. (1976) Scanning electron microscopy of a humic acid and of a fulvic acid and its metal and clay complexes, *Soil Science of America Journal* **40**, 682-686.

109

104. Rainville, D. P. and Weber, J. H. (1982) Complexing capacity of soil fulvic acid for Cu^{2+}, Cd^{2+}, Mn^{2+}, Ni^{2+}, and Zn^{2+} measured by dialysis titration: a model based on fulvic acid aggregation, *Canadian Journal of Chemistry* **60**, 1-5.
105. Underdown, A. W., Langford, C. H. and Gamble, D. S. (1985) Light scattering studies of the relationship between cation binding and aggregation of a fulvic acid, *Environmental Science & Technology* **19**, 132-136.
106. Engebretson, R.B. and von Wandruszka, R. (1998) Kinetic aspects of cation enhanced aggregation in aqueous humic acids, *Environmental Science & Technology* **32**, 488-493.
107. Bryan, N.D., Jones, M.N., Birkett, J. and Livens, F.R. (2001) Aggregation of humic substances by metal ions measured by ultracentrifugation, *Analytica Chimica Acta* **437**, 291-308.
108. Bryan, N.D., Jones, M.N., Birkett, J. and Livens, F.R. (2001) Application of a new method of analysis of ultracentrifugation data to the aggregation of a humic acid by copper(II) ions, *Analytica Chimica Acta* **437**, 281-289.
109. Cheam, V. and Gamble, D. S. (1974) Metal-fulvic acid chelation equilibrium in aqueous $NaNO_3$ solution. Hg, Cd, and Cu fulvate complexes, *Canadian Journal of Soil Science* **54**, 413-417.
110. Matsuda, K. and Ito, S. (1970) Adsorption strength of Zinc for soil humus: III. Relationship between stability constants of zinc-humic and -fulvic complexes, and the degree of humification, *Soil Science And Plant Nutrition* **16**, 1-10.
111. Truitt, R. E. and Weber, J. H. (1981) Determination of complexing capacity of fulvic acid for copper(II) and cadmium(II) by dialysis titration, *Analytical Chemistry* **53**, 337-342.
112. Soares, H. and Vasconcelos, M. T. (1994) Study of the lability of copper(II)-fulvic acid complexes by ion selective electrodes and potentiometric stripping analysis, *Analytica Chimica Acta* **293**, 261-270.
113. Stevenson, F.J. and Chen, Y. (1991) Stability constants of copper(II)-humate complexes determined by modified potentiometric titration, *Soil Science Society of America Journal* **55**, 1586-1591.
114. Ricca, Pastorelli and Severini (1997) Humic acid from leonardite: structural investigations and complexes with metal ions, in Drozd, Gonet, Senesi, and Weber (eds.), *The role of humic substances in the Ecosystems and in environmental protection,* IHSS, Wroclaw, Poland, pp. 175-181.
115. Evangelou, V.P., Marsi, M. and Vandiviere, M.M. (1999) Stability of Ca^{2+}-, Cd^{2+}-, Cu^{2+}-[illite-humic] complexes and pH influence, *Plant and Soil* **213**, 63-74.
116. Gao, K.Z., Pearce, J., Jones, J. and Taylor, C. (1999) Interaction between peat, humic acid and aqueous metal ions, *Environmental Geochemistry and Health* **21**, 13-26.
117. Dzombak, D. A., Fish, W. and Morel, F. M. M. (1986) Metal-humate interactions. 1. Discrete ligand and continuous distribution models, *Environmental Science & Technology* **20**, 669-674.
118. McBride, M.B., Richards, B.K., Steenhuis, T. and Spiers, G. (1999) Long-term leaching of trace elements in a heavily sludge-amended silty clay loam soil, *Soil Science* **164**, 613-623.
119. Kaschl, A. (2001) *Trace metal binding by organic matter from municipal solid waste compost and consequences for mobility in compost-amended soils under semiarid conditions (Gaza Strip)*, Ph.D. Thesis, Hohenheim University, Stuttgart, Germany.
120. Giusquiani, P.L., Concezzi, L., Businelli, M. and Macchioni, A. (1998) Fate of Pig Sludge Liquid Fraction in Calcareous Soil: Agricultural and Environmental Implications, *Journal Of Environmental Quality* **27**, 364-371.
121. Naidu, R. and Harter, R.D. (1998) Effect of different organic ligands on cadmium sorption by and extractibility from soils, *Soil Science Society of America Journal* **62**, 644-650.
122. Al-Wabel, M.A., Heil, D.M., Westfall, D.G. and Barbarick, K.A. (2002) Solution chemistry influence on metal mobility in biosolids-amended soils, *Journal Of Environmental Quality* **31**, 1157-1165.
123. Kaschl, A., Romheld, V. and Chen, Y. (2002) The influence of soluble organic matter from municipal solid waste compost on trace metal leaching in calcareous soils, *Science of the Total Environment* **291**, 45-57.
124. Holtzclaw, K. M., Keech, D. A., Page, A. L., Sposito, G., Ganje, T. J. and Ball, N. B. (1978) Trace metal distribution among the humic acid, the fulvic acid, and preccipitable fractions extracted with NaOH from sewage sludges, *Journal of Environmental Quality* **7**, 124-127.
125. Senesi, N., Sposito, G., Holtzclaw, K.M. and Bradford (1989) Chemical properties of metal-humic acid fractions of a sewage sludge-amended aridisol, *Journal Of Environmental Quality* **18**, 186-194.

110

126. Petruzzelli, G., Guidi, G. and Lubrano, L. (1980) Chormatographic fractionation of heavy metals bound to organic matter of two Italian composts, *Environ. Technol. Lett.* **1**, 201-208.

127. Davis, W.M., Erickson, C.L., Johnston, C.T., Delfino, J.J. and Porter, J.E. (1999) Quantitative Fourier Transform Infrared spectroscopic investigation of humic substance functional group composition, *Chemosphere* **38**, 2913-2928.

128. Evangelou, V.P., Marsi, M. and Chappell, M.A. (2002) Potentiometric-spectroscopic evaluation of metal-ion complexes by humic fractions extracted from corn tissue, *Spectrochimica Acta Part A- Molecular and Biomolecular Spectroscopy* **58**, 2159-2175.

129. Hatcher, P.G., Schnitzer, M., Dennis, L.W. and Maciel, G.E. (1981) Aromaticity of humic substances in soils, *Soil Science Society of America Journal* **45**, 1089-1094.

130. Hatcher, P.G., Schnitzer, M., Dennis, L.W. and Maciel, G.E. (1983) Solid state ^{13}C-NMR of sedimentary humic substances: New revelations on their chemical composition, in R.F. Christman and E. T. Gjessing (eds.), *Aquatic and terrestrial humic materials,* Ann Arbor Science Publ., MI, pp. 37-81.

131. Chefetz, B., Hatcher, P.G., Hadar, Y. and Chen, Y. (1996) Chemical and biological characterization of organic matter during composting of municipal solid waste, *Journal Of Environmental Quality* **25**, 776-785.

132. Kawahigashi, M., Fujitake, N. and Takahashi, T. (1996) Structural information obtained from spectral analysis (UV-VIS, IR, H-1 NMR) of particle size fractions in two humic acids, *Soil Science & Plant Nutrition* **42**, 355-360.

133. Howe, R.F., Lu, X.Q., Hook, J. and Johnson, W.D. (1997) Reaction of aquatic humic substances with aluminium: a Al-27 NMR study, *Marine and Freshwater Research* **48**, 377-383.

134. Schnitzer, M. and Gosh, K. (1981) Characteristics of water-soluble fulvic acid-copper and fulvic acid-iron complexes, *Soil Science* **134**, 354-363.

135. Martin-Neto, L., Saab, S.d.C., Ferreira, J.A., Nascimento, O.R., Bayer, C., Novotny, E.H. and Mielniczuk, J. (2000) Recent applications of ESR spectroscopy in soil humic substances research, in *10th International Meeting of the International Humic Substances Society,* July 24th-28th, 2000, Toulouse, France, pp. 45-48.

136. Ryan, J. and Weber, J.H. (1982) Fluorescence quenching titration for determination of complexing capacities and stability constants of fulvic acid, *Analytical Chemistry* **54**, 986-990.

137. Gregor, J.E., Powell, H.K.J. and Town, R.M. (1989) Metal-fulvic acid complexing: evidence supporting an aliphatic carboxylate mode of coordination, *The Science of the total Environment* **81/82**, 597-606.

138. Saar, R.A. and Weber, J.H. (1980) Comparison of spectrofluorometry and ion-selective electrode potentiometry for determination of complexes between fulvic acid and heavy-metal ions, *Analytical Chemistry* **52**, 2095-2100.

139. Fründ, R., Guggenberger, G., Haider, K., Knicker, H., Kögel-Knabner, I., Lüdemann, H.-D., Luster, J., Zech, W. and Spiteller, M. (1994) Recent advances in the spectroscopic characterization of soil humic substances and their ecological relevance, *Zeitschrift fur Pflanzenernaehrung und Bodenkunde* **157**, 175-186.

140. Bidoglio, G., Ferrari, D., Selli, E., Sena, F. and Tamborini, G. (1997) Humic acid binding of trivalent Tl and Cr studied by synchronous and time-resolved fluorescence, *Environmental Science & Technology* **31**, 3536-3543.

141. Tiseanu, C.-D., Kumke, M.U., Frimmel, F.H., Klenze, R. and Kim, J.I. (1998) Time-resolved fluorescence spectroscopy of fulvic acid and fulvic acid complexed with Eu^{3+} - a comparative study, *Journal Of Photochemistry and Photobiology A: Chemistry* **117**, 175-184.

142. Takahashi, Y., Kimura, T. and Minai, Y. (2002) Direct observation of Cm(III)-fulvate species on fulvic acid- montmorillonite hybrid by laser-induced fluorescence spectroscopy, *Geochimica et Cosmochimica Acta* **66**, 1-12.

143. Perdue, E. M. and Lytle, C. R. (1983) Distribution model for binding of protons and metal ions by humic substances, *Environmental Science & Technology* **17**, 654-660.

144. Fitch, A. and Stevenson, F.J. (1984) Comparison of models for determining stability constants of metal complexes with humic substances, *Soil Science Society of America Journal* **48**, 1044-1050.

111

145. Mandal, R., Sekaly, A.L.R., Murimboh, J., Hassan, N.M., Chakrabarti, C.L., Back, M.H., Gregoire, D.C. and Schroeder, W.H. (1999) Effect of the competition of copper and cobalt on the lability of Ni(II)-organic ligand complexes. Part I. In model solutions containing Ni(II) and a well-characterized fulvic acid, *Analytica Chimica Acta* **395**, 309-322.

146. Mandal, R., Hassan, N.M., Murimboh, J., Chakrabarti, C.L. and Back, M.H. (2002) Chemical speciation and toxicity of nickel species in natural waters from the Sudbury area (Canada), *Environmental Science & Technology* **36**, 1477-1484.

147. Pandeya, S.B. (1993) Ligand competition method for determining stability constants of fulvic acid iron complexes, *Geoderma* **58**, 219-231.

148. Stevenson, F.J. (1976) Stability constants of Cu, Pb, and Cd complexes with humic acids, *Soil Science Society of America Journal* **40**, 665-672.

149. Stevenson (1977) Nature of divalent transition metal complexes of humic acids as revealed by a modified potentionmetric titration method, *Soil Science* **123**, 10-17.

150. Pandeya, S.B. and Singh, A.K. (2000) Potentiometric measurement of stability constants of complexes between fulvic acid carboxylate and Fe^{3+}, *Plant and Soil* **223**, 13-21.

151. Santos, E.B.H., Esteves, V.I., Rodrigues, J.P.C. and Duarte, A.C. (1999) Humic substances' proton-binding equilibria: Assessment of errors and limitations of potentiometric data, *Analytica Chimica Acta* **392**, 333-341.

152. Bresnahan, W. T., Grant, C. L. and Weber, J. H (1978) Stability constants for the complexation of cu ions with water and soil fulvic acids measured by an ion selective electrode, *Analytical Chemistry* **50**, 1675-1679.

153. Buffle, J., Greter, F.-L. and Haerdi, W. (1977) Measurement of complexation properties of humic and fulvic acids in natural waters with lead and copper ion-selective electrodes, *Analytical Chemistry* **49**, 216-222.

154. Saar, R.A. and Weber, J.H. (1980) Comparison of sprectrofluorometry and ion-selective electrode potentiometry for determination of complexes between fulvic acid heavy-metal ions, *Analytical Chemistry* **52**, 2095-2100.

155. Candelaria L.M., Chang A.C. and Amrhein C. (1995) Measuring cadmium ion activities in sludge-amended soils, *Soil Science* **159**, 162-175.

156. Bhat, Saar, Smart and Weber, J.H. (1981) Titration of soil-derived fulvic acid by copper(II) and measurement of free copper(II) by anodic stripping voltammetry and copper(II) selective electrode, *Analytical Chemistry* **53**, 2275-2280.

157. Plavsic, M., Cosovic, B. and Miletic, S. (1991) Comparison of the behaviours of copper, cadmium and lead in the presence of humic acid in sodium chloride solutions, *Analytica Chimica Acta* **255**, 15-21.

158. vandenHoop, M.A.G.T. and vanLeeuwen, H.P. (1997) Influence of molar mass distribution on the complexation of heavy metals by humic material, *Colloids and Surfaces A-Physicochemical and Engineering Aspects* **120**, 235-242.

159. Town, R.M. and Filella, M. (2000) Determination of metal ion binding parameters for humic substances - Part 2. Utility of ASV pseudo-polarography, *Journal of Electroanalytical Chemistry* **488**, 1-16.

160. Celo, V., Murimboh, J., Salam, M.S.A. and Chakrabarti, C.L. (2001) A kinetic study of nickel complexation in model systems by adsorptive cathodic stripping voltammetry, *Environmental Science & Technology* **35**, 1084-1089.

161. Holm, P.E., Andersen, S. and Christensen, T.H. (1995) Speciation of dissolved cadmium: Interpretation of dialysis, ion exchange and computer (GEOCHEM) methods, *Water Research* **29**, 803-809.

162. Vasconcelos, M.T., Santos, A.P. and Machado, A.A. (1989) Evidence of conformational changes in fulvic acids from dialysis, *Science of the total Environment* **81/82**, 489-499.

163. Fish, W., Dzombak, D.A. and Morel, F.M.M. (1986) Metal-humate interactions. II. Application and comparison of models, *Environmental Science & Technology* **20**, 676-683.

164. Manunza, B., Gessa, C. and Rausa, R. (1992) A normal distribution model for the titration curves of humic acids, *Journal of Soil Science* **43**, 127-131.

165. Manunza, B., Deiana, S., Maddau, V., Gessa, C. and Seeber, R. (1995) Stability constants of metal-humate complexes: Titration data analyzed by bimodal Gaussian distribution, *Soil Science Society of America Journal* **59**, 1570-1574.

112

166. Kim, J.I. and Czerwinski, K.R. (1996) Complexation of metal ions with humic acid: Metal ion charge neutralization model, *Radiochimica Acta* **73**, 5-10.

167. Bryan, N.D., Hesketh, N., Livens, F.R., Tipping, E. and Jones, M.N. (1998) Metal ion-humic substance interaction - A thermodynamic study, *Journal of the Chemical Society-Faraday Transactions* **94**, 95-100.

168. Bryan, N.D., Jones, D.M., Appleton, M., Livens, F.R., Jones, M.N., Warwick, P., King, S. and Hall, A. (2000) A physicochemical model of metal-humate interactions, *Physical Chemistry Chemical Physics* **2**, 1291-1300.

169. Bryan, N.D., Robinson, V.J., Livens, F.R., Hesketh, N., Jones, M.N. and Lead, J.R. (1997) Metal-humic interactions: A random structural modelling approach, *Geochimica et Cosmochimica Acta* **61**, 805-820.

170. Shane Yu, Y., Bailey, G.W. and Xianchan, J. (1996) Application of a lumped, nonlinear kinetics model to metal sorption on humic substances, *Journal Of Environmental Quality* **25**, 552-561.

171. Bolton, K.A., Sjoberg, S. and Evans, L.J. (1996) Proton binding and cadmium complexation constants for a soil humic acid using a quasi-particle model, *Soil Science Society of America Journal* **60**, 1064-1072.

172. Woolard, C.D. and Linder, P.W. (1999) Modelling of the cation binding properties of fulvic acids: An extension of the RANDOM algorithm to include nitrogen and sulphur donor sites, *Science of the Total Environment* **226**, 35-46.

173. Benedetti, M.F., VanRiemsdijk, W.H., Koopal, L.K., Kinniburgh, D.G., Gooddy, D.C. and Milne, C.J. (1996) Metal ion binding by natural organic matter: From the model to the field, *Geochimica et Cosmochimica Acta* **60**, 2503-2513.

174. Kinniburgh, D.G., van Riemsdijk, W.H., Koopal, L.K., Borkovec, M., Benedetti, M.F. and Avena, M.J. (1999) Ion binding to natural organic matter: competition, heterogeneity, stoichiometry and thermodynamic consistency, *Colloids and Surfaces A-Physicochemical and Engineering Aspects* **151**, 147-166.

175. Gustafsson, J.P. (2001) Modeling the acid-base properties and metal complexation of humic substances with the Stockholm Humic Model, *Journal of Colloid and Interface Science* **244**, 102-112.

176. Koopal, L.K., van Riemsdijk, W.H. and Kinniburgh, D.G. (2001) Humic matter and contaminants. General aspects and modeling metal ion binding, *Pure and Applied Chemistry* **73**, 2005-2016.

177. van den Hoop, M.A.G.T., Porasso, R.D. and Benegas, J.C. (2002) Complexation of heavy metals by humic acids: analysis of voltammetric data by polyelectrolyte theory, *Colloids and Surfaces A-Physicochemical and Engineering Aspects* **203**, 105-116.

178. Thaer, A.D. (1808) *Grundriss der Chemie für Landwirte,* Berlin, Germany.

179. Grandeau, L. (1872) *Recherches sur le role des matieres organiques du sol dans les phenomenes de la nutrition des vegetaux,* Paris, France.

180. Liebig, J.V. (1841) *Organic chemistry in its applications to agriculture and physiology,* Translated by J.W.Webster and J.Owens, Cambridge, UK.

181. Liebig, J.V. (1856) On some points of agricultural chemistry, *Journal of the Royal Agricultural Society* **17**, 284-326.

182. Lawes, J.B. and Gilbert, J.H. (1905) Collected papers, in W.H. Hall (ed.), *The Book of the Rothamsted Experiments,* John Murray, London, UK.

183. Russell, E.J. (1921) *Soil conditions and plant growth,* London, UK.

184. Bottomley, W.B. (1914) The significance of certain food substances for plant growth, *Ann. Bot.* **28**, 531-540.

185. Bottomley, W.B. (1914) Some accessory factors in plant growth and nutrition, *Proceedings of the Royal Society* **88**, 237-247.

186. Bottomley, W.B. (1917) Some effects of organic growth-promotion substances (auximones) on the growth of *Lemna minor* in mineral cultural solutions, *Proceedings of the Royal Society* **89**, 481-505.

187. Bottomley W.B. (1920) The effect of organic matter on the growth of various plants in culture solutions, *Ann. Bot.* **34**, 353-365.

188. Olsen, C. (1930) On the influence of humus substances on the growth of green plants in water culture, *Comptes-rendus du Laboratoire Carlsberg* **18**, 1-16.

189. Burk, D., Lineweaver, H. and Horner, C.K. (1932) Iron in relation to the stimulation of growth by humic acid, *Soil Science* **33**, 413-435.

190. Vaughan, D. and Malcolm, R.E. (1985) Influence of Humic Substances on Growth and Physiological Processes., in D. Vaughan and et al. (eds.), *Soil Organic Matter and Biological Activity,* Martinus Nijhoff, Dordrecht, The Netherlands, pp. 37-75.

191. Chen, Y. and Aviad, T. (1990) Effects of humic substances on plant growth, in P. MacCarthy, C. E. Clapp, R. L. Malcolm and P. R. Bloom (eds.) *Humic substances in soil and crop sciences: Selected readings.,* SSSA, Madison, WI, USA, pp. 161-186.

192. Rauthan, B.S. and Schnitzer, M. (1981) Effects of a Soil Fulvic Acid on the Growth and Nutrient Content of Cucumber (*Cucumis Sativus*) Plants, *Plant and Soil* **63**, 491-495.

193. Tan, K.H. and Nopamornbodi, V. (1979) Effect of Different Levels of Humic Acids on Nutrient Content and Growth of Corn (*Zea Mays L.*), *Plant and Soil* **51**, 283-287.

194. Mylonas, V.A. and McCants, C.B. (1980) Effects of humic and fulvic acids on growth of tobacco: 1. Root initiation and elongation, *Plant and Soil* **54**, 485-490.

195. Dormaar J.F. (1975) Effects of humic substances from chenozemic Ah horizons on nutrient uptake by *Phaseolus vulgaris* and *Festuca scabrella, Can. J. Soil Sci.* **55**, 111-118.

196. Ernst, W.H.O., Kraak, M.H.S. and Stoots, L. (1987) Growth and Mineral Nutrition of *Scrophularia Nodos* With Various Combinations of Fulvic and Humic Acids, *J. Plant Physiol* **127**, 171-175.

197. Lee, Y.S. and Bartlett, R.J. (1976) Stimulation of Plant Growth by Humic Substances, *Soil Science Society of America Journal* **40**, 876-879.

198. Levesque, M. (1970) Fulvic acid and fulvometallic complexes in mineral nutrition of plants, *Can. J. Soil Sci.* **50**, 385-390.

199. Fagbenro, J.A. and Agboola, A.A. (1993) Effect of different levels of humic-acid on the growth and nutrient uptake of teak seedlings, *J. Plant Nutr.* **16**, 1465-1483.

200. Irintoto, B., Tan, K.H. and Sommer, H.E. (1993) Effect of humic-acid on callus-culture of slash pine (*Pinnus-elliotti engelm*), *J. Plant Nutr.* **16**, 1109-1118.

201. Chen, Y., Magen, H. and Riov (1994) Humic substances originating from rapidly decomposing organic matter: properties and effects on plant growth, in Senesi and Miano (eds.), *Humic substances in the global environment and implications on human health,* Elsevier, The Netherlands, pp. 427-443.

202. Chen, Y. (1996) Organic Matter Reactions Involving Micronutrients in Soils and Their Effect on Plants, in A. Piccolo (ed.), *Humic Substances in Terrestrial Ecosystems,* Elsevier, Oxford, UK, pp. 507-529.

203. Marschner, H., Römheld, V., Horst, W.J. and Martin, P. (1986) Root induced changes in the rhizosphere: importance for the mineral nutrition of plants, *Z. Pflanzenernaehr. Bodenk* **149**, 441-456.

204. Marschner, H., Römheld, V. and Kissel, M. (1986) Different strategies in the higher plants in mobilization and uptake of iron, *J. Plant Nutr.* **9**, 695-713.

205. Chen, Y., Magen, H. and Clapp, C.E. (2001) Plant growth stimulation by humic substances and their complexes with iron, in *Proceedings of The Dalia Greidinger Symposium,* The International Fertiliser Society.

206. Bar-Ness, E. and Chen, Y. (2003) Manure and Peat Based Fe-Organo Complexes. I. Characterization and Enrichment, *Plant and Soil* **130**, 35-43.

207. Bar-Ness, E. and Chen, Y. (1991) Manure and peat based Fe-enriched complexes: II. Transport in soils, *Plant and Soil* **130**, 45-50.

208. Yehuda, Z., Shenker, M., Römheld, V., Marschner, H., Hadar, Y. and Chen, Y. (1996) The role of ligand exchange in the uptake of iron from microbial siderophores by graminaceous plants, *Plant Physiol.* **112**, 1273-1280.

209. Nardi, S., Pizzeghello, D., Muscolo, A. and Vianello, A. (2002) Physiological effects of humic substances on higher plants, *Soil Biol. Biochem.* **34**, 1527-1536.

Chapter 5

INFLUENCE OF UV-OXIDATION ON THE METAL COMPLEXING PROPERTIES OF NOM

F.H. FRIMMEL, K. VERCAMMEN, D. SCHMITT
*Engler-Bunte Institute, Department of Water Chemistry, University of
Karlsruhe, Postfach 6980, D-76128 Karlsruhe, Germany
<fritz.frimmel@ciw.uni-karlsruhe.de>*

Abstract

The interaction of metals and natural organic matter (NOM) is of influence on the metal transport in aquatic systems and soil. In the presented study, NOM of a brown water lake was oxidized with UV irradiation to elucidate the influence of oxidative degradation on the metal complexing properties of NOM. UV/Vis and fluorescence spectroscopy, as well as size-exclusion chromatography (SEC) with online detection of UV absorption, fluorescence and metal concentration were used to investigate the property changes of NOM caused by oxidation. After oxidation, the fluorescence intensity of NOM increased considerably despite a decrease in the UV absorption. The SEC experiments showed a shift towards smaller molecular sizes in the oxidized NOM samples and a decrease in the stability constants of corresponding Al-NOM complexes. For Pb no such effect could be determined. The studies on the dissociation kinetics of metal-NOM complexes revealed a slower dissociation of Al and Pb complexes with original NOM compared to oxidized NOM. The determined dissociation rate constants were used to predict the migration of metal ion-NOM complexes in column experiments. The experimental data and model predictions were in good agreement for the divalent metals Pb and Zn, but differed substantially for Al.

The stability of the metal complexes with original NOM decreased in the order Al > Pb > Zn. After UV-oxidation, the complexing ability of NOM towards Al- ions was decreased. In NOM equilibrated quartz columns, Pb and Zn were immobilized in the presence of all types of NOM, whereas Al was readily eluted. The original NOM was more effective for complexing Al than the oxidized form. Colloid formation however can stimulate the Al migration.

1. Introduction

The mobility of metals in aquatic systems is strongly influenced by natural organic matter (NOM) [1, 2]. NOM can act as an electron-donor ligand for metal ions in dissolved state as well as being immobilized on surfaces of solid phases. Since metals are common pollutants in dumping areas and industrialized grounds [3], the role of NOM in remediation of polluted environments is of special interest [4]. From water

I. V. Perminova et al. (eds.),
Use of Humic Substances to Remediate Polluted Environments: From Theory to Practice, 115–133.
© 2005 *Springer. Printed in the Netherlands.*

treatment it is known that oxidation reactions can alter the structure of NOM and its technical behaviour [5, 6, 7]. NOM is also known to be among the most important sunlight-absorbing components of soil surfaces and aquatic environments with the consequence of photochemical transformations [8, 9]. Bleaching of the yellow colour, decrease of molecular size, introduction of oxygen-containing functional groups and increased biodegradability are the most obvious results of NOM oxidation [10]. Some of these changes should also influence the interaction of NOM with metals. Whereas complex formation of NOM and metals has been studied extensively [11, 12], little is known about how oxidation alters the NOM complexing properties toward metal ions [13].

Aerated UV-irradiation turned out to be a promising tool for investigating the changes of NOM and its complexation capacity. In that clean oxidation there is no application of chemicals and hence no formation of oxidation by-products. Therefore UV-irradiation is not only of increasing interest for the photic zone of surface waters due to the increasing depletion of ozone in the earth's atmosphere, the irradiation method is also well suited for the production of substances which can be expected as products of NOM after aerobic degradation or technical oxidation. In addition to the UV-absorbance measurements, fluorescence spectra have shown that they are well suited for the characterization of NOM and its oxidation products [14, 15].

The objectives of the work were to 1) show the influence of partial oxidation on the structure of NOM by fluorescence spectroscopy and gel chromatography, 2) quantify the complexation of Al^{3+}, Pb^{2+} and Zn^{2+} by the differently oxidized NOM, 3) compare the transport of the NOM-complexed metals in sand columns, and 4) conclude on the role of NOM-metal complex formation in remediation.

The NOM-metal dissociation rates were studied using a cation exchange resin technique and the estimated rate constants were used to predict the breakthrough of metal cations through a sand column in the presence and absence of oxidized and untreated NOM. The results were compared with the calculations obtained using a kinetic model for estimating the NOM-facilitated transport of metal ions.

For a better understanding of the results, the theoretical backgrounds of the methods applied are given.

2. Theory

2.1. SIZE-EXCLUSION CHROMATOGRAPHY

Molecular size, or more precisely molecular hydrodynamic volume, governs the separation process of size-exclusion chromatography (SEC). As a mixture of solutes of different size passes through a column packed with porous particles, the molecules that are too large to penetrate the pores of the packing elute first. Smaller molecules, however, which can penetrate or diffuse into the pores, elute at a later time or larger elution volume. The retention of molecules in the column can be described by the dimensionless distribution coefficient K_d:

$$K_d = \frac{V_e - V_0}{V_p - V_0}, \tag{1}$$

where V_e is the elution volume of the molecule, V_0 is the void volume, and V_p is the permeation volume. Besides diffusion of the molecules into the porous gel, two other mechanisms influence the separation of the solutes, i.e. hydrophobic adsorption of solutes onto the gel and electrostatic repulsion of the analytes [16, 17]. Both disturb the purely size-based separation.

2.2. DISSOCIATION KINETICS OF METAL-NOM COMPLEXES

Experiments with cation-exchange resins were used for the determination of dissociation rates of metal-NOM complexes in solution. This method is applicable for slowly dissociating complexes (on the time scale of minutes and hours) [18, 19]. Free metal ions sorb onto the resin, whereas NOM and metal-NOM complexes may not. By binding to the resin, the free metal ion has to dissociate first from the metal-NOM complex. The dissociation rate can be determined by adding resin to the Me-NOM system and measuring the concentration of remaining metal ion in solution as a function of time. The reaction scheme can be described as follows (charges are omitted):

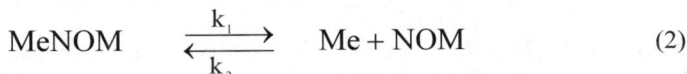

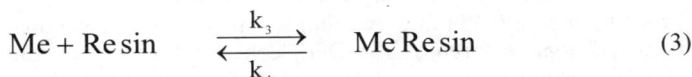

$$MeNOM \quad \underset{k_2}{\overset{k_1}{\rightleftarrows}} \quad Me + NOM \tag{2}$$

$$Me + Resin \quad \underset{k_4}{\overset{k_3}{\rightleftarrows}} \quad MeResin \tag{3}$$

Equation (2) describes the formation of metal-NOM complexes (MeNOM) from free ions (Me) and free ligands (NOM) with k_1 and k_2 being the rate constants for the dissociation and the association of the complexes. Equation (3) accounts for the interaction of free metal ions with the solid phase (Resin) leading to surface complexes (MeResin). k_3 and k_4 are the rate constants for the adsorption of metal ions on and the desorption of metal ions from the solid phase. Assuming first order rate laws for these reactions, the total concentration of metal species in solution $[Me]_{tot,aq}$ is defined as:

$$[Me]_{tot,aq} = [Me] \times e^{-t/t_3} + [MeNOM] \times e^{-t/t_1} \tag{4}$$

where $t_3 = 1/k_3$ is the time constant for the adsorption of metal ion on the resin, and $t_1 = 1/k_1$ is the time constant for dissociation of the metal-NOM complexes. For a detailed description, the reader is referred to Schmitt et al. [20]. If the dissociation of the metal ion from the metal-NOM complex is much slower than the cation-exchange of the metal ion with the resin (i.e. $k_1 \ll k_3$), equation (4) can be simplified to:

$$[Me]_{tot,aq} = [MeNOM] \times e^{-t/t_1} \tag{5}$$

2.3. TRANSPORT MODELING OF METAL-NOM COMPLEXES

Most transport models describing metal-NOM systems are based on the assumption that all chemical processes are fast compared to the rate of flow of the solution phase through porous media. However, a certain fraction of the metal ions bound to NOM may dissociate very slowly from the metal-NOM complex, which results in a much faster transport of the metal ion than would be expected from a system at equilibrium. In this study, a transport model which is able to account for slow chemical reactions is used. The model is based on the following equations:

$$\frac{d[MeNOM]}{dt} = k_{ass}[Me][NOM] - k_{diss}[MeNOM] \tag{6}$$

$$\frac{d[MeSurface]}{dt} = k_{ads}[Me][Surface] - k_{des}[MeSurface] \tag{7}$$

where [MeNOM] is the concentration of metal-NOM complexes, [Me] is the free metal concentration, [NOM] is the concentration of NOM, k_{ass} and k_{diss} are the rate constants of association and dissociation of the metal-NOM complexes, respectively. [MeSurface] is the concentration of metal ions adsorbed onto the stationary phase, [Surface] is the concentration of adsorption sites, k_{ads} and k_{des} are the rate constants for adsorption of the metal ion on and desorption of the metal ion from the stationary phase, respectively. For a more detailed description the reader is referred to Schmitt et al. [20] and Bryan et al. [21].

3. Experimental Section

NOM sample. The NOM sample originates from an environmentally protected brown water lake (lake Hohloh, Northern Black Forest, Germany). The water sample was filtered through a 0.45 μm polycarbonate filter before use and contained 21 mg/l DOC (dissolved organic carbon). The DOC concentration was measured using a Shimadzu 5000 TOC analyser. The proton capacity of the NOM was 15 μmol/mg DOC, and the complexation capacity for copper was 3.5 μmol/mg DOC. Metal concentrations of the brown water were $\beta(Al) = 111$ μg/l, $\beta(Fe) = 335$ μg/l, $\beta(Na) = 1$ mg/l, $\beta(Ca) = 1.5$ mg/l, $\beta(Pb) = 4$ μg/l, $\beta(Zn) = 37$ μg/l, $\beta(Cu) = 3.6$ μg/l.

Me-NOM samples. Fifteen different samples containing NOM and/or metal ions were studied: *NOM original* ($\beta(DOC) = 9.7$ mg/l), *NOM 30min* ($\beta(DOC) = 9.1$ mg/l) and *NOM 4h* ($\beta(DOC) = 8.6$ mg/l) each in the absence and presence of the following metal ions: Al ($0.7 \cdot 10^{-5}$ M), Zn ($0.9 \cdot 10^{-5}$ M) and Pb ($1.1 \cdot 10^{-5}$ M). Three samples contained only the metal ion of interest. The samples were made from 10^{-3} M Al, Pb or Zn stock solutions in 1% HNO_3. For these, ICP standards of the metal ions ($\beta(Me) = 1$ g/l, Merck) were used. All samples were made in 10 mM $NaClO_4$ and brought to pH 5. The samples were shaken head-over-head for one week. The pH was examined and corrected, if necessary.

UV irradiation. In the UV irradiation experiments, a low pressure mercury lamp (24W, type AR-14D, Katadyn GmbH, Germany) was used at the maximum irradiation wavelength of 254 nm. The photon flux was 7.3 μEinstein/s as determined by actinometry with $K_3Fe(C_2O_4)_3$. The radiation power equalled 3.4 W. The photoreactor was made up of two concentric quartz tubes with the inner tube containing the lamp. Between both tubes, one litre of sample solution was circulated at a flow rate of 50 ml/s. A plexiglass tube surrounded the whole reactor. The NOM sample was irradiated for 30 min (denoted as *NOM 30min*) and for 4 h (denoted as *NOM 4h*) under air saturated conditions. The non-irradiated NOM sample is denoted as *NOM original*.

UV/Vis absorption and fluorescence measurements. The absorption measurements were carried out with a Varian Cary 50 spectrophotometer in the wavelength range of 190-600 nm.

Steady-state fluorescence measurements were performed using a FL900CDT fluorescence spectrometer (Edinburgh Analytical Instruments, UK). A 450 W Xenon arc lamp was used for excitation. The spectral bandpass of the monochromators was 2.2 nm in the excitation and emission paths. The emission spectra were recorded for excitation wavelengths in the range of 275-401 nm ($\Delta\lambda = 3$ nm) in the emission range of 281-545 nm ($\Delta\lambda = 1$ nm). Each fluorescence spectrum was quantum-corrected in terms of the detector sensitivity. The emission spectra were corrected for inner-filter effects [22] and combined in an excitation-emission matrix (EEM). In the data evaluation, the ratio of the fluorescence intensities before and after oxidation was calculated.

Size-exclusion chromatography measurements. Chromatographic separation was performed by a metal-free HPLC-system (Sykam GmbH) with on-line UV absorption detection ($\lambda = 254$ nm, Linear Instruments, UVIS-206), fluorescence detection ($\lambda_{ex} = 254$ nm, $\lambda_{em} = 450$ nm, Linear Instruments, LC-305) and metal detection by ICP-MS (Elan 6000, PE Sciex). A Superdex 75 HR 10/30 column (Amersham Pharmacia) filled with dextran-agarose gel was used (length = 30 cm, inside diameter = 10 mm). Determination of the void volume V_0 with Blue Dextran (2000 kDa) and of the permeation volume V_p with acetone resulted in: $V_0 = 6.75$ ml and $V_p = 16.3$ ml. The eluent was 10 mM $NaClO_4$ at pH 5. The injection volume was 400 µl and the flow rate was 1 ml/min. Recovery calculation was done by determining the ratio between column and bypass (direct sample injection into the detector) measurements. The NOM samples were injected in a DOC concentration range of 10 mg/l.

Cation-exchange batch experiments for determining the dissociation rate constants of metal-NOM complexes. Two kinds of solutions were prepared for the cation-exchange experiments. The first series of solutions, containing only the metal ion of interest, was used to determine the adsorption kinetics of the metal ions onto the resin. The second series of solutions, containing the metal ion and NOM, were equilibrated for two weeks before use to allow for metal-NOM complexation. These solutions were used for the determination of the dissociation rate of the metal-NOM complexes. All solutions were at pH 5 and contained 10 mM $NaClO_4$, the metal ion of interest ($0.7 \cdot 10^{-5}$ M Al, $1.1 \cdot 10^{-5}$ M Pb or $0.9 \cdot 10^{-5}$ M Zn) and, if desired NOM (β(DOC) = 10 mg/L). 1g of cation-exchange resin (Analytical Grade Chelex 100 Resin Na^+-form, Bio-Rad) was added to 250 ml of each solution. The resin was brought to pH 5 beforehand. The suspension was stirred vigorously. For sampling at defined time intervals, the suspension was allowed to settle before withdrawing small aliquots of the supernatant. The samples were stabilized with 1% HNO_3 and analysed for metal concentration.

Metal transport experiments. Column experiments were performed to study the metal transport. The experimental set-up consisted of a reservoir for the carrier solution, a peristaltic pump, a six port injection valve, a glass column filled with quartz sand (length: 22.6 cm, diameter: 2.0 cm) and a fraction collector (Frac 100, Pharmacia) for sample collection every 10 min. In these experiments, the breakthrough of unretarded metal ions (at one bed volume) was of special interest. The injected samples contained the metal ion of interest (ca. $1 \cdot 10^{-5}$ M), 10 mM $NaClO_4$, β(DOC) = 10 mg/l) (*NOM original* or *NOM 4h*) and were at pH 5. The samples were equilibrated for at least one week before injection to allow for metal-NOM complexation.

Prior to the breakthrough experiments, the packing material in the column was equilibrated with NOM solution until the outflow DOC concentration reached that of the feed solution. The feed solution was of the same ionic strength, pH value and DOC concentration as the injected samples. This loading process took at least one week and assured no significant adsorption of NOM from the injected sample, containing metal ion and NOM, to the packing material. The flow velocity, the effective porosity and the longitudinal dispersion coefficient of the column were characterized by inert tracer experiments with Cl⁻. The characteristic parameters of the quartz columns are summarized in Table 1.

Table 1. Characteristic parameters of the quartz columns used for metal transport experiments.

	column loaded with NOM original	column loaded with NOM 4 h
mass quartz [g]	120	111
flow rate [ml/min]	0.095	0.1
bed volume [ml]	31.6	35.0
flow velocity [cm/min]	0.0692	0.0674
effective porosity	0.42	0.47
longitudinal dispersion coefficient [cm²/min]	0.0015	0.0025

Transport modelling. The computer code k1D which is explained elsewhere [21] was used for estimating the direct breakthrough of metal ions through the quartz columns loaded with NOM. The dissociation rate constants of the metal-NOM complexes were determined in the cation-exchange experiments ($k_1 = k_{diss}$) and used for transport calculations. The other rate constants (k_{ass}, k_{ads}, k_{des}) were set to such values that dissociation of the metal-NOM complex was rate limiting (exact values are given in Table 4). The hydraulic characteristics of the column were also taken into account in the calculations.

4. Results and Discussion

The changes in NOM content, structure and complexing properties brought about by exposure to UV-irradiation were followed using DOC-measurements, UV/Vis spectrophotometry, fluorescence spectroscopy, and SEC.

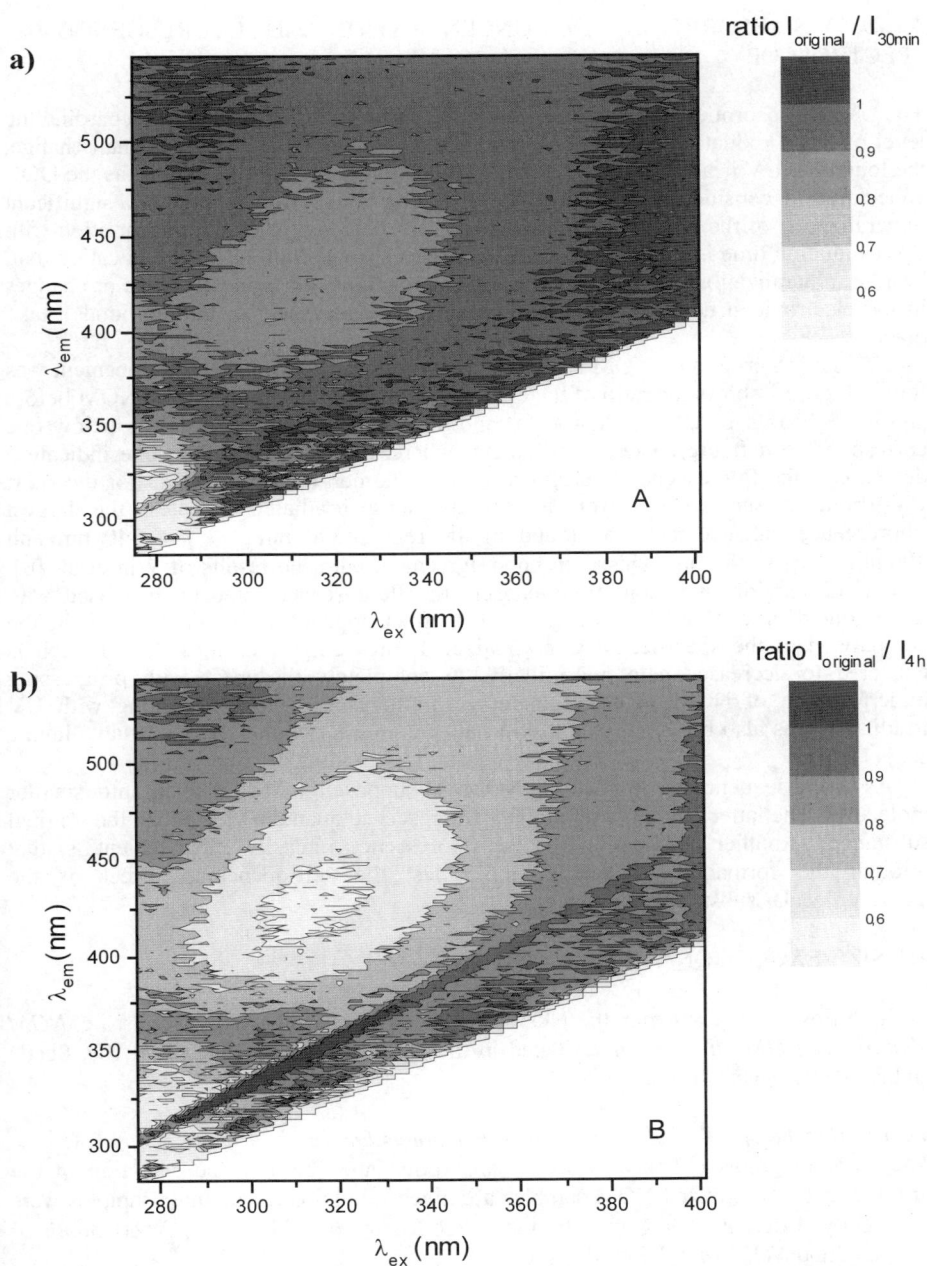

Figure 1. The excitation-emission-matrix of the ratio of the fluorescence intensities of NOM before and after (a) 30 min of UV irradiation and (b) 4h UV irradiation (DOC = 15 mg/l).

4.1. UV/VIS ABSORPTION, DOC CONCENTRATION AND FLUORESCENCE SPECTROSCOPY

The UV/Vis absorption and DOC concentration showed similar tendencies regarding the level of NOM oxidation. While short time UV irradiation induced only a small change, the long time UV irradiation decreased the absorption of the sample, as well as the DOC concentration, considerably. This indicates that long UV irradiation leads to a significant mineralization of the NOM sample. Moreover, the decrease in DOC concentration with UV irradiation time is smaller than the decrease in absorption. This is an indication that, besides mineralization of NOM, photochemical reactions, like bond cleavage or changes in the electronic structure of the chromophores (e. g. decrease of double bonds), take place.

For the UV-irradiated NOM samples, a non-uniform fluorescence enhancement was found. Figure 1 shows the ratio of the excitation-emission matrix (EEM) of NOM before and after 30 min of UV irradiation (a) and 4h UV irradiation (b). Values below one correspond to a fluorescence enhancement, whereas values larger than one indicate a decrease in the fluorescence intensity relative to the non-oxidized NOM. For the short UV-irradiated samples, as well as for the long irradiated samples, the largest fluorescence enhancement was found in the region 290 nm $< \lambda_{ex} <$ 340 nm and 390 nm $< \lambda_{em} <$ 480 nm. This is in good agreement with the results of Win et al. [6]. Under the experimental conditions chosen, the fluorescence intensity increased with increasing degree of NOM oxidation. This effect appears even more clearly in the comparison of the specific, DOC normalized fluorescence intensities. However, it is expected to decrease again and, finally, to completely disappear with progressing mineralization of NOM. A slight increase in fluorescence intensity along with UV irradiation was also observed by Lipski et al. for several commercially available humic acids [15].

As a consequence of oxidation, NOM is supposed to be cleaved into smaller molecules. The latter might possess higher fluorescence quantum yields than the original substances. Another explanation for the fluorescence intensity enhancement is, that through the formation of smaller molecules, the self-quenching effect of the macromolecular substances is reduced.

4.2. SIZE-EXCLUSION CHROMATOGRAPHY

In the following experiments, the NOM sample and its two irradiated forms, i.e. *NOM 30 min* and *NOM 4h*, were investigated in the absence and presence of Al(III), Pb(II) and Zn(II) by means of SEC.

4.2.1. Influence of UV-oxidation on optic and molecular weight properties of NOM
The chromatograms obtained with UV absorption and fluorescence detection of the original and UV-oxidized NOM samples are given in Figure 2. The three samples were of slightly different DOC-concentrations: *NOM original* = 9.7 mg/l, *NOM 30min* = 9.1 mg/l, and *NOM 4h* = 8.6 mg/l.

The elution of all samples was mostly between the exclusion and permeation volume, i.e. 6.75 and 16.3 ml, respectively. This is an indication that the separation is

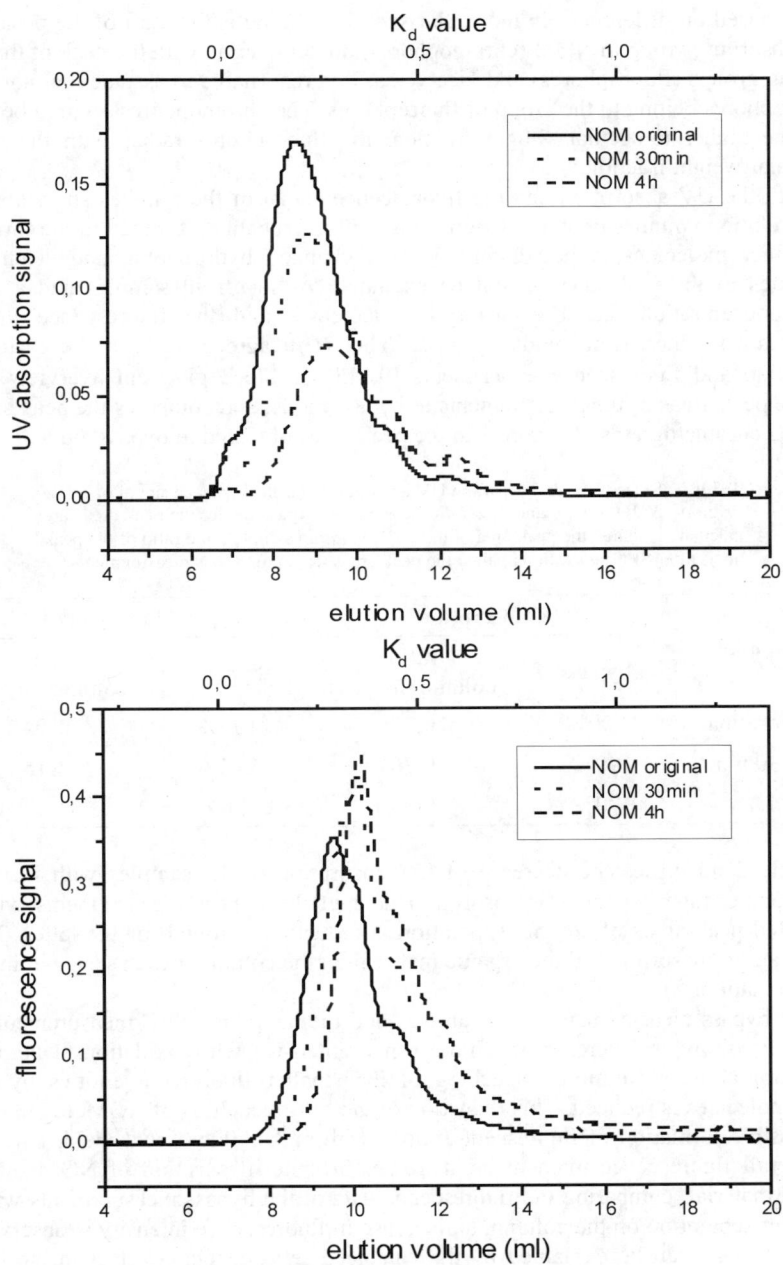

Figure 2. SEC chromatograms of NOM original, NOM 30min and NOM 4h with UV absorption detection (left) and fluorescence detection (right).

mainly based on differences in molecular size. Interesting is that part of the peak of the light-absorbing group in NOM (chromophores) do not overlap with the peak of the light-emitting groups (fluorophores) and vice versa. It is reasonable to assume that not all the chromophores belong to the group of fluorophores. The chromophores seem to be linked with the higher molecular weight fraction, the fluorophores rather with the smaller molecular weight fraction.

Both the UV absorption and the fluorescence signal of the samples show a shift to higher elution volumes upon oxidation. This shift is probably attributed to the formation of smaller molecules, although an effect of changed hydrophobic and electrostatic properties of the molecules cannot be excluded. Although all samples had a similar DOC concentration, the UV absorption decreases and the fluorescence intensity increases with increasing oxidation time. This is in agreement with the earlier UV absorption and fluorescence experiments [9, 13]. Table 2 gives an overview of the relative peak areas of the measurements in bypass mode, and compares the peak areas of the SEC chromatograms of Figure 2 to the peak areas obtained in bypass mode.

Table 2. Relative peak area of the UV absorption and fluorescence signal of *NOM original*, *NOM 30min*, and *NOM 4h* measured in bypass mode (the peak areas are calculated against the peak area of the *NOM original* sample), and ratio of the peak areas after SEC separation against the peak areas from the bypass measurements.

Sample	UV absorption		Fluorescence	
	bypass	ratio column / bypass	bypass	ratio column / bypass
NOM original	100%	0.77	100%	1.04
NOM 30 min	90%	0.76	112%	1.18
NOM 4 h	70%	0.68	136%	1.12

Table 2 quantifies the decrease of UV absorption of the samples with increasing oxidation. Comparing the UV absorption data of the bypass measurements with the integrated peak-areas of column separation, a reduction is found for the latter. This is likely caused by sorption of the organic material on the column, which seems to increase upon oxidation.

The bypass measurements (see Table 2) and the integrated SEC measurements (see Figure 2) show an increase in fluorescence intensity with oxidation. Due to the chromatographic separation, quenching of the smaller fluorescent entities by larger NOM molecules is reduced. This reduction of the self-quenching of NOM, together with the possible formation of fluorescent entities with higher fluorescent yield, may cause the overall fluorescence intensity to increase. In spite of sorption of NOM onto the column material, comparing the fluorescence data of the bypass measurements with the data after separation on the column, an increase in fluorescence intensity is observed for the latter. This can be explained by the enhanced self-quenching effect in the bypass measurement due to high NOM concentration passing the fluorescence detector, whereas in the SEC measurement, the NOM sample is more diluted. Inner-filter effects, i.e. reabsorption of emitted light, presumably play only a minor role in these samples of DOC of ca. 9 mg/l. Other concentration and pH-dependent molecular aggregation of humic substances in solution and on mineral surfaces was extensively discussed by Wershaw [23].

4.2.2. Influence of UV-oxidation on the complexing properties of NOM with respect to al

In Figure 3, the chromatograms of Al in the presence of (a) *NOM original*, (b) *NOM 30 min* and (c) *NOM 4 h* are shown. They demonstrate that Al elutes simultaneously with the UV-absorbing components of NOM. The injection of Al in the absence of NOM yielded no Al signal. This leads to the conclusion that free Al is retained by the column. Therefore, it can be assumed that the Al signal in Figures 3 (A) – (C) represents Al which is bound to NOM [24].

Two kinds of UV-absorbing compounds can be distinguished: those with an elution volume of 9 ml, i.e. with smaller molecular weight, and those eluting at the exclusion volume ($K_d = 0$), i.e. larger molecular weight compounds. After oxidation, the UV peak area of the low molecular weight fraction decreased. Simultaneously, the amount of Al bound to this fraction has also decreased. On the other hand, the UV peak with an elution volume at $K_d = 0$ increased after oxidation. The above results contradict the trends shown in Figure 2, where only a decrease of the fraction at $K_d = 0$ was observed. The amount of Al in the fraction at $K_d = 0$ is also increased. Therefore, it is hypothesized that this peak represents colloidal Al-NOM, which causes light scattering at the UV absorption detector. Thus, decreasing complexation of Al by lower molecular size NOM due to oxidation leads to increasing Al hydroxide formation, as pictured in the following reaction scheme:

$$NOM - Al \rightleftarrows Al^{3+} \rightleftarrows Al(OH)_3 \qquad (8)$$

The presence of oxidized NOM may prevent the fast agglomeration of the $Al(OH)_3$-colloids due to complexation with colloid surface. Indications for the formation of Al-NOM colloids are also given in Schmitt et al. [20].

It is also striking that the fluorescence signals show a fairly constant intensity and appear at much higher elution volumes than the UV-absorbance. The first Al-fraction between 6 and 7 ml elution volume has practically no fluorescence at all. This also supports the hypothesis on Al-colloid formation. Sharpless and McGown [25] reported on the effect of Al-induced aggregation on the fluorescence of humic substances (HS) and concluded that fluorescence quenching was mainly the result of decreased HS concentration due to precipitation by Al. The intensities of the fluorescence signals in Fig. 3 indicate that most of the fluorophores obviously remained mobile and the maxima appeared at higher elution volumes than those of the UV absorbing regions and of Al. This supports the finding that the highest fluorescence intensity of NOM is intrinsic to its smaller molecules (Fig. 2).

The strength of interaction between Al and NOM can be quantitatively given as stability constants. Lambert et al. have determined conditional stability constants for Al/HS complexes to be 2.0 and 4.3, depending on the $c(Al_{complex})/\beta(DOC)$ ratio and pH value [26]. In order to determine how stable the Al-NOM complexes are, the recovery of Al during SEC separation was measured. Stable complexes are able to pass the column, whereas labile complexes tend to dissociate during passage, and the free metal

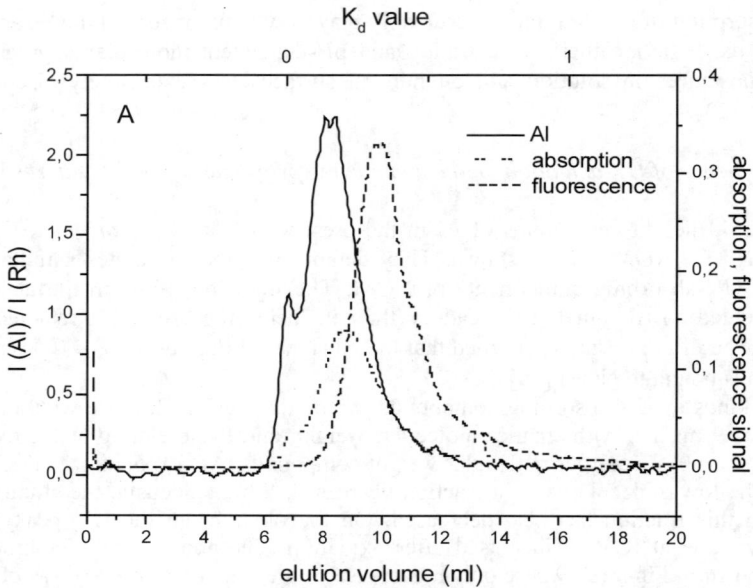

Figure 3A. SEC-Chromatograms of *NOM original* (9.7 mg DOC/l) in the presence of $0.7 \cdot 10^{-5}$ M Al with UV absorption and fluorescence detection, and ICP-MS detection of Al. The Al signal is corrected against the internal Rh standard.

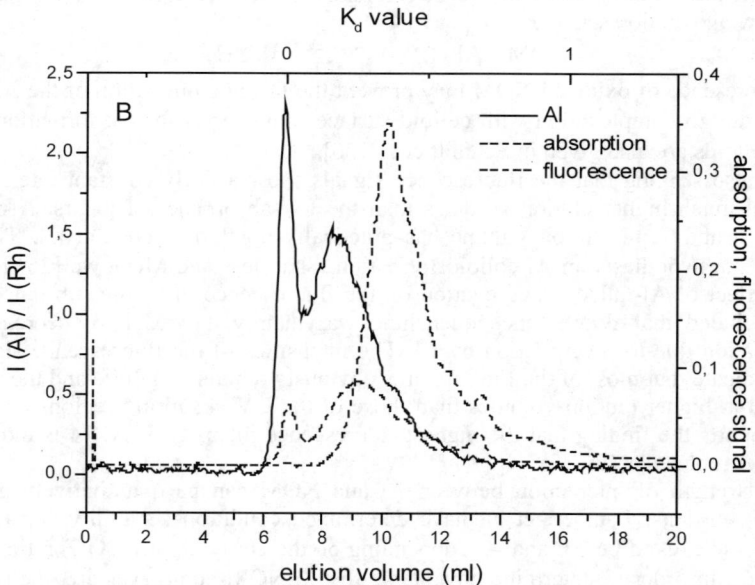

Figure 3B. SEC-Chromatograms of *NOM 30min* (9.1 mg DOC/l) in the presence of $0.7 \cdot 10^{-5}$ M Al with UV absorption and fluorescence detection, and ICP-MS detection of Al. The Al signal is corrected against the internal Rh standard.

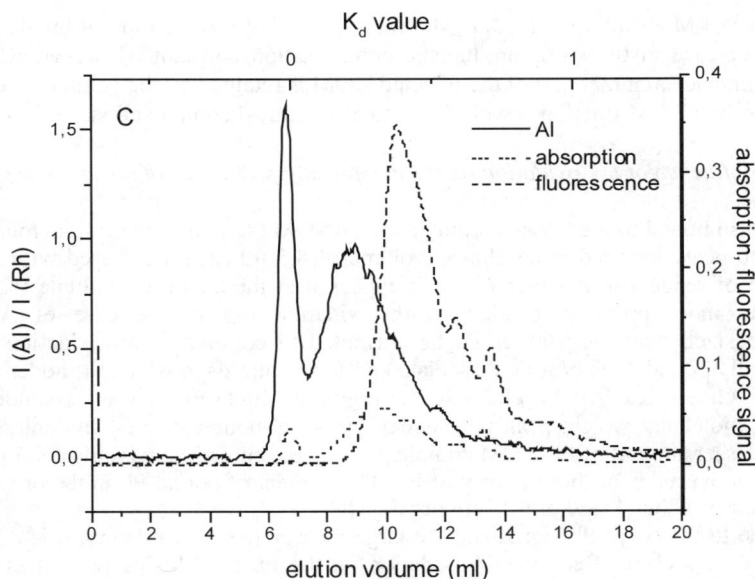

Figure 3C. SEC-Chromatograms of *NOM 4h* (8.6 mg DOC/l) in the presence of $0.7 \cdot 10^{-5}$ M Al with UV absorption and fluorescence detection, and ICP-MS detection of Al. The Al signal is corrected against the internal Rh standard.

ion is retained. Table 3 gives an overview of the recovery of Al. The results are based on experiments where (i) Al was added to *NOM original, NOM 30min* and *NOM 4h*, (ii) Zn or Pb were added to *NOM original, NOM 30min* and *NOM 4h* and (iii) no metal ion was added to *NOM original, NOM 30min and NOM 4h*. In the last two types of experiments the originally bound Al present in the NOM samples was observed, whereas in the first type of experiments the sum of the original and the additional Al was measured. The Al-concentration in the original NOM sample was 100 µg/l which means ca. 4 µg Al per mg OC. The Fe-concentrations were about twice as high.

Table 3. Recovery of Al and Pb for different NOM samples in percent of the injected amount of metal ion.

Sample	Recovery of Al % (Average ± std. dev.)	Recovery of Pb % (Average ± std. dev.)
NOM original	80 ± 7	0.8 ± 0.5
NOM 30 min	72 ± 9	1.4 ± 0.6
NOM 4 h	61 ± 15	1.3 ± 0.7

The results in Table 3 show that in the presence of UV-oxidized NOM, lesser amount of Al passes through the column than with *NOM original*. Given that this effect is more pronounced with higher degree of oxidation, it was reasonable to assume the formation of labile Al complexes with oxidized NOM. This is also in agreement with the findings of Monsallier et al. [27], who found that photodegradation of humic acid leads to a decrease in the europium-humate complexation constant. However, it has to be taken into account that part of the injected NOM is retained by the column (see Table 2). This sorbed NOM fraction is well suited to also bind Al complexation.

4.2.3. Influence of UV oxidation on the complexing properties of nom with respect to Pb and Zn

Pb is also bound to the UV-absorbing parts of NOM (similar to the results found for Al). One Pb peak occurred at an elution volume of 8.5 mL, and decreased with increasing degree of oxidation. Another Pb peak appeared at the exclusion volume ($K_d = 0$) and became more pronounced along with oxidation. As in the case of Al, it was hypothesized that this peak could be attributed to colloids: upon oxidation of NOM, originally bound Al and/or Fe dissociate and form colloids to which the added Pb can be bound. Given that free Pb elutes in the region of 10-11 ml, it was concluded that its higher molecular weight peak is provided by NOM-bound species. In contrast to Al, a much lesser amount of Pb-NOM complexes is stable enough to pass the column. Table 3 gives an overview of the recovery of Pb. The content of bound Pb in the original NOM sample was about 4 µg/l or 0.13 µg per mg OC.

Due to the small Pb signal and the large measurement uncertainty, it is not possible to observe a clear effect of UV-oxidation on the Pb complexing properties of NOM. However, it can be concluded that the Al-NOM complexes are more stable than the Pb-NOM complexes.

Completely different elution behaviour was observed for Zn. In contrast to Al and Pb, Zn was totally retained by the column. This means that Zn-NOM complexes are very unstable and dissociate completely during discharge through the column.

4.3. DETERMINATION OF THE DISSOCIATION KINETICS OF THE METAL-NOM COMPLEXES

Prior to the cation exchange experiments, it was verified that no adsorption of *NOM original*, *NOM 30 min* or *NOM 4 h* onto the resin took place.

Figure 4 shows the cation exchange kinetics of Al on the resin in the presence and absence of NOM. In the presence of NOM, the cation exchange proceeded more slowly than in solutions without NOM. Moreover, in the presence of oxidized NOM, the complexes dissociated faster than with non-oxidized NOM. It can, therefore, be concluded that the Al complexes of *NOM original* are more stable than those with oxidized NOM. This is in good agreement with the above SEC results reflecting adsorption of Al on the column in the presence of oxidized NOM. However, a distinctive effect of the UV-irradiation on the dissociation kinetics of NOM-complexes was not observed in this experiment.

Note that Al concentration at t = 0 includes original-NOM bound Al ($4 \cdot 10^{-6}$ M) and the added Al ($7 \cdot 10^{-6}$ M). Figures 4B and 4C show the results of similar experiments for Pb and Zn, respectively. Despite the small differences in Pb exchange behaviour in the presence and absence of NOM, it is still possible to observe a slower exchange kinetic in the NOM presence. In case of Zn, no differences were observed in the presence and absence of NOM. This is indicative of formation of weak and very labile Zn-NOM complexes. The similar indications were obtained in the above SEC experiments.

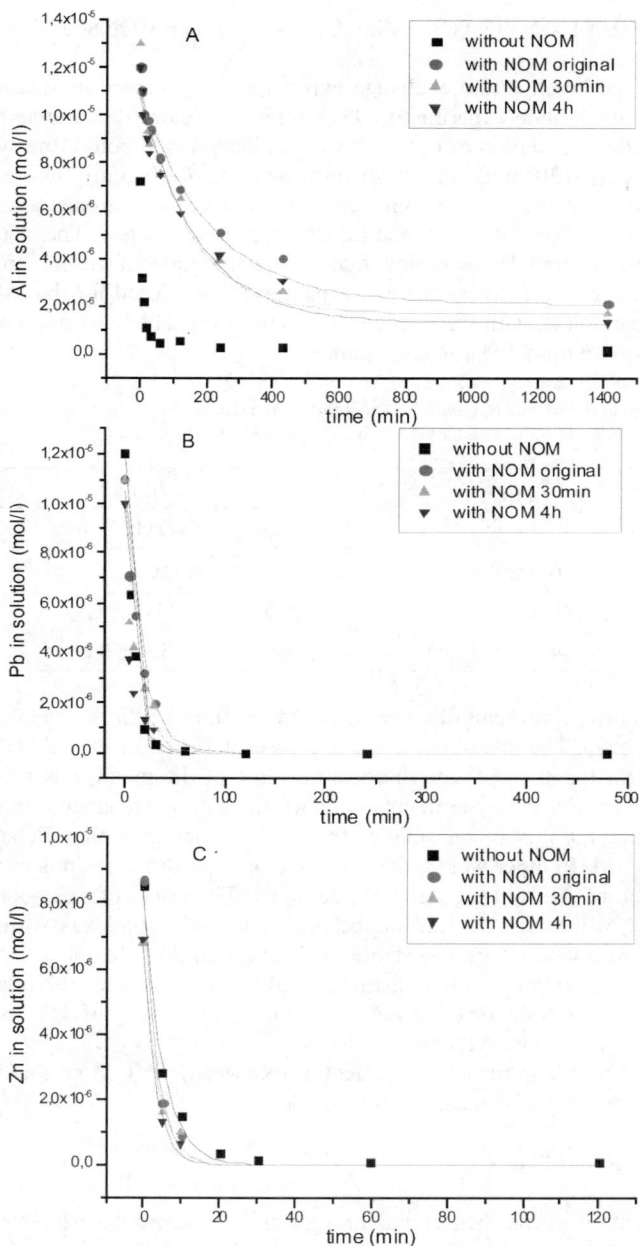

Figure 4. Cation exchange kinetics of (a) Al, (b) Pb, and (c) Zn in the metal --resin systems in the presence and absence of different types of NOM. (The lines are given for visualization of the obtained trends).

4.4. DETERMINATION OF THE DISSOCIATION RATE CONSTANTS

In contrast to the batch cation-exchange experiments, no equilibrium conditions could be reached in the column experiments. Therefore, only those data points from Figure 4 were used for the calculation of t_1, which were collected well before the system reached equilibrium: Al (0-120 min), Pb (0-60 min) and Zn (0-10 min). The exchange time constants t_3 and the dissociation time constants t_1 were determined according to the Equation (4) (with [MeNOM] = 0) and Equation (5), respectively. The calculated values for t_1 and t_3 are given in Table 4. Note that for the experiments with Pb and Zn, t_3 is not negligible compared to t_1. Nevertheless, Equation (5) was used for the calculation of t_1 instead of Equation (4), with fixed values for t_3 and [Me]. The latter approach resulted in values for t_1 with extremely large uncertainties.

Table 4. Exchange (t_3) and dissociation (t_1) time constants for the metal ion-resin systems and for the metal ion-NOM-resin systems, respectively.

Metal ion	t_3, min without NOM	t_1, min		
		NOM orig.	NOM 30 min	NOM 4 h
Al	6.5 ± 0.7	222 ± 37	169 ± 42	163 ± 29
Pb	8.4 ± 0.3	15.8 ± 1.5	11.7 ± 2.9	6.4 ± 1.2
Zn	4.7 ± 0.3	3.5 ± 0.5	3.9 ± 0.8	3.2 ± 0.5

The constant t_3 represents the cation exchange time of "free" metal ions with the surface of the resin. The rate of exchange decreases in the order Pb > Al > Zn.

The constant t_1 represents the dissociation time of the metal ions from the metal - NOM complexes. For the experiments with Al, the cation exchange rate determined in the absence of NOM is much smaller than the dissociation rates determined in the presence of NOM. In other words, the exchange of Al with the resin is much faster than the dissociation of Al from the Al-NOM complex. The effect of oxidation is expressed by the smaller values for t_1. This indicates that the Al complexes with UV-oxidized NOM dissociate faster and are less stable than those with *NOM original*.

For the experiments with Pb, a distinctive effect of the degree of oxidation of NOM on the dissociation time is observed. The Pb complexes f of UV-oxidized NOM dissociate faster than those with original NOM.

It was not possible to measure an effect of oxidation of NOM on the dissociation of the Zn-NOM complexes by means of this method.

4.5. METAL TRANSPORT EXPERIMENTS

For the calculation of the metal breakthrough in the column experiments, the value of k_{diss} (Eq. 6) was set equal to the rate constant k_1 (Eq. 2). The values of $k_1 = 1/t_1$ were calculated from Table 4. For the other rate constants, assumptions were made. The values for k_{ass} and k_{des} were set equal to 10^{-10} s^{-1}. This was based on the assumption that the formation of metal-NOM complexes and the desorption of metal ions from the quartz are slow compared to the dissociation rate of the metal-NOM complexes. Furthermore, it was assumed that the dissociation of the metal-NOM complexes was the rate-determining step. Therefore, k_{ads} was set to a high value compared to k_{diss}. The value of k_{ads} was set equal to $1.5 \cdot 10^{-2}$ s^{-1} for Al and $5 \cdot 10^{-2}$ s^{-1} for Pb and Zn to enable fast model calculations and accurate numerical results. Under these assumptions, the metal ions dissociating from the metal-NOM complexes during passage through the column are supposed to be retained and not to elute at one bed volume after injection.

Comparison of calculated and experimentally determined recoveries shows a good agreement for the divalent ions. Zn was not found to elute after one bed volume either with *NOM original* or with *NOM 4h*. For Pb, no breakthrough was found or predicted in the presence of *NOM 4h*. With *NOM original*, a very small breakthrough was found experimentally, which was not predicted by the model. Higher recoveries for the trivalent Al were found compared to the divalent ions. The model was able to predict the order of magnitude of the recovery. However, considerable differences were obtained for Al. The model underestimated Al breakthrough for the experiments with *NOM 4h*, whereas it was overestimated for the *NOM original*. A possible explanation for these differences is again the formation of colloids, which are only partially filtered through the column and not taken into account by the transport model. Furthermore, the metal-NOM solutions used in the cation-exchange experiments and the transport experiments were equilibrated before use to allow for metal-NOM complexation. However, for both types of experiments, this equilibration time was slightly different (on the order of a few days). If in both experiments a different degree of complexation was achieved, then this could explain the differences between the experimental results and the calculations using the dissociation time constants from the cation-exchange experiments. Finally, it should be noted that the column experiments with *NOM original* are performed with NOM of a different sampling time than the NOM used in the column experiments with *NOM 4h* and in the cation-exchange experiments.

5. Conclusions

The distribution and transport of metals in polluted sites is mainly governed by the hydraulic situation. In addition, migration of metals is strongly influenced by the speciation. NOM is ubiquitous in aquatic systems and soils and can act as ligand in metal complexes. Due to the broad variety of metals and electron donor sites of ligands, there is a wide distribution of stabilities of the complexes and their solubility. They can be systematically used for mobilization or immobilization of metals and, hence, for remediation of polluted sites. The NOM complexes of Al, Pb and Zn were investigated as typical examples for environmentally relevant tri- and divalent metals. Complexing properties of NOM were altered by aerated UV- irradiation leading to soft oxidation and to a decrease in molecular size. This change is typical for aerobic degradation. The UV-oxidized NOM showed a decrease in UV-absorbance and an increase in fluorescence intensity which is indicative of a higher reactivity of the excited electron states. The chromatographic experiments have shown that the oxidized NOM formed less stable metal complexes. This suggests a higher availability of metal species with increasing degree of oxidative degradation of the NOM-ligand. In case of Al, this could lead to formation of highly mobile colloids in porous media.

A general tendency towards better mobilization of metals with a defined and stable oxidation state being the reason for scarcely soluble species can be postulated for aerobic media. The divalent metals – Pb and Zn – were shown to form only labile complexes with NOM independently on the level of its oxidation. This can be indicative of negligible influence of NOM on the behaviour of those metals in the polluted sites. The developed procedure can be extended to the other metals to assess their fate in polluted sites in the presence of NOM and its UV-oxidation products.

6. Acknowledgements

The financial support by the Deutsche Forschungsgemeinschaft (research project "MetalOM", Grant FR 536/21-2) is greatly appreciated.

7. References

1. Buffle, J. and De Vitre, R.R. (1994) *Chemical and biological regulation of aquatic systems*, Lewis Publishers, New York.
2. Weber, J.H. (1988) Binding and transport of metals by humic substances, in F.H. Frimmel, R.F. Christman (eds.), *Humic Substances and Their Role in the Environment*, John Wiley & Sons, Chichester.
3. Förstner, U. (1993) *Umweltschutztechnik*, Springer Verlag, Berlin.
4. Yates III, L.M. and von Wandruszka, R. (1999) Decontamination of Polluted Water by Treatment with a Crude Humic Acid Blend, *Environ. Sci. Technol.* **33(12)**, 2076-2080.
5. Suffet, I.H. and MacCarthy, P. (1989) *Aquatic Humic Substances. Influence on Fate and Treatment of Pollutants*, American Chemical Society, Washington D.C.
6. Win, Y.Y., Kumke, M.U., Specht, C.H., Schindelin, A.J., Kolliopoulos, G., Ohlenbusch, G., Kleiser, G., Hesse, S. and Frimmel, F.H. (2000) Influence of oxidation of dissolved organic matter (DOM) on subsequent water treatment processes, *Wat. Res.* **34(7)**, 2098-2104.
7. Hesse, S., Kleiser, G. and Frimmel, F.H. (1999) Characterization of Refractory Organic Substances (ROS) in Water Treatment, *Wat. Sci. Tech.* **40(9)**, 1-7.
8. Zafiriou, O.C., Joussot-Dubien, J., Zepp, R.G. and Zika, R.G. (1984) Photochemistry of natural waters, *Env. Sci. Tech.* **18**, 358A-371A.
9. Korshin, G.V., Kumke, M.U., Li, C. and Frimmel, F.H. (1999) Influence of Chlorination on Chromophores and Flurophores in Humic Substances, *Environ. Sci. Technol.* **33(8)**, 1207-1212.
10. Frimmel, F.H. (1998) Impact of Light on the Properties of Aquatic Natural Organic Matter, *Environ. Int.* **24 (5/6)**, 559-571.
11. Christensen, J.B., Jensen, D.L. and Christensen, T.H. (1996) Effect of Dissolved Organic Carbon on the Mobility of Cadmium, Nickel and Zinc in Leachate Polluted Groundwater, *Wat. Res.* **30(12)**, 3037-3049.
12. Stumm, W., and Morgan, J.J. (1996) *Aquatic Chemistry: chemical equilibria and rates in natural waters*, John Wiley & Sons, New York.
13. Kumke, M.U., Tiseanu, G., Abbt-Braun, G., Frimmel, F.H. (1989) Fluorescence Decay of Natural Organic Matter (NOM) – Influence of Fractionation, Oxidation and Metal Ion Complexation, *J. Fluorescence* **8(4)**, 309-318.
14. Kumke, M.U., Abbt-Braun, G. and Frimmel, F.H. (1998) Time-resolved Fluorescence Measurements of Aquatic Natural Organic Matter (NOM.), *Acta hydrochim. hydrobiol.* **26(2)**, 73-81.
15. Lipski, M., Slawinski, J. and Zych, D. (1999) Changes in the Luminescent Propereties of Humic Acids Induced by UV-Radiation, *Journal of Fluorescence* **9(2)**, 133-138.

16. Perminova, I.V., Frimmel, F.H., Kovalevskii, D.V., Abbt-Braun, G., Kudryavtsev, A.V. and Hesse, S. (1998) Development of predictive model for calculation of molecular weight of humic substances, *Wat. Res.* **32(3)**, 872-881.

17. Swift, R.S. (1999) Macromolecular Properties of Soil Humic Substances: Fact, Fiction, and Opinion, *Soil Sci.* **164(11)**, 790-802.

18. Muller, F.L.L. and Kester, D.R. (1990) Kinetic Approach to Trace Metal Complexation in Seawater: Application to Zinc and Cadmium, *Environ. Sci. Technol.* **24(2)**, 234-242.

19. Shuman, M.S. and Michael, L.C. (1978) Application of the Rotating Disk Electrode to Measurement of Copper Complex Dissociation Rate Constants in Marine Costal Samples, *Environ. Sci. Technol.* **12(9)**, 1069-1072.

20. Schmitt, D., Saravia, F., Frimmel, F.H. and Schuessler, W. (2002) NOM-Facilitated Transport of Metal Ions in Aquifers: Importance of Complex-Dissociation Kinetics and Colloid Formation, *Wat. Res.* (accepted for publication).

21. Bryan, N.D., Jones, D., Griffin, D., Regan, L., King, S., Warwick, P., Carlsen, L. and Bo, P. (1999) in G. Buckau (ed.), *Effects of humic substances on the migration of radionuclides: Complexation and transport of the actinides*, Second Technical Progress Report, Forschungszentrum Karlsruhe GmbH, Karlsruhe, pp. 303-338.

22. Gauthier, Th.D., Shane, E.C., Guerin, W.F., Seitz, W.R. and Grant, C.L. (1986) Fluorescence quenching Method for Determining Equilibrium Constants for Polycyclic Aromatic Hydrocarbons Binding to Dissolved Humic Materials, *Environ. Sci. Technol.* **20**, 1162-1166.

23. Wershaw, R.L. (1999) Molecular Aggregation of Humic Substances, *Soil Science* **164 (11)**, 803-813.

24. Vogl, J. and Heumann, K.G. (1997) Determination of Heavy Metal Complexes with Humic Substances by HPLC/ICP-MS Coupling Using Isotope Dilution Technique, *Fresenius J. Anal. Chem.* **359**, 438-441.

25. Sharpless, C.M. and McGown, L.B. (1999) Effects of Aluminum-Induced Aggregation on the Fluorescence of Humic Substances, *Environmental Sci. Technol.* **33(18)**, 3264-3270.

26. Lambert, J., Buddrus, J. and Burba, P. (1995) Evaluation of conditional stability constants of dissolved aluminum / humic substances complexes by means of ^{27}Al nuclear magnetic resonance, *Fresenius J. Anal. Chem.* **351**, 83-87.

27. Monsallier, J., Scherbaum, F.J., Buckau, G., Kim, J., Kumke, M.U., Specht, C.H. and Frimmel, F.H. (2001) Influence of photochemical reactions on the complexation of humic acid with europium(III), *J. Photochem. Photobiol. A* **138**, 55-63.

Use of Humic Substances to Remediate Polluted Environments: From Theory to Practice, 135–154.

Chapter 6

ROLE OF HUMIC SUBSTANCES IN THE COMPLEXATION AND DETOXIFICATION OF HEAVY METALS: CASE STUDY OF THE DNIEPER RESERVOIRS

P.N. LINNIK, T.A.VASILCHUK

Department of Hydrochemistry, Institute of Hydrobiology, National Academy of Sciences, 04210 Kiev, Ukraine <mvasilchuk@mail.ru>

Abstract

Results of long-term investigations of the heavy metal content and speciation in the Dnieper water bodies (Ukraine) are presented. The particular importance was given to the study of dissolved organic matter (DOM) composition, to binding affinity of metals such as Fe, Mn, Cu, Zn, Pb, Cr, and Cd for DOM, and to molecular weight distribution and chemical nature of metal-DOM complexes. The chemiluminescence and anodic stripping voltammetry in combination with membrane filtration, gel permeation chromatography, and ion-exchange chromatography were used for this purpose. It is shown that complexation with natural organic ligands plays the major role in the fate of HM in fresh water systems. The prevailing fraction of dissolved metals is found to present as complexes with humic substances, mainly, with fulvic acids. For the evaluation of the stability of aquatic ecosystems to the toxic action of HM, the potential complexing ability of DOM was investigated. The results on copper toxicity are presented obtained in long-term experiments with distilled and tap waters as well as with the natural water from the Kanev reservoir containing additives of copper ions and humic acids (HA). The toxic effect of free Cu^{2+} ions is determined using biotests with *Daphnia magna*. It was established that a decrease in integral water toxicity correlated with a reduction in free Cu^{2+} concentration in water medium. Hence, Cu^{2+} ions present in aquatic media in the form of non-toxic complexes with organic ligands. The maximal decrease in toxicity was observed in natural water where the complexation occurred with participation of both DOM and added HA.

1. Introduction

Heavy metals (HM) are obligatory components of surface waters. They essentially influence the quality of the water environment and the functioning of aquatic ecosystems. Many of them are mutagenic and carcinogenic [12, 14, 27]. HM may exist in fresh water bodies as hydrated ions or as inorganic and organic complexes of different molecular weight. They may also be sorbed onto inorganic and organic particulate matter.

135

I. V. Perminova et al. (eds.),
Use of Humic Substances to Remediate Polluted Environments: From Theory to Practice, 135–154.
© 2005 *Springer. Printed in the Netherlands.*

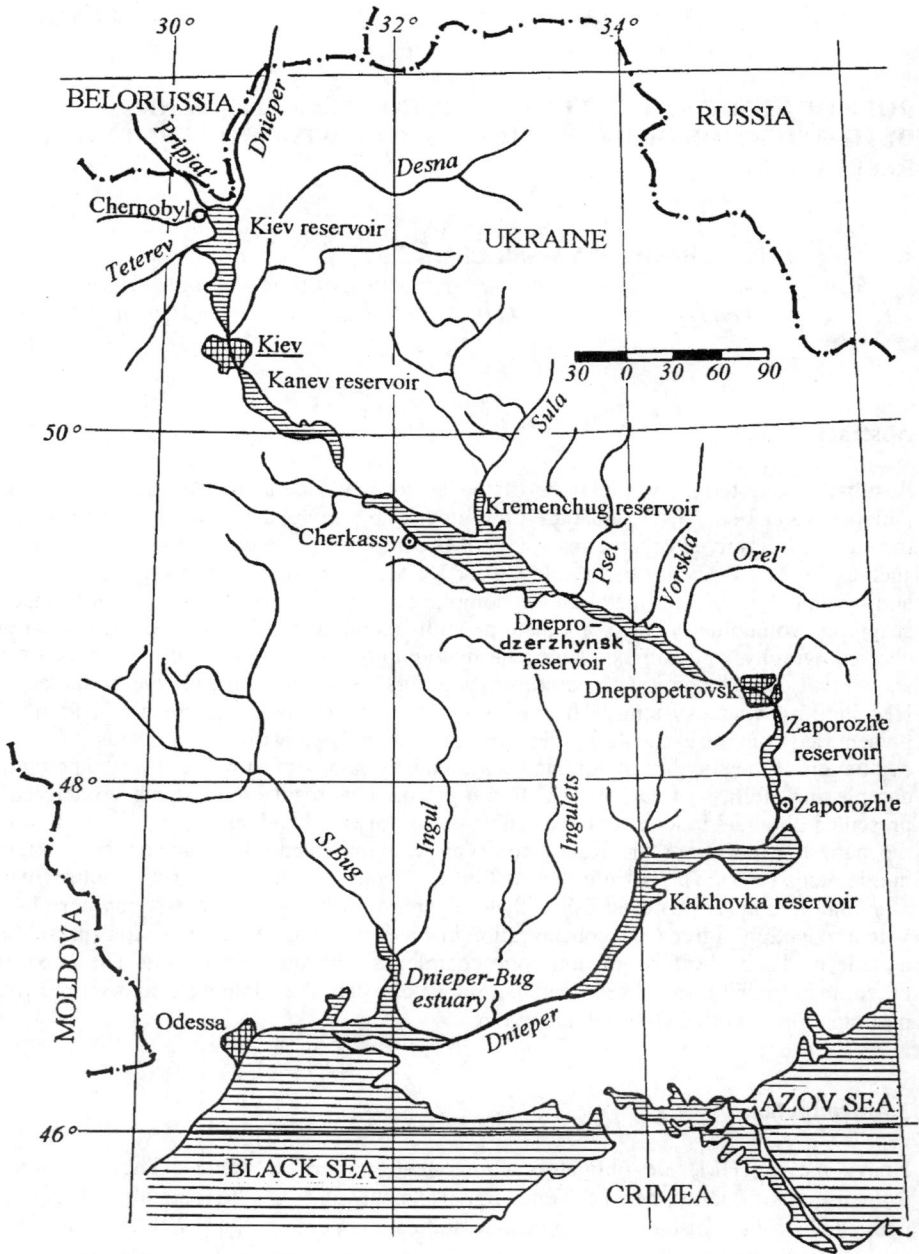

Figure 1. Map of the Dnieper reservoirs.

The speciation of HM in natural waters as well their total content, has an important impact on the water quality. It is widely studied because of the importance for prediction of transport, bioavailability, and fate of metals [5, 6, 12, 21, 27, 29, 32, 40]. Knowledge of chemical speciation of HM in natural waters is essential for the interpretation of their biological and geochemical cycling in water ecosystems. It is well known that HM, due to their toxicity and potential to accumulate in biological systems, constitute a risk for the health of living organisms [14]. Toxicity of the aquatic environment for living organisms, and finally for humans, depends primarily not on total content, but the speciation of HM in water. Freely dissolved (hydrated) metal ions are regarded as the most toxic form. The HM bound into complexes with inorganic and; particularly, with naturally occurring organic matter, expose much less toxicity [2, 7, 11, 21, 24, 27, 29, 31, 39, 44].

The aquatic ecosystem has to cope with the continuously increasing levels of HM. In most cases, a reduction in HM toxicity is a result of transformation of toxic forms into less active forms. The degree of detoxification depends first on the intensity of processes within a water body, which favour reducing the concentration of free HM ions.

The Dnieper reservoirs (Ukraine, Figure 1) are regarded as highly productive water bodies [9]. The concentration of dissolved organic matter (C_{org}) in reservoirs fluctuates widely and accounts in average, for 7.4-15.8 mg·L^{-1}. This explains the intensive complexation processes that dominate the fate of metals studied.

The objective of this paper was to present the results of long-term investigation of HM speciation in the Dnieper reservoirs, and to highlight the dominant role of DOM, and of humic substances (HS) in particular, in metal speciation and toxicity in fresh water systems.

2. Materials and Methods

Water sample preparation. The water samples were taken during boat expeditions along the Dnieper reservoirs in various seasons and filtered using membrane filters ("Synpor", Czech Republic) with pore diameter of 0.4 µm.

Metal analysis. The heavy metals were determined using chemiluminescence (Fe, Mn, Cu, Cr) [28] and the anodic stripping voltammetry (Zn, Pb, Cd) [20, 22]. The concentration of free (hydrated) metal ions was determined *in situ*. Determination of the total concentration of dissolved metals was preceded by photochemical destruction of DOM in water samples during 4-5 h by using a UV-irradiation from DRT-1000 mercury lamp (Ekaterinburg, Russia). The solutions were acidified with sulphuric acid to pH 1.0.

DOM fractionation. Ion-exchange chromatography (columns with ion-exchange cellulose derivatives – diethylaminoethyl cellulose (DEAE) and carboxymethyl cellulose (CM) (Reanal, Budapest, Hungary), were used to investigate the component composition of DOM and the chemical nature of metal complexes [38, 43]. This allowed to separate the DOM pool into three fractions: anionic, cationic and neutral.

Molecular weight distribution (MWD) determination. MWD of DOM fractions was determined using size-exclusion chromatography (SEC). The gels used were Sephadex G-25, G-50, and G-75 (Pharmacia, Sweden); and Toyopearl HW-50 (Japan). The void

138

volume of the gel-filled columns was determined with Blue Dextran (2000 kDa). Calibration of columns was carried out using standard substances with a known MW: albumin (45.0, 70.0 kDa), carboanhydrase (29.0 kDa), chymotrypsin (25.0 kDa), myoglobine (17.8 kDa), lysozyme (14.8 kDa), cytochrome (12.3 kDa), insulin (5.6 kDa), polyethylene glycol (1.0, 0.6, and 0.3 kDa) and glucose (0.18 kDa). The elution rate was 1 ml·min^{-1}; the eluents were 0.025 mol·L^{-1} NaCl and 0.05 mol·L^{-1} KNO$_3$.

Determination of humic and fulvic acids (HA and FA, respectively) was carried out spectrophotometrically measuring their characteristic absorption at 400 nm and by their reaction with diazotised 4-nitroaniline after chromatographic separation [33]. Calibration plots for the determination of HA and FA were constructed with HS extracted from the water of the Kiev reservoir as described elsewhere. Proteins were determined spectrophotometrically by the Loury reaction [8], carbohydrates – by reaction with anthrone [37].

Copper toxicity assessment. To investigate the impact of HA on integral toxicity of water medium, the complexation of Cu (II) with HA including its kinetic characteristics was studied. Experiments were carried out in three variants: (1) in distilled water saturated with oxygen; (2) in tap water; (3) in natural water from the Kanev reservoir. *Daphnia magna* Straus was used as the test object; the test response was mortality of test organisms. Three series of experiments were performed at pH 8.0, and an oxygen concentration of 8.7 mg·L^{-1}, t = 26 ± 0,5 °C. In the first case, the investigated concentrations of Cu (II) were 100.0 and 200.0 µg·L^{-1}, HA, 2.5 and 5.0 mg·L^{-1}; in the second case, concentrations of Cu (II) were 50.0 and 100.0 µg·L^{-1}, HA, 5.0 and 10.0 mg·L^{-1}; in the third experiment, these concentrations were 50.0, 100.0, 250.0, and 500.0 µg·L^{-1} for Cu (II) and 2.5, 5.0 and 10.0 mg·L^{-1} for HA. The investigations of water toxicity were done for Cu (II) or HA separately and as the combination of these substances. The distilled water was not used because it does not have the nutrients needed for test-organisms.

3. Results and Discussion

3.1. CONTENT OF DISSOLVED ORGANIC MATTER IN THE DNIEPER RESERVOIRS

The concentration of DOM in the Dnieper reservoirs varies widely (Table 1), depending on intensity of biological processes within a water body. DOM is considered to play the main role in complexation and detoxification of heavy metals in natural water. HS, mainly FA, are the predominant component of the DOM pool in the Dnieper reservoirs. HS content reaches 70-90%, 55-68% and 45-60% of the total DOM in the Kiev, Kremenchug and Kakhovka reservoirs, respectively. The FA concentration is 20 to 40 times higher than that of HA (Table 2). In the successive reservoirs, the concentration of HS falls. HS are mainly of the allochthonous origin brought into reservoirs with surface water. The maximum content of HS is usually observed in spring, owing to the high-water season. The concentration of HS gradually declines through the summer, autumn, and winter (Figure 2). This appears not to be completely true for HA, which seems to have an equivalent mid-summer/mid-fall peak.

The MWD of HS in the reservoirs studied depends on the various factors such as the hydrological regime, the cascade arrangement, the seasons, and others. Data on the MWD of FA showed that this DOM fraction is dominated with the low MW compounds (< 1 kDa). The particular feature of the MWD of HA is a decrease in the content of the above low MW fractions from North to South. So, the content of fractions with MW < 1 kDa accounted for 89% of the total FA pool and 63% of HA pool in the Kanev reservoir in summer, 2000 (Figure 3). At the same time, in summer 1992, the content of fractions with MW > 1 kDa comprised 35%, 24% and 20% of the total FA pool in the Kiev, Kremenchug, and Kakhovka reservoirs, respectively (Figure 4). It maybe due to the different hydrological conditions in that summer and the cascade arrangement of the reservoirs. The more drastic transformation of FA to the low MW compounds is shown to take place from the North to the South in these reservoirs.

Table 1. Concentration of DOM (mg $C_{org} \cdot L^{-1}$) in the waters of the Dnieper reservoirs in different seasons.

Reservoir	Spring	Summer	Autumn
Kiev	7.5-47.0 (13.2)[a]	8.8-38.6 (15.8)	5.1-15.2 (11.4)
Kanev	8.7-23.2 (11.6)	4.0-18.3 (12.7)	6.2-16.9 (10.3)
Kremenchug	4.3-12.0 (8.7)	6.6-22.4 (10.3)	4.9-13.5 (8.8)
Kakhovka	4.1-12.3 (7.4)	4.9-13.5 (8.8)	5.1-14.0 (8.0)

[a] The average values are given in parentheses.

Table 2. Concentrations of HS ($mg \cdot L^{-1}$) in waters of the Dnieper reservoirs in different seasons.

Reservoir	Spring		Summer		Autumn	
	HA	FA	HA	FA	HA	FA
Kiev	0.75-2.17[a] 1.46	24.30-45.80 35.7	0.32-1.96 1.05	20.60-38.54 30.50	0.50-1.12 0.76	14.05-31.72 27.75
Kremenchug	0.16-0.58 0.35	11.22-23.20 16.80	0.25-1.12 0.59	8.85-20.45 13.70	0.25-0.87 0.48	13.45-17.20 14.36
Kakhovka	0.24-0.39 0.31	1.30-12.45 10.95	0.17-0.82 0.40	5.32-20.17 12.90	0.19-0.44 0.27	5.14-15.94 8.85

[a] The limits of fluctuations are given above bar, the average values are presented below bar.

The contribution of high MW FA to the total DOM increases in the spring (Figure 5). In the Kiev reservoir, this fraction is still present in summer and autumn, but in the Kakhovka reservoir, it is almost absent in these seasons, indicating that FA undergo a deeper transformation in the water of the south reservoir.

140

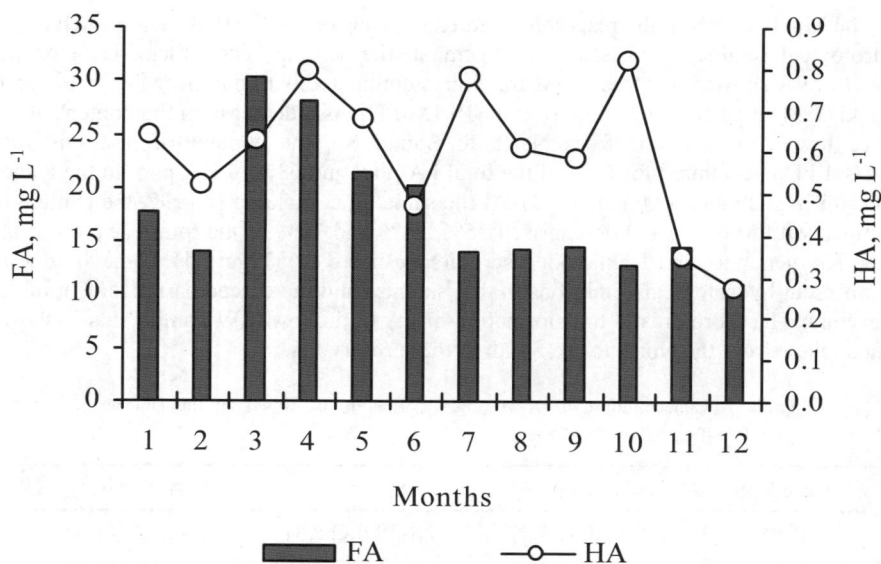

Figure 2. Seasonal dynamics of the concentration of FA and HA in the lower section of the Kiev reservoir in 1992.

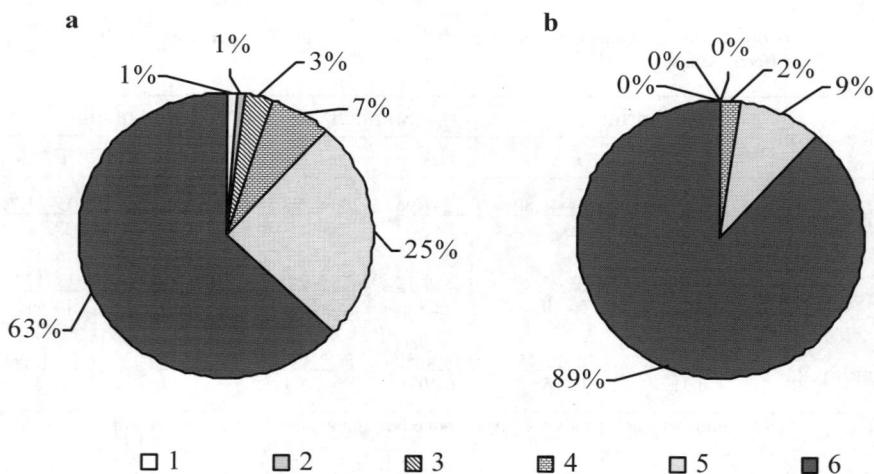

Figure 3. Molecular weight distribution of HA (a) and FA (b) in the water of the upper section of the Kanev reservoir (summer 2000). 1 – > 70 kDa, 2 – 70–40 kDa, 3 – 40–20 kDa, 4 – 20–5 kDa, 5 – 5–1 kDa, 6 – < 1 kDa.

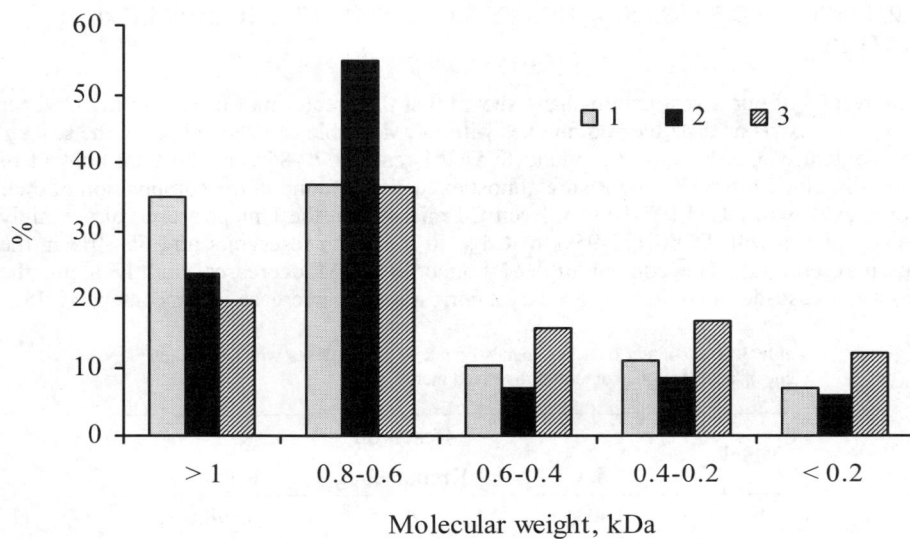

Figure 4. Molecular weight distribution of FA in the water of the Kiev (1), Kremenchug (2), and Kakhovka (3) reservoirs (summer 1992).

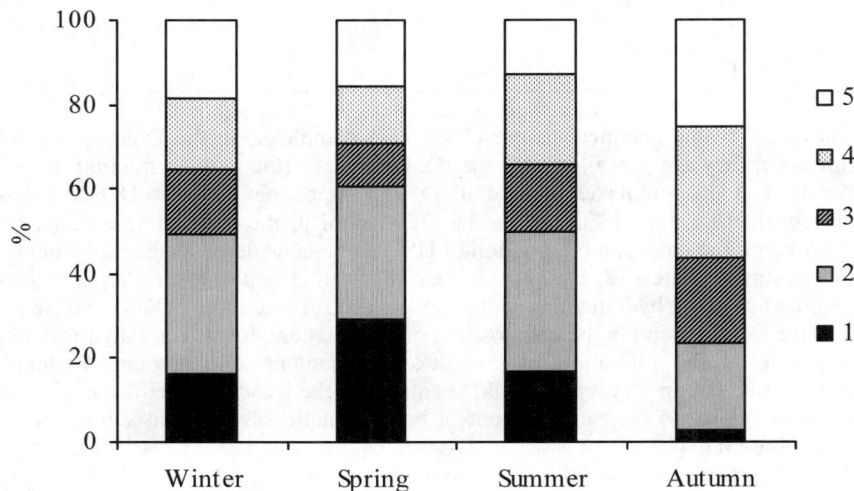

Figure 5. Molecular weight distribution of FA in the water of the lower section of the Kiev reservoir in different seasons. 1 – > 1 kDa, 2 – 0.8–0.6 kDa, 3 – 0.6–0.4 kDa, 4 – 0.4–0.2 kDa, 5 – < 0.2 kDa.

3.2. DISSOLVED SPECIES OF HEAVY METALS AND THEIR CHEMICAL NATURE

The results of our investigations have shown that the greater part of HM in the Dnieper reservoirs exists as dissolved complexes with DOM (Table 3). Depending on the season, the content of Cu, Zn, and Cr bound to DOM reaches 70-85% of the total content of their dissolved forms. Fe and Pb are almost exclusively found in the composition of their complexes with DOM (95-100%). Even Cd migrates in the Dnieper reservoirs, mainly as complexes with DOM (75-95% of Cd_{diss} in the north reservoirs and 65-80% in the south reservoirs). The content of HM bound to DOM decreases slightly along the reservoir cascade. It could be caused by a corresponding decrease in the content of HS.

Table 3. The content of heavy metals bound into complexes with DOM (average values), % of the total content of dissolved metals.

Metal	Reservoir		
	Kiev	Kremenchug	Kakhovka
Fe	100.0	98.5	98.0
Mn	59.4	41.5	32.5
Cu	85.8	74.9	71.0
Zn	84.3	71.7	72.5
Pb	97.6	94.5	91.6
Cr	77.8	76.5	71.4
Cd	78.9	79.6	77.2

The study of the chemical nature of the HM complexes in the Dnieper reservoirs revealed that they are prevailed with the HS complexes (the anionic fraction of DOM) (Table 4). The HS complexes include 40-74% of the metals bound to DOM. It was to expect considering that HS dominate the DOM pool in the Dnieper reservoirs. In its turn, FA being the most soluble fraction of HS, can be considered as the most important ligands in surface waters [2, 10, 21]. Another important group of natural organic ligands is presented by carbohydrates that comprise the neutral fraction of DOM and bind 12-49% of the metals found in the composition of organic complexes. Carbohydrates play a leading role in the formation of complexes in summer and autumn. Protein-like substances (the cationic group of DOM) contributes the least in the HM complexation. This can be related to their minute content in the aquatic environment and rather rapid decomposition rate [19].

The seasonal dynamics of the DOM composition influences the MWD of HM-DOM complexes . In general, the tendency toward the domination of the complexes with MW < 5 kDa is observed (Figure 6). These complexes included nearly 60% of the bound Fe, 70% of Cu, 51% of Cd, 67% of Zn, 51% of Pb, and 48% of Cr. The most probable that it is FA fraction that binds the above metals. The given data corroborate well the findings of the other researchers on the predominance of bound metals in the low MW fractions of HS (< 10 kDa) [3, 40].

Table 4. Distribution of heavy metals among DOM fractions of different chemical nature in waters of the Dnieper reservoirs (average values), summer 1994–1995; % of the bound metal concentrations.

Metal	Kiev reservoir			Kakhovka reservoir		
	anionic	cationic	neutral	anionic	cationic	neutral
Fe	57.0	4.5	38.5	52.5	14.7	32.8
Zn	50.5	6.8	11.9	40.5	10.5	49.0
Pb	61.0	8.5	30.5	62.6	15.4	22.0
Cr	54.0	16.0	30.0	43.5	33.5	23.0
Cd	73.5	10.0	16.5	60.5	9.6	29.9
Cu	68.0	9.8	22.2	61.5	14.0	24.5

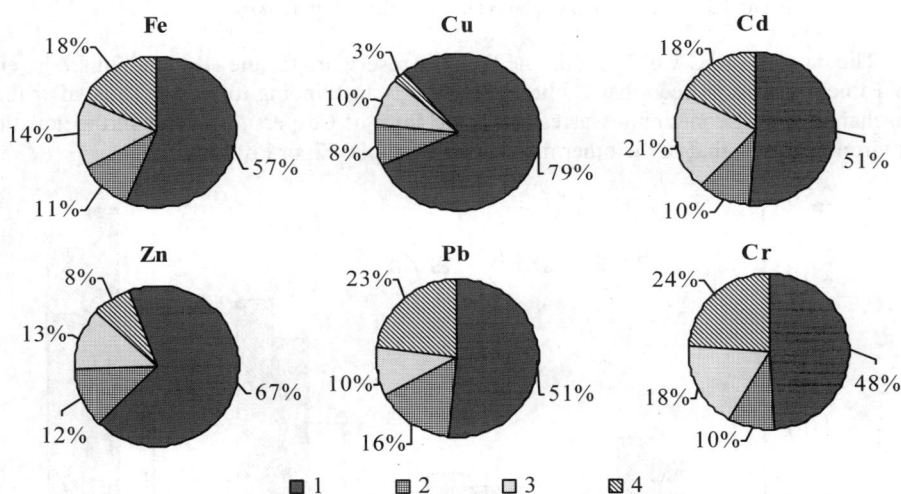

Figure 6. Distribution of the metals among the complex compounds with DOM of different molecular weight in the water of the lower section of the Kiev reservoir in spring. 1 – 0.25–5 kDa, 2 – 5–20 kDa, 3 – 20–60 kDa, 4 – > 60 kDa.

3.3. COMPLEXING ABILITY OF DISSOLVED ORGANIC MATTER

Complexing ability (CA) is the maximum total content of the coordinating sites of the organic ligands that can be occupied with metal ions. Consequently, the dimensional units of CA is μmol of metal per litre. Our investigations [19, 23] as well as the findings of the other authors [1, 42] show that the CA of DOM differs greatly from metal to metal. These differences can be governed by (1) the complex-forming affinity of the metal; (2) the strength of the complexes formed; and (3) the composition of the DOM. Therefore, CA should be determined for every metal taking into account its chemical properties and affinity for DOM.

144

Table 5. Complexing ability[a] of dissolved organic matter in the Dnieper reservoirs in relation to different metals, $\mu g \cdot L^{-1}$.

Metal	Reservoir		
	Kiev	Kremenchug	Kakhovka
Fe	580.0-1800.0	510.0-1250.0	440.0-1050.0
Mn	115.0-340.0	70.0-225.0	45.5-105.0
Cu	64.0-390.0	52.0-340.0	56.0-280.0
Zn	124.0-319.0	101.5-226.0	110.0-220.0
Pb	66.5-240.0	256.7-382.9	88.5-180.0
Cr	176.0-525.0	115.0-430.0	105.0-320.0
Cd	38.0-132.5	32.0-115.0	27.6-88.5

[a] Expressed as quantity of metal bound into complexes with DOM.

The data on the CA of DOM in the Dnieper reservoirs for the HM studied are given in Table 5. It can be seen that Cd has least affinity for binding to DOM. Thus, after the discharge of HM-containing wastes, the toxic form of Cd persists longer in the aquatic environment than that of the other metals (for example, Zn or Pb).

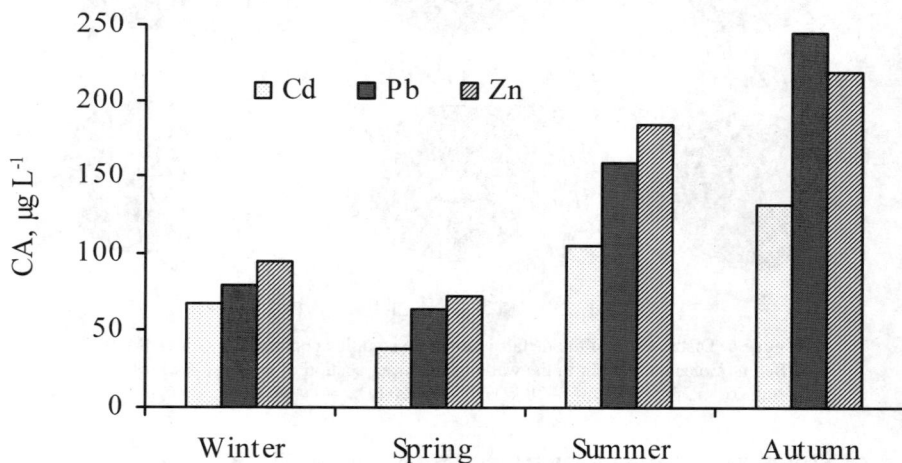

Figure 7. Mean complexing ability (CA) of DOM in the Kiev reservoir in different seasons.

CA also undergoes seasonal changes (Figure 7) connected to the changes in the DOM composition. Minimal quantities of metals are bound in late spring and early summer, even though the concentration of DOM at that time is usually high. This suggests that only a small fraction of effective ligands is present in water body. Of importance is, that HS are dominant in the water during flood-time (late spring). However, the CA of HS differ greatly with the season. So, it was established that in spring the high MW HS of allochthonous origin prevail the total pool of DOM. High MW HA have a lower CA than FA, which are generally of lower MW [4]. The low CA of HA-rich water has been also reported by other researchers [13]. The decrease in CA usually occurs in late spring-early summer and may, therefore, have an adverse effect on water quality because at that very time an increase in the concentration of the highly toxic metals is usually observed. In the late summer-autumn, CA reaches maximum values. At that time, organic compounds – products of metabolism – take part in complexation together with humic compounds (mainly FA). The above data suggest that the Dnieper reservoirs are more resistant to HM pollution and can more easily detoxify wastes containing HM in summer-autumn. In spring, when CA is minimal, the aquatic environment is most vulnerable to HM pollution.

The binding rate of metals in complexes with DOM should be considered the important factor from the ecological point of view [3, 5, 6, 15, 16]. The faster complexation occurs, the more probable the metals toxicity will decrease because of decreasing free ion concentration, which is the most toxic form of metals. The results of our long-term investigations have shown that the binding of metals in complexes with DOM, first of all, HS occurs sufficiently slowly. The complexation rate depends on the metal concentration brought into natural water and on the component composition of DOM. The higher the concentration of metals in water samples, the more metals bind. Earlier, we suggested [23] that equilibrium in a system under such conditions was reached in a few days. But the more detailed investigation showed that the equilibrium was not reached within a few days, even with a high concentration of metals.

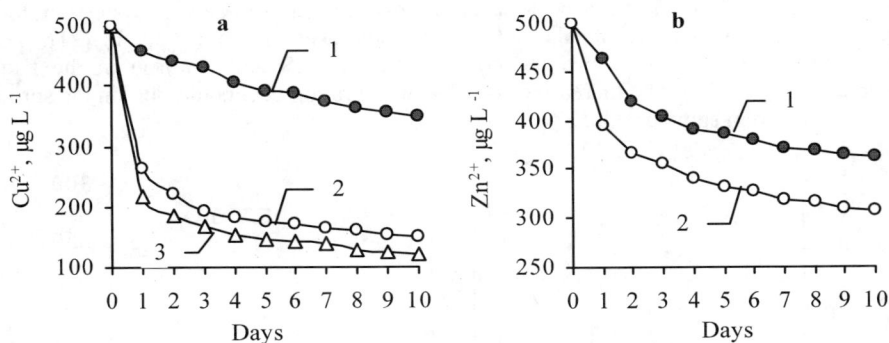

Figure 8. Complexation kinetics of copper(II) and zinc(II) with DOM of the upper section of the Kanev reservoir. a: 1 – July, 2 – August, 3 – November; b: 1 – April, 2 – October.

The kinetic data on complexation of Cu and Zn during various seasons support the above conclusion on the slowest rate of metal binding to DOM in spring and early summer (Figure 8). So, in the sample of the Kanev reservoir water taken in June 1998, the complete binding of Cu^{2+} ions to DOM was reached only in three months (Figure 9). It is difficult to explain this phenomenon because the concentration of HS was rather high (20-30 mg·L^{-1}) and considerably exceeded the metal concentration in the system. In summer 1993, when it was raining, and HS concentration increased by a factor of 3-4 (about 114 mg·L^{-1}), the binding of Cu^{2+} ions was also minimal.

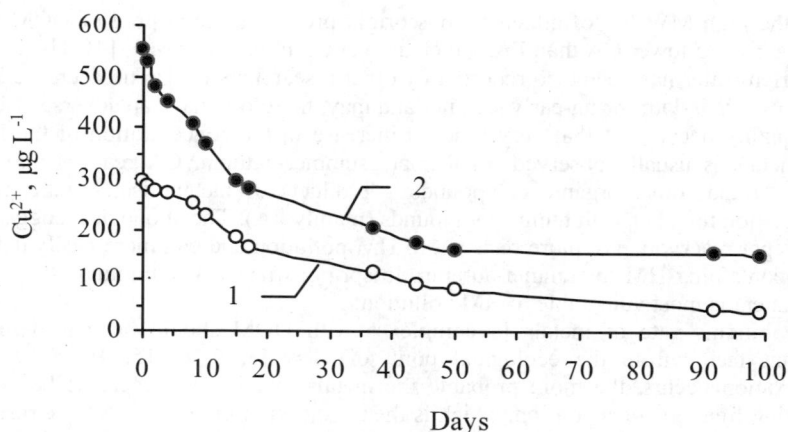

Figure 9. Complexation kinetics of copper(II) with DOM in the Kanev reservoir, June 1998. C_{Cu}: 1 – 300 µg·L^{-1}, 2 – 550 µg·L^{-1}.

Insignificant complexation under an increasing concentration of HS in the aquatic environment has been insufficiently studied until recently. It is possible that the increase of HS concentration results in some conformational changes in the macromolecules, thereby making their active centres unattainable for complexation [29]. Moreover, some of the active centres are already occupied by other metals, e.g., Fe, Al, Cu [25, 26, 36]. Fe is the first to bind with HS because the Fe concentration in surface water is rather high compared with other metals and its HS-complexes are rather stable [21]. The correlation is observed between the concentration of HS and the iron in the Kiev reservoirs (Figure 10). The increase of the Fe concentration occurs not only in the spring freshet but also in summer rain.

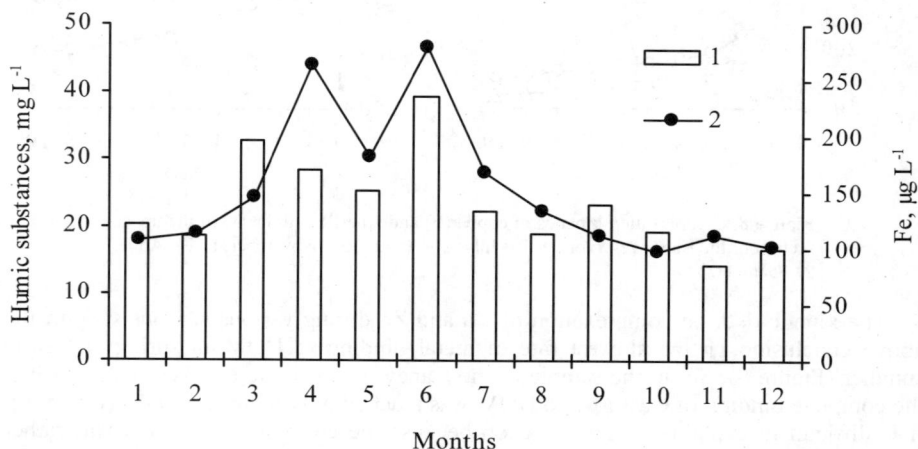

Figure 10. Seasonal dynamics of the content of HS (1) and Fe (2) in the water of the Kiev reservoir, 1992.

The binding of Cu to aquatic HS can take place as a result of the concurrent complexation with iron [25, 26]. But the release of iron is complicated because the Fe-HS complexes have high stability that was proven studies on isotope exchange between[59]Fe and Fe incorporated in aquatic humus [35]. It was established that only 6.5-10.1% of iron was exchanged after 1 h; whereas the rest was not exchanged even after 24 h at pH 3-5. The complete release of iron from the HS complexes occurs after photochemical oxidation of HS or ultrasonic treatment of samples [30].

The kinetics of copper (II) complexation with FA was studied also. FA were isolated from the natural water of the Kanev reservoir and purified according to the existing procedure [17]. The conditions of the experiment were as follows: pH 8, concentration of FA was 27 $mg \cdot L^{-1}$, concentration of Cu was 500 $\mu g \cdot L^{-1}$. It was established that under the chosen experimental conditions copper binds with FA faster than in natural water. However, in this case, complexation occurs over several days and equilibrium is reached only in 5 or 6 days (Figure 11).

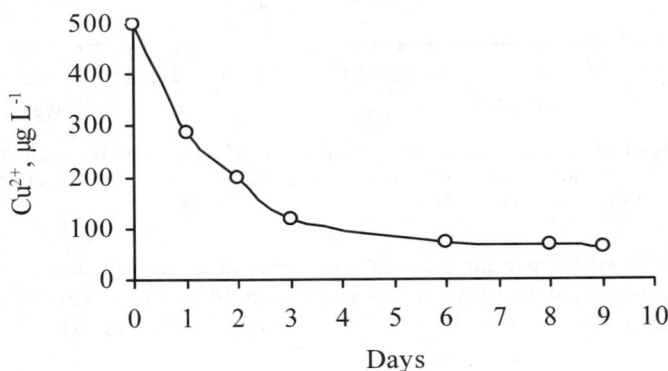

Figure 11. Complexation kinetics of copper (II) with the fraction of FA of molecular weight of 0.8-0.4 kDa.

3.4. BIOAVAILABILITY OF HEAVY METALS IN THE AQUATIC ENVIRONMENT

Complexation with DOM as a rule, leads to a decrease in HM toxicity up to its complete detoxification [12, 18, 21, 24, 27]. This is caused by strong complexation of HM with HS that was discussed above. The HM-HS complexes dissociate very slowly having inert nature [25, 26, 30, 42]. The latter is true of the compounds with relatively low MW (1–10 kDa). Detoxification can be also caused by the formation of high MW complexes, which cannot penetrate through the cell membrane [12, 34].

Bioassays with distilled water showed that when the HA concentration was 2.5 and 5.0 $mg \cdot L^{-1}$, mortality of the test organisms in 7 days was 56.7 and 66.7%, respectively,. It is possible that HS caused the oppression of the vital activity of the organisms. Reservoirs having high concentrations of HS are characterized by lower bioproductivity than the ones with low concentrations of HS [41]. At the Cu^{2+} concentration of 100.0 and 200.0 $\mu g \cdot L^{-1}$, the mortality of all test objects was observed in the first day.

The combined impact of Cu^{2+} ions and HA on *D. magna* indicates a decrease in water toxicity. This is due to the complexation of Cu^{2+} ions with HA. However, it should be noted that under these conditions (100.0 and 200.0 $\mu g \cdot L^{-1}$ of Cu^{2+} ions) the toxicity of the water medium was rather high even in 15 days. The residual concentration of free Cu^{2+} ions was 28-34 and 58-66 $\mu g \cdot L^{-1}$ (Figure 12a, b) respectively. 100% mortality of test objects occurred within 48 h.

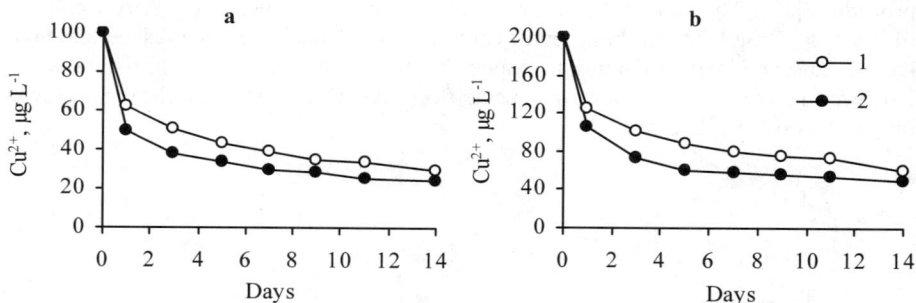

Figure 12. Kinetics of the binding of Cu^{2+}ions in complexes with HA in distilled water in the biotesting experiments. C_{Cu} ($\mu g \cdot L^{-1}$): 100.0 (a), 200.0 (b); C_{HA} ($mg \cdot L^{-1}$): 2.5 (1), 5.0 (2).

Experiments with tap water showed a decrease in toxicity greater than in distilled water. The complete detoxification was reached in 10-15 days. After this period, the concentration of free Cu^{2+} ions did not exceed 5-10 $\mu g \cdot L^{-1}$ (Figure 13).

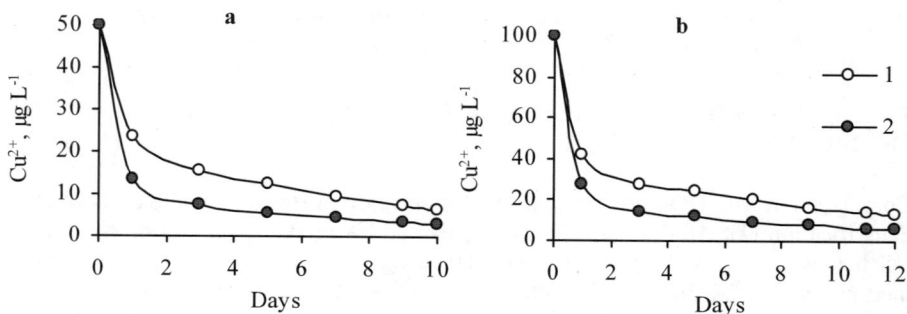

Figure 13. Kinetics of the binding of Cu^{2+} to HA in tap water used in the bioassays. C_{Cu} ($\mu g \cdot L^{-1}$): 50.0 (a), 100.0 (b); C_{HA} ($mg \cdot L^{-1}$): 5.0 (1), 10.0 (2).

Bioassays with the natural water samples occurred under conditions when the Cu complexation was provided both by DOM and by the added HA. At smaller concentrations, Cu^{2+} ions (50.0 and 100.0 $\mu g \cdot L^{-1}$) showed almost complete binding, and the complete detoxification was reached within 96 h (Figure 14a, b). At the 250.0 $\mu g \cdot L^{-1}$ concentration of Cu^{2+} ions, complete detoxification was reached within 14 days under maximal concentration of HA. The residual concentration of free Cu^{2+} ions was about 28-30 $\mu g \cdot L^{-1}$ (Figure 14c). At the 500.0 $\mu g \cdot L^{-1}$ concentration of Cu^{2+} ions, the toxicity of the water sample maintained for a longer time because only 350.0 $\mu g \cdot L^{-1}$ Cu (II) was bound into complexes (Figure 14d).

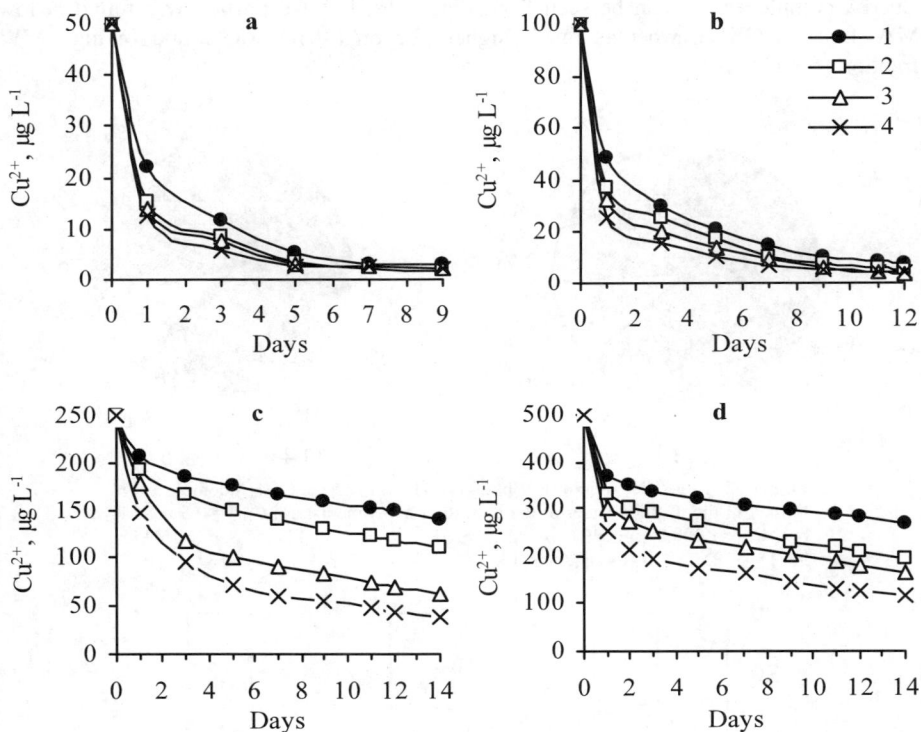

Figure 14. Kinetics of the binding of Cu^{2+} to DOM in natural water from the lower section of the Kanev reservoir (July 1999) used in the bioassays and to added HA. C_{Cu} ($\mu g \cdot L^{-1}$): 50.0 (a), 100.0 (b), 250.0 (c), 500.0 (d); 1 – natural water, 2, 3, 4 – natural water + HA at the concentrations of 2.5, 5.0 and 10.0 $mg \cdot L^{-1}$, accordingly.

In addition to the complexation kinetics, we studied the distribution of Cu(II) among the complexes formed. In distilled water, HA and their complexes with Cu(II) showed similar MWD (Figure 15). The main fraction of Cu(II) was found in the low MW complexes (< 1 kDa). It should be noted that in natural water, 52% of copper was bound to FA (Figure 16). At the same time, the concentration of Cu-HA complexes did not exceed 8%. With increasing concentrations of Cu^{2+} ions, the MWD of Cu-HA complexes had changed. When the concentration of Cu^{2+} ions and HA was 100.0 $\mu g \cdot L^{-1}$

and 10.0 mg·L^{-1}, respectively, the share of Cu-HA complexes became 22%, that is 3 times higher than in natural water. An increase in Cu(II) concentration up to 250.0 µg·L^{-1} resulted in an increase of the share of Cu-HA complexes up to 37%. The obtained data on Cu(II) distribution among complexes with different MW fractions of HA showed that under 100.0 µg·L^{-1} Cu^{2+} and 10.0 mg·L^{-1} HA, the low MW complexes (< 1 kDa) predominated (54.6%), and 30% of Cu (II) was bound to high MW fraction (> 30 kDa). An increase in Cu(II) concentration up to 250.0 µg·L^{-1} shifted the MWD of Cu-HA complexes. As can be seen from Figure 17, less Cu portion was found in low MW fraction (25%), whereas much higher portion (40%) was found in high MW fraction (> 30 kDa).

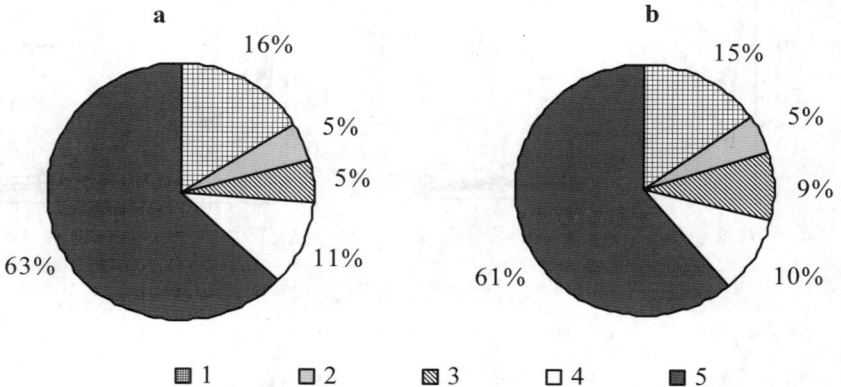

Figure 15. Molecular weight distribution of HA (a) and the Cu-HA complexes (b) formed in the distilled water in the biotesting experiments. a – C_{HA} = 0.5 mg·L^{-1}; b – C_{Cu} = 200.0 µg·L^{-1}, C_{HA} = 5.0 mg·L^{-1}; 1 – > 30 kDa, 2 – 30–15 kDa, 3 – 15–5 kDa, 4 – 5–1 kDa, 5 – < 1 kDa.

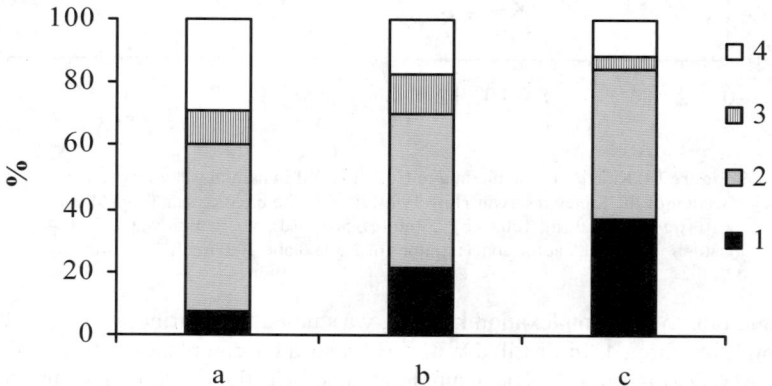

Figure 16. Distribution of Cu(II) among the complexes with DOM of different chemical nature in the natural water from the Kanev reservoir and with the added HA in the bioassay experiments. a – natural water, b – natural water + Cu^{2+} (100.0 µg·L^{-1}) + HA (10.0 mg·L^{-1}), c – natural water + Cu^{2+} (250.0 µg·L^{-1}) + HA (10.0 mg·L^{-1}); 1, 2 – Cu-HA and Cu-FA complexes, respectively, 3, 4 – complexes of Cu (II) with protein-like compounds and carbohydrates, respectively.

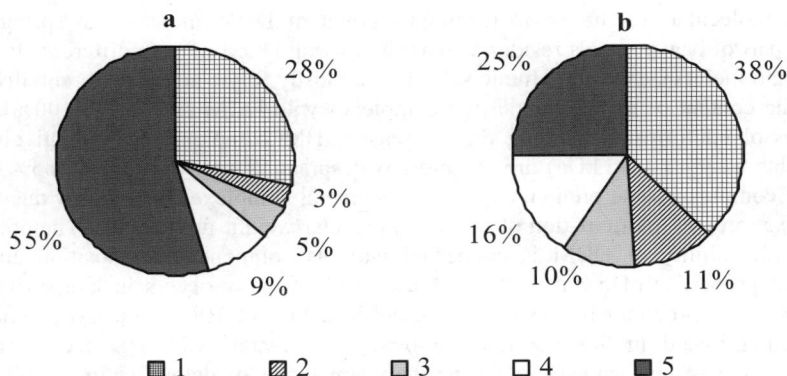

Figure 17. Distribution of Cu (II) among the complexes with HA from the natural water of the Kanev reservoir and HA added after separation of HA and FA in acidic medium (pH = 1). C_{Cu} (µg L^{-1}): 100.0 (a), 250.0 (b); 1 – > 30 kDa, 2 – 30–15 kDa, 3 – 15–5 kDa, 4 – 5–1 kDa, 5 – < 1 kDa.

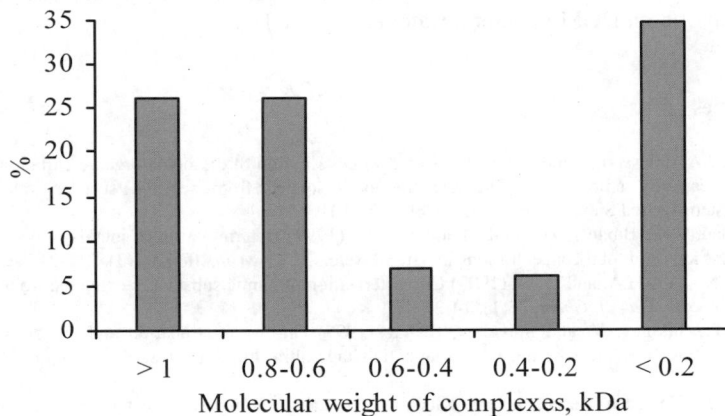

Figure 18. Distribution of Cu(II) among complexes with FA of different molecular weight in natural water from the lower section of the Kanev reservoir (July 1999).

The results of the corresponding study on the MWD of Cu(II)-FA complexes are shown in Figure 18. At the concentration of Cu^{2+} of 250 µg·L^{-1}, Cu(II) was bound mainly by fractions of FA with MW of 0.8-0.6 and > 1 kDa (about 50%), whereas 35% was found as a low MW fraction (< 0.25 kDa).

4. Conclusions

The results of our research show that HS are the main component of DOM in the Dnieper reservoirs. The concentration of FA is 20-40 times higher than that of HA. The content of HS decreases from the north to south in these reservoirs. FA in the Dnieper reservoirs are mostly compounds of low molecular weight, up to 1kDa. The contribution of high molecular weight FA to the total amount of DOM increases in spring. The greater part of heavy metals resides as complexes with DOM of both different chemical nature and molecular weight. Humic substances, mostly fulvic acids, preferentially bind HM. The content of metals bound into complexes with DOM reaches 70-100% of the total dissolved forms, depending on the season. HM complexes with relatively low molecular weight (0.5-5 kDa) are the most widespread. They make up 40-65% of all organic complexes. The predominance of these metal complexes is probably due to the complexation with participation of the low molecular weight fulvic acids. The potential complexing ability of DOM is connected with its component composition and the chemical properties of metals. The decrease in CA usually occurs in late spring and early summer, probably because of the specific nature of HS as the most important complexing ligand in this period. The binding of metals with HS occurs slowly. Complexation of HM with DOM is the dominant factor of detoxification resulting in disappearance of free metal ions, which are the most toxic forms. The results of experiments on the impact of HA on the toxicity of Cu (II) in aquatic environments (distilled, tap, and natural water) show that the maximum decrease of toxicity was observed in natural water because the binding of Cu (II) into complexes occurred with participation of both DOM in natural water and added HA.

5. References

1. Alberts, J.A., Giesy, J.P. and Evans, D.M. (1984) Distribution of dissolved organic carbon and metal-binding capacity among ultra-filterable fractions isolated from selected surface waters of the southeastern United States, *Environ. Geol.* **8**(2), 91-101.
2. Appelblad, P.K., Baxter, D.C. and Thunberg, J.O. (1999) Determination of metal-humic complexes, free metal ions and total concentrations in natural waters, *J. Environ. Monit.* **1**(3), 211-217.
3. Belzile, N., Joly, H.A. and Li, H. (1997) Characterization of humic substances extracted from Canadian lake sediments, *Can. J. Chem.* **75**(1), 14-27.
4. Burba, P., Shkinev, V. and Spivakov, B. Ya. (1995) On-line fractionation and characterization of aquatic humic substances by means of sequential-stage ultrafiltration, *Fresenius' J. Anal. Chem.* **351**, 74-82.
5. Burba, P., Van den Bergh, J. and Klockow, D. (2001) On-site characterization of humic-rich hydrocolloids and their metal loading by means of mobile size-fractionation and exchange techniques, *Fresenius' J. Anal. Chem.* **371**, 660-669.
6. Campanella, L., Pyrzyñska, K. and Trojanowicz, M. (1996) Chemical speciation by flow-injection analysis. A review, *Talanta* **43**, 825-838.
7. Campbell, J.H. and Evans, R.D. (1987) Inorganic and organic ligands binding of lead and cadmium and resultant implications for bioavailability, *Sci. Total Environ.* **62**, 219-227.
8. Debeyko, E.V., Ryabov, A.K. and Nabivanets, B.I. (1973) Direct photometric determination of dissolved proteins in natural waters, *Gidrobiologicheskiy Zhurnal (Hydrobiol. J.)* **9**(6), 109-113 (in Russian).
9. Denisova, A.I., Timchenko, V.M., Nakhshyna, Ye.P., Ryabov, A.K. and Novikov, B.I. (1989) *Gidrologiya i gidrokhimiya Dnepra i yego vodokhranilishch (Hydrology and Hydrochemistry of the Dnieper and Its Reservoirs)*, Naukova Dumka Publishers, Kiev (in Russian).
10. Esteves da Silva, J.C.G., Machado, A.A.S.C., Oliveira, C.J.S. and Pinto, M.S.S.D.S. (1998) Fluorescence quenching of anthropogenic fulvic acids by Cu (II), Fe (III) and UO_2^{2+}, *Talanta* **45**, 1155-1165.
11. Flemming, C.A. and Trevors, J.T. (1989) Copper toxicity and chemistry in the environment: a review, *Water, Air, Soil Pollut.* **44**, 143-158.

12. Förstner, U. and Wittman, G.T.W. (1983) *Metal Pollution in the Aquatic Environment*, 2nd ed., Springer-Verlag, New York.

13. Frimmel, F.H., Immerz, A. and Niedermann, H. (1983) Heavy metal interaction with aquatic humus, *Intern. J. Environ. Anal. Chem.* **14**(2), 105-11.

14. Goyer, R.A. (1991) Toxic effects of metals, in M.O. Amdur, J. Doull and C.D. Klaassen (eds.), *Toxicology: The Basic Science of Poisons*, Pergamon Press, New York, pp. 623-860.

15. Imai, A., Fukushima, T., Matsushige, K. and Kim, Y. H. (2001) Fractionation and characterization of dissolved organic matter in a shallow eutrophic lake, its inflowing rivers, and other organic matter sources, *Wat. Res.* **35**(17), 4019-4028.

16. Itoh, A., Kimata, C., Miwa, H, Sawatari, H. and Haraguchi, H. (1996) Speciation of trace metals in pond water as studied by liquid chromatography/inductively coupled plasma mass spectrometry, *Bull. Chem. Soc. Jpn.* **69**, 3469-3473.

17. Ivanova, E., Pershina, I.V., Polenova, T.V. and Chernyak, C.M. (1986) Fluorescence method of determination of fulvic acids in sea water, *Zhurnal Analiticheskoy Khimii (J. Anal. Chem.)* **41**(7), 1256-1259 (in Russian).

18. Khangarot, B.S., Ray, P.K. and Chandra, H. (1987) Preventive effects of amino acids on the toxicity of copper to *Daphnia magna*, *Water, Air, Soil Pollut.* **32**, 379-387.

19. Linnik, P.N. (1999) Heavy metals in the surface waters of the Ukraine: Content and migration forms, *Gidrobiologicheskiy Zhurnal (Hydrobiol. J.)* **35**(1), 22-42 (in Russian).

20. Linnik, P.N. and Iskra, I.V. (1994) Application of anodic stripping voltammetry to the investigation of the physicochemical state of cadmium in surface water in the Ukraine, *Microchem. J.* **50**, 184-190.

21. Linnik, P.N. and Nabivanets, B.I. (1986) *Migrating Forms of Metals in Fresh Surface Waters*, Gidrometeoizdat Publishers, Leningrad (in Russian).

22. Linnik, P.N. and Nabivanets, Yu.B. (1988) Application of anodic stripping voltammetry to the determination free and bound in complexes ions of zinc and lead in natural waters, *Gidrobiologicheskiy Zhurnal (Hydrobiol. J.)* **24**(1), 68-71 (in Russian).

23. Linnik, P.N., Nabivanets, Yu.B., Iskra, I.V. and Chubar, N.I. (1994) Complexing ability of dissolved organic matter and the buffering capacity of aquatic ecosystems, *Gidrobiologicheskiy Zhurnal (Hydrobiol. J.)* **30**(5), 87-99 (in Russian).

24. Luoma, S.N. (1983) Bioavailability of trace metals to aquatic organisms – a review, *Sci. Total Environ.* **28**, 1-22.

25. Mandal, R., Sekaly, A.L.R., Murimboh, J., Hassan, N.M., Chakrabarti, C.L., Back, M.H., Gregoire, D.C. and Schroeder, W.H. (1999) Effect of the competition of copper and cobalt on the lability of Ni (II)-organic ligand complexes. Part I. In model solutions containing Ni (II) and a well-characterized fulvic acid, *Anal. Chim. Acta* **395**, 309-322.

26. Mandal, R., Sekaly, A.L.R., Murimboh, J., Hassan, N.M., Chakrabarti, C.L., Back, M.H., Gregoire, D.C. and Schroeder, W.H. (1999) Effect of the competition of copper and cobalt on the lability of Ni (II)-organic ligand complexes, Part II: In freshwaters (Rideau River surface waters), *Anal. Chim. Acta* **395**, 323-334.

27. Moore, J.W. and Ramamoorthy, S. (1987) *Heavy Metals in Natural Waters*, Mir Publishers, Moscow (Russian ed.).

28. Nabivanets, B.I., Linnik, P.N. and Kalabina, L.V. (1981) *Catalytic Methods for Natural Water Analysis*, Naukova Dumka Publishers, Kiev (in Russian).

29. Nifant'eva, T.I., Shkinev, V.M., Spivakov, B. Ya. and Burba, P. (1999) Membrane filtration studies of aquatic humic substances and their metal species: a concise overview. Part 2. Evaluation of conditional stability constants by using ultrafiltration, *Talanta* **48**, 257-267.

30. Nomizu, T., Sanji, M. and Mizuike, A. (1988) Removal of iron for the spectrophotometric determination of humic substances in fresh waters, *Anal. Chim. Acta* **211**(1-2), 293-297.

31. Perminova I.V. (2000) *Analysis, Classification and Prediction of Properties of Humic Substances*, Dr. Sci. Thesis. (in Chemistry), Lomonosov Moscow State University, Moscow, Russia (in Russian).

32. Peters, A.J., Hamilton-Taylor, J. and Tipping, E. (2001) Americium binding to humic acid, *Environ. Sci. Technol.* **35**, 3495-3500.

33. Popovich, G.M. (1990) *Sorption concentration and spectrophotometric determination of humic and fulvic acids in waters*, Abstract of Ph.D. Thesis, Kiev (in Russian).

34. Salomons, W. and Förstner, U. (1984) *Metals in the Hydrocycle*, Springer-Verlag, Berlin.

154

35. Sedláček, J., Gjessing E. and Rambaek, J.P. (1987) Isotope exchange between inorganic iron and iron naturally complexed by aquatic humus, *Sci. Total Environ.* **62**, 275-279.

36. Sekaly, A.L.R., Mandal, R., Hassan, N.M., Murimboh, J., Chakrabarti, C.L., Back, M.H., Gregoire, D.C. and Schroeder, W.H. (1999) Effect of metal/fulvic acid mole ratios on the binding of Ni (II), Pb (II), Cu (II), Cd (II), and Al (III) by two well-characterized fulvic acids in aqueous model solutions, *Anal. Chim. Acta* **402**, 211-221.

37. Semenov, A.D. (Ed.) (1977) *Manual on the Chemical Analyses of Surface Waters of Land,* Gidrometeoizdat Publishers, Leningrad (in Russian).

38. Sirotkina, I.S., Varshal, G.M., Lur'ye, Yu.Yu. and Stepanova, N.P. (1974) Application of cellulose sorbents and sephadexes in systematic analysis of natural water organic substances, *Zhurnal Analiticheskoy Khimii (J. Anal. Chem.)* **29**(8), 1626-1633 (in Russian).

39. Steinberg, C.E.W., Haitzer, M., Brueggemann, R., Perminova, I.V., Yashchenko, N. Yu. and Petrosyan, V.S. (2000) Towards a quantitative structure activity relationship (QSAR) of dissolved humic substances as detoxifying agents in freshwaters, *Int. Rev. Hydrobiol.* **85**(2-3), 253-266.

40. Town, R.M. and Filella, M. (2000) A comprehensive systematic compilation of complexation parameters reported for trace metals in natural waters, *Aquat. Sci.* **62**, 252-295.

41. Tseeb, Ya. Ya. and Maistrenko, Yu. G. (eds.) (1972) *Kiev Reservoir: Hydrochemistry, Biology, Productivity,* Naukova Dumka Publishers, Kiev (in Russian).

42. Van den Bergh, J., Jakubowski, B. and Burba, P. (2001) Investigations on the conditional kinetic and thermodynamic stability of aquatic humic substance–metal complexes by means of EDTA exchange, ultrafiltration and atomic spectrometry, *Talanta* **55**, 587-593.

43. Varshal, G.M. (1977) On the state of mineral components in surface waters, in M.M. Senyavin (Ed.), *Methods of Analysis of Natural Waters and Effluent: Problems of Analytical Chemistry,* Vol. 5, Nauka Publishers, Moscow, 94-107 (in Russian).

44. Zhilin, D.M. and Perminova, I.V. (2000) Approach to assessment of the buffer capacity of an aquatic ecosystem on the example of Hg (II), *Priroda (Nature)* **11**, 43-50 (in Russian).

Chapter 7

COMPLEXATION OF RADIONUCLIDES WITH HUMIC SUBSTANCES

V. MOULIN
*Commissariat à l'Energie Atomique, CEA, Nuclear Energy Division &
UMR CEA-CNRS-UEVE, Laboratory of Analysis and Environment,
91191 Gif–sur-Yvette, France <valerie.moulin@cea.fr>*

Abstract

Humic substances are ubiquitous compounds presenting specific properties with respect to complexation, sorption and transport. Because they may enhance metal solubilities due to their strong complexing properties, as well as increase or decrease metal retention on mineral surfaces due to their affinity for mineral substrates and their scavenging properties, humic substances may impact metal speciation in the environment, and in particular radionuclides speciation. Various examples based on actinides, lanthanides or iodine will be presented to illustrate the complexation properties as well as speciation calculations with emphasis on the thermodynamic constants used and the techniques used to obtain them.

1. Introduction

Humic substances (HS) are ubiquitous in natural environments, namely in aquatic and pedologic media, and they present specific properties with respect to complexation, sorption and transport. Because HS may enhance metal solubilities due to their strong complexing properties, and either increase or decrease metal retention on mineral surfaces due to affinity for mineral substrates and scavenging properties, they may impact the mobility and speciation of inorganic xenobiotics in the environment. Because the speciation governs the mobility, toxicity, and risk resulting from the presence of trace elements in natural systems [1, 2], its determination contributes to a better understanding of the fate (retention, transport, bioavailability) of radionuclides, in the framework of environmental and waste management purposes. Moreover, the consequences on remediation purposes are of prime importance because the knowledge of speciation constitutes a key point to this aspect.

Here, the term "speciation" is defined by the chemical form of the element (i.e., its oxidation state, charge, and complexing properties) but also includes the bearing phase (physical speciation). Hence, the dissolved species are distinguished from those bound to solid phases (colloids, particles, and mineral substrates), the various complexes formed with the ligands (inorganic, organic) present in solution, and the various oxidation states [2]. Templeton et al. [3] detailed the various terms, based on IUPAC recommendations, to define the chemical speciation from the fractionation of elements, including speciation analysis.

I. V. Perminova et al. (eds.),
Use of Humic Substances to Remediate Polluted Environments: From Theory to Practice, 155–173.

The purpose of this chapter is to present all information (thermodynamic, chemical, and spectroscopic) to determine the speciation of a given radionuclide in the presence of HS. Hence, complexation studies were performed to characterize the stoichiometry of the complexes formed as well as the associated interaction constants. Data on actinides (thorium, uranium, etc.), on fission products (iodine), and on activation products (lanthanides) will be presented with a focus on the possibilities of analytical techniques to study the systems, *relying on CEA* (Commissariat à l'Energie Atomique) *works.* Moreover, the influence of HS on trace element retention onto mineral surfaces will also be illustrated. The results presented here have been obtained with Aldrich humic acid (HA), and aquatic fulvic acid (extracted from a clayey water) (FA).

2. Ligand Description

It is important to define the description of the specific ligand constituted by HS. Recent CEA works performed by electrospray ionisation high resolution mass spectrometry (Q-TOF) show that FA are supramolecules constituted by aggregates of many small molecules (200-800 atomic mass unit (amu)) linked by hydrogen or other weak bonds [4-5]. Figure 1 illustrates the proposed structure of FA monomers (with possible existence of isomers) obtained after analysis of MS-MS spectra for various aquatic FA.

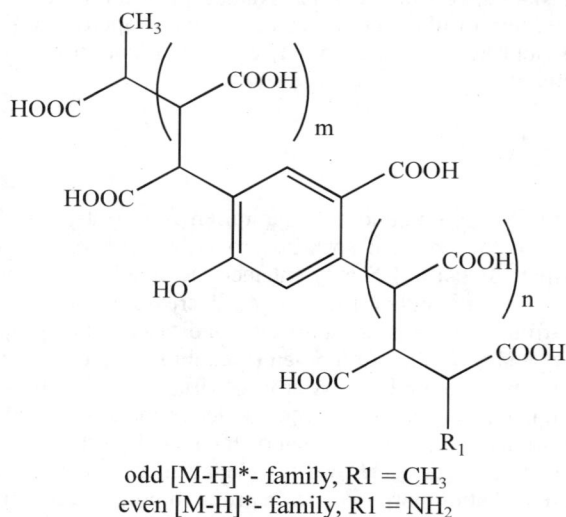

odd [M-H]*- family, R1 = CH_3
even [M-H]*- family, R1 = NH_2

Figure 1. Proposed structure for FA obtained by ES-MS-MS [4]. *deprotonation of the ligand M within the ionization process.

As can be seen, the structure presents phenolic cores together with carboxylic moieties as well as amino groups (it was not possible to observe S atoms). It is also important to underline that this structure is in agreement with the elemental analysis of the FA studied. Moreover, it is in agreement with literature data [6]. ES-MS-MS data obtained on HA present the same kind of molecules composed of the aggregates.

Hence, HS can be considered as a complex mixture of ligands with various functional groups (mainly carboxylic), the structure of which is constituted by these monomers (Figure 1) forming the molecular structure of HS.

2.1. SINGLE SITE MODEL

The *single site model* is retained to describe the binding of a cation to the humic molecules. Here the ligand is defined as a site (A) with no particular assumption about its chemical nature, even though these sites are mainly carboxylic groups (see also Figure 1) [7-10]. All sites are thus considered identical and independent, and no differences between the size of HS fractions are taken into account. Another important feature concerning the structure of these sites is the number of ligands involved in the interaction with a cation. In a first approach, only monodentate (or monoprotic) sites are considered, and 1:1 complexes (metal:site) are supposed to be formed according to the equilibrium:

$$M^{(z+)} + A^{(-)} \Leftrightarrow MA^{(z-1)+}$$

This system is described by the mass action law:

$$\beta = \frac{[MA]}{[M] \times [A]} \quad \text{(for simplification, charges are omitted)} \tag{1}$$

where β is the complexation constant associated with the interaction between the cation and a complexing site, M is the metal ion in solution, and A is one site. β is an apparent constant and a global interaction constant.

The site concentration [A] used in equation (1) will govern the dimensional unit of the interaction constant. This concentration is derived from the complexing capacity W, which is defined as the number of equivalents, or moles, of metal bound to one gram of HS. W is accessible from experiments (e.g., titration of the metal ion by the HS ligand). In this system, the interaction constant has dimensional units of L/eq (or L/mol). This point is particularly important because the comparison between literature data will be valid only if β units are identical, or if they can be easily recalculated from the known technique of [A] determination.

Other models are proposed in the literature for the description of the binding between metal ion and HS, as detailed in Section 2.2.

2.2. OTHER MODELS

Other approaches to description of metal-HS interactions, based on discrete sites models, are used in the literature as detailed in [10, 11]. Among them are the "conservative roof" model [11], the charge neutralization model [12], the polyelectrolyte model [13], and other models presented in [10-11, 14] with references therein.

Hummel *et al.* [11] propose the "conservative roof" model, which allows estimation of the maximum effect of a various number of parameters, the objective of which is the application of such a model for performance assessment purposes.

The charge neutralization model used by Kim and Czerwinski [12] considers that a metal M^{z+} occupies z sites, leading to an operational concentration HA(z) of humic sites for every metallic species. The value of W in that case can be considered as W/z. The number of accessible sites can increase with pH. Hence, equation (1) is the same, except for the definition of [A], and the dimensional units of β are also expressed as L/eq (or L/mol).

The double site model developed by Choppin and co-workers [13, 14] considers β as a function of the ionisation degree α_{HA} of the acid sites of a polyelectrolyte (related to the acidity constant pK_{HA}), referring to Henderson-Hasselbach treatment. The complexation constant can, hence, be described by a different approach, in which:

$$\log \beta = a\, \alpha_{HA} + b \qquad (2)$$

where a and b are two characteristic factors of the complexation of the metal by HA.

The α_{HA} and pK_{HA} values are determined experimentally from HA titration curve. Operational values of pK_{HA} are thus determined for $\alpha_{HA} = 0.5$. With this description, the second reaction is also introduced considering two neighbouring complexing sites:

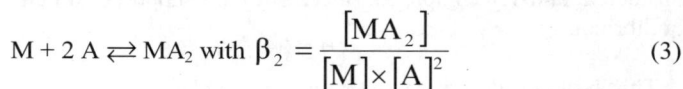

$$M + 2\,A \rightleftarrows MA_2 \text{ with } \beta_2 = \frac{[MA_2]}{[M] \times [A]^2} \qquad (3)$$

In this model, the ligand concentration is equal to the number of ionised sites at the pH of the study. Hence, it is equal to the proton exchange capacity (eq/g) multiplied by the ionisation degree α_{HA} and the ligand concentration (g/L).

The models [12-14] developed by Kim et al. and by Choppin et al. are very similar but cannot be compared directly [14]. The dimensional units obtained for the interaction constants are identical, but based on different assumptions for the sites.

Other models are used for the description of radionuclides by HS, namely NICA-Donnan [15] and Model V [16]. Their major advantage is the representation of the competition with alkaline earth elements. Their main drawback is the lack of application in the case of actinides for NICA-Donnan and the multiple adjusted parameters for Model V.

2.3. COMPETITION WITH MAJOR CATIONS

HS are strong complexing agents of cations, and the affinity order conforms to the "classical" Irving-Williams series (see development in [10]) :
$Fe^{3+} > Al^{3+} > Pb^{2+} > Cu^{2+} > Cd^{2+} > Ni^{2+} > Zn^{2+} > Fe^{2+} > Co^{2+} , Mn^{2+} > Ca^{2+}, Mg^{2+}, Ba^{2+}$

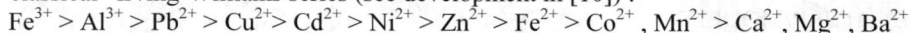

In natural waters, the concentration of alkali earth elements (e.g., Ca and Mg) can be high, and thus, can induce competitive complexation reactions with HS. In the case of Ca and Mg, controversy remains on their complexation by HS and on the values of interaction constants. In fact, some studies show that these divalent cations do not compete with other divalent or trivalent cations because the binding sites are different or a difference exists in the nature of the binding of the metals to the humic material (see citations in [10] and [17]). As pointed out by Reiller et al. [10] and Moulin et al. [17], in the case of Ca and Mg, the HS interaction constants have not yet been determined unambiguously, and for Al and Fe, scarce data exist. It is also to be noted that the real oxidation state of Fe in the presence of HS is a subject of controversy because reduction may occur (see [10] and citations in it).

3. Techniques for Studying Complexation with HS

Various techniques may be used to study the interaction of metal ions with HS, allowing the determination of an interaction constant. Investigations are classically performed either by separation or non-separation techniques:

Separation techniques include ultrafiltration (UF), ion exchange resins (IE), solvent extraction (SE), equilibrium dialysis (ED), and chromatography [18]

Non separation techniques are either spectroscopic, such as time-resolved laser-induced fluorescence (TRLIF) or spectrophotometry (SP), or spectrometric, such as mass spectrometry with electrospray ionisation (ES-MS) [19].

If the separation techniques allow determination of the fraction of either bound or free metal, thus permitting calculation of the interaction constant, the spectroscopic techniques are more informative due to their non destructive character. In addition to the calculation of interaction constants, important features can be deduced from spectroscopic data, such as the stoichiometry and type of the complex.

The following section will present the principles of some of the analytical tools used at CEA for metal-HS studies [20-22].

3.1. TIME-RESOLVED LASER-INDUCED FLUORESCENCE AND SPECTROPHOTOMETRY

The TRLIF technique (Figure 2) allows us i) to work at very low metal concentrations (less than 1 µM) and ii) to identify spectrally and temporally the various species, *i.e.,* to directly determine speciation of the cation under study. Besides its selectivity, sensitivity, dynamic nature (for analysis purposes larger than 10^6: from ng to mg) and

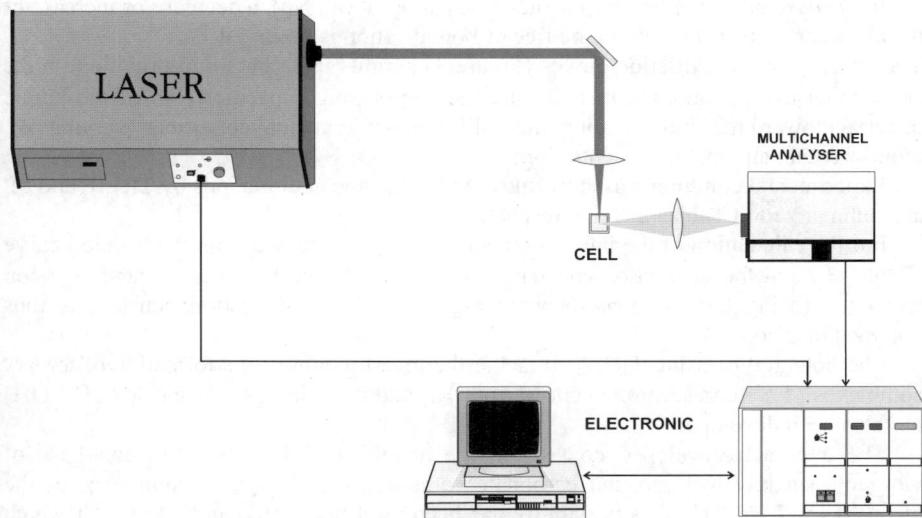

Figure 2. TRLIF set-up.

remote measurement, TRLIF's main drawback is its applicability only to fluorescent elements, which include in the case of lanthanides/actinides U, Cm, Am, Eu, Dy, Tb, Sm, and Gd. Moreover, a great advantage of this technique is that it may also be used to perform speciation studies [23-25]. The speciation limit in these cases is at least three orders of magnitude less than the analysis detection limits (see Table 1).

Table 1. Detection limits for TRLIF [25].

Species	Limit of detection (μmoles/L)	Speciation limit (μmoles/L)
U(VI)	5×10^{-7}	0.01
Am(III)	0.001	1
Cm(III)	5×10^{-7}	0.01
Eu(III)*	5×10^{-7}	0.01

*Sm, Gd, Tb, Dy, Tm, Ce, and Nd can also be determined at low level.

The principle of TRLIF and SP for the study of complexation is based on the *titration of the cation* (in the nanomolar to micromolar range for TRLIF, and in the micromolar range for SP) *by the organic ligand* (HS) at a constant pH and ionic strength and on the measurement of the signal characteristic of each technique, namely fluorescent signal for TRLIF, and absorbance for SP.

In the case of TRLIF applied for lanthanides (Eu, Dy) and actinide (Cm) cations, an increase of the fluorescence (at the wavelength of cation emission) is observed when adding the ligand, whereas for a hexavalent cation such as U, a decrease of the fluorescence signal is obtained.

In the case of SP applied for two actinide cations (Am, Np), a decrease or increase of the absorbance, respectively, to the free or bound cation is observed.

The analysis of such titration curves (Figure 3) permits us to obtain information on the conditional interaction constants (β) and the complexing capacities (W) of the humic materials toward the studied cation under different experimental conditions (varying pH, ionic strength, and metal concentration).

Experimental conditions used to study lanthanide and actinide ions by TRLIF and SP are summarized in Tables 2 and 3, respectively.

For the calculation of the interaction constant, W is obtained from the titration curve (Figure 3), and the interaction constant (log β) is evaluated by a non linear regression technique (using the Marquadt-Newton algorithm). The interaction constant is thus obtained in L/eq.

The potential of using TRLIF to tackle the investigation of the formation of ternary complexes of a fluorescent element M with HS and main inorganic ligands (CO_3, OH) has also been developed.

The approach developed consists in the acquisition of a spectrum data-base of inorganic (mainly hydroxo and carbonate complexes), and organic complexes of the metal M [5, 7, 26-31]. This possibility has been used for Eu(III) and U(VI), for which such data bases have been constituted (for carbonato and hydroxo Eu complexes, and for only U hydroxo complexes because carbonate complexes are not fluorescent), before studying the interaction with HS. The main spectral and temporal characteristics of Eu and U species are detailed in [26-31].

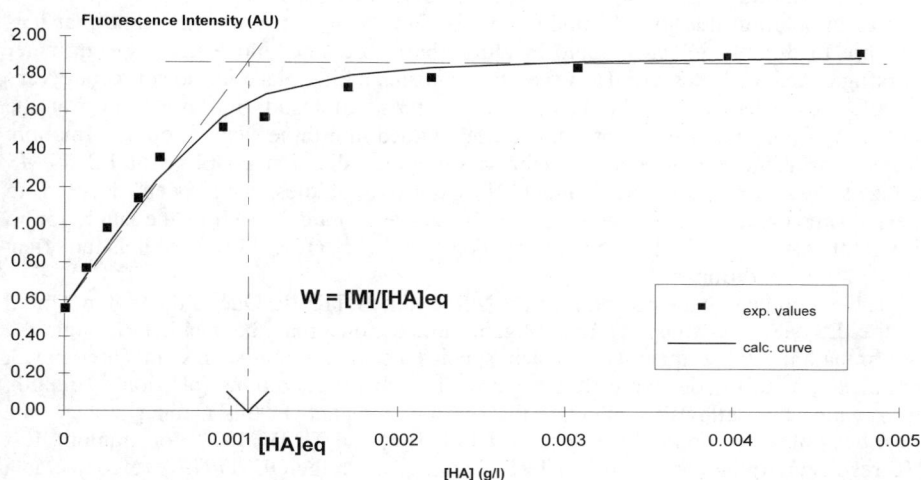

Figure 3. Typical titration curve obtained for a system M-HA.

Table 2. Experimental conditions used for TRLIF measurements.

TRLIF	Eu(III)	Cm(III)	Dy(III)	U(VI)
$\lambda_{ex,}$ nm	266 (Nd-YAG laser) or OPO at 395	337 (N_2 laser) or 355 (Nd-YAG laser)	355 (Nd-YAG laser)	266 (Nd-YAG laser)
$\lambda_{em,}$ nm	593-617	601	576	514
[M], μM	1-70	0.03-10	2	0.04-0.4

Table 3. Experimental conditions used for SP measurements.

SP conditions	Am(III)	Np(V)
λ_{abs} for M_{free}, nm	503.1	981
λ_{abs} for M_{bound}, nm	504.5-505	991
[M], μM	30	60

3.2. ELECTROSPRAY MASS SPECTROMETRY

Complementary to the above photon-based methods, new ion detection methods such as electrospray-mass spectrometry (ES-MS) are very promising for speciation studies. Initially developed for biological applications, since it allows to analyse very high mass by making multicharged species, ES-MS is also very suitable for direct speciation studies in solution due to soft mode of ionisation. Hence, it is the first time that it is possible to directly inject a liquid at atmospheric pressure into a mass spectrometer working at reduced pressure. However, mechanisms taking place in the source are very complex and special care should be taken in terms of results interpretation. For the purposes of convenience, the process can be divided into three steps: droplet formation, droplet shrinkage and gaseous ion formation. Since the first coupling of ES to MS realized by Yamashita and Fenn [32], extensive studies have been devoted to mechanisms taking place in the source with the use of models such as the single ion in droplet theory (SIDT)[33] and the ion evaporation theory (IET) [34] as well as the effect of operating conditions [35].

Table 4 compares the potential of ES-MS to that of TRLIF. Due to the high potential of the ES-MS, the study of metal-ligand interactions may be tackled through the determination of the quantity of each species occurring in the system (free metal, complexed metal) under the different physical-chemical conditions (pH, ionic strength, etc.), which then allows to calculate the stability constant. In this manner, recent data have been obtained on the hydrolysis of U(VI) [36] and Th(IV) [37]. For uranium, ES-MS results are quite comparable to TRLIF data given in the OECD/NEA reference data base for U hydrolysis. In the case of thorium, it has been possible to observe the apparition of the first thorium hydroxo-complexes at pH 1 as well as the other species at higher pH. Complexing constants have been determined and are in fair agreement with literature data. Moreover, ES-MS has also been applied for complexation studies with organic molecules [38]. The feasibility of these studies was well demonstrated on the example of Eu(III)-glycolic acid system (as a model of functional groups occurring in HS) as detailed in [27].

Table 4. Comparison of spectrometric methods with regard to speciation characteristics.

	Oxidation State	Chemical species	Quantitative analysis	Sensitivity	In situ analysis
TRLIF	yes	yes	yes	very sensitive	yes
ES-MS	yes	yes	yes	yes	no

4. Formation of Organic Complexes

The complexation of trace metals with HS modifies metal speciation and solubility in dependence on both the HS concentration and their binding strength [9-11, 13, 39-40]. Many studies have examined complexation reactions of radioisotopes with HS in a pH range in which no competing reactions exist (hydrolysis or carbonate complexation).

The characterization of these simple complexes formed between HS and actinide ions has been mainly performed by means of spectroscopic methods (TRLIF or SP) for M(III), M(V), M(VI) actinide ions in the case of Am, Cm, Np, and U, and for lanthanide ions (Dy, Eu, and Tb). At the same time for tetravalent elements such as Th (which is an analogue for other tetravalent actinides such as Pu or Np), not spectroscopic, but separation method has been used, based on the competition between the organic ligand and a colloidal phase.

The characterization of ternary complexes (or mixed complexes such as M-L-L', where L can be hydroxide or carbonate, and L' can be organic ligands (HA or FA) has been studied by TRLIF for Eu(III) and U(VI).

4.1. COMPLEXATION OF TRIVALENT ELEMENTS

In the case of trivalent elements (Eu^{3+}, Am^{3+}, Dy^{3+}, Cm^{3+}), the use of TRLIF has allowed determination of apparent interaction constants through titration of the metal solution by the organic ligand, as descried in details in [5, 7-9, 21-22].

The studies on the influence of various parameters, in particular pH and metal concentration, showed no effect of pH on the interaction constants in the pH range where no hydrolysis occurred, whereas the metal concentration influenced greatly the interaction constants, in particular, in the case of trivalent ions (Figure 4). This effect (observed for Tb, Eu, Am, and Cm [5, 7, 41]) is attributed to the presence of strong and weak sites in humic macromolecule. Different sites have also been revealed by TRLIF in the case of Eu(III) [42]. Nevertheless, the influence of metal concentration remains a controversial question, because no similar effect was observed in the studies of Kim and Czerwinski [12]. This controversy is discussed in [43].

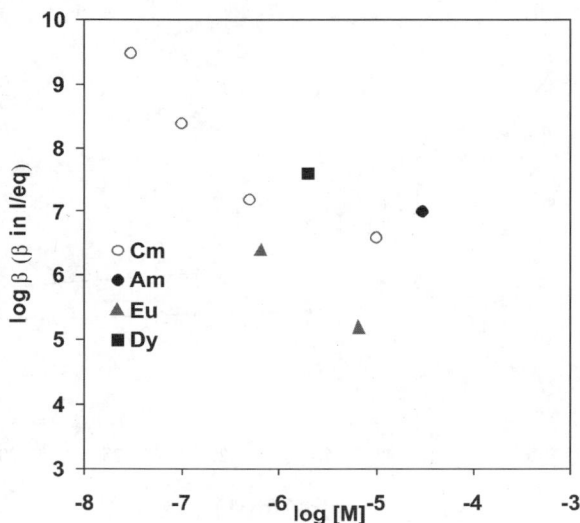

Figure 4. Effect of M concentration on log β for the systems constituted by humic acid and trivalent elements (actinides/lanthanides) – I = 0.1 M NaClO$_4$.

Moreover, through the analysis of lifetimes obtained by TRLIF, it has been possible to characterize the formation of outer-sphere complexes formed between Eu^{3+} and HA [27-28].

In order to tackle the study of ternary complexes, as mentioned in Section 4.1, an approach has been developed based on the acquisition of a spectrum data-base of the inorganic and organic species, including spectral and temporal characteristics. Then the global system is studied, which has allowed in the case of trivalent elements, particularly Eu(III), the direct evidence by TRLIF, of the formation of mixed carbonate-humate complexes [29], the formation kinetics of which is rapid. The formation of such complexes have already been proposed with the associated apparent interaction constants, but without direct evidence [45] or by spectral deconvolution for Cm(III) [44]. In this later case, hydroxo-humate complexes have also been studied.

4.2. COMPLEXATION OF TETRAVALENT ELEMENTS

In the case of actinides at the tetravalent oxidation state, which is the major one along with the trivalent oxidation state under reducing conditions, Th(IV) is the most studied element as an analogue of tetravalent actinides. Because the Th chemistry is relatively complex (strong hydrolysis, low solubility, easy formation of polynuclear species or colloidal species), leading to an extremely narrow domain of free thorium ions, the formation of mixed or ternary species should be seriously taken into consideration when organic ligands such as HS are present in the reaction system.

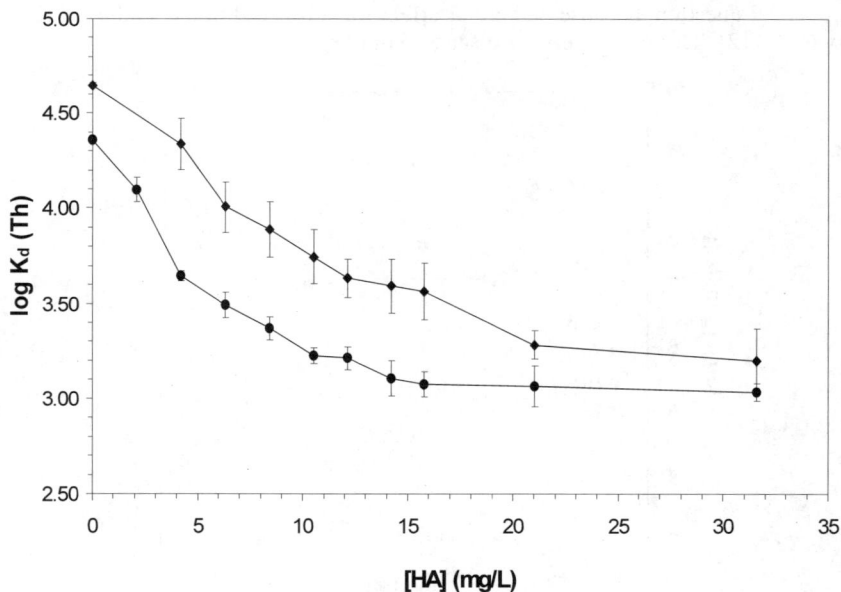

Figure 5. Dependence of Th(IV) distribution coefficient (log K_d, mL/g) on Aldrich HA concentration at pH 7 (dots) and 8 (squares). $[SiO_2]$=250 mg/l, $[Th]$=10^{-12} M, $[NaClO_4]$=0.1 M.

Studies on Th(IV) complexation were performed using an extraction technique based on a competition between the organic ligand and silica colloids. They have shown that HA act as strong competing ligand for Th. Figure 5 shows the effect of HA on the distribution coefficient of Th in the system silica colloids-HA-Th at pH 7 and 8. The retention of Th decreases strongly with an increase in HA concentration indicating the formation of organic species that govern Th behaviour [46]. Global interaction constants can be determined from experimental data obtained at various pH in the range from 6 to 8 (Figure 5) [47]. These interaction constants are very high, indicating very strong organic complexes with Th(IV). The formation of hydroxo-humates of Th(IV) is assumed in [46-47]. Figure 6 illustrates the effect of formation of such organic complexes on Th speciation as a function of HA concentration.

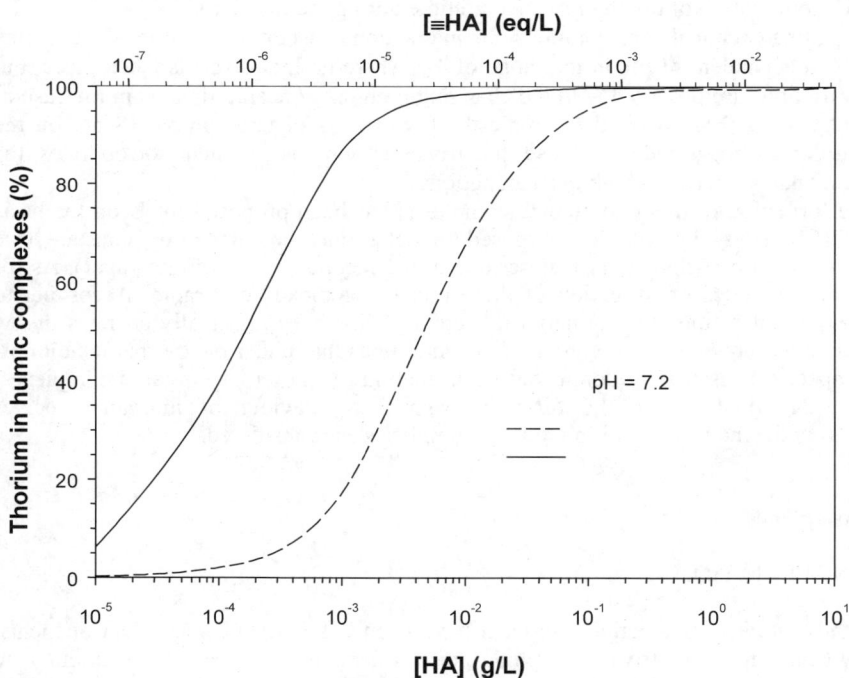

Figure 6. Percentage of Th(IV) in HA complexes as a function of HA concentration at pH 7.2: solid line is given for log β 17.3 [47], dotted line – for log β 15.8 [13] (same Th hydrolysis data set), (β in L/eq).

4.3. COMPLEXATION OF PENTAVALENT ACTINIDES

Neptunium is chosen as a model for pentavalent actinides. The study on this element present in solution as the neptunyl ion NpO_2^+ has been performed by spectrophotometry [8, 20, 48]. From spectroscopic data, the presence of one isobestic point indicates the existence of a single type of complex with the HS ligand and under our experimental conditions (at the laboratory time scale), no reduction of Np(V) to Np(IV) has been observed. The magnitude of conditional interaction constant values obtained from the titration curves (log β = 4.6 with β expressed in L/eq) shows a relatively low affinity of the neptunyl cation for HA, as it could be expected from the low charge of the ion.

4.4. COMPLEXATION OF HEXAVALENT ACTINIDES

The main features arising from the fluorescence study [7, 22, 26] on the system uranyl ion-HS, are the following: no reduction of U(VI) is induced by HA under the experimental conditions (pH 4-5, ionic strength of 0.1 M), but the interaction of the uranyl ion with HA induces a static quenching (which explains a decrease of the signal with no modification of the fluorescence lifetime during the titration).

Regarding chemical considerations, the interaction constant obtained for the complex U-HA is independent of pH in the range of 4-5 where no hydrolysis is taken place, but some variations are observed with the changing uranium concentration, as in the case of trivalent cations. Moreover, the complexation constants of uranium to HS are on the same order of magnitude as those for trivalent actinides, which corroborates the chemical analogy between both types of cations.

The formation of hydroxo-humate complexes has been proposed for U on the basis of a TRLIF study [7, 22, 26] purposed to determine the interaction constant (the principle used for such a study is described in the second section of this paper). As for Eu(III), the kinetics of formation of these mixed complexes was rapid. According to these important results, the formation of ternary complexes drastically changes the M speciation, as show below (Figure 7). This underlines that under neutral pH conditions, U(VI) speciation is mainly dominated by formation of the ternary organic complexes, the fate of which should be different from the behaviour of inorganic species constituting the main species if no ternary complexes are considered.

5. Consequences

5.1. ON SPECIATION

From these studies, interaction constants have been determined for trivalent elements, namely lanthanides, and trivalent actinides, pentavalent elements, namely neptunium and hexavalent elements namely uranium according to a model retained for the description of the organic/metal system. The model describes metal-humate interactions, namely 1:1 stoechiometry, ligand assumed to be a site, equivalent sites.

These studies lead to the following affinity order of actinides (An) for humic acid ligands: An(V) < An(VI) ~ An(III) < An(IV)).

Table 5 summarizes the apparent interaction constants determined by the techniques presented in this paper. It should be noticed that the kinetics of formation of these complexes is rapid under our experimental conditions.

From the interaction constants here determined, sensitivity studies can be performed underlining the conditions under which HS may have a strong impact on M speciation.

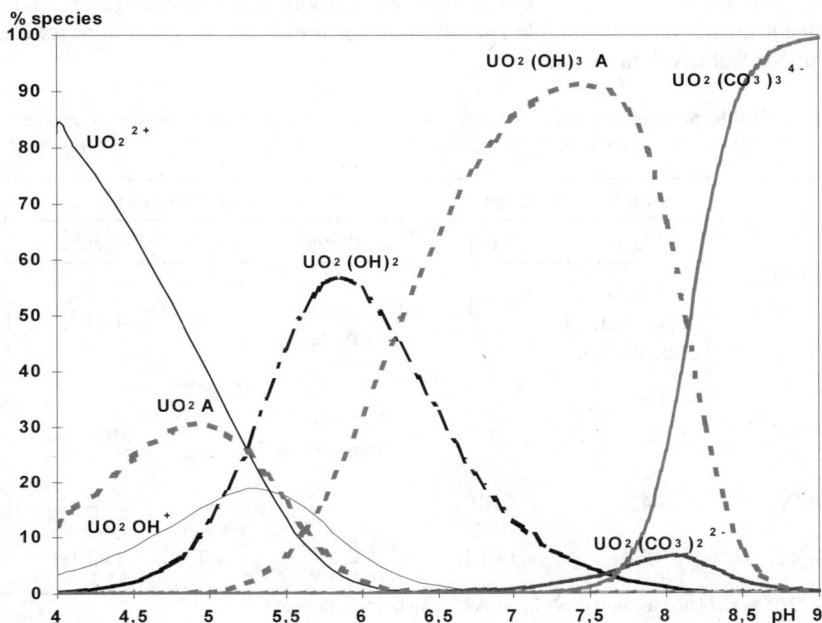

Figure 7. Impact of ternary complexes on U(VI) speciation ([U]=1 mg/L) in the presence of HA (5 mg/L), ionic strength 0.1 M $NaClO_4$, $pCO_2=10^{-3.5}$ atm; log β used for calculations:

Hydrolysis reactions
$$UO_2^{2+} + nH_2O \Leftrightarrow UO_2(OH)_n^{(2-n)} + nH^+$$
with n=1 to 3 ; log β_n = -5.2 (n=1), -10.3 (n=2), -19.2 (n=3)
$$pUO_2^{2+} + rH_2O \Leftrightarrow (UO_2)_p(OH)_r^{(2p-r)} + rH^+$$
log $\beta_{p,r}$ = -5.62 (p=2, r=2), -15.55 (p=3, r=5)
Complexation with carbonate ions
$$UO_2^{2+} + mCO_3^= \Leftrightarrow UO_2(CO_3)_m^{(2-2m)}$$
with m= 1 to 3 ; log β_m = 9.7 (m=1), 16.9 (m=2), 21.6 (m=3)
Complexation with HA (for simplification, charges are omitted)
$$UO_2 + A \Leftrightarrow UO_2A \text{ log } \beta = 5.4$$
$$UO_2(OH)_3 + A \Leftrightarrow UO_2(OH)_3A \text{ log } \beta = 6.7$$

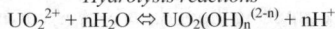

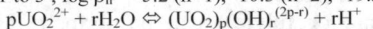

In general, trivalent elements are relatively sensitive to the presence of HS even at low concentrations (less than 1 mg/L), leading to the predominance of organic complexes [7-10]. The study [10] indicates which radionuclides among the actinides (Th, Pa, U, Np, Pu, Am, and Cm), and activation or fission products (Ni, Co, and Cd), can have their speciation modified by the presence of humic ligands, and for which conditions (pH, presence of competing actions, humic acids concentration range) HS may govern their speciation under natural water conditions (6 < pH < 9, reducing conditions, presence of competing cations). A large part of this study [10] is also devoted to the role of competing cations such as Ca and Mg in complexation behaviour of trace elements toward humic ligands. On the contrary, in another study [11] under specific conditions of Swiss granitic deep waters (with an imposed carbon content, namely, few mg/L as a maximum), similar calculations have shown that HS have no significant impact on radionuclide speciation under these specific conditions, except for lanthanides and trivalent actinides.

Table 5. Apparent interaction constants obtained for different M-humic acids systems (β in L/eq; ionic strength of 0.1 M).

	Simple complexes		Mixed complexes			Ref
	log β	Tech.	evidence	log β	Tech.	
An(III) with An = Dy, Eu, Am, Cm	5 to 9 (depend on M concentration)	TRLIF, SP	Carbonate-humate	–	TRLIF	[7] [29] [48]
Th(IV)			Hydroxo-humate		extraction	[47]
Np(V)	4.6	SP	–	–	–	[48]
U(VI)	5.4	TRLIF	Hydroxo-humate	6.7	TRLIF	[7]

* calculated using Baes & Messmer Th hydrolysis data set.

Most of the above experimental works did not take into consideration competing reactions of hydrolysis or of carbonate complexation. The question that is emerging is the validity of extrapolating the interaction constants at higher pH (pH range of natural waters 6 <pH < 9) where competing ligands are present and where ternary complexes (or mixed complexes such as $M-L_1-L_2$, where L_1 can be hydroxide or carbonate, L_2 can be organic ligands (humic or fulvic)) can be formed. Thus the speciation can be completely modified by the presence of such species. They are not considered as a general manner in the sensitivity studies mentioned above (except in [10]). Hence, an important question to be addressed is the possible existence of such complexes in the system of humic acids/hydroxide ions/carbonate ions (representing the most important ligands present in deep waters, for example) in the case of actinides.

The above mentioned indications on formation of missed complexes are summarized below:
- the formation of carbonate-humate complexes has been observed spectroscopically by TRLIF in the case of Eu [29],
- interaction constants for carbonate-humate in the case of Cm(III) and Eu(III) have been determined [44, 45] as well as for hydroxo-humate of Cm(III) [44],
- the formation of hydroxo-humate complexes has been proposed in the case of U basing on a TRLIF study [7, 22, 26],
- the formation of hydroxo-humate are also assumed in the case of Th(IV).

From the above results follows that the formation of ternary complexes drastically changes the M speciation, as pointed out in the case of uranium in Figure 7.

In the present paper, the effects of HS on actinide speciation have been discussed for various elements except for plutonium. This element has been particularly studied by

Choppin and co-workers [49, 50, and references therein]. Plutonium exists at various oxidation states (III, IV, V, and VI). According to the aquatic chemistry, it will be mainly present as Pu(III) and Pu(IV) in anoxic waters, whereas under oxic water conditions, Pu(V) will be the major species. In the presence of HS, reduction and complexation reactions can occur: in oxic waters, reduction of Pu(V) to Pu(IV) is observed, and hydrolysis and complexation take place. André and Choppin [50] have recently shown that this reduction (tracer concentration of Pu) occurs for low HA concentration (at pH values of 6 and 8, for brine solutions). Moreover, an enhancement of Pu(V) reduction is observed in the presence of light and divalent cations, compared to experiments conducted in the dark without divalent cations [50]. Pu reduction to the III- and IV-valent states has also been observed under estuarine conditions in the presence of colloidal and dissolved species of HA [51]. For anoxic waters, complexation by HS will be the predominant phenomenon for both Pu(III) and Pu(IV). In fact, even if the complexation data (log β) on the systems Pu(III) and Pu(IV)-HA are very scarce, chemical analogies with lanthanide or trivalent actinides, and Th(IV), respectively, can be made allowing then speciation calculations. For these oxidation states, organic complexes are very strong as shown in the previous sections.

5.2. ON RETENTION

In the presence of mineral surfaces, HS can retard trace element migration when fixed on mineral surfaces or by filtration into the porous medium. Hence, the behaviour of HS toward mineral phases is important, and particularly the conditions of formation of an organic film, as well as the mechanisms governing the formation of this coating should be known [5, 7-8]. The knowledge of the functional groups involved in such processes is quite important because they govern the complexing behaviour of the free residual sites and, hence, they influence the behaviour of trace elements in solution. The strong influence of HS on the sorption of tri- and tetravalent actinides, and to a lesser extent, on hexavalent actinides onto oxides has been shown in the case of silica [52, 53] and iron oxides [54], implying that each binary system (radionuclide/surface and humic acids/surface) has been thoroughly characterized. The determination of interaction constants relevant to the systems using surface complexation models has allowed us to describe systems independently from the systems being used for their quantification [52, 54]. From the ternary system study, it appears that HA may increase radionuclide availability because they are oversaturated compared to the surface capacity and organic complexes are of high binding strength. On the contrary, if conditions favour the formation of an organic film, it could induce a higher retention of radionuclides. In such systems, the sequence order of constituents is of primary importance in order to tackle the reversibility aspect of such interactions.

6. Reactivity of humic substances toward anions

If the reactivity of HS toward cations is implicit, their reactivity toward anions should also be taken into account, although the mechanisms are somewhat different.
The presence of relatively high contents of iodine in HS extracted from various environments (up to some % mass) [55] has led to more precise study of the reactivity of

HS toward iodine directed on elucidation of the binding mechanism, the functional groups involved, and the type of bonding. For this purpose, various analytical techniques (photon electron spectroscopy (XPS) and ES-MS) have been applied [5].

Information on the chemical environment of iodine in HS has been obtained by XPS by comparing the characteristic signal, namely, the bonding energy (BE) of the core electrons ejected from I, for a series of iodinated model samples (with ionic or covalent bonding) with the signal obtained on natural HS enriched with iodine. Using this approach [55, 56], it has been possible to show that iodine in HS does not present as an anion (iodide or iodate), but is associated probably through covalent bonding.

Because the technique used does not allow to elucidate the nature of the functional groups involved in the binding, electrospray with quadrupole/time-of-flight mass spectrometry (Q-TOF) has been used to provide more details on the functional groups involved as well as on the mechanism of iodine bonding to HS. The study of iodinated FA (reaction of a I_2-I^- mixture with FA) by MS-MS has showed an enrichment of a certain population compared to the reference sample due to the scavenging of iodine by FA, and the key result is a similar structure of the iodinated FA showing that iodine is covalently bound to aromatic moieties [57]. Further investigations are in progress directed toward deeper understanding of the mechanism involved in iodine reactivity toward HA and FA (role of oxidants and light), as well as on the kinetics of reactivity.

7. Conclusions and Perspectives

HS are ubiquitous in various environments and may strongly affect radionuclide speciation due to the formation of simple or mixed complexes, and, hence, may increase their solubility. This is of particular importance in relation to environmental threat of radionuclides. Speciation is directly connected to the fate of a toxic element (predicted through modelling) in the biosphere, that is why the role of these natural ligands whose structure has been recently elucidated by direct measurements, could not be neglected in speciation codes or in geochemical modelling. Further investigations are needed to reach deeper understanding of the role of HS. Among them are the studies on the role of competing natural cations, on kinetics and reversibility, and the study of redox properties as well as the properties of adsorbed HS.

The implication for bioremediation technologies appears to be straightforward because the characterization of the system is necessary before remediation. The knowledge of the speciation of the elements in the polluted environment will contribute to the choice and applications of the selected technologies.

8. Acknowledgments

Drs. Amekraz (CEA), Mercier (UMR CEA-CNRS-UEVE 8587), Plancque (CEA), Reiller (CEA), and Toulhoat (CEA) are acknowledged for the participation in these works, as well as N. Barré (UMR CEA-CNRS-UEVE 8587), Dr Laszak and Pr. Moulin (CEA). The presented results are a part of EU projects, as well as of a CEA project 98T13 "Transport, transfer of radionuclides in the life".

9. References

1. Tessier, A., Turner, D.R. (1995) *Metal Speciation and Bioavailability in Aquatic systems*, John Wiley and Sons, Chichester.
2. *Contamination des Sols par les éléments en traces : les risques et leur gestion.* Rapport N°42 (Août 1998). Académie des Sciences (Lavoisier TECDOC).
3. Templeton, D.M., Freek, A., Cornelis, R., Danielsson, L. , Muntau, H., van Leeuwen, H., and Lobinski, R. (2000) Guidelines for terms related to chemical speciation and fractionation of elements. Definitions, structural aspects and methodological approaches, *Pure Appl. Chem.* **72**, 1453-1470.
4. Plancque, G., Amekraz, B., Moulin, V., Toulhoat, P., and Moulin, C. (2001) Molecular structure of fulvic acids by electrospray with quadrupole/time-of-flight mass spectrometry, *Rapid Communications in Mass Spectrometry* **15 (10),** 827-835.
5. Moulin, V., Amekraz, B., Barre, N., Plancque, G., Mercier, F., Reiller, P., and Moulin, C. (in press) The role of humic substances on trace element mobility in natural environments and applications for radionuclides, in E. A. Ghabbour and G. Davies (eds.), *Humic Substances: Nature's Most Versatile Materials*, Taylor and Francis, Inc. New York.
6. Conte, P., and Piccolo, A. (1999) Conformational arrangement of dissolved humic substances. Influence of solution composition on association of humic molecules, *Environ. Sci. Techn.* **33**, 1682-1690.
7. Moulin, V., and Moulin, C. (2001) Radionuclide speciation in the environment: a review, *Radiochim. Acta*, **89**, 773-778.
8. Moulin, V., and Moulin, C. (1995) Fate of actinides in the presence of humic substances under conditions relevant to nuclear waste disposal, *Applied Geochem.* **10**, 573-580.
9. Buckau, G. (ed.) (2000) *Effects of Humic Substances on the Migration of Radionuclides: Complexation and Transport of Actinides*, EUR Report 19610 EN.
10. Reiller, P., Moulin, V., and Giffaut, E. (in press) On the influence of humic substances upon radionuclide speciation. A sensitivity study. *Applied Geochem.*
11. Hummel, W., Glaus, M.A., and van Loon, L.R. (2000) Trace metal-humate interactions. II. The conservative roof model and its application, *Applied Geochem.* **15**, 975-1001.
12. Kim, J.I., and Czerwinski, K. (1996) Complexation of metal ions with humic acid: metal ion neutralization model, *Radiochim. Acta* **73**, 5-10.
13. Choppin, G.R. and Allard, B. (1985) Complexes of actinides with naturally occurring organic coumpounds, in A.J. Freeman and C. Keller (eds.), *Handbook on the Physics and Chemistry of the Actinides*, Chapter 11, Elsevier, Amsterdam.
14. Choppin, G.R. and Labonne-Wall, N. (1997) Comparison of two models for metal-humic interactions, *J. Radioanal. Nucl. Chem.* **221**, 67-71.
15. Kinniburgh, D.G., van Riemsdijk, W.H., Koopal, L.K., Borkovec, M., Benedetti, M.F. and Avena, M.J. (1999) Ion binding to natural organic matter: competition, heterogeneity, stoichiometry and thermodynamic consistency, *Colloids Surf. A* **151**, 147-166.
16. Tipping, E., and Hurley, M. A. (1992) A unifying model of cation binding by humic substances, *Geochim. Cosmochim. Acta* **56**, 3627-3641.
17. Moulin, V., Tits, J. and Ouzounian, G. (1992) Actinide speciation in the presence of humic substances in natural water conditions, *Radiochim. Acta* **58/59**, 179-190.
18. Buffle, J. (1988) *Complexation Reactions in Aquatic Systems,* Wiley & Son, New-York.
19. OECD Proceedings of the Workshop on *Evaluation of Speciation Technology* held in Japan/October 1999. Code 662001041P1 OECD May 2001.
20. Moulin, V., Moulin, C. and Dran, J.C. (1996) Role of humic substances and colloids on the behaviour of radiotoxic elements in relation with nuclear waste disposals: confinement or enhancement of migration? In J.S. Gaffney, N.A. Marley and S.B. Clark (eds.), *ACS Symposium Series 651 on Humic and Fulvic Acids and Organic Colloidal Materials in the Environment*, Chapter 16, ACS, Washington, pp. 259-271.
21. Plancque, G., Moulin, C., Moulin, V. and Toulhoat, P. (2000) Complexation of Eu (III) by humic substances: Eu speciation determined by time-resolved laser-induced fluorescence, In G. Buckau (ed.) *Third Technical Progress Report FZKA 6524 Report*, October, Research Center Karlsruhe.

172

22. Moulin, V., Reiller, P., Dautel, C., Plancque, G., Laszak, I. and Moulin, C. (1999) Complexation of Eu(III), Th(IV) and U(VI) by humic substances. G. Buckau (ed.), *Second Technical Progress Report FZKA 6324 Report*, June, Research Center Karlsruhe.
23. Moulin, C. (2001) Speciation, from photon to ion detection. Experience in CEA, Speciation from photon to ion detection. In G.R. Choppin, J. Fuger, and Z. Yoshida (eds.), *Proceedings of a Workshop on Evaluation of Speciation technology*, (JAERI, Nov 99) code 662001041P1 OECD May.
24. Moulin, C., Mauchien, P. and Decambox, P. (1991) Analytical applications of Time-Resolved Laser-Induced Fluorescence in the nuclear fuel cycle, *Journal de Physique* 1, Sup. 3, C7, 677-680 .
25. Moulin, C., Decambox, P. and Mauchien, P. (1997) State of the art in TRLIF for actinides analysis: applications and trends, *J. Radioanalyt. Nucl.* **226**, 135.
26. Laszak, I. (1997) Etude des interactions entre colloïdes naturels et éléments radiotoxiques par Spectrofluorimïtrie Laser a Rïsolution Temporelle. *Etude Spectroscopique et chimique.* Thèse de l'Universitï Pierre et Marie Curie-Paris VI, France.
27. Plancque, G. (2001) Etude des interactions entre colloïdes organiques et polluants inorganiques: structure et réactivitï des substances humiques. Thèse de l'Universitï d'Evry, France.
28. Plancque, G., Moulin, V., Toulhoat, P. and Moulin, C. (2003) Europium speciation by time-resolved laser induced fluorescence, *Anal. Chim. Acta* **478**, 11-22.
29. Moulin, C., Wei, J., Van Iseghem, P., Laszak, I., Plancque, G. and Moulin, V. (1999) Europium complexes investigations in natural waters by time-resolved laser-induced fluorescence, *Anal. Chim. Acta*, **396**, 253-261.
30. Moulin, C., Laszak, I., Moulin, V. and Tondre, C. (1998) Time-resolved laser-induced fluorescence as a unique tool for low-level uranium speciation, *Appl. Spectroscopy* **52**, 528.
31. Moulin, C., Decambox, P., Moulin, V. and Decaillon, J.G. (1995) Uranium speciation in solution by time-resolved laser-induced fluorescence, *Anal. Chem.* **34**, 348-353.
32. Yamashita, M. and Fenn, J.B. (1984) Negative ion production with the electrospray ion source, *J. Phys. Chem.* **88**, 4671-4675.
33. Dole, M., Mack, L. L., Hines, R. L., Mobley, R. C., Ferguson, L. D. and Alice, M. B. (1968) Molecular beams of macroions, *J. Chem. Phys.* **49**, 2240-2249.
34. Thomson, B.A. and Iribarne, J.V. (1979) Field-induced ion evaporation from liquid surfaces at atmospheric pressure, *J. Chem. Phys.* **71**, 4451-4463.
35. Wang, G. and Cole, R.B. (1997) *Electrospray Ionization Mass Spectrometry*, J. Wiley&Sons, New-York, 137-174.
36. Moulin, C., Charron, N., Plancque, G. and Virelizier, H. (2000) Speciation of uranium by ES-MS: comparison with TRLIF, *Appl. Spectroscopy* **54**, 843.
37. Moulin, C., Amekraz, B., Hubert, S. and Moulin, V. (2001) Speciation of thorium hydrolysis species by ES-MS, *Anal. Chim. Acta* **441**, 269-279.
38. Lamouroux, C., Moulin, C., Tabet, J.C. and Jankowski, C. (2000) Characterization of Zr complexes of interest in spent nuclear fuel reprocessing by ESI-MS, *Rapid. Comm. Mass Spectrom.* **14**, 1869.
39. Glaus, M.A., Hummel, W. and van Loon, L. (2000) Trace metal-humate interactions. I. Experimental determination of conditional stability constants, *Appl. Geochem.* **15**, 953-973.
40. Choppin, G.R. (1992) The role of natural organics in radionuclide migration in natural aquifer systems, *Radiochim. Acta* **58/59**, 113-120.
41. Bidoglio, G., Grenthe, I., Robouch, P. and Omenetto, N. (1991) Complexation of Eu and Tb with humic substances by time-resolved laser-induced fluorescence, *Talanta* **9**, 999-1003.
42. Shin, H.S. and Choppin, G.R. (1999) A Study of Eu(III)-humate complexation using Eu(III) luminescence spectroscopy, *Radiochim. Acta* **86**, 167-174.
43. Hummel, W., Glaus, M.A. and van Loon, L. R. (1999) Complexation of radionuclides with humic substances: the metal concentration effect, *Radiochim. Acta* **84**, 111-114.
44. Panak, P., Klenze, R. and Kim, J.I. (1996) A study of ternary complexes of Cm(III) with humic acid and hydroxide or carbonate in neutral pH range by time-resolved laser-induced fluorescence spectroscopy, *Radiochim. Acta* **74**, 141-146.
45. Diercks, A. (1995) Complexation of europium with humic acids – Influence of cations and competing ligands, Ph-D, Leuven University, Belgium.
46. Reiller, P., Moulin, V., Dautel, C. and Casanova, F. (2000) Complexation of Th(IV) by humic substances G. Buckau (ed.), *Third Technical Progress Report FZKA 6524*, October, Research Center Karlsruhe.

47. Reiller, P., Moulin, V., Casanova, F. and Dautel, C. (in press) On the study of Th(IV)–humic acids interactions by competition towards sorption onto silica and determination of global interaction constants, *Radiochim. Acta.*

48. Moulin, V., Tits, J., Laszak, I., Moulin, C., Decambox, P. and de Ruty, O. (1995) Complexation behaviour of actinides with humic substances studied by time-resolved laser-induced fluorescence and spectrophotometry, in J.I. Kim and G. Buckau (eds.), *Effects of Humic Substances on the Migration of Radionuclides: Complexation of Actinides with Humic Substances*, RCM 01394 (Sixth Progress Report CEC-Contract FI2W-CT91-0083), March, Institut für Radiochemie, Technische Univ. München.

49. Choppin, G.R. and Morgenstern, A. (2001) Distribution and movement of environnemental plutonium, in A. Kudo (ed.), *Plutonium in the Environment*, Elsevier ScienceLtd, Oxford.

50. Andre, C. and Choppin, G.R. (2000) Reduction of Pu(V) by humic acid, *Radiochim. Acta* **88**, 613-616.

51. Garcia, K., Boust, D., Moulin, V. Douville, E., Fourest, B. and Guillaumont R. (1996) Multiparametric investigation of the reactions of plutonium under estuarine conditions, *Radiochim. Acta*, **74**, 165-170.

52. Labonne-Wall, N., Moulin, V. and Vilarem, J.P. (1997) Retention properties of humic substances onto amorphous silica: consequences for the sorption of cations, *Radiochim. Acta* **79**, 37-49.

53. Mercier, F., Moulin, V., Barré, N., Trocellier, P. and Toulhoat, P.(2001) Study of a ternary system silica/humic acids/iodine: capabilities of the nuclear microprobe, *Nuclear Instr. Methods* **B181**, 628-633.

54. Reiller, P., Moulin, V., Casanova, F. and Dautel, C. (2002) Retention behaviour of humic substances onto mineral surfaces and consequences upon Th(IV) mobility: case of iron oxides, *Appl. Geochem.* **17**, 1551-1562.

55. Mercier, F., Moulin, V., Guittet, M.J, Barré, N., Toulhoat, N., Gautier-Soyer, M. and Toulhoat, P. (2000) Applications of different analytical techniques such as NAA, PIXE and XPS for the evidence and characterization of the humic substances/iodine associations, *Radiochim. Acta* **88**, 779-785.

56. Mercier, F., Moulin, V., Guittet, MJ., Barré, N., Gautier-Soyer, M., Trocellier, P. and Toulhoat, P. (2002) Applications of new surface analysis techniques (NMA and XPS) to humic substances, *Org. Geochem.* **33**, 247-255.

57. Moulin, V., Reiller, P., Amekraz, B. and Moulin, C. (2001) Direct characterization of covalently bound iodine to fulvic acids by electrospray mass spectrometry, *Rapid Comm. in Mass Spectrometry* **15**, 2488-2496.

Chapter 8

HUMIC ACIDS AS BARRIERS IN ACTINIDE MIGRATION IN THE ENVIRONMENT

S.N. KALMYKOV[1,2], A.P. NOVIKOV[2], A.B. KHASANOVA[1],
N.S. SCHERBINA[1], YU.A. SAPOZHNIKOV[1]

[1] *Department of Chemistry, Lomonosov Moscow State University, 119992 Moscow, Russia <stepan@radio.chem.msu.ru>*
[2] *Vernadsky Institute of Geochemistry and Analytical Chemistry, 119991 Moscow, Russia*

1. Abstract

Humic acids (HAs) and fulvic acids (FAs) are the main components of natural waters that can influence the migration ability of cations. They form stable complex compounds with actinide elements that can change their solubility. In addition, HAs could be responsible for migration of colloidal species. The role of HAs in formation of geochemical barriers in actinide migration in the geosphere is discussed on the basis of results of laboratory experiments and speciation of plutonium and neptunium in field samples collected from the regions contaminated by human-made radionuclides.

2. Introduction

The concept of deep geological repositories for disposal of nuclear wastes (NW) has been adopted in many countries, including Russia and the United States. This concept is based on multi-barrier system that includes both engineered (artificial) and natural barriers. Actinides are considered the most toxic components of NW because most of them are long-lived alpha-emitters. If engineered barriers fail to block the release of radionuclides from NW canisters, the surrounding geological environment should provide adequate isolation properties.

HAs are natural polyelectrolytes that are the major components of dissolved organic carbon (DOC). Moulin et al. [1] studied the role of HA in actinide speciation and pointed out that HAs govern actinide behaviour in natural waters of pH 7. Because actinides are hard acids, they form rather stable complexes with carboxylic and phenolic groups of HAs. Choppin [2] summarized the values of stability constants for actinide-HAs complexes, as presented in Table 1.

However, it is still questionable whether HAs can act as geochemical barriers in actinide migration in the geosphere. According to the observations of Fujikawa et al. [3], the mobile plutonium fraction in Nagasaki soil was associated with HAs. At the same time, it is known that immobilization of natural uranium in the environment is due to the presence of HAs [4].

I. V. Perminova et al. (eds.),
Use of Humic Substances to Remediate Polluted Environments: From Theory to Practice, 175–184.
© 2005 *Springer. Printed in the Netherlands.*

Table 1. Stability of actinide humates [2], T = 25°C, I = 0.1 M, α = 0.5.

Element	logβ₁	logβ₂
Am(III)	8.9 ± 0.2	12.9 ± 0.2
Th(IV)	11.8 ± 0.1	17.3 ± 0.2
Pu(IV)	9.6	18.3
Np(V)[a]	2.34 ± 0.01	
U(VI)	6.7	11.5

^{a} - from [5], T = 25°C, I = 0.1 M, α = 0.94.

In this paper the results of several model laboratory experiments and field observations on the possible role of HAs in actinide migration are discussed.

3. Laboratory Experiments

The HA that was used in laboratory experiments was obtained from Aldrich Co. (USA) and purified by several repetitive precipitations with hydrochloric acid and dissolution with sodium hydroxide as described in [5]. The ash content was less than 0.05%. The HAs sample was characterized by ^{1}H, ^{13}C-NMR and elemental analysis was also determined [5, 6]. The ^{1}H–NMR spectrum as well as proton distribution between various functional groups are presented in Fig. 1 and Table 2. The pKa value for α = 0.5 (α is the degree of deprotonation) was 4.68 as determined by potentiometric titration. The results of elemental analysis are presented in Table 2 as well.

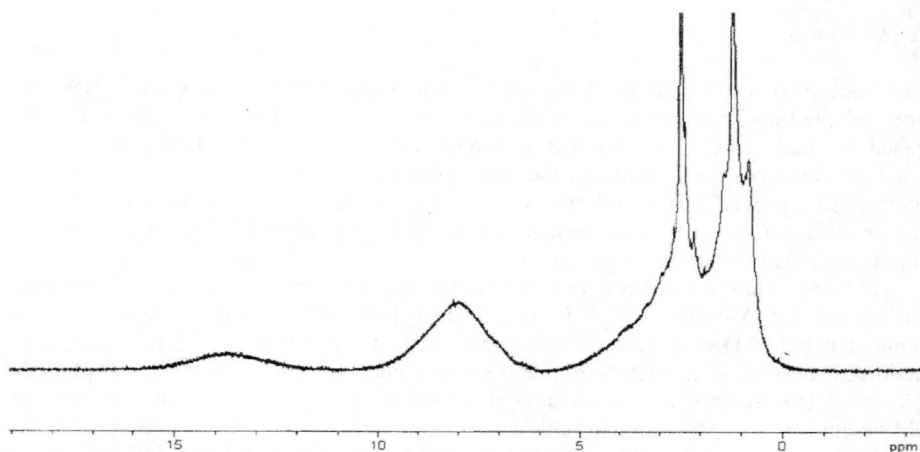

Figure 1. ^{1}H –NMR spectra of Aldrich Has (in DMSO-d6).

Table 2A. Proton distribution (in %) between various functional groups as determined by ^{1}H-NMR of purified Aldrich HAs.

COOH	Ar-OH	Ar-H	Alk-OH	O-Alk-H	a-Alk-H	Alk-H	Ar-H/ Alk-H	Ar-OH / Ar-H
13	18	3	9	0	20	37	0.273	0.17

Table 2B. Elemental composition of purified Aldrich HAs.

C	H	N	S	O
56.2	6.2	4.3	0.5	32.8

All the laboratory experiments were performed under nitrogen atmosphere to prevent carbonate complexation of actinides.

3.1. PENTAVALENT ACTINIDE REDUCTION BY HA

Pentavalent actinides possess the highest migration ability in the environment among actinides in different oxidation states. In contrast, tri-, tetra- and to a lesser extent, hexavalent actinides show high sorption affinity and low solubility. Therefore, if HAs can reduce actinides from pentavalent to tetravalent states, this will diminish their migration.

The reduction potential of HAs was estimated earlier to be 0.5-0.7 V in laboratory experiments [7], which is possibly due to the presence of hydroquinone groups. In addition, HAs undergo photolysis by sunlight with production of hydrogen peroxide. Therefore the redox properties of HAs can affect the migration of actinides in the environment. A positive correlation of Pu(III+IV) to Pu(V+VI) ratio to DOC was found for a lake environment [8]. At pH's higher than 6 Pu(VI) was reduced to Pu(IV) in carbonate solutions [9]. Andre and Choppin [10] studied the influence of sunlight, divalent cations and HAs concentration on reduction of Pu(V) by HAs. They indicated that the rate of reduction is increasing upon addition of Ca^{2+} and Mg^{2+}. However the fact data on pentavalent actinides are rather scare and laboratory experiments are necessary.

The Pu(V) and Np(V) reduction by HAs was studied in batch kinetic experiments in the dark at various pH values. The pentavalent Pu was prepared in trace concentrations following the method described by Saito et al. [11]. The total metal concentrations were $1.8 \cdot 10^{-8}$ M and $5.6 \cdot 10^{-8}$ M for Pu and Np, respectively, while the HAs concentration was 10 ppm. The ionic strength of solutions was 0.1 mol/L ($NaClO_4$). The separation of tetravalent and pentavalent states was performed by solvent extraction with tenoyl-trifluoroacetone (TTA) after acidification to pH 2. It has been shown that such a procedure enables the complete destruction of humates and quantitative separation of tetravalent actinides to TTA solution [12].

Figure 2 demonstrates the kinetics of Pu(V) reduction by HAs at various pH values. The reduction is pH dependent. The rates of reduction were determined for pH 8 and pH 9 to be 0.05 h^{-1} and 0.03 h^{-1}, respectively, and the order of reaction was close to 1. In the case of Np(V), no reduction occurred during 1000 hours. This contradicts data obtained by Artinger and co-authors [12], who found that the main mechanism of Np migration is colloid transport of tetravalent Np humates.

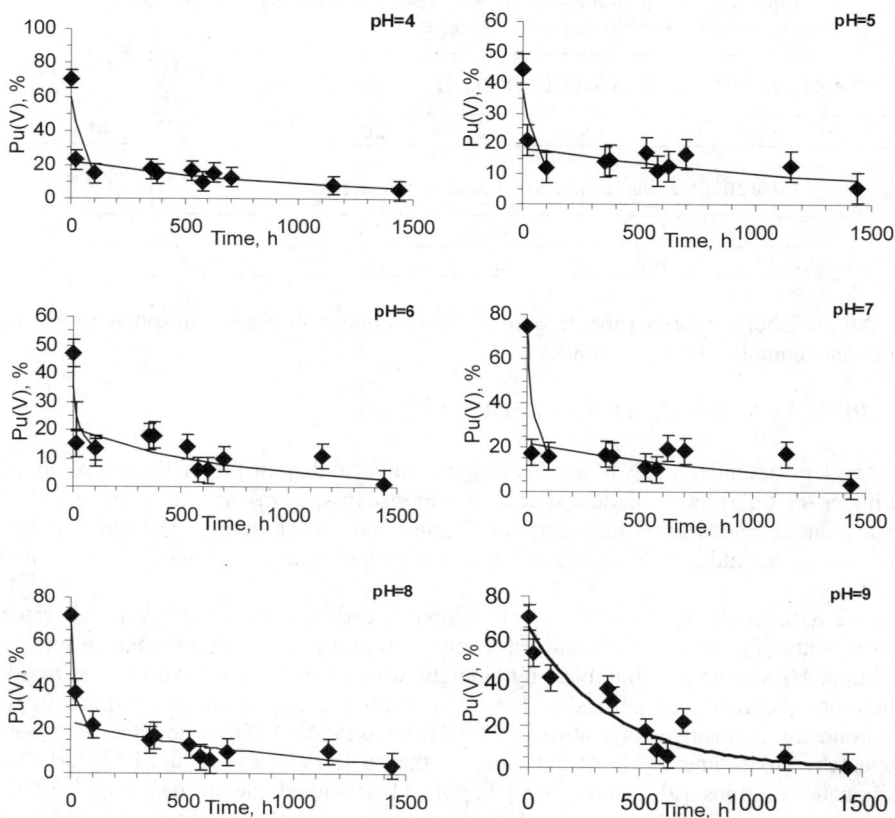

Figure 2. Kinetics of Pu(V) reduction by Aldrich HAs as a function of pH ($C_{Pu}=1.8\cdot10^{-8}$ M, $C_{HA}=10$ ppm).

3.2. SOLUBILITY OF ACTINIDES IN THE PRESENCE OF HAs

The solubility of actinide valence state analogues (Nd(III), Th(IV), and U(VI)) was studied in solutions of high ionic strength at constant pcH value of 9.5 in oversaturated mode (pcH is negative logarithm of H^+ ion concentration. At high ionic strengths it differs from pHr, which is the negative logarithm of H^+ activity in solution). The brine solutions were close to that at the Waste Isolation Pilot Plant (WIPP, New Mexico, USA) and consisted of sodium, potassium, calcium, and magnesium salts (I = 5.08 mole/L). The influence of Co(II), Ni(II), Pb(II), Cu(II), and Al(III) that can compete for binding to HA (was studied as well. Co(II) was taken as an analogue to Fe(II), and Ni(II), Pb(II), and Cu(II) could be present in nuclear wastes at WIPP. Al(III) was taken as a cation commonly present in the surrounding environment.

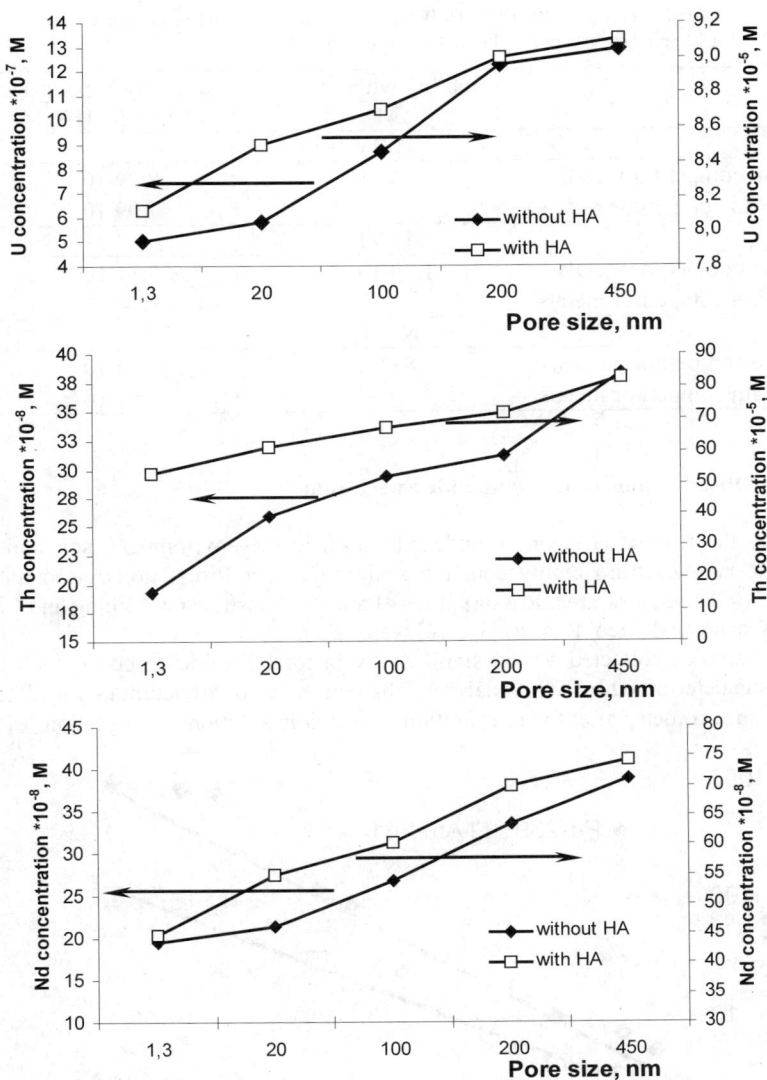

Figure 3. Size distribution of U, Th, and Nd hydroxocolloids and humic colloids.

The colloid formation was studied using different micro- and ultrafiltration (5 kD – 0.45 μm) with subsequent metal determination using ICP-MS. The solubility of Nd(III), Th(VI), and U(VI) is presented in Table 3, and the distribution of colloids according to their size is presented in Fig. 3. The increase in solubility of f-elements in the presence of HAs is well observed with less or no influence of d-metals. The size of humic colloids is significantly than the size of actinide hydroxocolloids that can increase the migration ability of actinides in the environment.

Table 3. Solubility of Nd(III), Th(IV) and U(VI) in brine solutions in the presence of 10 ppm HAs (0,45 μm filtration was applied).

	Solubility without HA, M	Solubility in presence of 10 ppm HA, M
Th(IV)		
No competitor metals	$3.83 \cdot 10^{-8}$	$8.29 \cdot 10^{-5}$
With competitor metals		$7.09 \cdot 10^{-5}$
U(VI)		
No competitor metals	$1.29 \cdot 10^{-6}$	$9.11 \cdot 10^{-5}$
With competitor metals		$9.14 \cdot 10^{-5}$
Nd(III)		
No competitor metals	$3.88 \cdot 10^{-8}$	$7.44 \cdot 10^{-8}$
With competitor metals		$7.57 \cdot 10^{-8}$

4. Field Observations of Radionuclide Speciation

To study the role of HAs on actinide migration in the environment, soil and bottom sediment samples from highly contaminated zones near Production Association (PA) "Mayak" (Chelyabinsk area, Russia) [13, 14] and the Krasnoyarsk Mining and Chemical Plant (Krasnoyarsk area, Russia) [15, 16] were taken.

The samples collected varied significantly in total organic carbon (TOC) content, which was determined by TOC-analyser. The concentration of actinides was determined using α-spectrometry after their separation from acidic solutions by digestion.

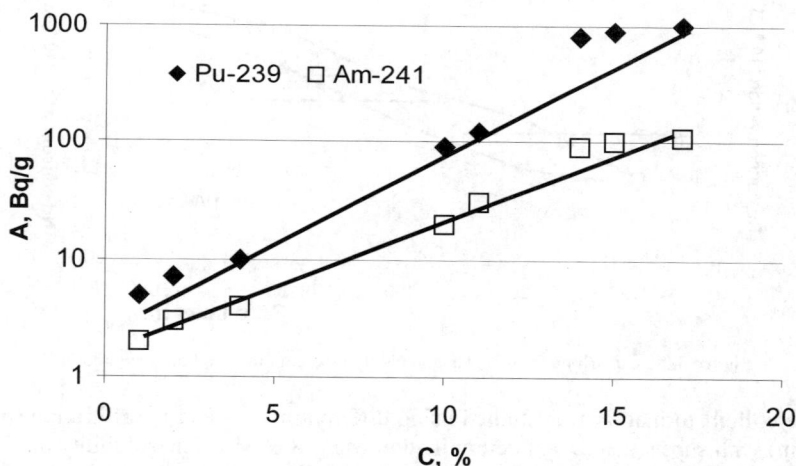

Figure 4. Correlation of ^{241}Am and ^{239}Pu content with total organic carbon concentration in bottom sediments from V-10 reservoir at PA "Mayak".

The positive correlation of Pu and Am content with TOC concentration was obtained for bottom sediments from the V-10 reservoir at Pa Mayak, as presented in Fig. 4. The higher slope for Pu dependence than for Am may have been due to significantly higher interaction constants for tetravalent actinides than for trivalent.

The correlation of migration abilities of various radionuclides (^{239}Pu, ^{90}Sr, and ^{137}Cs) with HAs concentration was studied for a variety of soil samples of high clay content from the Eastern Ural Radioactive Trace zone [14]. The vertical profiles of ^{239}Pu, ^{90}Sr, and ^{137}Cs were obtained, and migration abilities of radionuclides as presented by diffusion resistance of soils is presented in Fig. 5.

The higher the HA content, the lower was the migration ability of plutonium. The opposite dependence was observed for ^{137}Cs, which had low affinity toward HAs but high affinity toward clays. No correlation was observed for ^{90}Sr. HAs may be present as mineral coatings that prevent radionuclides from penetrating to exchange sites of clays.

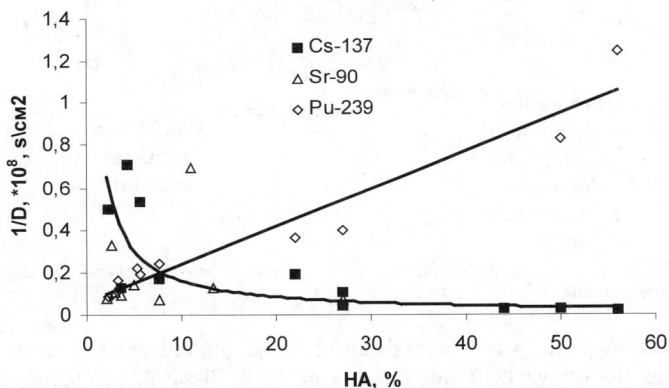

Figure 5. The dependence of diffusion resistance of various soils to ^{137}Cs, ^{90}Sr, and ^{239}Pu vertical migration upon HAs concentration.

The fractionation of organic matter from bottom sediments of the Ynisey River (Russia) was performed using the method described by Myasoedov and Novikov [13]. This multistep method is based on differences in solubility of HA and FA in acidic and basic solutions.

The fraction of FAs and low-molecular weight organic acids that is bound to mobile metal hydroxides was separated by treatment with 0.05 M sulphuric acid solutions (the sample to solution ratio was 1:5) during 4-5 hours with continuous mixing. The separation was performed by centrifuging the vessel at 3000-6000 RPM. To separate HAs and FAs that are bound to metal oxides, the sample was treated with 0.1 M NaOH solution (the sample to solution ratio was 1:20) with continuous mixing overnight. The separation of solution was also performed using centrifugation. The separation of HAs from FAs was performed by lowering the pH to 2 using sulphuric acid. After the solution turn cloudy, it was heated to 80^0C in a water bath. The HAs fraction in precipitate was collected by centrifugation, while FAs stayed in solution. DOC fraction bound to minerals was separated from the subsamples after separation of low molecular weight organics, HA, and FA by treatment with 0.02 M NaOH during 6 hours at 80°C.

182

The relative content of plutonium and americium in different fractions of natural organic matter is presented in Fig. 6. The plutonium is associated with the low mobile fraction, and americium is bound to mobile fulvic acids or low molecular weight organic substances.

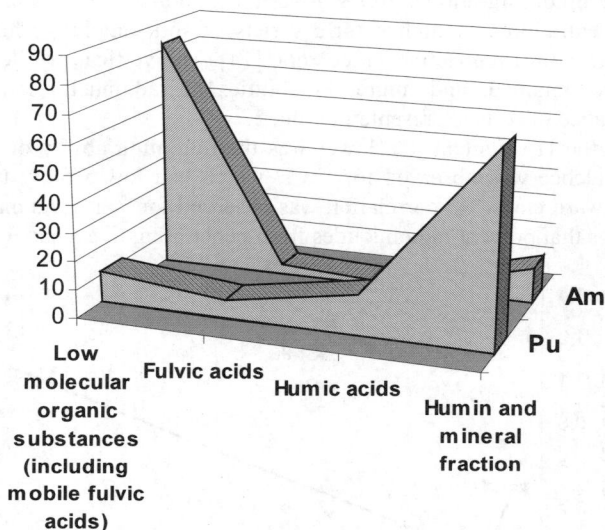

Figure 6. Relative distribution of plutonium and americium among fractions of organic matter.

This could explain why the concentration of plutonium decreased greatly with distance from Krasnayarsk Mining and Chemical Plant if compared to americium concentrations (Fig. 7).

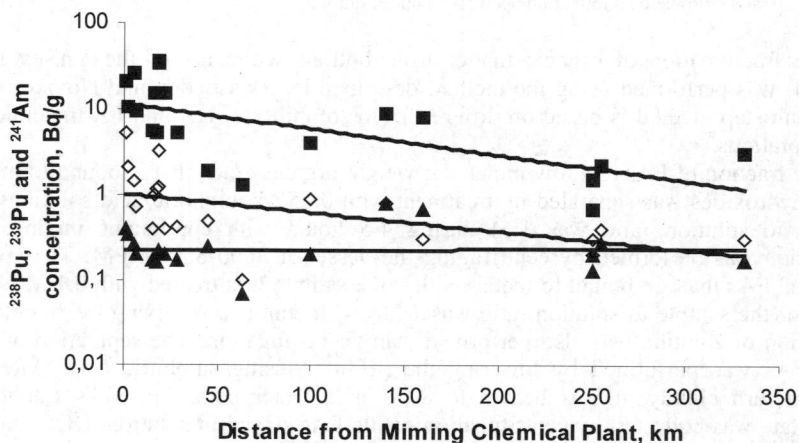

Figure 7. The concentration of ^{238}Pu (open symbols), ^{239}Pu (closed squares), and ^{241}Am (closed triangle) with distance from the mining plant.

5. Conclusions

The role of HAs in actinide migration is discussed. It is shown that HAs reduce Pu(V) to Pu(IV) with no effect for Np(V). The rate of reduction is pH dependent. However, HAs increase the solubility of valence state analogues of Pu (Nd(III), Th(IV), and U(VI)), and the size of humic colloids is relatively smaller than the size of actinide hydroxocolloids.

Data on actinide speciation in highly contaminated soils and bottom sediments from Russia is presented. It is shown that HAs and FAs play the major role in actinide migration in the environment. The strong correlation of Pu and Am activity with HAs content is observed for a number of samples. It is shown that Pu is mainly associated with the humin and mineral fraction of bottom sediments and Am is associated with low molecular weight organic substances and FAs.

6. Acknowledgments

This work was supported by the Joint U.S. Department of Energy–Russian Academy of Sciences Program (grant M6RAS0008 and grant M6RAS0006).

7. References

1. Moulin, V., Tits, J. and Ouzounian G. (1992) Actinide speciation in the presence of humic substances in natural water conditions, *Radiochim Acta* **58/59**, 179-185.
2. Choppin, G.R. (1999) Role of humics in actinide behavior in ecosystems, in G.R. Choppin and M. Kh. Khankhasayev (eds.), *Chemical Separation Technologies and Related Methods of Nuclear Waste Management*, Kluwer Academic Publishers, Dordrecht, Boston, London, pp. 247-260.
3. Fujikawa, Y., Zheng, J., Cayer, I., Sugahara, M., Takigami, H. and Kudo, A. (1999) Strong association of fallout plutonium with humic and fulvic acids as compared to uranium and ^{137}Cs in Nishiyama soil from Nagasaki, *Japan. J. Radioanal. Nucl. Chem.* **240** (1), 69-74.
4. Perelman, A.I. (1989) *Geochemistry*, Moscow, High School, Moscow (in Russian).
5. Rao, L., and Choppin, G.R. (1995) Thermodynamic study of the complexation of neptunium(V) with humic acids, *Radiochim. Acta* **69**, 87-95.
6. Kalmykov St.N., Boldesko A.S., Sapozhnikov Yu.A., Badun G.A. (2000) Migration of Np in clay minerals. Influence of humic and fulvic acids, in *Proceedings of 3rd Russian conference in Radiochemistry*, Sankt-Petersburg, p. 197.
7. Skogerboe, R.K. and Wilson, S.A. (1981) Reduction of ionic species by fulvic acids, *Anal. Chem.* **53**, 228-232.
8. Wahlgren, M.A. and Orlandini, K.A. (1982) *Environmental migration of long-lived radionuclides*, IAEA, Vienna.
9. Bondietti E.A., Reynolds S.A. and Shanks M.N. (1976) *Transuranic nuclides in the environment*, IAEA, Vienna.
10. Andre C. and Choppin G.R. (2000) Reduction of Pu(V) by humic acid, *Radiochim. Acta* **88**, 613-616.
11. Saito A., Roberts R.A. and Choppin G.R. (1985) Preparation of solution of tracer level plutonium(V), *Anal. Chem.* **57**, 390-391.
12. Artinger, R., Marquardt, C.M., Kim, J.I., Seibert, A., Trautmann, N. and Kratz, J.V. (2000) Humic colloid-borne Np migration: influence of the oxidation state, *Radiochim. Acta* **88**, 609-612.
13. Myasoedov, B.F. and Novikov A.P. (1998) Main sources of radioactive contamination in Russia and methods for their determination and speciation, *J. Radioanalyt. Nucl. Chem.*, **229** (1-2), 33-38.

14. Novikov, A.P., Pavlotskaya, F.I., Goryachenkova, T.A., Smagin, A.I., Kazinskaya, I.E., Emelyanov, V.V., Kyzovkina, E.V., Barsykova, K.B., Lavrinovich, E.A., Korovaykov, P.A., Drozko, E.G., Rovniy, S.I., Posokhov, A.K. and Myasoedov, B.F. (1998) The content and distribution of radionuclides in waters and bottom sediments from some industrial reservoirs of PA "Mayak", *Russian Radiochem.*, **40** (5) 453-461.

15. Myasoedov, B.F. and Novikov, A.P. (1999) Radiochemical procedures for speciation of actinides in the environment. Methodology and data obtained in contaminated by radionuclides regions of Russia, in *Proceedings of Speciation Workshop*, Tokai-Mura, pp. 3-21.

16. Kyznetsov, Yu.V., Legin, V.K., Strykov, I.R. and Novikov, A.P. (2000) Transuranium elements in bottom land soils from Enisey river, *Russian Radiochem.* 42 (6), 519-529.

Chapter 9

THE USE OF HUMATES FOR THE DETOXIFICATION OF SOILS CONTAMINATED WITH HEAVY METALS

O.S. BEZUGLOVA, A.V. SHESTOPALOV
Soil Science and Agrochemistry Department, Rostov State University,
B. Sadovaya St., 105, 344006 Rostov-on-Don, Russia
<bezuglov@rndavia.ru>

Abstract

Effective remediation technologies for heavy metal polluted soils have not been developed until now. The currently used techniques, such as flushing or immobilizing washing, binding heavy metals to non-toxic forms, and others, have substantial limitations: they are very costly and are applicable to only limited soil types.

The presented research was devoted to a use of brown coal to remediate soils polluted with heavy metals. Brown coal contains 46% (mass) of humic acids that can bind heavy metals into complexes. In addition, brown coal can bind heavy metals by physical absorption.

Sorption-desorption behaviour of heavy metals onto brown coal was studied in several model experiments. The sample taken from the upper horizon of zonal soil was mixed first with different forms of metals. After a month of incubation in vessels, the sample was mixed with various amounts of brown coal. In another experiment, the samples taken from the upper layers of polluted urban soils were mixed with various amounts of brown coal. These soils were put into glass vessels for long-term incubation under optimum temperature and humidity conditions. The soil samples were taken out after a week, 1 month, or 6 months. The amounts of heavy metals in mobile forms dissolved in ammonium-acetate extract with pH 4.8 were measured using atomic absorption spectrometry and the polarographic method. The results were checked under field conditions, and the results showed that brown coal is an effective ameliorant of soils polluted by heavy metals.

1. Introduction

Humic substances (HS) occupy a special niche among natural biologically active substances due to the peculiar mechanism of their formation. They are not synthesized by living organisms, but they are formed as a result of chemical-microbiological transformation of the debris of living organisms. This provides HS with a high biothermodynamic stability.

The main HS reserves are organic ores such as peat and brown coal. These humic materials can be used for production of organic fertilizers with high biological activity.

I. V. Perminova et al. (eds.),
Use of Humic Substances to Remediate Polluted Environments: From Theory to Practice, 185–200.

The latter can be used for purposes such as agriculture, biotechnology, medicine, and other fields. The humic fertilizers are a subject of growing interest today. It is caused by accumulation of data on the positive influence of HS on the growth and development of plants and the quality of crop. Another area of HS application could be their use for immobilizing heavy metals in the polluted soils. So, Piccolo et al. [20] have shown that an introduction of HS solution into the soil promoted immobilization of metals and caused a decrease in their availability to plants. The given effects depended on the stability constant of the metal-humate complex.

The problem of heavy metal pollution belongs to severe environmental problems today. The soil pollution by heavy metals is mainly local in nature. Highly contaminated soils occupy, as a rule, territories surrounding industrial enterprises and highways. In addition, elevated concentrations of heavy metals – far above the maximum permissible limit – are frequently found in soils and plants of large cities. The elevated concentrations of heavy metals influence greatly the structure and functioning of microbial communities in the polluted soils. This, in turn, has an impact of the properties of HS formed in the polluted soil. For example, we have found a greater hydrolysing ability of HS formed in the polluted soil [10]. Of particular importance is that even 6 months after the pollution incident occurs, the humus system was still not balanced [10]. In the end, this results in decreasing soil fertility.

Addition of lime-stone is the most widely used method of chemical amelioration of soils polluted by heavy metals [1, 14]. However, its application is limited to acidic soils. Binding of heavy metals into stable complexes with a use of artificial chelating agents such as ethylenediaminetetraacetic acid (EDTA) is hardly feasible because those reagents are expensive and their application can cause a secondary pollution of the soil [1]. The latter limitation is also valid for application of ion-exchange resins as an adsorbent to reduce the mobility of heavy metals [18].

The application of natural organic compounds such as decomposed leaves, manure, peat, sphagnum moss, and river sludge as ameliorants for anthropogenically polluted soils is relatively simple, universal, and promising [2]. However, if the degree of pollution is high, the binding of heavy metals by these materials can be not sufficient to provide the necessary detoxification level.

In our view, nature itself offers an inexpensive material that can be used to solve the problem. It is brown coal–the source of natural HS, which is capable of binding heavy metals to stable complexes. In addition, the brown coal, possess a high sorption affinity to heavy metals.

According to the reported studies, brown coal is ecologically pure material [13]. It contains heavy metals, but not in mobile forms, and their total quantity is insignificant. In our previous studies it was shown that brown coal improved soil fertility and optimised plant nutrition [6, 8, 9]. So, the brown coal amendment improved soil properties such as humus content and water-stability of the structure (Figures 1 and 2). The effects obtained were dependent on the dose of the fertilizer applied. Low doses of brown coal and coal-humic fertilizers provided an increase in the content of fulvic acids. High doses of brown coal resulted in a stable increase in the degree of humification of soil organic matter. The proportion of agronomically valuable fractions of organic matter increased, thus causing a more than double increase in the content of water-stable aggregates. So the more brown coal or humic fertilizer was added, the more effective was their structuring action on the soil.

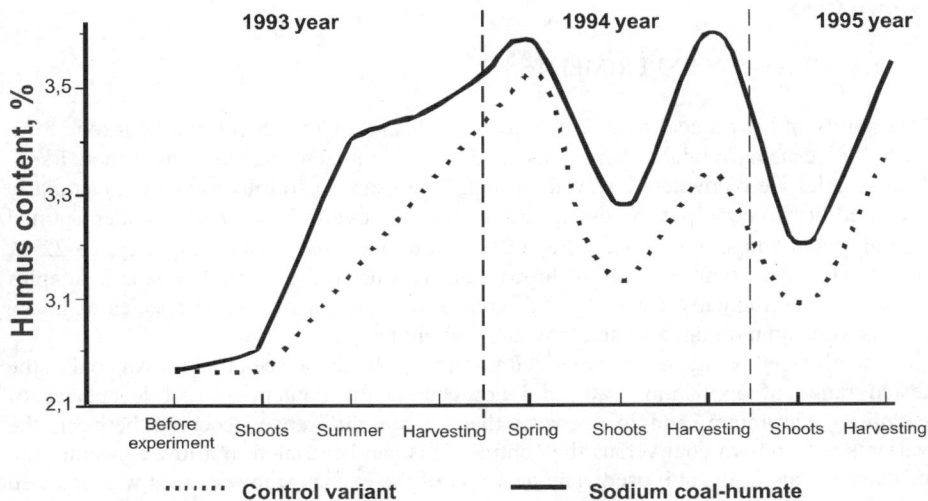

Figure 1. The effect of sodium coal-humate on humus content in ordinary chernozem.

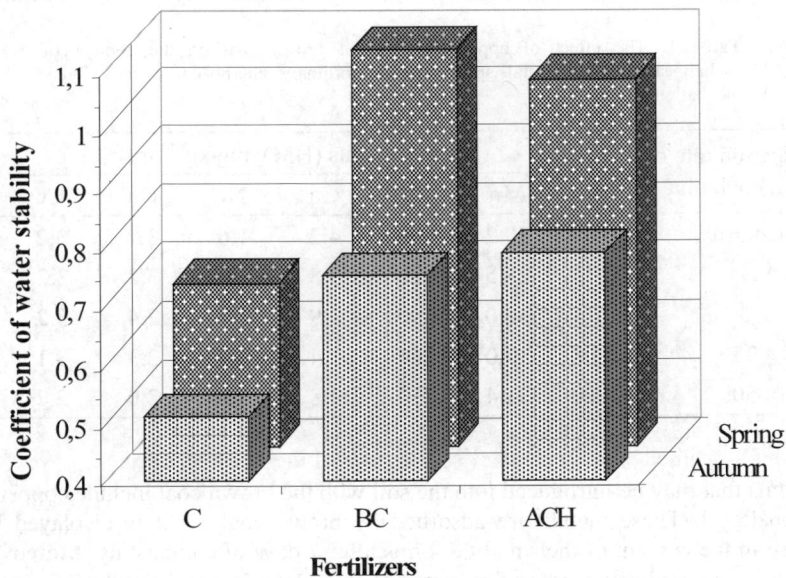

Figure 2. The effect of brown coal (BC) and ammonium coal-humate (ACH) on the water-stability of aggregates in ordinary chernozem (C – control variant).

2. Studies on the Heavy Metal Adsorption in the Polluted Soils Amended with Brown Coal

2.1. LABORATORY EXPERIMENTS

The ability of brown coal to adsorb heavy metals and to keep them in a non-toxic form was studied using model experiments. Ordinary chernozem was used as a model soil. Soil samples were powdered, sieved through 2 mm sieve, carefully mixed with similarly prepared brown coal, put in vessels, and composted over a 7 day period under optimal humidity and temperature conditions: 60% of total moisture-holding capacity, 18-22°C. The same soil without addition of brown coal was used as a control. The soil samples were taken periodically for analysis to control the content of mobile species of heavy metals using ammonium acetate extraction technique.

The obtained results are summarized in Table 1. It can be seen that brown coal in the tested range of application rates did not change the content of mobile species of cadmium, chromium, and manganese: there is no difference observed between the variants with brown coal versus the control. This can be related to a low concentration of these metals in the soil used. The decrease of the exchangeable species was detected only for cobalt in variants with high application rates of brown coal: 5, 25, and 50 t·ha^{-1}. At the same time, for manganese and nickel, increase was observed in the variant with the maximum amount of brown coal versus the control variant. The latter metals (Mn and Ni) probably got into soil from the brown coal, and although the increase is insignificant, it reminds us to be cautious with high amounts of brown coal amendment.

Table 1. The effect of application rate of brown coal on the content of exchangeable heavy metal species in the ordinary chernozem over a 7-day incubation period.

Application rate of brown coal, t·ha^{-1}	Heavy metals (HM), mg·kg^{-1} of soil							
	As	Mn	Zn	Cu	Ni	Co	Cr	Cd
Control	0.09	30.2	800	4.2	1.6	3.2	3.2	2.2
1	0.15	28.5	630	3.8	1.0	3.2	3.2	2.2
5	0.11	30.0	690	2.9	2.2	2.4	3.2	2.2
25	0.08	30.0	700	3.4	3.4	2.3	3.1	2.2
50	0.20	31.4	730	3.8	2.8	2.0	2.7	2.3

Metals that may be introduced into the soil with the brown coal include copper, zinc, and cobalt [7]. These metals are adsorbed by brown coal, as it is displayed by the decrease in the content of their mobile forms after 7 days of composting. Moreover, the increase in the application rate of brown coal from 1 to 50 t·ha^{-1} hardly influenced the quantity of adsorbed copper, zinc, and cobalt. This allows us to conclude that it is not reasonable to use high amounts of brown coal for amelioration.

Table 2. The effect of different application rates of brown coal on adsorption of zinc and copper in ordinary chernozem over a 7-day incubation period.

Application rate of brown coal, t·ha^{-1}	Content of Zn		t*	Content of Cu		t
	mg·kg^{-1}	% of control		mg·kg^{-1}	% of control	
Control	800	100	–	4.2	100	–
1	630	79	3.21	3.8	90	0.50
5	690	86	3.11	2.9	69	1.46
25	700	88	2.83	3.4	80	1.00
50	730	91	1.98	3.8	90	0.50

* t is a Student's coefficient.

The studied soil appeared to be polluted only with zinc: it exceeded the maximum permissible concentration by almost 35 times. The Zn adsorption appeared to be most effective in the variant with 1 t·ha^{-1} of brown coal (Table 2). There is a high statistical confidence (P = 0.95) in the decrease in concentration of zinc's mobile form. The maximal decrease of the concentration of mobile form of copper is found in the variant 3 (5 t·ha^{-1} of brown coal). Nevertheless, it should be noted that the concentration of the mobile species of zinc did not drop to a permissible level. A longer composting with brown coal would probably be more effective because zinc is characterized with slow reaction kinetics with organic substances in comparison with other metals. This assumption is confirmed in the further experiments. The majority of the above mentioned elements in the sampled soil are below the MAC. It is possible to hypothesize that brown coal will not immobilize microelements and negatively influence soil fertility.

In another model experiment, pollution was simulated by the addition of various forms of copper, such as oxide, acetate, and sulphate of zinc (Table 3).

In general, the obtained results are in agreement with the data of the previous experiment. Of importance is that the introduced amount of zinc acetate provided the concentration of its mobile species equal to 4 maximum allowable concentration (MAC) (100 mg·kg^{-1}). Composting of the soil without brown coal (control variant) for one month resulted in a sharp decrease in the mobile zinc content. This shows the high sorption properties of chernozem. The decrease is so significant that the ameliorating action of brown coal almost has lost its importance. The determined content of mobile zinc can be considered being below MAC (22.2 mg·kg^{-1}) because in the given soil, taking into account a background content (9 mg·kg^{-1}), the amount of zinc corresponding to 1 MAC would have been 32 mg·kg^{-1}. It is known that the mobile forms of zinc in soils are chiefly bound by hydroxides of iron and aluminium and immobilized by carbonate of limestone [17, 12]. The latter is especially important in our case because of the high concentration of calcium carbonate in this soil. However, the largest portion of zinc in the soil is present in a residual fraction, where the metals are included in a crystal lattice of primary and secondary minerals [16]. Chernozems are characterized by high adsorption capacity. After addition of salts, chernozem was composted under optimum temperature and humidity conditions for one month, that brought about immobilization of most of the zinc. Nevertheless, the results show that the 50 t·ha^{-1} brown coal treatment (variant 10) with a 1-day incubation is capable of immobilizing about one-fourth of the zinc remaining in an exchangeable form.

Table 3. Content of zinc and copper introduced as different salts in the ordinary chernozem mixed with brown coal after 1 day of composting.

№	Salt-pollutant	Content of Zn		Content of Cu	
		mg·kg^{-1}	% of control	mg·kg^{-1}	% of control
1	Control-1 (without brown coal): C1	9.0	–	0.9	–
2	Control-2 (with brown coal): C2	5.0	–	1.0	–
3	C1+Cu(CH$_3$COO)$_2$	4.6	–	20.0	100
4	C2+Cu(CH$_3$COO)$_2$	4.2	–	11.8	59
5	C1+CuSO$_4$	3.2	–	16.2	100
6	C2+CuSO$_4$	4.2	–	10.6	65
7	C1+Zn(CH$_3$COO)$_2$	22.2	100	3.6	–
8	C2+Zn(CH$_3$COO)$_2$	17.0	76.6	0.7	–
9	C1+CuO	2.5	–	5.6	100
10	C2+CuO	3.0	–	4.8	86

Of a great interest are the data on variants in which soil was polluted by addition of copper acetate 100 mg·kg^{-1} (var. 3, 4), copper sulphate 100 mg·kg^{-1} (var. 5, 6), and copper oxide 500 mg·kg^{-1} (var. 9, 10). As well as in case of zinc, the concentration of copper after a month of composting of soil got much lower as it was expected. Nevertheless, the copper concentration has exceeded 6 MAC for the acetate, and 5 MAC for the sulphate. Thus, the high immobilizing ability of chernozem in relation to copper is demonstrated.

The soil amendment with brown coal caused an essential decrease in the content of mobile copper: up to 41% (var. 4) and 35% (var. 6), respectively. It is known that more than 90% of copper is bound to the most conservative fraction (trioctahedral and dioctahedral minerals), and the fraction bound to organic substances is the next most abundant [16]. Copper has been reported to have high affinity for forming chelates with humic acids that are characterized with the high stability [27, 24, 28, 15, 12]. So, it can be suggested that addition of brown coal increases an amount of the humic acids in the soil, and this explains the decrease of the content of mobile forms of copper.

In the third model experiment, we studied influence of brown coal amendment on the content of heavy metals in the soil sampled at the most heavy metal-polluted site in the city of Rostov-on-Don (Russia) [5]. The concentration of mobile forms of lead accounted for 16 MAC. The mobile forms of heavy metals were determined using extraction with ammonium-acetate solution at pH 4.8. Samples prior to analyses were incubated for 7 days, 1 month, and 6 months (Table 4).

Table 4. The effect of brown coal amendment on adsorption dynamics of zinc and lead in ordinary chernozem.

Brown coal, t·ha^{-1}	Heavy metals, mg·kg^{-1} of soil					
	7 days		1 month		6 months	
	Zn	Pb	Zn	Pb	Zn	Pb
Control	44.0	100	44.5	106	45.3	100
1	43.0	100	42.7	90	44.2	80
5	42.0	83	43.2	69	45.8	53
10	41.4	71	42.9	49	37.6	31

A decrease of lead concentration was observed in all variants with brown coal (Figure 3). The lead sorption in the variants with 5 and 10 t·ha^{-1} of brown coal was rather high. After the 7-day incubation period, there was a statistically significant distinction observed only between the control and the variant with the maximum amount of brown coal added (t = 2.46, tabulated value is 2.23). The lead sorption was increasing with an increase in incubation time. The statistically significant distinctions were found between the control and the variants with 5 and 10 t·ha^{-1} after one month of incubation (t = 2.75 and t = 5.25, tabulated value is 2.23), and after six months of incubation (t = 6.08 and t = 11.00, tabulated value is 2.26).

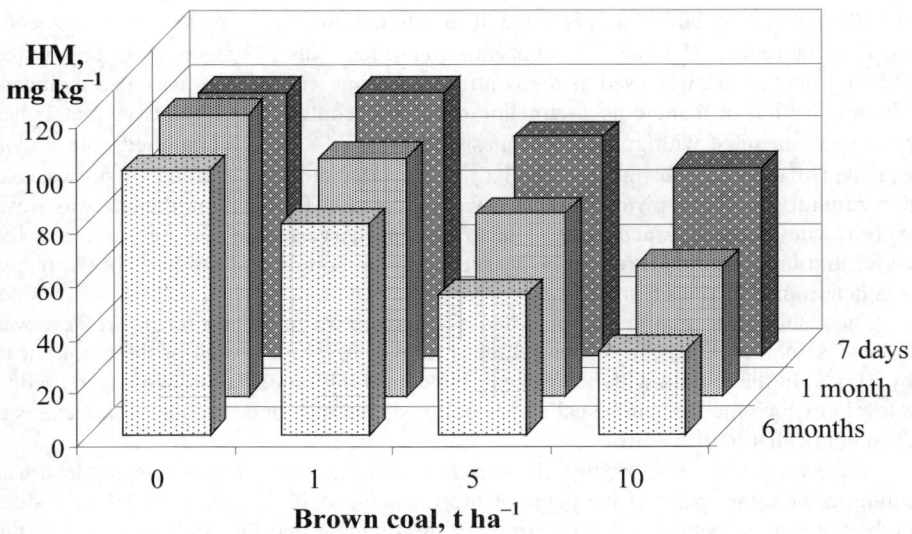

Figuge 3. The effect of brown coal on the adsorption dynamics of mobile form of lead from contaminated chernozem, mg·kg^{-1}.

The adsorption of zinc was found to be significant only in the variant with 10 t·ha^{-1} of brown coal added after six months of incubation (t = 3.31, tabulated value is 2.57). In this soil, the content of zinc was equal to MAC, so, it was present as a nutritional element rather than a pollutant. The insignificant adsorption of zinc confirms the above assumption that brown coal does not produce a negative influence on the plant nutrition.

The amount of brown coal that should be added for remediation of soil contaminated with heavy metals can be calculated using the following formula [25]:

$$D = 0.0001 \times S \times H \times d \times \sum_{i=1}^{n} \left(C_i \times K_i \right)$$

where D is an amount of brown coal added (t), S is the polluted area (m^2), H is a depth of the polluted layer (cm), d is the soil bulk density (g·m·cm^{-3}), C is the concentration of mobile forms of heavy metals extracted into ammonium-acetate solution at pH 4.8 (mg·kg^{-1}) measured using atomic absorption spectrometry, and K is an empiric coefficient. It was shown, that for Pb and Cu, K = 0.0044 and for Zn, K = 0.0052.

If the brown coal with other properties is used, such as the content of humic acids, ash percentage, etc., the K-coefficient will change; however, the ideal amount of brown coal can be calculated according to the above formula.

2.2. FIELD EXPERIMENT

The results of laboratory experiments were verified under the field conditions. The experiment was carried out on soil extremely polluted by zinc. The site is near a chemical factory where zinc oxide is produced. The concentration of mobile forms of zinc was determined prior to the beginning of the experiment using extraction into ammonium-acetate buffer at pH 4.8 and accounted for 874.5 mg·kg^{-1}. This exceeds MAC by more than 35 times. The concentration of lead was 109 mg·kg^{-1}. An unpolluted ordinary chernozem was used as the control. A variant with manure was also included. The application of manure as an ameliorant of soils contaminated by heavy metals has been recommended with application rates of 20-80 t·ha^{-1} [22]. We applied both brown coal and manure at the rates of 50 t·ha^{-1}, after calculating an amount of brown coal according to the above given formula. Two months after fertilization, the site was sown with fescue grass *(Festuca spp)*. The grass was harvested in the autumn, and the concentration of mobile forms of heavy metals in the bottom and top parts of the plants was determined (Tables 5 and 6).

The results show that in the soil where the manure was used as a fertilizer, there was a decrease observed for the concentration of mobile zinc by 9% and for the mobile lead by 21.4%. In the variant with brown coal, the concentration of mobile zinc decreased to a level at which the soil can be called not polluted, and the concentration of the lead was even below that in the control.

In the variant with the urban soil a very high concentration of both zinc and lead was found in the upper parts of the plants. Unfortunately, MAC for heavy metals in fodder herbs has not yet been developed. However, it is known that the content of zinc in the upper parts of fescue grass on average is 16 mg·kg^{-1} of dry weight [23]. So, an almost 10-times higher value of this parameter sharply reduces the fodder quality.

Table 5. The effect of brown coal amendment on the content of the mobile forms of heavy metals in the urban soil (field experiment, 6 months after fertilization).

Variant	Zn			Pb		
	mg·kg^{-1}	% of control		mg·kg^{-1}	% of control	
		1*	2*		1	2
Ordinary chernozem (Control 1)	16.8	100	–	1.9	100	–
Urban soil (Control 2)	740.3	4407	100	87.2	4590	100
Urban soil + manure	674.7	4016	91	68.5	3463	79
Urban soil + brown coal	24.9	148	3.3	1.2	63	1.4

* 1 and 2 are given for the per cent calculated in relation to control 1 and 2, respectively.

Table 6. The effect of brown coal amendment on the content of heavy metals in the fescue grass (aboveground part).

Variant	ash, %	Zn			Pb		
		mg·kg^{-1}	% of control		mg·kg^{-1}	% of control	
			1	2		1	2
Chernozem ordinary (Control 1)	4.8	24.5	100	–	8.6	100	–
Urban soil (Control 2)	6.2	155.0	419	100	48.4	433	100
Urban soil + manure	5.8	116.0	392	80	19.7	189	44
Urban soil + brown coal	4.9	27.9	112	23	5.4	61	14

*1 and 2 are given for the per cent calculated in relation to control 1 and 2, respectively.

A use of manure as an ameliorant has caused a significant decrease in the heavy metal content in the plants; however, it remained very high. In the variant with brown coal, the content of zinc in plants was reduced almost to the level of the control. The content of lead in the grass grown on the variant with brown coal was even lower than in the control variant.

2.3. MECHANISTIC STUDIES

To study the mechanism of binding of heavy metals by brown coal, we performed the following experiment. It was carried out with the unpolluted soil - ordinary chernozem. The content of the mobile copper in the model soil was on the background level, while the content of mobile lead slightly exceeded it. The soil was composted with brown coal for two months under optimal temperature and soil moisture conditions. Brown coal was applied at the rate of 50 t·ha^{-1}.

Preparation of soil extracts was made according to the Rin'kis method [21] using the acetate buffer solution and hydrochloric acid. Copper and lead were determined using polarography technique that allows to determine both concentration and speciation of metals. Hydrochloric acid was used as a media for polarographic measurements. The high reversibility of reduction processes in this media provided signals with good reproducibility, high sensitivity, and high resolution, and allowed us to determine both copper and lead in the same sample. All solutions were measured against 3 M HCl.

Voltammetric determinations were made using Oscillofoloro TsLA, model O2A equipped with a temperature-controlled ($25 \pm 0.2°C$) three-electrode cell. The working electrode was a mercury stripping electrode ($\tau^{2/3}m^{2/3} = 6.5$ mg$^{2/3}$). The reference electrode was a saturated calomel electrode or mercury-pool; the auxiliary electrode was a platinum wire. Oxygen was removed from polarographic solutions with an argon flow. The polarograms obtained with an LP-7 polarograph and cyclic voltammeter were used. The results obtained are shown in Table 7.

Table 7. The effect of brown coal amendment on the concentration of mobile copper and lead in ordinary chernozem.

Variant	Composting duration, days	Concentration, mg·kg^{-1}	
		Cu	Pb
Soil (Control)	–	4.18 ± 0.177	3.9 ± 0.105
Brown coal (without soil)	–	0.22 ± 0.087	not found
Soil + Brown coal	–	0.98 ± 0.073	–"–
Soil + Brown coal	15	0.61 ± 0.062	–"–
Soil + Brown coal	30	0.57 ± 0.022	–"–
Soil + Brown coal	60	0.34 ± 0.018	–"–

± is given for standard deviation.

The data show that the introduction of brown coal promoted the transformation of heavy metals into relatively immobile forms. The initial concentration of lead in the soil was on average 3.9 mg·kg^{-1}, and of copper 4.2 mg·kg^{-1}. Brown coal promoted a sharp decrease in copper content down to 0.98 mg·kg^{-1} immediately after its introduction (variant 3). Further composting resulted in a progressive decrease in the mobility of copper compounds during two months to minimal values, which was similar to the content of mobile copper in the brown coal. Due to the introduction of brown coal, the mobile lead vanished because of its rapid and complete binding into relatively immobile complexes.

The character of the polarography curves suggests that the mechanism of metal binding by brown coal is rather complex. Firstly, the formation of complexes between copper and lead and anions of humic acids of brown coal is probable. This is supported by a shift in the reduction potential of these metal ions to the cathode zone accompanied by kinetic constraints of the peak current.

Secondly, the adsorption of complexed metals by the active surface of brown coal cannot be excluded. The shapes of oscillopolarograms and values of diagnostic c-,v-, τ-, ψ-criteria are indicative of the ability of the complexed metals to adsorb. In the absence of the brown coal (control), the contribution of adsorptive compounds to the current is insignificant. This can be explained by a considerably lower adsorptive capacity of soil organic matter compared to the humic components of brown coal.

3. The Effect of Brown Coal on Humus Conditions

In the above model experiments, the humus state of soil was also controlled. The content of the alkaline-soluble forms of humus was measured as well as the absorptivity coefficients of humic acid-1 (HA-1) (the E4/E6 index). The data show that addition of brown coal considerably increased the content of mobile humic acids (Table 8). It also increased the content of fulvic acid, which should be favourable for plant nutrition.

Table 8. The effect of application rate of brown coal on the humus state of ordinary chernozem over a 7-day incubation period.

Brown coal, $t \cdot ha^{-1}$	C_{soil}, %	$C_{extract}$, %	%		in the ratio of C_{soil}, %		E4:E6	t
			C_{HA}	C_{FA}	C_{HA}	C_{FA}		
0 (control)	2.81	0.38	0.17	0.21	6	7	0.47	–
1	2.64	0.45	0.19	0.26	7	10	0.64	3.87
5	2.81	0.43	0.27	0.16	10	6	0.82	4.32
25	3.48	1.97	0.61	1.36	18	39	3.91	2.34
50	3.75	1.94	0.72	1.22	26	44	4.01	3.12

The results show that the content of mobile fulvic acid (FA-1.1a) and of mobile humic acid (HA-1) in variants with addition of brown coal is rather high. The concentration of mobile fulvic acid on average is 1.5-2 times more than the concentration of mobile humic acid. The character of changes in the E4/E6 index indicates that heavy metals were bound to peripheral part of brown coal's humic acids.

The same findings were observed on variants with high amounts of brown coal. The increase of fulvic acid concentration seems strange. On the one hand, in the chernozems, there is no such quantity of fulvic acid, especially of the mobile fractions: according to our data [4], their average concentration is about 18% of the total content of soil organic carbon. Brown coal contains almost no fulvic acid. A question arises where such a high content of these compounds comes from. Brown coal transformation in the soil seems to cause the formation of fulvic acid. It is possible that humic acids of brown coal bound to heavy metals were subjected to transformation and that the fulvication is a first stage of this process. Micro-organisms play a primary role in the process of removing various fragments from specific soil and brown coal organic substances. It is possible that brown coal humic acid mineralization proceeds under the influence of many biological and abiotic factors and is accompanied with formation of fulvic acid.

In another model experiment we studied the processes of brown coal transformation. The fraction and groups of organic substances of brown coal were analysed by Tyurin's method (modified by Ponomareva and Plotnikova); organic carbon was determined by wet ashing technique. It was established that during composting of brown coal under optimum conditions of humidity and temperature, the structure of organic substances changes. It should be noted that humic acids of brown coal are characterized with the high content of aromatic structures [19]. Hence, their reactivity to oxidation is rather low. So, during the composting, the amount of mobile fulvic acid increased more than 2 times. The same tendency is observed with other fulvic acid fractions. During a 12-month incubation, the total amount of fulvic acid increased almost six times (Table 9).

Table 9. The changes in the content of humus acids in the brown coal during composting.

C_{org}, %	% of total C_{org}								Humin*
	Fractions of humic acids				Fraction of fulvic acids				
	1	2	3	the sum	1+1a	2	3	the sum	
Initial brown coal (before composting)									
27.3	0.18	1.6	9.88	11.6	0.12	0.18	0.77	1.07	87.4
1 month composting									
32.44	1.26	5.36	8.26	14.88	1.20	1.12	0.81	3.13	81.99
2 months composting									
36.88	1.31	5.47	9.56	15.36	1.25	1.20	1.02	3.47	80.17
12 months composting									
46.4	2.13	8.62	9.96	20.71	2.35	2.50	1.50	6.35	73.48

* Humin is a non-extractable fraction of humus.

The fractional composition of humic acids have also changed during composting. Increase in composting time has caused an increase in their mobility: the content of the mobile HA-1 fraction increased almost 12 times, the content of HA-2 – about 5 times. The contribution of the third fraction of humic acids increased up to 8%. During 12 months of composting, the total content of humic acid in the organic matter of brown coal increased almost twice (1.78 times). The analysis of humic acid fractions shows that the structure of their molecules changed. The quantity of acidic functional groups increased from 600 up to 713 meq per 100 g of a preparation during the time of experiment. Simultaneously, the portion of the humin fraction decreased, and the ratio $C_{HA}:C_{FA}$ narrowed (from 10.4 up to 4.3 after 12 months of composting).

The sharp increase in a fulvic acid fraction in the brown coal humus can be explained by the fact that under new hydrothermal conditions the presence of mobile components is more preferable energetically. The biological factor plays a significant role in the processes of brown coal humic compound destruction. The major source of nutrition are mobile organic compounds, composing only a small portion of brown coal organic matter. Further, in the process of transformation, microorganisms utilize a hard component of brown coal. This explains the decrease of the humin in the structure of the brown coal organic substance.

So, at the beginning of composting, the stimulation of microorganisms by optimising the humidity is observed. Our studies have shown that the brown coal amendment increases an amount of microorganisms in soil, such as *Azotobacter* and actinomycetes. The amount of micromycetes, on the opposite, was decreased in the dose-independent way, regardless, how much of brown coal was applied (Figure 4). The observed differences in the amount of actinomycetes and micromycetes are statistically significant with a level of confidence of 0.9 and higher.

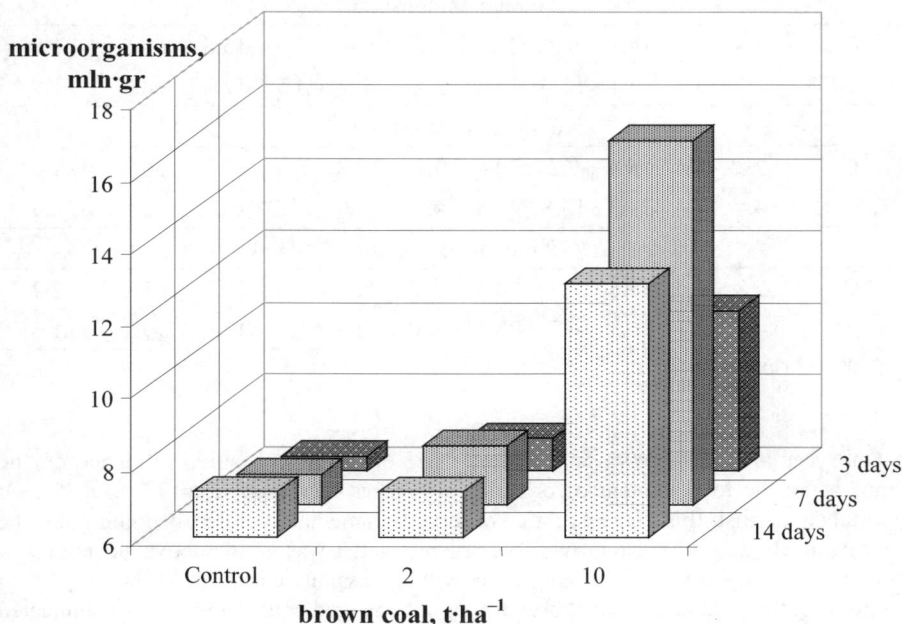

Figure 4. The effect of brown coal on the amount of actinomycetes in dark-chestnut soils.

198

The large increase in soil enzymatic activity (almost on 10%) also is indicative to stimulation of microflora.

The above processes observed for brown coal can also take place in soil. However, the results of model experiment on composting the ordinary chernozem amended with brown coal showed the different tendencies in transformation of the humic components (Table 10). This was despite the fact that the brown coal addition stabilized redox conditions: during all experiments the value of the redox potential remained at a constant level of 500-515 mV.

Table 10. The effect of brown coal on the humus state of ordinary chernozem during various periods of composting.

Variant	C_{org}, %	Fractions of humic acids				Fractions of fulvic acids				Humin	$\dfrac{C_{HA}}{C_{FA}}$
		1	2	3	the sum	1+1a	2	3	the sum		
Initial soil (before composting)											
Control (C)	2.8	2.1	13.6	9.3	25.0	3.5	5.7	4.3	13.5	61.5	1.9
2 month composting											
C	2.4	2.1	14.3	9.3	25.7	2.8	6.4	4.3	13.5	60.8	1.9
C+ BC*	4.1	6.1	17.1	11.0	34.2	7.7	5.5	4.5	17.7	48.1	1.9
6 month composting											
C	2.3	5.4	15.4	9.2	30.0	3.1	10.8	5.4	19.3	50.7	1.6
C+ BC	4.4	7.4	24.8	14.8	46.5	4.7	1.3	11.7	17.7	35.8	2.6
12 month composting											
C	2.8	1.4	18.6	6.3	26.3	1.4	8.0	2.8	12.2	61.4	2.2
C+BC	4.3	2.1	16.3	4.7	23.1	3.3	12.9	1.7	17.9	59.1	1.3

* BC – brown coal.

One year after the brown coal addition, we observed a positive influence on the humus state: the total content of organic carbon has increased from 2.7 to 4.3%. Of importance is that this increase was observed immediately after beginning of the experiment. By that time, the brown coal organic matter was in non-activated state, so it could not be determined by combustion with potassium chromate. After a year of composting, the organic matter of brown coal has changed considerably, this is indicated by a steady increase in the amount of organic carbon in the composted soil, as well as by an increase in humic and fulvic acid fractions during the first 6 months. During the next 6 months, the content of mobile humic and fulvic fractions has decreased. This might be a result of the high rate of microbiological activity because these very forms are the major source of plant and microorganism nutrition [3, 11].

After a year of incubation, a decrease was observed in the content of all humic fractions. The HA-1 content was down to 71.4%, HA-2 to 34.5%, and HA-3 to 68.2%. The carbon content decreased two times and we can hypothesize that the system of humus is stabilized. Therefore, we hypothesize that brown coal humic substance in the soil will be actively utilized only if there is thermodynamically more stable matter present. Hence, brown coal can have a prolonged impact on the soil fertility. The heavy metals bound by brown coal will be in an immobilized state for a long time. This makes it possible to use brown coal as an ameliorant for heavy metal polluted soils.

4. Conclusion

Research was carried out on brown coal transformation. Microorganisms were decomposing humic substances of brown coal during the composting of various mixtures of soil and brown coal under optimal temperature and humidity. As a result, the mobility of humic and fulvic acids was increased.

The formation and properties of heavy metal complexes with humic substances of brown coal depend on the nature of metals and on their influence on soil microorganisms. The immobilization of zinc and lead in the soils amended with brown coal was established. The immobile forms accounted for 30-80% of the initial content of heavy metals.

The complexation with humic substances was assumed to be the main mechanism underlying immobilization of zinc, copper, and lead in the soils amended with brown coal. It was indirectly confirmed by an increase of mobile humic fractions characterized by a higher complexing affinity (e.g., HA-1, FA-1, FA-1a) The content of these fractions increased 6–10 times compared to the non-amended soil.

5. References

1. Alekseev, J. V. (1987) *Heavy Metals in Soils and Plants*, Agropromizdat; Leningrad.
2. Bassuk, N.L. (1986) Reducing lead uptake in lettuce, *Hortic. Sci.* **21(4)**, 993-995.
3. Bezuglova, O.S. (1982) The effect of the humus quality on the crops' yield, *Pochvovedenie* **2**, 134-137.
4. Bezuglova, O.S. (2001) *The humus state of soils of the South Russia*, SKNCVSH-Publisher, Rostov-on-Don.
5. Bezuglova, O., Gorbov, S., Morozov, I. and Privalenko, V. (2001) The soil pollution with heavy metals in Rostov-on-Don, *Abstract Book of the 11th International Symposium on Environmental Pollution and its Impact on Life in the Mediterranean Region*, Limassol, Cyprus. p. D6.
6. Bezuglova, O.S., Etsenkova, E.V. (1992) The application of brown coal and coal-humate fertilizers on some soils of the Rostov region, *Pochvovedenie* **1**, 139-143.
7. Bezuglova, O.S., Ignatenko, E.L., Morozov, I.V. and Shevchenko, I.D. (1996) The effect of brown coal on the decrease in the content of mobile forms of copper and lead in ordinary calcareous Chernozem, *Eurasian Soil Sci.* **29(9)**, 1103-1106.
8. Bezuglova, O.S., Kogan, I.B., Morozov, I.V., Ponomarenko, A.V. and Shevchenko, I.D. (1997) Coal-humic fertilizer, *The Resolution of All-Russian Scientific Research Institute of State Patent Examination № 96111989/13 (018511) from 19.05.1997* (Russia).

200

9. Bezuglova, O.S., Morozov, I.V. and Stepovoi, V.I. (1996) Effect of some biologically active substances on the humus of Ordinary Chernozem and Dark Chestnut soil, *Eurasian Soil Sci.* **28(10)**, 100-110.
10. Bezuglova, O.S., Val'kov, V.F., Kazeev, K. Sh., Kolesnikov, S.I. and Morozov, I.V. (1999) The effect of high concentration of heavy metals on the humus state and biological activity of calcareous chernozems, *Izvestiya VUZ'ov. North Caucasus region. Natural Sciences* **2**, 66-71.
11. Kleinhempel, D. (1970) Beziehungen einiger Huminstoffkennzahlen zum Ertrag auf Schwarzerdstandorten, *Thaer-Arch.* **14(1)**.
12. Ladonin, D.V. and Margolina, S.E. (1997) Interaction between humic acids and heavy metals, *Pochvovedenie* **7**, 806-811.
13. Lozanovskaya, I.N. and Luganskaya, I.A. (1991) The ecological estimation of the opportunity of brown coal use as humic fertilizers, *Biologicheskie Nauki.* **10**, 155-160.
14. Obukhov, A.I. and Efremova, L.L. (1988) Protection and remediation of soils polluted by heavy metals, *Proceedings of the 2nd All-Union Conf.: Heavy Metals in the Environment and Nature Protection,* Moscow, vol. 1, pp. 23-36.
15. Orlov, D.S. (1992) *Soil Chemistry*, Moscow State University Publisher, Moscow.
16. Pampura, T.V., Pinskiy, D.L., Ostroumov, V.G., Gershevich, V.D. and Bashkin, V.N. (1993) Experimental research on Chernozem buffer capacity to copper and zinc pollution, *Pochvovedenie* **2**, 104-110.
17. Pampura, T.V. (1996) *Absorption of copper and zinc by typical chernozem under conditions of model experiments*, Ph.D. Thesis. Department of Soil Science, Lomonosov Moscow State University, Moscow.
18. Pavlikhina, G.P. (1990) Methods of purification of polluted soils and ground water, *Problemy okruzhayushcei sredy i prirodnykh vod: obzor inf. (Problems of the Environment and Natural Resources: Information Review)* **7**, 73-83.
19. Perminova, I.V. (2000) *Analysis, classification and prediction of properties of humic substances*, Dr. Sc. Thesis, Department of Chemistry, Lomonosov Moscow State University, Moscow.
20. Piccolo, A. (1989) Reactivity of added humic substances towards plant available heavy metals in soils, *Sci. Total Environ.* **81-82**, 607-614.
21. Rin'kis, G.Y. (1963) *Methods for rapid colorimetric determination of microelements in biological objects*, Riga.
22. Samoilova, T.S. (1988) The effect of fertilizers on fixation of mobile forms of lead and copper in soils, *Proceedings of the 2nd All-Union Conf.: Heavy Metals in the Environment and Nature Protection,* Moscow, vol. 1, p. 224.
23. Scerling, V.V. (1990) *Diagnostics of nutrition of agricultural cultures*, Agropromizdat, Moscow.
24. Schnitzer, M. and Kerndorf, H. (1981) Reactions of fulvic acid with metal ions, *Water, Air and Soil Pollut.* **5**, 97-108.
25. Shestopalov, A.V. and Bezuglova, O.S. (1999) Brown coal as detoxicant of soils polluted with heavy metals, *Ninth annual Goldschmidt Conference, www.lpi.usra.edu/meetings/gold99/pdf,* p. 7217.
26. Shestopalov, A. and Bezuglova, O. (2000) The remediation technique for heavy metal polluted soils, *Proceedings of the First International Conference on Soil of Urban, Industrial, Traffic and Mining Areas, Vol. 3, The Soil Quality and Problems: What Shall We Do?* University of Essen, Germany, pp. 795-798.
27. Stevenson, F.J. (1976) Stability constants of Cu, Pb and Cd complexes with humic acids, *Soil Sci. Soc. Am. J.,* **40**, 665-672.
28. Zhorobekova, S. Z. (1987) *Macroligand Properties of Humic Acid*, Ilim-Publisher, Frunze.

Part 3

Sorptive-partitioning interactions of humic substances
with organic ecotoxicants and their implementation
for remediation technologies

UTILIZATION OF IMMOBILIZED HUMIC ORGANIC MATTER FOR *IN-SITU* SUBSURFACE REMEDIATION

G.U. BALCKE, A. GEORGI, S. WOSZIDLO, F.-D. KOPINKE,
J. POERSCHMANN
*UFZ Center for Environmental Research Leipzig-Halle, Permoserstr. 15,
04318 Leipzig, Germany <balcke@hdg.ufz.de>*

Abstract

HOM enriched areas within aquifers retain or bind organic pollutants by sorptive or specific interactions. This contribution briefly reviews some established concepts on *in situ* remediation for removal of dissolved organic contaminants from groundwater using reactive and sorptive barriers. In particular, we present a new concept for use of HOM (humic organic matter) as sorptive barrier. According to this concept HOM is sorbed to metal salt surface precipitates on aquifer minerals. Induced by sorptive interactions HOC (hydrophobic organic carbon) groundwater pollutants are retarded while passing the HOM barrier. An extended residence time within the aquifer subjects the pollutant plume to extensive microbial degradation and diminished spreading. We give information on pertinent laboratory experiments and field considerations describing design and possible ways to construct HOM barriers *in situ* in the subsurface by infiltration techniques. With regard to sorptive pollutant removal we discuss the mechanisms that drive the binding of HOM to mineral surfaces and those that govern the interactions between HOC and HOM. Operational approaches to determine sorption coefficients of HOC on HOM are conferred. The sorption of HOC on HOM in flocculated, surface-bound, and the freely dissolved state is compared.

1. Introduction

1.1. ESTABLISHED CONCEPTS FOR *IN SITU* SUBSURFACE REMEDIATION WITH SORPTIVE AND REACTIVE BARRIERS

Contamination of groundwater by chlorinated solvents, gasoline components, diesel fuel residues, and gas plant wastes is a worldwide problem. Contaminant plumes up to several kilometres in length downstream of contamination sources have been reported [1-3]. Although many contaminants of such plumes are known to be biodegradable, they sometimes appear to be refractory under natural groundwater conditions [1, 4, 5]. These

I. V. Perminova et al. (eds.),
Use of Humic Substances to Remediate Polluted Environments: From Theory to Practice, 203–232.
© 2005 *Springer. Printed in the Netherlands.*

findings are frequently attributed to the lack of electron acceptors, nutrients, or suitable microorganisms to support the subsurface bioremediation process [6-9].

Available technologies for groundwater remediation can be classified as above-ground and below-ground (or "*ex situ*" and "*in situ*").

Traditional technologies to clean up groundwater involve pumping the contaminated water to the surface and then treating it. However, in many cases *ex situ* field scale remediation applications result in unsatisfactorily high costs and indefinite operation time assessments [10-12]. An approach to increase the viability of pump and treat technologies at contaminated sites has been to solubilize and mobilize HOC by various agents including surfactants and dissolved and colloidal [aqueous] HOM. This approach is described in Chapter 11 of this book.

A potentially more cost-efficient alternative to pump & treat systems is the implementation of permeable reactive barriers (PRB) into the subsurface. They are often designed as a "funnel and gate" configuration in order to focus the plume towards the permeable sorption or reaction barrier by means of impermeable walls [13, 14]. During the passage through the barrier, the contaminants are bound, precipitated or chemically degraded, often merely resulting in less hazardous products rather than in complete mineralization.

PRB can be classified into three types:
– oxidation barriers,
– reduction barriers, and
– sorption barriers.

Oxidation barriers are based on the introduction of an electron acceptor into the groundwater flow (usually NO_3^-, O_2, peroxides) in order to provide either biodegrading microorganisms with conditions suitable for their reproduction, or to facilitate direct chemical oxidation (e.g., by permanganate) [12, 15, 16]. If oxygen is introduced as a gas, i.e., by air sparging, the dominant pollutant removal mechanism is often volatilisation rather than on the establishment of a biobarrier [17].

Oxygen-based oxidation barriers are most amenable for contaminants having chemical structures that are susceptible for initial oxygenase enzyme attacks. These include BTEX, phenols, PAHs, halogenated aromatics with up to three ring-bound halogens, or halogenated olefines and alkanes possessing up to 2 or 3 halogen atoms. Nitrate amendments may assist in anaerobic biodegradation of oxygenated metabolites of aromatic structures [18].

Higher oxidized organic compounds such as polyhalogenated hydrocarbons can act as oxidants (that is, electron acceptors) at high dissolved hydrogen levels and conditions lacking natural electron acceptors provided that bacteria capable of an initial reductive biotransformation are present. Hydrogen or enzymatic bound forms of hydrogen can be produced *in situ* by fermentative processes [19] or by **reduction barriers** based on zero valent iron (ZVI) while carbon dioxide serves as the carbon source for bacterial growth [20-22].

Sorption barriers are designed to retard dissolved pollutants in relation to the groundwater flux. The sorption approach can be regarded as the opposite strategy to the above-mentioned "mobilization/solubilization" strategy. The pollutant mass flux throughout the passage of the barrier does not necessarily have to be reduced to zero (complete sorption). In some cases it may also be desirable to achieve longer residence times in order to achieve better biodegradation within a segment of an aquifer.

In several pilot plant and full scale cases permeable walls were constructed using particulate sorbent materials such as peat, coal-derived HOM, or activated carbon [23].

Another promising strategy to establish permeable sorption walls has been demonstrated by means of subsurface injections of surfactants. After injection, attempts were made to immobilize the surfactants onto the aquifer matrix, thus serving as a hydrophobic mineral coating capable of HOC retardation [24-26].

1.2. THE NEW CONCEPT

Several features of humic organic matter (HOM) cause it to have a high potential in subsurface remediation.

The utilization of HOM as a means to increase the organic carbon (OC) content of the aquifer matrix results in higher retardation of HOC within the aquifer. In contrast to solid OC-rich materials placed in the aquifer, such as activated carbon, HOM liquid injections are not in need of soil excavations which potentially makes the approach less cost-intensive, especially at deep aquifer contamination. In contrast to surfactants, HOM does not only sorb HOC, it can also undergoes specific interactions with functionalised organic pollutants [27-29], it promotes biodegradation processes [30-32] (see also Chapter 17 of this book for details), and it may result in irreversible binding in the long term [33]. The ultimate goal should be therefore the formation of so called bound residues, humification by formation of covalent bonds or other very strong interactions between pollutant and the humic matrix.

In succeeding to establish a stable but permeable HOM barrier by injection of dissolved humates and their precipitation onto the surfaces of aquifer materials the sorption potential of coarse sand poor in organic carbon (OC) and gravel aquifer mineral matrices is improved. However, since several parameters limit the increase in the OC content by HOM precipitation (low specific surface area of aquifer sands, limited porosity) we consider extended sorption zones with moderately increased OC content in the subsurface rather than permeable high OC sorption walls.

Further advantages of the HOM based approach are that HOM is not known to be toxic as, for instance, described for cationic surfactants [34] and is hardly biodegradable in the absence of oxygen and light. At last, HOM is commercially available at technical scale by alkaline extraction of lignites and weathered lignites, so called oxygen-coals.

Sorption by hydrophobic partitioning of HOC between the humic and the water phase is only little effected by the present environmental conditions (e.g., pH, ionic strength, redox potential). These features make HOM a powerful and robust tool to remove HOC from a polluted groundwater. Besides its hydrophobic moieties HOM carries a variety of functional groups that can specifically interact with pollutants. However, sorption governed by specific interactions can be constricted by the amount of active functional groups or units (e.g., carboxylic groups for heavy metal sorption [29].

Disadvantageous can be that by the suggested method only a relatively small OC enrichment of the aquifer material can be achieved, the risk of aquifer clogging by precipitated HOM is given, and the contaminants are not removed from groundwater flow by degradation.

Figure 1 elucidates the general idea of the suggested procedure. Subsequent to an extended hydrogeological survey on site, dissolved (coal-derived) HOM is infiltrated

206

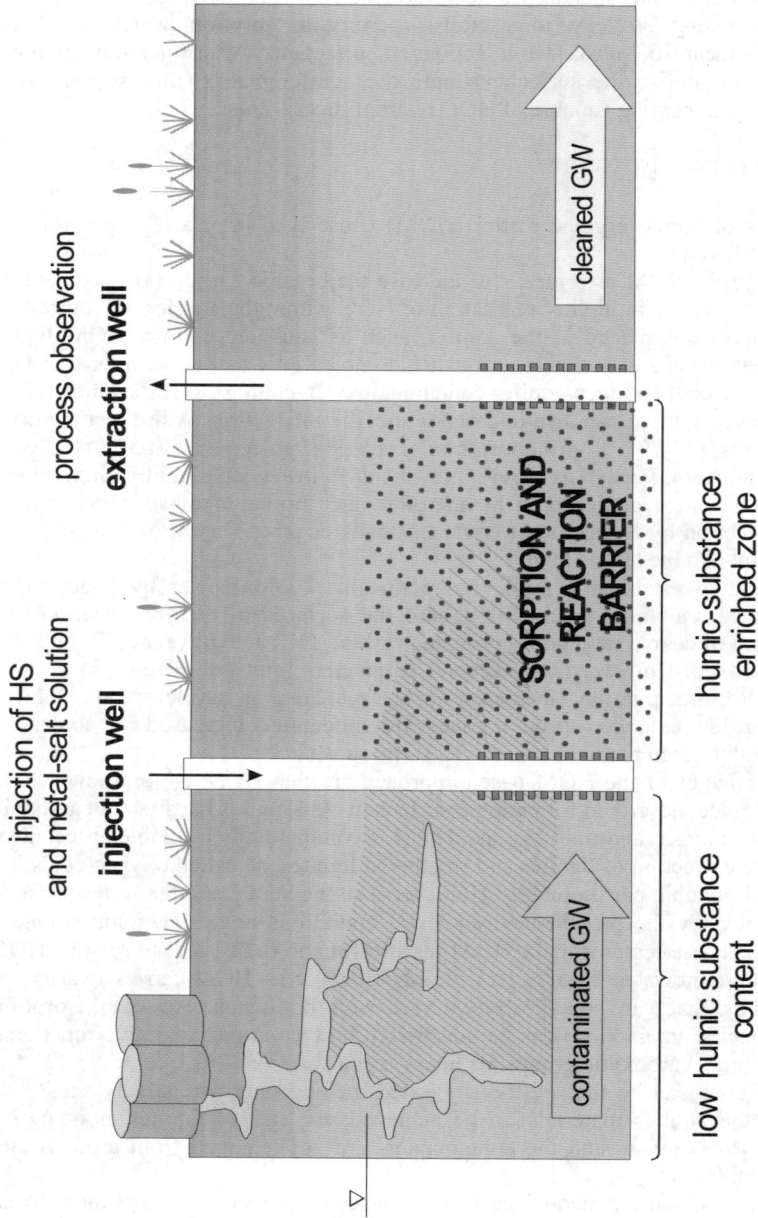

Figure 1. Conceptualisation of the procedure how to build up a permeable HOM barrier.

via injection wells to the subsurface. To promote the infiltration velocity the groundwater table is lowered by a net of pump wells. These wells serve to set the position of the HOM-barrier and can be used later as observation wells. Immobilization of the injected HOM is achieved by direct surface precipitation on clay coatings of aquifer sand grains, or calcium- or iron-mediated HOM precipitation facilitates the binding to the aquifer matrix (see section 5 of this chapter). The humic enriched zone serves as a sorptive/reactive barrier. While sorptive processes prolong the time during which HOC are exposed to natural attenuation processes, reactive binding to the HOM barrier removes HOC from the groundwater.

2. Humic Binding to Mineral Surfaces

The sorption of HOM onto oxide surfaces has been explained in the literature by mechanisms of ligand exchange of surface hydroxyl for humic carboxyl and phenolic groups [35-40]. The strength of sorption is determined mainly by the degree of dissociation of carboxyl and phenolic groups of the HOM and by the position of these groups relative to each other. Particularly ortho-positioned phenolic groups in neighbourhood to carboxylic groups promote sorption of HOM onto metal oxide surfaces [38]. Also hydrophobic moieties of HOM support sorption to mineral surfaces [39-44].

At pH values common for groundwater conditions (pH about 6.0 to 7.5) both, the dissolved HOM and most mineral surfaces, carry negative net surface charges [37, 45]. While a negative net surface charge for HOM is caused by dissociation of carboxylic moieties [46], negative surface charges of mineral surfaces may be of various origins.

The charge on mineral surfaces is attributed to
- isomorphic substitution in tetrahedrons of silicates [47] (e.g., Al^{3+} for Si^{4+}, i.e., the replacement of cations of lower total positive charge results in a permanent "charge deficiency" that needs to be balanced by monovalent cations, which can be replaced easily),
- the protonation/deprotonation equilibrium of hydrolysed surface hydroxyl groups: $SS\equiv Si$ (Fe, Al)-OH2+ $\leftrightarrow$ $SS\equiv Si$ (Fe, Al)-OH $\leftrightarrow$ $SS\equiv Si$ (Fe, Al)-O-, (SS-sorbent surface, $\equiv$ refers to inner crystal bonds),
- the adsorption of charged species on the mineral-water interface, e.g., humic substances or biofilms or competition with other surface complex forming anions [45]. The state of the surface protonation equilibrium at a given pH is mainly determined by the electron deficiency of the central ion in the surface complex. Whilst aluminium or ferric ion surface complexes exhibit typical Lewis acid character, silicate surfaces are described rather by weak Broenstedt acid behaviour. Accordingly, the first are good and the latter poor sorbents for HOM [37].

Moreover, typical coarse materials of unconsolidated aquifers exhibit small specific surface areas (<2 m^2g^{-1}) [3]. As a result dissolved HOM is only poorly retarded on materials like quartz sand or gravel [48].

To avoid repulsion forces between negatively charged HOM and mineral surfaces with negative charges these surfaces need to be modified.

208

Looking at the nature we observe that podzolization of soils results in ferric iron coatings of silica surfaces. In principle, this also might be accomplished by means of precipitation of iron or aluminium salts as a "mediator" between HOM and mineral surface.

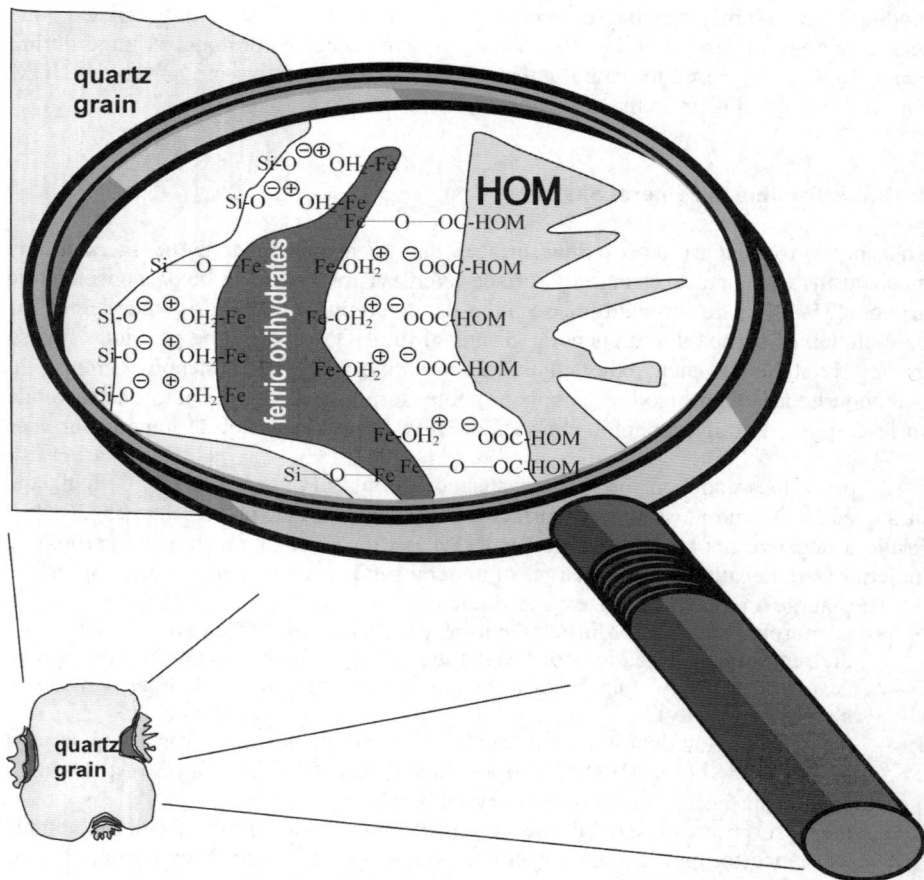

Figure 2. HOM coatings on ferric iron precipitates in niches of a quartz sand surface.

Figure 2 schematically illustrates HOM sandwich coatings on iron surface precipitates on silicate materials. Microscopic magnifications reveal that iron and HOM enrichments on quartz surfaces occur predominantly in glyphs and cavities [3].

Despite their high binding affinity to mineral surfaces, hydroxides of Fe(III) and Al(III) have drawbacks. Aluminium salts are known to display a potential toxicity under acidic conditions [49, 50]. Even though at neutral pH values aluminium is almost completely insoluble and non-toxic, Al(III)-salt solution injections may not be tolerated by regulatory authorities.

Table 1. Adsorption of different metal salt precipitates on quartz sand after precipitation of the metal salt at neutral pH ($c_{Me} = 0.25$ mol L⁻¹) in the presence of the sand (specific surface = 0.4 m² g⁻¹, 500 g L⁻¹).

sand / metal salt		
metal salt	$X_{adsorbed}$ [mg/g]	degree of interaction
Fe(OH)$_3$	4	
FeCO$_3$	2.6	
Fe$_3$(PO$_4$)$_2$	1.6	
FeS	1.2	
Ca$_3$(PO$_4$)$_2$	0.1	

Table 2. Adsorption of HA (Roth humic acid, Germany) by different metal salt precipitates ($c_{Me} = 0.5$ mmol L⁻¹) from a 200 mg L⁻¹ HA-solution (pH 7) and degree of HA desorption from HA-coated minerals after a single replacement of the supernatant by 0.1 M NaCl.

HA / metal salt			stability under anaerobic conditions
metal salt	% HA adsorbed	% HA desorbed	
Fe(OH)$_3$	99.8	0.6	-
Fe$_3$(PO$_4$)$_2$	96.9	3.0	+
Ca$_3$(PO$_4$)$_2$	87.8	n.d.	+
FeCO$_3$	24.2	81.7	+
FeS	4.5	n.d.	+

The triangle symbolically depicts the degree of interaction between either quartz mineral surface and metal ion precipitate or metal ion precipitate and HOM.

Although the placement of $Fe(OH)_3$ as a precipitate on silicate surfaces that can also bind HOM is relatively straightforward, attention should be paid to anoxic groundwater conditions. Ferric hydroxide shows poor redox stability under reducing conditions [51]. Moreover, it was shown that dissolved HOM facilitates the biodegradation of organic pollutants with ferric iron as terminal electron acceptor [31]. A remobilisation of dissolved ferrous iron could lead to non-stable HOM- Fe(III) oxide barriers. Therefore, we tested the sorption behaviour of several environmentally important precipitates, including poorly soluble Fe(II) salts, with respect to their binding abilities towards a quartz surface (Table 1), and towards dissolved HOM (Table 2). Details of these tests were reported earlier [3]. Next we investigated the stability of a HOM barrier under simulated ferric iron reducing conditions (Figure 3A and 3B) [3].

A set of metal ion precipitates representative of mineral surfaces in a polluted aquifer were formed (partially under exclusion of oxygen) in presence of quartz sand (200 – 630 μm) at pH 7. As can be seen from table 1, surface coatings of silicates can be achieved by several metal ion surface precipitates. Besides $Fe(OH)_3$, several Fe(II)-coated mineral surfaces show a strong tendency to sorb to quartz (e.g., $Fe_3(PO_4)_2$). Ferrous iron sulphide, on the other hand, shows neither a high degree of interaction towards the quartz surface nor towards HOM. For some salts like $Ca_3(PO_4)_2$ and $FeCO_3$ complexes with HOM are only weak.

In combining the results of tables 1 and 2 it becomes clear that not all ferrous iron species may work as mediators for sorption of HOM onto silicate surfaces. Some precipitates, like $Ca_3(PO_4)_2$, do not form stable surface complexes with the quartz surface. Others, like $FeCO_3$, redissolve when the supernatant is exchanged against a 0.1 M NaCl solution (maintaining the same ionic strength and pH conditions). Under reducing conditions, the ferrous compound $Fe_3(PO_4)_2$ may serve as an alternative to ferric species such as $Fe(OH)_3/FeOOH$ for it forms stable complexes with HOM and silicate surfaces.

In another experiment [3] we subjected a ferric iron hydroxide/HOM-coated quartz sand, that has been produced by batch $Fe(OH)_3$-precipitation prior to the experiment (Fe: 1.5 g/kg, OC: 0.1 g/kg), to anaerobic respiration under flow through conditions in order to investigate the stability of ferric humate coatings against reductive dissolution. As carbon source we used a lactate solution (500 mg/L). A groundwater bacteria consortium from the Rositz site, a former tar factory site south of Leipzig, Germany, was inoculated to grow on lactate in the absence of oxygen. 0.33 L of a slightly modified Brunner nutrient solution containing the bacteria have been recycled through the coated quartz sand column for several days maintaining a flow rate of 10 mL/min (equivalent Darcy velocity: 35 m/d). System parameters such as redox potential, pH, dissolved ferrous iron, and the disappearance of lactate were measured.

Figures 3A and 3B represent the reductive dissolution of ferric iron humates in the presence and in the absence of phosphate.

In both experiments lactate was consumed rapidly. Further, sulphate was rapidly consumed by the microbial community, and the redox potential dropped slowly. The marked difference between Figures 3A and 3B is the appearance of ferrous iron in the effluent only in the absence of phosphate.

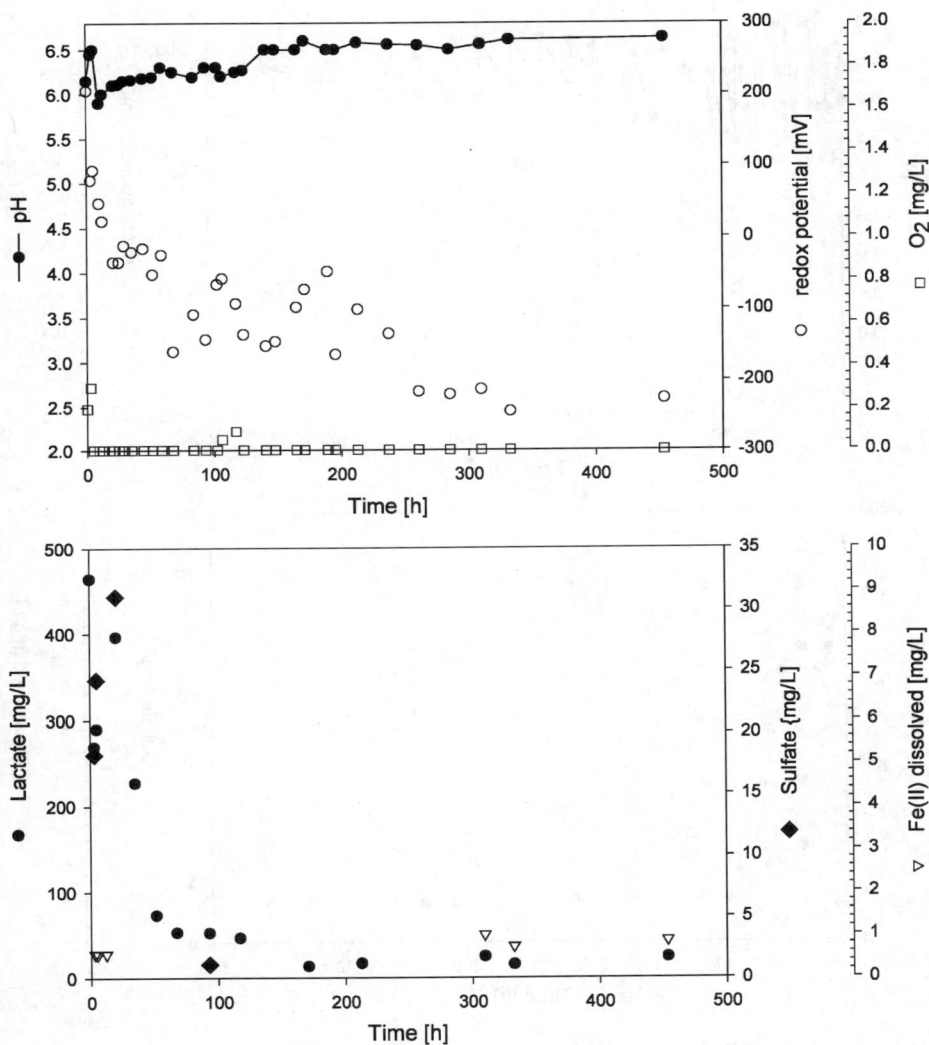

Figure 3A. Anaerobic respiration of lactate in presence of a Fe(III)-HOM coated sand and of a phosphate buffer. Even though ferric iron is reduced to ferrous iron no resolubilization of Fe(II) and of the sorbed HOM is observed.

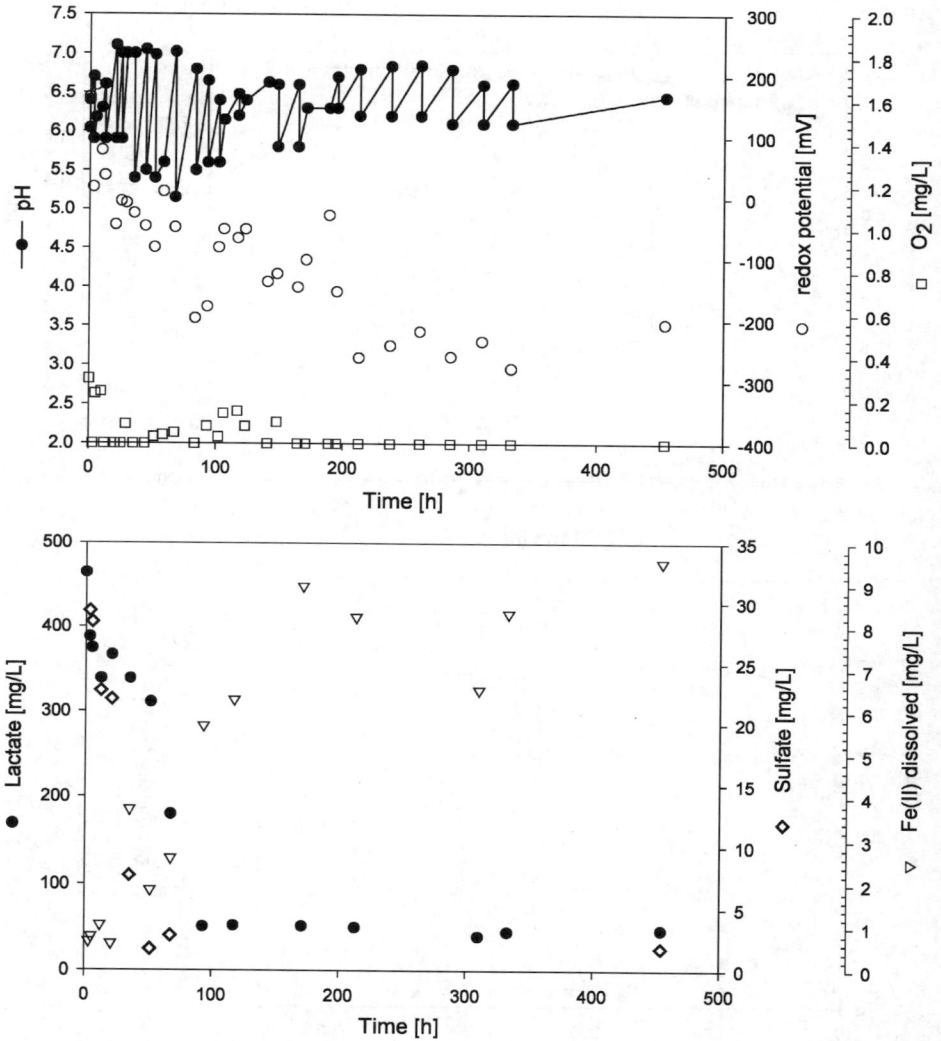

Figure 3B. Anaerobic respiration of lactate in presence of Fe(III)-HOM coated sand but in absence of phosphate (only Redfield formula conserving amounts). Iron is partially dissolved by reduction. Despite this no mobilization of sorbed HOM could be observed. The pH value was artificially forced to a range between 6.0 and 6.5 by successive amendments of 0.1 M NaOH solution.

In the presence of phosphate, ferric iron was completely reduced to ferrous iron. After the experiment was finished the washed sand surface contained high amounts of phosphate and iron, most likely indicating the formation of $Fe_3(PO_4)_2$. No ferrous iron formation could be detected in the effluent. A high phosphate background obviously facilitates ferric iron reduction. In the absence of phosphate only about one quarter of the initial surface-bound iron was reduced. During the course of the experiment (450 hours) it was only partly mobilized. The level of dissolved Fe did not decline or approach zero at the apparent end of the experiment, about 450 h. This suggests that ongoing ferrous iron mobilization would occur at a longer time scale. Further studies are required to examine long term stability/instability of the iron precipitates. A quantitative evaluation of the HOM in solution in the presence of bacteria, lactate and extracellular polymeric substances (EPS) is difficult. We solved this problem by sterile filtration over 0.2 μm cellulose acetate and subsequent UV spectroscopy. While bacterial cells are retained, HOM is small enough in size to pass the filter. Preliminary investigations showed that bacterial EPS suspensions possess much lower extinction coefficients at 270 nm than the HOM. By introduction of one dilution step before UV-absorption analysis the contribution of EPS to the total absorption signal could be neglected. By this method the HOM concentration in the column effluent at the last day of the recirculation was below the detection limit (ca. 2 mg/L DOC), the effluent was turbid from washed-off cells but not brown coloured. With progressing iron reduction the colour of all coated sand columns turned from ochre to deeply brown, which is typical for humic substances.

Moreover, all HOM-OC precipitated as an HOM coating was recovered at the end of the experiment by elution of the columns with 0.1 N NaOH, readjustment of the pH to 6.5, and UV measurement as described above. From these indications we derive that HOM is not significantly redissolved despite the reduction of ferric iron. Mobilization following a reductive dissolution of ferric iron at pH 6.5 is prevented by the presence of phosphate that forms slightly soluble ferrous iron species under these conditions. As shown in the above mentioned batch experiments, ferrous phosphate also strongly binds HOM hence not releasing any HOM to the effluent. In the absence of high concentrations of phosphate not all of the ferric iron was reduced. We may conclude that HOM is not redissolved as long as still enough ferric iron is available to retain the HOM on the sand grain surfaces. Future experiments will show whether anions that are more prevalent under natural conditions, such as sulphide or carbonate, also function as scavengers for ferrous iron, and whether these precipitates also inhibit a redissolution of the surface-HOM complexes.

3. Interactions between HOM and Hydrophobic Organic Compounds

Interactions between HOM and HOC range from reversible sorption equilibrium to the formation of bound residues by chemical reactions. While the latter process has been described to occur with specific functionalised compounds such as aromatic amines and reactive phenols [27, 28, 52, 53] and some very reactive PAHs [54] most of the hydrophobic xenobiotics interact with HOM only by reversible sorption or partitioning unless they are (biologically) transformed into more reactive metabolites. The degree of sorption between HOC and HOM is determined by the specific properties of both partners. The main driving force for sorption of HOC on HOM is the incompatibility of the HOC with the water phase, that is their hydrophobicity. Therefore, K_{OC} values for sorption of HOC on HOM can be roughly predicted by linear correlations with the octanol-water partition coefficients K_{OW} of the HOC [55-57]. However, these empirical K_{OC}-K_{OW} correlations are specific for certain compound classes. The predictive power of K_{OC}-K_{OW} correlations for grouped data including different compound classes is considerably worse [57]. This is due to the fact that the degree of sorption is not only influenced by the hydrophobicity of the HOC but also on the affinity between HOC and HOM, i.e., their ability to form intermolecular interactions. In case of reversible sorption these intermolecular interactions can comprise non-specific intermolecular interactions, hydrogen bonding or charge-transfer interactions. Several more sophisticated models have been developed, based for example on Flory-Huggins theory [58-61] or linear solvation energy relationships (LSERs) [62, 63]. These models that consider at least partly these intermolecular interactions between HOC and HOM and can, therefore, be applied to a broader range of compounds.

With respect to our concept of a HOM-based sorption barrier we were also interested in the question which structural properties of the HOM provide a high sorption potential for HOC. One approach to answer this question is to correlate sorption coefficients of HOM to a variety of structural parameters that can be experimentally determined. In order to consider a broad range of structurally different HOM we chose HOM of various origins. The model sorbate applied was pyrene representing the class of PAHs.

The 11 humic and fulvic acids investigated (table 3) were extracted and isolated according to the methods proposed by the IHSS. They included:

- aquatic humic acid (HA) as well as fulvic acid (FA) from Lake Hohloh, Germany (BW-HA and BW-FA)
- terrestrial HAs from the A_h horizon and the B_v horizon of a forest soil in Bitterfeld, Germany (SoilA-HA and SoilB-HA, respectively) and a moor (Moor-HA, Kleiner Kranichsee, Ore Mountains, Germany)
- humic substance-like polymers isolated from the water body and the sediments of a former lignite wastewater pond (Lake Schwelvollert, Germany) [64], referred to as anthropogenic HOM (Anth-DW-FA, Anth-DW-HA, Anth-Sed-FA, Anth-SW-FA and Anth-SW-HA; where DW, Sed and SW stand for deep water, sediments and surface water respectively as the sources of the HOM), and
- a commercially available, coal-derived HA from Roth Ltd., Germany (Com-HA).

Structural parameters of the HOM investigated were atomic ratios of O/H, O/C, (N+S+O)/C determined by elemental analysis, content of carboxylic groups (f_{COOH}) and total content of acidic groups ($f_{COOH+Ph-OH}$) determined by acid-base titration (both in mval g^{-1} HOM), UV absorptivity at 280 nm (ε_{280} in cm^2 mg$_{DOC}$$^{-1}$) as well as the following properties obtained from ^{1}H-NMR spectra: fraction of aromatic protons (f_{arom}, 6.3-8.1 ppm) and fraction of non-polar aliphatic protons (f_{NPAl}, 0.4-1.7 ppm) of total protons in ^{1}H-NMR spectra as well as the ratio of aromatic and non-polar aliphatic protons f_{arom}/f_{NPAl}.

The sorption experiments were performed using solid phase micro-extraction (SPME) [60, 65, 66].

In SPME a tiny silica fibre coated with a thin polymeric film samples only from the fraction of an analyte that is freely dissolved and not associated with the HOM. If the conditions are adjusted properly, the sampled amount is so little that the sorption equilibrium between HOC and HOM is practically not affected. Since this procedure does not significantly disturb the sorption equilibrium, SPME belongs to the so-called "non-invasive" methods. One must be aware that SPME, like the dialysis or solubility enhancement approaches, supplies activity related K_{OC} values rather than concentration-based partition coefficients [54, 65, 67]. We will later see in our discussion that this point may explain possible differences in K_{OC} values obtained by different methods. The main advantages of the use of SPME include
- the applicability to investigate both dissolved and particulate HOM,
- that the sorption equilibrium on HOM is not disturbed (thus also allowing multiple extractions), and
- that this approach can be applied for multicomponent systems and it is not limited to fluorophores as the fluorescent quenching technique (FQT) is.

A disadvantage of the SPME method is the slow equilibration between solute and fibre (especially in the headspace mode). Furthermore, one has to consider that since SPME is an activity based method it is sensitive to the conditions of the aqueous phase (e.g., ionic strength).

For sorption experiments, pyrene as a highly hydrophobic compound and a representative of the PAH family, was spiked as ethanolic solution into the various aqueous solutions (c_{HOM} = 50 to 1000 mg L^{-1}; pH = 7; 4 mM NaN$_3$; IS = 10 mM, NaN$_3$ + NaCl). The final concentration was 15 μg L^{-1} pyrene in reference solutions and 30 μg L^{-1} in solutions containing HOM. The HOM solutions were prepared by dissolving the solid HOM in dilute NaOH. After further dilution with deionised water, the final pH of the solution was adjusted by adding dilute HCl. The sorption and SPME experiments were conducted in 100 mL Erlenmeyer flasks equipped with PTFE-lined septa and glass-coated magnetic stirring bars. Sorption was allowed to come to equilibrium for at least 12 hours. The SPME was performed either in the solution extraction (direct) mode or in the headspace mode according to the procedure described in [65]. For the direct mode, the fibre was equilibrated with the HOM/pyrene solution over 6 h. For the headspace mode, the fibre was placed above the solution for 18 h. The loaded fibre was then transferred into a GC injector (splitless injection at 290°C for 3 min) for pyrene desorption. In order to simplify the quantitative treatment of the SPME data, we used conditions where the depletion of the analytes from the sample and their vapour fractions were insignificant: 7 μm polydimethylsiloxane (PDMS) fibre from Supelco, 80-100 mL sample volume, external calibration.

Tables 3 and 4 contain the sorption coefficients of pyrene as well as the results for the structural parameters of all HOM samples studied, see also [54].

The sorption coefficients of pyrene with the various HOM samples differ by about one order of magnitude. Similar differences between sorption coefficients of one and the same sorbate with HOM of various origins have also been reported by other authors [68-70]. They are in the expected range of variation. Fulvic acids (FAs) generally give lower sorption coefficients than the HAs of the same origin. Within the natural HAs the soil-derived HAs possess the highest sorption potential in accordance with literature data [68, 71]. The highest sorption coefficients within all HOM samples studied were determined for the anthropogenic HA isolated from 15 m depth (anaerobic zone) of a lignite wastewater pond and the commercially available coal-derived HA. Recently, Kulikova and Perminova [72] studied the sorption of atrazine to HOM of various origins and also reported that a coal-derived humic acid showed the highest sorption potential.

Table 3. Sorption coefficients of pyrene (log K_{DOC}) and structural parameters expressing the polarity of the HOM.

Sample	log K_{DOC}	O/H	O/C	(N+S+O)/C	f_{COOH}[1] [mmol g^{-1}]	$f_{COOH+OH}$[2] [mmol g^{-1}]
Anth-DW	5.02±0.07	0.22	0.21	0.307	2.17	3.67
Com-HA	4.95±0.07	0.36	0.37	0.407	3.41	5.40
Anth-Sed-HA	4.80±0.08	0.36	0.35	0.449	2.27	3.41
Moor-HA	4.77±0.08	0.32	0.38	0.413	2.45	3.07
SoilA-HA	4.70±0.08	0.41	0.50	0.58	2.60	3.16
SoilB-HA	4.47±0.10	0.36	0.50	0.602	n.d.[3]	n.d.
Anth-SW-HA	4.41±0.11	0.41	0.38	0.487	2.49	4.42
Anth-DW-FA	4.35±0.11	0.44	0.49	0.596	2.17	3.67
Anth-SW-FA	4.23±0.13	0.48	0.50	0.603	3.22	5.06
BW-HA	4.21±0.13	0.65	0.52	0.543	3.05	5.27
BW-FA	4.01±0.18	0.68	0.57	0.589	3.83	5.74

[1] mass related carboxyl acidity, calculated from the proton capacity of the HOM in the range of pH 3 to pH 7.5.

[2] mass related phenol and carboxyl acidity, calculated from the proton capacity of the HOM in the range of pH 3 to pH 10.

[3] not determined.

Table 4. Structural parameters expressing the aromaticity or aliphaticity of the HOM.

Sample	ε_{280} [cm^{2}mg$_{DOC}^{-1}$]	f_{arom}	f_{NPAl}	f_{arom}/f_{NPAl}
Anth-DW	65.8	0.179	0.453	0.395
Com-HA	83.0	0.287	0.433	0.663
Anth-Sed-HA	71.9	0.174	0.437	0.398
Moor-HA	50.8	0.185	0.288	0.642
SoilA-HA	57.8	0.112	0.371	0.302
SoilB-HA	45.6	0.095	0.515	0.184
Anth-SW-HA	44.2	0.136	0.343	0.396
Anth-DW-FA	46.4	0.128	0.470	0.272
Anth-SW-FA	43.3	0.143	0.395	0.362
BW-HA	46.3	0.133	0.230	0.578
BW-FA	39.2	0.096	0.276	0.35

In general, one can consider two cases for the correlation between structural parameters and sorption affinities. If one regards the sorption affinity of the various building blocks i to be more or less independent of their environment in the macromolecule, then the overall sorption coefficient K_{DOC} can be approximated as a linear combination of single $K_{DOC,i}$ values (eq. 1)

$$K_{DOC} = \sum_i x_i \cdot K_{DOC,i} \tag{1}$$

(with x_i being the proportion of e.g. aromatic, aliphatic, polar... units)

In the second scenario, a change in the relative amount, size or arrangement of the structural units may result in a significant change of their sorption affinities. As an example one can consider an increase in aromatic centres which leads to the formation of larger aromatic regions with a higher affinity towards the sorbate than that of all isolated aromatic units combined. In such a case, we have to consider a more complex correlation between structural parameters and the overall sorption coefficient, possibly an exponential correlation (eq. 2) based on the exponential relationship between K_{DOC} and the free enthalpy of interaction $G_{interaction}$.

$$K_{DOC} \sim e^{-\Delta G\ interaction} \rightarrow log\ K_{DOC} \sim x_i \tag{2}$$

This relation assumes a linear dependence between $G_{interaction}$ and the proportion x_i of a certain constituent i in the sorbent, which is not expected to be completely fulfilled in reality. For the complex and heterogeneous humic substances, a superposition of both effects (symbolized by eqs 1 and 2) is to be expected.

Table 4 shows the results for correlations between the sorption coefficients and the various structural parameters according to both approaches $K_{DOC} = f(x)$ and $log\ K_{DOC} = f(x)$. For several of the structural parameters studied significant correlations with the sorption coefficient of pyrene of the HOM were obtained. Both approaches, the linear regression according to $K_{DOC} = f(x)$ or $log\ K_{DOC} = f(x)$ differ only slightly in their r^2 values for each of the parameters. In general, all structural parameters which are a measure of the content of functional groups in the HOM and consequently reflect their polarity, such as the ratios O/H, O/C, (N+S+O)/C as well as the content of acidic groups, show correlation equations with a negative slope. This is in accordance to the results of other authors who also found a negative effect of the polarity of HOM (calculated from atomic ratios) on the sorption potential towards HOC [73,74]. The weak correlation with the content of acidic groups (Table 5) indicates that the polarity of the HOM is not only determined by carboxylic and/or phenolic groups but also to a large extent by the fraction of carbohydrate structures (parameters O/C and (N+O+S)/C).

Furthermore, significant correlations were also found between the sorption potential of the HOM samples and their aromaticity measured as absorptivity at 280 nm or by ^{1}H-NMR spectra. A strong positive relationship between aromaticity of HOM and sorption potential toward PAHs was also found by other authors [69, 70, 72, 75]. Chin et al. [69] reported a negative effect of the content of non-polar aliphatic carbon (determined by ^{13}C-NMR) on the sorption coefficient of pyrene as determined with five aquatic humic substances, which seems to underline the importance of aromatic structures. In contrast, Chefetz et al. [76, 77] found a positive trend between the K_{OC} level of pyrene and the aliphaticity calculated from ^{13}C-NMR spectra for natural organic matter of various origins. In our study no significant correlation between pyrene sorption and the content of highly shielded aliphatic protons in the ^{1}H-NMR spectrum as a measure of the content of non-polar aliphatic structures was observed.

Table 5. Results of the linear regression according to $K_{DOC}=f(x)$ or log $K_{DOC}=f(x)$ for the correlations between the sorption coefficients of pyrene and different structural parameters (x) of HOM of various origins.

structural parameter (x)	[number of samples]	$K_{DOC} = f(x)$ r^2; F-Test	log $K_{DOC} = f(x)$ r^2; F-Test	slope
O/H	[11]	0.73; s[1]	0.72; s	positive
O/C	[11]	0.74; s	0.69; s	negative
(N+S+O)/C	[11]	0.75; s	0.64; s	negative
f_{COOH}	[9]	0.31; ns[2]	0.,44; s	negative
$f_{COOH+Ph-OH}$	[9]	0.28; ns	0.45; s	negative
ε_{280}	[11]	0.75; s	0.72; s	positive
f_{Ar}	[11]	0.54; s	0.48; s	positive
f_{NPAl}	[11]	0.13; ns	0.18; ns	positive
f_{Ar}/f_{NPAl}	[11]	0.15; ns	0.11; ns	positive

[1] significant correlation according to F-test (P = 0.95).
[2] correlation is not significant according to F-test (P = 0.95).

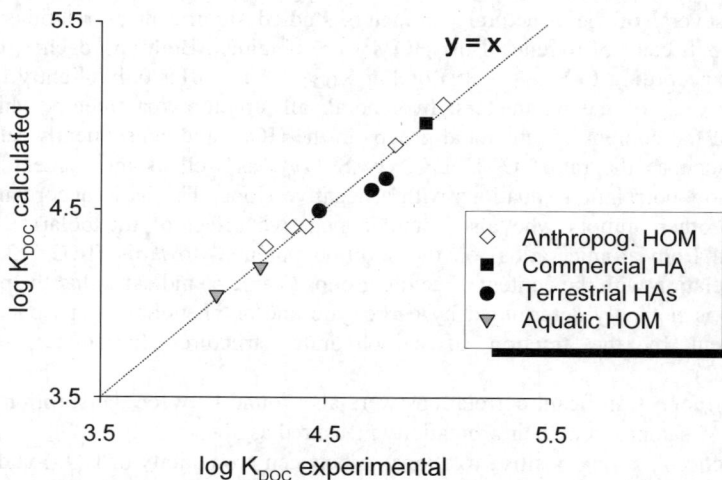

Figure 4. Calculated vs. experimental log K_{DOC} values for pyrene with HOM of various origins (Table 3); calculation according to equation 3.

However, even with the most significant parameters (e.g. O/H, ε_{280}) the predictive power of the correlations is rather poor ($r^2 \leq 0.75$). Obviously, a single parameter alone is not able to sufficiently describe the sorption behaviour of HOM of various origins.

Taking into account *two* characteristic properties of HOM, their polarity *and* aromaticity, we obtained an excellent fit ($r^2 = 0.97$) for our data between log K_{DOC} of pyrene, the O/H atomic ratio and the UV absorptivity at 280 nm [cm^2 mg_{DOC}^{-1}] using eq. 3 (see figure 4):

$$log\,K_{DOC} = 1.02 - 1.31\,log\,[O/H] + 1.75\,log\,\varepsilon_{280} \qquad (3)$$

Because of possible intercorrelations between several HOM properties (e.g., aromaticity and molecular weight [69], mechanistic interpretations of empirical correlations between sorption potential and specific structural parameters have to be treated with caution. However, the correlation described above might be a guideline for the selection of a HOM that is best suited for establishing a sorption barrier for PAHs. For the various HOM samples studied coal-derived HOM best incorporate both favourable properties: low polarity and high aromaticity.

4. Sorption Potential of Mineral-Associated, Flocculated and Dissolved HOM

The sorption capability of HOM in the dissolved and the mineral-associated state has been a matter of interest both for methodological and practical reasons. In several studies non-invasive methods like the fluorescence quenching technique (FQT) were applied to compare K_{OC} values of HOM in dissolved and solid state. K_{OC} values obtained for the association of phenanthrene to dissolved organic matter (DOM) were found to be about one order of magnitude higher than the corresponding K_{OC} data for mineral-associated HOM [78-80].

With respect to our intention to generate a sorptive/reactive barrier made of mineral-associated HOM coatings, these findings prompted us to scrutinize these phenomena.

The question arises, if the decrease in K_{OC} data measured for mineral-associated HOM is a real phenomenon (e.g., due to steric effects), or if it is an artefact of the analytical method, FQT. There is a lot of debate in the literature about the validity of FQT for determining sorption coefficients [82]. Problems can arise from several effects that must be taken into account: inner filter effects caused by DOM, cross-quenching by other solutes such as oxygen, and degradation of the analyte by UV light. Unlike the references cited above, the determination of K_{OC} data in the present study was conducted using SPME (see paragraph 3).

The sorption behaviour of flocculated HOM was also investigated within this study. The experimental determination of sorption coefficients is depicted schematically in figure 5 and shall be described briefly: Three 250 mL flasks were each filled with a DOM solution ($c_{DOM} = 200$ mg L^{-1}, DOM: commercial humic acid, purchased from Roth Ltd., Germany), leaving a headspace volume of 75 mL. 200 mg L^{-1} sodium azide was added to suppress microbial activity. Sodium chloride was also added to give a concentration of 0.1 M ensuring constant ionic strength. To investigate mineral-associated HOM (denoted as SOM for sorbed organic matter), the mineral hematite (Fe_2O_3), which was prepared according to [81], was suspended in the DOM solution by rigorous stirring. The pH was set to pH 4.5 for the dissolved and hematite-associated HOM. A flocculated HOM (denoted as FOM) was simulated by preparation of a ferric iron-humate. For this purpose a 0.1 M $FeCl_3$-solution was added to the same humic material and adjusted to pH 5.5. Both the flocculation and the sorption on hematite resulted in an almost complete removal of DOM. A mixture consisting of various PAHs

Table 6. Sorption coefficients of dissolved (K_{DOC}), sorbed (K_{SOC}), and flocculated (K_{FOC}) HOM normalized to the OC contents. $\bar{s}$ standard deviation of 20 replicate measurements (5 different PDMS fibres in fourfold replicate measurement). K_{Min} denotes the sorption to the solid material in absence of HOM.

HOM	Tetrahydro-thiophene		2,4,4-Trimethyl-cyclopentanone		Naphthalene		Acenaphthylene		Phenanthrene	
	$\log \overline{K_{OC}}$	$\bar{s}$	$\log \overline{K_{OC}}$	$\bar{s}$	$\log \overline{K_{OC}}$	$\bar{s}$	$\log \overline{K_{OC}}$	$\bar{s}$	$\log \overline{K_{OC}}$	$\bar{s}$
sorbed to Fe_2O_3 (pH 4.5)	3.78	0.22	3.46	0.40	3.56	0.43	4.21	0.26	4.78	0.07
freely dissolved (pH 4.5)	3.82	0.20	3.68	0.27	3.37	0.57	4.23	0.15	4.64	0.06
flocculated (pH 5.5)	3.76	0.22	3.66	0.24	3.45	0.30	4.18	0.18	4.68	0.05
freely dissolved (pH 5.5)	3.97	0.17	3.97	0.20	3.57	0.32	3.88	0.10	4.49	0.07
	$\log \overline{K_{Min}}$	$\bar{s}$	$\log \overline{K_{Min}}$	$\bar{s}$	$\log \overline{K_{Min}}$	$\bar{s}$	$\log \overline{K_{Min}}$	$\bar{s}$	$\log \overline{K_{Min}}$	$\bar{s}$
Fe_2O_3 blanc (pH 4.5)	rcn. with matrix		0.98	0.71	no sorption		rcn. with matrix		0.78	1.06
$Fe(OH)_3$-flocculate blanc	2.22	1.08	2.12	1.07	no sorption		rcn. with matrix		1.99	0.78

1) Tetrahydrothiophene and acenaphthylene reacted with the blanc hematite surfaces, no acceptable standard deviations could be obtained, data where omitted.

(naphthalene through phenanthrene) as well as tetrahydrothiophene and 2,2,4-trimethylcyclopentanone was spiked to each of the flasks from acetonic solution to give a concentration of 100 µg L^{-1} per compound. After equilibrating the multiphase systems for one day, SPME fibres (7 µm PDMS coating) were pierced through an PTFE-coated septum and placed in the headspace above the solution. An SPME sampling time of 18 h was maintained (typical overnight procedure to ensure equilibrium conditions in the headspace SPME; see [65]. Each measurement was carried out in four replicates.

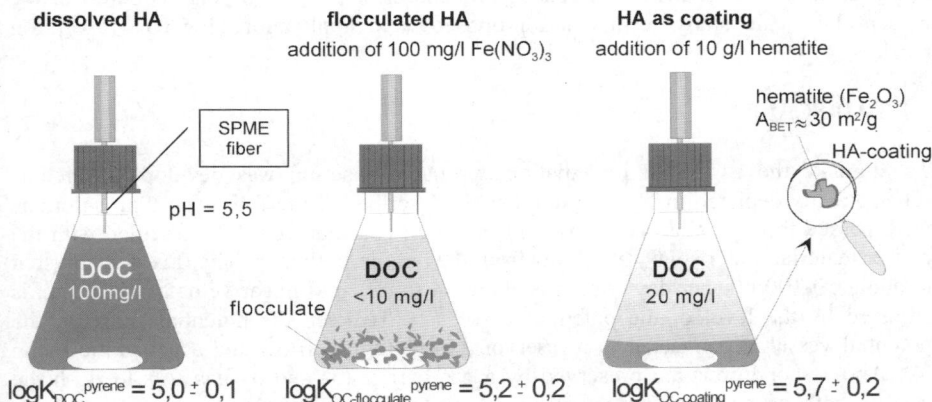

Figure 5. Determination of sorption coefficients by means of headspace SPME on DOM, flocculated HOM, and mineral-associated HOM.

The results, which are summarized in Table 6, give strong evidence that there is no significant deviation in sorption data for a given solute dependent on the state of HOM "aggregation". Therefore, the findings in [79] and [80] are assumed to be associated with peculiarities of the fluorescence quenching technique. A methodological comparison of this technique and our SPME approach [54, 67, 83] confirmed that K_{DOC} data of different PAHs measured by FQT were up to one order of magnitude higher than data obtained by using SPME. These methods differ in their definition of the sorbed state. In sorption experiments with dissolved sorbents, FQT includes all solute molecules which are sufficiently close to the sorbent that their fluorescence signal is quenched. When the sorbent is in the particulate state, the sorbed fraction of the solute and the particular sorbent are separated from the solution phase. The concentration of the freely-dissolved fraction of the solute is determined by a fluorescence measurement. With the FQT approach the sorbed state is method defined [67]. In comparison, SPME is an activity-based method, where all effects caused by the sorbent which contribute to a decrease in the thermodynamic activity of the solute are defined as sorption or sorptive interaction, regardless of the mechanism, strength or life-time of binding. Whether the sorbent is in the dissolved or particulate state does not change this principle. Therefore, we evaluate the SPME approach described above to be more reliable for a comparison of the sorption potential of OM in its different states (dissolved versus particulate).

Consequently, considering our concept of generating a mineral-associated HOM barrier, we expect that the sorption potential of the HOM remains completely preserved after the immobilization of HOM.

5. Engineering Procedures for Immobilization of HOM on Selected Aquifer Materials

In this section we describe HOM-coating procedures under flow-through conditions. We present lab scale study results and proposed field applications for HOM sorption barriers.

5.1. LAB-SCALE

To simulate the suggested remediation approach, a set-up was developed which is schematically depicted in figure 6 and briefly described in the following: Salt solutions were passed through glass columns (5) (length 20 cm, diameter 2.2 cm) filled with the carrier material under study, with flow from the bottom to the top. The flux was adjusted to about 50-100 mL per day (which is about 34 to 68 cm/d linear velocity). The flux is adjusted by the level of the potential vessel (2). To keep the potential constant, the potential vessel is connected to a reservoir (1) via an overflow and a peristaltic pump (4). Anoxic conditions are preserved by a gas barrier (3) and by gas frits to flush the vessels with argon (7). The set-up in figure 6 is thought to be quite universal for investigation of the immobilization of the HOM as well as the retardation of the HOC on the immobilized coatings.

In the latter case, a conventional solid phase extraction (SPE) cartridge (6) filled with 200 mg C18 and 200 mg SDB-2 material (both from Baker) allows to monitor the breakthrough of organic analytes of a wide range of polarity. For studying the HOC breakthrough the peristaltic pump is not activated to prevent sorption of HOC onto the tubing. 200 mg L^{-1} NaN_3 was added to the influent in order to suppress biodegradation of the HOC under investigation.

To demonstrate the principle functioning of an HOM-barrier under flow-through conditions we first had to build up a ferric iron barrier. Lacking representative aquifer materials from polluted sites in native condition (e.g., ferrous iron surface species, exclusion of oxygen, undisturbed texture), we injected ferric iron solutions to precipitate ferric hydroxide onto mineral surfaces.

The suggested procedure depicted schematically in figure 1 includes the following steps
  – injection of ferric iron salt solutions ($FeCl_3$) and precipitation as ferric hydroxide as a result of increased pH or reaction with phosphate buffer (alternative to ferrous iron oxidation)
  – injection of dissolved HOM resulting in its immobilization
  – examination of the retardation of HOC on the sorptive/reactive barrier.

In succeeding to build up a ferric iron coating we passed a 0.1 M $FeCl_3$-solution (pre-titrated to pH 2.5) alternating with either 0.1 mM NaOH (A) or 0.5 M phosphate buffer (pH 6.5) (B) through the columns. Several experiments with a pre-cleaned quartz

223

sand (fraction 200-630 μm, BET(N₂)-surface = 0.4 m²/g, porosity = 0.356) led to 0.7 mg Fe/g sand when precipitated by NaOH and 2.2 mg Fe/g sand when phosphate buffer was added.

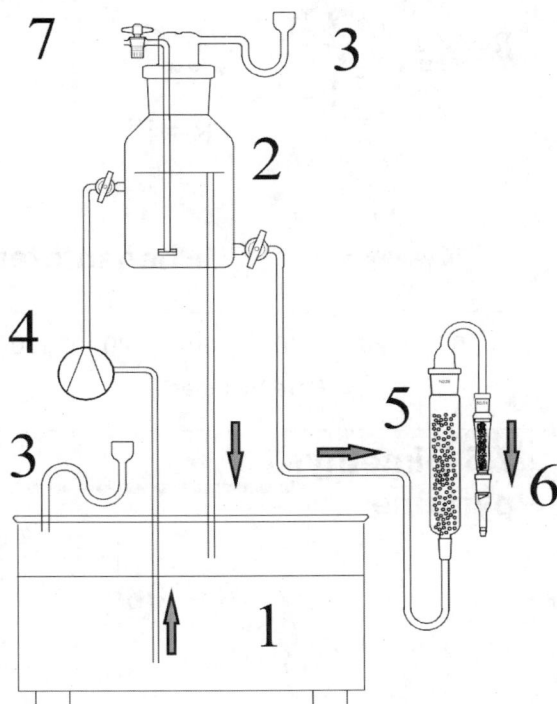

Figure 6. Apparatus for immobilization of HOM under bench scale flow through conditions. Numbered components are explained in the text.

According to these results the immobilization of a reference HOM (200 mg L^{-1} commercial humic acid, Roth Ltd., Germany) resulted in 0.26 mg OC/g sand (A) and 0.35 mg OC/g sand (B). The iron and HOM precipitates dyed the sand homogeneously and did not lead to clogging effects, most likely due to a high portion of voids and very little dispersivity of the applied sand grains. Yet, batch coating experiments confirmed that iron covers the quartz surface only on "hot spots" in cleavages and surface cracks. XPS investigations also revealed less than 5% iron at the quartz surface on an atomic scale.

The HOC mixture applied in the retardation studies represented typical coal pyrolysis contaminants as they occur e.g. downstream the above mentioned tar factory site in Rositz, Germany [3]. The compounds which were used cover a log K$_{OW}$-range from 1.79 to 4.46 and include compound classes like ketones, phenols, pyridines, quinolines, and PAHs. Each compound was spiked into a 0.1 M NaCl feed solution (pH 6.5) to give a concentration of 100 μg L^{-1}. Analysis of the solid phase extracts at the outlet of the columns was performed by GC/MS in SIM mode.

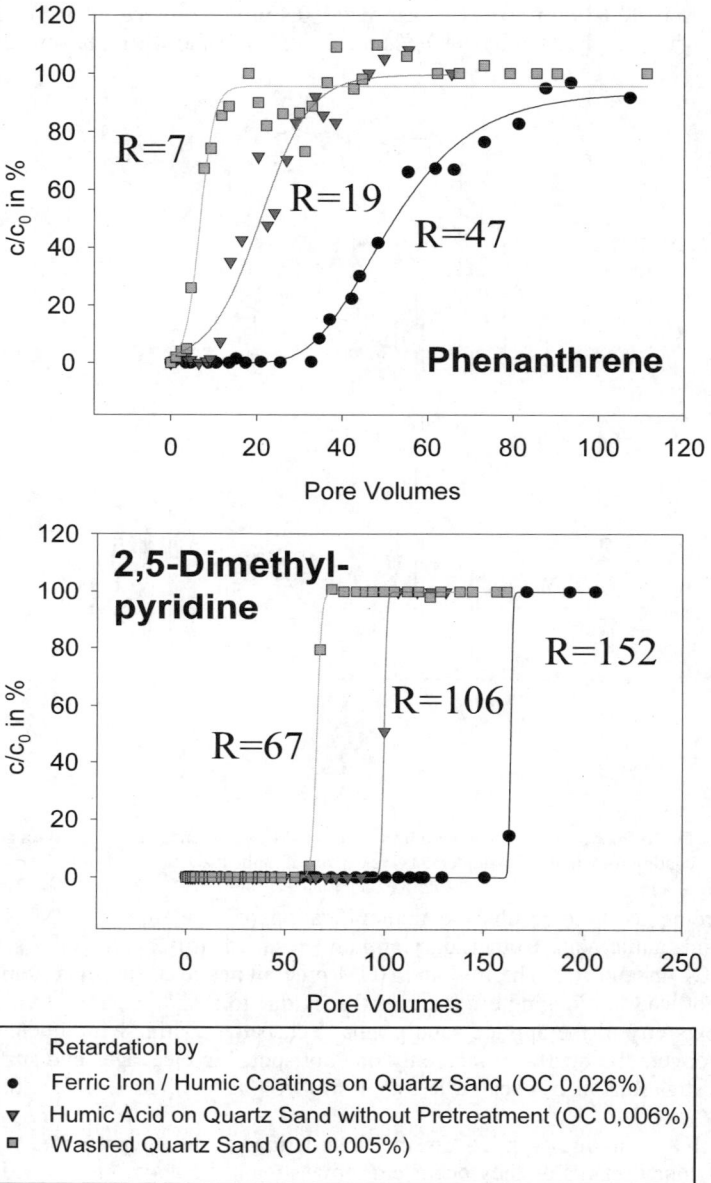

Figure 7. Breakthrough curves of two HOC in columns of quartz sand coated with Roth humic acid. c/c_0 denotes the ratio of the HOC concentration at the outlet of the column and its initial concentration in the reservoir. The total pore volume of a column was determined to be 21 mL.

Taking hydrophobic partitioning as a basis to describe sorption of HOC on OC-coated aquifer materials retardation factors (R) can be calculated according to equation 4 [84].

$$R = 1 + \frac{\rho}{\Theta} K_{OC} f_{OC} = \frac{\overline{u}}{\overline{u}_{HOC}} \qquad (4)$$

Where ρ is the density of the wet aquifer matrix (in our case 1.695 g cm^{-3}), Θ – the porosity of the sand (in our case 0.356), and K_{OC} – the distribution coefficient of the HOC between HOM and the aqueous phase, normalized to the OC contents of the HOM, and f_{OC} – a factor representing the OC content in the water saturated sand. For calculation of R we used the K_{FOM}-values listed in the paragraph above. R is also experimentally accessible by relating the average velocity of an inert tracer ($\overline{u}$) to the migration velocity of an HOC ($\overline{u}_{HOC}$).

The breakthrough curves of two selected compounds are shown in figure 7.

The environmental behaviour of phenanthrene as an representative of the class of PAHs is mainly determined by hydrophobic partitioning (log K_{OW} = 4.46). When comparing the breakthrough curves for this compound on pure quartz sand and HOM-coated sand with the breakthrough of an inert tracer (KNO$_3$, not shown) a significant sorption enhancement on HOM coated sand is noticed (R_{exp} = 47, see table 7). However, a small retardation effect is observed also for a "clean" quartz sand surface (R_{exp} = 7).

2,5-Dimethylpyridine (2,5-DMPy), a more polar compound of the Rositz plume (log K_{OW} = 1.8), is even more retarded (R_{exp} up to 150) by the barrier than the much more hydrophobic phenanthrene. The reason for this surprising observation cannot be explained only by hydrophobic interactions. Specific interactions of the N-base with mineral surfaces may account for this effect. The fact that the columns prepared by the phosphate method completely loose their ability to retard 2,5-DMPy as compared with a clean quartz also supports this idea. Phosphate sorbed on the surface of ferric hydroxide may block reactive Fe-binding sites, which are assumed to be responsible for strong interactions with 2,5-DMPy.

Tables 7 and 8 represent measured and calculated retardation factors for several HOC studied. Table 8 outlines considerable decreases in pollutant mobility for highly hydrophobic or specifically interacting pollutants when quartz sand is coated by HOM albeit the immobilized OC is fairly low (f_{OC} = 0.001). Consequently, measured R-values smaller than those calculated according to eq. 4 based on established partitioning equilibrium may imply restricted accessibility of immobilized HOM for the permeating HOC (dead end voids, stagnant water, pore exclusions by clogged ferric iron precipitates etc.).

Moreover, the iron/phosphate-HOM barrier showed significantly less HOC retardation even if the OC-contents were higher than in HOM-barriers prepared by FeCl$_3$/NaOH. These results suggest a dependency of HOC-sorption on the accessibility of humic organic carbon [85].

Table 7. Breakthrough behaviour of 5 HOC on HOM-coated quartz sand columns – Experimentally determined retardation factors.

Coating Procedure	Naphthalene	Acenaphthylene	Phenanthrene	2,5-Dimethyl-pyridine	2,6-Dimethyl-quinoline
Precipitation of $Fe(OH)_3$ by OH^-	2.9	6.7	43	152	>189
Precipitation of $Fe(OH)_3$ by OH^-	n.d.	7.6	47	123	>189
Precipitation of $Fe(OH)_3$ by phosphate	2.0	4.0	35	ca.100	189
Precipitation of $Fe(OH)_3$ by phosphate	1.7	3.1	18	75	>189
No coating (pure quartz)	1.6	1.7	6.6	67	>189

Table 8. Calculated retardation factors according to pure hydrophobic partitioning based on the OC values given in the text.

Coating Procedure	Naphthalene	Acenaphthylene	Phenanthrene
Precipitation of $Fe(OH)_3$ by OH^-	4.5	19.7	60.3
Precipitation of $Fe(OH)_3$ by OH^-	4.2	18.3	55.7
Precipitation of $Fe(OH)_3$ by phosphate	5.7	26.2	80.8
Precipitation of $Fe(OH)_3$ by phosphate	4.0	16.9	51.1
No coating (pure quartz)	1.7	4.6	12.4

5.2. FIELD SCALE

The *in situ* construction of permeable HOM barriers to retain dissolved contaminants comprises several considerations that need to be addressed here.

HOM barriers can be created on native or artificially coated mineral surfaces of the aquifer. One general recipe to build up an HOM barrier cannot be given. The way the barrier needs to be constructed is rather site dependent. Firstly; an extended geological survey of the polluted aquifer is required. This survey should include geohydraulic tests, information about the grain size distribution, clay, chalk, OC and iron content, dissolved ferrous iron concentration, and mineral surface textures. If clays are present, they frequently form indigenous surface coatings with increased HOM retention potential.

Laboratory results implied (see above) that metal ion surface precipitates must be generated in case the majority of the aquifer material is made of low surface quartz and other silicate materials.

If the groundwater is rich in dissolved ferrous iron, aeration technologies may attribute ferric iron coatings in advance to provide for a better HOM sorption onto the aquifer material. Field tests are needed to confirm the stability of such barriers at shutoff of the oxygen supply.

As with the mediator, the HOM-coating must be long-term stable and withstand leaching in the specific anoxic groundwater environment. For aluminium hydroxide coatings may not be allowed by the local authorities, iron coatings may be more suitable instead. Here the redox instability of iron and the potentially associated barrier leaching at reduction of ferric iron must be taken into consideration. Assuming the pH is kept in the neutral range, the supply of phosphate may assist in generating milieu stable ferrous phosphate surface coatings which in turn are highly susceptible to bind HOM.

In order to avoid bypassing of the pollutant plume, the original hydrodynamic conditions have to be preserved as best as possible. In this respect, thin and uniform mediator and HOM coatings on the aquifer matrix are to be pursued. Hence, we would like to suggest to create sorptive/reactive areas of moderate OC content within an aquifer rather than barriers of spatially high HOM content. The production of tailored HOM for this specific field of application might be considered; e.g. the introduction of chelating groups into the HOM backbone to retard heavy metal contaminations or the enhancement of the HOM's hydrophobicity when considering HOC as targets.

A locally controlled precipitation of the injected HOM can potentially be achieved in the field by two methods.

First, we will consider the case where HOM shows non-conservative flow behaviour, as occurring especially in aquifers with significant contents of inorganic surface precipitates and clays [3, 86, 87]. The concept in this case is either to directly flush the HOM solutions into the aquifer, if the aquifer grains are already coated with ferric iron or clay precipitates, or to use solutions of polyvalent cations (such as Ca^{2+}), that show almost conservative flow behaviour, subsequent to HOM injections which is retarded due to interactions with the mineral surfaces of the aquifer material. As a consequence of the different flow velocities the HOM becomes caught up by the polyvalent cations and flocculated. Studying the velocities of both reactants in the aquifer the barrier can be placed that way at a desired spot [88]. While the latter method should result in high local

OC spots, uncontrolled flocculation of the HOM in the voids of the aquifer may result in an undesirable decrease in hydraulic conductivity [Balcke, unpublished].

The approach preferred in this study is to coat the mineral surfaces of the aquifer with polyvalent ion precipitates so the sorption affinity towards HOM increases. Ferric iron surface coatings can be established in practice by temporary groundwater oxygenation if the aquifer carries enough dissolved ferrous iron or ferrous precipitates, such as siderite or pyrite. In chalk-rich aquifers or those of high alkalinity dilute acidic ferric iron solution injections will be neutralized and dispersed finally precipitating ferrous oxyhydrates onto the aquifer mineral surfaces rendering them more susceptible for sorption of HOM. If the aquifer contains neither chalk nor a sufficient clay content, based on an intensive geological survey the utilization of transverse mixing of two spatially separated injections into the aquifer may be possible. These could include mixing of solutions of different density (e.g., high salinity ferric iron solution, low salinity dilute sodium hydroxide solution) or mixing of solutions by transverse dispersion with precipitation in the contact mixing zone.

In a future project we intend to investigate the construction and effectiveness of a HOM barrier adapted to the specific conditions of a contaminated site. The field study site will be the grounds of a former wood preserving plant that is contaminated by tar oil. The main groundwater contaminants consist of HOC suitable to be retarded/fixed by a HOM barrier (PAHs and their metabolites as well as heterocyclic compounds). Within this pilot scale field test the HOM barrier will be created according to both previously mentioned strategies.

In summary, we may conclude that HOM sorption/ reactive barriers possess a high potential to retard dissolved environmental pollutants in comparison to the groundwater flux. Even small effective retardation coefficients (e.g., order of magnitude 10), caused by low OC enrichment, will give rise to relevantly longer exposure to microbial attacks and covalent fixation to surface bound HOM.

6. References

1. Pankow, J.F., Cherry, J.A. (1996) *Dense Chlorinated Solvents & Other DNAPLs in Groundwater. History, Behavior, and Remediation*, Waterloo Press, Rockwood, Canada.
2. Schiedek, T. (1997) *Literaturstudie zum natürlichen Rückhalt / Abbau von Schadstoffen im Grundwasser*, Eberhard-Karls-Universtität Tübingen, Lehrstuhl für Angewandte Geologie, Tübingen.
3. Balcke, G.U. (2000) *Anthropogene Huminstoffe als Sorbenzien und Reagenzien zur Immobilisierung von organischen Schadstoffen in Grundwässern*, PhD Thesis. Centre for Environmental Research, Dep. of Remediation Research, Leipzig, University of Leipzig.
4. Dermietzel, J., Vieth, A. (2002) Chloroaromatics in groundwater: Chances of bioremediation, *Environ. Geol.* **41**, 683-689.
5. Rieger, P.G., Meier, H.M., Gerle, M., et al. (2002) Xenobiotics in the environment: present and future strategies to obviate the problem of biological persistence, *J. Biotechnol.* **94**, 101-123.
6. Leahy, J.G., Olsen, R.H. (1997) Kinetics of toluene degradation by toluene-oxidizing bacteria as a function of oxygen concentration, and the effect of nitrate, *FEMS Microbiol. Ecol.* **23**, 23-30.
7. Kostka, J.E., Viehweger, R., Stucki, J.W. (1999) Respiration and dissolution of iron(III) containing clay minerals by bacteria, *Environ. Sci. Technol.* **33** (18), 3127-3133.
8. Durant, N.D., Jonkers, C.A.A., Bouwer, E.J. (1997) *Spatial variability in the naphthalene mineralization response to oxygen, nitrate, and orthophosphate amendments in MGP aquifer sediments*, Biodegradation **8**, 77-86.

229

9. Dybas, M.J., Hyndman, D.W., Heine, R., Tiedje, J., Linning, K., Wiggert, D., Voice, T., Zhao, X., Dybas, L., Criddle, C.S. (2002) Development, operation, and long-term performance of a full-scale biocurtain utilizing bioaugmentation, *Environ. Sci. Technol.* **36**, 3635-3644.
10. Renner, R. (1997) Pump-and-treat enters the supercomputer age, *Environ. Sci. Technol./News* **31**, 30A-31A.
11. *Guide to Documenting and Managing Cost and Performance Information for Remediation Projects - Revised Version* (1998) EPA 542-B-98-007, http://www.frtr.gov/cost/guide.pdf.
12. *Treatment Technologies for Site Cleanup: Annual Status Report (ASR)* (2001) Tenth Edition, EPA 542-R-01-004, http://clu-in.org/products/asr/Home.htm.
13. McGovern, T., Guerin, T.F., Horner, S., Davey, B. (2002) Design, construction and operation of a funnel and gate in-situ permeable reactive barrier for remediation of petroleum hydrocarbons in groundwater, *Water, Air, Soil Pollut.* **136**, 11-31.
14. Guerin, T.F., Horner, S., McGovern, T., Davey, B. (2002) An application of permeable reactive barrier technology to petroleum hydrocarbon contaminated groundwater, *Water Res.* **36**, 15-24.
15. Yerushalmi, L., Manuel, M.F., Guiot, S.R. (1999) Biodegradation of gasoline and BTEX in a microaerophilic biobarrier, *Biodegradation* **10**, 341-352.
16. Siegrist, R.L., Lowe, K.S., Murdoch, L.C., Case, T.L., Pickering, D.A. (1999) In situ oxidation by fracture emplaced reactive solids, *J. Environ. Eng.-ASCE* **125**, 429-440.
17. Aelion, C.M., Kirtland, B.C. (2000) Physical versus biological hydrocarbon removal during air sparging and soil vapor extraction, *Environ. Sci. Technol.* **34**, 3167-3173.
18. Durant, L.P.W., D'Adamo, P.C., Bouwer, E.J. (1999) Aromatic hydrocarbon biodegradation with mixtures of O_2 and NO_3^- as electron acceptors, *Environ. Engin. Sci.* **16**, 487-500.
19. Nielsen, A.T., Amandusson, H., Bjorklund, R., Dannetun, H., Ejlertsson, J., Ekedahl, L.G., Lundstrom, I., Svensson, B.H. (2001) Hydrogen production from organic waste, *Int. J. Hydrogen Energy* **26**, 547-550.
20. Gillham, R.W., o Hannesin, S.F. (1994) Enhanced degradation of halogenated aliphatics by zero-valent iron, *Groundwater* **32**, 958-967.
21. Orth, W. S., Gillham, R. W. (1996) Dechlorination of trichloroethene in aqueous solution using Fe^0, *Environ. Sci. Technol.* **30**, 66-71.
22. Richardson, J.P., Nicklow, J.W. (2002) In situ permeable reactive barriers for groundwater contamination, *Soil Sed. Cont.* **11**, 241-268.
23. Sanjay, H.-G., Fataftah, A.K., Walia, D., Srivastava, K. (1999) Humsaorb CS: A humic acid-based adsorbent to remove organic and inorganic contaminants, in G. Davies and E.A. Ghabbour (eds.), *Understanding humic substances, advanced methods, properties and application*, Royal Society of Chemistry, Cambridge, pp. 241-254.
24. Wang, Y.L., Banziger, J., Dubin, P.L., Filippelli, G., Nuraje, N. (2001) Adsorptive partitioning of an organic compound onto polyelectrolyte-immobilized micelles on porous glass and sand, *Environ. Sci. Technol.* **35**(12), 2608-2611.
25. Sun, S., Jaffe, P.R. (1996) Sorption of phenanthrene from water onto alumina coated with dianionic surfactants, *Environ. Sci. Technol.* **30**, 2906-2913.
26. Smith, J.A., Sahoo, D. McLellan, H.M. Imbrigiotta, T.E. (1997) Surfactant-enhanced remediation of a trichloroethene-contaminated aquifer: 1. Transport of Triton x-100, *Environ. Sci. Technol.* **31**, 3565-3572.
27. Achtnich, C., Fernandes, E., Bollag, J.-M., Knackmuss, H.-J., Lenke, H. (1999) Covalent binding of reduced metabolites of [^{15}N3]TNT to soil organic matter during a bioremediation process analyzed by ^{15}N NMR spectroscopy, *Environ. Sci. Technol.* **33**, 4448-4456.
28. Bollag, J.-M. (1999) Effect of humic constituents on the transformation of chlorinated phenols and anilines in the presence of oxidoreductive enzymes or birnessite, *Environ. Sci. Technol.* **33**, 2028-2034.
29. Verstraete, W., Devliegher, W. (1996) Formation of non-bioavailable organic residues in soil: Perspectives for site remediation, *Biodegradation* **7**, 471-485.
30. Dunnivant, F.M., Schwarzenbach, R., Macalady, D.L. (1992) Reduction of substituted nitrobenzenes in aqueous solutions containing NOM, *Environ. Sci. Technol.* **26**, 2133-2145.
31. Bradley, P.M., Chapelle, F.R., Lovley, D.R. (1998) Humic acids as electron acceptors for anaerobic microbial oxidation of vinyl chloride and dichlorethene, *Appl. Envir. Microbiol.* **64**, 3102-3105.
32. Lovley, D.R., Fraga, J.L., Coates, J.D., Blunt-Harris, E.L. (1999) Humics as an electron donor for anaerobic respiration, *Environ. Microbiol.* **1**, 89-98.

230

33. Kästner, M., Lotter, S., Heerenklage, J., Breuer-Jammali, M., Stegmann, R., Mahro, B. (1995) Fate of ^{14}C-labeled anthracene and hexadecane in compost manured soil, *Appl. Microbiol. Biotechnol.* **43**, 1128-1135.
34. Valles, B., Riva, M.C., Sanpera, C., Canela, M., Griful, E. (2000) Toxicity values for cationic surfactants - Comparison of EC50/LC50 values obtained from different toxicity bioassays for cationic surfactants in emulsions and microemulsions, *Tens. Surfac. Deterg.* **37**, 290-296.
35. Sposito, G. (1984) *The Surface Chemistry of Soils*, Oxford University Press, New York.
36. Murphy, E.M., Zachara, J.M., Smith, S.C. (1990) Influence of mineral-bound humic substances on the sorption of hydrophobic organic compounds, *Environ. Sci. Technol.* **24**, 1507-1516.
37. Spark, K. M., Wells, J.D., Johnson, B.B. (1997) Characteristics of the sorption of humic acid by soil minerals, *Austral. J. Soil Res.* **35**, 103-112.
38. Evanko, C.R., Dzombak, D.A. (1998) Influence of structural features on sorption of NOM-analogue organic acids to goethite, *Environ. Sci. Technol.* **32**, 2846-2855.
39. Vermeer, A.W.P., Koopal, L.K. (1998) Adsorption of humic acids to mineral particles. 2. Polydispersity effects with polyelectrolyte adsorption, *Langmuir* **14**, 4210-4216.
40. Vermeer, A.W.P., van Riemsdijk, W.H., Koopal, L.K. (1998) Adsorption of humic acid to mineral particles. 1. Specific and electrostatic interactions, *Langmuir* **14**, 2810-2819.
41. Gu, B., Schmitt, J., Chen, Z., Liang, L., McCarthy, J.F. (1994) Adsorption and desorption of natural organic matter on iron oxide: Mechanisms and models, *Environ. Sci. Technol.* **28**, 38-46.
42. Balcke, G.U., Kulikova, N.A., Hesse, S., Kopinke, F.-D., Perminova, I.V., Frimmel, F.H. (2002) Adsorption of humic substances onto kaolin clay related to their structural features, *Soil Sci. Soc. Am. J.* **66**, 1805-1812.
43. Murphy, E.M., Zachara, J.M., Smith, S.C., Phillips, J.L., Wietsma, T.W. (1994) Interaction of hydrophobic organic compounds with mineral-bound humic substances, *Environ. Sci. Technol.* **28**, 1291-1299.
44. Kaiser, K., Zech, W. (1997) Structure-Dependent Sorption of Dissolved Organic Matter on Soils and Related Minerals, in J. Drozd., S.S. Gonet, N. Senesi, W.J. Weber (eds.), *The Role of Humic Substances in the Ecosystems and in Environmental Protection*, Polish Society of Humic Substances, Wroclaw, pp. 385-390.
45. Kretzschmar, R., Hesterberg, D., Sticher, H. (1997) Effects of adsorbed humic acids on surface charge and flocculation of kaolinite, *Soil Sci. Soc. Am. J.* **61**, 101-108.
46. Perdue, E.M. (1989) Effects of humic substances on metal speciation, *Am. Chem. Soc.*, 281-295.
47. Scheffer, N.N., Schachtschabel, P. (1998) *Lehrbuch der Bodenkunde*, Ferdinand Enke Verlag, Stuttgart.
48. Kögel-Knabner, I., Totsche, K.U. (1998) Influence of dissolved and colloidal phase humic substances on the transport of hydrophobic organic contaminants in soils, *Phys. Chem. Earth* **23**, 179-185.
49. Walker, C.H., et. al. (1996) *Principles of Ecotoxicology*, Taylor & Francis Ltd.
50. Marquardt, H., Schäfer, S.G. (1997) *Lehrbuch der Toxikologie*, BI-Wissenschaftsverlag, Mannheim.
51. Sigg, L., Stumm, W. (1996) *Aquatische Chemie*, B.G. Teubner, Stuttgart.
52. Morimoto, K., Tatsumi, K., Kuroda, K.I. (2000) Peroxidase catalyzed co-polymerization of pentachlorophenol and a potential humic precursor, *Soil Biol. Biochem.* **32**, 1071-1077.
53. Bollag, J.-M., Dec, J, Huang, P.M. (1997) Formation mechanism of complex organic structures in soil habitats, *Adv. Agron.* **63**, 237-266.
54. Kopinke, F-D., Georgi, A., Mackenzie, K., Kumke, M. (2002) Sorption and chemical reactions of PAHs with dissolved humic substances and related model polymers, in F.H. Frimmel (ed.), *Refractory organic substances in the environment*, John Wiley, Heidelberg, Germany, pp. 475-515.
55. Karickhoff, S.W. (1984) Organic pollutant sorption in aquatic systems, *J. Hydraulic Engineering* **110**, 707-734.
56. Seth, R., Mackay, D., Muncke, J. (1999) Estimating the organic carbon partition coefficient and its variability for hydrophobic chemicals, *Environ. Sci. Technol.* **33**, 2390-2394.
57. Burkhard, L. P., (2000) Estimating dissolved organic carbon partition coefficients for non-ionic organic chemicals, *Environ. Sci. Technol.* **22**, 4663-4668.
58. Chiou, C.T., Porter, P.E., Schmedding, D.W. (1983) Partition equilibria of nonionic organic compounds betwen soil organic matter and water, *Environ. Sci. Technol.* **17**, 227-231.

59. Chin, Y.P., Weber, W.J. (1989) Estimating the effects of dispersed organic polymers on the sorption of contaminants by natural solids. 1. A predicitive thermodynamic humic substance-organic solute interaction model, *Environ. Sci. Technol.* **23**, 978-984.

60. Kopinke, F.-D., Poerschmann, J., Stottmeister, U. (1995) Sorption of organic pollutants on anthropogenic humic matter, *Environ. Sci. Technol.* **29**, 941-950.

61. Georgi, A., Kopinke, F.-D. (2002) Validation of a modified Flory-Huggins concept for description of hydrophobic organic compound sorption on dissolved humic substances, *Environ. Toxicol. Chem.* **21**, 1766-1774.

62. Poole, S.K., Poole, C.E. (1999) Chromatographic models for the sorption of neutral organic compounds by soil from water and air, *J. Chromatogr.* **A845**, 381-400.

63. Ohlenbusch, G., Frimmel, F.H. (2001) Investigations on the sorption of phenols to dissolved organic matter by a QSAR study, *Chemosphere* **45**, 323-327.

64. Poerschmann, J, Kopinke, F.D., Remmler, M., Mackenzie, K., Geyer, W., Mothes, S. (1996) Hyphenated techniques for characterizing coal wastewaters and associated sediments, *J. Chromatogr.* **A750**, 287-301.

65. Kopinke, F.-D., Poerschmann, J., Georgi, A. (1999) Application of SPME to study sorption phenomena on dissolved humic organic matter, in J. Pawliszyn (ed.), *Applications of SPME*, RSC Chromatographic Monographs, Cambridge, UK, pp 111-128.

66. Pörschmann, J., Zhang, Z., Kopinke, F.-D., Pawliszyn, J. (1997) Solid phase microextraction versus liquid-liquid extraction for determining organic pollutants in contaminated water rich in humic organic matter, *Anal. Chem.* **69**, 597-600.

67. Mackenzie, K., Georgi, A., Kumke, M., Kopinke, F.-D. (2002) Sorption of pyrene to dissolved humic substances and related model polymers. 2. SPME and FQT as analytical methods, *Environ. Sci. Technol.* **36**, 4403-4409.

68. Chiou, C.T., Kile, D.E., Brinton, T.I., Malcolm, R.L., Leenheer, J.A. (1987) A comparison of water solubility enhancement of organic solutes by aquatic humic materials and commercial humic acids, *Environ. Sci. Technol.* **21**, 1231-1234.

69. Chin, Y.-P., Aiken, G.-R., Danielsen, K.M. (1997) Binding of pyrene to aquatic and commercial humic substances: The role of molecular weight and aromaticity, *Environ. Sci. Technol.* **31**, 1630-1635.

70. Perminova, I.V., Grechishcheva, N. Yu., Petrosyan, V.S. (1999) Relationships between structure and binding affinity of humic substances for polycyclic aromatic hydrocarbons: Relevance of molecular descriptors, *Environ. Sci. Technol.* **33**, 3781-3787.

71. Chiou, T., Malcom, R. C., Brinton, T. I., Kile, D. E. (1986) Water solubility enhancement of some organic pollutants and pesticides by dissolved humic and fulvic acids, *Environ. Sci. Technol.* **20**, 502-508.

72. Kulikova, N.A., Perminova, I.V. (2002) Binding of atrazine to humic substances from soil, peat, and coal related to their structure, *Environ. Sci. Technol.* **36**, 3720-3724.

73. Xing, B., McGill, W.B., Dudas, M.J. (1994) Sorption of a-naphthol onto organic sorbents varying in polarity and aromaticity, *Chemosphere* **28**, 145-153.

74. Grathwohl, P. (1990) Influence of organic matter from soils and sediments from various origins on the sorption of some chlorinated aliphatic hydrocarbons. Implications on KOC correlations, *Environ. Sci. Technol.* **24**, 1687-1639.

75. Chiou, C.T., McGrody, S.E., Kile, D.E. (1998) Partition characteristics of polycyclic aromatic hydrocarbons on soils and sediments, *Environ. Sci. Technol.* **32**, 264-269.

76. Chefetz, B., Deshmukh, A.P., Hatcher, P.G., Guthrie, E.A. (2000) Pyrene sorption by natural organic matter, *Environ. Sci. Technol.* **34**, 2925-2930.

77. Salloum, M., Chefetz, B., Hatcher, P.G. (2002) Phenanthrene sorption by aliphatic-rich natural organic matter, *Environ. Sci. Technol.* **36**, 1953-1958.

78. Laor, Y. (1995) *Sorption of phenanthrene to dissolved and mineral associated humic acids and its effect on phenanthrene bioavailability*, PhD thesis, Rutgers State University of New Jersey, New Brunswick.

79. Laor, Y., Farmer, W.J., Aochi, Y., Strom, P.F. (1998) Phenanthrene Binding and Sorption to Dissolved and Mineral-Associated Humic Acid, *Water Res.* **32**, 1923-1931.

80. Jones, K.D., Tiller, C.L. (1999) Effect of solution chemistry on the extent of binding of phenanthrene by a soil humic acid: A comparison of dissolved and clay bound humic, *Environ. Sci. Technol.* **33**, 580-587.

232

81. Schwertmann, U., Cornell, R.M. (1991) *Iron Oxides in the Laboratory*, VCH, Weinheim.
82. Tiller, C.L., Jones, K.D. (1997) Effects of dissolved oxygen and light exposure on determination of KOC values for PAHs using fluorescence quenching, *Environ. Sci. Technol.* **31**, 424-429.
83. Doll, T.E., Frimmel, F.H., Kumke, M.U., Ohlenbusch, G. (1999) Interaction between natural organic matter (NOM) and polycyclic aromatic compounds (PAC) - comparison of fluorescence quenching and solid phase microextraction (SPME), *Fres. J. Anal. Chem.* **364**, 313-319.
84. Schwarzenbach, R.P., Gschwend, P.M., Imboden, D.M. (1993) *Environmental Organic Chemistry*, John Wiley & Sons, New York.
85. Luthy, R.G., Aiken, G.R., Brusseau, M.L., Cunningham, S.D., Gschwend, P.M., Pignatello, J.J., Reinhard, M., Traina, S.J., Weber, Jr., W.J., Westall, J.C. (1997) Sequestration of hydrophobic organic contaminants by geosorbents, *Environ. Sci. Technol.* **31**, 3341-3347.
86. Oeste, F.D., Kempfert, J. (1995) *Huminstoffbarriere und Verfahren zu ihrer Herstellung*, Patent: DE 4443828A1.
87. Totsche, K.U., Weigand, H. (1998) Flow and reactivity on dissolved organic matter transport in soil columns, *Soil Sci. Soc. Am. J.* **62**, 1268-1274.
88. Oeste, F., et al. (1994) *Verockerungsbarriere und Verfahren zu ihrer Herstellung*, Patent: DE 4242682A1.

Chapter 11

THE USE OF AQUEOUS HUMIC SUBSTANCES FOR *IN-SITU* REMEDIATION OF CONTAMINATED AQUIFERS

D.R. VAN STEMPVOORT[1], S. LESAGE[1], J. MOLSON[2]
[1]*National Water Research Institute, P.O. Box 5050, Burlington, Ontario, Canada L7R 4A6 <dale.vanstempvoort@ec.gc.ca>*
[2]*Département des génies civil, géologique et des mines, École Polytechnique de Montréal, C.P. 6079, Succ. Centre-ville, Montréal, Québec, Canada H3C 3A7*

Abstract

This chapter provides a review of the literature on binding of organic contaminants by aqueous humic substances (AHSs). Colloidal dispersions of AHSs are shown to be potential carriers (flushing agents) for enhanced removal of hydrophobic organic contaminants from aquifers. The process involves binding of contaminants by AHSs which can enhance the apparent solubility and mobility of contaminants. Binding, often modelled as linear partitioning, may vary with aqueous concentrations of contaminants and/or AHSs, and other parameters. Evidence is mixed whether aggregation of AHSs at high concentrations increases or decreases their capacity to carry organic contaminants. Sorption of contaminants and AHSs to solid aquifer particles and co-aggregation of AHSs with inorganic colloids/clays are also important, potentially clogging pores and reducing aquifer permeability. Advanced numerical models (e.g., BIONAPL/3D), which include binding/sorption kinetics and *in situ* biodegradation, can now be used to simulate carrier-assisted transport of contaminants in aquifers.

This chapter includes a discussion of a case study: a unique 5-year laboratory test, in which diesel fuel within a pilot-scale model sand aquifer was flushed with water containing 0.8 g/L AHSs (Aldrich® humic acid). AHS flushing increased aqueous concentrations of methylated naphthalenes from diesel two to ten fold. As a direct consequence, *in situ* biodegradation of the methylated naphthalenes increased. As hydrocarbons were depleted from the diesel, the contaminant plume shrank and disappeared. Numerical simulations using BIONAPL/3D indicated that without AHS flushing, complete diesel dissolution would have taken about 6 times longer.

Practical recommendations on use of AHS as flushing agents are given. The use of AHSs at levels > 1 g/L would most effectively flush hydrophobic contaminants (e.g., PAHs). Inexpensive, naturally-derived, non-toxic commercial humic products may offer significant advantages compared to other chemical flushing agents (e.g., surfactants). It may be possible to use AHSs for a combination of flushing, enhanced bioremediation and/or sequestation of organic contaminants in aquifers.

I. V. Perminova et al. (eds.),
Use of Humic Substances to Remediate Polluted Environments: From Theory to Practice, 233–256.
© 2005 *Canadian Crown. Printed in the Netherlands.*

1. Introduction

Over the past decade, interest has grown in potential applications of commercial humic products for remediation of contaminated soils, aquifers and sediments. Typically the studies are with colloidal, aqueous humic substances (AHSs), often referred to as humic acids, or humates. In this paper, AHSs are not limited to naturally occurring, waterborne humics, but also include concentrated solutions and colloidal suspensions of humic substances in water, whether as commercial products, as preparations in scientific laboratory experiments, or as a result of any other human activities. This chapter considers potential applications of AHSs as flushing agents for *in situ* remediation of aquifers contaminated by organics. Other chapters in this volume discuss alternative potential remediation applications of humic substances: to sequester (immobilize) contaminants [5], or to enhance the biodegradation of organic contaminants ([22], see also [54]).

AHSs are polydisperse mixtures of macromolecules that aggregate with various other organic compounds, metals, inorganic colloids and microorganisms. The structures and chemical properties of AHSs vary as a function of cation exchange, complexation with metals, binding with inorganic colloids, and affiliation with various organic compounds [109]. Isolation and concentration procedures that are used to analyse AHSs change their properties, and this greatly restricts our ability to determine how AHSs interact with other components in aquatic systems [23].

Recent reviews of research on molecular structures of humic substances are available [24, 31], however, after decades of research, there is still no consensus on the basic structural components of humic substances [85]. Various methods have been used to investigate the size distribution and structures of AHSs [49], and in recent years, many new analytical techniques have been introduced [30, 69].

In water, AHSs are polydisperse mixtures, which are more accurately described as colloidal dispersions or hydrosols, rather than as solutions [67]. AHSs in such mixtures are predominately colloidal in size (0.001 to 1 μm) but include some dissolved compounds (< 0.001 μm) and suspended particles (> 1 μm) [78]. Tombácz and Rice [89] have provided a useful multi-perspective conceptual model of AHSs as colloids, surfactants, "macroions" or polyelectrolytes, and amphiphilic units that tend to form fractal polymer-like networks (self-assemblages, micelles or aggregates).

The currently available commercial humic products that can be used to generate hydrosols (colloidal dispersions) of AHSs are generally derived from weathered coal deposits, typically oxidized lignite, sometimes referred to as leonardite [32, 67]. Commercial humic substances that have been most commonly used in laboratory experiments include those marketed by Aldrich® as dry form "humic acid" or "sodium humate". A number of other commercial humic products are marketed in bulk quantities as "water-soluble" powders or pellets. These products are typically used in agricultural and horticultural applications [32]. Over the past decade, several groups of researchers have examined the potential use of commercial humic products as AHSs (hydrosols or aqueous "solutions") in subsurface remediation, as chemical carriers to flush (solubilize and mobilize) hydrophobic organic contaminants [1, 6, 17, 38, 50, 51, 94, 103].

2. Experimental Methods to Evaluate Use of AHSs as Flushing Agents in Aquifer Remediation

Of the experiments pertinent to this topic, almost all have been conducted in laboratories at the bench-top scale. These have included tests to measure the binding of organic contaminants to AHSs, the sorption of AHSs to aquifer materials, the partitioning of organic contaminants between water and aquifer solids, and the facilitated transport of contaminants in aquifer materials in the presence of flowing water with AHSs. One experiment has examined the AHS-facilitated flushing of contaminants in a model aquifer at the pilot-scale in the laboratory (Section 5). A few field tests have examined the sorption of AHSs to soil or aquifer materials [61]. The methods used in such experiments are outlined in this section, along with pertinent references.

2.1. LABORATORY METHODS TO MEASURE BINDING OF ORGANIC CONTAMINANTS BY AHSs

Various methods are available for measuring the binding of organic contaminants to AHSs [7, 46]. Researchers typically use batches containing both AHSs and organic contaminants as dissolved phases in water, under static or actively mixed conditions. At inferred equilibrium, the free dissolved or total aqueous concentrations of contaminants in these batches are measured either *in situ* (e.g., fluorescence) or *ex-situ* (e.g., gas chromatography, HPLC). Techniques to measure the binding coefficients usually compare results for several batches with different concentrations of AHSs, including controls without AHSs. The equilibrium dialysis technique [52] is sometimes used, in which dissolved contaminants are partitioned between water with AHSs on one side of a dialysis membrane, and water without AHSs on the other side. An alternative technique is to measure quenching of contaminant fluorescence due to binding with AHSs [18, 79]. This is only applicable for contaminants that fluoresce, and for batches with relatively low AHS concentrations (< 50 mg/L), where blocking of fluorescence of the unbound contaminant phase by the bulk AHS phase is negligible. Yet another common technique involves the measurement of increases in apparent solubility of contaminants in the presence of AHSs [10]. Here one measures the total aqueous concentrations of contaminants in batches that contain excess contaminant in a non-aqueous form, such as a non-miscible liquid. Apparent solubility tests are relatively prone to artefacts, such as formation of colloids or emulsions of the excess contaminant phase [93].

Recently, solid phase microextraction (SPME) [18, 71, 93] and headspace sampling techniques [95, 106] have been adapted to examine contaminant-AHS binding. The main advantage for the latter is that the headspace samples that are analysed (*ex situ*) do not contain AHSs. In other techniques, samples that contain high concentrations of AHSs may foul surfaces of analytical devices (HPLC columns, SPME fibres). However, headspace techniques are only applicable for volatile or semi-volatile organic contaminants. In some cases, the application of head space sampling and SPME is combined [108].

"Reverse-phase separation" is a column-based technique that has been used to measure contaminant-AHS binding [47]. In this method, the free dissolved contaminant phase in water sorbs to a stationary phase in a column, while the AHS-bound contaminant phase that passes through the column in the water is analysed.

Sorption of strongly hydrophobic contaminants to glassware, dialysis membranes or other testing materials generally poses a problem, regardless of the technique. Binding tests are generally conducted at various contaminant and AHS concentrations, to determine any significant changes in binding strength. Other variables such as ionic strength and pH have also been examined, whereas studies on the effects of variations from room temperature (20-25°C) have been rare.

2.2. METHODS TO EXAMINE SORPTION OF AHSs AND CONTAMINANTS TO AQUIFER MATERIALS

Laboratory experiments are used to examine the sorption of AHSs to surfaces of soil and aquifer solids. In most cases, these involve bench scale tests with batches of AHSs in water mixed with aquifer materials or pure mineral phases, under various pH and ionic strength conditions [3, 27, 97]. Alternatively, water containing AHSs is passed through a test column containing the solid material [20]. Residual AHS concentrations in the supernatant or effluent water are analysed, generally *ex situ* by UV/Vis spectrophotometry [97] or total organic carbon analyses [27], to determine sorption of AHSs to the solid phase. In column tests, a time series of AHS concentrations is generated, and a "conservative" (non-sorbing) tracer, such as chloride or bromide is monitored for comparison [1, 20]. Numerical models are used to interpret the transport of various aqueous species in column tests, including AHS colloids and contaminants, as discussed in Section 4.3.

Some sorption studies have shown that the ratio of water to aquifer solids has a large effect on the resulting sorption coefficient. Thus, batch experiments should be designed such that the water to solid ratio closely resembles the subsurface conditions [96]. The sorption kinetics can be investigated by collecting a time series of samples from the batch supernatant or column effluent. Failure to account for sorption kinetics or hysteresis (differences between sorption and desorption processes) may result in poor simulation and/or underestimation of the sorption (see Section 4.2).

Field or pilot scale experiments to examine sorption of AHSs to aquifer solids are rare [61, 96]. Water containing AHSs is injected into an aquifer, and the AHS concentrations in down gradient wells are monitored.

Typically, AHS analyses (lab or field experiments) do not account for the polydispersivity of the AHSs. This is a drawback, given that different size fractions of AHS colloids have different properties, including their UV/Vis absorbance and sorption behaviour [98]. For this reason, a few AHS sorption studies have monitored the dynamic size distribution of the AHSs, for example, by size exclusion filtration [61]. Other techniques that can be used to measure size and/or size distribution of AHS colloids include dynamic light scattering, flow field flow fractionation, and others [49].

Batch or column tests to examine partitioning of organic contaminants between water and aquifer materials are similar to those for investigating sorption of AHSs. Typically, the tests are bench top batch experiments. Dynamic and equilibrium aqueous concentrations of contaminants are analysed *ex situ* (e.g., gas chromatography, HPLC).

The typical approach to examine the facilitated transport of organic contaminants by AHS carriers in aquifers is the use of bench top column tests. In such tests, water containing AHSs is passed through columns containing samples of the aquifer materials [1, 6, 19, 37, 53, 58, 90, 100]. The contaminants of interest are added as dissolved phases to the influent water, or as a nonaqueous phase ("immobile" liquid or sorbed phase) to the aquifer material prior to the column test. The *ex situ* analyses of concentrations of AHSs and contaminants along the flow path or in the column effluent are by the same methods described in preceding paragraphs. The methods for a unique pilot scale laboratory test are reported in Section 5.

3. Binding of Organic Contaminants by Aqueous Humic Substances: Quantification and Mechanisms

The binding of organic contaminants by AHSs is often modelled as partitioning (i.e., solid-phase distribution), following a linear isotherm [10], defined for each pair of contaminant and type of AHS:

$$C_s = K_p \cdot C_w \qquad (1)$$

where C_s is the concentration of contaminant bound/sorbed to AHSs; K_p is the partitioning coefficient; C_w is the concentration of the "freely" dissolved contaminant in water.

Aldrich® humic acid and other commercial humic products tend to be relatively strong binders of organic contaminants [11, 68]. This is apparently related to their relatively low polarity (high carbon, low oxygen contents) [11] or their high aromatic C content [68]. Binding is generally enhanced as the water solubility of the contaminant decreases, or conversely as its hydrophobicity (often measured as K_{ow}) increases [7, 10, 100]. Accurate prediction of K_p as a function of K_{ow} may be limited to individual compound "classes" or "families", such as PAHs, PCBs or alkanes [25, 71].

With changes in aqueous organic contaminant concentrations, non-linear sorption of these contaminants to AHSs has been observed. This behaviour has generally been modelled using the Freundlich isotherm [12, 102, 107]:

$$C_s = K_F \cdot C_w^N \qquad (2)$$

where K_F = Freundlich sorption constant, and N = Freundlich exponent. Note that if $N = 1$, the Freundlich isotherm reduces to the linear isotherm (Equation 1).

Non-linear isotherms based on equation 2 generally indicate lower binding at higher contaminant concentrations, and/or that binding is competitive [65, 102]. Competitive and/or non-linear binding cannot be explained by the partitioning model, but indicates that hydrophobic adsorption is important and may dominate. Xing and co-workers [102] proposed that the non-linear sorption behaviour may be a result of "dual mode" sorption: dissolution (i.e., partitioning) and an adsorption-like "hole-filling" mechanism. Sorption linearity reflects the properties of both the host sorbent and the sorbate [12, 91, 107].

With varying AHS concentrations, some researchers have reported negligible changes in partitioning coefficients (K_p in Equations 1) for binding of organic contaminants to AHSs [38, 46, 80, 103] (Figure 1, curve A). Other experiments have indicated decreases in binding strength with increases in AHS concentrations [18, 52,

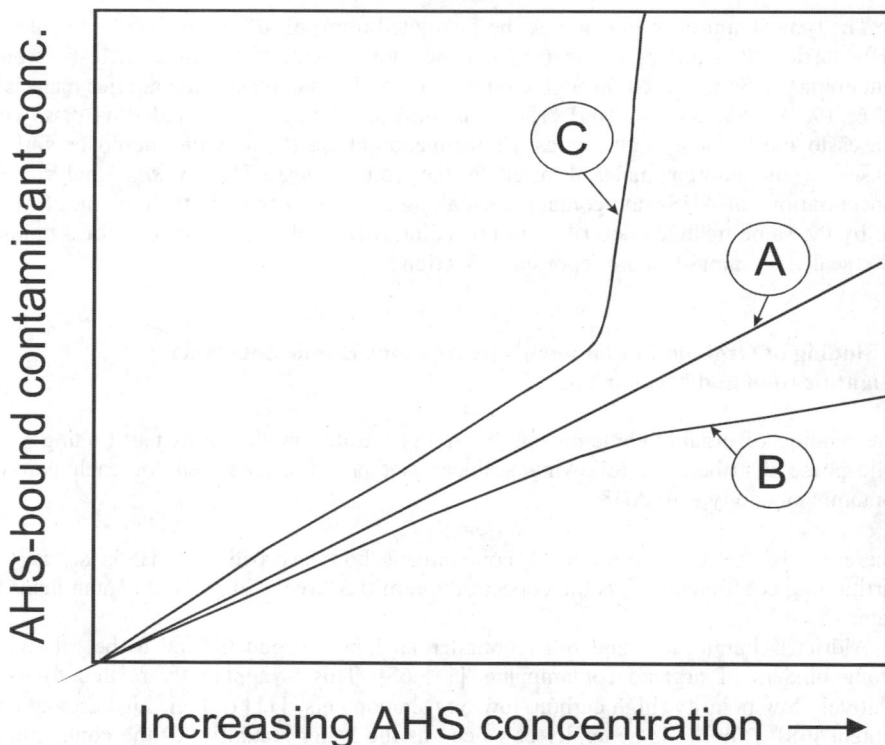

Figure 1. Schematic of different behaviors of binding of organic contaminants to aqueous humic substances (AHSs), with increasing AHS concentrations. Curve (A) is expected linear behavior if the binding coefficient, K_p, is constant. Curve (B) indicates a decrease in binding strength, perhaps related to aggregation. Curve (C) indicates an abrupt increase in binding at a "critical micellar concentration", associated with formation of micelles or "pseudomicelles" of AHSs.

80, 93, 106] (Figure 1, curve B). Still other studies have reported sharp increases in binding of contaminants at relatively high levels of AHSs [9, 28] (Figure 1, curve C). The latter type of trend has been attributed to enhanced binding of hydrophobic organics by AHSs above a "critical micellar concentration" (cmc) of the AHSs [28] (Figure 1). Evidence related to this inferred micelle-like behaviour of AHSs is mixed. Evidence in support was reported by Guetzloff and Rice: an abrupt increase in binding of DDT at their inferred cmc of 7.4 g/L of Aldrich® humic acid (HA) [28]. In contrast, the same authors [29] found no abrupt increase in binding for other sorbents above this inferred cmc. Subsequent studies of contaminant binding to concentrated Aldrich® HA [38, 93, 95] found no evidence for a cmc for ranges in Aldrich® HA concentrations up to 10 g/L as carbon (or approx. 20 g/L as AHSs).

The binding of organic contaminants to AHSs also changes with variations in the bulk aqueous chemistry, including pH and ionic strength [42, 46, 77]. Such changes are probably largely related to dynamic aggregation and conformation (expansion-collapse of molecular structure) of AHSs [107].

Typically, investigators have reported that binding/sorption of organic contaminants to AHSs in batch tests reaches equilibrium in minutes to a few hours [18, 42, 77]. However, slow reaction between some organic contaminants and AHSs may result in a slow phase of binding [18]. The binding is not always reversible; in some cases, a fraction of the contaminants that bind to AHSs is not extractable by conventional methods [101]. Numerical terms can account for binding kinetics [99].

The mechanisms of the binding of organic contaminants to AHSs may include cation exchange, hydrogen bonding, charge-transfer, covalent bonding, van der Waals forces, ligand exchange, and hydrophobic adsorption and partitioning [70, 81]. There is a growing perspective that an understanding of the molecular structures of contaminant compounds and/or the nature of their molecular interactions with AHSs are required in order to predict accurately how strongly they bind to AHSs [25, 70, 71, 77, 91]. Given the current gaps and limitations in the theory and in predictive models of the binding of organic contaminants to AHSs, an empirical approach is generally suitable for environmental remediation applications. This approach relies on laboratory tests conducted over the range of conditions of interest, and the resulting data can be modelled as either linear (Equation 1) or non-linear binding (Equation 2), with kinetic terms included as appropriate (see Section 4.3).

4. Binding of Organic Contaminants by Aqueous Humic Substances: Implications for Aquifer Flushing Technologies

The efficiency of conventional groundwater remediation technologies, such as "pump and treat", are often limited by the low aqueous solubility of hydrophobic organic compounds, and the large retardation factors for the transport of these compounds in the subsurface environment [92]. For this reason, chemical flushing agents, including surfactants and/or cosolvents are sometimes added to water, singly or as mixtures, in order to enhance the efficiency of the flushing of the contaminants from the source zone. This technology has been referred to as agent-enhanced soil flushing [35] or *in situ* flushing [76]. However, the latter term has also been applied when plain water was injected for contaminant flushing [76].

Agent-enhanced *in situ* flushing is a very new technology. The first documented field tests were conducted in the 1990s [35]. Various chemical flushing agents have been tested, but in the majority of cases, synthetic surfactants were selected [76]. Commercial humic substances in aqueous form (AHSs) have been considered by some investigators as flushing agents for remediation of soil and groundwater. To date, the published studies with AHSs have generally been laboratory batch and column experiments at the bench scale. A unique pilot-scale laboratory test is described in Section 5. To date, there are apparently no published field tests in which AHSs have been used as groundwater flushing agents.

4.1. ENHANCING APPARENT SOLUBILITY AND MOBILITY OF CONTAMINANTS

The binding of organic contaminants by AHSs (Section 3) results in an enhancement of the apparent solubility of these contaminants in water [10]. Although this has generally been documented in laboratory studies, one field study indicated concentrations of PAHs in groundwater at a coal tar site that were 3 to 50 times higher than purely dissolved concentrations, and the elevated concentrations were inferred to be due to binding of PAHs to humic-like organic colloids [56].

For strongly hydrophobic compounds (i.e., those with solubility in water < 0.01 mg/L) the enhancement in apparent solubility in the presence of AHSs is striking, even at relatively low AHS concentrations. For example, a greater than ten-fold enhancement in the apparent solubility of p,p'-DDT in the presence of 30 mg/L commercial AHSs was reported [11], as well as a similar increase in the apparent solubility of chlordane in the presence of 40 to 60 mg/L AHSs [41]. At higher concentrations of AHSs, the enhanced apparent solubility of strongly hydrophobic organic contaminants may be several hundred fold [95]. For organic contaminants that have moderate solubility in water (> 0.01 mg/L and < 100 mg/L), such as naphthalene, phenanthrene and methylated naphthalenes, an enhancement in apparent solubility becomes important at elevated concentrations of AHSs (e.g., above 1 g/L) [93, 103]. For relatively hydrophilic organics (i.e., those with solubility in water > 100 mg/L) such as BTEX [1], PCE [38] and TCE [6], the enhancement in apparent solubility is usually marginal at relatively low concentrations of AHSs (i.e. < 1 g/L). However, at high AHS concentrations, the enhanced solubilization of such compounds may be significant, even without evidence for micellar behaviour [6, 38]. A two fold increase of the aqueous concentration of PCE in 10 g/L humic acid [38], and a seven fold increase in the aqueous concentration of TCE in 5 % humic acid [6] have been reported.

Assuming linear hydrophobic partitioning (Equation 1), and ignoring kinetic considerations, the relationships between hydrophobicity, AHS concentration and apparent aqueous solubility can be predicted by Equation 3 [10]:

$$S_w^* = S_w(1+X \cdot K_{p,HS})$$ (3)

where S_w^* is the apparent solubility of a contaminant in water containing AHSs; S_w is the solubility of the same contaminant in pure water; X is the concentration of the AHSs; $K_{p,HS}$ is the contaminant partitioning coefficient between AHSs and water.

Applications of AHSs for the remediation of organic contaminants in groundwater must account for the sorption of the contaminants by the solid-phase aquifer material (sediment). When water flows through the aquifer, the movement of the contaminants is retarded, i.e., slower than the flow of water. Assuming linear, instantaneous sorption (Equation 1), and neglecting the role of carriers, this behaviour can be quantified as a retardation factor, R:

$$R = v_w/v_c = 1 + K_{p,s} \cdot \rho_s/n$$ (4)

where v_w is the average linear velocity of the groundwater; v_c is the velocity of the center of mass of a contaminant; $K_{p,s}$ is the coefficient for the partitioning the contaminant between the aquifer solids and water; ρ_s is the bulk mass density of the aquifer solids; n is the effective porosity of the aquifer.

The presence of natural AHSs or other colloidal carriers may enhance the mobility of the contaminants. Various equations have been introduced to account for the resulting change in the retardation factor, R, including Equation 5 [33]:

$$R = 1 + K_{p,s}/(1+K_{p,HS} \cdot C_{AHS})(1-n)(\rho_s/n) \qquad (5)$$

where $K_{p,HS}$ is the coefficient for partitioning between water and carrier (AHS or colloid); C_{AHS} is the concentration of the aqueous carrier.

West [100] conducted the pioneering laboratory column tests that demonstrated the role of AHSs as "mediators" of the transport of organic contaminants in the subsurface. The presence of 20 mg/L humics (extracted from groundwater) "assisted" the aqueous transport of hexachlorobenzene and anthracene through columns of sandy clay loam soil. Several years later, a key review article was written by McCarthy and Zachara [62] on the role of colloids, including humic substances, as carriers of contaminants. This article highlighted the potential negative impact that such carriers may have, by increasing the mobility of contaminants in the subsurface environment. But it also noted that there was potential for the manipulation of such carriers, as a positive tool in the remediation of subsurface contaminants.

The cogent review article by McCarthy and Zachara set the stage for a number of detailed laboratory investigations by various researchers in the 1990s on the potential role of AHSs as contaminant-mobilizing agents. Some of these studies examined the potential role played by low levels of aqueous natural organic matter (< 100 mg/L as organic carbon), largely AHSs, in mobilizing PAHs or PCBs through columns of sand or quartz [19, 37, 53, 58, 90]. Similarly, up to 25 mg/L of a commercial humic acid (Aldrich®) was added to water to simulate facilitated transport of chlordane through sand in the presence of natural AHSs ("aqueous organic matter") [41]. All but one of the above studies found that adding a few mg of AHSs per L of water increased the mobility of the contaminants and decreased the retardation factor, R. In the one exception [90], the addition of up to 35 mg/L (as organic carbon) AHSs to water retarded the transport of PAHs through a low pH soil. The increase in retardation was apparently related to the retention of 70 % of the added AHSs by the soil, apparently due to the low pH conditions. Pertinent to such negative results, the sorption of AHSs to solid surfaces is discussed in the following Section 4.2.

In some of the above facilitated transport studies, additional tests were conducted to study desorption of the contaminants from the columns in the presence or absence of AHSs. One study generally found no enhancement of the desorption of PAHs from quartz in the presence of AHSs [53], while another found that AHSs increased the rate of desorption of PAHs from sand [40]. Similarly, the presence of dissolved organic matter derived from composts and organic wastes increased the rate-limited desorption of PAHs from soil [44].

Similar to the above studies, a number of investigators have examined the potential beneficial use of Aldrich® humic acid as a carrier to increase the efficiency of the flushing of hydrophobic contaminants from subsurface sources. These have included several bench-scale column studies [1,6, 17, 38, 103] and one pilot-scale laboratory test (see Case Study - Section 5).

Abdul et al. used low concentrations of Aldrich® humic acid (29 mg/L), and found small enhancements in the mobility of the more hydrophobic alkyl-benzenes that were tested [1]. In later experiments, other researchers have used much higher concentrations of Aldrich® humic acid, ranging from around 1 g/L [17, 50, 51, 94, 103] to 10 g/L [38], and ~ 50 g/L (5 % by weight) in one experiment [6]. Ding and Wu demonstrated that DDT and aldrin were much more mobile when 1 g/L Aldrich® HA was added to water, such that large fractions of these pesticides could be translocated downward in soil columns [17]. Johnson and John found that the flushing of relatively soluble PCE from a NAPL phase in a sand column was enhanced by approximately 2.5 times in the presence of 10 g/L Aldrich® HA [38]. In a similar sand column study, in the presence of 5 weight % aqueous Aldrich® HA, Boving and Brusseau reported a 7 to 8 fold increase in aqueous concentrations and mobility for TCE derived from a NAPL source [6].

Collectively, most of the above studies have indicated that AHSs increase the subsurface mobility of contaminants, particularly at AHS concentrations above 1 g/L. Despite these findings, some investigators have downplayed the importance of this mobilizing potential. For example, using hypothetical calculations based on Equation 5, Gounaris et al. concluded that facilitated transport of organic contaminants in the subsurface by organic-rich colloids in landfill leachate (2.5 to 25 mg/L) would only be important for relatively hydrophobic contaminants [26]. In their example, the mobility of such compounds would still be very limited, increasing from 0.5 mm to 0.5 cm/a in groundwater flowing at 1 m/a.

The recent study of Boving and Brusseau indicates that it may be possible to utilize commercial AHSs as carriers at concentrations up to 5 % by weight (approx. 50 g/L) [6]. Such concentrations are several orders of magnitude greater than natural AHS concentrations. For the hypothetical system described by Guonaris et al. [26], using the same partitioning coefficient for the carrier and Equation 5, the mobility of a hydrophobic contaminant would increase from 0.0005 to 0.9 m/a, approaching the velocity of the groundwater.

Most of the laboratory studies of AHSs to date have consisted of batch tests. Dynamic column testing of AHS-facilitated transport of contaminants in aquifer materials should be expanded, given that relatively few such tests have been published to date [39], only one at the pilot scale (Section 5).

4.2. LIMITATIONS OF MOBILIZATION BY SORPTION AND CO-AGGREGATION

A key limitation to the mobilization of organic contaminants in groundwater by AHSs is the fact that a fraction of the AHSs will sorb to the surfaces of solid particles in the aquifer. As a consequence, binding of organic contaminants to AHSs does not always result in enhanced mobility in the subsurface. This sorption factor is illustrated in a study by Larsen et al. who observed that the transport of various organic contaminants in landfill leachate (355 mg/L TOC) through columns of aquifer material was sometimes retarded relative to transport in groundwater (5 mg/L TOC) [48]. Similarly, Totsche et al. found that, at low pH (4.5), the mobility of various PAHs in soil columns was retarded by the addition of low levels (10 to 20 mg/L) of aqueous, soil-derived organic matter (largely AHSs) [90]. In both of these studies, the authors explained the retardation behaviour by noting that portions of the aqueous phase organics (largely AHSs) became sorbed to soil or aquifer materials, affecting the sorption of the contaminants to these materials as well.

A number of laboratory studies have investigated the sorption of AHSs to mineral/solid surfaces [3, 20, 27, 66, 83, 96, 97, 98]. A few studies have examined the sorption of AHSs to soils and aquifer materials at the pilot scale in the laboratory [96] or in the field [2,61].

The sorption of AHSs onto mineral and sediment surfaces is non-linear. Some authors have reported that as the AHS concentration increases, the concentration of sorbed humic substances (HS) tends to reach a plateau, suggestive of site saturation [27, 65, 66, 72, 96]. This behaviour has sometimes been modelled using the Langmuir isotherm, as shown in Equation 6 [65, 96]:

$$C_{HS} = (K_L \cdot Q \cdot C_w)/(1 + K_L \cdot C_w) \tag{6}$$

where K_L = Langmuir adsorption constant, and Q = sorption maximum. A modification of this model that accounts for hysteresis has also been used [27].

Sorption of AHSs onto iron oxides [3, 27, 88, 98] and soil [87] decreases with increasing pH. The above Langmuir-type, pH-dependent behaviour suggests adsorption of relatively polar and hydrophilic AHSs onto the surfaces of hydrous oxides (including colloids), dominantly by ligand exchange [27, 65]. Recently, several authors have used polyelectrolyte models to simulate the sorption of AHSs to mineral surfaces [3, 97, 98]. Sorption of AHSs by mineral surfaces is particularly enhanced by the presence of polyvalent cations, such as Ca^{2+}, Cu^{2+}, and Fe^{3+} [87, 88]. AHSs also occur in polydisperse mixtures, and several studies have shown that the larger macromolecules are preferentially adsorbed [66, 83].

Sorption of the HS often appears to be largely "irreversible" [27]. Gu et al. [27] explained such behaviour in terms of "time-dependent adsorption and displacement processes" between different aqueous HS components. In contrast, Avena and Koopal [4] found that the HS sorbed onto iron oxide surfaces could be rapidly desorbed (reversibly) by changing the pH.

When AHSs co-aggregate with colloid-sized mineral grains (e.g., iron oxides), this can enhance the aqueous concentration and mobility of the mineral colloids [16, 45]. However, the same studies have indicated that the stability and mobility of the AHS-colloid complexes are reduced at low pH or at elevated concentrations of Ca^{2+} and other cations. The net effect of the presence of humic substances (aqueous and solid phase) on either the net aggregation or dispersion of colloids/clays is complex [60]. The presence of AHSs may lead to dispersal [60] and swelling of clays, which, combined with the sorption of AHSs, may lead to a substantial reduction in the size of pores and a reduction in the hydraulic conductivity of soils or aquifers [8]. Such a mechanism may be responsible for large reductions in permeability observed during soil leaching tests with solutions containing concentrated AHSs [17, author's unpublished data].

4.3. NUMERICAL MODELING OF MOBILIZATION OF CONTAMINANTS BY AHS CARRIERS

Laboratory studies have indicated that, in addition to the partitioning of contaminants between water, AHSs and aquifer solids (e.g., Equation 5), and the sorption of AHSs to aquifer solids (e.g., Equation 6), the kinetics of the sorption and/or desorption of

244

contaminants to AHSs [99], and of AHSs to the aquifer solids [96] must also be considered. If the contaminant is present as one component of an immobile, non-aqueous liquid phase (e.g., oil spill), the effective solubilities in water of the various components and the associated rates of dissolution must be calculated [57]. Biodegradation and/or abiotic transformations of the aqueous phase contaminants also have to be taken into account. Given all these factors, the facilitated transport of contaminants by AHS carriers is a complex process (Figure 2) that has required the development of novel numerical modelling approaches.

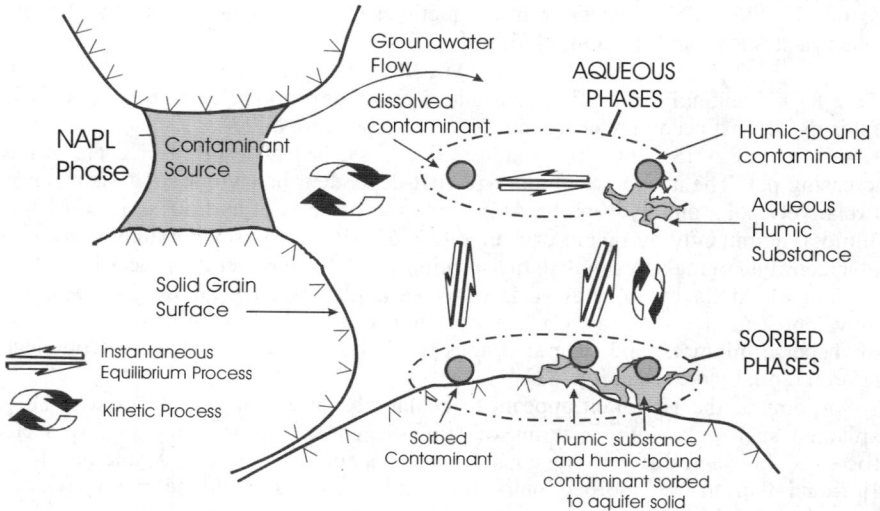

Figure 2. Conceptual model of processes included in the BIONAPL/3D simulation of the pilot scale case study (Section 5). Modified after Molson et al. [64].

In recent decades, various groups of researchers have developed numerical models that can simulate the transport and fate of contaminants in the groundwater environment. Some contaminant transport models are available in the public domain, such as BIOMOC [21], BIOPLUME III [75], HBGC123D [105], MT3D99 (S.S. Papadopulos & Associates, Bethesda, Maryland) and BIOMOD 3-D (Draper Aden Environmental Modeling, Inc., Blacksburg, VA). However, few if any of the existing public domain models incorporate carrier-assisted transport (flushing) of contaminants.

Over the past decade, various research models have been developed to account for carrier-assisted (-facilitated) transport of contaminants in groundwater [13, 14, 34, 36, 39, 40, 43, 58, 64, 73, 86]. Others have focused specifically on the aqueous transport of colloids (i.e., analogues of humic carriers) in soils and aquifers [84]. All of the carrier models simulate partitioning of contaminants between a stationary solid matrix, a mobile fluid (aqueous) phase, and a mobile carrier (colloid or organic matter) phase, under water-saturated conditions (i.e., groundwater). Most assume one dimensional transport and a homogenous (on macroscopic scale) porous solid matrix. Some are limited to equilibrium (instantaneous mass transfer), linear partitioning between phases, as in Equation 1 [58, 73], and one assumes negligible sorption of AHSs to the stationary solid matrix [73].

The majority of the above cited research models can incorporate kinetic expressions to simulate mass transfer between phases, and can simulate non-linear sorption (Langmuir or Freundlich) of the carrier or contaminant to the solid matrix, and/or of the contaminant to the carrier (cf. Sections 3 and 5). A few [36, 64, 86] can simulate the dissolution of contaminants from an immiscible, non-aqueous liquid phase. The comprehensive research model BIONAPL/3D, developed by Molson et al. [64], can be used to simulate carrier-assisted transport of multiple, biodegrading contaminants in a heterogeneous, 3-dimensional aquifer.

In the future, AHS-facilitated contaminant transport modelling may be further developed to include expanded and additional numerical terms. These would incorporate the rapidly expanding pool of information on complex relationships between the chemistry, conformation and/or aggregation of AHSs and their binding/sorption properties, including kinetic factors. For example, in some applications, it may become necessary to incorporate interactions between AHSs and inorganic colloids and/or attendant changes in the hydraulic properties of soils/aquifers (Section 4.2).

5. Case Study: Pilot-Scale Laboratory Experiment with Diesel

The pilot scale experiment reported in this section was conducted by the authors and co-workers. This section provides a brief summary of the experimental methods and results; details have been published elsewhere [50, 51, 64]. In this unique pilot-scale test, a stationary diesel fuel contaminant source was placed below the water table in a model sand aquifer (1.2 m x 5.5 m x 1.8 m deep) and flushed with water at a flow rate of 2 cm/h, for a period of 5 years. The aquifer material was a medium to very coarse, carbonate-rich sand, obtained from a local aggregate supplier. The model sand aquifer had a water-filled porosity of 28 to 32 %. At 51 d, Aldrich® humic acid (as sodium salt, technical product # H16752) was added as an AHS flushing agent to the water and maintained at a level of approximately 0.8 g/L for the balance of the experiment.

The methods and conditions for this experiment are summarized in Table 1. During the experiment, hydrocarbons were dissolving from the diesel source and their concentrations were monitored throughout the model aquifer at regular time intervals. The addition of the AHSs had a significant effect on the flushing of polycyclic aromatic hydrocarbons (PAHs), including methylated naphthalenes, from the diesel source. Binding to the AHSs enhanced the solubilization of these PAHs two to ten fold. This was detected as two to ten fold increases in total aqueous concentrations of the methylated naphthalenes immediately down gradient of the diesel source (Figures 3, 4). In contrast, the addition of the AHSs had only a small impact on the aqueous concentrations of the relatively hydrophilic BTEX components (benzene, toluene, ethylbenzene and xylenes), which were rapidly dissolved from the diesel.

During this flushing experiment, biodegradation of the BTEX and PAHs limited the lateral and longitudinal extent of the diesel contaminant plume. It appears that as a

Table 1. Experimental set-up and methods used for the pilot scale laboratory study.

Artificial groundwater flow (2 cm/hour)

---

model aquifer subdivided into two side-by-side, hydraulically isolated cells, one with diesel source, other (control) without; water introduced by gravity flow at 400 mL/min, from head tank, upgradient of both cells

tapwater (51 days), followed by 1 g/L Aldrich® humic acid in tapwater for duration

withdrawal at same rate, opposite end of model aquifer by two peristaltic pumps

tracers for defining flow regime (prior to diesel source emplacement): fluorescent lissamine (2mg/L), sodium bromide (100 mg/L)

Conditions

---

15 to 27°C, pH 6.5 (tapwater) to 7.9 (0.8 to 1 g/L Aldrich® humic acid)

water filled porosity: 28-32%

Chemical analytical techniques

---

humic acid concentration: UV/Vis spectrophotometry: (Varian CARY3) at 400 nm

methylated naphthalenes: dilution/extraction in methanol, HPLC (Waters 600E)

volatile aromatics: headspace sampling, GC-PID (Photovac 10S)

Dissolved oxygen: Orion 830 meter with galvanic D.O., probe (#83010)

Supporting batch tests with methylated naphthalenes

---

anaerobic conditions, measured binding by enhanced apparent solubility in presence of AHSs immersion SPME (non-equilibrium); partitioning between aqueous and aquifer solid phases: static and mixed batches, analyses by HPLC (Waters 600E)

Microbial analytical techniques

---

plate counts of water samples (APHA Standard Method 9050C)

phospholipid fatty acid analyses of sediment samples by Microbial Insights (Rockford, TN)

Confocal laser scanning microscopy (Bio-Rad MRC 1024, Nikon Microphot SA)

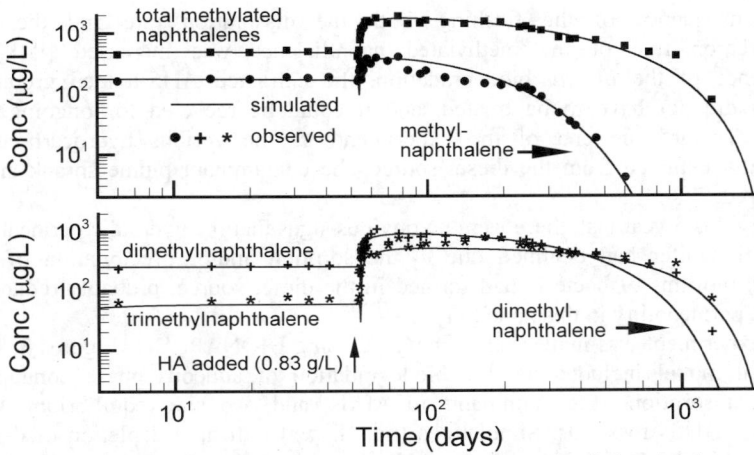

Figure 3. Observed and simulated trends in concentrations of methyl-naphthalenes immediately down gradient of the diesel source over the course of the experiment. Modified after Molson et al. [64].

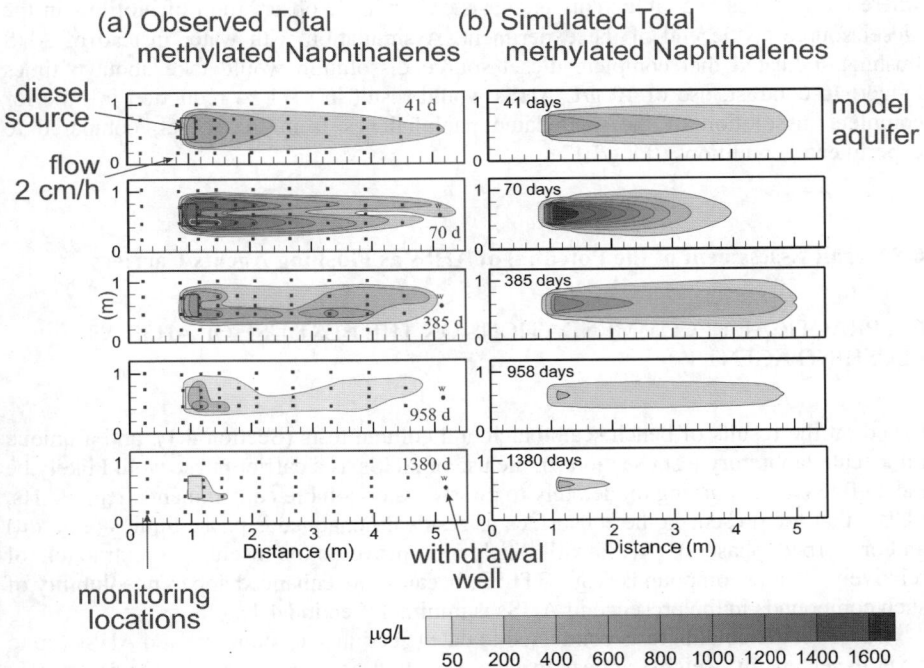

Figure 4. "Snapshot" plan views of model aquifer showing sequential changes in the dissolved diesel contaminant plume (total methylated naphthalenes) over time. Observed trend is shown on left, top to bottom; simulated trend on right. Note abrupt increase in dissolved concentrations between 41 and 70 d, corresponding to time when AHS carrier was added (51 d). Modified After Molson et al. [64].

direct consequence of the flushing, both the dissolution rate and the overall biodegradation rate of the methylated naphthalenes was increased [94]. As a consequence of the *in situ* biodegradation, the extracted AHS-loaded groundwater generally did not have to be treated, and it could be recycled for ongoing *in situ* flushing. Toward the end of the experiment, as the various hydrocarbons were sequentially depleted from the diesel source, the contaminant plume shrank and then disappeared.

During this 5 year test, there were no obvious signs that the hydraulic permeability of the model aquifer had declined due to the sorption and/or precipitation of AHSs. However, biofilms of bacteria had formed in the diesel source, probably reducing the hydraulic permeability in this zone [94].

The experiment was numerically simulated using BIONAPL/3D (Figures 3, 4). The conceptual model included aerobic biodegradation of aqueous phase contaminants, equilibrium sorption of contaminants to AHSs, and two site (equilibrium, kinetic) sorption of AHSs to solids. Simulations (not shown) indicated depletion of dissolved oxygen within the contaminant plume, with almost complete loss of oxygen within the core of the plume, as observed. The simulations also suggested that build-up of bacterial biomass would occur at the source and down gradient, along the margins of the plume, where oxygen was available. This is consistent with the observation of biofilms in the diesel source at the end of the experiment. A simulation with water instead of AHS flushing indicated that complete diesel source dissolution would take about 6 times longer. In contrast, use of 10 g/L AHSs would result in a 71 % reduction in time for complete dissolution of the methylated naphthalenes from the diesel, compared to experimental conditions (0.8 g/L).

6. Overall Assessment of the Potential of AHSs as Flushing Agents/Carriers

6.1. PRACTICAL RECOMMENDATIONS ON THE BEST USE OF AHSS AS FLUSHING AGENTS

Based on the results of bench scale batch and column tests (Section 4.1), and a unique pilot scale laboratory test (Section 5), the use of AHSs as a carrier phase would likely be most effective for purging moderately to poorly water-soluble contaminants (e.g., PAHs, PCBs) that are present in the subsurface as residual nonaqueous phase liquids (e.g., oil) and/or sorbed phase(s). AHSs will likely not prove to be useful for extraction of relatively soluble compounds (e.g., BTEX), because the enhanced apparent solubility of such compounds in the presence of AHSs is minimal (Section 4.1).

Based on the column tests conducted to date (Section 4.1), concentrated AHSs (up to several percent by weight) appear to have potential as carriers to flush some organic contaminants from aquifers. Barring problems such as hydraulic clogging (Section 4.2), AHS concentrations should be maximized for enhanced mobilization applications. Successful column tests with Aldrich® humic acid at 5 weight % in water supports this conclusion [6]. In spite of some evidence that increasing the concentration of AHSs may sometimes lead to reduced strength in the binding of contaminants to the AHSs (Section 3), the overall efficiency of solubilization/mobilization increases with an increase in AHS concentrations. Further, there is some evidence that AHSs may exhibit micellar-like behaviour, potentially resulting in a more efficient binding of some organic contaminants at higher AHS concentrations (Section 3).

High AHS concentrations may provide more effective buffering of the groundwater chemistry, including the pH and dissolved cation concentrations. This may be particularly important where the groundwater considered for remediation has a relatively low pH and/or a relatively high divalent cation (Ca^{2+}, Mg^{2+}, Fe^{2+}) chemistry, which can induce sorption or precipitation of AHSs [27, 87, 88].

The sorption of AHSs to mineral surfaces is a surface area-limited process, following Langmuir-like behaviour (Section 4.2). As the AHS concentration is increased, the percentage of total added AHSs that becomes sorbed is reduced. In a dynamic system, where the aquifer is continuously flushed with AHS-loaded groundwater, net co-sorption of contaminants and AHSs becomes less significant over time, and the overall retardation of the aqueous transport of the contaminant by the carrier is reduced.

Due to the potential for hydraulic clogging of aquifers by AHSs (Section 4.2), the use of these products as flushing agents may be restricted to relatively coarse grained materials (e.g., clean coarse sands and gravels). It may be possible to use AHSs as flushing agents in fractured rock aquifers, although we are not aware of any such tests.

To date, marginal attention has been given to final treatment options for contaminants flushed by concentrated AHS carriers from aquifers. As one option, the contaminants could potentially be removed from the water by inducing precipitation of the AHSs by adding lime, alum or ferric chloride [74, 104]. In other applications, it may be feasible to treat extracted AHS-loaded groundwater using a biofilter [15] or batch bioreactor, under either aerated or anaerobic conditions, as appropriate, to biodegrade the contaminants that have been flushed from the subsurface. In this case, it might be feasible to recover most or some of the AHS-loaded groundwater for subsequent reuse (cf. Section 5).

The results of our pilot scale experiment with a diesel source (Section 5) suggest that final treatment of extracted groundwater during *in situ* AHS-flushing may not be a concern in cases where this technology can be combined with *in situ* enhanced biodegradation. It may also be possible to use AHSs for the remediation of organic contaminants in aquifers in a process that combines both flushing and sequestration/immobilization [55, 82]. During the flushing of an aquifer with a hydrosol of concentrated AHSs, a fraction of the AHSs would be (inadvertently) sorbed to the solid phase and/or precipitate in the presence of dissolved metals or other colloids (Section 4.2). This immobilized humic phase could act as a sorptive sink for aqueous organic contaminants. Depending on the fraction of AHSs that was immobilized, this phase could act as an important sequestering agent, resulting in a significant reduction of the dissolved concentrations of the organic contaminants in the aquifer, and an attendant reduction in their mobility [87, 90]. Some of the sequestered organic contaminants might become "irreversibly" bound to the solid phase humics [9, 101]. Further, sequestration-lowered concentrations of dissolved organic contaminants in aquifers, in source areas and down gradient, might meet the targets required for "clean-up" or risk-based management.

6.2. COMPARISON OF AHSs TO ALTERNATIVE FLUSHING AGENTS

There is little comparative information available on the potential advantages of commercial AHSs compared to other flushing agents that are currently being tested and used at contaminated sites. At this stage, it can be argued that the selection of a chemical flushing agent for full-scale field applications should be made on a case-by-case basis [63].

There is some potential that the use of AHSs may prove to offer significant advantages over other chemical flushing agents, such as chemical surfactants and cosolvents. Compared to some other agents, such as specialty surfactant products or cyclodextrins, commercial humic products may offer a relatively inexpensive alternative. Humic substances are generally non-toxic, and compared with many organic substrates, including some of the other organic compounds that have been tested as flushing agents (e.g. surfactants), AHSs are generally less biodegradable and thus relatively weak competitors for oxygen and other electron acceptors [59]. Consequently, the addition of AHSs may have a negligible inhibition effect on concurrent bioremediation. In fact, the presence of AHSs may, in some cases, enhance the biodegradation process, in ways not afforded by other flushing agents. For example, AHSs sometimes may act as an electron shuttle, or supply Fe^{3+} or other cations as electron acceptors [54].

Very few published studies provide direct comparisons of the use of AHSs as flushing agents compared to use of surfactants, or other agents, even at a bench scale. The recent study by Boving and Brusseau provides a unique and useful comparison of laboratory tests with several different types of flushing agents [6]. Boving and Brusseau tested the various agents separately at concentrations of 5 % each, by weight, in water flowing through a sand column containing residual phase trichloroethene (TCE). Under the experimental conditions, they found that the surfactant DOWFAX 8390 was less effective as a flushing agent than a commercial preparation of AHSs, Aldrich® humic acid. In contrast, the surfactant sodium dodecyl sulphate (SDS) was more effective than Aldrich® humic acid on a mass basis, requiring 25 % less total flushing volume to achieve final recovery (contaminant decline to non-detectable levels). However, the efficiency of flushing by SDS declined over time, whereas the efficiency of Aldrich® humic acid remained relatively constant over 40 pore volumes. SDS may be readily biodegraded [59], which would render it non-recyclable, which could be a disadvantage. It appears that on a cost basis alone, a relatively inexpensive commercial humic product, with properties similar to the Aldrich® product tested, may offer a distinct advantage in some cases, compared to the surfactants that were tested.

Boving and Brusseau found that two biosynthesized cyclodextrins, also tested at 5 %, were not as effective as Aldrich® humic acid in flushing the TCE from the columns. They also tested the use of ethanol as a cosolvent flushing agent. They found that 5 % ethanol had a negligible affect on TCE mobilization in their test column compared to water, whereas 50 % ethanol was the most effective mobilizing solution that they tested. However, such a high concentration of ethanol might not be practical for subsurface remediation applications, given cost and environmental impact considerations.

7. Topics Requiring Further Investigation

Given the current controversies and uncertainties regarding the molecular structures of humic substances, more information is required on the aggregation and conformation of AHSs, and their behaviour in the environment. Further data are required on the effects of various factors, including inorganic colloids and various ions on the aggregation of AHSs, and the resulting effects on binding of organic contaminants. This would include a closer look at the "micellar" or "pseudomicellar" behaviour inferred by some investigators under some conditions. In particular, studies are required to confirm the conditions in which aggregation causes either formation of "pseudomicelles" that enhance binding, or reduces binding, perhaps related to less surface area available.

More information is required on interaction of AHSs with other colloids, including clay and oxide minerals. For example, the presence of AHSs tends to enhance the concentration and mobilization of colloidal clay minerals and oxides, and this may cause significant reductions in aquifer permeability (Section 4.2). Such a response is a potential barrier to the use of AHSs for *in situ* flushing.

The earliest tests of AHSs as flushing agents in subsurface remediation tended to consider these substances as stable macromolecules, having no significant changes in size or structure over time. There is a need to further examine the conformation/aggregation behavior of AHSs under various hydrochemical conditions representative of subsurface environments, and how this affects the mobility of the AHSs and their binding of organic contaminants. In particular, there are very few published data on the mobility of AHSs in aquifers (or soils) at high levels (> 1 g/L) that might be used in contaminant flushing applications.

Most investigators of AHSs as "carriers" of organic contaminants have ignored ion exchange processes. Currently, there is little information on the role that ion exchange could have on the dynamic behaviour of AHSs in soil and groundwater environments, and particularly on the conformational changes and aggregation of AHSs. Batch tests to date have often had very low solids/solutions ratios. As a consequence, researchers may have underestimated the role that cation exchange between humic substances and solid phases play in subsurface environments. Mineral surfaces in aquifers contain large pools of exchangeable cations. Thus, aquifers tend to buffer the concentrations of cations in groundwater and on the surface exchange sites of AHSs. This buffering effect would tend to be magnified at the leading edge of an injected volume of AHS-loaded water, as it is transported through the aquifer, thus being progressively exposed to more cation exchange sites. It is unknown whether ongoing flushing of an aquifer with AHSs would eventually exert a measurable impact on the cation exchange population on mineral surfaces within the aquifer.

Further modelling, matched with column studies, is required to provide further insights into mechanisms and kinetics of the binding of organic contaminants to AHSs, and the sorption/desorption of AHSs on aquifer materials. Future numerical modelling may need to incorporate kinetics and non-linear processes (e.g., sorption of AHSs to aquifer materials). Complex interactions of AHSs with inorganic colloids/aggregates may have to be considered in these models. Given the divergence in batch test results to date, there is a need to further test, explain and predict the role of various interactive

252

factors, such as the effect of the concentration of AHSs on the binding of contaminants, or on the rate of *in situ* biodegradation of organic contaminants in the subsurface.

The possibilities of the use of AHSs for a combination of *in situ* flushing, enhanced *in situ* biodegradation, and/or sequestration should be explored further. In particular, further pilot scale tests and field experiments would provide vital information at the application scale for this array of emerging technologies.

8. References

1. Abdul, A. S., Gibson, T. L. and Rai, D. N. (1990) Use of humic acid solution to remove organic contaminants from hydrogeologic systems, *Environ. Sci. Technol.* **24**, 328-333.
2. Alborzfar, M., Villumsen, A. and Grøn, C. (2001) Artificial recharge of humic ground water, *J. Environ. Qual.* **30**, 200-209.
3. Au, K.-K., Penisson, A. C., Yang, S. and O'Melia, C. R. (1999) Natural organic matter at oxide/water interfaces: Complexation and conformation, *Geochim. Cosmochim. Acta* **63**, 2903-2917.
4. Avena, M. J. and Koopal, L. K. (1998) Desorption of humic acids from an iron oxide surface, *Environ. Sci. Technol.* **32**, 2572-2577.
5. Balcke G. U., Georgi, A., Woszidlo, S., Kopinke, F.-D. and Poerschmann, J. (2005) Utilization of immobilized humic organic matter for in situ subsurface remediation, in I.V. Perminova, N. Hertkorn, K. Hatfield, *Use of humic substances to remediate polluted environments: from theory to practice*, Chapter 10, pp. 203-232 (this volume).
6. Boving, T. B. and Brusseau, M. L. (2000) Solubilization and removal of residual trichloroethene from porous media: comparisoin of several solubilization agents, *J. Contam. Hydrol.* **42**, 51-67.
7. Burkhard, L.P. (2000) Estimating dissolved organic carbon partition coefficients for nonionic organic chemicals, *Environ. Sci. Technol.* **34**, 4663-4668.
8. Chen, Y. and Tarchitzky, J. (1999) Humic substances and pH effects on clay dispersion and the hydraulic conductivity of soils. Program with Abstracts, *Humic Substances Seminar III*, March 22-23, 1999, Northeastern University, Boston, p. 12.
9. Chien, Y.-Y., Kim, E.-G. and Bleam, W. F. (1997) Paramagnetic relaxation of atrazine solubilized by humic micellar solutions, *Environ. Sci. Technol.* **31**, 3204-3208.
10. Chiou , C. T., Malcolm, R. L., Brinton, T. I. and Kile, D. E. (1986) Water solubility enhancement of some organic pollutants and pesticides by dissolved humic and fulvic acids, *Envir. Sci. Technol.* **20**, 502-508.
11. Chiou, C. T., Kile, D. E., Brinton, T. I., Malcolm, R. L. and Leenheer, J. A. (1987) A comparison of the water solubility enhancements of organic solutes by aquatic humic materials and commercial humic acids, *Environ. Sci. Technol.* **21**, 1231-1234.
12. Chiou, C. T., Kile, D. E., Rutherford, D. W., Sheng, G. and Boyd, S. A. (2000) Sorption of selected organic compounds from water to a peat soil and its humic-acid and humin fractions: Potential sources of the sorption nonlinearity, *Environ. Sci. Technol.* **34**, 1254-1258.
13. Corapcioglu, M. Y. and Jiang, S. (1993) Colloid-facilitated groundwater contaminant transport, *Water Resour. Res.* **29**, 2215-2226.
14. Corapcioglu, M. Y. and Wang, S. (1999) Dual-porosity groundwater contaminant transport in the presence of colloids, *Water Resour. Res.* **35**, 3261-3273.
15. Couillard, D. (1994) The use of peat in wastewater treatment, *Water Res.* **28**, 1261-1274.
16. Degueldre, C., Triay, I., Kim, J.-I., Vilks, P., Laaksoharju, M. and Miekeley, N. (2000) Groundwater colloid properties: a global approach, *Appl. Geochem.* **15**, 1043-1051.
17. Ding, J.-Y. and Wu, S.-C. (1997) Transport of organochlorine pesticides in soil columns enhanced by dissolved organic carbon, *Wat. Sci. Tech.* **35**, 139-145.
18. Doll, T. E., Frimmel, F. H., Kumke, M. U. and Ohlenbusch, G. (1999) Interaction between natural organic matter (NOM) and polycyclic aromatic compounds (PAC) - comparison of fluorescence quenching and solid phase micro extraction (SPME), *Fresenius J. Anal. Chem.* **362**, 313-319.

19. Dunnivant, F. M., Jardine, P. M., Taylor, D. L. and McCarthy, J. F. (1992) Cotransport of cadmium and hexachlorobiphenyl by dissolved organic carbon through columns containing aquifer material, *Environ. Sci. Technol.* **26**, 360-368.

20. Dunnivant, F. M., Jardine, P. M., Taylor, D. L. and McCarthy, J. F. (1992) Transport of naturally occurring dissolved organic carbon in laboratory columns containing aquifer material, *Soil Sci. Soc. Am. J.* **56**, 437-444.

21. Essaid, H.I., and Bekins, B. (1997) *BIOMOC, A Multispecies Solute - Transport Model with Biodegradation,* U.S. Geological Survey, Water-Resources Investigations Report 97-4022, Menlo Park, CA.

22. Field, J.A. and Cervantes, F.J. (2005) Microbial redox reactions mediated by humus and structurally related quinones, in I.V. Perminova, N. Hertkorn, K. Hatfield, *Use of humic substances to remediate polluted environments: from theory to practice,* Chapter 17, pp. 343-352 (this volume).

23. Frimmel, F. H. (1998) Characterization of natural organic matter as major constituents in aquatic systems, *J. Contam. Hydrol.* **35**, 201-216.

24. Gaffney, J. S., Marley, N. A. and Clark, S. B. (eds.) (1996) *Humic and Fulvic Acids, Isolation, Structure and Environmental Role,* ACS Symposium Series 651, American Chemical Society, Washington, DC.

25. Goss, K.-U. and Schwarzenbach, R.P. (2001) Linear free energy relationships used to evaluate equilibrium partitioning of organic compounds, *Environ. Sci. Technol.* **35**, 1-9.

26. Gounaris, V., Anderson, P.R. and Holsen, T.M. (1993) Characteristics and environmental significance of colloids in landfill leachate, *Environ. Sci. Technol.* **27**, 1381-1387.

27. Gu, B., Schmitt, J., Chen, Z., Liang, L. and McCarthy, J.F. (1994) Adsorption and desorption of natural organic matter on iron oxide: Mechanisms and models, *Environ. Sci. Technol.* **28**, 38-46.

28. Guetzloff, T.F., and Rice, J.A. (1994) Does humic adic form a micelle? *Sci. Total Environ.* **152**, 31-35.

29. Guetzloff, T.F. and Rice, J.A. (1996). Micellar nature of humic colloids, in J.S. Gaffney, N.A. Marley and S.B. Clark (eds.), *Humic and Fulvic Acids, Isolation, Structure and Environmental Role,* ACS Symposium Series 651, American Chemical Society, Washington, DC, pp. 18-25.

30. Haberhauer, G., Bednar, W., Gerzabek, M.H. and Rosenberg, E. (1999) MALDI-TOF-MS analysis of humic substances - A new approach to obtain additional structural information? in E.A. Ghabbour and G. Davies (eds.), *Understanding Humic Substances, Advanced Methods, Properties and Applications,* Royal Society of Chemistry, Cambridge, UK, pp. 121-128.

31. Hayes, M.H.B. (1998) Humic substances: Progress towards more realistic concepts of structures, in G. Davies, and E.A. Ghabbour (eds.), *Humic Substances, Structures, Properties and Uses,* Royal Soc. Chem., Cambridge, pp. 1-27.

32. Hoffman G.L., Nikols D.J., Stuhec S. and Wilson, R.A. (1993) *Evaluation of leonardite (humalite) resources of Alberta,* Open File Report 93-18, Alberta Geological Survey, Edmonton, AB. Prepared by Retread Resources Ltd. (Calgary, AB) for Energy, Mines and Resources Canada.

33. Hutchins, S.R., Tomson, M.B., Bedient, P.B. and Ward, C.H. (1985) Fate of trace organics during land application of municipal wastewater, *CRC Crit. Rev. Environ. Control* **15**, 355-416.

34. Ibaraki, M., and Sudicky, E.A. (1995) Colloid-facilitated contaminant transport in discretely fractured porous media. 1. Numerical formulation and sensitivity analysis, *Wat. Resour. Res.* **31**, 2945-2960.

35. Jalvert, C.T. (1996) *Surfactants/cosolvents,* Technology Evaluation Report TE-96-02, Ground-Water Remediation Technologies Analysis Centre, Pittsburg, PA. (http:/www.gwrtac.org).

36. Ji, W. and Brusseau, M. L. (1998) A general mathematical model for chemical-enhanced flushing of soil contaminated by organic compounds, *Water Resour. Res.* **34**, 1635-1648.

37. Johnson, W.P. and Amy, G.L. (1995) Facilitated transport and enhanced desorption of polycyclic aromatic hydrocarbons by natural organic matter in aquifer sediments, *Environ. Sci. Technol.* **29**, 807-817.

38. Johnson, W.P., John, W.W. (1999) PCE solubilization by commercial humic acid. *J. Contam. Hydrol.* **35**, 343-362.

39. Johnson, W.P. (2000) Sediment control of facilitated transport and enhanced desorption, *J. Environ. Eng.* **126**, 47-56.

40. Johnson, W.P., Amy, G.L. and Chapra, S.C. (1995) Modeling of NOM-facilitated PAH transport through low-foc sediment, *J. Environ. Eng.* **121**, 438-446.

41. Johnson-Logan, L.R., Broshears, R.E. and Klaine, S.J. (1992) Partitioning behavior and the mobility of chlordane in groundwater, *Environ. Sci. Technol.* **26**, 2234-2239.

254

42. Jones, K.D. and Tiller, C.L. (1999) Effect of solution chemistry on the extent of binding of phenanthrene by a soil humic acid: A comparison of dissolved and clay bound humic, *Environ. Sci. Technol.* **33**, 580-587.
43. Knabner, P., Totsche, K.U. and Kögel-Knabner, I. (1996) The modeling of reactive solute transport with sorption to mobile and immobile sorbents, 1. Experimental evidence and model development, *Water Resour. Res.* **32**, 1611-1622.
44. Kögel-Knabner, I., Totsche, K.U. and Raber, B. (2000) Desorption of polycyclic aromatic hydrocarbons from soil in the presence of dissolved organic matter: Effect of solution composition and aging, *J. Environ. Qual.* **29**, 906-916.
45. Kretzschmar, R., and Sticher, H. (1997) Transport of humic-coated iron oxide colloids in a sandy soil: Influence of Ca^{2+} and trace metals, *Environ. Sci. Technol.* **31**, 3497-3504.
46. Krop, H.B., van Noort, P.C.M. and Govers, H.A.J. (2001) Determination and theoretical aspects of the equilibrium between dissolved organic matter and hydrophobic organic micropollutants in water (Kdoc), *Reviews Environ. Contam. Toxicol.* **169**, 1-122.
47. Landrum, P.F., Nihart, S.R., Eadie, B.J. and Gardner, W.S. (1984) Reverse-phase separation method for determining pollutant binding to Aldrich humic acid and dissolved organic carbon of natural waters, *Environ. Sci. Technol.* **18**, 187-192.
48. Larsen, T., Christensen, T.H., Pfeffer, F.M. and Enfield, C.G. (1992) Landfill leachate effects on sorption of organic micropollutants onto aquifer materials, *J. Contam. Hydrol.* **9(4)**, 307-324.
49. Leppard, G.G. and Buffle, J. (1998) Aquatic colloids and macromolecules: Effects on analyses, in R.A. Meyers (ed.), *Encylcopedia of Environmental Analysis and Remediation,* John Wiley and Sons, Inc., New York, pp. 349-377.
50. Lesage, S., Novakowski, K.S., Xu, H., Bickerton, G., Durham, L. and Brown, S. (1995) A large scale aquifer model to study the removal of aromatic hydrocarbons from the saturated zone, Proceedings, *Solutions '95*, International Association of Hydrogeologists Congress, June 4-10, Edmonton, AB.
51. Lesage, S., Brown, S., Millar, K. and Novakowski, K. S. (2001) Humic acids enhanced removal of aromatic hydrocarbons from contaminated aquifers: Developing a sustainable technology, *J. Environ. Sci. Health* **A36(8)**, 1515-1533.
52. Li, A.Z., Marx, K.A., Walker, J. and Kaplan, D.L. (1997) Trinitrotoluene and metabolites binding to humic acid, *Environ Sci Technol.* **31**, 584-589.
53. Liu, H., and Amy, G. (1993) Modeling partitioning and transport interactions between natural organic matter and polynuclear aromatic hydrocarbons in groundwater, *Environ. Sci. Technol.* **27**, 1553-1562.
54. Lovley, D.R., Woodward, J.C. and Chapelle, F.H. (1996) Rapid anaerobic benzene oxidation with a variety of chelated Fe(III) forms, *Appl. Environ. Microbiol.* **62**, 288-291.
55. Lueking, A.D., Huang, W., Soderholm-Schwarz, S., Kim, M. and Weber, W.J., Jr. (2000) Relationship of soil organic matter characteristics to organic contaminant sequestration and bioavailability, *J. Environ. Qual.* **29**, 317-323.
56. MacKay, A.A., and Gschwend, P.M. (2001) Enhanced concentrations of PAHs in groundwater at a coal tar site, *Environ. Sci. Technol.* **35**, 1320-1328.
57. Mackay, D., Shiu, W.Y., Maijanen, A., and Feenstra, S. (1991) Dissolution of non-aqueous phase liquids in groundwater, *J. Contam. Hydrol.* **8**, 23-42.
58. Magee, B.R., Lion, L.W., and Lemley, A.T. (1991) Transport of dissolved organic macromolecules and their effect on the transport of phenanthrene in porous media, *Environ. Sci. Technol.* **25**, 323-331.
59. Margesin, R. and Schinner, F. (1999) Biodegradation of diesel oil by cold-adapted microorganisms in presence of sodium dodecyl sulfate, *Chemosphere* **38**, 3463-3472.
60. Mbagwu, J.S.C., Piccolo, A., and Mbila, M.O. (1993) Water-stability of aggregates of some tropical soils treated with humic substances, *Pedologie* **43**, 269-284.
61. McCarthy, J.F., Gu, B., Liang, L. and Mas-Pla, J. (1996) Field tracer tests on the mobility of natural organic matter in a sandy aquifer, *Water Resour. Res.* **32**, 1223-1238.
62. McCarthy, J.F. and Zachara, J. M. (1989) Subsurface transport of contaminants: Binding to mobile and immobile phases in groundwater aquifers, *Environ. Sci. Technol.* **23**, 496-504.
63. McCray, J.E. and Brusseau, M.L. (1998) Cyclodextrin-enhanced in situ flushing of multiple-component immiscible organic liquid contamination at the field scale: Mass-removal effectiveness, *Environ. Sci. Technol.* **32**, 1285-1293.

64. Molson, J.W., Frind, E.O., Van Stempvoort, D.R. and Lesage, S. (2002) Humic acid enhanced remediation of an emplaced diesel source in groundwater: 2. Numerical model development and application, *J. Contam. Hydrol.* **54**, 277-305.

65. Murphy, E.M., Zachara, J.M. and Smith, S.C. (1990) Influence of mineral-bound humic substances on the sorption of hydrophobic organic compounds, *Environ. Sci. Technol.* **24**, 1507-1516.

66. Namjesnik-Dejanovic, K., Maurice, P.A., Aiken, G.R., Cabaniss, S., Chin, Y.-P. and Pullin, M.J (2000) Adsorption and fractionation of muck fulvic acid on kaolinite and goethite at pH 3.7, 6, and 8 *Soil Sci.* **165**, 545-559.

67. Olson, E.S., Diehl, J.W. and Froelich, M.L. (1988) Hydrosols from low rank coals. 1. Preparation and properties, *Fuel* **67**, 1053-1061.

68. Perminova, I.V., Grechischeva, N.Y. and Petrosyan, V.S. (1999) Relationship between structure and binding affinity of humic substances for polycyclic aromatic hydrocarbons: Relevance of molecular descriptors, *Environ. Sci. Technol.* **33**, 3781-3787.

69. Pfiefer, T., Klaus, U., Hoffmann, R. and Spiteller, M. (2001) Characterization of humic substances using atmospheric pressure chemical ionization and electrospray ionisation mass spectrometry combined with size-exclusion chromatography, *J. Chromat.* **A 926**, 151-159.

70. Piccolo, A. (1994) Interactions betwen organic pollutants and humic substances in the environment, in N. Senesi and T.H. Miano (eds.), *Humic Substances in the Global Environment and Implications or Human Health*, Elsevier, Amsterdam, Netherlands, pp. 961-980.

71. Poerschmann, J. and Kopinke, F.-D. (2001) Sorption of very hydrophobic organic compounds (VHOCs) on dissolved humic organic matter (DOM). 2. Measurement of sorption and application of a Flory-Huggins concept to interpret the data, *Environ. Sci. Technol.* **35**, 1142-1148.

72. Rav-Acha, C. and Rebhun, M. (1992) Binding of organic solutes to dissolved humic substances and its effects on adsorption and transport in the aquatic environment, *Wat. Res.* **26**, 1645-1654.

73. Rebhun, M., de Smet, F. and Rwetabula, J. (1996) Dissolved humic substances for remediation of sites contaminated by organic pollutants, *Wat. Res.* **30**, 2027-2038.

74. Rebhun, M., Meir, S. and Laor, Y. (1998) Using dissolved humic acid to remove hydrophobic contaminants from water by complexation-flocculation process, *Environ. Sci. Technol.* **32**, 981-986.

75. Rifai, H.S., Newell, C.J., Gonzales, J.R., Dendrou, S., Kennedy, L. and Wilson, J. (1997) *BIOPLUME III. Natural Attenuation Decision Support System Version 1.0, Users Manual,* Prepared for the U.S Air Force Centre for Environmental Excellence, Brooks Air Force Base, San Antonio, TX.

76. Roote, D.S. (1998) *In Situ Flushing. Technology Status Report, TS-98-01*, Ground-Water Remediation Technologies Analysis Centre, Pittsburg, PA.

77. Sabljic, A. (2001) QSAR models for estimating properties of persistent organic pollutants required in evaluation of their environmental fate and risk, *Chemosphere* **43**, 363-375.

78. Schimpf, M.E. and Petteys, M.P. (1997) Characterization of humic materials by flow field-flow fractionation, *Colloids and Surfaces, A: Physicochem. and Eng. Aspects* **120**, 87-100.

79. Schlautman, M.A. and Morgan, J.J. (1993) Effects of aqueous chemistry on the binding of polycyclic aromatic hydrocarbons by dissolved humic materials, *Environ. Sci. Technol.* **27**, 961-969.

80. Schmitt, P., Freitag, D., Trapp, I., Garrison, A., Schiavon, M. and Kettrup, A. (1997) Binding of s-triazines to dissolved humic substances: Electrophoretic approaches using affinity capillary electrophoresis (ACE) and micellar electrokinetic chromatography (MEKC), *Chemosphere* **35**, 55-75.

81. Senesi, N. (1992) Binding mechanisms of pesticides to soil humic substances, *Sci. Total Environ.* **123/124**, 63-76.

82. Shimizu, Y., Sogabe, H. and Terashima, Y. (1998) The effects of colloidal humic substances on the movement of non-ionic hydrophobic organic contaminants in groundwater, *Water Sci. Tech.* **38**, 159-167.

83. Specht, C.H., Kumke, M.U. and Frimmel, F.H. (2000) Characterization of NOM adsorption to clay minerals by size exclusion chromatography, *Wat. Res.* **34**, 4063-4069.

84. Sun, N., Sun, N.-Z., Elimelech, M. and Ryan, J.N. (2001) Sensitivity analysis and parameter identifiability for colloid transport in geochemically heterogeneous porous media, *Wat. Resour. Res.* **37**, 209-222.

85. Swift, R.S. (1999) Macromolecular properties of soil humic substances: fact, fiction and opinion, *Soil Sci.* **164**, 790-802.

86. Tatalovich, M.E., Lee, K.Y. and Chrysikopoulos, C.V. (2000) Modeling the transport of contaminants originating from the dissolution of DNAPL pools in aquifers in the presence of dissolved humic substances, *Transport in Porous Media* **38**(1-2), 93-115.
87. Temminghoff, E.J.M., van der Zee, S.E.A.T.M. and de Haan, F.A.M. (1997) Copper mobility in copper-contaminated sandy soil as affected by pH and solid and dissolved organic matter, *Environ. Sci. Technol.* **31**, 1109-1115.
88. Tipping, E. (1986) Some aspects of the interactions between particulate oxides and aquatic humic substances, *Marine Chem.* **18**, 161-169.
89. Tombácz, E. and Rice, J.A. (1999) Changes of colloidal state in aqueous systems of humic acids, in E.A. Ghabbour and G. Davies (eds.), *Understanding Humic Substances, Advanced Methods, Properties and Applications,* Royal Society of Chemistry, Cambridge, UK, pp. 69-78.
90. Totsche, K.U., Danzer, J. and Kögel-Knabner, I. (1997) Dissolved organic matter-enhanced retention of polycyclic aromatic hydrocarbons in soil miscible displacement experiments, *J. Environ. Qual.* **26**, 1090-1100.
91. Uhle, M.E., Chin, Y.-P., Aiken, G.R. and McKnight, D.M. (1999) Binding of polychlorinated biphenyls to aquatic humic substances: the role of substrate and sorbate properties on partitioning, *Environ. Sci. Technol.* **33**, 2715-2718.
92. (1999) *Treatment Technologies: Annual Status Report*, Ninth Edition, United States Environmental Protection Agency (USEPA), Washington, DC.
93. Van Stempvoort, D.R. and Lesage, S. (2002) Binding of methylated naphthalenes to concentrated aqueous humic acid, *Adv. Environ. Res.* **6**, 495-504.
94. Van Stempvoort, D.R., Lesage, S., Novakowski, K S., Millar, K., Brown, S. and Lawrence, J.R. (2002) Humic acid enhanced remediation of an emplaced diesel source in groundwater: 1. Laboratory-based pilot scale test, *J. Contam. Hydrol.* **54**, 249-276.
95. Van Stempvoort, D.R., Lesage, S., and Steer, H. (2002) Binding of hydrophobic organic contaminants to humalite-derived aqueous humic products, with implications for remediation, *Water Qual. Res. J. Can.* **38**, 267-281.
96. Van Stempvoort, D.R., Molson, J.W., Lesage, S. and Brown, S. (2000) Sorption of Aldrich humic acid to a test aquifer material and implications for subsurface remediation, in E.A. Ghabbour and G. Davies (eds.), *Humic Substances: Versatile Components of Plants, Soils and Water*, Royal Society of Chemistry, Cambridge, UK, pp. 153-163.
97. Vermeer, A.W.P., van Riemsdijk W.H. and Koopal, L.K. (1998) Adsorption of humic acid to mineral particles. 1. Specific and electrostatic interactions, *Langmuir* **14**, 2810-2819.
98. Vermeer, A.W.P. and Koopal, L.K. (1998) Adsorption of humic acids to mineral particles. 2. Polydispersity effects with polyelectrolyte adsorption, *Langmuir* **14**, 4210-4216.
99. Weber, E.J., Spidle, D.L. and Thorn, K.A. (1996) Covalent binding of aniline to humic substances. 1. Kinetic studies, *Environ. Sci. Technol.* **30**, 2755-2763.
100. West, C.C. (1984) *Dissolved Organic Carbon Facilitated Transport of Neutral Organic Compounds in Subsurface Systems*, Ph.D. Thesis, Rice University, Houston, TX.
101. Xie, H., Guetzloff, T.F. and Rice, J.A. (1997) Fractionation of pesticide residues bound to humin, *Soil Sci.* **162**, 421-428.
102. Xing, B. and Pignatello, J.J. (1997) Dual-mode sorption of low-polarity compounds in glassy poly(vinyl chloride) and soil organic matter, *Environ. Sci. Technol.* **31**, 792-799.
103. Xu H., Lesage S. and Durham, L. (1994) The use of humic acids to enhance removal of aromatic hydrocarbons from contaminated aquifers, in *Proceedings, 4th Annual Symposium on Groundwater & Soil Remediation*, Calgary, AB, Canada, pp. 635-645.
104. Yates, L.M. III and von Wandruszka, R. (1999) Decontamination of polluted water by treatment with a crude humic acid blend, *Environ. Sci. Technol.* **33**, 2076-2080.
105. Yeh, G.-T., Salvage, K.M., Gwo, J.P., Zachara, J.M. and Szecsody, J.E. (1998) HydroBioGeoChem: A Coupled Model of Hydrologic Transport and Mixed Biogeochemical Kinetic/Equilibrium Reactions in Saturated-Unsaturated Media, *Report ORNL/TM-13668*, Oak Ridge National Laboratory, Oak Ridge, TN.
106. Yin, C. and Hassett, J. P. (1986) Gas-partitioning approach for laboratory and field studies of mirex fugacity in water, *Environ. Sci. Technol.* **20**, 1213-1217.
107. Yuan, G. and Xing, B. (2001) Effects of metal cations on sorption and desorption of organic compounds in humic acids, *Soil Sci.* **166**, 107-115.
108. Zhang, Z. and Pawliszyn, J. (1993) Headspace solid phase microextraction, *Anal. Chem.* **65**, 1843-1852.
109. Zwiener, C. Kumke, M.U., Abbt-Braun, G. and Frimmel, F. (1999) Adsorbed and bound residues in fulvic acid fractions of a contaminated groundwater: Isolation, chromatographic and spectrographic characterization, *Acta Hydrochim. Hydrobiol.* **27**, 208-213.

Chapter 12

ADVANTAGES OF *IN-SITU* REMEDIATION OF POLLUTED SOIL AND PRACTICAL PROBLEMS ENCOUNTERED DURING ITS PERFORMANCE

J.F. DE KREUK
*BioSoil R&D, Nijverheidsweg 27, 3341 LJ Hendrik Ido Ambacht,
the Netherlands <j.f.dekreuk@wxs.nl>*

Abstract

For many centuries soil has become contaminated by human activities, but only recently has soil pollution was considered a threat for man and environment based on increasing knowledge about the effects pollutants may have and on ways to detect them. Now, society is faced with an increasing number of polluted sites that are discovered and considered to need treatment. Although excavation and treatment of the excavated soil seems the most definitive measure, costs of such an operation and damage to buildings and infra structure prevents such an approach. For that reason (biological) *in situ* techniques have been developed and finally accepted. When carried out properly, *in situ* remediation leads to environmentally sound solutions in accordance with (the Netherlands) regulations. This does not mean that problems are not encountered. The principles of *in situ* treatment are discussed in this paper as well as the problems which are associated with certain types of contaminants and possible solutions for these problems. The paper is based on personal experience in the field of the development of *in situ* techniques, the performance of a large number of *in situ* projects and the problems to be solved in carrying out this task.

1. Introduction

As a result of human activities the soil of a large number of sites has become contaminated with the materials which were used in labour and personal activities. Carelessness with the environment is something of all ages. Part of it is welcomed as a human heritage, but the more recent form is considered a burden for present and future generations. It must be taken into account, however, that the problems associated with the environment only became part of the public interest about 20 years ago. Politicians considered it an issue in the mid eighties and lost interest in the mid nineties or a just little later.

Water pollution was considered a problem for a long period of time already leading to an increasing number of wastewater treatment installations for both industries and communities. Abatement of air pollution also was an important issue. Once emissions

I. V. Perminova et al. (eds.),
Use of Humic Substances to Remediate Polluted Environments: From Theory to Practice, 257–265.
© 2005 *Springer. Printed in the Netherlands.*

are controlled in these media a rapid improvement may occur at least locally, because residues are swept away and they are degraded or diluted in the environment. In soil movement of contaminants is slow leading to very long residence times of contaminants and sometimes to extensive areas with polluted groundwater. In addition natural attenuation alone in most cases does not yield a final solution within a reasonable time frame due to unfavourable conditions for microbial growth and, therefore, to low degradation rates. The measures needed, consequently, are of quite another nature than those for surface waters (not their sediments) and air. In this respect it must also be noted that good quality groundwater, that can be used as drinking water straight away is getting a scarce commodity and its protection by prevention of pollution and the uncontrolled spreading thereof should have high priority.

After the beginning of the eighties, in the Netherlands alone, the number of contaminated sites discovered grew from around 200 in 1980 to about 200.000 by the end of the century. Although it was assumed at the start of that period that excavation of all contaminated soil in the Netherlands would solve the problem, over the years it became clear that the country could not afford such an approach.

The problems of financing the decontamination of contaminated sites caused a growing acceptance during that period of *in situ* techniques. This was boosted by the NOBIS-programme (Netherlands Research into Biological *In situ* Remediation), which started around 1995 and is still active under the name SKB.

Although for smaller sites excavation and treatment of the excavated soil still is the most economical solution, this rapidly changes when buildings have to be demolished or when infra structures are damaged or put out of order for a longer period of time. For larger sites or when the contaminants have penetrated the soil to larger depth *in situ* remediation in almost all cases is the most economic one.

The further discussion on *in situ* techniques and the problems encountered is by no means meant to be a literature review etc.; it is based on personal experience in the fields of research into the development of techniques for environmental pollution abatement and of the *in situ* remediation of a large number of sites in the Netherlands and abroad.

2. Soil Decontamination

In deciding which technique is to be used, the nature of the contaminants play an important role. Based on their properties and the approach needed the following division can be made:
- mineral oil (petrol stations, tank parks, solvents)
- chlorinated hydrocarbons (dry cleaning, solvents)
- PAH – polynuclear aromatic hydrocarbons (wood preservation, gas works)
- pesticides
- POP's - persistent organic pollutants (PCB's, dioxines, etc.)

The order of this list more or less also indicates the order in which the it was "discovered" that biotreatment was feasible. Consequently mineral oils for a very long period of time were already considered treatable leading to activities like sludge farming in which oil sludges were brought on the soil after which biodegradation was enhanced by farming like techniques. This was followed by so called landfarming of contaminated soils in roughly the same manner. Later *in situ* techniques were developed starting with pump and treat and soil vapour extraction and finally bioremediation.

In practice the following properties of mineral oils determine its behaviour in the environment:

– penetrates soils when spilled;
– floats on groundwater;
– low water solubility apart from the light aromatics (BTEX);
– high retardation factor in soil

The size of the areas to be decontaminated after an oil spillage is, even after relative long periods between the spill and the remedial action, in most cases within manageable limits either by excavation, bioremediation or a combination of both (see also problems in *in situ* remediation).

Chlorinated hydrocarbons, however, differ in their behaviour with mineral oil in this respect that:

– they are heavier than water and may penetrate soil to depth of 30 m or more;
– the most common ones are from an environmental standpoint soluble in groundwater;
– they form DNAPL's (dense non aqueous phase liquids);
– their retardation is relatively low;
– they may be carcinogenic leading to low target values for remediation.

These properties result in large areas to handle even after limited spillages. At least in the Netherlands problems with chlorinated hydrocarbons were more or less ignored, because a solution did not seem to be available and polluters could not afford long term pump and treat measures. In addition pumping of large volumes of groundwater would also damage the environment by increasing withering of soils leading to loss of specific plants and animals.

When it was found, that dehalogenation of chlorinated hydrocarbons took place under anaerobic conditions in the presence of a suitable electron donor, this opened the way to translate these findings to actual remediation techniques. (see *in situ* remediation).

PAH containing contaminants such as creosote equal the behaviour of both mineral oil and chlorinated hydrocarbons.

The mixture fractionates after entering the soil in one floating on groundwater and in one heavier than groundwater, of which the latter contains a large portion of PAH and which also forms DNAPL's. A compound such as naphthalene is rather soluble in water, which may, therefore, result in large areas with contaminated groundwater.

Apart from mineral oil and PAH considerable levels of phenolic compounds are found.

The feasibility of the biodegradation of PAH was shown already in the eighties. This and later laboratory tests were translated into field work resulting on a full scale *in situ* remediation of a site in the Netherlands in the city of Rotterdam. The actual efficiency of the biodegradation depends on the both the composition of the mixture of contaminants and the nature of the soil (see also problems in *in situ* remediation – Table 1).

Bioremediation of pesticide containing soils is mainly limited to sites were pesticide residues were dumped and those where spillages have occurred such as production and blend plants. In the latter cases the pesticide contamination very often is associated with solvents, emulsifiers and other auxiliary compounds. The distribution of the pesticide itself in the soil is then determined by the behaviour and transport of these auxiliaries. Once their degradation took place the mobility of the pesticide residue in general is strongly reduced.

In the Netherlands, much attention is paid to the fate of pesticides in normal agricultural use. In addition the bioremediation was investigated of residues of the lindane production. So far it did not result in further practical developments.

Although the POP's are important much developmental work has to be done before bioremediation becomes feasible for solving these problems (see also problems in *in situ* remediation). On the other hand, fundamental research as carried out at universities showed the possibilities of (partial) dehalogenation of many chlorinated hydrocarbons even of these which are still considered persistent so restoring the faith in microbial infallibility.

3. Principles of Biological *In Situ* Remediation

As indicated above an increasing number of chemicals is found to be biodegradable. Mainly because more attention is paid to biodegradation and other degradation routes but not in the least because micro-organisms seems to induce the ability to degrade chemicals after a prolonged exposure to these substances. So the necessary micro-organisms are in almost all cases present in the soils to be treated and extra measures to introduce degraders are, therefore, very seldom needed.

It now seems strange that huge quantities of inherently biodegradable materials pollute soils for almost infinitive times. The reason is that one or more of the prerequisites for microbial growth are not met. Active biological *in situ* decontamination, therefore aims at changing the soil conditions in such a manner that microbial growth is feasible and biodegradation of the pollutants can take place.

In general this can be obtained by:
– improving the nutrient status of the soil;
– supplying oxygen or an electron donor in sufficient quantities;
– controlling the soil humidity;
– increasing the bio-availability of the contaminants.

Since the soil is not disturbed during *in situ* remediation, the only way to change the conditions for the available microflora is to use water as the medium for transport of nutrients, microbial growth factors, oxygen sources, micro-organisms, etc. In addition air can be used as an oxygen supply. To achieve this, groundwater is extracted from wells or drains and used for reinfiltration after purification and after the addition of nutrients, etc. as indicated schematically below.

groundwater purification, dosing of nutrients, etc.

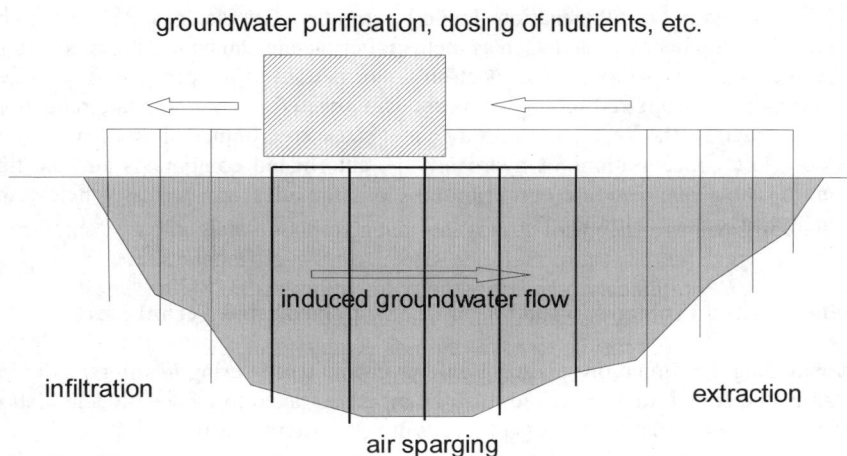

for clarity the systems for extraction, infitration and sparging are given separately ; in practice these are alternated

Figure 1. Groundwater purification, dosing of nutrients, etc.

In situ treatment, however, is by no means a soil washing process, because the water solubility of most contaminants is low and they bind effectively to soil organic matter. It is, therefore, in most cases impossible to obtain a significant reduction of contamination by washing (pump and treat) alone. Consequently, biodegradation must for the greater part take place in the soil itself; the soil is the bioreactor.

It must be noted, that these techniques are limited to the saturated zone of the soil.

In general the following approach is followed.

– Based on a soil and contamination inventory the *in situ* treatment system is designed. When the system is actually made and wells are installed, the distribution of the contamination is checked with respect to the earlier inventories and the system is adapted if necessary.

– Soil properties and groundwater flow determine to a great extent the design of the *in situ* treatment system. The distance between infiltration and extraction wells is governed by the permeability of the soil for water and when horizontal clay layers are present the treatment system must be designed and laid out for treating the soil both below and above that layer.

– The progress of the decontamination is monitored and the way the system is managed is adapted to the monitoring results. For that reason each well and sparging point is connected separately to a manifold in such a manner, that the flows can be controlled and extraction and infiltration and thus the artificial groundwater flow may be reversed.

– Apart from inorganic nutrients which are by no means strange (xenobiotic) to the environment and are consumed by the degrading micro-organisms no chemicals are added in the process of the aerobic biodegradation of contaminants. In addition the microflora naturally present is stimulated and "constructed" micro-organisms are not applied.

- When anaerobic dehalogenation should be stimulated an electron donor has to be added. A single compound such as methanol or lactate can be used or mixtures of natural organic materials. The electron donor is used in the process. Any excess normally is transferred by methane producing bacteria or for sulphate reduction. Consequently, the treatment in itself cannot cause environmental problems.
- Very low concentrations of hydrocarbons, chlorinated compounds and the like are feasible and residual concentrations of aromatics are normally below the analytical detection limits.

4. Limitations and Problems in the Use of *In Situ* Remediation Techniques

In understanding the limitations and/or problems encountered during *in situ* remediation of contaminated soils it must be noted, that the measures taken to induce biodegradation and the actual biodegradation process proceeds at very different scales.

Biodegradation runs at the scale of micro-organisms, because they need for growth at about the same moment and at the same place nutrients, the contaminant and oxygen (or another electron acceptor) for aerobic processes or an electron donor for anaerobic dehalogenation.

In order to influence the conditions in the soil and to turn it into a bioreactor a system of infiltration and extraction wells and when necessary air sparging points are used. The distances between wells are in the order of several metres or more. Transport at the m^3-scale is used to stimulate degradation at the μ-scale. Where transport fails or is slow, biodegradation rates are low. In general it can be stated, that the biodegradation processes in soil as such are always governed by mass transfer rates.

Transport limitations can then be caused by:
– inhibition of groundwater flow;
– irregular soil structures;
– binding of contaminants to organic matter or in porous structures;
– binding of essential nutrients in the soil;
– presence of contaminants as DNAPL's.

These limitations are different in nature for the groups of contaminants mentioned above.

Groundwater transport is greatly influenced by the presence of hydrophobic materials such as mineral oil and PAH-containing contaminants. When concentrations are high (above a few thousand mg/kg of soil) preferential pathways are formed around the zones with high contamination levels. Consequently nutrients which are transported in groundwater do not reach the areas where they are needed most and decontamination does not take place at a reasonable rate. At the end of the day pockets of soil can be found where the contamination can still be found at the original levels.

The soil structure in the Netherlands results from the presence of the large rivers and the geological processes from the ice ages on. Although soil profiles seem to be uniform over a location they often contain thin layers of fine sediments or clay which inhibit horizontal movement of water and air even if their thickness is less than 10 or 20 mm.

Depending on the nature of a contaminant its bio-availability is greatly influenced by the presence of soil organic matter leading to biodegradation rates, which are limited by the actual desorption rates. Also adsorption into soil or porous media may play an important role.

Table 1 shows the results of two tests regarding the degradation of creosote. One test was run with soil and the other one with broken rubble in which the creosote had penetrated. The conditions of the test were such, that nutrients were present in sufficient quantities and the system was completely aerobic.

The results presented show that in soil all PAH analysed with the exception of benzo(ghi)perylene show a distinct reduction in concentration. Since the soil particles were more or less covered in creosote at these levels a steady decrease in film thickness occurred till the availability of the heavier residue became that low, that no further visible degradation took place within a reasonable time frame. In broken rubble, however, only naphthalene, which has a relatively high water solubility and is easily biodegradable, has been removed completely. In this case diffusion plays an important role and the low solubility prevents a significant mass transfer. The larger particles between 2 and 5 mm even showed less reduction in concentrations than the smaller ones below 2 mm most likely owing to the longer diffusion path.

Another problem associated with binding of contaminants in soil is the underestimation of the actual load of the contamination in a certain case. This is not really important in the case that the contaminated soil is excavated, but to design a proper *in situ* remediation system the distribution of the contaminants and the levels to be expected should be known. It is not uncommon that the actual load is double of the one calculated at the start of the remediation. This is in particular the case with chlorinated hydrocarbons, because it is very often assumed that the level in the aqueous phase are representative for the total quantities present in soil and groundwater. The Koc for PCE is 760 and in a soil with 1 or 2 % organic matter when an equilibrium is assumed between soil and groundwater, more than 50 % of the total mass of PCE present, is associated with the organic matter in the soil and it is only set free when the concentrations in the aqueous phase drop as a result of treatment or groundwater flow.

Further to the above it can be stated, that this bio-availability problem is certainly associated with the remediation of pesticide residues and POP's. Handling these compounds within survivable time frames requires means to overcome these problems.

The presence of nutrients such as nitrogen and phosphorous are essential for allowing a level of microbial growth, which is sufficient for obtaining a measurable decontamination of the soil. It seems never to be a problem to distribute sufficient nitrogen containing compounds like nitrate over the soil to be treated. A large number of soil types, however, posses high phosphate binding capacities, which inhibit its even distribution over the soil leading to phosphate limitation. Since soil tend to contain naturally reasonable levels of phosphate, this is generally speaking only a problem with concentrations of oil above 2,000 tot 3,000 mg/kg. If this is the case like in the creosote example given above care must be taken to dose levels of phosphate that overcome the binding problem.

Table 1. Results of biodegradation tests with creosote in soil and broken rubble.

Component	Concentration (mg/kg dm)				
	soil		broken rubble		
	start	end	start	end (fraction < 2 mm)	end (fraction 2-5 mm)
naphthalene	780	< 0.05	3700	< 1	< 1
acenaphthylene	12	0.09	150	28	21
acenaphthene	1200	0.10	2000	960	800
fluorene	1200	0.25	1700	760	670
fenanthrene	2900	0.09	7300	650	600
anthracene	480	0.35	760	390	370
fluoranthene	1700	1.8	4200	2600	3000
pyrene	840	2.2	2200	2300	2500
benzo(a)anthracene	210	0.89	520	400	500
chrysene	170	1.8	320	300	310
benzo(b)fluoranthene	62	8.5	170	180	240
dibenzo(a,h)anthracene	3.3	1.5	84	71	74
benzo(k)fluoranthene	28	2.8	120	60	80
benzo(a)pyrene	46	5.4	2,1	6	6
benzo(ghi)perylene	4.1	3.8	67	17	15
indeno(1,2,3-c,d)pyrene	11	3.4	47	27	33
PAH-sum	9700	33	23000	9000	9200

DNAPL's are very often not found in the initial inventories. The penetration for instance of chlorinated hydrocarbons in soil is very irregular and free phase product stays behind in fine sandy inclusion in courser material, on clay lenses, etc. The scale of these phenomena in many cases is no more than a few meters and it is, therefore, very unlikely that a monitoring well is installed in such a spot during the investigations to determine the size of the contamination. This also will lead to a gross underestimation of the problem and the mass of contaminants to be removed or degraded.

Once the remediation starts many more wells are installed and these spots are very likely to be found either directly or because the levels in a few wells stay high in the course of the remediation with respect to those of other wells. Even then it takes time to remove such a DNAPL in particular when it is caught in fine sands, because these are difficult to penetrate by the induced groundwater flow and for the electron donor which is dosed to obtain the anaerobic dehalogenation of that chlorinated hydrocarbon.

5. Possible Solutions

When the sketch of the development of *in situ* remediation and the problems associated are summarized it boils down to:

- The hydrophobic nature of soils contaminated with hydrocarbons leading to insufficient transport of nutrients into zones with higher levels, where they are needed most;
- The binding of organic contaminants in soil organic matter;
- The binding of phosphate to soil minerals;
- The presence of DNAPL's of chlorinated hydrocarbons in for example inclusions of fine sands in coarser sand layers, where they are difficult to reach.

The hydrophobic nature of soil layers containing hydrocarbons may be influenced by the use of surface active agents. In principle micro-organisms are able to produce their own in which case the oil is emulsified and it is in that case mobile enough to pump a yoghurt like oil in water mixture from these layers. In practice this has been noticed only in two out of a 130 cases when conditions were optimal (high levels of mineral oil, high microbial activity and more than sufficient levels of nutrients and oxygen).

The use of surfactants in *in situ* remediation cannot be compared with soil washing techniques, for *in situ* processes mixing energy is limited to only groundwater flow. Consequently the changes to be obtained depend on the diffusion of the surfactant into the oil. These processes also are slow and require sufficient contact time. In *in situ* experiments this contact time was found to be in the order of one tot two weeks for a surfactant like Tween 80 and its concentrations in this period shall remain above the critical micelle concentration (CMC) to act on the oil. Several products commercially available were tested and found not to be effective *in situ* (effectiveness was high in shake flasks and significant in column experiments, the former because mixing energy could be applied, the latter because the solution is forced through the contaminated soil, while in practice this is not the case).

In principle it is not necessary to mobilize the oil at all, when the penetration of groundwater with nutrients and oxygen in the oily layers is facilitated, because the micro-organisms present are then able to start the biodegradation of the contamination. Once degradation processes occur, the hydrophobic nature of the soil is lost and normal *in situ* practices can be applied. There is a need of low cost materials, which can be used for this purpose.

Binding of contaminants to soil organic matter retards their biodegradation. It was found that once degradation starts, concentrations in the groundwater are very much reduced, which will draw the solubility equilibrium in the direction of groundwater and finally will result in the complete removal of a contaminant. In the case of compounds that may bind strongly to the soil organic matter like pesticides and POP's this process will take a long period of time. The transport rate may be increased when the compound under consideration could be linked into for instance a soluble humate/contaminant complex, by which aqueous concentrations would drop and the driving force would increase. This complex should then be checked for its biodegradability to solve the complete problem.

Phosphate fixation is a problem in calcium or iron rich soils leading to nutrient shortages at the spots where they are needed. Organic phosphates which are water soluble do exist and can be used for this type of problems. A cost effective alternative would, however, be welcomed.

The DNAPL-type of problem may be approved by a surfactant like solution. On the other hand by adapting the *in situ* system in general this problem can be solved.

Chapter 13

WASTEWATER TREATMENT USING MODIFIED NATURAL ZEOLITES

P. PRINCZ[1], J. OLAH[1], S.E. SMITH[2], K. HATFIELD[2],
M.E. LITRICO[2]
[1] Living Planet Environmental Research Ltd., H-2040 Budaors,
Szivarvany u. 10, Hungary <lplanet@hungary.net>
[2] Department of Civil and Coastal Engineering, University of Florida,
Gainesville, Florida USA

Abstract

Current wastewater treatment processes generally improve the quality of effluent water by enhancing the efficiency of the pollutant reduction process or by increasing the retention time of the wastewater in the treatment facility. The latter approach, however, requires a larger aeration basin and higher operation cost.

One of the most promising approaches to improve the efficiency and increase the capacity of wastewater treatment plants without increasing size is based upon application of natural zeolites in the aeration basin. Zeolite particles are good carriers of bacteria, which adsorb on the zeolite surface resulting in increased sludge activity. There is a significant drawback to the application of the zeolite additive. Formation of the bacteria layer on the zeolite surface is a slow process and becomes effective only after approximately a week.

A new zeolite modification method accelerates the interaction between zeolites and activated sludge, which further increases the sludge activity. The effects of modified zeolites on organic degradation rates have been tested in laboratory and full-scale experiments. Based on the experiments, new industrial technologies for zeolite modification and wastewater treatment have been developed.

1. Introduction

In Hungary, as in many other countries, industrial and domestic wastewaters usually feed into a common public-sewer system and are treated together. And the most common wastewater purification method is aerobic biological treatment in which bacteria in the presence of oxygen decompose oxidizable contaminants. The bacteria biomass represents a living activated sludge.

However, a variety of organic and inorganic compounds, at some concentration level, inhibit the activated sludge process. The majority of these toxic organic compounds are biodegradable, albeit some degrade very slowly. Efficient operation of

I. V. Perminova et al. (eds.),
Use of Humic Substances to Remediate Polluted Environments: From Theory to Practice, 267–282.

the activated sludge process in the presence of recalcitrant pollutants, such as oils and pesticides, requires effective control of influent loads. Consequently, a number of modifications of the activated sludge process have been developed to accommodate specific wastewater characteristics and operational needs. To mention a few of them, these are *Selector Activated Sludge* when the growth of floc-forming organisms is stimulated to maintain sludge quality control; *Extended Aeration when a* longer hydraulic retention time (18 to 24 hours) is used for treatment of poorly degradable organic wastes; *High Purity Oxygen Systems* when waste influent, recycled sludge and oxygen gas are introduced at an initial stage of treatment promoting degradation by high dissolved oxygen operating levels, which are achieved using pure oxygen; *Powdered Activated Carbon Treatment (PACT)* when the powdered activated carbon is mixed with the influent wastewater or fed directly into the aeration basin to achieve the sorption of both degradable and non-degradable organic compounds, reducing the effect of inhibition caused by toxic constituents by reducing their "free" concentration; *Two-Stage Activated Sludge (TSAS)* when the more readily degradable organic components are removed on the first stage, and on the second stage, a longer retention time of eight to sixteen hours is used to effect adequate treatment of any recalcitrant pollutants.

The given modifications improve the quality of effluent water by enhancing the efficiency of the pollutant reduction process or by increasing the retention time of the wastewater in the treatment facility. The latter approach, however, requires a larger aeration basin and higher operation cost. One of the most promising approaches to improve the efficiency, and simultaneously increase the capacity of wastewater treatment plants without increasing size is based upon application of natural zeolites in the aeration basin. Zeolite particles are good carriers of bacteria, which adsorb on the zeolite surface resulting in increased sludge activity.

Wastewater treatment using zeolites has been developed in Hungary since 1970s [1-6]. The most advanced technology got the name "Zeoflocc" [7]. It was proven that addition of unmodified clinoptilolite improves the quality of effluents causing greater reduction of chemical oxygen demand (COD), better oxidation of ammonia, more effective phosphorus removal, and faster settling of the sludge. The Zeoflocc process was tested in WTPs of differing capacities and influent characteristics in Hungary. Further development of Zeoflocc process implied a use of modified zeolites. The beneficial effects of zeolite modification or activation are summarized in [8]. The modification means, e.g., dealumination, increase of pore size by acid treatments, transformation of polycationic natural zeolites containing mainly Na^+, K^+, Ca^{2+}, Mg^{2+} exchangeable cations into monocationic ions, e.g., into Na-form, by exhaustive ion exchange and follow up treatment with Fe^{3+} ions to reverse the charge of the zeolite particle. Horváthova [9] found that the ion exchange capacity of clinoptilolite increased by fifteen percent after ion exchange for sodium, and by 30 percent after acid-treatments and following heat-treatments.

Another kind of treatment that can effectively change the charge and surface of natural zeolites was that of introducing methylamine hydrochloride into cationic positions on the zeolites in order to adsorb chlorinated hydrocarbons (trichloro-ethylene, dichloro-ethane, dichloro-ethylene, chloroform) [10]. After modification, the sorption capacity increased by 35-40 percent. This shows that organic cations can be bound electrostatically (e.g., they can be bound rather strongly to the zeolite) as known from the literature for quaternary amines. Of particular importance is that this demonstrated that surface properties of zeolites can be customized (e.g., organophillic sites can be created) thus enlarging the field of potential applications of natural zeolite in wastewater treatment.

The above developments lead to a new modification of zeolite using cation active polyelectrolytes (CAP). In the experimental phase of the work, zeolites of different structures and origins were pre-modified with HCl and NH_4Cl solutions at various temperatures and times of treatment. For modification quaterner ammonium groups containing CAPs of different molecular weight were used in various concentrations. By using CAP it is then feasible to overcome a significant drawback of using the zeolite as an additive in wastewater treatment, namely, the slow formation of a bacterial layer on the zeolite surface. Via the CAP-chains attached to the external surface of the zeolite particle, bacteria can rapidly adhere to the zeolites. This significantly reduces the time for the displaying positive effects of zeolite addition, and increases the decomposition activity of the zeolite-bacteria system. The effect of modified zeolites on the decomposition rate of organic compounds was tested first in laboratory experiments. Based on the successful laboratory tests, bench and full-scale experiments were carried out at several Hungarian water treatment plants. This new process is named ZEORAP. Its implementation was shown to increase the capacity of WTPs and to improve the quality of the treated wastewater.

The objective of this paper is to describe the development of the new waste water treatment technology based on a use of CAP-modified zeolites, and to demonstrate its successful application in full-scale, functioning water treatment plant in Szob (Hungary).

2. Development of ZEORAP Technology

2.1. THEORETICAL BACKGROUND FOR USING CAP-MODIFIED ZEOLITES IN ACTIVATED SLUDGE

The biochemical activity of activated sludge can be increased with the addition of porous additives, such as those with high surface area to which bacteria adhere and are thus immobilized. The bacteria are then accessible in high concentrations for the utilization of substrates and oxygen. The specific gravity of bacteria flocs adhered to porous solids, bacterial/solid flocs, is higher than that of simple bacterial flocs alone. Therefore, the settling rate of the bacterical/solid flocs increases and the effluent contains less suspended matter.

Porous materials such as basalt, lava stone calcined clay, activated carbon, anthracite, brown coal coke, pumice, plastic beats of polyethylene, polystyrene, polyamides, etc. are used. Most of these substances have large pores, rather low surface area of less favourable polarity and no ion exchange ability, which play important roles in the effectiveness of sewage decontamination. In order to have more effective additive, application of zeolitic tuff was suggested and patented [11] for increasing the biochemical activity and the settling rate.

The tuff consists of different minerals such as zeolite clinoptilolite, mordenite, chabazite, phillipsite, quartz, cristobalite, feldspar, montmorillonite and volcanic glass [12]. The crystallites of rock building minerals set up a porous structure ranging from several Angstroms which is the size of zeolitic pores, to micro-, meso- and macropores up to 75,000 Å, accessible for large molecules, viruses and even bacteria [13, 14]. The tuff has ion exchange properties from the zeolite content. In zeolites, trivalent aluminium substitutes isomorphously for the tetravalent silicon in the SiO_2 lattice, bringing about a negative lattice charge that is compensated by exchangeable inorganic or organic cations [15].

The theory of the new method is based on the concept that natural zeolites contain easily removable Na^+, K^+, Ca^{2+} and Mg^{2+} ions. In aqueous media, as a result of dissociation, zeolites have negative lattice charge. Bacteria and bacterial flocs in the biological reactor also have a negative surface charge. Although particles of negative charges repulse each other, bacteria can adsorb on the zeolite surface. This can be explained by the presence of polyelectrolytes produced by the bacteria. The polyelectrolytes production, however, is a slow process, which is generally inhibited by the toxic components of wastewaters.

The CAP's organic compounds of high molecular weight have a large number of positive charges after dissociation. When zeolite is modified properly, CAPs fixed on the zeolite surface still have a number of free positive charges that facilitate the accumulation of bacteria on the zeolite surface. Consequently, the formation of a bacterial layer on the zeolite surface accelerates regardless of the polyelectrolyte production of bacteria.

CAP containing quaternary ammonium groups are bound electrostatically to the external surface of zeolite crystals embedded in rock (comprising the "natural zeolite"). The "free" ammonium groups of CAP bind with the negatively charged sites of bacteria active in biological degradation. The ion exchange capacity of the zeolite is therefore not exhausted. Most cationic sites are occupied by H^+, Na^+, K^+, Mg^{2+}, Ca^{2+}, or NH_4^+, depending on the zeolite pre-treatment. When the pH of the wastewater is above 3.5, these latter cations can be selectively exchanged for transition or heavy metal cations present in the sewage. At lower pH values, protons occupy cationic sites of zeolites instead of the metal cations [16].

Interactions between zeolite particles and CAP molecules depend on the form of particles, the specific surface area, the Si/Al ratio (e.g., specific lattice charge of zeolites), the zeolite-to-CAP ratio, and the type and molecular weight of CAP used for modification. The zeolite-CAP bonding formation rate is faster if natural zeolites (zeolites of Na^+, K^+, Ca^{2+} and Mg^{2+} form) are converted to H^+- or NH_4^+-form. With larger specific surfaces and lattice charge, there is a corresponding increase in the specific CAP concentration on CAP-modified zeolites. Thus, there is an optimal molecular weight range for modification. The shorter CAP molecules can adsorb only lower numbers of bacteria, but the longer ones can easily be remobilised from the zeolite surface.

The zeolite-to-CAP ratio or, more exactly, the ratio of the equivalent weights (ew) $zeolite_{ew}$ to CAP_{ew} should be less than one and may have an optimal value. If this ratio is greater than one, i.e., the number of negative charges on the zeolite lattice is larger than the number of the positive charges associated with the CAP molecules, there will not be sufficient cationic character associated with the bacteria to attract. If the ratio is too small (<<1) the bond between zeolites and CAP will not be strong enough. To ensure high bond-strength between zeolite and CAP, a polymer containing quaternary ammonium or amino groups should be used for modification. Creating an effective zeolite-CAP bond resistant to the physical and chemical effects extant in a biological reactor of a water treatment plants was one of the most critical objectives of the project.

2.2. EXPERIMENTAL DESIGN

To convert the natural zeolites to H^+-forms, treatment with aqueous HCl solution at room temperature was applied. Conversion to NH_4^+-form was carried out with aqueous NH_4Cl solution at room temperature. To convert the natural zeolite to H^+-form and increase its specific surface simultaneously, treatment with aqueous HCl at the boiling point was applied. Strong acids and high temperature affect not only the specific surface, but will increase the Si/Al ratio as well. Since the higher Si/Al ratio represents lower specific surface charge, which is disadvantageous, conditions of acidic treatment used for increasing the specific surface were also optimised.

To quantify the amount of CAP bound to the zeolite surface, a total organic carbon (TOC) was determined. First, the easily remobilisable part of the CAP was removed by washing the modified zeolite with TOC-free water. After this treatment, zeolite has only "unwashable" CAP bound chemically. The quantity of this CAP was determined using a thermal TOC method. The bond-strength between zeolite and CAP was determined by X-ray photoelectron spectroscopy.

Modified zeolite having chemically bound CAP was tested further by the respirometric method. Experiments were carried out with a continuous respirometer set to the operation parameters of an average Hungarian water treatment plant. The effect of modified zeolite application on the decomposition rate of organic compounds and the quality of treated water was determined. Based on the results, modified zeolite was ranked. Respirometric examinations of the best modified zeolite was continued in respirometers set to the operation parameters of the water treatment plant where the bench and full-scale experiments have been performed.

2.3. RESULTS OF PRELIMINARY EXPERIMENTS

Clinoptilolite containing rhyolite tuff was powdered and then converted to hydrogen or ammonium form using a respective acidic or NH_4Cl. 1 kg of the powdered zeolite, of the fraction particle size smaller than 200 μm, was mixed with 1-liter aqueous solution of CAP of acrylic acid-amide type containing 1 g/L of Praestol 444K or Cytec C573. The homogenized paste was dried at 25°C.

40 mg of zeolite coated with Praestol 444K or Cytec C573 was added to 1 litre of municipal wastewater originating from the South Budapest Sewage Treatment Plant in Hungary. The concentration of the activated sludge was 3 g/L. Similar experiments were carried out with zeolite modified with iron ions according to the "Zeoflocc" method. The obtained results are summarized in Table 1.

Table 1. Comparison of the Zeoflocc process and the new method using zeolite modified with Praestol 444K and Cytec C573.

Time of treatment (hour)	Zeoflocc process	Modified zeolite	
		Praestol 444K	Cytec C573
COD (initial value of 850 mg/L)			
1	250	135	260
5	95	58	80
10	72	35	75
48	76	27	75
NO_3 N (initial value of 30 mg/L)			
1	30	27	27
5	17	6	20
10	11	3	10
48	9	2	10
Total-N (initial value of 87 mg/L)			
1	75	60	78
5	55	34	54
10	31	19	35
48	17	8	19

The data given in Table 1 illustrate the beneficial effect of the natural clinoptilolite modified with Praestol 444K. No improvement in decomposition rate was detected in zeolite modified with Cytec C573, probably, due to the higher molecular weight of the polymer.

3. Implementation of ZEORAP-Technology at the Water Treatment Plant of Szob

3.1. ORGANIZATION OF THE FULL-SCALE EXPERIMENT AT THE WATER TREATMENT PLANT OF SZOB

The water treatment plant (WTP) for Szob is located on the east bank of the Danube. The total hydraulic capacity of the WTP of Szob is 1,000 m^3/day. Influents of the WTP frequently cause water and sludge quality problems in summer, autumn and winter due to fruit processing. The effluent COD generally exceeds the standard value (75 mg/L) and the sludge volume index (SVI) is usually higher than 100 indicating poor settling properties. The communal rate of the influent is so low that its nitrogen concentration usually does not reach the optimal value necessary for biochemical processes (C: N: P = 100: 3: 0.5). This suggests that the WTP is not suitable for demonstrating potential improvements in nitrification related to the use of modified zeolite (MZ). Because there is no denitrification activity at the water treatment plant, the MZ effect on denitrification is not evaluated.

Based on the laboratory-scale experiments, the technology using modified zeolite (ZEORAP) was installed and implemented at the Szob WTP of the Danube-Valley Regional Waterworks (DMRV). To ensure an accurate assessment of the ZEORAP effect, wastewater flows were divided between a "zeolite line" that applied ZEORAP technology, and a reference line that applied standard technology. Water and sludge quality data of the ZEORAP line were compared to the reference line. Clinoptilolite content of the natural zeolite used after modification in the experiments was 32 to 36 percent.

Concentration of MZ to be applied was five percent between 30 March and 18 April, eight percent between 19 April and 25 September. Since 25 September the applied zeolite concentration has been ten percent. (Percentile values are expressed in $g_{zeolite}/g_{dry\ sludge}$ x 100 units.) Wastewater and sludge samples were taken twice a week. Daily average water samples were collected from both the influent wastewater and the effluent water for the "zeolite" and reference lines. Sludge samples were taken from both the aeration tanks and the recirculation systems.

Water samples were analyzed for pH, COD, filtered COD (COD_f), BOD_5, NH_4-N, Kjeldahl-N, NO_3–N, total-P and suspended solids. The monitored quality parameters of sludge were as follows: (1) sludge concentration and sedimentation in the aeration tank; (2) sludge concentration and sedimentation in the recirculation system; (3) excess sludge concentration and organic content; (4) daily quantity of excess sludge.

3.2. RESULTS OF FULL-SCALE EXPERIMENTS AND THEIR CRITICAL ASSESSMENT

Changes in wastewater quality parameters are shown on the example of chemical oxygen demand (COD) (Figure 1) for seasons with the moderate (regular) organic matter load (March to October) and high organic matter load characteristic for the period of fruit processing (October to December). SVI measurements are shown in Figure 2. Analytical data of the experiments with 5%, 8% and 10% modified zeolite are summarized in Table 2.

Figure 1a shows that there was a 37 percent decrease in COD concentrations in the effluent from the zeolite line compared to the reference line observed for the period of moderate organic loading. During the fruit-processing season (October to December) the organic loads increased, therefore more significant differences were detected between the zeolite and the reference treatment lines (Figure 1b). In general, it could be seen that the addition of MZ considerably reduced the impact of industrial shock-loads on the effluent COD (13 and 30 August, October 17 – November 12).

BOD measurements confirmed the above tendency that the zeolitic line was not as sensitive to shock-loads as the reference line. BODs measured in the zeolite line were always lower. The average BOD concentration in the effluent of zeolite line was 11.9 mg/L, whereas in the reference line it was 16.9 mg/L. The average decrease in BOD concentrations was 30% in the zeolitic line with respect to the reference line.

274

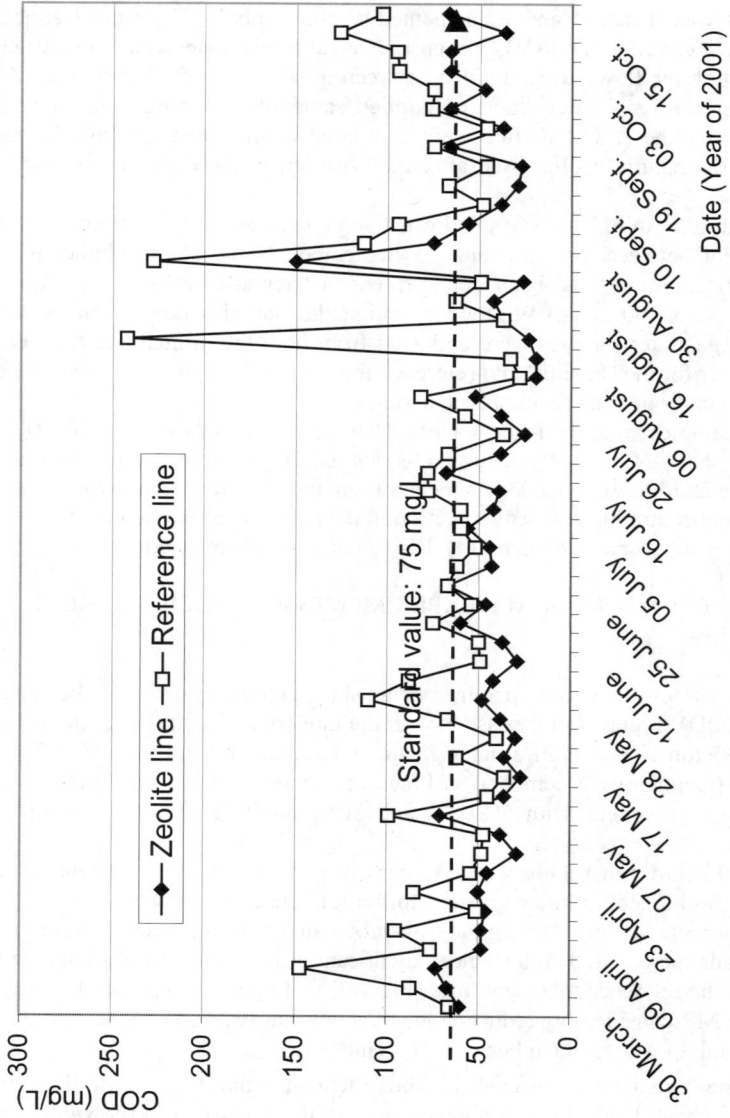

Figure 1a. COD concentrations measured in the effluents of WTP of Szob: season of the regular organic matter load.

275

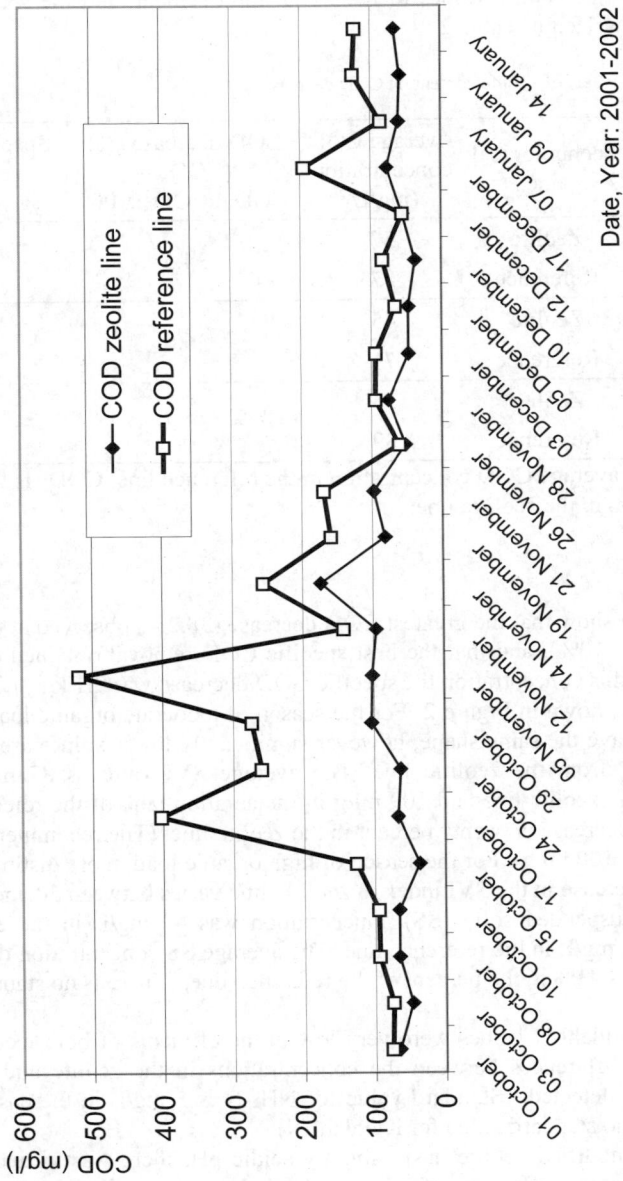

Figure 1b. COD concentrations measured in the effluents of WTP of Szob: season of the high organic matter load during fruit-processing.

276

The relationship between the applied MZ concentration and the water quality improvement of the effluent was examined using COD as an example. The average COD concentrations, the percentile values of the COD improvement, and the specific COD removal are summarized in Table 2.

Table 2. Effect of Zeolite content on COD decrease.

Zeolite content, percent of dry sludge	Technological line	Average COD concentration (mg/L)	COD decrease (%) $(COD_R - COD_Z)/COD_R \times 100^*$	Specific COD removal (g_{COD}/g_{zeolit})
5%	Zeolite	57	34.5	2.1
	Reference	87		
8%	Zeolite	45	36.6	1.2
	Reference	71		
10%	Zeolite	55	38.2	1.6
	Reference	89		

* COD_R is the average COD concentration in the reference line; COD_Z is the average COD concentration in the zeolitic line.

The given data show that the greatest COD decrease (38.2%) observed at the highest MZ concentration (10%), and that the best specific COD removal obtained at 5% MZ concentration. At this concentration the specific COD decrease was 2.1 kg_{COD}/kg_{MZ}.

SVI values are shown in Figure 2. For the season of moderate organic loads (Figure 2a), both curves have the same shape, however significantly lower values are seen with sludge originating from the zeolitic line. The average SVI value is 87 ml/g in the aeration tank of the zeolite line and 109 ml/g in the aeration tank of the reference line. The average SVI decrease is twenty percent in the zeolite line. (The recommended value of SVI is less than 100 ml/g). For the season of high organic load, more distinct changes are observed: a decrease in the SVI index in zeolitic line varies between 20 and 70%.

The average suspended solid (SS) concentration was 49 mg/L in the effluent of zeolite line and 83 mg/L in the reference line. The average SS concentration decrease in the zeolitic line was 41% in the percent of the reference line. (There is no standard value for SS.)

NH_4-N and Kjeldahl-N values were very low in the effluents of both technological lines. Significant differences between the concentrations in the zeolite and reference lines could not be detected. (Standard value for NH_4-N is 5 mg/L in the case of Szob WTP and there is no standard value for Kjeldahl-N).

Although the modified zeolite has a slightly acidic pH, there was virtually no pH difference between the effluents of the zeolite and reference lines observed. The measured pH values varied between 6.8 and 7.7. (The standard pH range is between 6.5 and 9.0 for the Szob WTP.)

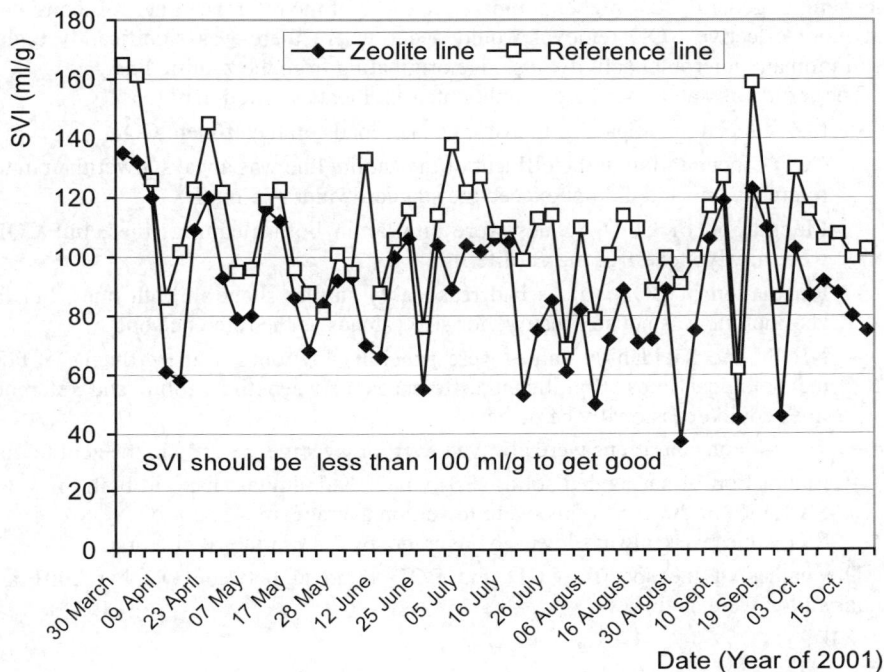

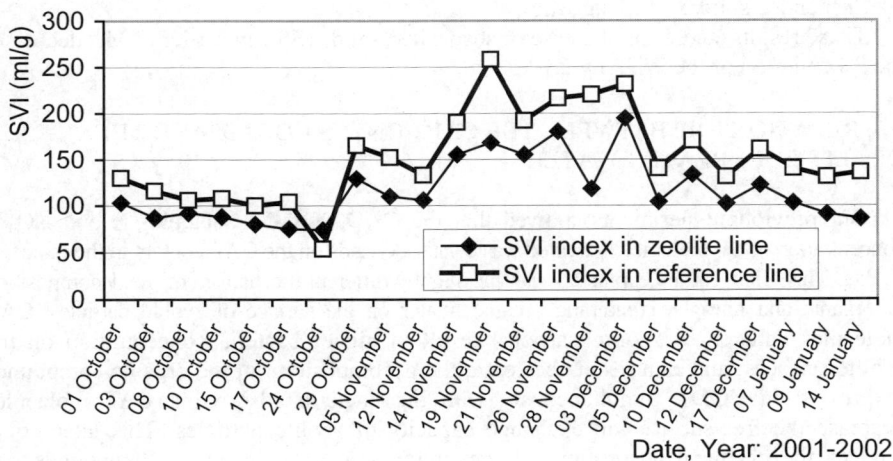

Figure 2. Sludge volume index values of sludges originating from the effluents of the WTP of Szob.

The values of both endogenic respiration and activity of sludge found for the zeolitic line were, in general, 30% higher compared to those of the reference line. It is consistent that more effective COD removal would occur where there was significantly higher respiration and microbial activity in sludge originating from the zeolitic line.

The results of water and sludge quality measurements showed that:

- MZ reduced the impact of industrial shock-loads on the effluent COD
- COD concentration in the effluent of the zeolite line was always lower than in the reference one and did not exceed the standard value (75 mg/L)
- Fluctuation of COD_f values were similar in both cleaning lines, but COD_f were always lower in the zeolite line
- Fluctuation of BOD values had reasonably similar shape in both lines, but the zeolitic line was not so sensitive for shock-loads as the reference one
- NH_4-N and Kjeldahl-N values were generally low ones in the effluents of both technological lines. Significant differences between the zeolite and reference lines could occasionally be detected.
- Total-P concentrations were always lower, on average, by 20% in the zeolite line
- Fluctuation of suspended solids (SS) values had similar shape in both lines, but SS values of the zeolite line were lower, on average, by 41%.
- SVI values were always lower, on average, by 20% in the zeolite line

The values of the specific COD and BOD removal between October 2001 and January 2002 were as follows:

COD: 3.4 kg_{COD}/kg_{MZ}
COD_f: 0.8 kg_{CODf}/kg_{MZ}
BOD: 0.4 kg_{BOD}/kg_{MZ}
Suspended solids: 2.7 kg_{SS}/kg_{MZ}

These results suggest in the case of suspended solids (SS), that 1 kg of MZ decreases the SS emission of the WTP by 2.7 kg.

3.3. RELATIONSHIP BETWEEN THE CAP CONTENT OF MZ AND THE QUALITY OF TREATED WATER

Our previous experiments showed that the COD, NH_4-N, Kjeldahl- N and NO_3-N concentration of the biologically treated wastewater depends on the CAP content in the modified zeolite. This effect can be explained with the slightly different mechanism of the decomposition of organic and nitrogen containing organic matter on the surface of zeolite particles. CAP molecules adhered to zeolite particles result in higher bacteria concentration on the zeolite surfaces, and as a result accelerate the decomposition of the organic compounds (expressed in COD). Simultaneously with the bacteria adsorption, CAP molecules decrease the free-surface ion exchange capacity of zeolite particles. This later effect reduces the chemical adsorption of ammonium ions and organic-N compounds and consequently decreases the decomposition rate of the molecules containing nitrogen. The experiments were carried out in a continuous respirometer. The obtained data are summarized in Table 3.

Table 3. Biological degradability test of the wastewater of Szob WTP using modified zeolites of different CAP content.

CAP content of modified zeolite (mg$_{TOC}$/g$_{zeolite}$)	COD (mg/L)	NH$_4$-N (mg/L)	Kjeldahl-N (mg/L)	NO$_3$-N (mg/L)	Total-P (mg/L)	SVI (ml/g)
influent	1,100	15.6	45.2	1.0	15.3	–
No zeolite addition	226	12.3	40,9	5.1	12.6	275
0.0 (natural zeolite)	158	10.8	32.1	18.2	11.2	126
0.5	147	10.8	31.3	18.4	11.0	122
1.0	135	10.8	30.5	18.3	10.8	118
1.5	129	10.9	29.3	18.0	10.7	114
2.0	121	10.9	28.7	18.0	10.6	111
2.5	111	10.9	30.0	17.9	10.6	105
3.0	102	10.0	31.7	17.5	10.5	101
3.5	91	11.1	33.0	17.3	10.3	97
4.0	85	11.1	33.6	17.0	10.0	94
4.9	81	11.2	34.1	16.0	10.0	88

It can be seen that:
- MZ of higher CAP concentration results in better COD, total-P and sludge volume index (SVI) values in the effluents. The best results are found with the MZ of highest CAP content (4.9 mg$_{TOC}$/g$_{zeolite}$).
- In the case of Kjeldahl-N and NH$_4$-N there is an optimal CAP concentration between 0.5 – 2.5 mg$_{TOC}$/g$_{zeolite}$.
- The optimal CAP concentration range depends on the composition of the wastewater to be treated.
- CAP content of MZ does not appear to have an effect on the NO$_3$-N concentration of the effluents examined

It can be concluded that the effective COD removal and the decomposition of the nitrogen containing organic, as well as inorganic compounds require MZ of different CAP content. Therefore, in a decision to use MZ consideration must given to the composition of the wastewater to be treated, the effluent water quality parameters of concern, the applicable water quality standards and the cost of CAP.

3.4. EXAMINATION OF THE ZEOLITE-CAP BOND-STRENGTH

To evaluate the possible release of CAP from its zeolite complexes during the process of wastewater treatment, the physical-chemical processes that take place in the course of modification were considered as follows:

a. Dissociation of CAP

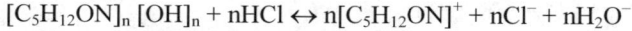

$$[C_5H_{12}ON]_n [OH]_n + nHCl \leftrightarrow n[C_5H_{12}ON]^+ + nCl^- + nH_2O^-$$

where n = the number of the monomers.

b. Creation of Zeolite CAP bond

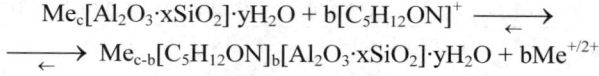

$$Me_c[Al_2O_3 \cdot xSiO_2] \cdot yH_2O + b[C_5H_{12}ON]^+ \longrightarrow$$

$$\xrightarrow{\longleftarrow} Me_{c-b}[C_5H_{12}ON]_b[Al_2O_3 \cdot xSiO_2] \cdot yH_2O + bMe^{+/2+}$$

where Me: Na, K, Mg, Ca; c: number of cations; x: >2 (depend on zeolite type); y: 1-8 (depend on zeolite type); b: c/30 (depend on the applied zeolite-CAP ratio).

The above reaction is shifted to the right, because the CAP-zeolite product has lower solubility than the reactants, i.e. the CAP-zeolite bonds must be stable at higher pHs as well. In order to verify this statement, 100g of modified zeolite was resuspensed in 1 litre of TOC-free water, the suspension was intensively stirred and the TOC concentration in the aqueous phase was measured as a function of pH and time. The corresponding data of TOC measurements are shown in Figure 3.

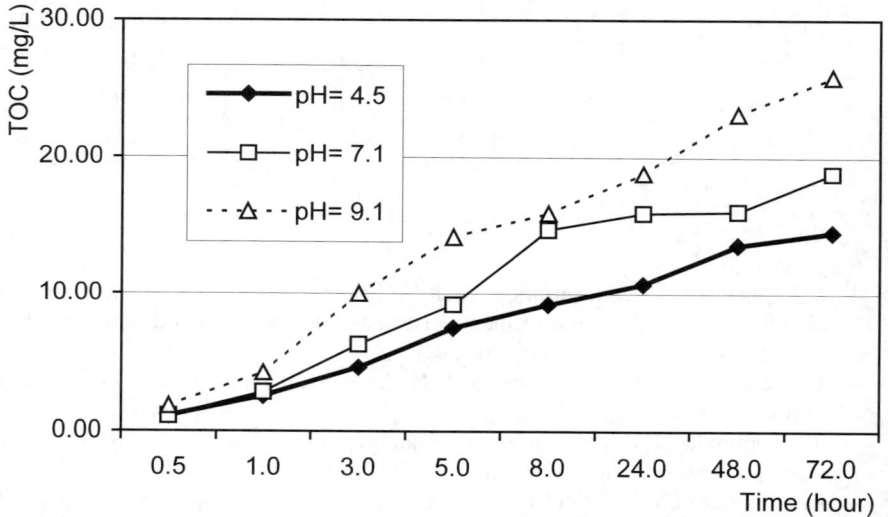

Figure 3. Remobilisation of CAP content expressed in TOC as a function of pH and time.

To evaluate the data, it should be noted that for the full-scale experiments the "dry" modification of natural zeolite was used. For this purpose, 10.40 litres of CAP solution was stirred in 32 litres of 1 M HCl and made up with tap water to 48 litres. The obtained solution was sprayed onto 800 kg of natural zeolite that equated to the content of the added total organic carbon (TOC) in the modified zeolite of 4.0 mg/L. Hence, the total remobilisation of CAP attached to zeolite should increase the TOC concentration of filtered samples up to 400 mg/L. From the data shown it can be seen that more than 93 percent of the CAP-zeolite bonds remained stable even at pH 9 after three days. There, since the pH of communal wastes is always below 9, the CAP bonding to zeolite can be considered stable in the process of wastewater treatment.

4. Environmental and Economic Benefits of the ZEORAP Technology

It was shown that implementation of the ZEORAP technology increased the capacity of WTPs and improved the quality of the treated wastewater. These benefits can be attained with negligibly low investment costs at the existing WTPs and with reduced operation costs the total cost of wastewater treatment decreases. For new water treatment plants, ZEORAP can expedite plant start-up and attenuate fluctuations in treatment system performance. Due to increased treatment efficiency the cost of investing in ZEORAP technology can be reduced by approximately 25 percent.

The enhancement of biological treatment capacity results in a decrease of environmental loading, i.e., less ground and surface water contamination. Since a significant portion of nutrient loading to surface waters is derived from inadequate wastewater treatment, it is expected that decreasing these loads through the use of ZEORAP technology would translate into reduced costs of environmental remediation.

Excess sludge will contain zeolite and heavy metals in the form of oxides and hydroxides. When this sludge is dewatered, stabilized and composted, the metals form organic chelates. This type of sludge treatment may decrease the pH, and result in the dissolution of heavy metals that subsequently become fixed to the cationic sites of the zeolite.

Heavy metal cations occupying a small fraction of cationic sites of the zeolite are strongly bound, e.g., these harmful cations are retarded under conditions of plant cultivation. After composting, excess sludge produced with ZEORAP technology, can be safely and economically utilized as fertilizer. This environmentally sound method of sludge disposal has been used in a number of successful experiments around the world. In zeolite-free sludge, toxic cations are not fixed or immobilized and, consequently, are taken up by plants.

The amount of zeolite in excess sludge is approximately 8-wt % in the dry substance. The zeolite additive contains $5 \cdot 10^{-2}$-wt % CAP, and remobilisation is slow. Since the treatment (dewatering) of surplus sludge containing zeolites requires 10 percent less quantity of CAP, the CAP content of the ZEORAP sludge is smaller and, consequently, less harmful than sludge of conventional wastewater treatment.

5. Conclusions

ZEORAP represents a significant advance in wastewater treatment technology. Effluent water quality standards can be met with minimum infrastructure investment and cost. Using the ZEORAP process the following goals can be achieved:
- increase in the decomposition of organic compounds (measured as COD) by approximately 20%;
- increase in nitrification activity by 50-100%;
- increase in the degradation of nitrogen compounds (proteins and protein derivatives) by 30-50%; and,
- increase of sludge settling rate by 100%.

In spite of these achievements, opportunities exist to advance the current ZEORAP technology. Thus far, research has focused on the effectiveness of various synthetic cation active polyelectrolytes. As an alternative to synthetic polyelectrolytes, natural or modified humics could be used. Zeolites modified with humics have not been tested in wastewater systems.

6. Acknowledgements

This research was supported by the NATO Science for Peace Sub-Programme, Project Ref. Number SfP-972494.

7. References

1. Kalló, D., Papp, J. and Valyon, J. (1982) Adsorption and catalytic properties of sedimentary clinoptilolite and mordenite from Tokaj Hill/Hungary, *Zeolites* **2**, 13-16.
2. Kalló, D. (1995) Wastewater purification in Hungary using natural zeolites, in D. W. Ming, F. A. Mumpton (eds.), *Natural Zeolites '93 Occurrence, Properties, Use*, Intern. Committee on NaturalZeolites, Brockport, New York, pp. 437-445.
3. Mucsy, G. (1992) Grosstechnischer Einsatz von Zeolith auf Kläranlagen in Ungarn, *Abwassertechnik* **43(2)**, 48-54.
4. Oláh, J., Papp, J., Mészáros-Kiss, Á., Mucsy, G. and Kalló, D. (1988) Removal of suspended solids, phosphate ions from communal sewage using clinoptilolite derivatives, in D. Kalló, H.S. Sherry (eds.), *Occurrence, Properties and Utilisation of Natural Zeolite*, Akadémiai Press, Budapest, pp. 511-520.
5. Oláh, J., Papp, J., Mészáros-Kiss, Á., Mucsy, G. and Kalló, D. (1989) Simultaneous separation of suspended solids, ammonium and phosphate ions from wastewater by modified clinoptilolite, *Stud. Surf. Sci Catal.* **46**, 711-719.
6. Oláh, J., Papp, J. and Kalló, D. (1991) Upgrading the efficiency of biological sewage treatment with zeolites, *Hidrológiai Közlöny* **71(2)**, 70-76 (in Hungarian).
7. Charuckyj, L. (1997) *Brisbane Water Zeoflocc Performance Report*, Zeoflocc Process Selected by Queensland Government, Zeolite Australia Limited.
8. Vansant, E. F. (1990) *Pore Size Engineering in Zeolites*, J. Wiley & Sons, Chichester, New York, Toronto.
9. Horváthová, E. (1986) Ionselektiver Austausch an slowakischem Klinoptilolith-Tuffit, *Acta Hydrochim. Hydrobiol.* **14(6)**, 495-502.
10. Rustamov, S. M., Bashirova, Z. Z., Nasiri, F. M., Yagubov, A. I. and Muradova, S. A. (1988) Use of zeolites for purification wastewaters from toxic organic chlorine compounds, in D. Kalló, H.S. Sherry (eds.), *Occurrence, Properties and Utilisation of Natural Zeolites*, Press, Budapest, pp. 521-528.
11. Kiss, J., Hosszú, Á., Deák, B., Kalló, D., Papp, J., Mészárosné Kiss, Á., Mucsy, G., Oláh, J., Urbányi, G., Gál. T., Apró, I., Czepek, G., Törőcsik, F. and Lovas, A. (1984) *Process and equipment for removal of suspended material, biogenetic nutrients and dissolved metal compounds from sewage contaminated with organic and/or inorganic substances*, Hung. Patent 193-550, Europatent 0177-543; (1988) *Process for preparing an agricultural fertiliser from sewage*, US Patent 4, 772, 307.
12. Gottardi, G. and Galli., E. (1985) *Natural Zeolites*, Springer, Berlin, Heidelberg, New York, Tokyo.
13. Kalló, D. (1992) Natürliche Zeolithe – Herkunft und Wirkungsmechanismen, *Abwassertechnik* **43(2)**, 40-43.
14. Papp, J. (1992) Einsatzmöglichkeiten von Zeolith in der Abwassertechnik:, *Abwassertechnik* **43(2)**, 44-47.
15. Smith, J. V. (1976) Origin and structure of zeolites, in J. A. Rabo (ed.), *Zeolite Chemistry and Catalysis*, Am. Chem Soc., Washington D. C., pp. 1-79.
16. Blanchard, G., Maunaye, M. and Martin, G. (1984) Removal of heavy metals from water by means of natural zeolites, *Wat. Res.* **18(2)**, 1501-1507.

Part 4

Impact of humic substances on physiological functions
of living organisms and on microbial transformations
of ecotoxicants

Chapter 14

MITIGATING ACTIVITY OF HUMIC SUBSTANCES: DIRECT INFLUENCE ON BIOTA

N.A. KULIKOVA[1], E.V. STEPANOVA[2], O.V. KOROLEVA[2]
[1]*Department of Soil Science, Lomonosov Moscow State University, Leninskie Gory, 119992 Moscow, Russia <knat@soil.msu.ru>*
[2]*Bach Institute of Biochemistry of the Russian Academy of Sciences, Leninsky prospect 33, 119071 Moscow, Russia*

Abstract

Mitigating activity of HS can be defined as a phenomenon of lowering the adverse effects of contaminants toxicity and of those of abiotic stress factors such as unfavourable temperature, pH, salinity, etc. As a rule, it is related to the detoxifying properties of HS or to their beneficial effects on biota. This review focuses on the latter effects and considers the possible mechanisms of mitigating activity of humic materials in terms of biological activity of HS. The beneficial effects of HS on biota are segregated into the four categories according to the underlying mechanism of their action: an influence on the organism development; an enhancement in nutrient supply; catalysis of the biochemical reactions; and an antioxidant activity. The published data on the mentioned essential biological functions of HS are reviewed.

The given review shows that in spite of the numerous reports published recently to elucidate the mechanisms underlying biological effects of HS, the data reported are quite contradictory. The main reason is the complexity of HS structure. A wide variety of the effects observed can be explained by the polyfunctionality of HS. The general conclusion can be made – all biological effects of HS have been more pronounced under stress conditions independent of what stress factors are involved. Thus, HS can be considered as an environmental modulator mitigating the harmful consequences of stress factors.

1. Introduction

Humic substances (HS) are natural organic compounds comprising from 50 to 90% of the organic matter of peat, lignites, sapropels, as well as of the non-living organic matter of soil and water ecosystems [1-3]. According to the classical definition [4], HS are "a general category of naturally occurring heterogeneous organic substances that can generally be characterized as being yellow to black in color, of high molecular weight and refractory". They are a mixture of complex organic compounds that are usually

I. V. Perminova et al. (eds.),
Use of Humic Substances to Remediate Polluted Environments: From Theory to Practice, 285–309.
© 2005 *Springer. Printed in the Netherlands.*

separated into three fractions based on aqueous solubility: humic acids (HA), the fraction of HS that is not soluble in water under acidic conditions (pH < 2) but is soluble at higher pH values; fulvic acids (FA), the fraction of HS that is soluble in water under all pH conditions; humins, the fraction of HS that is not soluble in water at any pH value. HS cannot be described by unique, chemically defined molecular structures. They are operationally defined by a model structure constructed on the basis of available compositional, structural, functional, and behavioral data and containing the same basic structural units and the same types of reactive functional groups [5]. So, HS are very often supposed to be just an alkali extract from any organic substrate including sewage sludge, beer factory sludge, raw tea waste, composted grape marc, composted spend mushroom, composted bark, tobacco dust and others. Some authors use term "HS" to define materials which would be more correctly called as "humic-like" substances. This complicates interpretation of the reported data considerably. Despite of that, in this review, we kept the terminology used by the authors of the original papers.

HS have been the subject of numerous scientific studies due both to their mitigating effects on contaminants toxicity to biota [6-12], and to their anti-stress effects under abiotic stress conditions (unfavourable temperature, pH, salinity, et al.) [13-15]. The mitigating effects can be defined as a phenomenon of lowering the adverse effects of both specific and non-specific stress on biota. The detoxifying properties of HS are generally attributed to their capability for binding ecotoxicants. The binding of xenobiotics to HS causes formation of less bioavailable complexes and adducts followed by lowering their toxicity and bioaccumulation. Another phenomenon of importance is acceleration of biotic and abiotic mineralization of organic contaminants in the presence of HS [23-25]. In addition, the detoxifying ability of HS can be underlay by the beneficial effects exhibited onto living organisms. The latter are of particular importance for anti-stressor (or adaptogenic) activity of HS under conditions of the abiotic stress (water stress, unfavorable temperature).

The beneficial effects of HS on living organisms have been numerously reported (see Table 1). They can be provided by either indirect or direct impact of HS. The indirect effects are mostly provided by the HS-driven changes in environmental conditions such as bioavailability of some nutrients (due to increasing solubility), salt balance, physical and physico-chemical soil properties (soil structure, aeration, drainage, water retaining capacity, soil temperature, and others). On the other hand, HS are supposed to influence biota directly. The principal direct effects exhibited by HS onto living organisms include an increase in biomass accumulation, nutrient uptake, biosynthesis, antiviral activity, and others [13, 14, 27-53, 90-96].

In spite of numerous studies on the biological effects of HS, the mechanism of their action remains unclear. The main reason seems to be the stochastic nature of HS. In contrast to common biological macromolecules, which are synthesized by a living organism according to the information encoded in DNA (nuclear acids, proteins, enzymes, antibodies etc.), HS are the products of stochastic synthesis. They are characterized as polydisperse materials having elemental compositions that are non-stoichiometric, and structures, which are irregular and heterogeneous [16]. The above features hamper a use of common biological approaches to study biological activity of HS. This review represents the first attempt to address the possible mechanisms underlying mitigating activity of humic materials in terms of biological activity.

2. Principal components of mitigating activity of humic substances

The ability of HS to reduce toxicity can result from both their influence on metabolic pathways of ecotoxicants and beneficial effects on living organisms (Fig. 1).

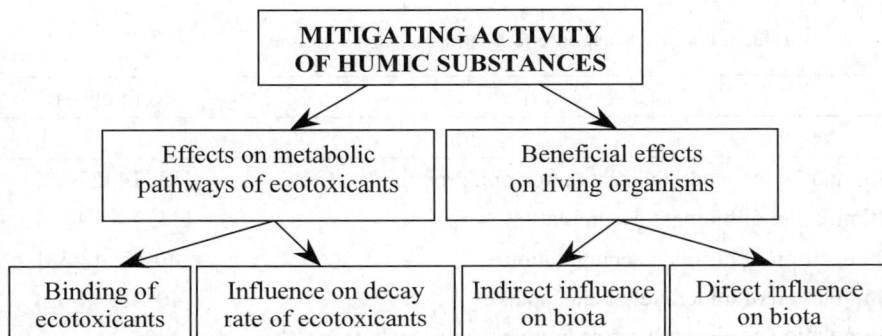

Figure 1. Principal components of mitigating activity of humic substances.

The effects on the metabolic pathway of the ecotoxicants are related to the chemical nature of humics that can be considered as irregular polymers of aromatic polyhydroxyl polycarbonic acids [3]. The peculiar structural feature of HS is a coexistence of both polar and hydrophobic environments in the same molecule. As a result, HS are able to bind both polar and hydrophobic xenobiotic organic compounds, and inorganic ions. The binding to HS causes a change in speciation of xenobiotics followed by a change in their toxicity and bioaccumulation. Based on the binding properties of HS, several remediation technologies were developed [17-22].

Humic materials can also facilitate degradation of some organic contaminants. For example, chemical hydrolysis, followed by degradation by soil microorganisms, could account for most of the atrazine breakdown in soil [23]. Addition of humic material was shown to increase the rate of atrazine hydrolysis. HS can also affect photodegradation, which is important for PAHs, chlorinated aromatic hydrocarbons, chlorinated phenols, and many other pesticides [162]. It was demonstrated that the photolysis rate of 1-aminopyrene could be enhanced by addition of HS [24]. The first order photolysis rate constant of 1-aminopyrene (10 µM) in the phosphate buffer (pH 7.0, 1 mM) containing HA (20-80 ppm) was enhanced by a factor of five compared to that in the absence of HA. Humic materials was hypothesized can transfer energy of oxygen facilitating formation of an excited state oxygen molecule. The single oxygen attacks the organic compound that decays due to indirect photolysis. The similar effect was observed for atrazine [24]. Of importance is, that the photolysis of atrazine was shown to be enhanced by an addition of HA, while the introduced FA caused no effect within 10 days. Recent study has shown that HS played a remarkable role in condensation reactions in the aquatic environment [25]. HS have been found to act as catalysts in Knoevenagel and Claisen-Schmidt reactions accelerating condensation of carbonyl compounds with aliphatic acids. Hence, they may influence the fate of organic pollutants in the environment.

It has been also suggested that these are rather humus-enzyme complexes than HS alone, that are essential for functioning of stable and resilient soil systems. These complexes are considered to be the "crossing-points" between mineral and organic reactions in soil [26]. On the other hand, HS are supposed to be able to influence biota directly. Principal observed direct effects of HS on living organisms are summarized in Table 1.

Table 1. Principal biological effects of HS on living organisms.

Observed effect	References
Plants	
Stimulation of seedling germination and growth	[13, 14, 27-38]
Stimulation of biomass accumulation	[13, 14, 39-46]
Stimulation of nitrogen accumulation	[40-42, 47-53]
Stimulation of mineral elements uptake	[40, 46, 54-61]
Stimulation of biosynthesis of proteins, carbohydrates, chlorophyll et al.	[13, 14, 46, 49, 62]
Stimulation of photosynthesis or respiration	[13, 14, 63-67]
Adaptogenic and detoxifying effects (against excess mineral nutrition, salinity, water stress, unfavourable pH, temperature etc., presence of xenobiotics)	[11, 13-15, 44, 68-73]
Animals	
Stimulation of biomass accumulation	[15, 74]
Immunomodulating activity	[15, 75]
Desmutagenic activity	[68, 76, 77]
Detoxifying ability	[9, 78, 79, 99]
Toxicity	[80, 81]
Microorganisms	
Stimulation of biomass growth and biosynthetic activity	[26, 82-87]
Detoxifying ability	[88, 89]
Viruses	
Antiviral activity	[90-96]

Table 1 shows that, first of all, the direct effects of HS on biota are very versatile; and second, there is no unified mechanism governing biological activity of HS. Of importance is that in contrast to the substances with a well-defined mode of action, HS effects drastically depend on the environmental conditions. David et al. [97] have shown that the most striking examples of HS beneficial effects on plant nutrition were observed under unfavorable growth conditions. Dunstone et al. [98] have found that the largest effects of fulvic acids on stomacal conductivity of wheat were observed under conditions of the sustained drought and hot, dry winds. At the same time, the authors were unable to demonstrate the similar effects of FA on grain yield in droughted plants either in the field or in glasshouse. This elucidates another peculiar feature of HS – instability and poor reproducibility of the biological effects.

Along with the numerous reports on HS detoxifying properties, some data on increasing toxicity of xenobiotics in the presence of HS have been reported as well [80, 81, 100-102]. HS was shown to suppress the immune and inflammatory reactions of cultured human umbilical vein endothelial cells (HUVECs) responsible for endotoxin lipopolysaccharide (LPS) detoxifying. Pretreatment of HUVECs with 100 mg/L HS for two days markedly suppressed the LPS induced expression of adhesion molecules and almost completely inhibited this process at a concentration of 200 mg/L. The authors suggested that HS could be a potential toxin causing blackfoot disease.

3. Beneficial effects of HS on the living organisms

Upon summarizing the discussed above beneficial effects of HS on biota, the following principal ways of HS action could be proposed:

1. HS affect the organism development. Being utilized as a substrate (a source of organic carbon) or nutrient source (N, P, trace elements and vitamins), HS can serve as a moiety of the biosynthesis chains. On the other hand, beneficial effects of HS on the plants are often attributed to hormone-like activity of HS.
2. HS enhance nutrient supply. HS can improve plant growth by increasing uptake of such nutrients as nitrogen, potassium, and some micronutrients.
3. HS catalyze some biochemical reactions. The mechanism of HS interaction with enzymes has not been established yet, - there are no direct evidences that HS can serve as co-oxidants or mediators of enzyme stimulating reactions. The participation of HS in catalysis of biochemical reactions should be thoroughly studied.
4. Antioxidant activity of HS. HS can reduce free radicals resulting from stress such as drought, heat, ultraviolet light and herbicide use. Free radicals are damaging because they are strong oxidizing agents, which damage lipids, proteins and DNA within plants cells.

The comprehensive and recent data on each of the essential biological functions of HS are described below.

3.1. INFLUENCE OF HS ON ORGANISM DEVELOPMENT

3.1.1. HS as a substrate or a nutrient source
The number of microorganisms which are reported to be able of decomposing HS is rather great (Table 2). Still, the information available on the mechanisms of HS transformation and utilization is very scarce.

Of interest is a study on the decomposition of HA isolated from two differently aged sanitary landfills by soil microflora under aerobic conditions [103]. With HA used as a supplementary nutrient source, the level of their utilization was 63.6% for a fresh refuse, and 88.5% - for the refuse disposed for 12 months. When HA was used as a sole source of both carbon and nitrogen, the complete utilization was observed. The decomposition process resulted in alteration of HS properties, namely, in a decrease of molecular weight of humic materials. FTIR spectroscopy has also revealed some changes in HS structure: the complete elimination of C=O band of COOH group, reducing or elimination of C-O stretching in polysaccharides, a strong decrease in absorption of CH_3 and CH_2 groups, and a removal of some aromatic structures [103].

The above results are in good agreement with those reported for HA samples extracted from a mixture of municipal waste and sewage sludge [104]. The yield of microbial biomass substantially increased in the cultures containing HA as a supplementary nutrient source and reached 195% in case of HA from the fresh refuse. At the same time, under the depleted nutrients and low energy source conditions, the same HA inhibited the production of microbial biomass. This effect was attributed to an inhibitory effect of HA decomposition products. The polymerization of low-molecular weight compounds with formation of more stable HS was hypothesized as well.

The numerous attempts to elucidate the mechanism of microbial decomposition of HS were made using a model system of fungi and coal of different ranks [82, 85, 87, 114-117]. These data allowed the authors of this review to propose the scheme of coal HA decomposition by *Basidiomycetes* and *Deuteromycetes* shown in Fig. 2. The effects of these white-rot fungi on the coal-derived HA are determined by the specific features of the fungal ligninolytic enzyme system. The latter consists of lignin peroxidase, manganese peroxidase, other peroxidases, laccase, and supporting enzymes generating H_2O_2. It was shown that depolymerization of lignite HA took place only after its methylation. It seems like lignin peroxidase catalyzed the bond cleavage in the methylated HA and oxidized non-phenolic aromatic groups to aryl cation radicals with further cleavage of C-C and C-O bonds. The predominant products of these enzymatic and enzyme-mediated reactions were methoxylated monoaromatic compounds. Mn peroxidase oxidized Mn(II) to Mn(III). The latter is stabilized due to complexation with organic acids (oxalate, malonate, malate, tartrate or lactate) produced by *Basidiomycetes*. The formed Mn(III) complexes act as high redox potential mediators. Such system – Mn peroxidase and chelated Mn(III) - can be supported by addition of appropriate redox-mediators such as thiols or lipids or unsaturated fatty acids. As a result, the degrading power of this system exceeds that of the enzyme itself. The formation of low molecular weight FA was observed both in the *in vivo* experiments with Mn(II) – amended fungal cultures and in the *in vitro* studies on the coal HA depolymerization using Mn-peroxidase. The correlation between laccase excretion and degradation of coal HA *in vivo* has been also observed for *Trametes versicolor* [110]. Moreover, system laccase and different redox mediators were successfully applied recently to lignin and kraft pulp depolymerization [115].

The ways of HS usage as an organic carbon or nitrogen source by higher organisms are even more sophisticated. Several studies on the uptake of HS by higher plants have been performed. Earlier work relied on the colour changes in the plant organs as an indication of HS uptake [118]. It was found that FA of lower molecular weight can be

291

Table 2. Some fungi and bacteria degrading humic substances.

Species[1]	Reference
Wood- and straw-degrading basidiomycetes[2]	
Basidiomycete strains RBS 1k, JF 596, i63-2	[105-107]
Clitocybula dusenii	[107, 108]
Fomitopsis pinicola, Gymnopilus sapineus, Hypholoma fasciculare, Hypholoma frowardii, Kuehneromyces mutabilis, Piptoporus betulinus, Stropharia rugosoannulata, Bjerkandera adusta	[86]
Lentinula endodes	[109, 110]
Nematoloma frowardii	[87, 108]
Phanerochaete chrysosporium	[82, 111]
Pleurotus ostreatus	[86, 110]
Poyiporus ciliatus, Pycnoporus cinnabarinus	[106]
Trametes versicolor	[109, 110, 112]
Terricolous basidiomycetes native to grassland and/or forest litter	
Agaricus arvensis, Agaricus bisporus, Agaricus porphyrizon, Clitocybe odora, Coprinus comatus, Lepista nebularis, Marasmius oreades	[86]
Collybia dryophila	[113]
Ectomycorrhizal fungi	
Amanita muscaria, Hebeloma crustuliniforme, Lactarius deliciosus, Lactarius deterrimus, Lactarius torminosus, Morchella conica, Morchella elata, Paxillus involutus, Suillus granulatus, Tricholoma lascivum	[86]
Soil-borne microfungi and plant pathogens	
Acremonium murorum, Botrytis cinerea, Chaetomium globosum, Cunninghamella elegans TM 1, Hyphomycetes G28.2&G28.30, *Rhizoctonia solani, Scytalidium lignicola, Trichoderma* sp.	[86]
Fusarium oxysporum, Trichoderma atroviride	[114]
Consortia of soil microorganisms	
Two consortia of predominantly gram-negative bacteria from black forest soil	[86]
Consortium of microorganisms from sandy brown earth	[103]
Bacteria	
Alcaligenes eutrophus, Alcaligenes faecalis, Bacillus brevis, Bacillus cereus, Pseudomonas fluorescens, Pseudomonas putida, Xanthomonas campestris	[86]

[1] Species were named as in original articles.
[2] Classification of fungi into groups was as described in [86].

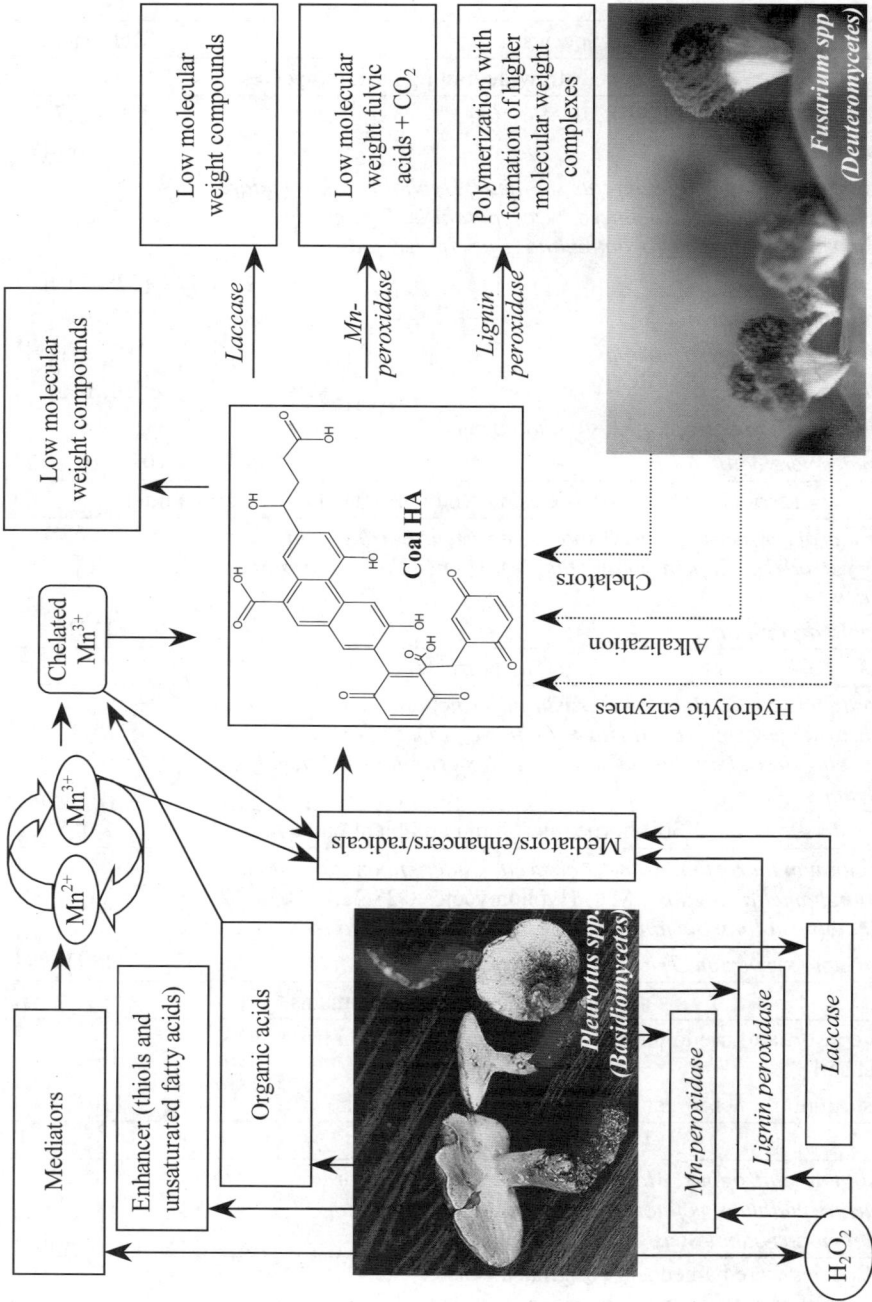

Figure 2. The proposed scheme of coal HA decomposition by *Basidiomycetes* and *Deuteromycetes*.

taken up to a large degree than humic acids. The more recent study used ^{14}C-labeled HS for this purpose [49] and also established that fulvic acids were taken by the plant cells. However, a particular care should be exercised on interpreting the results using the labeled HS. The latter are usually prepared from the substrate incubated with ^{14}C-labeled straw. The extracted HS are not identical to the naturally occurring materials. In addition, low molecular weight compounds from the humic extract can penetrate the plants.

Summarizing the data on HS as a nutrient source for microorganisms, the following conclusions can be made:

- the level of HS utilization by microorganisms under conditions of limited carbon or nitrogen supply is higher than that in a full nutrient solution;
- the efficiency of carbon conversion into microorganism biomass can be enhanced significantly if HS is a sole source of carbon, however, this effect has not been observed under conditions of low nutrient and energy sources;
- the microbial decomposition of HS includes non-enzymatic, enzymatic and enzyme-initiated reactions whose course depend on the specific cell metabolism;
- the long term microbial interaction with HS brings about both decomposition processes and synthesis of newly formed of humic like substances that hampers evaluation of decomposition processes.

3.1.2. Hormone-like activity of HS

Chemical substances identical to or mimicking the action of plant growth hormones have been investigated since the late 1910's. The main classes of those compounds are auxins, gibberellins, and cytokinins. The mode of their action is still not clearly understood. Although it is believed that the growth regulators act directly on the DNA and/or RNA of the cell affecting duplication, transpiration, and translation processes [34]. Interactions among the various groups of growth regulators are complex and complementary to their functions.

The plant hormones can be produced directly by the plant or by the soil microorganisms, especially by those residing in the rhizosphere. Since HS originate from the chemical and biological decomposition of the plant and animal residues and from metabolic activities of microorganisms, they might have a hormonal character. Indeed, in series of papers published since 1914 [27-31], it was shown that HS enhanced plant growth by exhibiting auxin-like activity. In 1917, Bottomley suggested that HS contained growth substances – so called "auximones" [27]. Hillitzer also claimed that HS could act as auxins [29]. Phuong and Tichy reported that humic and, in particular, fulvic acids showed some auxin, gibberellin or cytokinin-like activity [119]. The stimulation effects, however, were much lower compared to those of the real plant hormones. This brought the authors to a conclusion on a lack of phytohormone activity of HS. Cacco and Dell'Agnola detected a hormone-like behavior of HS from podzol using leaf disc senescence test [33]. The activity of HS was also much lower compared to the known plant hormones - indole-3-acetic acid (IAA) or N-6 benzyladenine.

The findings on auxin-like activity of HS have been supported and extended by experiments showing that some humic fractions have a high hormonal activity [33, 35-37]. In particular, it has been shown that low-weight molecular fraction induced morphological changes similar to those caused by IAA [66]. Moreover, this fraction increased IAA-oxidase activity [121]. In further experiments, it has been demonstrated that low molecular weight fraction of HS binds to the IAA cell membrane receptors [122]. However, there is still no direct chemical evidence that hormone-like activity of HS can be provided by the presence of plant growth regulators. Concentration of IAA in HS estimated using immunoassays varied from 0.5 to 3.7% (w/w) [123]. This shows that HS may contain different biologically active compounds – products of biosynthetic activity of microorganisms. For instance, the structure of HS can include polyamines that are plant regulators functioning similarly to the recognized plant hormones. The results of Young and Chen demonstrated that the content of putrescinene, spemidine, and spermine in HS from different sources ranged between 1.54-7.00, 0.39-3.88, 0.48-4.79 nM/g, respectively [38]. The authors concluded that polyamines might explain the hormone-like activity of HS.

Irrespective of principal structural fragments which are responsible for HS hormone-like activity, HS can be considered as a "storage" of hormone-like substances and transferring them to the cell membrane of a plant. Therefore, auxin-like activity of HS cannot be attributed only to their interaction with cell membrane receptors. An influence on the plasma membrane H^+-ATPase and an increase in microelement availability should be considered as well. It has been shown recently that extracts of soil root exudates exhibited a hormonal activity that was lacking in water extracts of soil and in the exudates used for extraction [124]. This confirms an important role of root exudates in decomposition of HS and in forming their hormone-like activity. Hence, the latter is determined by the specific features of soil-plant (root) interaction.

Our experiments with wheat coleoptiles and different fractions of coal-derived HA demonstrated their beneficial effect on coleoptiles growth at all the concentrations tested [120]. Fractionation of HA (CHA-GL02) included separation into hymatomelanic HA (CHM-GL02), a sum of brown and grey HA (CHR-GL02), and fractions of brown (CHB-GL02) and grey (CHG-GL02) HA. Auxin-like activity of different fractions of HA was estimated using bioassay with coleoptiles of wheat (*Triticum aestivum* L.) seedlings. Briefly, wheat seeds were germinated in the dark at 25°C for 72 hours. Then coleoptiles of similar length (about 5 mm) were cut and spited on glass capillaries (3 coleoptiles on a capillary). Upper part of coleoptiles of 4 mm length was preliminary removed to avoid influence of endogenous auxin. Then the length of coleoptiles was measured and capillaries with beaded coleoptiles were placed into the Petri plates added with 5mM K-phosphate buffer at pH 6.0 (blank) or with a solution of HA preparations in phosphate buffer. Then coleoptiles were grown at 25°C for 72 hours and the length of coleoptiles was measured again. The relative increase of coleoptiles length was used as a response. The concentration of HA varied in the range of 5 to 100 mg/L. Results are presented in Fig. 3.

All the HA samples tested showed the beneficial effect on coleoptiles growth at all the concentrations tested. This fact confirms the auxin-like activity of HA. Maximum stimulating activity was observed at the concentration of 10 mg/L and reached 156±9% of the control. The efficicency of different HA fractions was similar except for hymatomelanic and grey HA. The highest effect was observed for the grey HA characterized with the lowest contents of such fragments as carboxylic (COO^-) and aromatic (C_{Ar}) groups and with the highest molecular weight. On the other hand, hymatomelanic HA possessing the highest contents of alkoxy ($C_{Alk}O$) and methoxy (CH_3O) groups and the minimum value of molecular weight, exhibited the lowest beneficial effect on the coleoptiles growth. Statistical analysis of the data set on HA properties and their auxin-like activity have revealed a significant (P =95%) negative relationship between the contents of carboxylic groups in the HA preparations and the value of the beneficial effect (r = –0.96). Taking into consideration a determining role of carboxylic groups in formation of the surface negative charge of HA and the natural negative charge of cell surface, it could be suggested that this is adsorption of HA onto the coleoptiles surface that determines the auxin-like activity of HA.

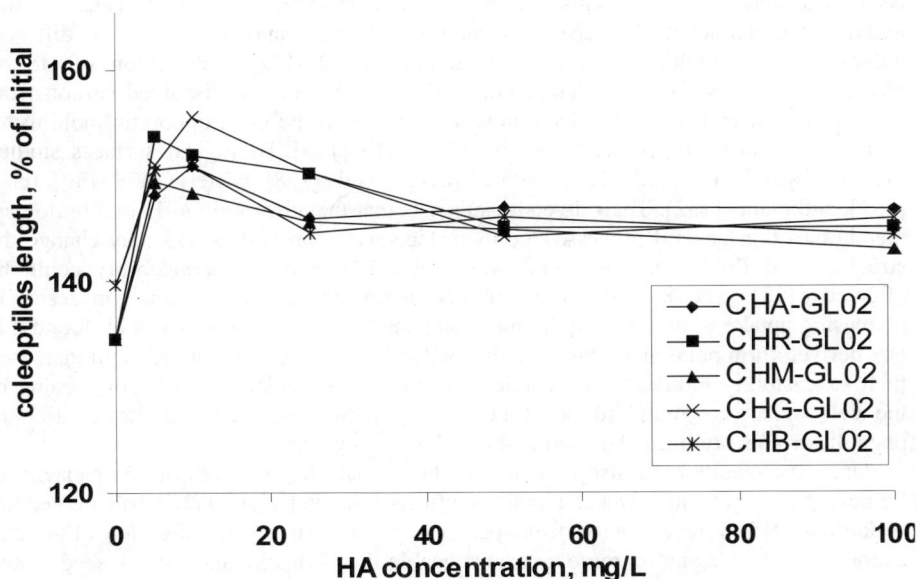

Figure 3. Influence of different fractions of coal HA on the wheat coleoptiles growth.

Summarizing the data on hormone-like activity of HS the following conclusions can be made:
- the level of hormone-like activity of HS is usually lower than that of plant hormones;
- the mechanism of hormone like activity of HS is still poorly understood and needs to be further investigated.

3.2. HS AS NUTRIENT CARRIERS

HS are generally considered to improve plant growth by enhancing nutrient supply [52, 55, 135, 136]. In contrary, some data indicate that there is no significant difference in biomass yield in the presence of HS [58, 70]. This contradiction could be provided by a use of supra-optimum conditions by the latter authors, whereas an impact of HS on the plant nutrition is the most pronounced under the conditions of limited nutrient availability [97]. For example, iron chlorosis, a symptom of iron deficiency in plants, can be corrected by applying HS which play a beneficial role in Fe acquisition by plants [61, 137- 139]. This beneficial effect of HS is attributed to the chelating activity of HS that provides the plant with Fe in easily assimilated form [140]. The presence of HS can also facilitate Fe uptake by plants as a result of HS participation in ion transport [141] and stimulation of membrane associated H^+-translocating enzymes [142, 143].

Humics have a surfactant-like structure containing both hydrophilic domains, such as carboxylic and phenolic groups, and alkylic and aromatic moieties. Because of this amphiphylic character, HS behave as natural surfactants and can adsorb on different natural surfaces including biological membranes [125-132]. Adsorption of HS on biological surfaces has been demonstrated directly, by loss of dissolved carbon from solution [125], and indirectly, by following changes in the electrophoretic mobility of individual cells in the presence or absence of HS [126]. Biological surfaces studied include phytoplankton [125-128], isolated fish gill cells [128], bacteria [129, 130], fungi [131], and plants [132]. Their diversity suggests that the adsorption of HS on biological membranes is a general process. Due to surface-active properties, HS may change the structure and fluidity of the membrane, and cell membrane permeability could be affected [133]. The increase in membrane permeability, in its turn, can result in increasing uptake of nutrients by living organisms [129, 132]. Ermakov et al. found that HA derived form potassium humates adsorbed to the surface of plant cells and increased their extensibility indirectly via reactions in the cell protoplasm [134]. Toughening of the cell walls was found to be stimulated by peroxidase-mediated dimerisation of phenolic groups present in HA and in the cell wall polymers.

Our experiments have also revealed the beneficial effects of HS on the capacity of Fe-deficient tomato and wheat plants to utilize iron under simulated calcareous soil conditions. HS were extracted from peat using standard alkali extraction [144] and cation exchange resin to remove exchangeable Fe. Tomato and wheat seeds were germinated on filter paper moistened with distilled water. After 10 or 5 days for tomato and wheat, respectively, the seedlings were transferred to plastic vessels containing aerated Hoagland nutrient solution either or without Fe supply, and grown for up 40 days. Fe, when added, was supplied at the concentration of 24 µM as $FeSO_4$, $FeSO_4$ in the presence of HS, and Fe-DTPA (an iron complex of diethylenetriaminepentaacetic acid). To simulate calcareous soil conditions, pH of the nutrition solution was adjusted to 8.0 using $CaCO_3$. After 20 days of the plant growing, photosynthesis efficiency in terms of electron transport rate (ETR) and effective quantum yield (Yield) were estimated using pulse amplitude modulation (PAM) fluorometer (PAM-2000, Walz, Germany). After 40 days the plants were harvested and used for length, weight, and chlorophyll content measurements. Data obtained are shown in Table 3.

Both the tomato and wheat plants grown in the nutrient solution without added Fe exhibited visual symptoms of Fe-deficiency (leaf chlorosis). The chlorophyll a/b ratio used as an indicator of light harvesting capacity of the plants under stress [145] was as low as 0.59 and 0.29 for the control tomato and wheat plants, respectively. As a consequence, the biometric parameters of the control plants – length and weight – were substantially suppressed as well.

Table 3. Length, weight, photosynthesis efficiency and chlorophyll content of tomato and wheat plants in the presence of different sources of iron and HS, % of blank .

Treatment	Weight	Length	Photosynthesis		Chlorophyll	
			ETR	Yield	Content	a/b
Tomato plants						
FeSO$_4$	90±12	97±12	104±11	103±17	142±15	189±14
Fe(III)-DTPA	4649±35	282±18	179±13	149±15	856±29	208±24
HS	110±11	105±10	200±11	154±21	107±11	156±12
FeSO$_4$+HS	4382±56	265±19	189±15	148±17	819±31	221±16
Wheat plants						
FeSO$_4$	101±5	105±12	97±7	100±8	101±11	95±10
Fe(III)-DTPA	172±12	146±12	218±15	197±12	200±11	205±5
HS	121±15	138±12	344±11	329±13	154±12	130±13
FeSO$_4$+HS	123±15	125±11	212±15	216±14	179±11	198±14

As it can be seen from Table 3, an addition of inorganic Fe (FeSO$_4$) alone did not increase iron-supply of the plants: the detected values of weight, length, photosynthesis efficiency and chlorophyll content were the same as for the control plants. This was to expect due to alkaline conditions (pH 8) of the nutrition media resulting in formation of insoluble Fe(III) hydroxide unavailable for plants uptake. At the same time, simultaneous addition of HS and FeSO$_4$ gave a substantial rise in the length, weight, photosynthesis efficiency and chlorophyll content of the plants. The photosynthetic parameters (ETR and Yield) even reached the values detected for the Fe-DTPA treated plants, whereas the weight, length and chlorophyll content were still lower than in the Fe-DTPA treated plants. The plants grown with addition of the synthetic iron chelate Fe-DTPA did not display any symptoms of chlorosis and were characterized with the highest values of biomass and length.

Of interest is that addition of HS solely led to a partial recovery of the Fe-deficient plants. In particular, the photosynthesis efficiency and chlorophyll content were significantly increased in humus-treated plants. Given that exchangeable Fe was removed during the extraction procedure, it seems like the plants could use constituent (endogenous) Fe of HS. Hence, the effect of enhancing nutrient supply can overlap a function of HS as a nutrient source. Another explanation is a direct influence of HS on the photosynthesis. The corresponding data are reported in the literature [65, 67] that claim the quinonoide structures of HS being able to support electron transport and, therefore, to increase the photosynthesis efficiency. In general, the obtained results have confirmed that treatment of iron-deficient plant with HS can result in increasing uptake of iron and, therefore, in increasing plant biomass. The effect observed seems to be extended for other micronutrients such as Zn and Mn.

Summarizing the data on enhancing nutrient supply in the presence of HS, the following conclusions can be made:

– HS show beneficial effects or no effects on the nutrient supply;
– the most pronounced beneficial effects of HS on the plant nutrition are observed under unfavourable growth conditions;
– the presence of HS under limited nutrition conditions can lead to a partial recovery of the plants.

3.3. HS AS CATALYSTS OF BIOCHEMICAL REACTIONS

The direct participation of HS in biochemical reactions has been observed only in few cases [26, 116]. However, HS can influence biochemical reactions indirectly, nominally, via formation of the HS-enzyme complexes, or as a substrate of enzymatic reaction. The principal processes that can occur in the HS-enzyme system are schematically represented in Fig. 4. Microorganisms and plants are the main sources of soil enzymes [146]. The latter should be segregated into existing in the soluble phase and in the solid phase. It is well known that extracellular soil enzymes can be associated with inorganic and organic colloids and, in particular, with HS. The essential soil enzymes such as urease, protease, phosphatase, hydrolases, laccase, and peroxidase have been detected in soil extracts as complexes with HS [26, 147]. The distribution of enzymes on the surface of organo-mineral particles depends strongly on the particle size as it was shown for invertase and xylanase [148, 149], phosphatase [150] and urease [151]. Moreover, various mineral and organo-mineral surfaces may have an inhibitory influence on the activity of enzymes as it was established for acid phosphatase [150].

The above data indicate that HS can play a role of enzyme carriers facilitating enzyme "immobilization" and improving their long-term stability. Due to the presence of carboxyl, carbonyl, hydroxyl and amide groups in HS structure [1, 2], they can serve as supports for the immobilized enzyme. Invertase was immobilized onto aminopropyl silica activated with soil HS. The immobilization technique yielded an increased amount of and an enhanced activity of the immobilized enzyme [152]. In the context of soil interactions, the formation mechanism of the HS - extracellular enzyme complexes can include an activation of inorganic particles with HS followed by a formation of inorganic particle-HS-enzyme complexes. As a result, upon extracting these complexes from soil, separation of inorganic particles occurs, and only HS-enzyme complexes can be obtained.

Reviewing the data on proteases-HS complexes, it can be noted that their formation depends on both the type of proteases and composition and structure of HS. Proteases able of hydrolyzing N-benzoyl-L-argininamide were shown to be associated with highly condensed humic materials, while N-benzoylcarbonyl-L-phenylalaninyl-L-leucine – hydrolyzing proteases are associated with less-condensed humic materials. Casein-hydrolyzing proteases are generally associated with non-humified organic matter [153]. The study on urease-HS complexes is in line with these data [154]. The effects of two different peat HA fractions, high molecular weight (HMW, 100-300 kDa) and low molecular weight (LMW, 10-20 kDa), on the activity and stability of urease were evaluated. It was shown that HMW fraction of HA significantly inhibited urease activity at pH 6.0, but caused no influence on its activity at pH 7.0 and 8.0. HMW fraction

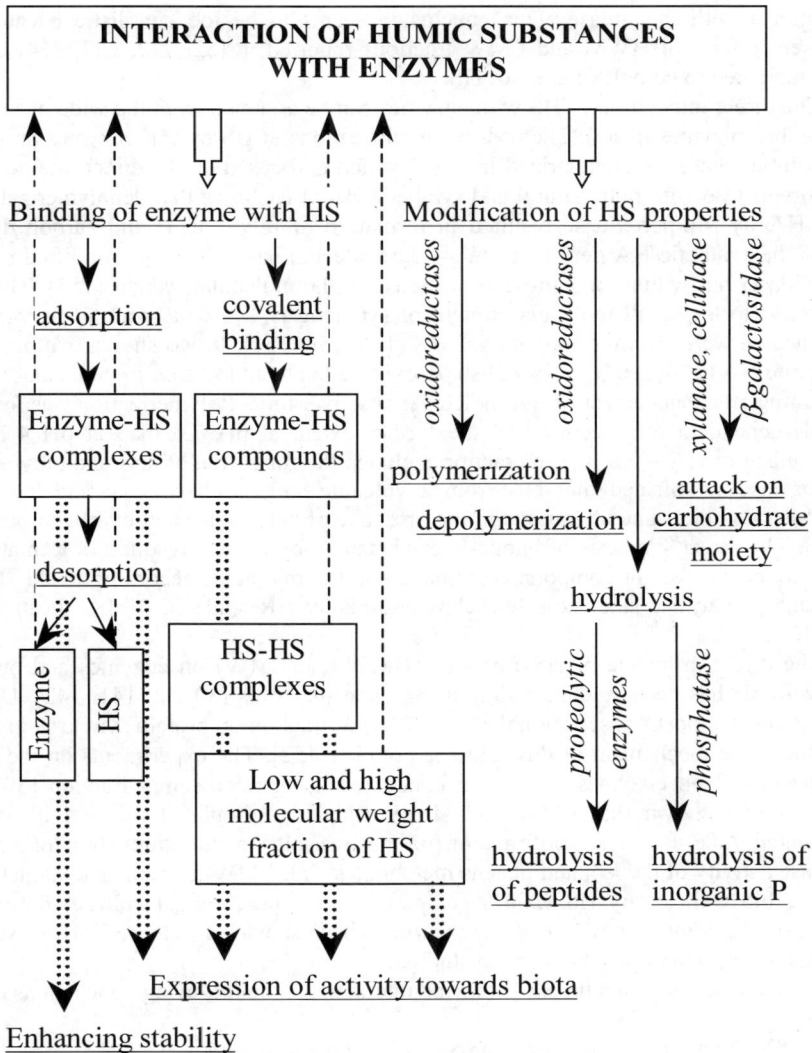

INTERACTION OF HUMIC SUBSTANCES WITH ENZYMES

Binding of enzyme with HS

adsorption

covalent binding

Enzyme-HS complexes

Enzyme-HS compounds

desorption

Enzyme

HS

HS-HS complexes

Low and high molecular weight fraction of HS

Modification of HS properties

oxidoreductases

oxidoreductases

xylanase, cellulase

β-galactosidase

polymerization

depolymerization

attack on carbohydrate moiety

hydrolysis

proteolytic enzymes

phosphatase

hydrolysis of peptides

hydrolysis of inorganic P

Expression of activity towards biota

Enhancing stability

Figure 4. Principal processes of interaction of HS with enzymes.

stabilized urease activity over a period of 11 days of treatments with protease and with Cu^{2+} and Hg^{2+} (two powerful inhibitors of soluble urease activity). On the contrary, the LMW fraction of HA inhibited urease activity at pH 6.0, 7.0 and 8.0, and did not stabilize urease activity in the presence of protease. These results show that the two HA fractions affected differently both the activity and stability of the urease. It has been proposed that the inhibition of the urease by HMW and LMW fractions is mainly due to the presence of two heavy metals which, although immobilized on the HA, are still able to interact with the urease. This conclusion seems to be too simplistic because the different effects of HWM and LMW fractions reported in [52, 122, 137, 154] can be also attributed to complex nature of HS.

The direct interaction of HS with enzymes can be assumed from the wide occurrence of soluble enzymes in soil. The possible modifications of HS by soil enzymes present in the soluble phase are summarized in Fig. 3 which is focused on the direct interaction of oxidoreductase with both natural and synthetic HS. It is shown that depolymerization of coal HA by Mn-peroxidase resulted in formation of fulvic acids and carbon dioxide [87]. The synthetic HA derived from 3-fluorocatechol have also been modified by Mn-peroxidase, but without a significant decrease in the molecular weight [155]. This was attributed to the small molecular weight of synthetic HA. On the contrary, peroxidase and laccase were shown to polymerize HS [116, 156, 157]. It was shown that oxidation of lignite HA catalyzed by horseradish peroxidase led to an increase in molecular weight indicating an enhancement of HA molecular size over time. Polymerization was found to be pH-dependent and occurred to the further extent at pH 7.0 than at pH 4.7. The mechanism of HA – enzyme interaction included the formation of aryl and alkyl ethers. It was also established that HA from a volcanic soil, oxidized coal and leonardite demonstrated the same alterations in the presence of peroxidase and hydrogen peroxide [116]. The direct synthesis of humic-like substances by laccase resulted in formation of high molecular weight compounds similar to soil humic acids according to the data of elemental analysis, size-exclusion chromatography, IR and ^{13}C-NMR spectroscopy [157].

The direct influence of HS fractions (HMW and LMW) on enzymes and on their biosynthesis has been studied mainly using plant materials [50, 52, 141, 142]. Despite the established post transcriptional effect of LMW fraction on protein synthesis in maize seedlings, the mechanism of this effect is not clear [52]. The experiments on the direct effects of HS on enzymes have been conducted using microsomal fraction [50, 141, 142]. It was shown that LMW and HMW fractions stimulated the activity of K^+-stimulated ATP-ase. The confirmation of these results is the stimulation of the H^+-ATPase activity of the isolated plasma membranes with LMW fraction. The induction of laccase biosynthesis by HS (200% compared to the pure fungal culture) during the growth of *Basidiomycetes Coriolus hirsutus* has been shown as well [157]. However, the mechanism of induction has not been studied.

Summarizing the data on HS – enzyme interactions, the following conclusions can be made:

- direct interaction of HS and enzymes is scarcely studied;
- polymerization/depolymerization transformations of HS as a result of intereactions with oxidoreductases are established.

3.4. ANTIOXIDANT ACTIVITY OF HS

Protection of cells, tissues and biological macromolecules against reactive oxygen species including free radicals is of great importance for living organisms. Oxidative stress arises in a biological system after an increased exposure to oxidants as a result of a decrease in the total antioxidant activity of the biological system. The main role in protection of the biological systems play specific compounds called antioxidants. The general description of these groups can be given in terms of their antioxidant activity – ability to interrupt radical-chain oxidation processes or to decrease the amount of free radicals (particularly, reactive oxygen species).

The study of antioxidant activity of HS is of a great interest. However, the measurement of the antioxidant activity of HS causes a lot of methodological difficulties. The first is HS interaction with system components producing free radicals, the second is the complexity of HS structure. The attempt to measure antioxidant activity of HS using voltammetry is described in [158]. The authors used method of electrochemical generation of oxygen superoxide radical and measured antioxidant activity of some standard antioxidants (vitamin C, glucose, resorcinol, and catechol) as well as of the plant extracts and of HS (water extracts from peat). The main problem in interpretation of their data is that no characteristics of HS were given, and correlation with standard units of antioxidant activity was not provided. However, among the samples of plant origin and of the standard antioxidants, HS demonstrated the highest value of antioxidant activity comparable to that of phthalocyanine Co, phthalocyanine Ni and phthalocyanine Cr. The data obtained allowed to propose the mechanism of interaction based on the dismutation of oxygen superoxide radical to dioxygen:

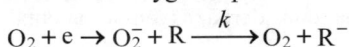

$$O_2 + e \rightarrow O_2^- + R \xrightarrow{k} O_2 + R^-$$

Therefore, HS can play protective function in ecosystems interrupting radical reaction and preventing damages of cell membranes and biological macromolecules.

Main interest to HS antioxidant activity is provoked by the role of HS in the environments exposed to the factors causing the formation of free radicals (OH·, O_2H·, ROO$_2$·), nominally, drought, heat, herbicide use, and ultraviolet radiation. Relationship between the structure of HS and their reactivity toward molecular ozone (O_3) and hydroxyl radicals (OH·) has been established [159-161]. The positive correlations of the scavenging activity of HS versus their concentration and the amount of C=C bonds in their structure were observed in radical reactions with O_3· and OH·. Ozone consumption and OH· scavenging rate constants were found to be positively correlated with the content of aromatic carbon, and inversely correlated with that of aliphatic carbon. The data obtained [159] allow to conclude that ozone reacts preferentially with aromatic constituents of HS and, specifically, with the electron enriched aromatics. In compliance with these findings, the statistically significant relationship was found between oxidation rate parameters and C/H ratio, - indicator of the unsaturation degree of the structure. So, the chemical nature of HS exerts a strong control over their reactions with ozone.

The study of MnO catalyzed ozonation of atrazine in the presence of HS revealed that HS at the concentration of 2-6 mg/L (DOC) might scavenge hydroxyl radicals produced during ozonation decreasing the oxidation efficiency of atrazine. However, when HS were present at low concentrations (1 mg/L DOC) both the manganese species and HS initiated and promoted the formation of hydroxyl radicals that enhanced the destruction of atrazine.

The following conclusions can be drawn:
- the radical scavenging effects are displayed at high concentrations of HS,
- the antioxidant activity of HS is strongly dependent on their structural features such as enrichment with unsaturated moieties.

4. Conclusions

Reviewing the data on mitigating activity of HS, the following principal ways of HS action were proposed:
- influence on the organism development;
- enhancing nutrient supply;
- interactions of HS with enzymes and catalysis of biochemical reactions;
- antioxidant activity.

The complexity of HS structure provides a wide variety of the observed biological effects. This is true for HS interactions with xenobiotics, nutrients, enzymes, membranes, etc. Therefore, the study of biological effects of HS neglecting either possible interaction of HS with components of biological system or their structural complexity leads to conflicting data and misinterpretation. In order to overcome the problem, one should standardize the experimental design and use characterized preparations of HS.

Another specific feature of biological activity exerted by HS is that the most pronounced effects on living organisms and on the components of the ecosystems can be observed under stress conditions induced by both specific and non-specific stress factors. Hence, HS can be considered as environmental modulators mitigating the adverse effects of stress factors.

5. Acknowledgements

The first author expresses her thanks for the financial support of the Research Center for Environment and Health (GSF, Neuherberg, Germany) FE 75184, BA 31/139166/02/U, and of the International Science and Technology Center (project KR-964).

6. References

1. Thurman, E. M. (1985) Organic geochemistry of natural waters, Martinus Nijhof/Dr. W. Junk Publishers, Dordrecht.
2. Orlov, D. S. (1990) *Soil humic acids and general theory of humification*, Moscow State University Publisher, Moscow.
3. Clapp, C.E., Hayes, M.H.B. and Swift, R.S. (1993) Isolation, fractionation, functionalities, and concepts of structure of soil organic macromolecules, in A J. Beck, K.C. Jones, M.B.H. Hayes, and U. Mingelgrin (eds.), *Organic substances in soil and water*, Royal Society of Chemistry, Cambridge.
4. Aiken, G.R., McKnight, D.M. and MacCarthy, P. (1985) Humic substances in soil, sediment, and water, Wiley, New York.
5. Senesi, N. (1993) Organic pollutant migration in soils as affected by soil organic matter. Molecular and Mechanistic Aspects, In D. Petruzzelli and F.G. Helfferich (eds), *Migration and fate of pollutants in soils and subsoils, NATO ASI Series*, Vol. G 32, Springer-Verlag, Berlin, p. 47.
6. Landrum, P.F., Reinhold, M.D., Nihart, S.R. and Eadie, B.J. (1985) Predicting the bioavailability of organic xenobiotics to Pontoporeia Hoyi in the presence of humic and fulvic materials and natural dissolved organic matter, *Environ. Toxicol. Chem.* 4, 459-467.
7. McCarthy, J.F. and Jimenez, B.D. (1985) Reduction in bioavailability to bluegills of polycyclic aromatic hydrocarbons bound to dissolved humic material, *Environ. Toxicol. Chem.* 4, 511-521.

8. Oris, J.T., Hall, A.T. and Tylka, J.D. (1990) Humic acids reduce the photo-induced toxicity of anthracene to fish and daphnia, *Environ. Toxicol. Chem.* **9**, 575-583.

9. Day, K.E. (1991) Effects of dissolved organic carbon on accumulation and acute toxicity of fenvalerate, deltamethrin and cyhalothrin to *Daphnia Magna* (Straus), *Environ. Toxicol. Chem.* **10**, 91-101.

10. Perminova, I.V., Kovalevsky, D.V., Yashchenko, N.Yu., Danchenko, N.N., Kudryavtsev, A.V., Zhilin, D.M., Petrosyan, V.S., Kulikova, N.A., Philippova, O.I., and Lebedeva, G.F. (1996) Humic substances as natural detoxicants, in C.E. Clapp, M.H.B. Hayes, N. Senesi, and S.M. Griffith (eds.), *Humic substances and organic matter in soil and water environments: characterization, transformations and interactions,* St. Paul, MN, USA, pp. 399-406.

11. Perminova, I.V., Grechishcheva, N.Yu., Kovalevskii, D.V., Kudryavtsev, A.V., Petrosyan, V.S. and Matorin, D.N. (2001) Quantification and prediction of detoxifying properties of humic substances to polycyclic aromatic hydrocarbons related to chemical binding, *Environ. Sci. Technol.* **35**, 3841-3848.

12. Misra, V., Pandey, S.D. and Viswanathan, P.N. (2000) Effect of humic acid on the bioavailability of γ-hexachlorocyclohexane in *Marsilea minuta* (L.), *Environ. Monitor. Assessment* **61**, 229-235.

13. Khristeva, L.A. (1953) The participation of humic acids and other organic substances in the nutrition of higher plants, *Pochvovedenie* **10**, 46-59.

14. Khristeva, L.A. (1970) Theory of humic fertilizers and their practical use in the Ukraine, in Robertson R.A. (ed.), *2-nd International Peat Congress*, Leningrad, HMSO, Edinburgh, 543-558.

15. Gorovaya, A.I., Orlov, D.S. and Shcherbenko, O.V. (1995) *Humic substances: structure, functions, mode of action, protective properties, role in the environment*, Naukova dumka, Kiev.

16. MacCarthy, P. and Rice, J. A. (1988) in *Proceedings of Chapman Conference on the Gaia Hypothesis*, San Diego, CA, March, 7-11.

17. Sawada, A., Tanaka, S., Fukushima, M. and Tatsumi, K. (2003) Electrokinetic remediation of clayey soils containing copper(II)-oxinate using humic acid as a surfactant, *J. Hazard. Mater.* **B96**, 145-154.

18. Fukushima, M. and Tatsumi, K. (2001) Functionalities of humic acid for the remedial processes of organic pollutants, *Analyt. Sci.* **17**, i821-i823.

19. Lesage, S., Novakowski, K.S., Brown, S. and Millar, K. (2001) Humic acids enhanced removal of aromatic hydrocarbons from contaminated aquifers: developing a sustainable technology, *J. Environ. Sci. Health* **A 36(8)**, 1515-1533.

20. Molson, J.W., Frind, E.O., Van Stempvoort, D.R. and Lesage, S. (2001) Humic acid enhanced remediation of an emplaced diesel source in groundwater - 2. Numerical model development and application, *J. Contam. Hydrol.* **54**, 277-305.

21. Schwartz, D.L. (1999) Coal-derived humic acid for removal of metals and organic contaminants, *Solid waste and emergency response (5102G), EPA 542-N-99-002.* **31**, 1.

22. Sanjay, H.G., Srivastava, K.C., Walia, D. S. (1997) *Mixed waste remediation using HUMASORB-CS^{TM} – an adsorbent to remove organic and inorganic contaminants,* ARCTECH Inc., Chantilly, Virginia.

23. Gamble, D.S, Khan, S.U. (1988) Atrazine hydrolysis in aqueous suspension of humic acid at 25.0°C *Can. J. Chem.* **66**, 2605-2617.

24. Zeng, K., Hwang, H.-M., Yu, H. (2002) Effect of dissolved humic substances on the photochemical degradation rate of 1-aminopyrene and atrazine, *Int. J. Mol. Sci.* **3**, 1048-1057.

25. Klavins, M., Dipane, J., Babre, K. (2001) Humic substances as catalysts in condensation reactions *Chemosphere* **44**, 737-742.

26. Masciandaro, G. and Ceccanti, B. (1999) Assessing soil quality in different agro-ecosystems through biochemical and chemico-structural properties of humic substances, *Soil Tillage Res.* **51**, 129-137.

27. Bottomley, W.B. (1914) Some accessory factors in plant growth and nutrition, *Proc. of the Royal Society of London (Biology)* **88**, 237-247.

28. Bottomley, W.B. (1914) The significance of certain food substances fro plant growth, *Annals of Botany (London)* **34**, 353-365.

29. Hillitzer, A. (1932) Uber den einfluss der humusstoffe auf das wurzelwachstum, *Beihefte zum Botanischen Zentralblatt* **49**, 467-480.

30. Paszewski, A., Trojanowski, J. and Lobarzewska, W. (1957) Influence of the humus fraction on the growth of oat coleoptiles, *Annales Universitatis Marie Curie Sklodowska, Sklodowska* **12**, 1-13.

31. O'Donnel, R.W. (1973) The auxin-like activity of humic preparations from leonardite, *Soil Sci.* **116**, 106-112.

32. Azam, F. and Malik, K.A. (1982) Effect of humic acids on seedling growth of wheat (*Triticum aestivum* L.) under different conditions, *Pak. J. Botany* **14**, 47-48.

304

33.   Cacco, G. and Dell'Agnola, G. (1984) Plant growth regulator activity of soluble humic complex, *Can. J. Soil Sci.* **62**, 306-310.
34.   Syltie, P.W. (1985) Effects of very small amaounts of highly active biological substances on plants growth, *Biol. Agricult. Horticult.* **2**, 245-269.
35.   Dell'Agnola, G. and Nardi, S. (1987) Hormone-like effect of enhanced nitrate uptake induced by depolycondenced humic fractions obtained from *Allobophora rosea* and *A. caliginosa* faeces, *Biol. Fertil. Soils* **4**, 115-118.
36.   Nardi, S., Arnoldi, G. and Dell'Agnola, G. (1988) Release of hormone-like activities from *Allobophora rosea* and *A. caliginosa* faeces, *Can. J. Soil Sci.* **68**, 563-567.
37.   Piccolo, A., Nardi, S. and Concheri, G. (1992) Structural characteristics of humus and biological activities, *Soil Biol. Biochem.* **24**, 273-380.
38.   Young, C.C. and Chen, Y. (1997) Polyamines in humic acid and their effect on radical growth of lettuce seedlings, *Plant Soil* **195**, 143-149.
39.   Niklewski, B. and Wojciechowski, J. (1937) Uber den Einfluss der wasserloeslichen Humusstoffe auf die Entwicklung einiger Kulturpflanzen, *Biochem. Z.* **271**, 11-122.
40.   Rauthan, B.S. and Schnitzer, M. (1981) Effects of a soil fulvic acid on the growth and nutrient content of cucumber (*Cucumis sativus*) plants, *Plant Soil* **63**, 491-495.
41.   Iswaran, V., Sen, A. and Vimal, O.P. (1973) Influence of humus spray on the yield of soybean *Glycine max* var, *Clark. Sci. Cult.* **39**, 143-144.
42.   Mishra, B. and Srivastava, L.L. (1988) Physiological properties of humic acids isolated from some major soil accosiation of Bihar, *J. Ind. Soc. Soil Sci.* **36**, 83-89.
43.   Solaiappan, U., Muthusankaranarayanan, A. and Muthusamy, P. (1995) Effect of humic acid on rainfed upland cotton (*Gossypium hirsitum*), *Ind. Agron.* **40**, 156-157.
44.   Loffredo, E., Senesi, N. and D'Orazio, V. (1997) Effect of humic acids and herbicides, and their combinations on the growth of tomato seedlings in hydroponics, *Z. Pflanzenernaehr. Bodenk.* **160**, 455-461.
45.   Sathiyabama, K. and Selvakumari, G. (2001) Effect of humic acid on growth, yield and nutrition of *Amaranthus*, *South Ind. Horticult.* **49**(Special): 155-156.
46.   Zachariakis, M., Tzorakakis, E., Kritsotakis, I., Siminis, C.I. and Manios, V. (2001) Humic substances stimulate plant growth and nutrient accumulation in grapevine rootstocks, *Acta Horticult.* (*ISHS*) **549**, 131-136.
47.   Prozorowskaya, A.A. (1936) The effect of humic acid and its derivatives on the uptake of nitrogen, phosphorus, potassium and iron by plants, in *Organo-mineral fertilizers, collected papers of Research Scientific Institute for Fertilizers, Insecticides, and Fungicides,* p. 127 (in Russian).
48.   Saalbach E. (1956) Einfluss von Huminstoffe auf den Stoffwechsel der Pflanzen, in Rapp. D. (ed), *Trans. 6 International Congr. Soil Science*, pp. 107-111.
49.   Vaughan, D., Malcom, R.E. and Ord, B.G. (1985) Influence of humic substances on biochemical processes in plants, in D. Vaughan and R.E. Malcom (eds.), *Soil organic matter and biological activity*, Martinus Nijhoff/Junk W Publishers, Dordrecht, pp. 77-108.
50.   Nardi, S., Concheri, G., Dell'Agnola, G. and Scrimin, P. (1991) Nitrate uptake and ATPase activity in oat seedlings in the presence of two humic fractions, *Soil Biol. Biochem.* **26**, 1341-1346.
51.   Pinton, R., Cesco, S., Iacolettig, G., Astolfi, S. and Varanini, Z. (1999) Modulation of NO_3^- uptake by water extractable humic substances: involvement of root plasma membrane H^+-ATPase, *Plant Soil* **215**, 155-161.
52.   Nardi, S., Gessa, C., Ferrarese, L., Trainotti, L., Casadoro, G., and Pizzeghello, D. (2000) A low molecular weight humic fraction on nitrate uptake and protein synthesis in maize seedlings, *Soil Biol. Biochem.* **32**, 415-419.
53.   Cacco, G., Attina, E., Gelsomino, A. and Sidari, M. (2000) Effect of nitrate and humic substances of different molecular size on kinetic parameters of nitrate uptake in wheat seedlings, *J. Plant Nutr. Soil Sci.* **163**, 313-320.
54.   Ram, N. and Verloo, M. (1983) Effect of natural complexants on the uptake of trace elements by barley and their extractable amounts in soil, *Agrochimica* **28**, 13-19.
55.   Fortun, C., Rapsch, S. and Ascaso, C. (1985) Action of humic acid preparations on leaf development, mineral elements contents and chloroplast ultrastructure of ryegrass plants, *Photosynthetica* **19**, 294-299.

56. Vaughan, D. and Malcom, R.E. (1985) Influence of humic substances on growth and physiological processes, in D. Vaughan and R.E. Malcom (eds.), *Soil organic matter and biological activity*, Martinus Nijhoff/Junk W Publishers, Dordrecht, pp. 37-76.

57. Chen, Y. and Avaid, T. (1990) Effect of humic substances on plant growth, in P. MacCarthy, C.E. Clapp, R.L. Malcom, and P.R. Bloom (eds.), *Humic substances in soils and crop science: selected readings*, Soil Sci. Soc. Am., Madison, pp.161-186.

58. Mackowiak, C.L., Grossl, P.R. and Bugbee, B.G. (2001) Beneficial effects of humic acid on micronutrient availability to wheat, *Soil Sci. Soc. Am. J.* **65**, 1744-1750.

59. Varanini, Z. and Pinton, R. (2001) Direct versus indirect effects of soil humic substances on plant growth and nutrition, in R. Pinton, Z. Varanini, and P. Nannipieri (eds.), *The Rizosphere*, Marcel Dekker, Basel, pp.141-158.

60. Clapp, C.E., Chen, Y., Hayes, M.H.B. and Cheng, H.H. (2001) Plant growth promoting activity of humic substances, in R.S. Swift and K.M. Sparks (eds.), *Understanding and managing organic matter in soils, sediments, and waters*, IHSS, Madison, pp. 243-255.

61. Sánchez-Sánchez, A., Sánchez-Andreu, J., Juárez, M., Jordá, J. and Bermúdez, D. (2002) Humic substances and amino acids improve effectiveness of chelate FeEDDHA in lemon trees, *J. of Plant Nutr.* **25**, 2433-2442.

62. Cincerová, A. (1964) The effect of humic acid on transamination in winter wheat plants, *Biol Plant* (Prague) **6**, 183-188.

63. Olsen, C. (1929) On the influence of humus substances on the growth of green plants in water culture, *C.R. trav. Lab. Carlsberg.* **18**, 1-16.

64. Sladký, Z. (1965) Die durch Blattduengung mit Humusstoffen bervorgerufen anatomischen une physiologischen Veraenderungen der Zuckerruebe, *Biol. Plant (Prague)* **7**, 251-260.

65. Visser S.A. (1986) Effects of humic substances on plant growth, in *Humic substances effect on soil and plants*, Italy, Reda, pp. 89-135.

66. Muscolo, A., Felici, M., Concheri, G. and Nardi, S. (1993) Effect of earthworm humic substances on esterase and peroxidase activity during growth of leaf explants of *Nicotiana plumbaginifolia, Biol. Fertil. Soils* **15**, 127-131.

67. Rea, E. and Pierandrei, F. (1994) Effects of fertilization with humic acids on soil and metabolism: a multidisciplinary approach, in N. Senesi, T. M. Miano (eds.), *Humic substances in the global environment and implications to human health*, Elsevier Science, Amsterdam, pp. 343-348.

68. Cozzi, R., Nicolai, M., Perticone, P., De Salvia, R. and Spuntarelli, F. (1993) Desmutagenic activity of natural humic acids: inhibition of mitomycin C and maleic hydrazide mutagenicity, *Mutat. Res.* **299**, 37-44.

69. Genevini, P.L., Saxxhi, G.A. and Borio, D. (1994) Herbicide effect of atrazine, diuron, linuron and prometon after interaction with humic acids from coal, in N. Senesi and T.M. Miano (eds.), *Humic substances in the global environment and implications on human health*, Elsevier Science, New-York, pp. 1291-1296.

70. Cooper, R.J., Liu, C. and Fisher, D.C. (1998) Influence of humic substances on rooting and nutrient content of creeping bentgrass, *Crop Sci.* **38**,1639-1644.

71. Ferrara, G., Loffredo, E., Simeone, R. and Senesi, N. (2000) Evaluation of antimutagenic and desmutagenic effects of humic and fulvic acids on root tips of *Vicia faba, Environ. Toxicol.* **15**, 513-517.

72. YuLing, C., Min, C., YunYin, Li and Xie, Z. (2000) Effect of fulvic acid on ABA, IAA and activities of superoxide dismutase and peridoxase in winter wheat seedling under drought conditions, *Plant Physiol. Com.* **36**, 311-314.

73. Haynes, R.J. and Mokolobate, M.S. (2001) Amelioration of Al toxicity and P deficiency in acid soils by additions of organic residues: a critical review of the phenomenon and the mechanisms involved, *Nutrient Cycling in Agroecosystems* **59**, 47–63.

74. Zhorina L. V. and Stepchenko L. M. (1991) The content of free amino acids in the tissues of broiler chicks administered sodium humate in the ration, *Nauchnye Dokl. Vyss. Shkoly Biol. Nauki* **10**, 147-150.

75. Lange, N., Golbs, S. and Kuhnert, M. (1987) Grundlagenuntersuchungen zu immunologishen Reaktionen an der Laboratoriumstratte unter dem Einflus von Huminsauren, *Arch. Exper. Veter.-Med.* **41**, 140-146.

76. Sato, T., Ose, Y. and Nagase, H. (1986) Desmutagenic effect of humic acid, *Mutat. Res.* **162**, 173-178.

306

77.	Sato, T., Ose, Y., Nagase, H. and Hayase, K. (1987) Mechanism of desmutagenic effect of humic acid, *Mutat. Res.* **176**, 199-204.
78.	Carlberg, G.E., Martinsen, K., Kringstad, A., Gjessing, E., Grande, M., Källqvist, T. and Skare, J.U. (1986) Influence of aquatic humus on the bioavailability of chlorinated micropollutants in Atlantic salmon, *Arch. Environ. Contam. Toxicol.* **15**, 543-548.
79.	*Humic acids and their sodium salts* (1999) Summary report, February, The European Agency for the evaluation of medical products, Committee for veterinary medical products, (EMEA/MRL) 554/99-FINAL.
80.	Ribas, G., Carbonell, E., Creus, A., Xamena, N. and Marcos, R. (1997) Genotoxicity of humic acids in cultured human lymphocytes and its interaction with the herbicides alachlor and maleic hydrazide, *Environ. Mol. Mutagen.* **29**, 272-276.
81.	Gau, R.J., Yang, H.L., Chow, S.N., Suen, J.L. and Lu F.J. (2000) Humic acid suppresses the LPS-induced expression of cell-surface adhesion proteins through the inhibition of NF-kappa B activation, *Toxicol. Appl. Pharmacol.* **166**, 59-67.
82.	Blondeau R. (1989) Biodegradation of natural and synthetic humic acids by the white rot fungus *Phanerochaete chrysosporium*, *Appl. Environ. Microbiol.* **55**, 1282-1285.
83.	Frimmel, F.H., Abbt-Braun, G., Hambsch, B., Huber, S., Scheck, S., and Schimiedel, U. (1994) Behaviour and functions of freshwater humic substances – some biological, physical and chemical aspects, in N. Senesi and T.M. Miano, (eds.), *Humic substances in the global environment and implications on human health*, Elsevier Science, New-York, pp. 735-755.
84.	Kirschner, R.A.Jr., Parker, B.C. and Falkinham, J.O.III. (1999) Humic and fulvic acids stimulate the growth of *Mycobacterium avium*, *FEMS Microbiol. Ecol.* **30**, 327-332.
85.	Dehorter, B. and Blondeau, R. (1992) Extracellular enzyme activities during humic acid degradation by the white rot fungi *Phanerochaete chrysosporium* and *Trametes versicolor*, *FEMS Microbiol. Letter* **94**, 209-216.
86.	Gramss, G., Ziegenhagen, D. and Sorge, S. (1999) Degradation of soil humic extract by wood- and soil-associated fungi, bacteria, and commercial enzymes, *Microbiol. Ecol.* **37**, 140-151.
87.	Hofrichter, D., Ziegenhagen, S., Sorge, R. U. and Bublitz, W. F. (1999) Degradation of lignite (low-rank coal) by ligninolytic basidiomycetes and their manganese peroxidase system, *Appl. Microbiol. Biotechnol.* **52**, 78-84.
88.	Morimoto, K., Tatsumi K., Kuroda, K.-I. (2000) Peroxidase catalyzed co-polymerization of pentachlorophenol and a potential humic precursor, *Soil Biol. Biochem.* **32**, 1071-1077.
89.	Balarezo, A.L., Jones, V.N., Yu, H. and Hwang, H-M. (2002) Influence of humic acid on 1-aminopyrene ecotoxicity during solar photolysis process, *Int. J. Mol. Sci.* **3**, 1133-1144.
90.	Schiller, F., Klocking, R., Wutzler, P. and Farber, I. (1979) Results of an oriented clinical trial of ammonium humate for the local treatment of herpesvirus hominis (HVH) infections, *Dermatoi Monatsschr* **165**, 505-509.
91.	Mentel, R., Helbig, B., Klocking, R., Dohner, L. and Sprossig, M. (1983) Effectiveness of phenol body polymers against influenza virus A/Krasnodar/101/59/H2N2, *Biochem. Acta* **42(10)**, 1353-1356.
92.	Schols, D., Wutzler, P., Klocking, R., Helbig, B. and De Clercq E. (1991) Selective inhibitory activity of polyhydroxcarboxylates derived from phenolic compounds agains human immunodeficiency virus replication, *Acquir Immune Defic Syndr* **4**, 677-685.
93.	Loya, S., Tal, R., Hizi, A., Issacs, S., Kashman, Y. and Loya, Y. (1993) Hexaprenoid hydroquinones, novel inhibitors of the reverse transcriptase of human iminunodeficiency virus type 1, *J. Nat. Products* **52**, 2120-2125.
94.	Polak, Z. and Pospisil, F. (1995) Alleviation of plant virus infection by humic acids, *Biologia Plantarum* **37**, 315-317.
95.	Laub, R. (1999) *Process for preparing synthetic soil-extract materials and medicament based thereon*, U. S. Patient No. 5 945 446.
96.	Laub, R. (2000) Developing humate with anti-HIV, HSV, HPV and other antiviral activity, *Antiviral Drug and Vaccine Development Information* **12**, Biotechnology Information Institute, p. 2.
97.	David, P.P., Nelson, P.V. and Sanders, D.C. (1994) A humic acid improves growth of tomato seedlings in solution culture, *J. Plant Nutr.* **17**, 173-184.
98.	Dunstone, R.L., Richards, R.A. and Rawson, H.M. (1988) Variable responses of stomatal conductance, growth, and yield to fulvic acid applications to wheat, *Aust. J. Agric. Res.* **39**, 547-553.

307

99. Leversee, G.J., Landrum, P.F., Giesy, J.P. and Fannin, T. (1983) Humic acids reduce bioaccumulation of some polycyclic aromatic hydrocarbons, *Can. J. Fish. Aquat. Sci.* **40**, 63-69.
100. Stewart, A.J. (1984) Interactions between dissolved humic materials and organic toxicants, in K.E. Cowser (ed.), *Synthetic fossil fuel technologies*, Butterworth Publisher, Boston, pp. 505-521.
101. Oikari, A., Kukkonen, J. and Virtanen, V. (1992) Acute toxicity of chemicals to *Daphnia magna* in humic waters, *Sci. Total Environ.* **117/118**, 367-377.
102. Steinberg, C., Haitzer, M., Hesse, S., Lorenz, R., Bueggemann, R. and Burnison, B.K. (1997) Change of bioconcentration and effect of pesticides in the presence of humic substances, *Umweltwissenschaften und schadstoff-forschung* **2**, 64-68.
103. Filip, Z. and Berthelin, J. (2001) Analytical determination of the microbial utilization and transformation of humic acids extracted from municipal refuse, *Fresenius J. Anal. Chem.* **371**, 675-681.
104. Filip, Z., Kanazawa, S. and Berthelin, J. (2000) Distribution of microorganisms, biomass ATP, and enzyme activities in organic and mineral particles of a long-term wastewater irrigated soil, *J. Plant Nutr. Soil Sci.* **163**, 143-150.
105. Willmann, G. and Fakoussa, R.M. (1997) Biological bleaching of water soluble coal macromole-cules by a basidiomycete strain, *Appl. Microbiol. Biotechnol.* **47**, 95-101.
106. Temp, U., Meyrahn, H. and Eggert, C. (1999) Extracellural phenol oxidase patterns during depolymerization of low-rank coal by three basidiomycetes, *Biotechnol. Letters* **21**; 281-287.
107. Scheel, T., Holker, U., Ludwig, S. and Hofer, M. (1999) Evidence for and expression of a laccase gene in three basidiomycetes degrading humic acids, *Appl. Microbiol. Biotechnol.* **52**, 66-69.
108. Hofrichter, M. and Fritsche, W. (1997) Depolymerization of low-rank coal by extracellular fungal enzyme systems. II. The ligninolytic enzymes of the coal-humic-acid-degrading fungus Nematoloma frowardii b19, *Appl. Microbiol. Biotechnol.* **47**, 419-424.
109. Götz, G. K. E. and Fakoussa, R. M. (1999) Fungal biosolubilization of Rhenish brown coal monitored by Curie-point pyrolysis/gas chromatography/mass spectrometry using tetraethylammonium hydroxide, *Appl. Microbiol. Biotechnol.* **52**, 41-48.
110 Fakoussa, R. M. and Frost, P. J. (1999) In vivo-decolorisation of coal-derived humic acids by laccase excreting fungus *Trametes versicolor*, *Appl. Microbiol. Biotechnol.* **52**, 60-65.
111. Ralph, J.P. and Catcheside, D.E.A. (1997) Transformations of low rank coal by *Phanerochaete chrysosporium* and other wood-rot fungi, *Fuel Process Technol.* **52**, 79-93.
112. Cohen, M.S., Bowers, W.C., Aronson, H. and Grey, E.T. (1987) Cell-free solubilization of coal by *Polyporus versicolor*. *Appl. Environ. Microbiol.* **53**; 2840-2844.
113. Steffen, K.T., Hatakka, A. and Hofrichter, M. (2002) Degradation of humic acids by the litter-decomposing *Basidiomycete Collybia dryophila*, *Appl. Environ. Microbiol.* **68**, 3442-3448.
114. Holker, U., Ludwig, S., Scheel, T. and Hofer, M. (1999) Mechanism of coal solubilization by the deuteromycetes *Trichoderma atroviride* and *Fusarium oxysporum*, *Appl. Microbiol. Biotechnol.* **52**, 57-59.
115. Call, H.-P. and Mücke, I. (1997) History, overview and applications of mediated lignolytic systems especially laccase-mediator-systems (lignozym(R)-process), *J. Biotechnol.* **53**,163–202.
116. Cozzolino, A. and Piccolo, A. (2002) Polymerization of dissolved humic substances catalyzed by peroxidase, Effects of pH and humic composition, *Org. Geochem.* **33**, 281-294.
117. Fakoussa, R. M. and Hofrichter, M. (1999) Biotechnology and microbiology of coal degradation, *Appl. Microbiol. Biotechnol.* **52**, 25-40.
118. Prát, S. (1963) Permeability of plant tissues to humic acids, *Biol. Plant (Prague)* **5**, 279-283.
119. Phoung, H.K. and Tichy, V. (1976) Activity of humic acids from peat as studied by means of some growth regulator bioassay, *Biol. Plant (Prague)* **18**, 195-199.
120. Kulikova, N.A., Dashitsyrenova, A.D., Perminova, I.V. and Lebedeva G.F. (2003) Auxin-like activity of different fractions of coal humic acids, *Bulgarian J. Ecolog. Sci.* **2(3-4)**, 55-56.
121. Nardi, S., Panuccio, M.R., Abenavoli, M.R. and Muscolo, A. (1994) Auxin-like effect of humic substances extracted from faeces of *Allobophora caliginosa* and *A. Rosea*, *Soil Biol. Biochem.* **26**, 1341-1346.
122. Muscolo, A., Bovalo, F., Gionfriddo, F. and Nardi, S. (1999) Earthworm humic matter produces auxin-like effects on Daucus carota cell growth and nitrate metabolism, *Soil Biol. Biochem.* **31**, 1303-1311.
123. Muscolo, A., Cutrupi, S. and Nardi, S. (1998) IAA detection in humic acids, *Soil Biol. Biochem.* **30**, 1199-1201.

308

124. Nardi, S., Pizzeghello, D., Muscolo, A. and Vianello, A. (2002) Physiological effects of humic substances on higher plants, *Soil Biol. Biochem.* **34**, 1527-1536.
125. Vigneault, B., Percot, A., Lafleur, M. and Campbell, P.G.C. (2000) Permeability changes in model and phytoplankton membranes in the presence of aquatic humic substances, *Environ. Sci. Technol.* **34**, 3907-3913.
126. Gerritsen, J. and Bradley, S.W. (1987) Electrophoretic mobility of natural particles and cultured organisms in fresh water, *Limnol. Oceanogr.* **32**, 1049-1058.
127. Parent, L., Twiss, M.R. and Campbell, P.G. (1996) Influences of natural dissolved organic matter on the interaction of aluminum with the microalga *Chlorella*: a test of the free-ion model of trace metal toxicity, *Environ. Sci. Technol.* **30**, 1713-1720.
128. Campbell, P.G., Twiss, M.R. and Wilkinson, K.J. (1997) Accumulation of natural organic matter on the surfaces of living – cells implications for the interaction of toxic solutes with aquatic biota, *Can. J. Fish. Aquat. Sci.* **54**, 2543-2554.
129. Visser, S.A. (1985) Physiological action of humic substances on microbial cells, *Soil Biol. Biochem.* **17**, 457-462.
130. Fein, J.B., Boily J.-F., Guclu, K. and Kaulbach, E. (1999) Experimental study of humic acid adsorption onto bacteria and Al-oxide mineral surfaces, *Chem. Geol.* **162**, 33-45.
131. Zhou J.L. and Banks C. J. (1993) Mechanism of humic acid colour removal from natural waters by fungal biomass biosorption, *Chemosphere* **27**, 607-620.
132. Samson, G. and Visser, S.A. (1989) Surface-active effects of humic acids on potato cell membrane properties, *Soil Biol. Biochem.* **21**, 343-347.
133. Visser, S.A. (1982) Surface active phenomena by humic substances of aquatic origin, *Rev. Fr. Sci. Eau.* **1**, 285-296.
134. Ermakov, E.I, Ktitorova, I.N. and Skobeleva, O.V. (2000) Effect of humus acids on the mechanical properties of cell walls, *Rus. J. Plant Physiol.* **47**, 518-525.
135. Malcom, R.E. and Vaughan, D. (1979) Humic substances and phosphatase activities in plant tissues, *Soil Biol. Biochem.* **11**, 65-72.
136. Xu X. (1986) The effect of foliar application of fulvic acid on water use, nutrient uptake and yield in wheat, *Aust. J. Agric. Res.* **37**, 343-350.
137. Pinton, R., Cesco, S., De Nobili, M., Santi, S., and Varanini, Z. (1998) Water- and pyrophosphate-extractable humic substances fractions as a source of iron for Fe-deficient cucumber plants, *Biol. Fertil. Soils* **26**, 23-27.
138. Santi, S., Pinton, R., Cesco, S., Agnolon, F. and Varanini, Z. (1999) Water-extractable humic substances enhance iron deficiency responses by Fe-deficient cucumber plants, *Plant Soil* **210**, 145-157.
139. Singh, A.K., Dhar, P. and Pandeya, S.B. (1998) Influence of fulvic acid on transport of iron in soils and uptake by paddy seedlings, *Plant Soil* **198**, 117-125.
140. Stevenson, F.J. (1991) Organic matter-micronutrient reactions in soil, in J.J. Morevedt, F.R. Cox, L.M. Shuman, and R.M. Welch (eds.), *Micronutrient in agriculture Soil Science,* Soc. of America, Madison, pp. 145-186.
141. Maggioni, A., Varanini, Z., Nardi, S. and Pinton, R. (1987) Action of soil humic matter on plant roots: stimulation of ion uptake and effects on $(Mg^{2+} K^+)$ ATPase activity, *Sci. Total Environ.* **62**, 355-363.
142. Pinton, R., Varanini, Z., Vizotto, G. and Maggioni, A. (1992) Soil humic substances affect transport properties of tonoplast vesicles isolated from oak roots, *Plant Soil* **142**, 203-210.
143. Varanini, Z., Pinton, R., De Biasi, M.G., Astolfi, S. and Maggioni, A. (1993) Low molecular weight humic substances stimulate H^+-ATPase activity of plasma membrane vesicles isolated from oat (*Avena sativa* L.) roots, *Plant Soil* **153**, 61-69.
144. Lowe, L.E. (1992) Studies on the nature of sulfur in peat humic acids from Froser river delta, British Columbia, *Sci. Total Environ.* **113**, 133-145.
145. Ramalho, J.C., Lauriano, J.A. and Nunes, M.A. (2000) Changes in photosynthetic performance of *Ceratonia siliqua* in summer, *Photosynthetica* **38**, 393-396.
146. Sollins, P., Homann, P. and Caldwell, B.A. (1996) Stabilization and destabilization of soil organic matter: mechanisms and controls, *Geoderma* **74**, 65-105.
147. Bentez, E., Melgar, R., Sainz, H., Gomez, M. and Nogales, R. (2000) Enzyme activities in the rhizosphere of pepper (*Capsicum annuum* L.) grown with olive cake mulches, *Soil Biol. Biochem.* **32**, 1829-1835.

148. Stemmer, M., Gerzabek, M. and Kandeler, E. (1998) Organic matter and enzyme activity of bulk soil and particle-size fractions of soils obtained after low-energy sonication, *Soil Biol. Biochem.* **30**, 9-17.

149. Stemmer, M., Gerzabek, M. and Kandeler, E. (1999) Invertase and xylanase activity of bulk soil and particle-size fractions during maize straw decomposition, *Soil Biol. Biochem.* **31**, 9-18.

150. Rao, M.A., Gianfreda, L., Palmiero, F. and Violante, A (1996) Interactions of acid phosphatase with clays, organic molecules and organo-miniral complexes, *Soil Sci.* **161**, 751-760.

151. Kandeler, E., Stemmer, M. and Klimanek, E-M. (1999) Response of soil microbial biomass, urease and xylanase within particle-size fractions to long-term soil management, *Soil Biol. Biochem.* **31**, 261-273.

152. Rosa, A.H., Vicente, A.A., Rocha, J.C. and Trevisan, H.C. (2000) A new application of humic substances: activation of supports for invertase immobilization, *Fresenius J. Anal. Chem.* **368**, 730–733. H

153. Nannipieri, P., Ceccanti, B. and Bonmati M. (1998) Protease extraction from soil by sodium pyrophosphate and chemical characterization of the extracts, *Soil Biol. Biochem.* **30**, 2113-2125.

154. Nemeth, K., Salchert, K., Putnoky, P., Bhalerao, R., Koncz-Kalman, Z., Stankovic-Stangeland, B., Bako, L., Mathur, J., Okresz, L., Stabel, S., Geigenberger, P., Stitt, M., Redei, G.P., Schell, J. and Koncz, C. (1998) Pleiotropic control of glucose and hormone responses by PRL1, a nuclear WD protein, in Arabidopsis, *Genes Development* **12**, 3059-3073.

155. Wunderwald, U., Kreisel, G., Braun, M., Schulz, M., Jager, C. and Hofrichter, M. (2000) Formation and degradation of a synthetic humic acid derived from 3-fluorocatechol, *Appl. Microbiol. Biotechnol.* **53**, 441-446.

156. Piccolo, A., Cozzolino, A., Conte P. and Spaccini, R. (2000) Polymerization of humic substances by an enzyme-catalyzed oxidative coupling, *Naturwissenschaften* **87**, 391-394.

157. Yavmetdinov, I.S., Stepanova, E.V., Gavrilova, V.P., Lokshin, B.V., Perminova, I.V. and Koroleva, O.V. (2003) Isolation and characterization of Humin-like substances produced by wood-degrading fungi causing white rot, *Appl. Biochem. Microbiol.* (Russia) **39**, 293-301.

158. Korotkova, E.I., Karbainov, Y.A. and Avramchik, O.A. (2003) Investigation of antioxidant and catalytic properties of some biologically active substances by voltammetry, *Analyt. Bioanalyt. Chem.* **375**, 465-468.

159. Westerhoff, P., Aiken, G., Amy, G. and Debroux, J. (1999) Relatioships between the structure of natural organic matter and its reactivity towards molecular ozone and hydroxyl radicals, *Wat. Res.* **33**, 2265-2276.

160. Danilov, R. and Ekelund, N.G.A. (2001) Effects of solar radiation, humic substances and nutrients on phytoplankton biomass in Lake Solumsjö, Sweden, *Hydrobiologia* **444**, 203-212.

161. Lipski, M., Slawinski, J. and Zych, D. (1999) Changes in the luminescent properties of humic acids induced by UV radiation, *J. Fluorescence* **9**, 133-138.

162. Fielding, M., D. Barcelo, A. Helweg, S. Galassi, L. Torstensson, P. Readman, J.W., T.A. Albanis, D. Barcelo, S. Galassi, J. Tronczynski, Van Zoonen, R. Wolter and G. Angeletti. (1992) Pesticides in ground water and drinking water, *Water Pollut. Res. Rep.* **27**, p. 1-136.

Chapter 15

CYTOGENETIC EFFECTS OF HUMIC SUBSTANCES AND THEIR USE FOR REMEDIATION OF POLLUTED ENVIRONMENTS

An experience of the Dnepropetrovsk School of Prof. Khristeva

A. GOROVA, T. SKVORTSOVA, I. KLIMKINA, A. PAVLICHENKO
*National Mining University, Ecology Department, Karl Marx Ave 19,
Dnepropetrovsk, 49027, Ukraine <gorovaa@nmuu.dp.ua>*

Abstract

Mechanisms of humic substances (HS) physiological activity in relation to different organisms at a molecular–cellular level were studied. Adaptation mechanisms of HS action under radiation and chemical injuries were examined. Antimutagenic and detoxifying properties of HS were revealed and a usage of physiologically active HS for remediation of both biotic and abiotic components of the environments and human's health recovery was proved. By the example of industrial area nearby the Dnepr River the HS applicability for remediation of disturbed soil was demonstrated. Experimental evidences of reduction of genetic damage caused by chemical and physical stress factors in the presence of HS were obtained using various bioassay techniques. A theoretical basis for HS application to retard uptake of contaminants by plants from soil and to improve biota resistance towards anthropogenic factors was developed. Recommendations on improving environmental conditions, recovering human health, and remediation of polluted areas were elaborated based on application of HS.

1. Introduction

Consequences of environment pollution caused by human activity is one of the highest priority global problem today. Accumulation of hazardous substances and radioactive elements in the biosphere threatens ecosystems, gene pool, public health, and limits opportunities for further development.

This becomes apparent first of all in the growth of mutagenic loads up to a level inducing doubling the human mutation frequency, which may lead to disastrous alterations in national gene pool. Considering the fact that a number pollutants can initiate mutations resulting in genetic pathologies, the issues of mutagenic contaminants monitoring and the development of new approaches providing environment protection from accumulation of mutagenic contaminants have to be solved. An understanding of the nature of the mutagen contamination should be sought and protective methods for live organisms should be investigated.

311

I. V. Perminova et al. (eds.),
Use of Humic Substances to Remediate Polluted Environments: From Theory to Practice, 311–328.
© 2005 *Springer. Printed in the Netherlands.*

A combined effect of genetic mechanisms and environment impurity has been established to results in 60 percent of the total gene pool public health injury, 10 percent increasing of a disablement level in Europe, reducing lifetime and fertility rate. Loads imposed on the gene pool by contaminants are enhanced by the Chernobyl disaster consequences and led to growth of ontological and endocrine diseases from deterioration of immunity protection [20]. Therefore, prediction and pollution consequences prevention are of great importance in Ukraine.

To decrease risk of gene damage caused by different contaminants, an issue of detoxification of the actual pollutants in the environments has to be solved. In this respect detoxifying and antimutagenic agents of natural origin, which HS belong to, seem to show considerable promise [16, 19].

Being components of ecosystems, HS influence various organisms. HS alter injurious effects of pollutants and thus regulate disturbed balance in the environment [16, 19]. Considering soil depletion in humus and its deterioration, a regulation of humus content in the environment and a usage of artificial physiologically active HS (PAHS) are of first rate importance. Wide spread and heavy stock of caustobioliths along with simple methods of PASH isolation from them make caustobioliths suitable for practical use. Principle effects of HS on living organisms were first defined by professor L.A. Khristeva and stated in the nine-volume transactions under the title ""Humic Fertilizers. Theory and Practice of Their Application". The study was further developed by A. Gorova, I. Yarchuk, L. Yepishina, T. Skvortsova, L. Bobyr, G. Batalkina and others. Numerous studies performed with plants, microorganisms, and laboratory animals provided evidence that HS mitigate damaging effects of radiation, pesticides, and excess of mineral fertilizers [6, 7, 11, 16, 18]. Using HS as natural antimutagens and detoxifying agents of wide activity appears to be a high priority point for the areas polluted with mutagens. However, antimutagen and detoxifying properties of HS under actual environmental conditions have not still adequately explored and need to be further investigated.

2. Physiological Activity of Humic Substances

2.1. NATURE AND PRINCIPLE MECHANISMS OF HUMIC SUBSTANCES PHYSIOLOGICAL ACTIVITY

At present two main hypotheses on the physiological activity of HS are accepted. Some researchers consider that humic compounds ameliorate physical and chemical soil properties, thus creating improved conditions for plants growth and development, whereas others suppose direct effects of HS on plants. Under natural conditions both mechanisms are seemingly actual.

Assuming polyphenol structure of HS, L.A. Khristeva developed a theory of their biochemical activity and supposed their direct influence on a protein synthesis in cells [2, 3]. A certain part of HS was hypothesized to enter the cells and then to become engaged in the plant metabolism. This hypothesis was later confirmed by bioassay experiments with plant and animals using labelled HS [4, 16, 22]. Besides, on the basis of results of experiments with natural and synthetic bio-membranes membranotropic properties of HS was revealed [1, 21].

Therefore two probable principle mechanisms of PAHS can be proposed:
1) entrance of HS into cells and their engagement in metabolism process;
2) membranotropic activity of HS.
In our opinion, both mechanisms can take place in nature in parallel.

2.2. INFLUENCE OF HUMIC SUBSTANCE ON THE PROTEIN SYNTHESIS AND MITOSIS

The first evidence of HS influence on the protein synthesis was obtained due to experiments with inhibitors of nucleic acids and protein synthesis [16]. The experiments showed that 50 to 60% decrease in growth processes caused by specific protein synthesis inhibitors can be completely recovered by PAHS. Based on the obtained results two following suppositions were made:
1) HS influence on protein synthesis;
2) HS can mitigate the impact of unfavourable environmental factors damaging the protein synthesis.

Cytophotometric studies on DNA showed that HS increased DNA synthesis rate in interphase nuclei of meristem cells of the plants. This phenomenon was proved by an established decrease of the number of cells in the G_1 phase and an increase of that in the S and the G_2 phases of the mitotic cycle. This finding was also confirmed by autoradiographic research which revealed increased rate of labelled DNA (^{3}H-thymidine) or RNA(^{3}H-uridine) precursors absorption by cells in the presence of HS. At that both concentration of tritium into the cells and a number of labelled cells increased significantly. A maximum value of the number of root meristem cells (96%) could be reached in 6 hours earlier in the presence of sodium salt of HS compared with the blank (no HS added) value [14, 16].

Contents of nuclear histon proteins in the plant meristem cells were determined using staining with methyl green according to Alfert and Geschwind and radioautographic data on ^{14}C-leucine uptake. Results of the study showed HS facilitated total protein synthesis and increased content of nuclear histons (Table 4). Translation rate was estimated on the ratio of polyribosomes to monoribosomes contents determined by differential centrifugation in a sucrose density gradient. Analysis of ribosomes sedimentation behaviour showed that HS introduction resulted in increase of portion of polysome, whereas that of monosome reduced (Table 4). Hence, HS facilitated monoribosomes association into polyribosomes, what led to increase of ribosome apparatus efficiency. In other words, HS effected beneficially on both basic molecular cell mechanisms and subcell structures.

The molecular processes taking place in the interphase nuclei are realized in mitosis. In research on various vegetation test systems, it was shown that HS extracted from chernozem, valley peat, and brown coal if taken in concentrations stimulating growth processes (50 to 100 mg/l) increased the value of mitotic indices by an average of two times. As for the time required for all phases of the mitotic cycle, it was 2.5 to 3 hours less, mainly due to reduce of the G_1 phase duration [16].

3. Nature of Mitigating Action of Humic Substances to Ionising Radiation and Chemical Damage

Taking into consideration that radiation and chemical sicknesses of the cells of various organisms are accompanied by damage to the protein synthesizing system, which is favourably influenced by HS, a supposition was made that HS can play a significant role in adaptation of the organisms to the impacts of unfavourable environmental factors. This was proven by bioassay experiments performed with various organisms inhibited by ionising radiation or pesticides [6, 7].

To cause the radiation injury of the plants, X and Y-radiation were used. Dry seeds were radiated in doses of 50 to 250 Gy, and germinated seeds in dose of 2 to 20 Gy. Chemical impact on the plants was caused by persistent pesticides such as hexachloran, triazines and others.

Based on data obtained a coefficient of relative biologic effect (RBE) and a dose reducing factor (DRF) for humic preparations were estimated. The calculated RBE coefficient for the humic preparations in the growing test (length of shoots and roots, weight of seedlings were used as responses) varied from 1.20 to 1.48 for unirradiated variants, whereas if radiated, that laid in the range from 1.38 to 2.07, i.e. the relative effect of HS was more pronounced under radiation stress. Calculated DRF values for brown coal HS were 1.40 to 2.00, those for peat preparations were 1.38 to 1.90, while DRF value for the conventionally used radio-protector cysteine was 2.00. This finding was evidence of high mitigating activity of HS in relation to radiation.

Mitigating activity of HS was also recorded in bioassay experiments with hexachloran. The calculated values of RBE and DRF varied form 1.50 to 2.00 and from 1.38 to 2.00, respectively. Hence, HS were sufficiently effective protectors for the plants suffered from pesticides.

3.1 CHANGING OF MITOTIC ACTIVITY OF MERISTEM TISSUES IN THE PRESENCE OF DAMAGING AGENTS AND HUMIC SUBSTANCES

This research was aimed to find out the influence of HS on mitotic activity under stress conditions. Data in Table 1 show that humic preparations of different origin eliminated the inhibiting action on mitosis of the ionising radiation and chlorinated organic pesticides. Note, that if the mitotic activity decreased to 50-60% of blank due to radiation or chemicals, HS could completely restore values of mitotic indexes. In case if the greater level of inhibition was observed, only minor increase of mitotic activity in the presence of humic preparations was recorded [16].

Analysis of distribution of the separated cell depending on the mitosis phase showed that radiation and chlorinated organic pesticides reduced amount of cells in all the phases of mitosis, whereas triazine herbicides caused metaphase delays, and therefore, the total mitotic index remained at the reference level or even slightly exceeded the latter. Increase of the value of mitotic index in that case, however, should be considered as a pathologic one as growing processes were slowed.

Table 1. Influence of HS of different origin on mitotic index of seedling root meristem injured by ionising radiation and hexachloran.

Variant	Growing medium	Mitotic index, ‰ (M±m)**	
		Triticum durum Desf.	*Zea mays* L.
Blank	Water	46.6±5.0	56.5±6.0
10 Gy	Water	18.0±2.0*	29.0±2.9*
	Na-humate (50 mg/l) from:		
10 Gy	brown coal	47.2±6.0	60.0±6.0
	valley peat	40.0±3.0	55.0±5.5
	chernozem	55.9±6.0	68.2±7.0
Blank	Water	85.5±8.0	86.9±7.0
Hexachloran, 7 mg/l	Water	40.5±4.1*	52.0±4.0*
	Na-humate (50 mg/l) from:		
Hexachloran, 7 mg/l	brown coal	91.2±9.1	145.0±7.7*
	valley peat	87.1±8.0	98.5±8.9*

* Indicates statistically significant difference from blank.
** Average ± standard deviation.

The mitosis pathologies represent a summary effect of the metabolic disturbances taking place in the interphase nucleus and can result from damage of the chromosomes (chromosome aberrations), mitotic apparatus (metaphase delays, chromosome scattering, etc.) or cytotomy process disorders. The stress factors in question damaged growing processes and cell division and, in addition, caused mitosis pathologies. The former two effects are non-specific responses of a plant for unfavourable factors of different origin, while a kind of mitosis pathologies reflects often a specific inhibitor effect. For example, radiation injury and long-term impact of chlorinated organic compounds (or their short-term impact at high concentrations) resulted mainly to chromosome damage, whereas triazines in phytotoxic doses led to metaphase delays connected to disorder of mitotic spindle functioning. The mutations induced by radiation included mainly circular chromosomes, dicentrics and complex damage, those induced by chemicals included bridges, fragments, retarded chromosomes, etc. [13].

When HS were introduced to the plants that have been damaged by radiation or hexachloran, a significant reduction of the mitosis pathology in the meristem tissues was observed (Table 2). This was evidence of the antimutagen activity of humic compounds.

Table 2. Influence of brown coal Na-humate on the amount of aberrant chromosomes in the meristem tissues of *Triticum durum* Desf. seedlings injured by radiation and hexachloran.

Test-object	Variant	Aberrant chromosomes in mitosis anaphase and telophase, % (M±m)*	
		No Na-humate	Na-humate, 50 mg/l
Root meristem	Blank	0.91±0.3	0.62±0.23*
	10 Gy	24.52±1.36	5.64±0.73
	20 Gy	34.2±1.5	11.7±1.02
Meristem of growing point	Blank	0.88±0.28	0.5±0.23*
	10 Gy	30.2±1.45	5.4±0.71
Root meristem	Blank	0.9±0.3	0.52±0.22*
	Hexachloran, 5 mg/l	10.01±0.94	4.5±0.21
	Hexachloran, 10 mg/l	18.5±1.22	6.0±0.75

* Average ± standard deviation.

3.2. DISTURBANCE OF PROTEIN SYNTHESIS INDUCED BY RADIATION AND CHEMICAL TOXICANTS AND MITIGATING ACTIVITY OF HUMIC SUBSTANCES

Typical symptom of the radiation and chemical induced injury of DNA synthesis in the interphase nuclei of the plant meristem cell is a reduction of the quantity of DNA in cells (Table 3). This is governed by the G_1-S block, which lead to the increase of number of the nuclei in the G_1 phase and decrease of number of nuclei in the G_1-S_2 cycle phases. Formation of hypodiploids can be occasionally observed because of the DNA degradation. HS mitigate above mentioned damage what is evidence of recovery of the DNA synthesis process disturbed by both radiation and pesticides.

Table 3. Recovery of DNA contents in the interphase nuclei of meristem cells of *Zea mays* L. plant damaged by radiation and hexachloran in the presence of HS.

Variant	DNA quantity, rel. units per 1 interphase nucleus, (M±m)*	
	No humate added	Humate Na, 50 mg/l
Blank	72.51±1.21	81.2±1.4
10 Gy	57.6±1.41	76.1±1.67
12 Gy	43.6±0.96	71.2±1.72*
Blank	60.0±0.5	69.1±0.9
Hexachloran, 10 mg/l	50.5±2.8	71.9±2.9

* Average ± standard deviation.

Mitigating activity of HS towards radiation damage was confirmed by study on a Chinese hamster fibroblast cell culture of B-11-a-ii-FAF-28 line. The study showed that radiation caused single-filament ruptures in the DNA that were repaired in the presence of HS into the cell culture both before and after exposure to radiation [12]. Similar results were obtained in experiments with plant cells [5]. Therefore, HS are capable of eliminating the single-filament ruptures of the DNA, obviously due to strengthening activities of the cell repairing systems.

Table 4. Influence of atrazine and sodium humate on the histon contents in the cell nucleus in the root meristem of *Hordeum sativum* Jess. (experiment were conducted in sand using Pryanishnikov nutrient solution).

Seed treatment	Variant	Quantity of histons, relative units per 1 interphase nuclei*
No treatment	No herbicide	57.5±3.1
Na-humate, 100 g per 1 ton of seeds	No herbicide	76.5±2.12
No treatment	No herbicide, plus Na-humate, 100 kg per hectare	77.7±3.0
No treatment	Atrazine, 0.5 kg	30.4±2.98
Na-humate, 100 g per 1 ton of seeds	Atrazine, 0.5 kg	53.8±1.84
No treatment	Atrazine, 0.5 kg, plus Na-humate, 100 kg per hectare	55.2±2.82

* Average ± standard deviation.

Table 5. Influence of hexachloran and sodium humate on relation of the plyribosome/monoribosome ratio in *Zea mays* L seedlings.

Seed treatment	Variant	Polyribosomes, %	Monoribosomes, %	Polyribosomes/ Monoribosomes
No treatment	Water (blank)	60.9	39.1	1.55
Na-humate, 25 mg/l	Water (blank)	71.9	28.1	2.56
No treatment	Hexachloran, 10 mg/ml	30.5	69.5	0.4
Na-humate, 25 mg/l	Hexachloran, 10 mg/ml	58.5	41.5	1.4

   The sodium humates were also demonstrated to repair the cell ribosome apparatus and to reactivate protein synthesis in the cells damaged by the DNA-tropic agents. This finding was confirmed by data on the intensity of inclusion of ^{14}C-leucine into the plant cells. In experiments with barley (*Hordeum sativum* Jess.) the blank trace intensity was 46.5+/-3.5 and 50.0+/-3.8 pcs of the traces per 1 cell in root meristem tissues and in the growing point of leaves, respectively. Atrazine at concentration 20 mg/l decreased the amount of ^{14}C-leucine to the level of 39.2+/-3.4 and 18.1+/-2.7 in the same tissues. HS at concentration 50 mg/l restored the trace index in the root meristem to 50.1+/-3.5, i.e., to the reference level, and in the leaves to only 31.6+/-3.3 pcs. This means that the repairing processes in the root system were more active due to the influence of the HS. In the root meristem cells, besides, after treatment of the barley seeds with HS and in the presence of HS in the nutrition solution, increase of histon content (Table 4) and recovery of ribosome apparatus (Table 5) were observed.

   Seed treatment with HS or introduction of HS into the nutrient solution appeared to be very effective in mitigation of inhibitory activity of the toxic doses of the chemical agents.

4. Physiologically Active Humic Substances as Promoters of Plant Ontogenesis Damaged by Ionising Radiation and Pesticides

Investigations of changes in plant ontogenesis in the presence of HS under optimum and stress conditions were performed in laboratory and field conditions using various plants [16]. As an example we show results of one of the laboratory experiment using *Lycopersicum esculentum* Mill. in sand with the nutrient mixture of Pryanishnikov. HS were added in the form of humophos.

   Under optimum conditions without humophos typical sigmoid curves of growth were obtained. When humophos was added, its promotional effect appeared mostly in initial and logarithm growth phases. The growth curves went up more steeply, achieving the stationary phase sooner. The maximum plant height in the presence of humophos was 80+/-5.8 cm, exceeding height of blank plants by 33 percent. Plant development observations showed that HS facilitated entrance of the plant into a phase of budding and fruiting. Appearance of the first ripe fruit was observed 5-7 days earlier as compared with the reference, and the number of fruits was twice as large compared with the blank. The average mass of one fruit among the treated plants was at the blank level. Productivity of the seed treated with HS increased by 37.3 percent, and the absolute mass of 1000 seeds increased by 48.5 percent. The same results were obtained under field experiments; humates facilitated productivity and development of plants *Zea mays* L. and *Hordeum sativum* Jess. These experiments also produced 7 to 8 percent increase in the protein contents in the grains.

   In the experiments performed with HS under stress conditions created by irradiation of the seeds or by adding pesticides to the growing medium, the mitigating effect for the damaged plants was observed. As a rule, the inhibiting factors slowed down the plant growth and development rates and reduced their productivity. The quality of the products became worse when the pesticides were introduced into the soil. When HS were introduced into the growing medium of the plants depressed by 50 percent, the recovery of the processes of growth, differentiation of the reproduction organs, and formation of the full value crops (Table 6).

All of the above appears to be theoretic grounds for practical usage of HS both under optimum and stress conditions.

Table 6. Influence of HS on recovery of the ontogenesis processes damaged by the impact of ionising radiation and hexachloran.

Variant	Growing medium	Height of plants, cm	Sterility of pollen, %	Yield, % of blank
Lycopersicum esculentum Mill.				
Blank	Pryanishnikov nutrient solution	75.5±6.1	1.6±0.22	100
Irradiation of seedlings, 15 Gy	–"–	60.0±5.0	35.3±0.87	58.5
irradiation of seedlings, 15 Gy	–"– + humophos, 20 t/hectare	72.3±5.8	4.4±0.37	97.2
Hordeum sativum Jess.				
Blank	Chernozem	24.4±3.0	6.4±0.96	100
Irradiation of dry seeds, 200 Gy	–"–	18.4±2.5	24.3±1.36	66.3
Irradiation of dry seeds, 200 Gy	–"– + humophos, 20 t/hectare	31.4±3.0	9.4±0.92	99.1
Hexachloran, 200 kg/hectare	–"–	12.0±1.8	21.8±1.3	70.2
Hexachloran, 200 kg/hectare	–"– + humophos, 20 t/hectare	25.1±4.6	5.9±0.74	109

5. Detoxifying Ability of Humic Substances towards Heavy Metals

Heavy metals that enter the environment together with wastes and emissions from manufacturing enterprises, railway and automobile transport, and air fleets are the priority contaminants in the industrial Dnepr region. Accumulation of heavy metals in soil means increasingly environmentally hazardous conditions, creating danger for human health and biological components of ecosystems. Unlike air and water, where purification processes are periodically possible with regard to the heavy metals, the soil is an active accumulator and soil purification capability is very poor. When the old industrial regions with developed mining and metallurgic complexes were investigated, maximum levels of metals were found in the ground near the manufacturing areas and at distances of 1 to 5 km from the sources of emission. The concentrations exceeded background levels by 10 or more times and only at distances of 15 to 20 km from the source of pollution did the metal contents in the ground reach the background level or a value close to it.

The soils of the Dnepr River region contain large quantities of heavy metals such as chromium (up to 64.3 mg/kg), cobalt (3.5 mg/kg), copper (3.6 mg/kg), zinc (16.1 mg/kg), lead (11.9 mg/kg), manganese (300 mg/kg), and iron (38600 mg/kg).

Being an essential component of ecosystem, heavy metals are necessary in the small doses for live organisms. In industrial regions where heavy metals are accumulated in the soil in high concentrations, they impact the plant organisms, resulting in reduced biological productivity and degradation of the natural and cultivated phytocenosis. Build up of heavy metals in the plant tissues is extremely hazardous if they may be eaten by humans or animals.

When consuming contaminated food products, gradual saturation of the human body with heavy metals occurs, resulting in a number of dangerous diseases. The most hazardous are the mutagenic, carcinogenic, and teratogenic impacts of the heavy metals, which may be aggravated if ionising radiation and other ecologically toxic agents influence the organism. The urgent problem then is to restrict entrance of the heavy metals from the soil into the plants and the human body.

It is known that HS interact with metals in the soil. When interacting with univalent metals, they form labile complexes that are easily soluble, and when interacting with bivalent metals or with those of higher valency, including heavy metals, poorly soluble compounds are formed, thus impeding their entrance into plants. Increase of HS content in the soil may therefore contribute to deceleration of heavy metal uptake by plants.

5.1. INFLUENCE OF HUMIC SUBSTANCES ON HEAVY METAL UPTAKE FROM SOILS BY PLANTS

The hypothesis on reduction of heavy metals uptake by plants in the presence of HS was confirmed by laboratory and field experiments. In the laboratory experiments with chernozem the growth of *Brassica capitana* L. was inhibited by lead, copper, and zinc at concentrations of 300, 277, and 281 mg/kg, respectively, which were 9.4, 9.2, and 12.2 times of maximum allowable concentrations (MAC). Sodium humate extracted from brown coal was used as a detoxifying agent at concentration 5 g/m^2. At the end of the experiment the contents of lead, copper, and zinc (with no humate when growing) equalled 2.4, 10.6, and 13.5 mg/kg, respectively, and where the humates were the background, these values were 0.5, 5.1, and 4.2 mg/kg, respectively. Apart from decreased rates of uptake of the heavy metals from soil by plants, sodium humate promoted a 20 percent increase in the crop, and the heavy metals showed a 33.4 percent drop as compared with the blank variant.

The field experiment was conducted on chernozem soils (Pavlograd region, Ukraine) polluted with radionuclides and heavy metals. To recover the soil the brown coal sodium humate was introduced into the soil at a rate of 150 kg per hectare. Pre-sowing treatment of the *Zea mays* L. seeds with humate "Christecol" (10 l of 30% solution for 1 t of grains) was performed.

The obtained results showed that with humus added, the amount of chromium entered the vegetative mass was 1.7 times less and the grain was 2.5 times less, as compared with the blank plant without humus. For lead that reduction was 2.5 and 3.0 times, respectively; for gallium, 1.0 and 5.0 times; for nickel, 1.3 and 1.0 times; for zirconium, 1.0 and 2.0 times; for cobalt, 1.0 and 1.7 times; for copper, 1.0 and 1.7 times; for vanadium, 1.0 and 1.6 times; for strontium, 2.1 and 2.1 times; for manganese, 1.7 times for both; for aluminium, 1.0 and 3.0 times; and for iron, 1.0 and 2.0 times.

The plant shoots was also observed to emerge two days earlier than the blank ones if the seeds were treated by the "Christecol" preparation and if sodium humate was introduced into the soil. At a four- or five-leaf stage non-treated and treated with humate plants differed neither in height nor in leaf surface area. The specific surface density of the leaves was, however, obviously greater than that of the blank plants. Chlorophyll contents of both a and b type was also higher in plants treated with HS. The increase in chlorophyll content led to increase in plant biomass. Calculated photosynthetic productivity rate of plants in blank variant was 3.9 g/m^2 per day, that value for HS-treated plants increased to 4.2 g/m^2 per day, i.e., by 8 percent more.

The beneficial effect of HS on development *Zea mays* L. was also evidenced by the pollen sterility features of the pollen cells of the plants. For the blank plants the pollen sterility was 18 percent, but only 8.5 percent for the treatment plants. That is, the vitality of the pollen cells of the treated plants was 2 times higher than that of the blank plants.

Thus, PAHS play an important role in the environment protecting plant from heavy metals and facilitating therefore plant development and biomass accumulation. HS can be considered as promising detoxifying agents for the areas polluted with heavy metals and radionuclids.

6. Ecological-Genetic Evaluation of Soils and Reduction of Genetic Consequences of Soil Pollution Using Humic Substances

Conventional physical and chemical methods of soil analysis do not reflect the biological effects of the combined influence of various damaging agents. This problem can be solved by using bioindicators; among them, the cytogenetic tests seems to be the most informative. They are extremely sensitive and sufficient for adequate risk evaluations [9, 10, 18]. So, an ecological-genetic monitoring system was developed to provide adequate evaluation of the genetic risk for populations due to environmental contamination. This system combined the bioindication of mutagen in various environments, the cytogenetic check-up of the population, and genetic monitoring in human populations. Studies of the total toxic-mutagen activity of the soils were also of special interest.

To evaluate ecological-genetic hazards, various indices were used:
a) indices of bioindicators, i.e. genetic mutations in hairs of stamen filaments of *Tradescantia poludosa* L. (clone 02), chromosome aberrations in *Allium cepa* L. root meristem cells, etc.,
b) indices at the level of the human population, i.e. neoplasm, congenital anomalies of development, child death rate up to 1 year, etc.

322

All the indices reflecting the genetic hazard were reduced to a single non-dimensional index (CI) using formula (1):

$$CI = \frac{|P_{comf} - P_{act}|}{|P_{comf} - P_{crit}|}.$$ (1)

Where P_{act} is the biological parameter value in the point of study; P_{comf} is the parameter value under favourable conditions; and P_{crit} is a value of parameter at maximum contamination level.

Integral conditional indices (ICI) of the damage rate were established using equation 2:

$$ICI_j = \frac{1}{n} \cdot \sum_{i=1}^{n} CI_i = \frac{1}{n} \cdot \sum_{i=1}^{n} \left[\frac{|(P_{comf} - P_{act})|}{|(P_{comf} - P_{crit})|} \right],$$ (2)

where ICI is the integral conditional index of biology system damage.

The basic index of the genetic hazard ($ICI_{gen risk}$) for the population with regard to the significance coefficient was established using formula 10:

$$ICI_{genrisk.} = 0.6 ICI_{bioind} + 0.4 ICI_{popul},$$ (3)

where $ICI_{bioind.}$ and ICI_{popul} are integral indices of the biology indicator damages and genetic alterations in the populations, respectively.

The significance of the conditional indices ranges from 0 (favourable conditions) to 1 (stress conditions).

To evaluate the environmental risk, a universal estimation scale (Table 7) was used.

Table 7. Estimation scale of environmental toxic and mutagen risk.

Range of ICI values[a]	Level of Bioindicator damage	Soil conditions
0 – 0.150	Low	Favorable
0.151 – 0.300	Below average	Disturbed
0.301 – 0.450	Average	Alarming
0.451 – 0.600	Above average	Risky
0.601 – 0.750	High	Critical
0.751 – 1.0	Maximum	Hazardous

[a]ICI = 0.300 we considered to be the maximum value of ICI when damaged environment can be recovered.

To detoxify polluted soils, preparations of PAHS derived form brown coal were used, i.e., the sodium humate and the "Christecol" preparation. The efficiency of the soil recovery by HS was estimated using bioindicators. As an object of study, the Dnepropetrovsk region soils were used. The region is recognized as an area of high environmental risk due to extremely high human-caused pollution loads imposed on the environment and populations.

6.1. ECOLOGICAL-GENETIC EVALUATION OF SOILS IN DNEPROPETROVSK REGION

In the framework of the program of comprehensive environmental monitoring of the Dnepropetrovsk region, environmental-genetic evaluation of the soils was performed. Bioindication of the mutagens in the soils of the cities and villages was undertaken. The soils from 12 plots in Dnepropetrovsk and 33 plots in four monitoring areas, Dnepropetrovsk-Dneprodzerzhinsk Agglomeration (DDA), Nikopol-Marganets-Ordzhonikidze region (NMO), Western Donbass (WD), and Pereshchepino-agdalinovka-Tsarichanka (PMT), were studied (Table 8).

Table 8. Toxic-mutagen activity of Dnepropetrovsk region soils according to "mitotic index" and "chromosome aberrations" bioassay with *Pisium sativum* L. cells.

Area	Mitotic Index			Chromosome Aberrations		
	min	max	average	min	max	average
DDA	78.5	147.3	109.9±11.1	9.3	26.4	17.7±2.5
NMO	104.5	173.2	113.5±12.5	7.8	23.1	14.6±2.2
WD	69.0	140.7	113.2±12.0	12.1	26.7	23.7±2.7
PMT	95.7	121.0	107.7±11.0	11.9	24.0	18.4±2.6
Dnepropetrovsk	72.0	130.1	98.8±10.5	7.2	17.2	14.9±2.2
Blank (distilled water)			143.8±7.2			1.3±0.4

The presented data evidence that the soils of the region differed significantly in their toxicity and mutagenicity. The most mutagen contaminated were the West Donbass soils and their soil conditions were evaluated as critical. High soil mutagenicity was also observed in Pereshchepino-Magdalinovka-Tsarichanka region, and the soil conditions were characterized as risky. The soil conditions of the Dnepropetrovsk-Dneprodzerzhinsk Agglomeration and the Nikopol-Marganets-Ordzhonikidze region was estimated as alarming. Fig. 1 shows results of the experiments on application of the "Christecol" preparation for the detoxification of polluted soils in Dnepropetrovsk region. Introduction of 0.01% water solution of the "Christecol" preparation into the polluted substrate led to statistically significant (P < 0.01) decrease of amount of chromosome aberrations in *Pisum sativum* L. meristem cells. Soil mutagenicity was reduced by 1.5 to 4 times. Application of the "Christecol" preparation decreased also soil toxicity, and the values of mitotic index increased significantly.

Similar results were obtained for the soils of Dnepropetrovsk were studied [8]. The antimutagen effect of the "Christecol" preparation was, however, more pronounced in that case. The number of chromosome aberrations for HS treated soils reduced by 2.9 to 12.4 times depending on city districts.

324

In the experiments with *Tradescantia poludosa* L. (clone 2), the ability of HS to reduce genotoxicity of soils was revealed. Sprinkling of contaminated soils with a 0.01 percent water solution of the "Christecol" preparation resulted in statistically significant (P < 0,01) reduction of the number of gene somatic mutations in the hairs of stamen filaments. The mutation frequency in various soils was reduced by 2.5 to 3 times (Fig. 2).

Hromosome abberations of *Pisum sativum*

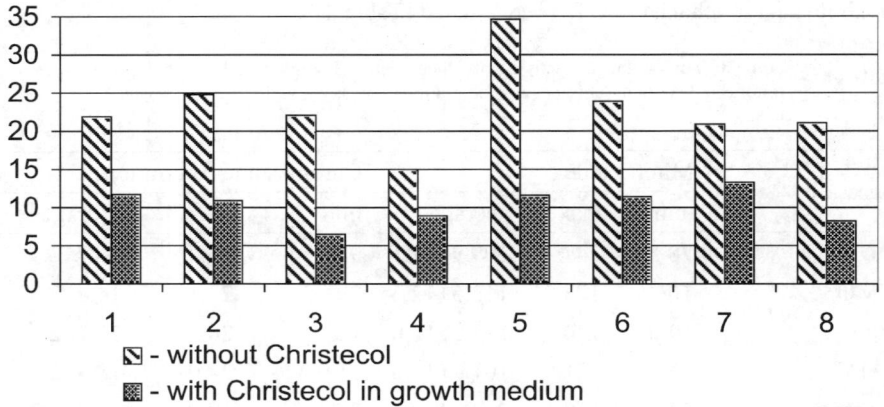

◩ - without Christecol

▦ - with Christecol in growth medium

Figure 1. Influence of the "Christecol" preparation on mutagenicity of Dnepropetrovsk region soils.

Gene mutation of *Tradescantia poludosa*

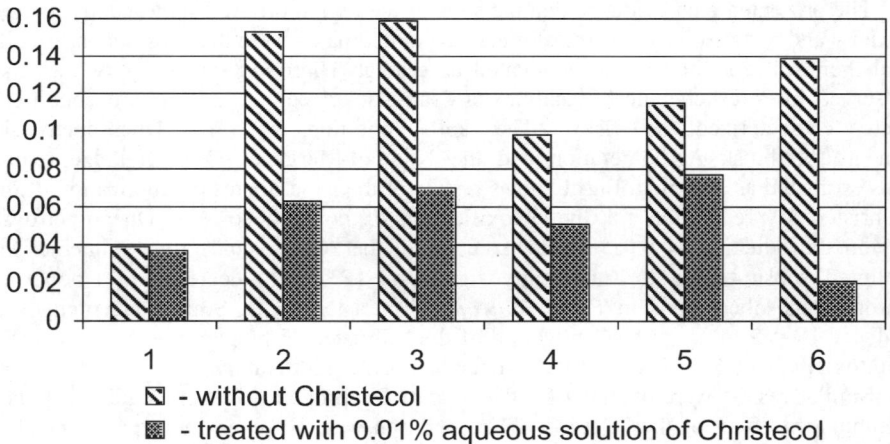

◩ - without Christecol

▦ - treated with 0.01% aqueous solution of Christecol

Figure 2. Influence of the "Christecol" preparation on toxicity of Dniepropetrovsk city soils.

Thus, the bioassays using higher plants as indicators proved an opportunity to reduce the mutagenicity and toxicity of the Dnepropetrovsk region soils by application of HS preparation.

6.2. CYTOGENETIC EVALUATION OF SOIL MUTAGENICITY IN DNEPROPETROVSK REGION AND THEIR RECOVERY BY HUMIC SUBSTANCES

Using Marganets Ore Mining and Processing Enterprise as an example, the soils condition was evaluated by cytological testing methods, and the efficiency of usage of the humates extracted from the brown coal of the Alexandria deposit was also studied. Obtained data (Figure 3 and Table 9) showed that the highest toxicity and mutageniciy of the soils were found in territories of the concentrating mill and the sludge depository; their soil condition was established as critical. Soil conditions in the immediate region of mine was estimated as risky, whereas that in the site 1000 m away from the mine was evaluated as alarming.

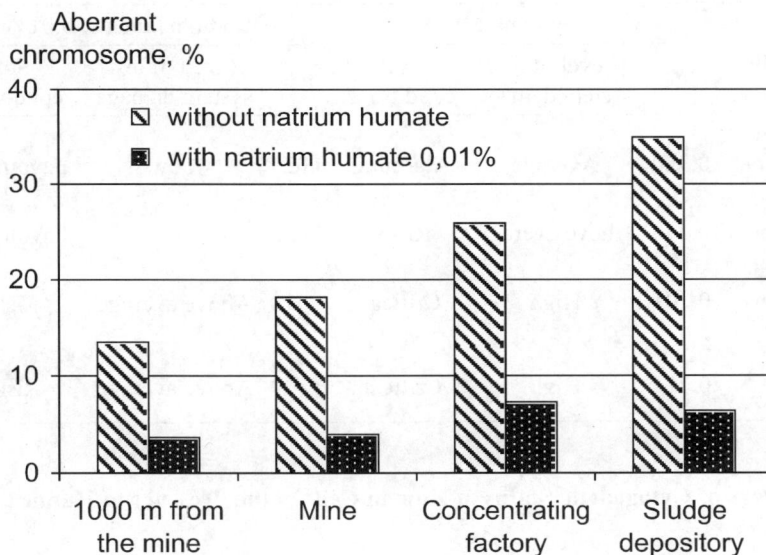

Figure 3. Quantity of chromosome aberrations in root meristem of *Allium cepa* L. seedlings grown on soil of Marganets Ore Mining with and without sodium humate.

Introduction of sodium humate into the soil at the level of 100 kg per hectare promoted normalization of the cell division processes and reduction of the amount of chromosome aberrations in the root meristem of the target plants.

The heavy metal contents was additionally determined in the soils under study. In soils sampled near the sludge depository, increased contents of manganese, nickel, barium, and molybdenum exceeding MAC was revealed. Manganese should be considered the main contaminant of the soils in the studied region. Its concentration near the sludge depository reaches 10050 to 15000 mg/kg, representing 6 to 10 times the MAC. Molybdenum contents in these soils amounted to 198 mg/kg (3.3 the MAC), barium was 195-300 mg/kg (1.3 to 2 times the MAC) and nickel was 200 mg/kg (2 times the MAC).

After the soils were treated with HS, the concentrations of heavy metal labile species significantly reduced, contributing to decrease of ICI values in that soils (Table 9).

Based on the data obtained, recommendations concerning use territories near industrial sites were formulated, i.e. only territories 1000 m away from the mining and processing enterprise may be used for economic purposes. When restoring the contaminated soils, usage of HS is recommended.

Table 9. Toxicity and mutagenicity of Soils of Marganets Ore Mining and Processing Enterprise before and after treatment with sodium humate.

Variant	No humate			Sodium humate, 0.01%		
	ICI	Level of test system damage	Soil conditions	ICI	Level of test system damage	Soil conditions
1000 m away from mine	0.320	Average	Alarming	0.029	Low	Favourable
Mine	0.470	Above average	Risky	0.038	Low	Favourable
Concen-tration factory	0.701	High	Critical	0.470	Above average	Risky
Sludge depository	0.714	High	Critical	0.520	Above average	Risky

7. Recovery of Cytogenetic Status of Human Cells in the Presence of Humic Substances

When experimenting with fibroblastic-like human cells, HS was found to reduce the level of spontaneous and radiation damage to the DNA. If HS were available in the cell-growing medium before, after, and during exposure to radiation, they promote repair of the DNA single fiber ruptures and recover functioning thereof [12, 16].

Experiments conducted at the Pulmonary Sanatorium of Dnepropetrovsk proved that children suffering recurring bronchitis after being treated with huminate (21-day course of treatment) displayed recovering of the immune system and the frequency of genetic damage in somatic cells reduced. Within the next autumn and winter period, the level of respiratory diseases of those children was reduced by 1.3 times.

Thus, high physiologic activity of HS and their antitoxic and antimutagenic properties are due to favourable influences on the protein synthesis, biological membranes, sub-cell structures, and the cell division processes. The unique structure and sorptive properties of HS provide for limitation of toxicant uptake from soil by plants, especially within industrial areas, and increase resistance rates of the plants and other organisms with regard to the variety of unfavourable environmental factors. This determines protective environmental functions of HS at the level of the ecosystem.

8. Conclusion

Information set forth in this article, of course, does not cover the accumulated data bank in full scale concerning the HS issue. For the first time, an attempt was made to summarize data concerning the nature of the HS physiological activity under normal and stress conditions, but the mechanisms of HS promotional and adaptive effects are so far not completely understood. There are grounds to consider that the role and significance of HS are not yet entirely revealed. If the view of V.I. Vernadskiy that there is no sharp line between live and dead mater is accepted, HS can easily be considered as an intermediate between them.

9. References

1. Batalkin, G.A., Koganov, M. and Makhno, M.Yu. (1980) *Membrane Activity of Certain Humus Substance Fractures. Humic Fertilizers. Theory and Practice of Their Usage*, Publishing House of Dnepropetrovsk Agriculture Institution, Dnepropetrovsk, V. 7, pp. 67-73.
2. Khristeva, L.A. (1948) Participation of Humic and Other Organic Substances in Higher Plant Nutrition, *Soil Science* 10, 46-59.
3. Khristeva, L.A., Gorovaya, A.I. (1967) *Physiologich active stoff derfesten ols wiekungsfaktor der Torfdunder*. Toft Koloquim DDR-VR, Polen Rostok.
4. Fokin, A.D., Bobyr, L.F., Yepishina, L.A. et al. (1975) Penetration of Humus Substances in Plant Cells, *Humic Fertilizers. Theory and Practice of Their Usage* 5, 38-56.
5. Golikova, O.P. (1976) *Repair of DNA in Plant Cells – Role in Radio Stability, Plant Radio Stability Mechanisms*, Kiev, Scientific Idea, pp. 59-97.
6. Gorovaya, A. I. (1988) Role of Physiologically Active Substances of Humus Origin in Increasing of Plant Stability to Influence of Pesticides, *Biological Sciences* 7 (295), 5-16.
7. Gorovaya, A.I. (1993) Role of Physiologically Active Substances in Adaptation to Influences of Ionizing Radiation and Pesticides, in *Humic Substances in Biosphere*, Moscow, Science, pp. 214-221.
8. Gorovaya, A.I. (2001) Methodological Aspects of Evaluation of Genetic Consequences of Technogenesis, in *Ecology and Usage of Nature*, Issue No 3, Dnepropetrovsk, pp. 143-151.
9. Gorovaya, A.I., Bobyr, L.F., Skvortsova, T.M., Klimkina, I.I. (1996) Methodological Aspects of Evaluation of Mutagen Background and Genetic Risk for Human and Biota Due to Influence of Mutagen Ecology Factors, *Cytology and Genetics* 30, 76-86.
10. Gorovaya, A.I., Digurko, V.M., Skvortsova, T.V. (1995) Cytological Evaluation of Mutagen Background in Industrial Dnepr Region, *Cytology and Genetics* 29, 16-22.
11. Gorovaya, A.I., Granovskiy, N.M., Kravtsova, L.V. (1977) Influence of Physiologically Active Substances of Humus Origin on Functional Activity of Plant, Animal and Germ Cells, in *Tissue Therapy According to V. P. Filatov*, Odessa.
12. Gorovaya, A.I., Khlyzina, N.V. (1987) Influence of Humic Preparations on Repair of Y-Induced Single Fiber Ruptures of DNA, in *Agricultural Radiobiology*, Kishinyov, pp. 76-79.

328

13.   Gorovaya, A.I., Kulik, A.F., Oginova, I.A. (1985) Role of Physiologically Active Humus Preparations in Regulation of Cellular Cycle Processes, in *Collection of scientific Works "Regulation of Plant Cellular Cycle"*, Kiev, Scientific Idea, pp. 101-110.
14.   Gorovaya, A.I., Kulik, A,F., Tkachenko, L.K. et al. (1981) Influence of Certain Biosphere Contaminants and Physiologically Active Substances of Humus Origin on Intercellular Nucleic Metabolism of Crops and Their Productivity, in *Collection of Scientific works "Nucleic Acids and Chromatin of Plants"*, Kiev, Scientific Idea, pp. 34-39.
15.   Gorovaya, A.I., Oginova, I.A. (1988) The Cytogenetic Effect of Peat Derived Sodium Humate Against Increasing Herbicide Application Rate Background, in *VIII International Peat Congress*, Leningrad, pp. 15-22.
16.   Gorovaya, A.I., Orlov, D.S, Shcherbenko, O.V. (1995) *Humic Substances*, Kiev, Scientific Idea.
17.   Gorovaya, A.I., Redko, Ye.S. Skvortsova, T.V. (1993) *Ecological Role of Humic Substances in Reducing of Aggressive Nature of Heavy Metals in Biogeocenosis*, Preprint IPPE ANU, Dnepropetrovsk.
18.   Gorovaya, A.I., Vasylieva, T.L. (1995) Cytogenetic Aspects of Mutagen Impact on Biota and Human Beings from Heavy Metals of Industrial Regions Environment and Rehabilitation Affect of Natural, in *Proceedings of the III International Conference on Biogeochemistry of Trace Elements*, Paris, France.
19.   Orlov, D. S. (1990) *Humus Acids of Soils and General Theory of Humification*, Publishing House of MGU, Moscow.
20.   Serdyuk, A. M. (1998) Natural Environment and Population Health of Ukraine, *Environment and Health* 4, 2-8.
21.   Visser, S.A. (1972) Physiological Action of Humic Acids on Living Cells, in *Proceedings of the IV International Peat Congress*, Finland, Ctaniemy.
22.   Yurin, V.M., Zhelyaeva, T.G., Kosobokova, R.V. (1982) Modification of Ion Penetrability of Protoplasm Cell Membrane. Nitelia under Influence of Physiologically Active Peat Compounds, in *III International Symposium IV and II MTO Committees*, Minsk, pp. 83-87.

Chapter 16

INFLUENCE OF METAL IONS ON THE ACTIVITY OF
SOIL HUMIC-ENZYME COMPLEXES

S. JOROBEKOVA, K. KYDRALIEVA, E. HUDAYBERGENOVA
Institute of Chemistry and Chemical Technology, NAS of Kyrgyz Republic, Chui prospect, 267, 720071 Bishkek, Kyrgyzstan
<jorobekova@mail.ru>

Abstract

In the present work, we attempted to elucidate the mechanisms of soil humic acids associations with some proteases. The stability of humic-enzyme complexes and the ways humic acids influence the activity of enzymes were determined. It was shown that humic acids are non-specific inhibitors of proteases. Enzyme-inhibitor complexes are formed by electrostatic, hydrophobic and other types of intermolecular interactions. The protease inhibiting ability of humic acids is influenced by their interaction with metal ions. The stability of enzyme-inhibitor complexes that form by association of proteases with humic-metal complexes depends on the nature of the metals and on the number of coordination centres on the polymeric chains of humic acids.

1. Introduction

It is known that extracellular enzymes, that have been produced by microorganisms and plants, influence the enrichment of soils with nutrients and the biotransformation of environmental pollutants [1-3]. Activity of such enzymes and their sensibility to the action of the inactivating factors changes by interaction with organic and mineral substances of soil. There are numerous references to clay-enzyme associations [1]. While many enzymes retain their activity after adsorption on clays, many others do not. Considerable research has been done on the association of soil enzymes with organic matter, in particular, with humic substances (HS) [5-8]. It has been reported that humics-immobilized enzymes can frequently retain their activity becoming stable against environmental stresses. There are many studies in the literature about the influences of humic acids (HA) on the activity of different enzymes [9-17]. Proteolytic enzymes are paid particular attention among other enzymes due to their important role in substances exchange in biological and environmental systems [18-19]. Early we reported that humics form stable complexes with proteolytic enzymes. It was shown that the activity of these enzymes generally decreased after complex formation [20-25]. However, the opposite effect was detected as well, nominally, an HA activated enzymatic reaction [26].

In spite of particular importance of the humics-enzymes interactions, the mechanisms governing the changes in enzyme activity are still poorly understood. The reported

I. V. Perminova et al. (eds.),
Use of Humic Substances to Remediate Polluted Environments: From Theory to Practice, 329–341.
© 2005 *Springer. Printed in the Netherlands.*

findings are very difficult to compare due to diversity of conditions and humic preparations used by the different authors. I addition, the interaction between humic acids and enzymes may be complicated by the presence of metal-ions in the biological systems. Metal-ions are able to form coordinative bonds with functional groups of both humic acids and enzymes. Consequently, they exert essential influence on the activity of enzymes and on the properties of the inhibitor.

The objectives of the present work were: to examine the influence of humic acids on the activity of different proteolytic enzymes; to investigate formation and stability of "humic acid - enzyme" complexes; to determine inhibition constants and their dependence on occupation of functional groups of humic acids with metal ions; and to surmise the mechanism of metal participation in humus-enzyme interactions.

2. Materials and Methods

Trypsin, subtilisin and α-chymotrypsin were used in this study. Trypsin had 33 % active molecules according to the results of titration by p-nitrophenyl ether of izanidinbenzoic acid, subtilisin and α-chymotrypsin had 78% and 77% active molecules, respectively, according to the results of titration by N-trans-cinnamoilimidazole [27]. Humic acids were isolated from soil .The fractions of humic acids possessing different molecular weights (M_w) were obtained using size exclusion chromatography fractionation.

The determination of enzyme's activity was conducted by pH-state method [22] under different reaction conditions (pH; ionic strength (μ); temperature) using ethyl ether of N-acetyl-L-tyrosine (ATEE) or methyl ether of N-benzoyl-L-leucine (BLME) as substrates.

The determination of kinetic parameters of mono-component, and joint inhibition of protease activity by humic acids and metal-ions was carried out using enzymatic catalytic techniques, implying selection of kinetic scheme of inhibition reactions and mathematical description of these schemes [28-30]. In some cases, inhibition constants were determined graphically using Dixon's or Henderson's methods [30]. Dissociation constants of "enzyme-inhibitor" complexes (K_i) are called inhibition constants, and are indicative of the degree of inhibition.

For the description of enzymatic processes affected by humic acids (inhibitor-1), by metal-ions (inhibitor-2), and by humic-metal complexes (inhibitor-3), as well as for calculation of theoretical values of the inhibition constants, the kinetic scheme and method were used as described in [31].

3. Results and Discussion

3.1. INFLUENCE OF HUMIC ACIDS ON THE ACTIVITY OF PROTEASES

The obtained results show that in the presence of humic acids the rate of enzymatic reaction decreases, and inhibition occurs reversibly. The latter follows from the linearity of the relationships of initial reaction rate versus inhibitor concentration (Figure 1).

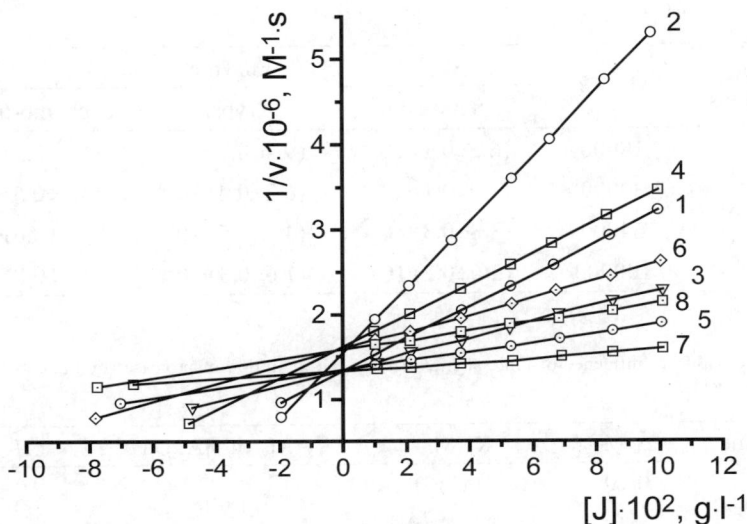

Figure 1. Dixon graph for inhibiting of trypsin activity by humic acids. Substrate – ethylether of N-acetyl-L-tyrosine, M_w of HA ~64 000 Da. Experimental conditions: pH 6.0; t = 25°C; $[E]_o = 1.98 \cdot 10^{-7}$ M; $\mu = 0.1$ M (lines 1 and 2); $\mu = 0.15$ M (lines 3 and 4); $\mu = 0.20$ M (lines 5 and 6); $\mu = 0.30$ M (lines 7 and 8); $[S]_o = 5.5 \cdot 10^{-3}$ M (lines 2, 4, 6 and 8); $[S]_o = 7.4 \cdot 10^{-3}$ M (lines 1, 3, 5 and 7) ($[E]_o$ – initial concentration of enzyme, $[S]_o$ – initial concentration of substrate in the reaction system).

Equilibrium was attained quickly in the reaction mixture (less than 1 minute). Preincubation (during 30 min) of the inhibitor both with enzyme and with substrate did not influence the value of enzymatic activity. The inhibition constants determined by graphic methods were on the order of 10^{-7}-10^{-6} M. These enzyme-inhibitor complexes may be classified as interpolymeric complexes, formed by the interaction of active centres on chemically complementary surfaces of inhibitor and protein globules. A lack of inhibition selectivity and the obtained values of the inhibition constants are indicative of the cooperative character of interaction between humic acid and the enzymes under study. The binding sites for protein catalysts seem to be located on the heteropolyfunctional surface of humic acids. This is consistent with the observed increase in inhibition along with an increase in molecular weight of humic acid (Table 1).

The interaction between protein and humic acid molecules has complicated nature. Nonetheless, the experimental data (Figure 1, Table 2) obtained at different ionic strengths for different HA samples reveal a significant role of electrostatic interactions between humic acid and enzymes.

Early it was shown that as solution's ionic strength increases, total electrostatic charge on HA macromolecules decreases [31]. Consequently, a decrease of inhibiting ability of humic acids at high ionic strengths may serve as an evidence of important role of electrostatic interactions in formation of enzyme-humic acid complexes.

Table 1. The dissociation constants of "humic acid-enzyme" complexes (inhibition constants, K_i, M).

Reaction conditions	M_w, Da	Enzyme		
		Subtilisin	Trypsin	α-chymo-trypsin
pH 7.8; μ = 0.1M; t = 25°C	18000	$(6.5\pm0.1)\cdot10^{-6}$	$(9.4\pm0.7)\cdot10^{-6}$	$(6.3\pm0.2)\cdot10^{-6}$
	30000	$(2.0\pm0.1)\cdot10^{-6}$	$(6.2\pm0.4)\cdot10^{-6}$	$(5.5\pm0.5)\cdot10^{-6}$
	64000	$(3.8\pm0.3)\cdot10^{-7}$	$(4.7\pm0.5)\cdot10^{-7}$	$(8.8\pm0.6)\cdot10^{-7}$
	100000	$(2.0\pm0.2)\cdot10^{-7}$	$(1.6\pm0.3)\cdot10^{-7}$	$(1.2\pm0.8)\cdot10^{-7}$

Table 2. Influence of ionic strength (μ) on the inhibition of proteases by humic acids.

Enzyme	μ, M	K_i, M	M_w of HA, Da	pH
Subtilisin	0.10	$(6.0\pm0.1)\cdot10^{-7}$	18000	8.0
	1.0	$(1.8\pm0.3)\cdot10^{-6}$		
Trypsin	0.10	$(2.4\pm0.4)\cdot10^{-7}$	30000	7.8
	0.15	$(6.0\pm0.1)\cdot10^{-7}$		
	0.20	$(1.2\pm0.8)\cdot10^{-6}$		
	0.3	$(1.6\pm0.4)\cdot10^{-6}$		

As it can be seen from Figure 2, the plots of experimental relationships of enzymatic reaction rate versus substrate concentration are linear in Lineweaver-Burk coordinates for all reaction systems studied. In most cases, the plots cross each other either in the upper left, or lower left quadrants, in the close proximity from the ordinate axis (Figure 2). This is intrinsic to a mixed type of inhibition. The data on inhibition are in good agreement with the calculated values of inhibition coefficients α and β obtained using kinetic scheme described by Webb [30]. The obtained values of α < 1 and β < 1 are indicative of a mixed type of inhibition. Consequently, humic acids react both with active centres and with other sections of surfaces of biological catalysts and change their activity.

A certain correlation was observed between the tendency of inhibitor molecules to associate and their ability to form a stable complex with enzymes [32]. The term "iso-association" was proposed to define the interaction of inhibitor molecules with each other, and "allo-association" – to define the interactions of inhibitor molecules with enzymes. The humic acids were able both to "isoassociation" and "alloassociation".

The data discussed confirm the previously reported results [20-25] and indicate that humic acids maybe considered as natural, non-specific inhibitors of proteolytic enzymes.

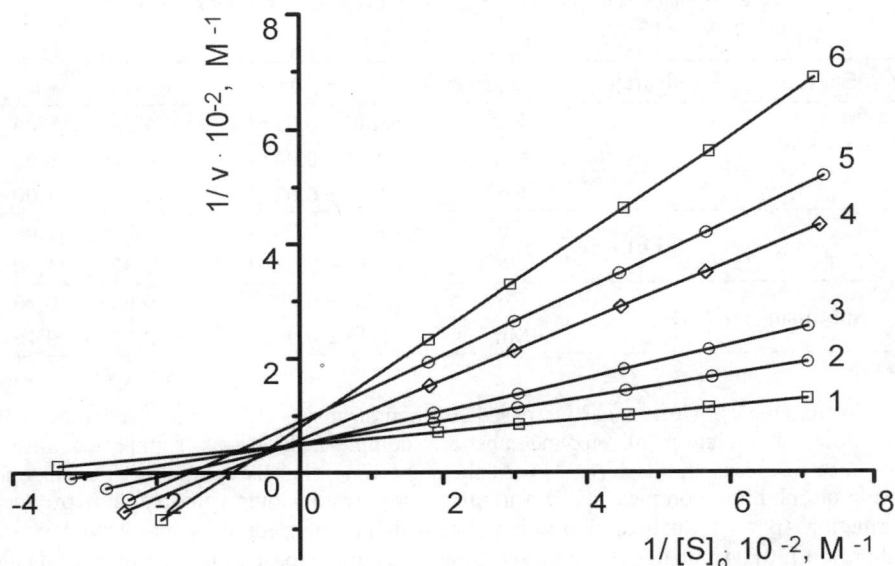

Figure 2. Lineweaver-Burk graph of humic acid influence on subtilisin activity. Substrate – ethylether of N-acetyl-L-tyrosine. Experimental conditions: $[E]_0 = 1.39 \cdot 10^{-8} M$; pH 8.0; $\mu = 0.1 M$; $t = 25°C$. Humic acid concentration: $[I] = 0 \ g \cdot l^{-1}$ (line 1); $[I] = 0.03 \ g \cdot l^{-1}$ (line 2); $[I] = 0.06 \ g \cdot l^{-1}$ (line 3); $[I] = 0.12 \ g \cdot l^{-1}$ (line 4); $[I] = 0.18 \ g \cdot l^{-1}$ (line 5); $[I] = 0.24 \ g \cdot l^{-1}$ (line 6); $M_w = 30\ 000$ Da.

3.2. METAL IONS IN ENZYMATIC PROCESSES

There are numerous literature data available on the influence of metal ions on enzymatic reactions [33-36]. It is known that metal-ions stabilize the configuration of some enzymes. Thus, Ca^{2+}, Mg^{2+} and Mn^{2+} may influence the equilibrium between native and reversibly inactivated protein [37]. High concentration of metal ions is believed to protect the enzyme from denaturing and, therefore, some bacteria can survive short treatment by a very high temperature [38]. Binding with metal ions can cause conformational changes of protein molecule as well as formation of active centre. Metal ions can catalyse redox-transformations of substrates, particularly, transition metal ions (Fe^{2+}, Cu^{2+}, Co^{2+}). Zn^{2+} may be included in this group as well [39]. However, the influence of metal ions on the catalytic activity of enzymes can be as positive, as negative. Formation of metal complexes may be considered as chemical modification of functional groups of proteins. As a result, a part of enzyme molecule can be altered and lacking activity.

The results of our investigations showed that metal ions reversibly inhibited the activity of proteolytic enzymes, such as trypsin, α-chymotrypsin and subtilisin [31, 40]. The values of inhibition constants and type of inhibition depended on the nature of the metal ions (Table 3).

Table 3. Parameters of inhibition of enzyme activity on the reactions of substrate hydrolysis.

Enzyme	Substrate	Metal ions	K_i, M	α	β
α-Chymo-trypsin	ATEE	Zn^{2+}	$(8.00 \pm 0.01)\cdot10^{-5}$	1.28	0.64
		Mn^{2+}	$(8.52 \pm 0.04)\cdot10^{-4}$	0.74	0.51
		Co^{2+}	$(3.08 \pm 0.02)\cdot10^{-5}$	1.22	1.00
Trypsin	ATEE	Cu^{2+}	$(0.84 \pm 0.06)\cdot10^{-4}$	28.4	1.00
		Co^{2+}	$(7.69 \pm 0.01)\cdot10^{-4}$	1.52	1.00
Subtilisin	BLME	Co^{2+}	$(8.20 \pm 0.09)\cdot10^{-5}$	0.65	0.16
		Mn^{2+}	$(1.75 \pm 0.04)\cdot10^{-5}$	2.69	0.15

For example, Mn^{2+} ions increased α-chymotrypsin affinity for substrate and, therefore, dissociation of enzyme-substrate complex (ES) was suppressed in the presence of these inhibitors ($\alpha < 1$). At the same time, the dissociation of the enzyme-substrate-inhibitor complex (ESI) was also suppressed along with reaction products formation ($\beta < 1$). Such an inhibition type can be referred to as non-trivial mixed inhibition [30]. An analogous analysis using calculations based on experimental data α and β coefficients of α-chymotrypsin inhibition by Zn^{2+} ions showed that this process occurred according to an incomplete mixed mechanism. Zn^{2+} ions decreased the enzyme affinity for the substrate ($\alpha > 1$), but ESI complexes dissociated along with product formation (as in the previous case) at a lower rate compared to the ES Michaelis complex ($\beta < 1$).

Inhibition of trypsin activity by Co^{2+} was characterized by the same mechanism. Partial competitive braking ($\infty > \alpha > 1$, $\beta = 1$) was observed at inhibition of α-chymotrypsin and trypsin activity by Co^{2+} ions, and of trypsin activity by Cu^{2+} ions. These inhibitors only partially hindered substrate binding, but prevented its conversion into reaction products. Incomplete mixed inhibition of subtilisin activity ($\alpha > 1$ and $\beta < 1$) was observed in the presence of Mn^{2+}-ions. Subtilisin activity was inhibited by Co^{2+} ions according to the non-trivial mixed type.

Thus, the kinetic studies of the influence of Mn^{2+}, Co^{2+}, Zn^{2+} and Cu^{2+} on the activity of proteolytic enzymes (trypsin, α-chymotrypsin, subtilisin) showed that it is formation of metal complexes with the binding sites of enzyme macromolecules that leads to reversible enzyme inactivation.

3.3. JOINT INHIBITION OF ENZYME ACTIVITY BY HUMIC ACIDS AND METAL IONS

The combined influence of humic acids and metal ions on enzyme activity was studied using a reaction of ATEE hydrolysis by α-chymotrypsin. The first set of the reactions was conducted at the constant concentration of metal (Co^{2+}) and increasing concentration of HS in the reaction system (Figure 3). Another set of the experiments was conducted for at the constant concentration of humic acid and increasing concentration of Mn^{2+} (Figure 4).

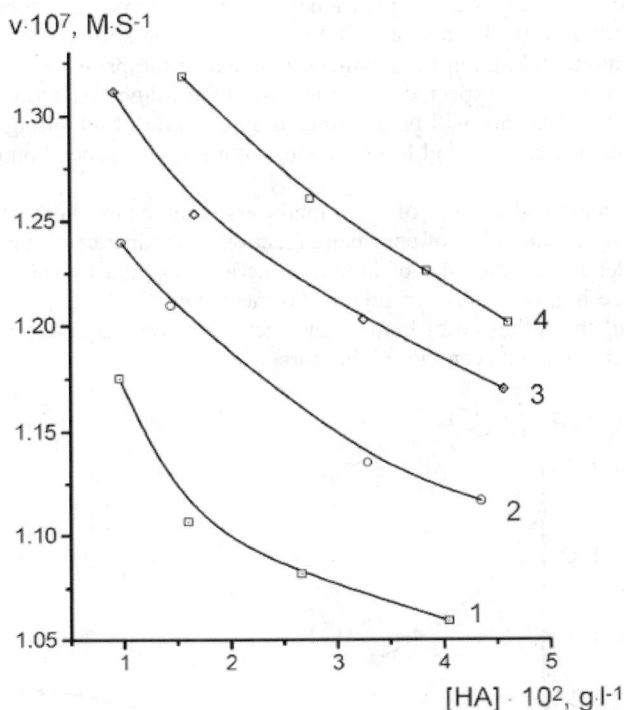

Figure 3. Dependence of ATEE hydrolysis rate by α-chymotrypsin on the concentration of humic acid ($M_w = 64\ 000$ D) in the presence of Co^{2+} ions. Experimental conditions: $t = 25°C$; $\mu = 0.1$ M; pH = 7.8. $[E]_o = 4 \cdot 10^{-9}$M; $[S]_o = 0.75 \cdot 10^{-3}$M; $[Co^{2+}]$: $0.25 \cdot 10^{-4}$ M. (line 1); $0.80 \cdot 10^{-4}$ M. (line 2); $1.25 \cdot 10^{-4}$ M. (line 3); $6.25 \cdot 10^{-4}$ M. (line 4).

As it can be seen, at the constant concentration of metal ions, the rate of enzymatic reaction monotonously decreased with increasing humic acid concentrations (Fig. 3). At the same time, at the constant concentration of humic acid, the dependence of the enzymatic reaction rate on metal concentration had more complicated character (Fig. 4) revealing a minimum at the low concentrations of metal ions in the range of $1\text{-}4 \cdot 10^{-4}$ M. At the increasing Mn^{2+} concentration, however, the summed inhibiting effect of both inhibitors (HS and Mn^{2+}) became weaker than the inhibition by the corresponding concentrations of the inhibitors in the mono-component systems.

To analyse the above system consisting of an enzyme, substrate and two inhibitors (I_1 and I_2), the assumptions were made that the inhibitors either compete for binding with the enzyme or associate with the enzyme independently. The first assumption excludes formation of the mixed complexes. The second assumption allows formation of I_1EI_2 complexes due to binding of the inhibitors to the active centres of the enzyme located far from each other. If the inhibitors react with each other ($I_1 + I_2 \rightarrow I_1I_2$), the

336

kinetics will be very complicated, in particular, when the formed I_1I_2 complexes expose inhibiting effects as well. In this case, if I_1I_2 possesses the stronger inhibiting activity than I_1 and I_2 alone, inhibition of enzymatic reaction in the presence of two inhibitors may be stronger than that expected on the basis of their individual action. The summed inhibiting effect in this case will be determined by the ratio of inhibiting activity of the I_1I_2 complex versus that of I_1 and I_2 alone. Most of all, it will depend on the stability of the I_1I_2 complex.

Thus, the combined action of two inhibitors may cause both weakening and strengthening of the inhibition of enzymatic reaction. The kinetics of the corresponding reaction will depend on the type of inhibiting activity exerted by each inhibitor [30]. Humic acids are highly reactive in relation to metal ions [31]. Hence, comprehensive consideration of the influence of humics and metal ions on enzymatic reaction should account for interaction between those inhibitors.

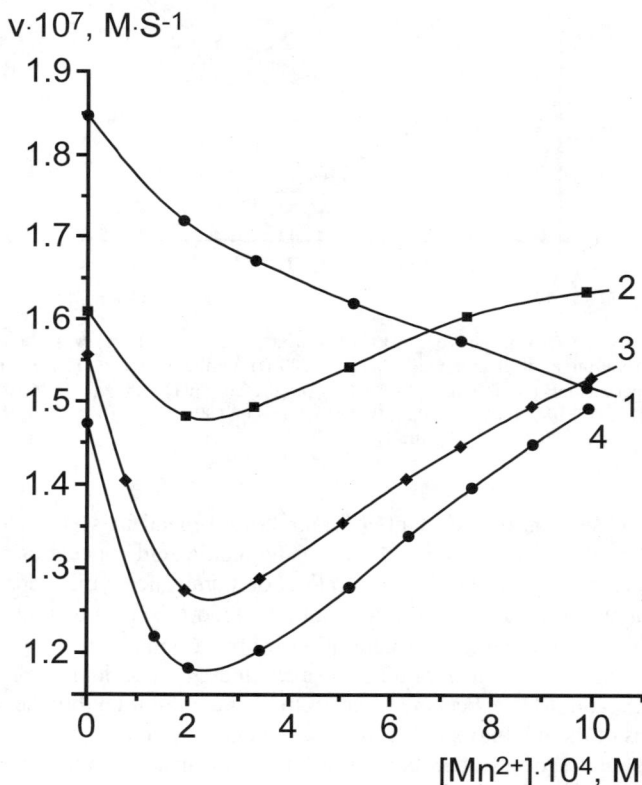

Figure 4. Dependence of ATEE hydrolysis rate by α-chymotrypsin on the concentration of Mn^{2+} ions in the presence of HA ($M_w = 18000$ Da). Experimental conditions: $t = 25°C$; $\mu = 0.1$ M; $pH = 7.8$, $[S]_o = 2.03\cdot10^{-4}$ M; $[E]_o = 1.0\cdot10^{-8}$ M; $[HA] = 0$ (line 1); $[HA] = 1.10\cdot10^{-5}$ M (line 2); $[HA] = 2.74\cdot10^{-5}$ M (line 3); $[HA] = 4.38\cdot10^{-5}$ M (line 4).

3.4. THE ROLE OF METAL IONS IN THE INTERACTION OF HUMIC ACIDS WITH ENZYMES

To evaluate the possibility of the participation of humic metal complexes in the process of enzyme activity inhibition, the equilibrium concentrations of free metal ions and free humic acids in the reaction system were calculated. The corresponding calculations were carried out on the basis of the stability constants values of humic metal complexes that were determined previously [31]. This allowed us to estimate the enzymatic reaction rate for two-component inhibition under assumption of their combined and independent action [41]. However, the obtained theoretical values did not agree with the experimentally determined reaction rates.

This motivated us to conduct the corresponding calculations under assumption that humic-metal complexes inhibit enzyme activity as well. Hence, there are three inhibitors in the system. The values of inhibition constants (K_{i3}) were calculated for humic-metal complexes formed at different concentrations of metal ions in reaction systems. The dependence of the calculated K_{i3} value on metal ion concentration for different divalent metals is shown in Figures 5 and 6 for enzymes trypsin and subtilisin, respectively.

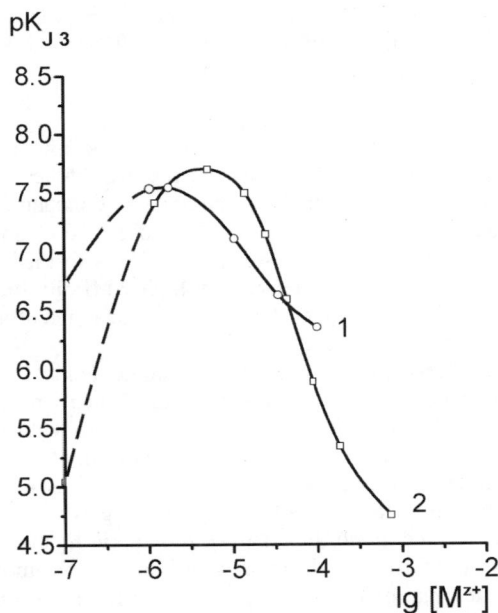

Figure 5. Dependence of pK_{i3} values on concentration of divalent metals: Cu^{2+} (line 1) and Co^{2+} (line 2) in the enzymatic reaction system. Substrate – ATEE; enzyme – trypsin; inhibitor (I_3) – humic- metal complexes. Experimental conditions: $t = 25°C$; $\mu = 0.1$ M. For Cu^{2+} experiments: $pH = 6.0$; $[E]_o = 3.09 \cdot 10^{-7}$ M; $[S]_o = 0.55 \cdot 10^{-3}$ M; $[HA] = 1.00 \cdot 10^{-6}$ M; M_w of HA ~100 000 Da. For Co^{2+} experiments: $pH = 7.8$; $[E]_o = 4.0 \cdot 10^{-9}$ M; $[S]_o = 0.75 \cdot 10^{-3}$ M; $[HA] = 3.47 \cdot 10^{-5}$ M; M_w of HA ~18 000 Da.

338

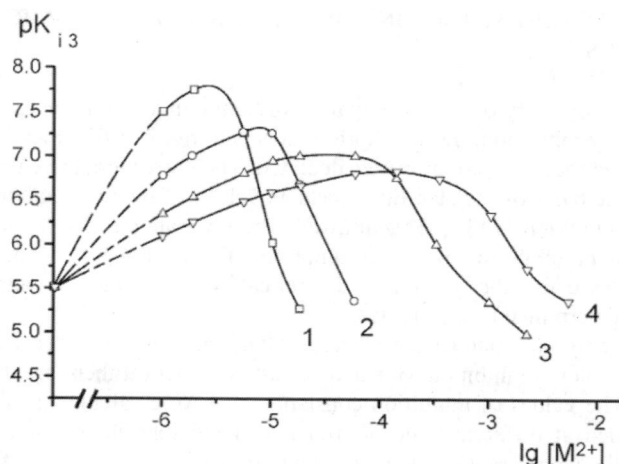

Figure 6. Dependence of pK_{i3} values on the concentration of divalent metals in the enzymatic reaction system. Metals: Ni^{2+} (line 1), Zn^{2+} (line 2), Co^{2+} (line 3) and Mn^{2+} (line 4); substrate – ATEE; enzyme – subtilisin; inhibitor (I_3) — humic-metal complexes. Experimental conditions: $t = 25°C$; $\mu = 0.1$ M; pH 7.8; $[E]_0 = 1.63 \cdot 10^{-3}$ M; $[HA] = 1.10 \cdot 10^{-5}$ M; M_w of HA ~30 000 Da; $[S]_0 = 2.5 \cdot 10^{-4}$, M.

The plots of the corresponding relationships for all the metals and both enzymes studied were characterized by a maximum: hence, the pK_{i3} values increased along with an increase in metal concentration in the low range of its concentrations and decreased with the further increase in metal concentration. The described character of the obtained plots can be interpreted in a way that the metal complexes formed in the range of low concentrations of metal ions are characterized by higher affinity to protein catalysts in comparison with HA. At larger values of divalent metal concentration, the inhibiting ability of metal complexes decreased.

Complex formation between metal ions and humic acids have some particular features while it involves both "microcoordination" and "macrocoordination" processes. "Microcoordination" is referred to the binding of metal ions with monomeric units of humic macromolecules and results in formation of the coordination centres. The number of coordination centres determines the conformational flexibility of the macromolecular ligand. "Macrocoordination" refers to polycentric binding when a high variety of the coordination centres is formed along the chain of polymeric ligand [31, 42]. Both types of complexation cause conformational changes in humic macromolecules [31, 43]. It should be noted that macromolecular metal complexes formed at low metal concentration are characterized by a small amount of the coordination centres. In this context, the inhibition efficiency of metal complexes will depend on the number of coordination centres formed by the polymeric chain of humic acids.

At low metal concentration, most of the functional groups of humic acids are ionised providing electrostatic charge of the macromolecule and, therefore, the unfolded conformations prevail. Under these conditions, enzyme-inhibitor complexes are formed at the expenses of electrostatic, hydrophobic and other types of intermolecular interactions. In addition, the formation of mixed coordination centres should be taken into account when functional groups of both humic acids and enzymes are bound with the came metal. This may strengthen the stability of enzyme humic complexes formed in the presence of metal ions.

At the metal concentration exceeding a critical value, the number of coordination centres on the macroligand chain increases, and the amount of ionised functional groups decreases. This causes folding of humic macromolecules with the formation of more densely packed macromolecular globules. As a result, the complementarity of the surfaces of humic macromolecules and enzyme is disturbed, and the formation of enzyme-humic complexes becomes hindered, that leads to the suppression of the inhibition process.

The metal ion concentration facilitating the most the formation of humic complexes depend on the nature of the metal. The metal influence on stability of enzyme-humic acid complexes decreases in the following order: $Cu^{2+} > Zn^{2+} > Co^{2+} > Mn^{2+}$.

4. Conclusions

Humic acids being macromolecular polyfunctional ligands display the ability to form interpolymeric complexes with proteolytic enzymes. This leads to inhibition of the activity of proteolytic enzymes. The inhibiting action of humic acids is sensitive to the presence of metal ions in the reaction system. Metal ions possess the ability to form coordination bonds with functional groups of protein molecule that renders significant influence on enzyme activity. The formation of humic-metal complexes also inhibits enzyme activity. Formation of enzyme-inhibitor complexes in this case may occur at the expenses of formation of mixed ligand coordination centres with participation of functional groups both of humic acids and of protein molecules. It also involves electrostatic and hydrophobic interactions of other parts of macromolecular chains.

The affinity of humic-metal complexes to an enzyme depends on the nature of the metal ion and on the amount of metal ions bound to humic acids. At small amounts of the bound metal, the electrostatic charge of humic macromolecule changes insignificantly, the unfolded conformation of macromolecules prevails which strengthens the stability of enzyme-humic complexes. A further increase in the extent of metal binding leads to denser packing of humic acids macromolecules, disturbing complementarity of surfaces of humic and protein globules. As a result, the stability of enzyme-inhibitor complexes decreases.

5. References

1. Burns, R.G. (ed) (1978) *Soil Enzymes*, Academic Press, London.
2. Skujn., J. (1976) Extracellular enzymes in soils, *Crit. Rev. Microbiol.* **6**, 383-421.
3. Frankenberg, W. and Tabatabai M.A. (1981) Amidase activity in soils. III. Stability and distribution, *Soil Sci. Soc. Am. J.* **45**, 333-338.
4. Burns R.G., Pukite, A.H. and Mc. Laren, A.D. (1972) Concerning the location and persistence of soil urease, *Soil Sci. Soc. Am. Proc.* **36**, 308-311.

340

5.   Ceccanti, B., Nannipieri, P., Cervelli, S. and Sequi, P. (1978) Fractionation of humus urease complex, *Soil Biol. Biochem.* **10**, 39-45.
6.   McLaren, A.D., Pukite, A.H. and Barshad, J. (1975) Isolation of humus with enzymatic activity from soil, *Soil Sci.* **102**, 85-93.
7.   Ladd, J.M. and Butler, J.H.A. (1975) Humus-enzyme systems and synthetic organic polymer-enzyme analogs, in E.A. Paul and A.D. McLaren. (eds.), *Soil Biochemistry,* Marcel Dekker, New York, **4**, pp. 143-194.
8.   Ruggiero, P. and Radogna, V.M. (1987) Tyrosinase activity on a humic-enzyme complex, *Sci. Total Environ.* **62**, 365-366.
9.   Mato, M.C., Olmedo, M.G. and Mendez, J. (1972) Inhibition of indoleacetic acid-oxidase by soil humic acids fractionated on Sephadex, *Soil. Biol. Biochem.* **4**, 469-473.
10.   Jahnell, J.B. and Frimmel, F.H. (1994) Comparison of the enzyme inhibition effect of different humic substances in aqueous solutions, *Chem. Eng. Process.* **33(5)**, 325-330.
11.   Malcolm, R.E. and Vaughan, D. (1979) Effects of humic acid fractions on invertase activities in plant tissues, *Soil Biol. Biochem.* **11**, 65-72.
12.   Pflug, W. and Ziechmann W. (1981) Inhibition of malate dehydrogenase by humic acids, *Soil Biol. Biochem.* **13**, 293-299.
13.   Vaughan, D. and Malcolm, R.E. (1979) Effect of soil organic matter on peroxidase activity of wheat roots, *Soil Biol. Biochem.* **11**, 57-63.
14.   Sarkar, J.M. and Bollag, J.-M (1987) Inhibitory effect of humic and fulvic acids on oxidoreductase as measured by the coupling of 2,4-dichlorophenol to humic substances, *Sci. Tot. Environ.* **62**, 367-378.
15.   Lu, F.J., Huang, T.S. and Shin, C.Y. (1994) Effects of humic acid-metal complexes on hepatic carnitine palmitoyl-transferase, carnitine ecetyltransferase and catalase activities, *Environ. Toxicol. Chem.* **13(3)**, 435-441.
16.   Jahnel, J.B., Mahlich, B. and Frimmel, F.H. (1994) Influence of humic substances on the activity of a protease, *Acta. Hydrochim, Hydrobiol.* **22(3)**, 109-116.
17.   Fujimura, V., Katayama, A. and Kuwatsuka, S. (1994) Inhibitory action of dissolved humic substances on the growth of soil bacteria degrading DDT, *Soil Sci. Plant Nutr.* **40(3)**, 525-530.
18.   Mosolov, W.W. (1971) *Proteolytic enzymes*, Nauka, Moscow.
19.   Ladd, J.N. (1972) Properties of proteolytic enzymes extracted from soil, *Soil Biol. Biochem.* **4**, 227-237.
20.   Jorobekova, Sh. (1983) Inhibition of activities of some proteases by humic acids, *2-nd Intern. Conference on Chemistry and Biotechnology of Biologicaly Active Natural Products*, Budapest, Hungary, p. 272.
21.   Jorobekova, Sh. (1995) Humic acids as inhibitors of proteases and possibility of their using in enzymatic therapy, *35-th IUPAC Congr. Istanbul, 14-19 Aug. 1995. Abstracts. Sec. 1-3*, Tyrkey, p. 64.
22.   Kazanskaya, N.F., Jorobekova, S. and Kost, O.A. (1986) Inhibition of activity of some proteases by humic acids, *Prikladnaya Biochymia i Microbiologya XXII* **4**, 480-485.
23.   Ladd, J.N. and Butler, J.H.A. (1969) Inhibitory effect of soil humic acids on the proteolytic enzyme pronase, *Aust. J. Soil Res.* **7**, 241-251.
24.   Ladd, J.N. and Butler, J.H.A. (1971) Inhibition by soil humic acids of native and acetylated proteolytic enzymes, *Soil Biol. Biochem.* **3**, 157-160.
25.   Butler, J.H.A. and Ladd, J.N. (1971) Importance of the molecular weight of humic and fulvic acids in determining their effects on protease activity, *Soil Biol. Biochem.* **3**, 249-257.
26.   Ladd, J.H. and Butler, J.H.A. (1969) Inhibition and stimulation of proteolytic enzyme activities by soil humic acids, *Austr. J. Soil Res.* **7**, 253-261.
27.   Bender, M.Z., Beque-Canton, M.L., Blakely, R.L., Brubacher, L.J., Feder, J., Gunter, C.R., Kezdy, F.J., Kilheffer, J.V., Marshall, T.H., Miller, C.G., Roeske, R.W. and Stoops, J.K. (1966) The determination of the concentration of hydrolytic enzyme in solutions: α-chymotrypsin, trypsin, papain, elastase, subtilisin and acetylcholinesterase, *J. Am. Chem. Soc.* **88, 24**, 5890-5913.
28.   Keleti, T. (1985) *Basic enzyme kinetics*, Akademiai Kiado, Budapest.
29.   Endrenyl, L. (ed.) (1981) *Kinetic Data Analysis*, Plenum Press, New York, London.
30.   Webb, L. (1963) Enzyme and Metabolic Inhibitors. *Academic Press,* New York and London.
31.   Jorobekoba, S. (1987) *Macroligand Properties of Humic Acids*, Ilim. Frunse, Kyrgyzstan.
32.   Mosolov, W.W. (1983) *Protein inhibitors as regulators of proteolyse process*, Bach readings XXXVI, Nauka, Moscow.

33. William, D.R (ed) (1976) *An Introduction to Bioinorganic Chemistry*, Ch. C. Thomas Publisher, Springfield USA.
34. Hill, H.A.O.(Ed) (1980-1982) *Inorganic Biochemistry*, Royal Soc. Chem. Burlington House, London.
35. Green, N.M. and Neurath, H. (1953) The effects of divalent cations on trypsin, *Biol. Chem.* **204**, 279-390.
36. Ladd, J.N. and Butter, J.H.A. (1970) The effect of inorganic cations on the inhibition and stimulation of protease activity by soil humic acids, *Soil. Biol. Biochem.* **2**, 33-40.
37. Hague, D.H. and Eigen, M. (1966) Role of metals in enzymatic reaction, *Trans. Far. Soc.* **62 (521)**, 1236-1248.
38. Gessly, G. and Jang, L. (1989) Interactions between metal ions and capsular polymers. In Metal Ions and Bacteria, *Academic Press, New York* **11**, 325-357.
39. Hughes, M.N. (1983) *The inorganic chemistry of biological processes*, John Willey and Sons Chichester, New York, Brisbane, Toronto.
40. Kydralieva, K.A. and Jorobekova, S. (2000) *Metal ions in enzyme-inhibitory systems*, Ilim. Bishkek, Kyrgyzstan.
41. Jorobekova, S. and Karabaev, S.O. (1986) Influence of Co^{2+} ions on the inhibition process of proteases activities by humic acids, in Mazalov L.N. (ed.), *Problems of Modern Bioinorganic Chemistry*, Nauka, Novosibirsk, 96-103.
42. Jorobekova, S., Kydralieva, K. and Khudaybergenova, E. (2001) Complex Formation Between Humic Acids and Metal Ions, in *9-th Intern. Symposium. On Macromolecule-Metal Complexes,* Polytechn. University, Brooklyn, New York, p. 42.
43. Jorobekova, S. and Maltceva G.M. (1987) About conformational changeability of humic acids, *Chim. tverdogo topliva* **3**, 34-37.
44. Lewis, J. and Wilkins, R.G. (eds) (1960) *Modern Coordination Chemistry Principles and Methods*, Internsciense Publishers Inc., New York, London.

Chapter 17

MICROBIAL REDOX REACTIONS MEDIATED BY HUMUS AND STRUCTURALLY RELATED QUINONES

J.A. FIELD[1], F.J. CERVANTES[2]

[1] *Department Chemical and Environmental Engineering, University of Arizona, P.O. Box 210011, Tucson, Arizona 85721-0011, USA <jimfield@email.arizona.edu>*
[2] *Departamento de Ciencias del Agua y del Medio Ambiente, Instituto Tecnologico de Sonora, 5 de Febrero 818 Sur, 85000 Cd. Obregon, Sonora, Mexico*

Abstract

Humic substances are natural organic compounds recalcitrant to biodegradation. Their structure is rich in quinone moieties that provides redox activity of humics. The evidence provided in this review indicates that quinonoid substructures of humics can play a number of significant roles in the anaerobic (bio)transformation of a wide variety of organic and inorganic compounds, including priority pollutants. Quinones can support the anaerobic microbial oxidation of different organic compounds by serving as terminal electron acceptors. Quinones, once being reduced, can also channel electrons to several distinct contaminants by acting as redox mediators. Reduced quinones can also serve as electron-donating substrates to bacteria supporting their growth.

1. Introduction

Humic substances have generally been regarded as a relatively non-biodegradable material, not readily subject to biotransformation by microorganisms. These highly condensed aromatic structures rich in quinone moieties are very recalcitrant to biodegradation as is evident from the mean residence time of humus in soil, which can vary from 250 to 1900 years [35]. The accumulation of natural organic matter (NOM) in anaerobic environments over geological time scales is the best testimony to the recalcitrance of humics in environments lacking elemental oxygen. There are also a few studies, which have documented the poor biodegradability of humus by anaerobic microbial communities. For example, peat samples were shown not to be degraded by anaerobic sludge, which was previously adapted to the degradation of lignocellulosic materials [29]. Likewise, a mixed bacterial community of heterotrophic bacteria was unable to degrade various samples of humic acid, synthetic polyphenols and melanoids under nitrate reducing conditions [42].

I. V. Perminova et al. (eds.),
Use of Humic Substances to Remediate Polluted Environments: From Theory to Practice, 343–352.

The evidence is compelling that humics are a poor source of carbon for microorganisms. Nonetheless, the quinoid substructures of humics have recently been shown to be susceptible to redox reactions catalysed by various microorganisms. New evidence is accumulating, indicating that humic substances can have multiple roles in the degradation and biotransformation of pollutants under anaerobic conditions. Humics and quinone substructure analogues have been shown to have three distinct roles as electron carriers to support biological biotransformation of priority pollutants:

1. electron acceptors for respiration;
2. redox mediators for reduction processes;
3. electron donors to microorganisms.

This paper reviews these roles of humics in the biodegradation and biotransformation of priority pollutants.

2. Humics as Electron Acceptor for Anaerobic Respiration

In 1996, Lovley and co-workers [25] reported the that the iron-reducing bacterium, *Geobacter metallireducens* was able to utilize purified humic substances as terminal electron acceptors supporting the anaerobic oxidation of acetate to CO_2. Similar results were also obtained with the humic model compound, anthraquinone-2,6-disulfonate (AQDS). During the process, cell growth was linked to the use of humic substances as electron acceptors. The reduced form of the quinone, anthrahydroquinone-2,6-disulfonate (AH_2QDS), accumulated in stoichiometric yields of 4 mol AH_2QDS per mol acetate oxidized. The results suggest that quinone substructures are the main moieties in humics responsible for the electron accepting properties. In further support of this hypothesis, experiments were conduced correlating humic substance quinone content and with their electron accepting capacity. Electron spin resonance (ESR), due to the unpaired electron in the semiquinone radicals, was used as an indicator of quinone content. This parameter was highly correlated with the microbial electron-accepting capacity of a wide variety of humic substances collected from sediments, soil and aquatic environments [33].

Since the initial observation, a large number of microorganisms capable of reducing quinones and humics have been identified (Table 1). These organisms come from diverse phylogenetic groups, ranging from *Archaea* to gram-negative and gram-positive bacteria. Taken as a whole, the present-day results suggest a ubiquitous capacity among microorganisms to reduce humics. Some of the studies demonstrate that microorganisms benefit from humic substance reduction with cell growth linked to the process. Table 1 indicates 11 species of bacteria from 6 genera for which cell growth was demonstrated to result from humic substance reduction. These bacteria also come from diverse phylogenetic groups, including iron-reducing [12, 25] and halorespiring bacteria [6]. To date, no *Archaea* has been found to benefit with cell growth from the reduction of humics or quinones.

The results shown in Table 1 refer to experiments utilizing simple substrates such as acetate, hydrogen and lactate. However, quinone-reducing microorganisms have also been shown to mineralise various organic pollutants to CO_2 under anoxic conditions at the expense of humics or AQDS reduction. Table 2 lists some priority pollutants, which have been shown to be degraded by different consortia with humic substances as the terminal electron acceptor.

3. Recycling of Reduced Humics

The electron accepting capacity of humics in soil and sediments, would quickly become exhausted if there was no mechanism of recycling reduced humic substances. However, reduced humics is chemically reactive and readily becomes oxidized by metal oxides of Fe(III) or Mn(IV) in soil. Direct reduction of Fe(III) by reduced humics was demonstrated in cell free systems [25]. Likewise it has been shown that Mn(IV) readily oxidizes hydroquinone to 1,4-benzoquinone [36]. In effect, low concentrations of humic substances can link quinone-reducing activity of microorganisms to dissimilatory metal reduction [6, 19, 23, 25].

Microorganims that are not typically metal-reducing organisms can indirectly utilize metals as electron acceptors via the reduction of humic substances. Non-iron-reducing microorganisms, such as the halorespiring bacterium, *Desulfitobacterium dehalogenans*, was shown to become an effective iron-reducer in the presence of sub-stoichiometric amounts of AQDS [6] by the mechanism shown in Figure 1.

4. Humics as a Redox Mediator for the Reduction of Priority Pollutants

Early on, NOM and quinone models were shown to shuttle electrons from sulfides, ferrous iron, and pyrite accelerating the abiotic reduction of nitroaromatic and organohalogens (Table 3). In many of the examples, the presence of the humic substances increased reduction rates by at least one-order of magnitude. The direct reduction of carbon tetrachloride (CT) and methyl parathion by chemically prepared hydroquinones was also demonstrated [14, 38].

Evidence is also emerging that quinones and humics can also play an important role in transferring electrons from microorganisms to support the reductive biotransformation of various priority pollutants. The initial observations were made with the reduction of azo dyes to colourless aromatic amines by aerobic bacteria incubated under anoxic conditions. Anthraquinone-2-sulfonate (AQS) was shown to greatly accelerate the reduction of the dyes [22, 24] and it was suggested that respiratory quinone-reductases might be implicated in the process. The reaction rate enhancements were shown to increase with quinones having lower redox potentials [31]. The application of redox mediators at sub-stoichiometric levels greatly accelerated the removal of azo dyes in continuous bioreactors [9, 39]. Molar ratios of AQDS/dye as low as 0.01 were shown to significantly improve dye removal rates.

The biotransformation of several organohalogens are also affected. Soil humics were shown to accelerate the reduction of CT to chloroform [13]. AQDS and humics were shown to greatly enhance the conversion of CT to inorganic chloride by anaerobic sludge (Figure 2). Also the reductive dechlorination of polychlorinated dibenzo-*p*-dioxins was shown to be improved in the presence of humic acids and even more so in the presence of 3,4-dihydroxybenzoate [21].

The reduction of soluble radionuclides to less soluble species could also be enhanced by humic substances. *Deinococcus radiodurans* was shown to reduce soluble U(VI) to insoluble U(IV) or soluble Tc(VII) to insoluble Tc(IV) in the presence of AQDS [19]. Biogenic AH_2QDS was also shown to directly reduce U(VI) in the absence of bacterial cells [20]. However, since AH_2QDS is more reactive with iron oxides than U(VI), the reaction can only occur if iron oxides are absent [16, 20].

Table 1. Phylogenetic diversity of quinone- or humics-reducing microorganisms reported in the literature.

Phylogeny	Strain	E-Acceptor*	Growth*	Reference
Archaea				
Methanococcales	*Methanococcus thermolithitrophicus*	AQDS	ND	[28]
	Methanococcus voltaei	AQDS	–	[2]
Methanobacteriales	*Methanobacterium thermoautotrophicum*	AQDS	ND	[28]
	Methanobacterium palustre	AQDS & HA	–	[2]
Methanosarcinales	*Methanosarcina barkeri*	AQDS & HA	–	[2]
	Methanolobus vulcani	AQDS	–	[2]
	Methanosphaera cuniculi	AQDS & HA	–	[2]
Methanomicrobiales	*Methanospirillum hungatei*	AQDS	–	[6]
Methanopyrales	*Methanopyrus kandleri*	AQDS	ND	[28]
Thermoproteales	*Pyrobaculum islandicum*	AQDS & HA	ND	[28]
	Pyrodictium abyssi	AQDS	ND	[28]
	Thermococcus celer	AQDS	ND	[28]
Thermococcales	*Pyrococcus furiosus*	AQDS	ND	[28]
Archaeoglobales	*Archaeoglobus fulgidus*	AQDS	ND	[28]
Bacteria				
γ-Proteobacteria	*Pantoea agglomerans*	AQDS	+	[18]
	Shewanella alga	AQDS & HA	+	[25]
	Shewanella putrefaciens	AQDS	ND	[26]
	Shewanella sacchrophila	AQDS	ND	[26]
	Aeromonas hydrophila	AQDS	ND	[26]
	Geospirillum barnseii	AQDS	ND	[26]
	Wolinella succinogenes	AQDS & HA	ND	[26]
	Escherichia coli K12	AQS & L	ND	[31]

*AQDS, anthraquinone-2,6-disulfonate; AQS, anthraquinone-2-sulfonate; ACNQ, 2-amino-3-carboxy-1,4-naphthoquinone; NQ, 1,4-naphthoquinone; DMBQ, 2,6-dimethyl-1,4-benzoquinone; L, lawsone (2-hydroxy-1,4-naphthoquinon); HA, humic acids; ND, not determined.

Table 1. Phylogenetic diversity of quinone- or humics-reducing microorganisms reported in the literature (continue).

Phylogeny	Strain	E-Acceptor*	Growth*	Reference
α-Proteobacteria	*Sphingomonas xenophaga BN6*	AQS	ND	[31]
β-Proteobacteria	*Ralstonia eutropha335*	AQS	ND	[31]
δ-Proteobacteria	*Geobacter metallireducens*	AQDS & HA	+	[25]
	Geobacter sulfurreducens	AQDS & HA	ND	[26]
	Geobacter humireducens	AQDS & HA	ND	[26]
	Geobacter JW-3	AQDS & HA	+	[12]
	Geobacter TC-4	AQDS & HA	+	[12]
	Geobacter grbiciae	AQDS	ND	[10]
	Desulfovibrio G11	AQDS	–	[6]
	Desulfovibrio vulgaris	DMBQ, NQ,	ND	[37]
	Desulfuromonas acetexigens	AQDS	ND	[26]
	Desulfuromonas SDB-1	AQDS	+	[12]
	Desulfuromonas FD-1	AQDS	+	[12]
Deinococci	*Deinococcus radiodurans*	AQDS	–	[19]
Thermotogales	*Thermotoga maritima*	AQDS	ND	[28]
Gram positives	*Thermoanaerobacter siderophilus*	AQDS	+	[34]
	Bacillus subtilis	AQS, AQDS & L	ND	[31]
	Propionibacterium freudenreichii	HA	–	[1]
	Enterococcus cecorum	HA	–	[1]
	Lactococcus lactis	HA	–	[1]
	Lactococcus lactis	ACNQ	ND	[41]
	Lactobacillus plantarum	ACNQ	ND	[41]
	Desulfitobacterium dehalogenans	AQDS & HA	+	[6]
	Desulfitobacterium PCE1	AQDS	+	[6]

*AQDS, anthraquinone-2,6-disulfonate; AQS, anthraquinone-2-sulfonate; ACNQ, 2-amino-3-carboxy-1,4-naphthoquinone; NQ, 1,4-naphthoquinone; DMBQ, 2,6-dimethyl-1,4-benzoquinone; L, lawsone (2-hydroxy-1,4-naphthoquinon); HA, humic acids; ND, not determined.

Table 2. Organic Priority Pollutants Mineralized under Anaerobic Conditions with Humic Substances Serving as the Terminal Electron Acceptor.

Pollutant	Electron Acceptor	Microbial Consortia	Reference
cis-dichloroethene	AQDS	organic rich stream sediments	[3]
vinyl chloride	AQDS & HA	organic rich stream sediments	[3]
toluene	AQDS & HA	enrichment culture from sediments	[7]
p-cresol	AQDS	anaerobic sludge	[8]
phenol	AQDS	anaerobic sludge	[8]
MTBE*	AQDS & HA	aquifer sediments	[17]

* MTBE, methyl tert-butyl ether.

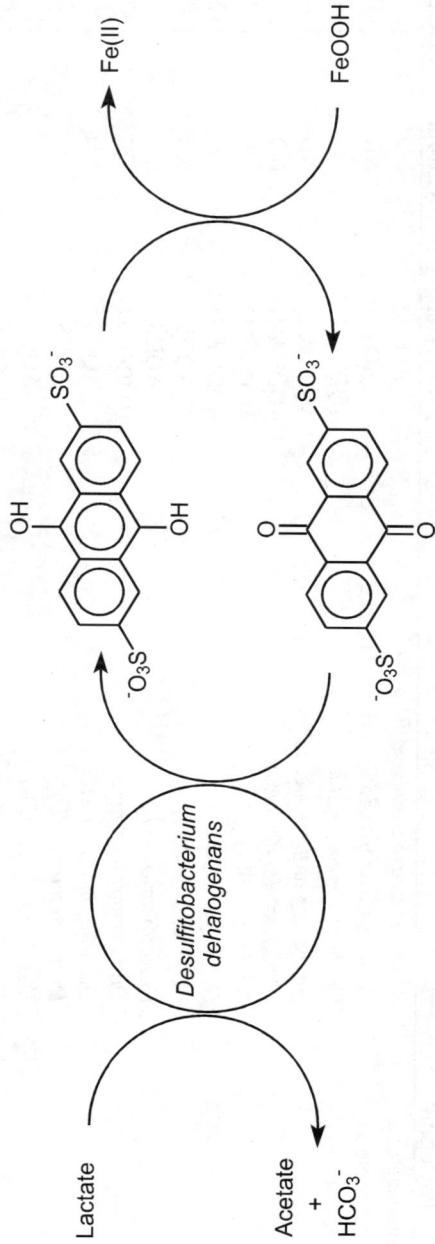

Figure 1. Quinone reduction by *D. dehalogenans* linked to iron oxide reduction [6].

Table 3. Humics and quinones as redox mediators for the abiotic reduction of priority pollutants by Fe(ii) and reduced sulphur.

Electron donor	Redox Mediator*	Reductive Reaction	Reference
		Organohalogens	
Fe(II) or HS⁻	HA	$C_2Cl_6 \rightarrow C_2Cl_4$	[14]
Fe(II) or HS⁻	HA	$CCl_4 \rightarrow CHCl_3$	[14]
Fe(II) or HS⁻	HA	$CHBr_3 \rightarrow$ unidentified	[14]
HS⁻ or S⁰	J	$C_2Cl_6 \rightarrow C_2Cl_4$	[30]
Azo Dyes			
HS⁻	AQDS	Acid orange 7 $\rightarrow$ aromatic amines	[40]
		Nitroaromatics	
HS⁻	J & L	various nitrobenzenes $\rightarrow$ anilines	[32]
HS⁻	NOM	various nitrobenzenes $\rightarrow$ anilines	[15]

* HA, humic acid; NOM, natural organic matter; AQDS, anthraquinone-2,6-disulfonate; J, juglone (5-hydroxy-1,4-naphthoquinone); L, lawsone (2-hydroxy-1,4-naphthoquinone).

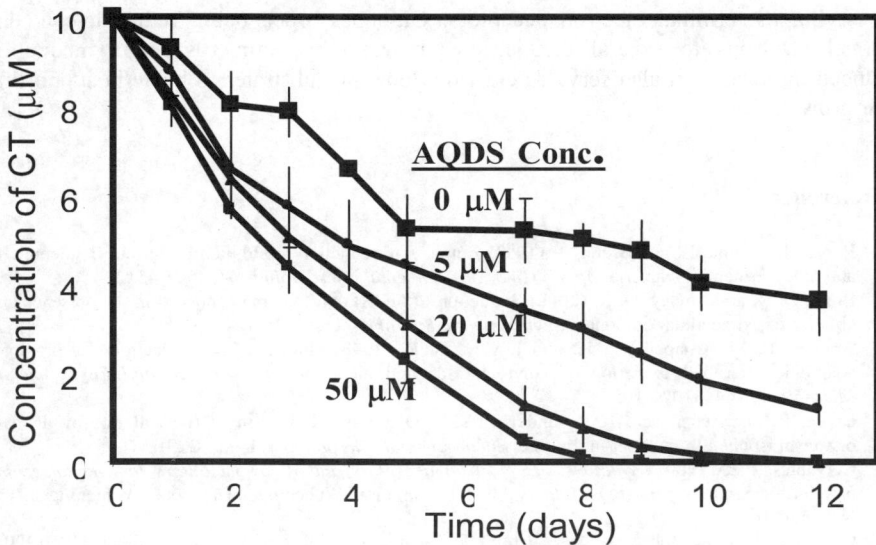

Figure 2. Effect of AQDS concentrations on the degradation of 100 μM CT by anaerobic granular sludge with 1 g acetate COD/l [5].

5. Humics as an Electron Donor to Microorganisms

Several studies have also shown that reduced humics or model hydroquinones are good electron donors to microorganisms, providing energy for metabolism. Several bacteria (*G. metallireducens, Geothrix fermentans* and *Wolinella succinogens*) were found that could link the anaerobic oxidation of AH_2QDS to the dissimilatory reduction of nitrate to ammonium (DRNA) [27]. Another bacterium, *Paracoccus denitrificans* was tested and found to link AH_2QDS oxidation or the oxidation of reduced humic acids to denitrification ($NO_3^- \rightarrow N_2$) [27]. A diversity of microorganisms were found in different sediments that were capable of coupling the oxidation of reduced humic substances to denitrification [11].

In addition to nitrate, *W. succinogens*, supported the oxidation of AH_2QDS with arsenate or selenate as electron acceptors [27]. *W. succinogens* and several other bacteria also linked AH_2QDS oxidation to the reduction of fumarate to succinate [27]. Finally, an unidentified bacterium, strain CKB isolated from paper mill waste, was shown to reduce chlorate (ClO_3^-) at the expense of AH_2QDS reduction [4].

6. Conclusions

The evidence provided in this review indicates that quinones, redox active groups in humics, can play a number of significant roles in the anaerobic (bio)transformation of a wide variety of organic and inorganic compounds, including priority pollutants. Quinones can support the anaerobic microbial oxidation of different organic compounds by serving as terminal electron acceptors. Quinones, once being reduced, can also channel electrons to several distinct contaminants by acting as redox mediators. Reduced quinones can also serve as electron-donating substrates to bacteria supporting their growth.

7. References

1. Benz, M., Schink, B. and Brune, A. (1998) Humic acid reduction by Propionibacterium freudenreichii and other fermenting bacteria, *Applied and Environmental Microbiology* **64**, 4507-4512.
2. Bond, D. R. and Lovley, D. R. (2002) Reduction of Fe(III) oxide by methanogens in the presence and absence of extracellular quinones, *Environmental Microbiology* **4**, 115-124.
3. Bradley, P. M., Chapelle, F. H. and Lovley, D. R. (1998) Humic acids as electron acceptors for anaerobic microbial oxidation of vinyl chloride and dichloroethene, *Applied and Environmental Microbiology* **64**, 3102-3105.
4. Bruce, R. A., Achenbach, L. A. and Coates, J. D. (1999) Reduction of (per)chlorate by a novel organism isolated from paper mill waste, *Environmental Microbiology* **1**, 319-329.
5. Cervantes, F. J. (2002) *Quinones as electron acceptors and redox mediators for the anaerobic biotransformation of priority pollutants*, Ph.D. Dissertation, Wageningen University, Wageningen, The Netherlands.
6. Cervantes, F. J., de Bok, F. A. M., Tuan, D. D., Stams, A. J. M., Lettinga, G. and Field, J. A. (2002) Reduction of humic substances by halorespiring, sulphate- reducing and methanogenic microorganisms, *Environmental Microbiology* **4**, 51-57.

7. Cervantes, F. J., Dijksma, W., Duong-Dac, T., Ivanova, A., Lettinga, G. and Field, J. A. (2001). Anaerobic mineralization of toluene by enriched sediments with quinones and humus as terminal electron acceptors, *Appl. Environ. Microbiol.* **67**, 4471-4478.

8. Cervantes, F. J., van der Velde, S., Lettinga, G. and Field, J. A. (2000) Quinones as terminal electron acceptors for anaerobic microbial oxidation of phenolic compounds, *Biodegradation* **11**, 313-321.

9. Cervantes, F. J., Van der Zee, F. P., Lettinga, G. and Field, J. A. (2001) Enhanced decolourisation of acid orange 7 in a continuous UASB reactor with quinones as redox mediators, *Wat. Sci. Technol.* **44**, 123-128.

10. Coates, J. D., Bhupathiraju, V. K., Achenbach, L. A., McInerney, M. J. and Lovley, D. R. (2001) Geobacter hydrogenophilus, Geobacter chapellei and Geobacter grbiciae, three new, strictly anaerobic, dissimilatory Fe(III)-reducers, *International Journal of Systematic and Evolutionary Microbiology* **51**, 581-588.

11. Coates, J. D., Cole, K. A., Chakraborty, R., O'Connor, S. M. and Achenbach, L. A. (2002) Diversity and ubiquity of bacteria capable of utilizing humic substances as electron donors for anaerobic respiration, *Applied and Environmental Microbiology* **68**, 2445-2452.

12. Coates, J. D., Ellis, D. J., Blunt-Harris, E. L., Gaw, C. V., Roden, E. E., and Lovley, D. R. (1998) Recovery of humic-reducing bacteria from a diversity of environments, *Applied and Environmental Microbiology* **64**, 1504-1509.

13. Collins, R. and Picardal, F. (1999) Enhanced anaerobic transformations of carbon tetrachloride by soil organic matter, *Environmental Toxicology and Chemistry* **18**, 2703-2710.

14. Curtis, G. P. and Reinhard, M. (1994) Reductive dehalogenation of hexachlorethane, carbon-tetrachloride, and bromoform by anthrahydroquinonedisulfonate and humic-acid, *Environmental Science & Technology* **28**, 2393-2401.

15. Dunnivant, F. M., Schwarzenbach, R. P. and Macalady, D. L. (1992) Reduction of Substituted Nitrobenzenes in Aqueous-Solutions Containing Natural Organic-Matter, *Environmental Science & Technology* **26**, 2133-2141.

16. Finneran, K. T., Anderson, R. T., Nevin, K. P. and Lovley, D. R. (2002) Potential for Bioremediation of uranium-contaminated aquifers with microbial U(VI) reduction, *Soil & Sediment Contamination* **11**, 339-357.

17. Finneran, K. T. and Lovley D. R. (2001) Anaerobic degradation of methyl tert-butyl ether (MTBE) and tert-butyl alcohol (TBA), *Environmental Science & Technology* **35**, 1785-1790.

18. Francis, C. A., Obraztsova, A. Y. and Tebo, B. M. (2000) Dissimilatory metal reduction by the facultative anaerobe Pantoea agglomerans SP1, *Applied and Environmental Microbiology* **66**, 543-548.

19. Fredrickson, J. K., Kostandarithes, H. M., Li, S. W., Plymale, A. E. and Daly, M. J. (2000) Reduction of Fe(III), Cr(VI), U(VI), and Tc(VII) by Deinococcus radiodurans R1, *Applied and Environmental Microbiology* **66**, 2006-2011.

20. Fredrickson, J. K., Zachara, J. M., Kennedy, D. W., Duff, M. C., Gorby, Y. A., Li, S. M. W. and Krupka, K. M. (2000) Reduction of U(VI) in goethite (alpha-FeOOH) suspensions by a dissimilatory metal-reducing bacterium, *Geochimica Et Cosmochimica Acta* **64**, 3085-3098.

21. Fu, Q. S., Barkovskii, A. L. and Adriaens, P. (1999) Reductive transformation of dioxins: An assessment of the contribution of dissolved organic matter to dechlorination reactions, *Environmental Science & Technology* **33**, 3837-3842.

22. Keck, A., Klein, J., Kudlich, M., Stolz, A., Knackmuss, H. J. and Mattes, R. (1997) Reduction of azo dyes by redox mediators originating in the naphthalenesulfonic acid degradation pathway of Sphingomonas sp. strain BN6, *Applied and Environmental Microbiology* **63**, 3684-3690.

23. Kieft, T. L., Fredrickson, J. K., Onstott, T. C., Gorby, Y. A., Kostandarithes, H. M., Bailey, T. J., Kennedy, D. W., Li, S. W., Plymale, A. E., Spadoni, C. M. and Gray, M. S. (1999) Dissimilatory reduction of Fe(III) and other electron acceptors by a Thermus isolate, *Applied and Environmental Microbiology* **65**, 1214-1221.

24. Kudlich, M., Keck, A., Klein, J. and Stolz, A. (1997) Localization of the enzyme system involves in anaerobic reduction of azo dyes by Sphingomonas sp. strain BN6 and effect of artificial redox mediators on the rate of azo dye reduction, *Appl. Environ. Microbiol.* **63**, 3691-3694.

25. Lovley, D. R., Coates, J. D., BluntHarris, E. L., Phillips, E. J. P. and Woodward, J. C. (1996) Humic substances as electron acceptors for microbial respiration, Nature **382**, 445-448.

352

26. Lovley, D. R., Fraga, J. L., Blunt-Harris, E. L., Hayes, L. A., Phillips, E. J. P. and Coates, J. D. (1998) Humic substances as a mediator for microbially catalyzed metal reduction, *Acta Hydrochimica Et Hydrobiologica* **26**, 152-157.

27. Lovley, D. R., Fraga, J. L., Coates, J. D. and Blunt-Harris, E. L. (1999) Humics as an electron donor for anaerobic respiration, *Environmental Microbiology* **1**, 89-98.

28. Lovley, D. R., Kashefi, K., Vargas, M., Tor, J. M. and Blunt-Harris, E. L. (2000) Reduction of humic substances and Fe(III) by hyperthermophilic microorganisms, *Chemical Geology* **169**, 289-298.

29. Owen, W. F., Stuckey, D. C., H.-J. J. B., Young, L. Y. and McCarty, P. L. (1979) Bioassays for monitoring biochemical methane potential and anaerobic toxicity, *Wat. Res.* **13**, 485–492.

30. Perlinger, J. A., Angst, W. and Schwarzenbach, R. P. (1996) Kinetics of the reduction of hexachloroethane by juglone in solutions containing hydrogen sulfide, *Environ. Sci. Technol.* **30**, 3408-3417.

31. Rau, J., Knackmuss, H. J. and Stolz, A. (2002) Effects of different quinoid redox mediators on the anaerobic reduction of azo dyes by bacteria, *Env. Sci. Technol.* **36**, 1497-1502.

32. Schwarzenbach, R. P., Stierli, R., Lanz, K. and Zeyer, J. (1990) Quinone and iron porphyrin mediated reduction of nitroaromatic compounds in homogeneous aqueous-solution, *Environmental Science & Technology* **24**, 1566-1574.

33. Scott, D. T., McKnight, D. M., Blunt-Harris, E. L., Kolesar, S. E. and Lovley, D. R. (1998) Quinone moieties act as electron acceptors in the reduction of humic substances by humics-reducing microorganisms, *Environmental Science & Technology* **32**, 2984-2989.

34. Slobodkin, A. I., Tourova, T. P., Kuznetsov, B. B., Kostrikina, N. A., Chernyh, N. A. and Bonch-Osmolovskaya, E. A. (1999) Thermoanaerobacter siderophilus sp nov., a novel dissimilatory Fe(III)-reducing, anaerobic, thermophilic bacterium, *International Journal of Systematic Bacteriology* **49**, 1471-1478.

35. Stevenson, F. J. (1994) *Humus Chemistry: Genesis, Composition, Reactions*, John Wiley and Sons, Inc., New York.

36. Stone, A. and Morgan, J. (1984) Reduction and dissolution of manganese(III) and manganese(IV) oxides by organics: 2. Survey of the reactivity of organics, *Environ. Sci. Technol.* **18**, 617-624.

37. Tatsumi, H., Kano, K. and Ikeda, T. (2000) Kinetic analysis of fast hydrogenase reaction of Desulfovibrio vulgaris cells in the presence of exogenous electron acceptors, *J. Phys. Chem.* **B 104**, 12079-12083.

38. Tratnyek, P. G. and Macalady, D. L. (1989) Abiotic reduction of nitro aromatic pesticides in anaerobic laboratory systems, *Journal of Agricultural and Food Chemistry* **37**, 248-254.

39. van der Zee, F. P., Bouwman, R. H. M., Strik, D. P. B. T. B., Lettinga, G. and Field, J. A. (2001) Application of redox mediators to accelerate the transformation of reactive azo dyes in anaerobic bioreactors, *Biotechnology and Bioengineering* **75**, 691-701.

40. van der Zee, F. P., Lettinga, G. and Field, J. A. (2000) The role of (auto)catalysis in the mechanism of an anaerobic azo reduction, *Water Science and Technology* **42**, 301-308.

41. Yamazaki, S., Kaneko, T., Taketomo, N., Kano, K. and Ikeda, T. (2002) Glucose metabolism of lactic acid bacteria changed by quinone- mediated extracellular electron transfer, *Bioscience Biotechnology and Biochemistry* **66**, 2100-2106.

42. Yanze, K. C. and Blondeau, R. (1990) Effect of heterotrophic bacteria on different humic substances in mixed batch cultures, *Can. J. Soil Sci.* **70**, 51–60.

Chapter 18

ENHANCED HUMIFICATION OF TNT BIOREMEDIATION OF CONTAMINATED
SOIL AND WATER IN PRACTICE

Bioremediation of Contaminated Soil and Water in Practice

H. THOMAS, A. GERTH
WASAG DECON GmbH, Deutscher Platz 5, 04103 Leipzig, Germany
<decon@wasag.de>

Abstract

The explosive 2,4,6-trinitrotoluene is a substance of environmental concern because of
its toxicity. In reducing environments TNT is reduced to amino-metabolites that are
immobilized into soil organic matter. The humification of TNT was tested with good
results for the cleanup of contaminated soil and water. Molasses proved to be a suitable
source of organic carbon for the biological reduction.

1. Introduction

2,4,6-trinitrotoluene (TNT) was the most widely used explosive during World Wars I
and II. During World War II, Germany produced approximately 800,000 tons of TNT
[1]. Environmental contamination with TNT is found at former TNT-producing plants,
ammunition factories, and storage sites due to improper waste disposal, accidents
(explosions), damage by bombing, and the destruction of the facilities after the war. It is
estimated that the explosives-contaminated areas in Germany cover more than
10,000 km² [2]. TNT forms slightly yellowish crystals at room temperature and is found
on contaminated sites either as dust or crystalline lumps of up to several decimetres.
Because of the low water solubility of TNT (130 mg/l), the soil contamination is very
slowly washed out into the ground- and surface water. Groundwater concentrations in
the range of one to several thousand μg/l are not unusual at explosives-contaminated
sites. TNT and other nitroaromatic compounds are of environmental concern because
they are toxic and suspected carcinogens [3, 4].

1.1. TRANSFORMATION OF TNT

The electron-withdrawing effect of the three nitro groups in the TNT molecule prevents
an oxidative (electrophilic) attack on the aromatic ring. The transformation usually starts
with the reduction of a nitro group to the corresponding hydroxylamino- or amino group.
The reduction can be initiated biologically by enzymes or by chemical reductants such
as elemental iron [5]. In the presence of a readily available carbon source, TNT will be

I. V. Perminova et al. (eds.),
Use of Humic Substances to Remediate Polluted Environments: From Theory to Practice, 353–364.
© 2005 *Springer. Printed in the Netherlands.*

transformed under aerobic conditions, but a reducing environment accelerates the reaction. The TNT reduction is co-metabolic, which means that external carbon and nitrogen sources are needed.

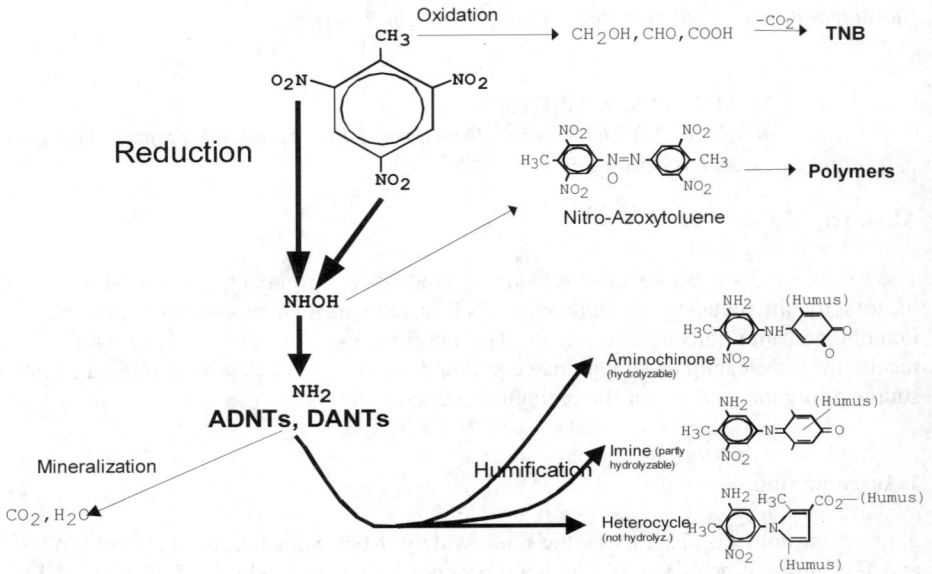

Figure 1. Transformations of TNT.

The most abundant products of the reduction are the compounds 2-amino-4,6-dinitrotoluene (2-ADNT) and 4-amino-2,6-dinitrotoluene (4-ADNT) and, after reduction of a second nitro group, the two isomeric diamino-nitrotoluenes (DANTs) [6] (see Figure 1). The reduction of all three nitro groups that leads to the very reactive compound triaminotoluene (TAT) is possible only under very strong reducing conditions (Eh $\leq$ -200 mV), which are not expected to occur in soil or groundwater, unless it is highly contaminated with easily degradable carbon sources.

The amino transformation products are reactive compounds that bind to the functional groups of the soil organic matter. This process is called humification. The products of this reaction are complex; besides physical and chemical sorption processes, different types of covalent bonds are formed. NMR studies have shown the types of bonds formed by reaction of reduced TNT metabolites with organic matter [7, 8, 9].

The bound transformation products of TNT undergo further reactions: in the first stage of a composting experiment about 20% of the transformed TNT could be re-

released through hydrolysis, but as the transformation process continued the products were no longer hydrolysable [10]. This means that additional bonds are formed and the transformation products were conjugated through multiple bonds.

Investigations of biologically treated TNT-contaminated soil proved that the bound transformation products are stable and non-toxic as remobilisation could not be observed [11].

Besides the reductive pathway that leads to humification, there are other possible reactions of the TNT. Of some importance is the oxidation of the methyl group to form polar metabolites and trinitrobenzene, which are found in soil and water of TNT-contaminated sites [12]. Under certain conditions, polymerisation of metabolites occurs. Mineralization of TNT has been observed under ligninolytic conditions with fungi [13].

The scheme of TNT transformations, compiled from various sources, is shown in Figure 1.

1.2. REMEDIATION OF TNT

1.2.1. Soil remediation

For the treatment of explosives-contaminated soils, the current practice is landfilling or incineration. Incineration is the most effective and widely used remediation alternative, but this method is expensive due to the costs of soil excavation, transport, and energy for incineration. Landfilling is cheap in some areas, but it is not a real solution, just a displacement of the contamination. Therefore, new remediation technologies are needed.

Biological removal of TNT and other explosives has been proven feasible (e.g., [14, 15, 16, 17]). At sites with large areas of contamination, *in situ* (bio-)remediation processes are economically and ecologically advantageous. Such processes are:

1. suitable for large areas,
2. environmentally friendly, and
3. cost-effective.

1.2.2. Constructed wetlands for water treatment

The currently available technology for treating explosives-containing water is activated carbon adsorption. Although effective, this treatment is expensive, especially if it has to be operated for decades, as in the case of a pump-and-treat groundwater remediation. As with soil remediation, new approaches are needed to deal with explosives in water. Constructed wetlands are especially promising due to their low long-term costs.

Constructed wetlands are an innovative and inexpensive treatment approach. They mimic the filtration processes of natural wetlands, effectively removing various contaminants. Wetlands are constructed as either surface flow or subsurface flow systems; subsurface flow can be horizontal or vertical. Subsurface flow systems consist of basins that are filled with a substrate such as sand or gravel. Contaminants are removed from the water through several mechanisms: sedimentation, microbial degradation, and plant uptake are the most important ones. Conditions are ideal for TNT transformation in a planted horizontal subsurface flow wetland: most of the wetland is anaerobic, supporting reduction, and the root zone of the plants provides some oxygen for the humification process.

2. 1st Case Study: Soil Cleanup

2.1. MATERIALS AND METHODS

2.1.1. Tests in a model system

Quartz sand was artificially contaminated with TNT dissolved in acetone and intensively homogenized in an industrial mixer. To 1 ton of the sand 1 kg of industrial-grade TNT, dissolved in 10 L of acetone, was added, so the nominal concentration was 1,000 mg/kg.

Soil from a contaminated site containing indigenous microorganisms or white-rot fungi (*Stropharia rugosoannulata*) precultivated on straw was used to inoculate the substrate. Molasses and rotting birchwood were used as the carbon sources. Additionally, *Medicago sativa* (lucerne) was planted in the substrate in one test variant for the purpose of recording the possible effects of plants on TNT attenuation in the soil.

The tests were carried out as pot experiments under field conditions, using vessels with a volume of 30 L. Soil samples were taken at fortnightly intervals by means of a rod-type probe, with eight individual samples being combined into one mixed sample for each vessel.

Because, in the case of the variant containing autochthonous microorganisms and molasses as the carbon source, the TNT concentration dropped very rapidly, this experiment was repeated. An additional experiment was set up with addition of molasses but without inoculation with microorganisms. The sampling interval was reduced to 2 days for this treatment.

2.1.2. Ecotoxicity test

Determination of the germination capacity and the root length growth of garden cress (*Lepidum sativum*) was used for ecotoxicological evaluation of the substrates of the model system. For this purpose, soil filtrates were prepared, and the seeds were subjected to this eluate. The germination capacity of the plant seeds and the lengths of the roots were evaluated.

2.1.3. Field trial in situ

A birch-covered area of about 250 m^2, exhibiting a mean TNT contamination of 6,000 mg per kg, was selected as the *in situ* field test area. The birch trees growing on this area were rooted out and crushed by means of a shredder. The shredded tree material, about 2 m^3, together with 0,5 m^3 of rotting birchwood chips, which were overgrown with fungi, was ploughed into the ground using a tiller. This mechanical treatment was supposed to homogenize the soil, mix it with the woodchips and provide oxygen for the humification process.

The area was irrigated at the beginning of the trial and during dry periods. After four weeks another 2 m^3 of rotting birchwood chips were applied and the tillage process was repeated. Another mechanical treatment was carried out after three months. Samples were taken at weekly intervals from the contaminated soil horizon (0-30 cm). The individual samples taken were combined to give representative mixed samples.

2.1.4. Container experiments to determine degradation exponents for original soil
Pot experiments similar to the sand systems were set up with soil taken from a contaminated site to determine the kinetic degradation exponents with various carbon sources and to calculate the time for a full scale remediation. The soil was sieved through a coarse sieve (about 10 mm) to separate stones and wood and approximately 10 L were placed into plastic pots for every experiment and mixed with the humification amendments. The pots were placed under a roof in the open air and secured with a net against animals. The following treatments were set up in duplicate (percentages are v/v):
- Control (without amendment)
- 10% wood chips
- 1% molasses
- 2.5% molasses
- 1% iron filings
- 1% molasses and 1% iron filings
- 1% molasses and 0.33% iron filings
- 2.5% sludge (from local water treatment plant) and 1% iron filings

After the set-up of the experiments and after 30 and 60 days the pots were intensively mechanically mixed to simulate a tillage. The pots were watered at least weekly depending on the air temperature to keep the soil at a constant humidity. Samples were taken at increasing intervals after 0, 2, 4 , 6, 8, 10, 12, 14, 21, 29, 35, 42, 46, 53, 60, 67, 75, 82 and 90 days. Six samples from each experiment, taken at random locations with a tube of 30 mm diameter, were mixed and then air-dried.

2.1.5. Analysis
The air-dried, screened soil samples were extracted with acetone in an ultrasonic bath and analysed by GC/ECD.

3. Results

3.1. TESTS IN THE MODEL SYSTEMS

The starting TNT concentration of the quartz sand was 800 mg/kg instead of 1,000 mg/kg, which was ascribed to degradation processes during the homogenisation in the mixer. In the control preparation (uninoculated and containing no C-source or plants), the TNT concentration remained constant for the duration of the trial. In the variant containing the white rot fungus (*Stropharia*) but no carbon, no reduction in the TNT concentration was detectable.

The addition of a carbon source brought about a reduction in the TNT concentration. In the variants containing wood chips as the carbon source, the TNT concentration dropped to a residual content of 350 mg/kg in about 100 days.

The most rapid attenuation of TNT was found in the variant containing autochthonous microorganisms and molasses; the initial concentration of 800 mg/kg dropped to 20 mg/kg within 14 days. Even the variant to which molasses had been added but not microorganisms clearly showed a TNT degradation.

358

In the control sample, total ADNT concentrations of 25-45 mg/kg were detectable. Considerably higher concentrations of primary metabolites were observed in the variants in which a reduction in TNT was detectable. While 2-ADNT was the predominant ADNT isomer in the test preparations containing no molasses, in those containing molasses, the 2- and 4-ADNT concentrations were nearly identical or showed slightly higher 4-ADNT levels. In some variants, the metabolite concentration initially increased before subsequently falling again.

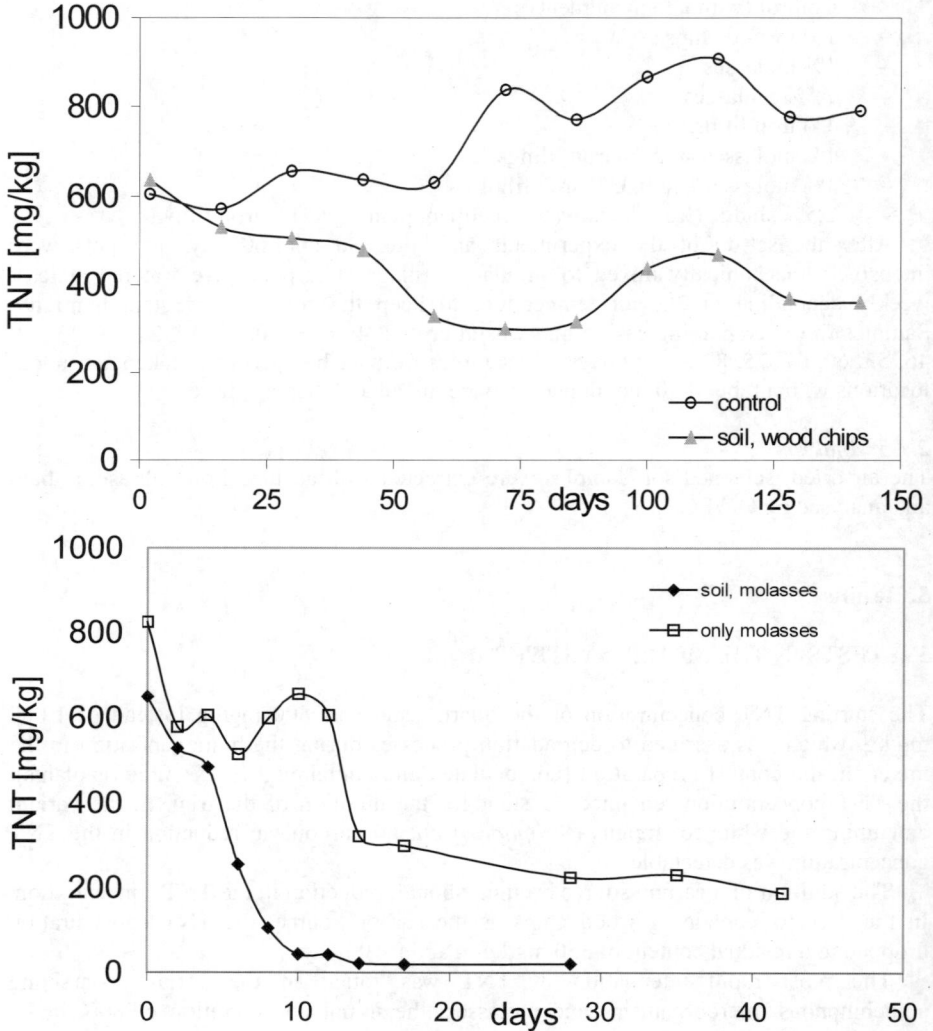

Figure 2. TNT-degradation in experiments with artificially contaminated quartz sand and various amendments.

3.1.1. Ecotoxicity test

The toxicity of the substrates from the pot experiments to cress seedlings diminished parallel to the TNT concentration. As expected, filtrates from test variants in which no TNT degradation was observable exhibited the highest ecotoxicity. Although the seedlings in the treatment with wood chips as the carbon source developed better, the seedlings' root growth was impaired. At the end of the trial, all the filtrates from variants in which the TNT concentration had dropped considerably exhibited only very slight phytotoxicity in the cress test. This suggests that no phytotoxicity emanates from metabolites or adducts formed during the TNT-transformation. The chemical fate of the TNT was not investigated in the described experiments.

Table 1. Phytotoxicity tests with *Lepidum sativum* of differently treated TNT-contaminated quartz sand experiments.

| Amendment | TNT transformation | pH | Ecotoxicity test with cress | |
			germination capacity	root length
Control without TNT	0%	7	optimal development of sprouts	9.5 cm
Molasses	97% in 20 d (650 to 20 mg/kg)	>8	optimal development of sprouts	6.5 to 7.0 cm
Rotting birch wood	50% in 86 d (636 to 317 mg/kg)	6	sprouts partially yellowed	0.5 to 1.0 cm

3.1.2. Field trial in situ

A roughly 80% reduction in the extractable TNT concentration of the soil was detectable over a 100-day period. The measured TNT concentrations of the soil samples exhibited great fluctuations. ADNT concentrations were in the range of 25-50 mg/kg, and the 4-ADNT content was somewhat higher, which, based on the results of the model tests, is indicative of active microbiological transformation. As a result of natural succession, a variety of grasses and herbs typical for the site were able to establish themselves.

3.1.3. Container experiments with original soil

Because original soil from a contaminated site was used, the determined concentrations showed the same high variability as is always observed in the field. The starting concentration was 1612 ± 637 mg/kg.

A degradation of TNT during the first 90 days was observed in the experiments that were amended with iron filings and/or an easily degradable organic carbon source. No decrease was observed in the control experiments or the wood chip-amended pots. After the first 90 days, no further reduction in TNT concentration was observed, and the experiment was stopped after 177 days. The halting of the TNT-degradation after the first 90 days of the experiment is ascribed to the cold winter conditions during the second phase of the experiment. Since the process used is not a composting operation (for composting much larger amounts of organic amendments up to 50% are utilized), there is no heat generated by the process and biological processes can be expected to stop at temperature around or below 0°C. From the regression of the experimental TNT –concentrations, the coefficients of a first order reaction were calculated. In Table 2 the coefficients of the equation

$$c = c_0 \, e^{-kt} \tag{1}$$

are tabulated. (c= concentration at time t, c_0 = initial concentration, k = degradation constant, t = time in days)

Table 2. Calculated coefficients for TNT degradation in pot experiments after 90 days.

Treatment	c_0 (mg/kg)	k (d^{-1})
molasses 2.5% (1)	1806	0.016
molasses 2.5% (2)	994	0.006
molasses 1% (1)	1950	0.011
molasses 1% (2)	1516	0.013
molasses 1%/ iron 1% (1)	1370	0.022
molasses 1%/ iron 1% (2)	1347	0.015
molasses 1%/ iron 0.33% (1)	1501	0.017
molasses 1%/ iron 0.33% (2)	1235	0.007
sludge / iron (1)	1723	0.010
sludge / iron (2)	1776	0.011
iron 1% (1)	1772	0.014
iron 1% (2)	calculation not possible	

From the results the following conclusions can be drawn:
- Without amendment of the soil no significant TNT degradation takes place.
- Iron used together with molasses leads to faster degradation than does either of the two alone.
- The usage of sludge together with iron filings gave lower degradation coefficients than iron alone, which means that the microorganisms from the sludge are not adapted to TNT–degradation, in contrast to the organisms indigenous to contaminated soil.

Based on these test results, a combination of iron and molasses is being used for a full-scale remediation of TNT-contaminated soil which started in August 2002.

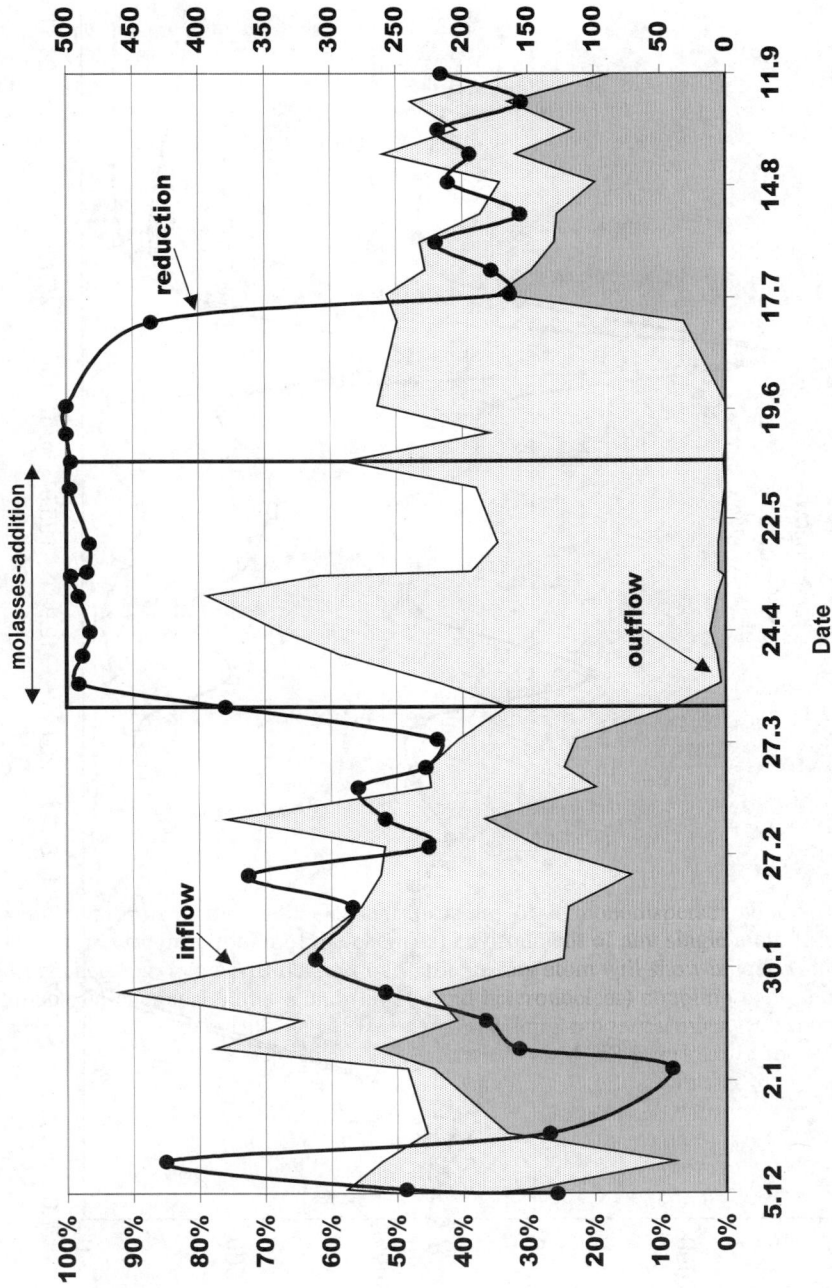

Figure 3. TNT removal in the pilot-scale wetland at Elsnig.

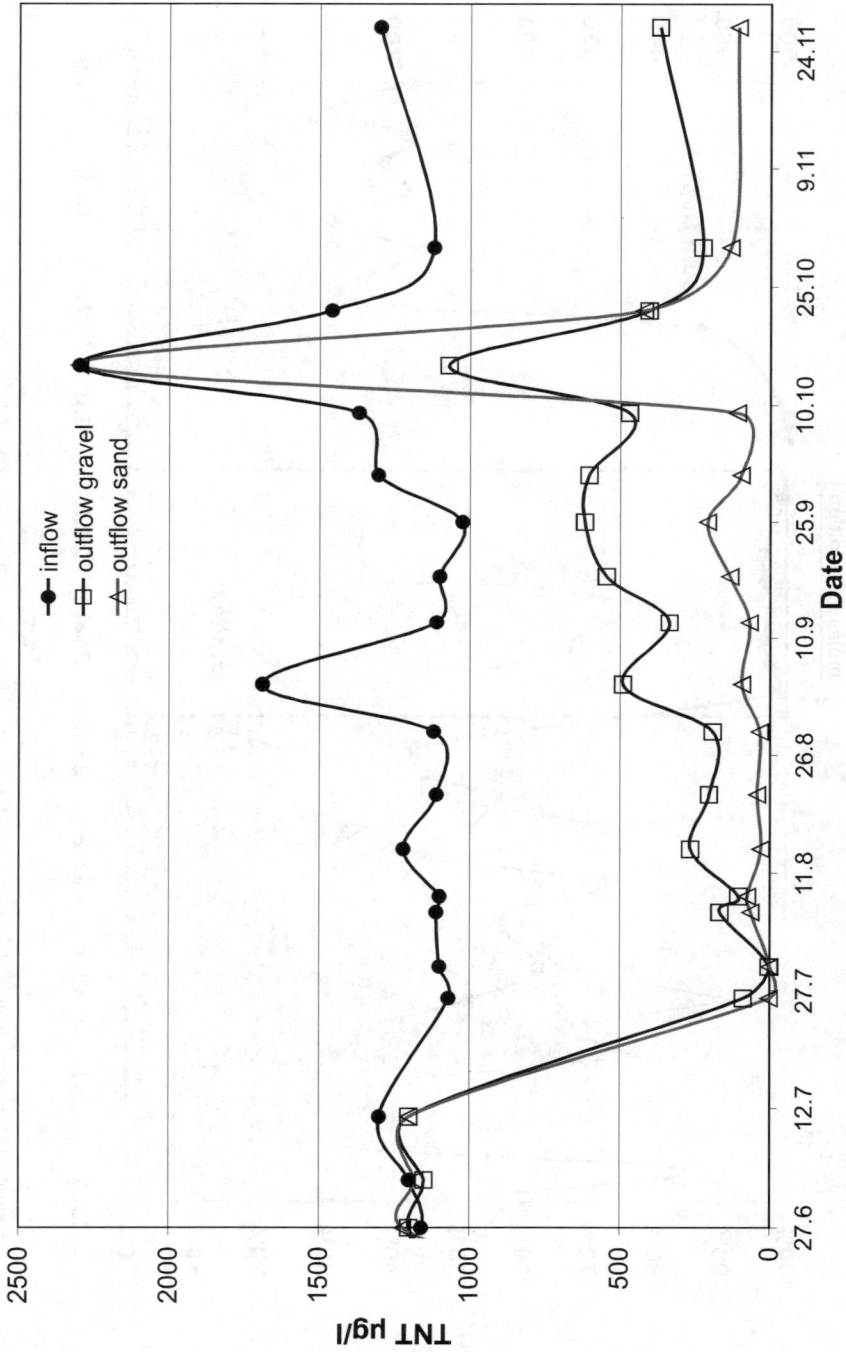

Figure 4. TNT removal in the pilot- scale wetlands at Unterlüß.

4. 2nd Case Study: Water Cleanup in Constructed Wetlands

4.1. MATERIALS AND METHODS

Pilot-scale constructed wetlands were operated on two explosives-contaminated sites. On one site, referred to as the Elsnig site, seepage water contaminated with explosives is collected in drainages. A partial stream of this water was fed into the wetland. On the second site at Unterlüß, groundwater contaminated with TNT by past dumping of wastewater is pumped from underground and fed into two pilot-scale wetlands.

The installations consist of containers of the size 1.2 x 1.0 x 0.76 m. The containers are filled with a sand or gravel substrate and planted with helophytes. Reducing conditions in the filter are established by the addition of molasses. These installations are operated in subsurface flow, which means that the water level is below the surface of the filling. The inflow of water is intermittent, with daily rates of 120 L and a hydraulic detention time of 24 h.

4.2. RESULTS

The in- and outflow concentrations of the wetland in Elsnig are shown in Figure 3. Concentrations and flows vary because the treated water is seepage water. The removal of TNT was 40-60% without a carbon source; after addition of molasses the removal was > 95% and the outflow concentration was less than 10 µg/L. The removal of 2,6-dinitrotoluene (2,6-DNT) was also measured (data not shown). It was initially 60-80% and after addition of molasses also reached > 95%. Adaptation phases for 2,6-DNT were longer than for TNT. The variation of the inflow concentration had no influence on the outflow in the phase of stable operation.

The performance of the two wetlands operated in Unterlüß is shown in Figure 4. There were some problems with these systems, because they suffered from occasional failures due to damage by wild animals and vandalism. For example the peak seen in Figure 4 on October 15th was a result of a power failure caused by martens. But it is clear that there is a good TNT removal when the systems are working properly. The mean inflow concentration was 1,240 µg/L TNT, and the mean outflow concentrations (only values for operational systems) were 356 µg/L for the gravel system and 116 µg/L for the sand system. The sand system was much more effective than the gravel system. The efficiency of the system in Elsnig could not be reached because the dosage of molasses was very low in this case (about 2.5 ml/day) to keep the chemical oxygen demand (COD) in the outflow as low as possible. If COD values in the outflow are not a concern, the efficiency could probably be greatly increased by larger additions of molasses.

5. Conclusion

From laboratory studies that different research groups conducted during the last decade, the mechanisms and products of the biological degradation and humification of TNT are quite good understood. Despite this theoretical understanding there has been little practical application of this process. We have shown in pilot-scale and field experiments, that it is possible to use a biological strategy for the remediation of soil and water. The results shown here are very encouraging and are presently used for full-scale remediation. Further optimisations from the experiences of the applications will provide site owners with a cost-effective and sustainable solution to the problem of explosives contamination.

364

6. References

1.   Preuss, J., Haas, R., Koss, G. (1988) Altstandorte - Altablagerungen - Altlasten. Das Beispiel eines ehemaligen Standortes der chemischen Industrie, *Geographische Rundschau* **40** (6), 31-38.
2.   Preuß, J. (1996) Alte Rüstungsstandorte – Erfassung und Bewertung von Umweltkontaminationen, *Forschungsmagazin Johannes-Gutenberg-Universität Mainz.* **1**, 35-51.
3.   Koss, G., Lommel, A., Ollroge, I., Tesseraux, I., Haas, R., Kappos, A.D. (1989) Zur Toxikologie der Nitrotoluole und weiterer Nitroaromaten aus rüstungsbedingten Altlasten, *Bundesgesundheitsblatt* **32**, 527-536.
4.   (1998) *Toxicity of military unique Ccompounds in Aaquatic Organisms: An annotated bibliography (studies published through 1996)*, Environmental Laboratory, Technical Report IRRP-98-4, U.S. Army Engineer Waterways Experiment Station, Vicksburg, MS. http://www.wes.army.mil/el/elpubs/pdf/trirrp98-4.pdf.
5.   Hundal, L.S., Singh, J., Bier, E.L., Shea, P.J., Comfort, S.D., Powers, W.L. (1997) Removal of TNT and RDX from Water and Soil using Iron Metal, *Environmental Pollution* **97**, 55-64.
6.   Vorbeck, C., Lenke, H., Fischer, P., Spain, J.C., Knackmuss, H.-J. (1998) Initial reductive Reactions in aerobic microbial Metabolism of 2,4,6-Trinitrotoluene, *Appl. Environ. Microbiol.* **64**, 246-252.
7.   Achtnich, C., Fernandes, E., Bollag, J.-M., Knackmuss, H.-J., and Lenke, H. (1999) Covalent Binding of reduced Metabolites of [^{15}N$_3$]TNT to Soil Organic Matter during a Bioremediation Process analyzed by ^{15}N NMR Spectroscopy, *Environmental Science & Technology* **33** 4448-4456.
8.   Achtnich, C., Lenke, H., Klaus, U., Spiteller, M., and Knackmuss, H.-J.(2000) Stability of immobilized TNT Derivatives in Soil as a Function of Nitro Group Reduction, *Environmental Science & Technology* **34**, 3698-3704.
9.   Thorn, K.A., Pennington, J.C., Hayes, C.A. (2002) ^{15}N NMR Investigation of the Reduction and Binding of TNT in an aerobic Bench Scale Reactor simulating Windrow Composting, *Environmental Science & Technology* **36**, 3797-3805.
10. Thorne, P. G., Leggett, D. C. (1998) Explosives Conjugation Products in Remediation Matrices: Interim Report 2, in J.C. Pennington, K.A. Thorn, D. Gunnison, V.A. McFarland, P.G. Thorne, L.S. Inouye, H. Fredrickson, D.C. Leggett, D. Ringleberg, A.S. Jarvis, D.R. Felt, C.H. Lutz, C.A. Hayes, J.U. Clarke, M. Richmond, B. O'Neal, B.E. Porter, (eds.), *Technical Report SERDP-98-12*, U.S. Army Engineer Waterways Experiment Station: Vicksburg, MS, pp. 38-52.
11. (2000) *Langzeit-Remobilisierungsverhalten von Schadstoffen bei der biologischen Bodensanierung*, Proceedings of the 2nd Statusseminar, Umweltbundesamt, Bremen.
12. Bruns-Nagel, D., Schmidt, T.C., Drzyzga, O., von Löw, E., Steinbach, K. (1999) Identification of oxidized TNT Metabolites in Ssoil Samples of a former Ammunition Plant, *Environmental Science and Pollution Research* **6**, 7-10.
13. Scheibner, K., Hofrichter, M., Herre, A., Michels, J., Fritsche, W. (1997) Screening for Fungi intensively mineralizing TNT, *Apl Microbiol Biotechnol* **47**, 452-457.
14. Boopathy, R. (2000) Bioremediation of Explosives contaminated Ssoil, *International Biodeterioration & Biodegradation* **46**, 29-36.
15. Boopathy, R., Manning, J., Kulpa, C.F. (1998) A Laboratory Study of the Bioremediation of TNT-contaminated soil using aerobic/anoxic soil slurry reactor, *Water Environment Research* **70**, 80-86.
16. Daun, G., Lenke, H., Reuss, M. and Knackmuss, H.-J. (1998) Biological Treatment of TNT-Contaminated Soil. 1. Anaerobic Cometabolic Reduction and Interaction of TNT and Metabolites with Soil Components, *Environmental Science & Technology* **32**, 1956-1963.
17. Dahn, A., Michels, J. (2002) Biologische Bodensanierung in der Praxis. Erfahrungen am Beispiel der maßstabsgerechten Erprobung biologischer Verfahren im ‚Werk Tanne', in: *Leitfaden – Biologische Verfahren zur Bodensanierung*, Umweltbundesamt PT AWAS, Berlin, pp. 445-472.

Chapter 19

COMMERCIAL HUMATES FROM COAL AND THEIR INFLUENCE ON SOIL PROPERTIES AND INITIAL PLANT DEVELOPMENT

O.S. IAKIMENKO
Soil Science Department, Lomonosov Moscow State University, 119992 Moscow, Russia <iakim@soil.msu.ru>

Abstract

Properties of three samples of coal-derived commercial humates (CH), produced as organic fertilizers by the same company, were examined. Despite the same origin, the samples possessed different properties even with respect to such basic parameters as content of C, N, and humic acids (HA). Two of the CH samples (CH-1 and CH-2) studied were very similar, but CH-3 differed significantly in having a lower content of C and HA; a higher content of N and FA-fraction; and in the character of molecular-weight distribution. The CH samples with similar properties (CH-1 and CH-2) exhibited different growth-stimulating effects; CH-2 was less effective. The least effective was CH-3 in spite of the highest N content. This brought us to the conclusion that either CH-bound N is unavailable for plants, or the amount and quality of HA are more important for growth-stimulating effects of CH than the total amount of nutrients. High application rates of CH inhibited plant development in spite of the higher nutritional value, which indirectly proves their physiological activity. Other possible reasons of inhibiting action of CH could be toxicity of their high doses, or inactivation of bound to HA nutrients.

1. Introduction

Commercial humates (CH), derived from various natural sources such as coal, peat, sediments, organic waste materials, and some other components, are reported to be effective soil conditioners. The phenomenon of positive influence of humic substances (HS) on plant growth and development was first discovered at the end of the 19th century and later was supported in classical works of Khristeva, Kononova, Tjurin, and Waksman [17, 18, 41, 47]. Since 1960, a lot of data on this subject have been accumulated [3, 4, 9, 16, 45, 46]. Commercial humates were shown to improve plant growth and development, increase plant resistance to unfavourable environmental conditions, and act similarly to natural soil HS [17, 24, 27]. In numerous field and laboratory experiments with different test-plants and crop species, it was demonstrated that commercial humates of sodium, potassium, and ammonia, no matter the sources of raw material for their production, when applied in optimum rates (50-100 $mg \cdot L^{-1}$ or 10-

365

I. V. Perminova et al. (eds.),
Use of Humic Substances to Remediate Polluted Environments: From Theory to Practice, 365–378.

100 kg·ha^{-1}), significantly stimulate the germination of agricultural seeds and plant growth [27, 42, 44], increase water absorption and respiration [25, 34] and the length and biomass of shoots and roots [20, 34, 45] of certain varieties, and decrease the uptake of heavy metals and radionuclides by plants [1, 12, 38, 37]. This effect is especially noticeable in the primary phase of plant development, but sometimes it can be realized during the entire ontogenesis of the plant, including the yield [17, 21, 22, 46]. Some potassium and ammonium humates as well as a number of other modified materials from coal were discovered to have growth stimulating effects exceeding their nutritional values, but they were also found to be toxic in certain relatively high rates [4, 5].

Higher efficiency of commercial humates in stimulating plant growth compared to *in vivo* soil HA was most often attributed either to their particular chemical structure [18, 4, 28] or to physiological influence [15, 17]. The former hypothesis refers to the alteration of HS from the initial raw materials during their production due to a number of processes such as breakdown of organo-mineral interactions, hydrolysis, and oxidation. Nevertheless, they still belong to the class of HA compounds, but the configuration of their molecules has been changed, enabling them to be activated. The latter hypothesis refers to the number of physiological mechanisms in plants influenced by HA, such as mitigating the effect of respiration inhibitors, accelerating the synthesis of proteins, affecting the metabolic reactions, or acting similarly to hormone-like substances [16, 17]. The strongest HS effect was revealed under a wide range of unfavourable environmental conditions such as insufficient or excessive moisture content, low temperatures, or insufficient elimination or contamination by heavy metals or radionuclides, which means that physiologically active HA increases plant resistance not only to certain factors of the environment, but also promote total plant resistance [17].

Many authors have reported that HS positively affect plant growth due to their indirect influence on the soil properties such as an increase in soil aggregation, aeration, permeability [34], and adsorption of soil nutrients [9, 35], and by improving distribution of metal ions within the plant due to formation of chelates [3, 18, 22].

The highly positive response of plants and soils to application of HS in laboratory experiments has aroused great interest in production of commercial humic fertilizers all over the world. Experience with using humic fertilizers in agriculture and horticulture allowed many companies to produce commercial humates from various organic raw materials, mostly, from peat, leonardite, and other type of lignites [14, 21]. Addition of nutrients, various conditioners, or special regimes of production resulted in a big variety of humates destined for different tasks: fertilization, reclamation of degraded soils, sorption of toxicants, and plant growth stimulators [20, 21, 29, 35, 38, 42]. Brown coals with low calorific values, but high content of humic acids, are the most widely used as raw materials for commercial production of HS. In Russia, production and application of coal humates nowadays is relatively widespread, and generally treated as a rational use of bioresources in modern agriculture. It is practiced most actively in the regions with large resources of low-quality brown coal, such as the Far East, Amur, and Altai regions, non-Chernozem areas of Russia. In Altai region, which is rich in supplies of oxidized brown coal, a number of companies produce a variety of commercial humic fertilizers using solid phase and liquid phase extractions and addition of nutrients. Scientific and practical conferences on application of humic fertilizers are regularly held there, mostly demonstrating data on CH influence on the decrease of heavy metals uptake by plants in contaminated soils, increasing yield, and the positive influence on soil structure and water-holding properties [1, 2, 10, 24, 25]. Similar activities take place in Amur region, where the total resources of brown coal were estimated as 6.4 billion tonnes. These coals are used for CH production because of their low heat values, low degree of coalification, and high content of HA (about 58-72%). It was found that the plants differ greatly in their response to application of CH. So, for Amur region plant species were divided into 4 groups, differing in their response to CH application: i) plants rich in carbohydrates that give up to 50% additional yield; ii) corns give up to 30% additional yield; iii) highly proteinaceous plants give up to 10%, and iv) plants that accumulate oils react weakly or negatively to addition of CH. CH, produced from local coal resources, are sold and applied in the Northwest of Russia and in the non-Cherozem and Chernozem areas, again being reported to give positive results predominantly on relatively low-fertility soils [1, 2, 22, 26, 31, 32].

Nevertheless, there are still some unanswered questions concerning the application of commercial humates in agriculture. Stevenson [39] stated that there is some doubt whether they have a real beneficial effect on plant yields in normally productive agricultural lands because unlike nutritional or sand media used for experiments, most mineral soils can inactivate HA. Today, as it was a decade ago, there is a lack of information about commercial humates with regard to their source, methods of isolations and pre-treatment, and evident chemical difference between natural and commercial humates. Some drawbacks and limitations on use of commercial humates have been stressed in the literature [21, 23]. The relationships between the structure of CH and their effect on soil and plants are still sparingly revealed, and many researchers have emphasized the need to establish a series of quality parameters that would enable the evaluation of the potential quality of a given commercial product [28, 14]. In spite of well-known general features of CH, due to the wide variety of commercial humates produced by worldwide companies, each CH sample of a certain brand and kind possesses some individual properties as well. Thus, Gonzales-Vila *et al.* reported [14] that a number of CH sold and recommended in Spain as soil amendments differed greatly from each other and from the parameters given by the producers not only in content and composition of carbohydrates, amino acids and some molecular parameters, but also in content of HA.

The above facts illustrate the need to create appropriate evaluation criteria for the quality of commercial humates produced by different companies. Generalization of properties of CH derived from similar raw materials or aimed at the same task as well as estimation of their individual properties and comparison with the parameters provided by producers could be the first step in solving this task.

The objective of this study was to investigate a number of CH samples produced by the Russian company "Technology Centre" from brown coal of the Moscow region with respect to their influence on soil properties and initial plant development under laboratory conditions; and to disclose a relationship between these effects and chemical characteristics of the humates.

2. Materials and Methods

The technological scheme of CH production includes oxidation of brown coal and addition of nutrients. Currently, the company is trying to determine the optimum technological regime, and different samples of CH vary remarkably. Therefore, three samples of CH (CH1, CH2, and CH3) were produced using three different technological schemes.

The CHs used in this study were declared by producers not as "humic fertilizers", but as "organo-mineral humate-containing soil amendments", derived from the oxidised brown coal. Therefore, the modification of the initial raw material was rather aimed on its minor alteration using addition of nutrients to increase nutritional value than on the production of the pure commercial humate.

2.1. ANALYSES OF COMMERCIAL HUMATES

Elemental analysis of CH–samples was conducted using dry combustion (total C) and Kjeldahl procedure (total N). pH in water suspension was determined using glass electrode. The amount of radioactive substances in CH was determined using the gamma-spectrometer. The yield of HA and acid-soluble humic fractions was determined in 0.1 M NaOH extract after acidification and precipitation of HA at pH 1.2.

The structure of CH was examined using IR-spectrometry. The IR spectra were registered in the range of 400-4000 cm^{-1} using pellets of 2 mg CH in 300 mg KBr. The molecular-weight distribution of CH-humic substances was determined using size-exclusion chromatography (SEC) in 0.1 M NaOH extract. The column dimensions were 1.6 × 65.9 cm (*Farmacia*). Sephadex G-100 was used as a fractionation gel; phosphate buffer (pH 8.2) containing 0.1% sodium dodecylsulphonate (DDS) was used as a mobile phase; elution rate was 10 mL/hr. The chromatograms were detected by UV-detector at 220 nm. Average molecular weights (MW) were calculated according to Determan for globular proteins [10] using the following equation:

$$\log MW = 5.941 - 0.847(V_e/V_o),$$

where V_e is the elution volume and V_o is the void volume of a column.

In CH samples alone as well as in their mixtures with soil, some microbiological properties, nominally, the total amount and taxonomic order of the microorganisms, presence of nitrate reduction, and redox conditions were examined.

2.2. MODEL EXPERIMENT

For the model experiment, air-dried CH were powdered and carefully mixed with crushed sandy loam soil in rates equivalent to 200, 400, 700, 1500, and 3000 kg·ha^{-1}, moistened, and placed into pots. Rates were calculated assuming that 1 ha of arable layer of sandy loam podzol soil having 20 cm depth and average bulk density of 1.5 g·cm^{-1} weighs 3000 ton. After one month of incubation, total C, total N, yield of HA and fulvic acid fraction (FAF) in 0.1 M NaOH extract of soil-CH mixtures were examined using the same procedures as described above for the CH samples alone.

Physiological activity of CH in soil was evaluated using the shoot length and root biomass as responses during the early ontogenesis of lettuce plants: 15 days after plants were seeded. The same test had been performed for the CH samples alone.

3. Results and Discussion

3.1. PROPERTIES OF COMMERCIAL COAL HUMATES

The CH samples studied were dark-brown dry granules of about 4 mm diameter with density 1.0-1.3 $kg \cdot L^{-1}$ and moisture content < 22%. According to the technical data of the producer, they contained up to 28 elements in amounts from 0.5 up to 2500 $g \cdot t^{-1}$; the content of nutrients varied around 2.6% K, 1.7% P and 3.5% N. They were recommended as organo-mineral soil amendments.

Due to recent environmental pollution by radionuclides, some open deposits of brown coal may be polluted; therefore, it was of interest to check the source materials for soil application for the radioactivity level. In the CH samples studied, the radioactivity did not exceed the global distribution level for unpolluted arable lands. The intensity of gamma-radiation of the samples studied ranged within the limits typical for natural unpolluted soils and averaged 8-15 $mkR \cdot h^{-1}$. The amount of cesium-137 was 1.2-3.9 $Bk \cdot kg^{-1}$; strontium-90 ranged between 0.4-0.7 $Bk \cdot kg^{-1}$; and potassium-40 amounted to 520-590 $Bk \cdot kg^{-1}$. The content of uranium-238 and radium-226 varied from 16-27 $Bk \cdot kg^{-1}$; thorium-232 - between 26 and 30 $Bk \cdot kg^{-1}$.

Basic parameters of CH are given in Table 1. The commercial humates studied had near neutral pH. CH-1 and CH-2 samples were very similar and contained about 1% N and 40% total C. The content of HA carbon in these samples reached > 10%, and that of acid-soluble FAF – > 3%, which corresponds to about 18% and 7% yield of HA and FAF, respectively. A high value of C_{HA}/C_{FA} ratio can be indicative of the high humus-forming potential of these humic materials.

Table 1. Some properties of commercial humates.

Sample	pH	N (%)	C (%)	C_{HA} (%)	C_{FA} (%)	C_{HA}/C_{FA}
CH-1	6.35	1.10	38.4	11.5	3.54	3.25
CH-2	6.38	0.86	40.2	10.4	3.87	2.70
CH- 3	6.32	4.26	22.2	4.86	3.40	1.43

The composition of sample CH-3 differed significantly from the others by the lower content of total C (about 20%) and enrichment with N (above 4%). This sample can be regarded as an intermediate product between slightly modified raw material and commercial humate. The content of HA carbon was also about a factor of two lower compared to CH-1 and CH-2, whereas the amount of FA and of non-humic materials remaining in the acid supernatant was approximately the same. As reported by Gonzales-Vila et al. [14], the obtained data are in accordance with average values of HS content in commercial humates (between 11 and 22%), whereas the content of other organic constituents (carbohydrates and amino acids) may vary on the order of several orders of magnitude. Unlike commercial humates, the content of HA in coal varies over a much wider range depending on its genesis and coalification degree. Thus, according to del Rio et al., the yield of HA in xylitic brown coal was about 8%, but in humic brown coal it reached 73% [8]. The brown coal of the Moscow basin HA contained about 20% of HA and minor amounts of FAF [35]. In general, the yield of HA in coals was found to be 17-80%, depending on coal type, coalification degree, weathering stage, and method of extraction [1, 13, 26, 31, 32].

IR-spectra (Figure 1) characterized the CH samples studied as complicated mixtures of organic and mineral compounds. Due to the high complexity of CH, the absorption bands are highly overlapped, which complicates their interpretation. Nevertheless, it can be seen that IR spectra for CH-1 and CH-2 were almost identical. The dominating bands can be assigned to clay minerals and kaolinite (3690, 3615, 1100, 1035, 1010, 915, 795, 595, 470, and 425 cm^{-1}), as well as to NO_3^- (2430, 1770, and 1380 cm^{-1}). The organic part of CH was characterized by the spectra similar to those of natural humates. The spectra of both samples (CH-1 and CH-2) showed the absorption bands in the range of 3400 cm^{-1}, which corresponds to vibration bands of OH-groups in alcohols, phenols, and acids. Weak absorption bands recorded in the range of 2940-2950 cm^{-1} and 2880-2890 cm^{-1} confirmed the presence of methylene and methyl aliphatic groups. Instead of an absorption peak at 1710-1720 cm^{-1} indicative of the presence of carbonyl groups in acids and ketones, typical for HA in protonated form, two absorption bands at 1590 and 1390 cm^{-1} of carboxylate ion were observed. Peaks in the range of 1605-1620 cm^{-1} revealed the presence of C=C bonds of the aromatic rings.

In the IR-spectra of the CH-3 sample, absorption bands of kaolinite and quartz were also recorded in the range of 1170, 695, 1100 and 795-780 cm^{-1}. The unique feature of the IR spectrum of this sample was that it contained more absorption bands characteristic of N-containing groups. A certain part of N was presented by amide: bands of primary and secondary amides were recorded at 1620 cm^{-1} (carbonyl of primary amide) and 3190 -3345 cm^{-1} (vibration of NH-groups). The aliphatic part of CH-3 was represented by -CH_2- and -CH_3 groups, whereas the absorption band for aromatic C=C bonds was shifted to 1520 cm^{-1}, most probably due to an intensive peak of amide at 1660 cm^{-1}. The IR data show that the CH samples studied contained many ash elements, both clay minerals and inorganic salts. This can probably also play a certain role in plant growth stimulation, either by nutrition or influencing osmotic pressure or ion force in soil solutions.

In general terms, the IR spectra of the CH samples studied are similar to those of humic materials isolated from peat, soils, and sediments [39, 40]. At the same time, some typical bands are missing whereas the bands of certain mineral compounds are present. This phenomenon has been reported in the literature, and IR-spectroscopy was recommended as a powerful tool to reveal differences between soil HAs and HAs of commercially available fertilizers [13, 14, 37].

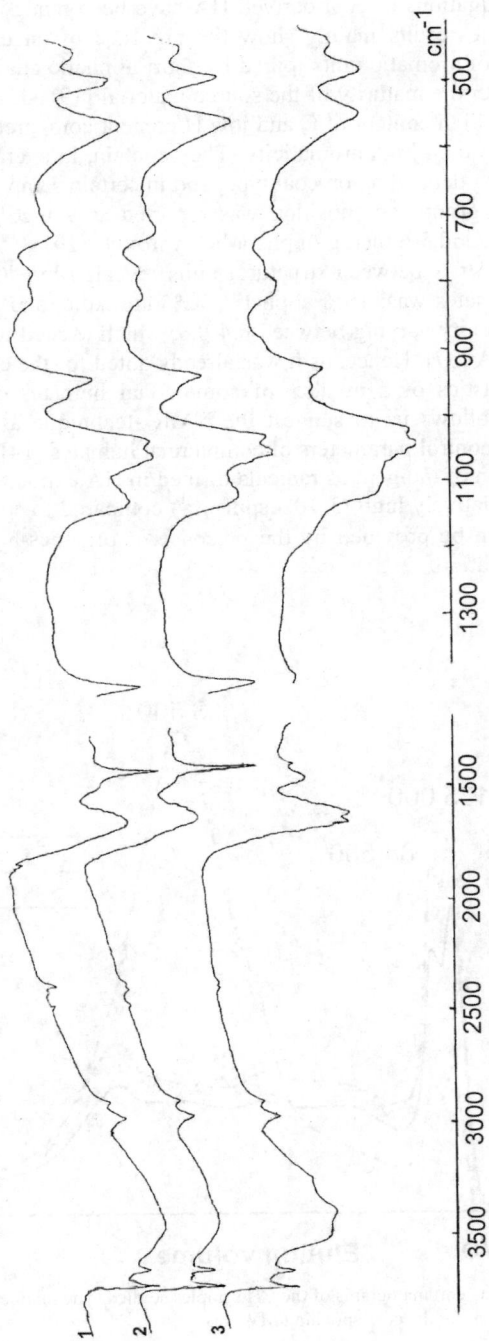

Figure 1. IR-spectra of the CH samples studied. Samples: 1- CH-1; 2 – CH-2; 3 – CH-3.

Structural investigations of coal-derived HA have been made in many studies [4, 8, 13, 26, 28, 37]. The results mainly show the presence of an extended hydrocarbon skeleton consisting of aromatic units joined by short aliphatic chains whose aromaticity degree increases with the maturity of the source material [19, 33, 36]. The coal-derived HAs have generally high content of C and low H content compared to an "average" HA [39] that agrees with their high aromaticity. They contain, in average, 62-64% C, 3-5% H, 0.9-2% N, varying depending on coal type, and in certain samples, about 5-6% S [13, 29, 36]. Functional groups composition was reported as 9 meq/g of total acidity, 3-4 meq/g of carboxyl, and 5-6 meq/g of phenolic hydroxyls [29].

The distribution of C between structural units revealed by ^{13}C-NMR spectroscopy for coal-derived humates was: 18% aliphatic, 72% aromatic, and 10% carboxyl C [27, 29], with the aromaticity varying between 0.4-0.6, which exceed those of commercially available Aldrich HA [37]. Hence, as it was already noted for the case of IR-spectra, the ^{13}C-NMR characteristics of a number of commercial humates differed greatly from natural HA, which allows us to suggest the NMR -technique as a powerful tool for establishing quality control parameters of commercial humates [14].

The concentration of indigenous radicals formed in HA extracted from brown coal is reported as being relatively high (3 10^{17} spins·g^{-1}) compared to soil, peat, and compost HAs [37], which can be provided by the presence of organics-bound iron, increasing free radical concentration.

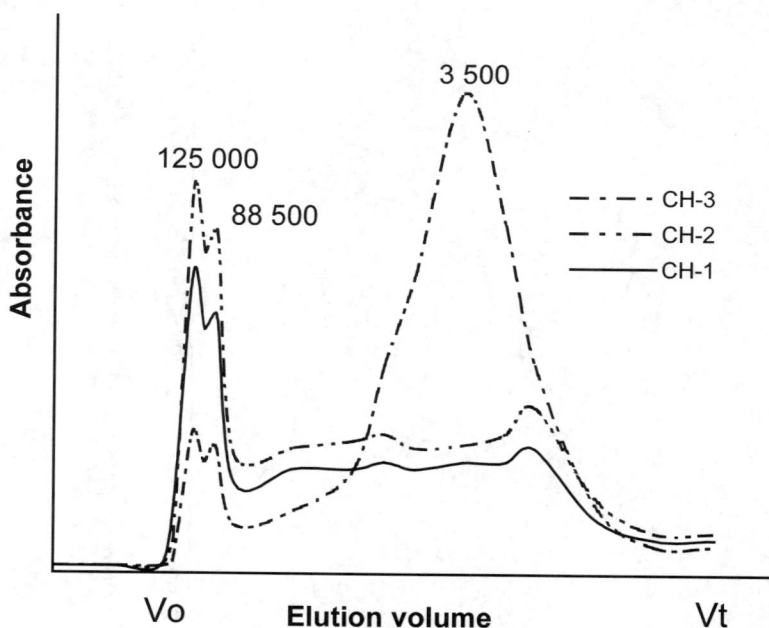

Figure 2. SEC chromatograms of the CH samples studied. The numbers above the peaks are given for the corresponding MW, Da.

HA derived from coal show considerable polydispersity yielding a wide range of variation in their molecular weight. In some experiments, HA have been fractionated by ultrafiltration into 6-8 weight ranges from 0.5 up to more than 500 kD [30, 36]. It was found that along with a decrease in MW values, the content of OH and carboxyl groups increases while the content of aliphatic and carbonyl groups decreases [37]. In our experiment, molecular sizes of the CH samples studied were evaluated using SEC. Despite some limitations of SEC application for HS studies [6, 7], this technique remains a useful tool for HS characterization. The obtained SEC-chromatograms allow us to treat CH samples as high-molecular weight polydisperse compounds consisting of 3 fractions (Figure 2). Similar to the other chemical properties, samples CH-1 and CH-2 demonstrated almost the same elution profile, whereas CH-3 showed distinct differences. The former two had well resolved high- and mid-molecular-weight fractions of about 125,000 D and 88,000 D and broad low-molecular-weight peaks of 1000-8000 D. In CH-3, the contribution of low-molecular-weight fraction is remarkably higher and exceeds the high-molecular-weight part. This fraction can consist of low-molecular-weight compounds of the acid-soluble fraction, probably N-containing compounds (amino acids or proteins) or FA.

3.2. INFLUENCE OF HUMATES ON SOIL PROPERTIES

3.2.1. Influence on Soil Chemical Properties

Many authors believe that the increase of yields due to CH application could be also related to improvement of soil chemical and physical properties [1, 2, 11, 25,29]. Soil application of coal-derived HA was reported to increase the saturated hydraulic conductivity, the amount of water-stable aggregates, and porosity; and to decrease bulk density of some degraded tropical soils [11]. It was also reported to improve aggregate stability and to reduce disaggregation effects of wetting and drying cycles in semi-arid and arid Mediterranean soils [29].

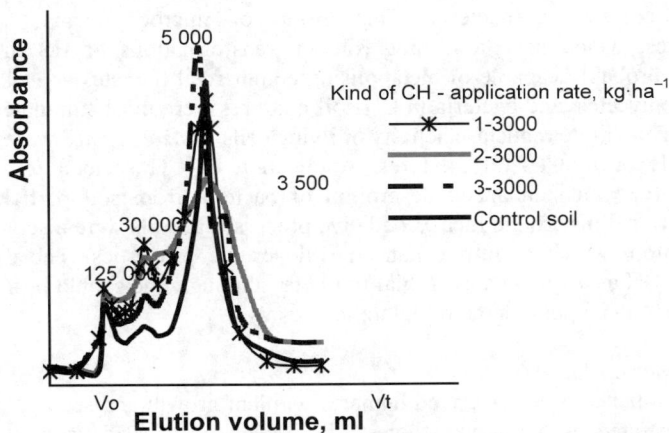

Figure 3. SEC-chromatograms of 0.1 M NaOH extracts from the soil-CH mixtures. The numbers above the peaks are given for the corresponding MW, Da.

Application of the CH samples studied in our model experiment did not cause a remarkable influence on the soil humus state. All the treatments resulted in a slight HA accumulation and a minor accumulation of FAF, which finally led to an increase in the C_{HA}/C_{FAF} ratio in soil. Similar results were observed by Mbagwu et al. [24], who reported that application of coal-derived HA in acidic, low-fertility tropical soils resulted in increased cation exchange capacity and organic matter accumulation and in an increase of pH by 0.5-1, which allows us to treat them as conditioners for reclaiming degraded soils.

The molecular-weight distribution of soil HS was not really influenced by CH application, even at the highest rates of their application (Figure 3). Elution profiles demonstrate 3 fractions with MW 125, 30, and 5 kDa. Application of CH-3, which was initially enriched with low-molecular-weight fraction, resulted in appearance of an additional peak with MW about 3.5 kDa.

3.2.2. Influence on Soil Microbiological Properties

In addition to their influence on physical and chemical soil properties, commercial humates, as well as other humified organic materials, are known to affect soil microbiological properties and to increase microbial growth and activity [43, 46]. Being added to a selective media, humates could increase the growth of a wide range of taxonomic and functional groups of soil bacteria [46]. As a possible mechanism, it was hypothesized that HS promote the modification of cellular activity and growth through the influence on cell membrane permeability or on nutrient availability [15, 17, 43]. Vallini et al. reported that the aerobic bacteria and actinomycetes were stimulated by rates of 1500 and 3000 mg·kg^{-1} HA, whereas the number of fungi living in the rhizosphere was not affected by any rates of HA [43].

In our experiment, the total amount and taxonomic distribution of microorganisms and presence of nitrate reduction and redox conditions were examined in CH and CH-soil mixtures. The CH samples, especially CH-3, were enriched with bacteria, which can live in soil and interact with fungi and actinomycetes. Application of CH promoted the growth of coryneform bacteria. This group of microorganisms is related to actinomycetes, which have a leading role in transformations of HS in soil. These bacteria are probably capable of metabolising commercial humates as well. In general, fungi, actinomycetes, and bacteria in CH-soil mixtures were most stimulated by the rate of 700 kg·ha^{-1}. Nitrate reductase activity or living cells of *Azotobacter* were not revealed either in CH or in CH-soil mixtures. Application of CH seemed to increase soil hygroscopicity, which enhanced desorption of bacteria from soil particles and their growth in water films around them. Oxidative processes in soils were also activated with CH application, which in turn enhanced soil aerobic properties. The data obtained characterize CH as a valuable microbial fertilizer, although one should bear in mind that at high rates CH can possess microbial toxicity as well.

3.2.3. Influence on Plant Growth

A positive influence of coal-derived humates on plant growth and seed germination has been demonstrated by numerous studies. The immersion of seeds in humate solutions has been shown to increase germination and the length and biomass of shoots and roots [9, 20, 27, 35, 44, 45].

In our experiment, the moistened CH samples studied turned out to be very toxic to plants: the germinated lettuce seeds did not give seedlings. After being mixed with soil, CH affected initial plant development. The length of shoots and roots of lettuce seedlings influenced by CH application in a pot experiment is presented in Figure 4. The positive influence of CH on lettuce plant growth was revealed with low rates of CH (200-700 kg/ha), whereas higher rates of CH inhibited plant development, in spite of the higher nutritional values. These results are in agreement with the reported data [4, 24, 27, 43] showing that application of CH was effective only at certain low rates, beyond which they become phytotoxic. The largest effect of CH application was observed for CH-1. Although the chemical properties of CH-1 and CH-2 were very similar, the latter gave a much lower positive effect. It could be caused by the presence of physiologically active low-molecular-weight organic acids or other factors.

Although CH-3 had higher nutritional value with respect to N content, it was characterized by the lowest beneficial effect on plant development. A possible reason may be the low content of HA in this sample: 2 times less than in the two other samples (Table 1). The obtained results allow us to conclude that the stimulating effect of CH can be attributed rather to the physiological activity of HS than to the nutritional value.

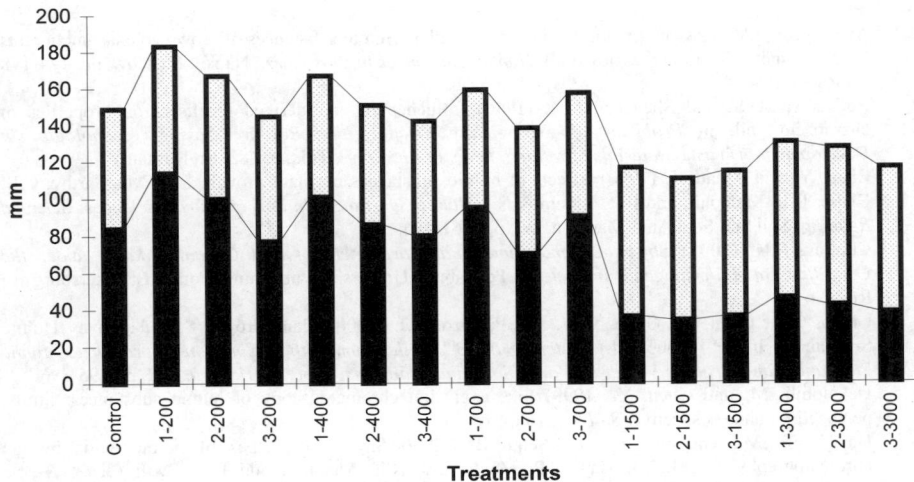

Figure 4. Influence of different application rates of CH samples on the growth of lettuce plants under conditions of pot experiment. Root length is shown by the black columns, shoot length – by the dotted columns; 1, 2, 3 stands for CH-1, CH-2 and CH-3, respectively; 200, 400, 700. 1500 and 3000 (kg/ha) are the application rates of humates.

4. Conclusions

Commercial humates from coal, produced by the same company, can possess different properties even with respect to their basic parameters such as content of C, N, and HA. Two kinds of examined CH (CH-1 and CH-2) were very similar, whereas sample CH-3 differed significantly by its lower content of C and HA, higher content of N and FA, and the character of molecular-weight distribution.

CH with similar properties (CH-1 and CH-2) exhibited different stimulating effects on plant growth; CH-2 was less effective. The lowest effects were detected for CH-3 in spite of its higher N content. This brought us to the conclusion that either CH-bound N is unavailable for plants or the physiological activity of HA is more important for the growth-stimulating effect of CH than the total amount of nutrients.

High rates of CH inhibited plant development in spite of the higher nutritional value, which indirectly proves physiologically active type of their influence.

5. Acknowledgments

The research was supported by RFBR, project 02-04-48016. We express deep appreciation to "Technology Centre" company for providing the CH-samples.

6. References

1. Alexandrov, I.V., Kossov, I.I. and Burkov, P.A. (1993) Humic substances of brown coal as ameliorants for solonchak soils, in D.S. Orlov (ed.), *Humic Substances in Biosphere*, Nauka, Moscow, pp. 174-178 (in Russian).
2. Bezuglova, O.S. and Shevchenko, I. (1996) Influence of coal-derived fertilisers on properties of chernozem soil, in *Problems of Ecologically Friendly and Resource-Saving Technologies in Development of Waste Materials of Mining. Proc. of Int. Conf.*, Tula, p. 182. (in Russian)
3. Chen, Y. and Aviad, T. (1990) Effects of humic substances on plant growth, in P. MacCarthy, C.E. Clapp, R.L. Malcolm, and P.R. Bloom (eds.), *Humic Substances in Soil and Crop Sciences: Selected Reading*, Soil Sci. Soc. Am., Madison, WI, pp. 161-186.
4. Chukov, S.N. (2001) *Structural and Functional Parameters of Soil Organic Matter under the Conditions of Anthropogenic Impact*, St. Petersburg University Publishing House, St Petersburg (in Russian).
5. Clapp, C.E., Chen, Y., Cline, V.W. and Palazzo, A.J. (2000) Plant Growth Stimulation by Humic Substances, in J.P. Croue (ed.), *Entering the 3rd Millennium with a Common Approach to Humic Substances and Organic Matter in Water, Soils and Sediments. Proc. of IHSS-10*, v. 2, pp.895-896.
6. De Nobili, M. and Chen, Y. (1999) Size exclusion chromatography of humic substances: limits, perspectives and prospective, *Soil Sci.* **164**, 825-833.
7. De Nobili, M., Gjessing, E. and Sequi P. (1989) Sizes and shapes of humic acids by gel chromatography, in M.H.B. Hayes, P. MacCarthy, R.L. Malcolm and R.S. Swift (eds.), *Humic Substances. II. In Search of Structure*, John Wiley & Sons, Chichester, pp. 561-591.
8. Del Rio, J.C., Czechowski, F., Gonzalez-Vila, F.J. and Martin, F. (1997) Molecular composition of humic acids from Xylitic and Humic brown coals as revealed by pyrolysis methods combined with gas chromatography-mass spectrometry, in J. Drozd, S.S. Gonet, N. Senesi, J. Weber (eds), *The Role of Humic Substances in Ecosystems and in Environmental Protection*, PTSH, Wroclaw, Poland, pp. 51-58.

9.   Dell'Agnola, N.S. (1986) News about biological effect of humic substances, in R.G. Burns, Dell'Agnola and S. Miele (eds.), *Humic Substances: Effects on Soil and Plants*, Reda, Rome, pp. 78-88.
10.   Determan, H. (1968) *Gel Chromatography*, Springer, NY.
11.   Ekeh, R.C., Mbagwu, J.S.C., Agbim, N.N. and Piccolo, A. (1997) Physical properties of two tropical soils amended with coal-derived humic substances, in J. Drozd, S.S. Gonet, N. Senesi and J. Weber (eds.), *The Role of Humic Substances in Ecosystems and in Environmental Protection*, PTSH, Wroclaw, Poland, pp. 329-333.
12.   Fecenko, J., Lozek, O. and Slamka, P. (2000) Utilization of sodium humate for inhibition of cadmium uptake by spring barley plants, *Toksyczne substancje w glebie - zrodla i wplyw na rosliny*, Warszawa **1**, 187-192.
13.   Gonet, S. and Kondratowicz-Maciejewska, K. (1999) Physico-chemical properties of humic acids extracted from brown coal, *Humic Substances in Ecosystems* **3**, 31-37.
14.   Gonzalez-Vila, F.J., Martin, F., del Rio, J.C., Castillo, R., Bautista, J.M. and Knicker, H. (1996) Assessment of the potential agronomic quality of commercial humic fertilizers by FTIR and CPMAS ^{13}C-NMR techniques, in C.E. Clapp, N. Senesi, M.H.B. Hayes, S.M. Griffit (eds.), *Humic Substances in Soil and Water Environments: Characterization, Transformation and Interactions*, pp. 227-233.
15.   Gorovaya, A.I., Orlov, D.S. and Scherbenko, O.V. (1995) *Humic Substances*, Naukova Dumka, Kiev.
16.   MacCarthy, P., Clapp, C.E., Malcolm, R.L., and Bloom, P.R. (eds), (1990) *Humic Substances in Soil and Crop Sciences: Selected Readings*, Soil Sci. Soc. Am., Madison, WI.
17.   Khristeva, L.A. (1970) Theory of humic fertilisers and their practical use in Ukraine, in R.A. Robertson (ed.), *Proc. of 2nd Int. Peat Congress*, Leningrad, HMSO, Edinburgh, pp. 543-550.
18.   Kononova, M.M. (1966) *Soil Organic Matter*, Pergamon, Oxford.
19.   Lawson, G.J. and Stewart, D. (1989) Coal humic acids, in M.H.B. Hayes, P. MacCarthy, R.L. Malcolm and R.S. Swift (eds.), *Humic Substances. II. In Search of Structure*, John Wiley & Sons, Chichester, pp. 641-680.
20.   Lee, Y.S. and Barlett, R.J. (1976) Stimulation of plant growth by humic substances, *Soil Sci. Soc. Am. J.*, **40**, 876-879.
21.   Lobartini, J.C. and Tan, K.H. (1992) The geochemical nature and agricultural importance of commercial humic matter, *Sci. Total Environ.* **1/3**, 1-5.
22.   Lozanovskaya, I.N., Luganskaya, I.A., Senchukov, G.A. and Gninenko, S.V. (1991) Effectiveness of coal-humate fertilizers on irrigated chernozems of Rostov area, in E.V. Poluektov (eds.), *Remediation and Utilization of Low-Fertile Soils*, pp. 58-64 (in Russian).
23.   Malcolm, R.L. and MacCarthy, P. (1986) Limitations in the use of commercial HA in water and soil research, *Environ. Sci. Technol.* **20**, 904-911.
24.   Mbagwu, J.S.C. and Piccolo, A. (1997) Effect of humic substances from oxidized coal on soil chemical properties and maize yield, in J. Drozd, S.S. Gonet, N. Senesi, and J. Weber (eds), *The Role of Humic Substances in Ecosystems and in Environmental Protection*, PTSH, Wroclaw, Poland, pp. 921-925.
25.   Noble A.D., Randall, P.J. and James, T.R. (1995) Evaluation of two coal-derived organic products in ameliorating surface and subsurface soil acidity, *Europ. J. Soil Sci.* **46**, 65-75.
26.   Orlov, D.S., Kulakov, V.V., Nikiforov, V.J. and Ammosova, Y.M. (1993) Humic substances from ashed brown coal of Moscow coal basin, in D.S. Orlov (eds.), *Humic Substances in Biosphere*, Nauka, Moscow, pp. 189-207 (in Russian).
27.   Piccolo, A., Celano, G. and Pietramellara, G. (1993) Influence of humic acids on laurel growth, associated rhizospheric microorganisms, and mycorrhizal fungi, *Biol. Fertil. Soils* **16**, 11-15.
28.   Piccolo, A., Nardi, S. and Concheri, G. (1992) Structural characteristics of humic substances as related to nitrate uptake and growth regulation in plant system, *Soil Biol. Biochem.* **24**, 373-380.
29.   Piccolo, A., Pietramellara, G. and Mbagwu, J.S.C. (1997) Use of humic acids substances as soil conditioners to increase aggregate stability, *Geoderma* **75**, 267-277.
30.   Piccolo, A., Rausa, R. and Celano, G. (1992) Characteristics of molecular size fractions of humic substances derived from oxidized coal, *Chemosphere* **24**, 1381-1387.
31.   Pokul, T.V., Paramonova, T.G., Kryukova, V.N. and Mitsuk, G.E. (1993) Humic substances from brown coal of Khandinsk deposit, in D.S. Orlov (eds.), *Humic Substances in Biosphere*, Nauka, Moscow, pp. 54-56 (in Russian).

32. Proskuryakov, B.A., Siroezko, A.M. and Sergeeva, S.A. (1996) Humic acids from coals of Kansk-Achinsk basin, in *Problems of Ecologically Friendly and Resource-Saving Technologies in Development of Waste Materials of Mining. Proc. of Int. Conf.*, Tula, p. 176 (in Russian).

33. Rausa, R., Calemma, V. and Girardi, E. (1989) Humic acids by dry oxidation of coal with air under pressure: Analytical and spectroscopical characteristics, in *Proc. Int. Conf. Coal Sci.*, Tokyo, v. 1, IAEA, Tokyo, pp. 237-240.

34. Rauthan, B.S. and Schnitzer, M. (1981) Effects of soil fulvic acid on the growth and nutrient content of cucumber (*Cucumis sativus)* plants, *Plant and Soil* **63**, 491-495.

35. Rode, V.V., Alyautdinova, R.H., Ekaterinina, L.N., Rizkov, O.G. and Motovilova, L.V. (1993) Plant growth stimulators from brown coal, in D.S. Orlov (eds.), *Humic Substances in Biosphere*, Nauka, Moscow, pp. 162-166 (in Russian).

36. Sebestova, E., Machovic, V. and Pavlikova, H. (1994) Isolation and characterisation of coal derived humates, in N. Senesi and T. Miano (eds.), *Humic Substances in the Global Environment and Implication on Human Health*, Elsevier Sci., Amsterdam, pp. 1359-1364.

37. Sebestova, E., Machovic, V., and Pavlikova, H. (1997) Structural characterization and heavy metal sorption properties of coal derived humates, in J. Drozd, S.S. Gonet, N. Senesi, J. Weber (eds.), *The Role of Humic Substances in Ecosystems and in Environmental Protection*, PTSH, Wroclaw, Poland, pp. 199-206.

38. Shalabey, O. and Bizik, J. (1998) Effects of manure, sodium humate and nitrogen on the content of some heavy metals in spring wheat, *Pol'nohospodarstvo*, **44**, 263-274

39. Stewenson, F.J. (1994) *Humus Chemistry: Genesis, Composition, Reactions*, John Wiley & Sons, NY.

40. Swift, R.S. (1996) Organic Matter Characterisation, in D.L. Sparks (eds.), *Methods of Soil Analysis. Part 3. Chemical Methods*, Soil Sci. Soc. Am. Book Series no. 5, pp. 1011-1069.

41. Tjurin, I.V. (1937) *Soil Organic Matter*, Nauka, Moscow (in Russian).

42. Ulanov, N.N. (1993) Possibilities of use of oxidized coal and humic substances in agriculture, in D.S. Orlov (eds.), *Humic Substances in Biosphere*, Nauka, Moscow, pp. 157-162 (in Russian).

43. Vallini, G., Pera, A., Avio, L., Valdrighi, M. and Giovannetti, M. (1993) Influence of humic acids on laurel growth, associated rhizospheric microorganisms, and mycorrhizal fungi, *Biol. Fertil. Soils* **16**, 1-4.

44. Van de Venter, H.A., Furter, M., Dekker, J. and Cronje, I.J. (1991) Stimulation of seedling root growth by coal-derived sodium humate, *Plant and Soil* **138**, 17-21.

45. Vaughan, D. and Linehan, D.J. (1976) The growth of wheat plants in humic acid solutions under axenic conditions, *Plant and Soil* **44**, 445-449.

46. Visser, S.A. (1986) Effects of humic substances on plant growth, in R.G. Burns, Dell'Agnola, and S. Miele (eds.), *Humic Substances: Effects on Soil and Plants*. Reda, Rome, pp. 89-135.

47. Waksman, S.A. (1932) *Humus*, Williams and Wilkins, Baltimore.

Chapter 20

IMPACT OF HUMIC SUBSTANCES ON PLANTS IN POLLUTED ENVIRONMENTS: IMPLICATIONS FOR PHYTOREMEDIATION TECHNOLOGIES

M.M. KHARYTONOV, R.A. KARBONYUK, M.P. BULGAKOVA
State Agrarian University, Voroshilov St. 25, Dnepropetrovsk, 49600, Ukraine <mykola_kh@yahoo.com>

Abstract

There is a great variety of soil types in Ukraine. Recently, some industrial regions of the country have encountered the problem of soil degradation due to technogenic pollution. Technogenic soil contamination with heavy metals has been found in the mining and metallurgical regions in the southeast part of Ukraine. Within Ukraine, 79% of all atmospheric emissions occur in the Donetsk-Pridneprovsky region. In the Dniepropetrovsk region, the concentration of mining and metallurgical production exceeds average indexes 7 to 10 times.

The zone of radioactive anomaly caused by the Chernobyl disaster is characterized by complex pollution of the environment. The objective of our work was to elaborate methods of soil detoxification with humates and to reduce the risk of plant contamination with heavy metals, radionuclides, and herbicides. Studies aimed to decrease the negative impact from industrial pollution of soils by heavy metals and radionuclides were conducted in the areas of industrial air and soil pollution in various areas of Ukraine. At the same time the risk of increased total toxicity from the combination of heavy metals and some herbicides presented in the local soils was also taken into account.

In the laboratory experiment a humic preparation "huminate" promoted benzoxazinones formation in the barley. Benzoxazinones formation further led to a 20% decrease of atrazine content in the seedlings. Introduction of humate preparation into the soil reduced also atrazine concentration to the trace concentrations. The protective action of peat, coal-derived humates and biohumus in relation to toxicity of heavy metals to plants was also studied.

A combination of two kinds of trials (introduction in soil and treatment of seeds before planting) for humic substance application led, as a rule, to increase of the barley productivity. The introduction of the vermicompost product (biohumus) into the soil reduced the negative heavy metals impact and the atrazine after-effect on barley growth. In particular, the introduction of biohumus and preliminarily processed poultry dung into the soil increased the barley yield by 12-27%.

I. V. Perminova et al. (eds.),
Use of Humic Substances to Remediate Polluted Environments: From Theory to Practice, 379–388.
© 2005 *Springer. Printed in the Netherlands.*

The field investigation to reduce radionuclide accumulation in plants through the application of humates was conducted in the Chernobyl District of Kiev region. In particular, the increase of fertilizer dose from 0.09 ton/ha to 0.27 ton/ha led to the reduction of radionuclide migration in the soil-plant system and to an increasing of crops yield. The highest yield of barley and decreases in the radionuclide uptake was observed in the trials as follows: $N_{90}P_{90}K_{270}$, $N_{90}P_{270}K_{270}$ and $N_{90}P_{270}K_{270}$ + humate of Na.

The results show that the introduction of the humic substances with other amendments provides for a reduction in total pollutant toxicity in case of industrial pollution. Model experiments showed distinct detoxification effects on soil depending on dose, soil buffering, microorganisms, and kinds of humic amendments.

1. Introduction

There is a great variety of soil types in Ukraine. Recently, some industrial regions of the country encountered the processes, connected with the soil degradation due to its aerotechnogenic pollution. Technogenic soil contamination with heavy metals has been found in the mining and metallurgical regions at the southeast part of Ukraine. Within Ukraine 79% of all atmospheric emissions occur in the Donetsk-Pridneprovsk Region. Major stationary emissions sources are centralized heat and power production (33%), metallurgical industries (25%), coal mining (23%), and chemical and petroleum industries 2% [1-3]. For example in Dniepropetrovsk region concentration of the mining and metallurgical production exceed average indexes 7-10 times. It was established that the zone of technogenic contamination of chernozem is 10-15 times lager than the area of land disturbed by mining [4].

During 16 years that have passed since the Chernobyl nuclear power plant catastrophe, the radiation situation in the country underwent considerable changes in the course of the following: natural effects or processes (natural disintegration of the radioactive nuclei, their fixation, and redistribution in the various components of the environment, other processes); implementation of the comprehensive countermeasures, which were aimed at the reduction of the radiation levels and the establishment of the radioecologically safe conditions for people living in the radioactively contaminated territories. The radioecological situation in the contaminated territories is monitored on the basis of data produced by radiation surveys. As a consequence of the Chernobyl nuclear power plant catastrophe, 8.4 million hectares of agricultural land were contaminated, out of which 3.5 million hectares were tillage land, nearly 400,000 hectares were natural pasturelands, and more than 3 million hectares were forest. Radiological contamination was the major reason for the abandonment of 180,000 hectares of arable land and 157,000 hectares of forest. The Ukrainian Law permits agricultural activities on the lands contaminated with Cs-137 at a level as high as 555 kBq/m² (15 Ci/km²). In 1998 the level of contamination with Cs-137 did not exceed 363 kBq/m² (~10 Ci/km²) [5].

The radionuclides in soil are mainly presented in fairly insoluble forms. The mobility of radionuclides can be however increased due to different local factors, among which the microbiological factors are believed to be the most important. Soil microorganisms are able to accumulate radioactive elements in their cells [6, 7]. It was shown that the number of bacteria in the soil samples contaminated with radionuclides was 7 times less as compared to the non-contaminated samples. The microorganisms responsible for nitrogen fixation in soils were found to be more sensitive to radioactive soil pollution. In particular, the population of *Azotobacterium* was 3 times, and that of *Clostridium* was 5 times lower than in the control [8]. The results obtained [9, 10] proved that most of the Cs-137 (47-93%) presented in topsoil was in a practically insoluble form. It is known the zone of radioactive anomaly caused by the Chernobyl disaster is characterized by complex pollution of the environment. In particular, it was established that the content of heavy metals in Chernobyl's soils had risen in comparison with control ones. The content of iron was 3-5 times higher than in control; those for cadmium, lead, magnesium, and calcium were 12-15, 2, 2-3, and 5-6 times, respectively [11].

The objective of our work was to elaborate methods of soil detoxification with humates and to reduce the risk of plant contamination with heavy metals, radionuclides, and herbicide traces.

2. Materials and Methods

The laboratory experiments were conducted using soil samples collected in the areas of soil pollution in Ukraine. They included common loamy chernozem and turf podzol soil. Humic preparations extracted form peat, brown coal, and other materials were used as detoxifying agents towards herbicides, heavy metals, and radionuclides. Barley (*Hordeum sativum L.*) and maize (*Zea mays L.*) were chosen as target plants.

The field experiments were conducted at the Samarsky farm of the Dniepropetrovsk State Agrarian University (DSAU). Humates and vermicompost were used as detoxifying agents.

In the first field experiment toxicity of Fe, Cu, Zn, Ni, and Pb in the selected soils was estimated tested. Heavy metal were introduced in the forms of salts at a rate of 30 kg/ha for $CuSO_4$, $ZnSO_4$, $Pb(OH)_2$, and at a rate of 3 kg/ha for $FeCl_3$, $Ni(NO_3)_2$. Plot area was 1 m^2, the experiments were conducted in 4 replications.

Toxicity of atrazine (after-effect in the next year after herbicide application) and 2,4-D (first year effect) was estimated in the laboratory and small-scale field experiments separately and with heavy metals pollution. Heavy metal salts were introduced into the soil at the doses indicated above. Manure (40 ton/ha) and $N_{60}P_{60}K_{60}$ were introduced into the soil in each experiment.

The application of humate preparations was tested in several field trials (A-introduction in soil, B-treatment of seeds before planting, C-surface treatment of plants). Vermicompost and preliminarily processed dung were tested in the same scheme at doses of 0.6-1.0 kg/m^2.

Laboratory and field experiments aimed to reduce radionuclide toxicity were conducted in the turf podzol soil. The field experiments aimed to reduce radionuclide accumulation through the application of humates were conducted in the Chernobyl District of Kiev region in 1988.

Experimental design included following variants: control (no treatment); $N_{90}P_{90}K_{90}$; $N_{90}P_{90}K_{90}$ + humate of Na (HNa) 0.3 t/ha; limestone 6 t/ha; limestone + HNa; manure 100 t/ha; $N_{90}P_{270}K_{90}$; $N_{90}P_{90}K_{270}$; $N_{90}P_{270}K_{270}$; $N_{90}P_{270}K_{270}$ +HNa. All the variants had 3 replications. Plot area for each variant was 2 m^2.

Turf podzol soil was characterized according unified methods [12] and was as follows: pH 7.1, humus content 1.11-1.16%, hydrolytic acidity 0.744 mg-eq./100 g of soil. Contents of exchangeable Ca and Mg was 3.2 and 0.8 mg-eq./100 g of soil, respectively. Specific radioactivity was $6 \cdot 10^{-6}$ Ci/kg.

Study of radionuclides accumulation in the barley plants in the presence of humate was conducted at the Novo-Zybkovsky field (Fertilizers & Agrochemistry Institute, Bryansk region, Russia). The experiments were conducted in 3 replications. Analysis of the obtained results included statistical data treatment using Least Significant Difference(LSD) index for the probability level 0.95.

The trace quantities of the herbicides were determined using high performance liquid chromatography methods. The total content of radionuclides was estimated using a radiometer.

3. Results and Discussion

The influence of humic-containing materials on the barley growth in the soil polluted with heavy metals was studied in the field trials. Results are presented in Table 1.

Table 1. Influence of humic-containing materials on barley growth in soils polluted with heavy metals.

№	Variant	Raw material & Country	Plant weight, g/m^2
1	Control (no treatment)	–	315
2	A		435
3	A+C	Brown coal, Kazakhstan	392
4	A+B		419
5	A+B+C		255
6	A+B	Brown coal, Ukraine	343
7	A+B+C		331
8	A+B	High-moor peat	348
9	A+B+C		300
10	A+B	Lowland peat	288
11	A+B+C		275
12		LSD = 46	

A – introduction in soil, B – treatment of seeds before planting, C – surface treatment of plants.

A combination of (A+B) treatment led, as a rule, to increases in the barley weight. This finding is evident for the detoxifying properties of humic-containing materials towards heavy metals.

In the laboratory experiment influence of preparation "Huminate" (sodium humate) on atrazine detoxification in barley plants was studied. Atrazine concentration was 5 mg/ml. Helrigel mixture was used as a background nutrition solution. Content of benzoxazinones, which facilitates atrazine hydrolysis, and atrazine in plants was used as a response.

The obtained results showed that "Huminate" promoted benzoxazinones formation in the barley plants (Table 2). That, in its turn, resulted in accelerated atrazine degradation in plants.

Table 2. Influence of preparation "Huminate" on benzoxazinones and atrazine content in 7-days seedlings of barley.

Variant	Atrazine content, mg/kg		Benzoxazinones content, mg/kg	
	Without treatment	Huminate treatment	Without treatment	Huminate treatment
Water			7.02±0.35	7.88±1.38
Water + atrazine	60.9±5.2	40.5±2.3	17.1±1.34	17.85±2.22
Helrigel mixture			7.2±1.16	10.26±1.21
Helrigel mixture + atrazine	48.0±1.3	38.2±1.0	19.4±3.18	29.45±4.23

"Huminate" application led to 20% decreasing of the atrazine content in the seedlings. Humate preparation introduction into the soil reduced also atrazine concentration to the traces (Table 3).

Table 3. Na-humate influence on atrazine degradation in chernozem.

Variant	Atrazine content, mg/kg	
	13.08	10.09
Atrazine, 4kg/ha, without plants	1.11±0.07	0.40±0.009
Atrazine, 4kg/ha, with plants	0.91±0.018	0.125±0.018
Atrazine + Na-humate, 100kg/ha, without plants	0.86±0.004	0.54±0.028
Atrazine + Na-humate, 100kg/ha, with plants	0.28±0.011	Traces

Atrazine after-effect and heavy metals toxicity in the presence of humic preparations was studied in the field experiments. Barley plants were used as a target object. The obtained results presented in Figure 1.

```
7
6
5
4
3
2
1
0
   Control      Heavy      Atrazine(A)   Biohumus+HM   Biohumus+A
              Metals(HM)
```

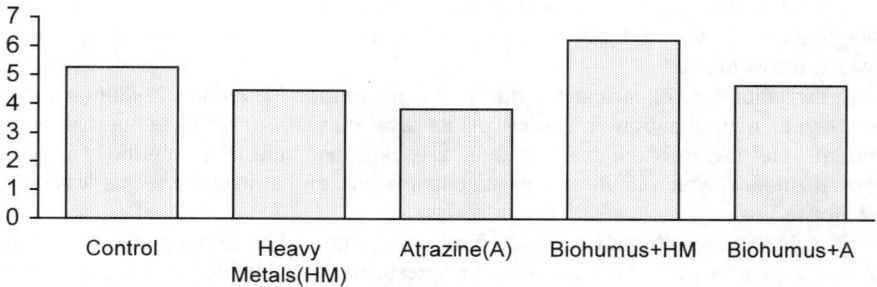

Figure 1. Influence of heavy metals and atrazine on weight of barley grain (ton/ha) in the presence and absence of humic preparation "Biohumus".

"Biohumus" (vermicompost product) introduction into the soil decreased toxicity of both heavy metals and atrazine. In particular, application of "Biohumus" and preliminary processed poultry dung into the soil provided the barley yield rising by 12-27%.

In the next field experiment "Biohumus" was additionally studied as a potential antidote in relation to combined toxicity of herbicide 2,4-D and heavy metals (Table 4).

Table 4. Influence of herbicide 2,4-D and heavy metals on barley weight (ton/ha) in the presence of Na-humates of different origin.

Herbicide	Fertilizers	Humate	Treatment	Without heavy metals	With heavy metals
–	$N_{60}P_{60}K_{60}$	–	–	5.06	
2,4-D	$N_{60}P_{60}K_{60}$	–	–	3.73	2.91
2,4-D	$N_{60}P_{60}K_{60}$	HNa	A (100 kg/ha)	3.86	3.35
2,4-D	$N_{60}P_{60}K_{60}$	HNa (brown coal)	B	4.2	4.26
2,4-D	$N_{60}P_{60}K_{60}$	HNa (lignite)	B	4.23	3.61
2,4-D	$N_{60}P_{60}K_{60}$	HNa (peat)	B	3.85	3.45
2,4-D	$N_{60}P_{60}K_{60}+$ "Biohumus"	–	–	5.96	5.76
2,4-D	$N_{60}P_{60}K_{60}+$ Dung	–	–	6.0	5.1
LSD			0.36		

A – introduction in soil, B – treatment of seeds before planting, C – surface treatment of plants.

2,4-D after-effect led to reduction of barley productivity up to 35.6% of control. Treatment of seeds with humates before planting provided a more pronounced detoxifying effect from heavy metals pollution. The greatest beneficial effect of humates was observed when preparation "Biohumus" was introduced into chernozem.

Influence of humates on sorption properties of soil and mobility of the radionuclides was determined in batch and column experiments. Before humate introduction the contents of both water-soluble and ion exchangeable forms of radionuclides were insignificant and were 0.5-2.4 and 3.6-6.2%, respectively [10, 13]. The data on humate influence on the desorption of radionuclides from soil are presented in the Table 5.

Table 5. Impact of the organic and inorganic amendments on the desorption of radionuclides from soil, % of the control.

Amendments to the soil sample, g/100g soil	Solution for desorption			
	H_2O	$CaCl_2$	KCl	Active
Control (without amendments)	100	100	100	100
Humate Na (brown coal), 0.05	86.6	47	98	84
Humate Na (brown coal), 0.25	140	115	96	105
Humate Na (brown coal), 1.0	275	77	96	106
Humate Na (peat), 0.05	100	107	92	103
Humate Na (peat), 0.25	189	100	103	106
Humate Na (peat), 1.0	350	81	78	102
Humic acid (peat), 0.25	63	105	90	92
Humate calcium (peat), 0.25	100	133	98	107
EDTA, 2.5	1508	28	91	259
Zeolite, 10	82	118	56	75
Limestone, 2.5	109	92	104	101
Limestone, 2.5 + humate Na, 0.25	106	36	87	99
Phosphate, 0.4	100	133	53	83
Phosphate, 0.4 + humate Na, 0.25	109	123	60	81
Natural phosphate meal, 0.6	92	113	101	103
Natural phosphate meal, 0.6 + humate Na, 0.25	108	71	95	88

The radionuclides sorption by turf podzol soil increased when zeolite and phosphates (separately and with humate) were added to soil samples.

The field study on the reduction of the radionuclide accumulation in plants in the presence of humates was conducted in the Chernobyl District of Kiev region. Different doses of phosphates, potassium fertilizers, and limestone were applied in together with humates (Table 6).

Table 6. Impact of the inorganic and organic fertilizers on the crop yield and radionuclide uptake by the crops, % of the control (Numerator-crop yield related to control, denominator-radionuclide uptake by the crops).

№	Variant	Barley		Lupine	Potato Tubers
		Grain	Straw		
1	Control	100	100	100	100
		100	100	100	100
2	$N_{90}P_{90}K_{90}$	146.7	132	77.8	122
		83.6	62.5	125.0	84.5
3	$N_{90}P_{90}K_{90}$ + humate of Na (HNa)	220	289	119	162
		88.1	73.0	119	90
4	Limestone	196	272	54	131
		89.2	58.0	106.0	81.2
5	Limestone + HNa	207.8	263.0	78.0	120
		127	45.0	144	84.0
6	Manure	191.2	270	118	209
		65.0	44.6	108	60.0
7	$N_{90}P_{270}K_{90}$	178	205	110	174
		80.0	55.5	127	76.0
8	$N_{90}P_{90}K_{270}$	249.4	310	106	229
		79.0	42.3	92	63
9	$N_{90}P_{270}K_{270}$	296	354	66	251
		63.0	48.6	100	69
10	$N_{90}P_{270}K_{270}$ + HNa	305.6	420	128	202
		86.0	70.8	104.0	72.0

In particular, the increasing of fertilizers dose from 0.09 to 0.27 ton/ha has led to the reduction of radionuclides migration in soil-plant system and the increasing of crops yield. The most pronounced effect was observed when for the following variants: $N_{90}P_{90}K_{270}$, $N_{90}P_{270}K_{270}$ and $N_{90}P_{270}K_{270}$ + humate of Na. In the case of lupine planting, the same effect was not seen. The triple dose of potassium, phosphorus, or potassium and phosphorus was enough to decrease the radionuclide migration in the soil-potato system.

These data are in closed relation with regulations, which were worked out with the Ukrainian Institute of Agricultural Radiology (UIAR) to decrease production of polluted agricultural crops in the districts affected after Chernobyl accident [14]. UIAR suggestions include application of peat, peaty sapropel and some composts.

Based on the obtained results the recommendations on agriculture and crop growing under radionuclide pollution in the Ukrainian territories for the period 1999-2002 were elaborated [15]. Unfortunately, humates were not taken in account for these recommendations as additional measure to decrease the radionuclide uptake in the soil-plant system.

Some decrease in radionuclide uptake was also found in another experiment with barley in 1988 for the variant $N_{90}P_{60}K_{60}+250$ kg/ha humate of Na in the Bryansk region of Russia. The field study on reduction of radionuclide accumulation in the soil-plant system due to humates application was conducted on the farm of the Novo-Zybkovsky filial of the Fertilizers & Agrochemistry Institute (Bryansk region, Russia). Turf podzol soil had following properties: pH 6.9, humus content 1.6-1.8%, hydrolytic acidity 0.85 mg-eq./100 g of soil, amount of exchangeable bases – 127 mg-eqv/100 g of soil, specific radioactivity $1.0 \cdot 10^{-7}$Bq/kg. The radionuclide accumulation in the barley grain decreased in the dose 0.25 ton/ha (Table 7).

Table 7. Impact of humate on barley yield and radionuclide uptake, % of the control.

№	Variant	Yield	Mass of 1000 seeds	Specific beta activity
1	Control (without fertilizers)	100	100	100
2	$N_{90}P_{60}K_{60}$	181	106	131
3	$N_{90}P_{60}K_{60}$+50kg/ha HNa	210	103	116
4	$N_{90}P_{60}K_{60}$+100 kg/ha HNa	207	110	141
5	$N_{90}P_{60}K_{60}$+150 kg/ha HNa	241	111	118
6	$N_{90}P_{60}K_{60}$+200 kg/ha HNa	264	112	105
7	$N_{90}P_{60}K_{60}$+250 kg/ha HNa	185	102	89

The data obtained in the field experiments were reflected in the DSAU recommendations on the humate application for crops[16]. Future prospects for humate application in cases of soil polluted with radionuclides seem to require additional experiments.

4. Conclusion

The results showed that the introduction of humates into polluted soil facilitated reduction of total soil toxicity. It was also shown that humic fertilizers and vermicompost were more efficient compared to ordinary fertilizers. Model experiments showed pronounced detoxifying effects of humates depending on dose, soil properties, and kinds of humic amendments.

5. Acknowledgments

Credit is due to people who provided the base for this investigation, including professors Lidia A. Khristeva and Ivan. I. Yarchuk.

6. References

1. Gritsan, N. and Babiy, A. (2000) Case study. Hazardous materials in the environment of Dnepropetrovsk Region (Ukraine), *Hazard. Mater.* **A76,** 59-70.
2. Kharytonov, M., Gritsan, N. and Anisimova, L. (2001) Environmental problems connected with air pollution in the industrial regions of Ukraine, in Abstract Book of NATO ARW: *Global Atmospheric Change and its Impact on Regional Air Quality,* Irkutsk, Lake Baikal, Russian Federation.
3. Babiy, A., Kharytonov, M. and Gritsan, N. (2002) Connection between emissions and concentrations of atmospheric pollutants, in Abstract Book of NATO ARW: *Air Pollution Processes in Regional Scale,* Kallithea, Halkidiki, Greece.
4. Kharitonov, N. (1996) Ecotoxicological problems under mining at the Ukrainian steppe, *30th International Geological Congress,* Beijing, China, **3,** 599.
5. National report on the environmental conditions in Ukraine for 1998 prepared by the Ministry for Environment Protection and Nuclear Safety of Ukraine (CD-ROM publication, Web pages created by the Ministry for Environmental Protection and Nuclear Safety of Ukraine, www.grida.no).
6. Haselwandter, K. and Berrect, M. (1988) Fungi as bioindicators of radiocesium contamination: pre- and post- Chernobyl activities, *Trans. Brit. Mycol. Soc.* **90 (2),** 171-174.
7. Sidorenko, L.P. and Kljonus, V.G. (1989) Fungi ability to accumulate strontium-90 and cesium-137, in *Int. Radiobiological Congress,* Pushchino, p. 986 (in Russian).
8. Korbanyuk, R. and Kharitonov, N. (1996) Treatment of contaminated soil from Chernobyl zone, in Abstracts Proceedings of *Second Int. Congress of the European Society for Soil Conservation,* Munchen, Germany.
9. Kravettz, A.P., Grodzinsky, D.M., Zhdanova, N.N., Vasilevskaya, A.J. and Sinyavskaya, O.L. (1992) Interaction of soil micromycetes with "hot" particles in the simple model system and in system soil-plant, in *Proc. Int. Symp. on Radioecology. Chemical speciation-Hot Particle,* Znojmo. October 12-16.
10. Korobova E.M., Uraletz T.M., and Ermakov A.I.(1995) Microbiological factor in mobilization of Cs-137 and Sr-90 in soils contaminated after the Chernobyl accident, in *Third Int. Conf. on the Biogeochemistry of Trace elements,* Paris, France. Theme A 3.
11. Filonik, I. and Guscha, N. (1995) Study of heavy metals accumulation in soils and cereals from Chernobyl zone, in *Third Int. Conf. on the Biogeochemistry of Trace Elements,* Paris, France. Theme A 1.
12. Agrochemical methods of soils investigation (1965) Nauka, Moscow (in Russian).
13. Korbanyuk, R., Kharitonov, N. and Bulgakova, M. (1997) Rehabilitation effects of the humic preparations, in *Natural and Agricultural Ecosystems in Peatlands and their Management,* Saint Malo, France, **1,** 38.
14. Prister, B.S., Perepelyatnikova, L.V. and Perepelyatnikov, G.P. (1991) Effective measures which are oriented on plant husbandry production pollution decreasing in the districts affected after Chernobyl accident, *Agricultural Radiology Problems,* 141-153.
15. Conducting of agriculture in conditions of radiological pollution for the territories of Ukraine after Chernobyl nuclear power station accident for the period 1999 -2002 (1998) in *Recommendations,* Kiev (in Ukrainian).
16. *Recommendation on the humate sodium application for crops* (1991) Dnepropetrovsk State Agrarian University (in Russian).

Part 5

Quantifying structure and properties of humic
substances and example studies on design of humic
materials of the desired properties

MOLECULAR LEVEL STRUCTURAL ANALYSIS OF NATURAL ORGANIC MATTER AND OF HUMIC SUBSTANCES BY MULTINUCLEAR AND HIGHER DIMENSIONAL NMR SPECTROSCOPY

N. HERTKORN, A. KETTRUP
GSF Institute of Ecological Chemistry, Ingolstaedter Landstrasse 1, D-85758 Neuherberg, Germany

Abstract

NMR spectroscopy offers uniquely versatile options to obtain molecular level structural characterization of natural organic matter (NOM) and humic substances (HS). By defining the relative amounts and some structural detail of fundamental building blocks, multinuclear quantitative one dimensional NMR spectroscopy provides the key margin of any structural model of NOM/HS. A large array of higher dimensional NMR spectra serves to enhance the reliability of NMR resonance assignments and allows to define rather extended substructures of NOM/HS.

Owing to the nature of NOM/HS as a mixture of small and large molecules, related by a continuous range of weak and strong interactions, any exhaustive analysis of the NMR spectra of NOM/HS requires understanding of the physical processes initiated by the pulse sequences. The NMR spectral analysis has to be corroborated at all levels by comparison with NMR spectra of model compounds, data base search and adjusted schemes of calculations of NMR parameters.

1. Role and significance of NOM

Natural organic matter (NOM) is an operationally defined, enormously complex mixture of organic, and some inorganic, constituents occurring in terrestrial, limnic and marine ecosystems [1-7]. On a global basis, NOM on average greatly outweighs biochemicals in the living organisms from which they derive. The origins, reactions and fates of these ubiquitous and inconspicuous materials are relatively obscure, in large part because the rich vein of geochemical information that typically derives from detailed structural and stereochemical analysis is yet to be tapped [8]. In recent years, the perception of NOM has evolved from an emphasis of a largely separate pool of remarkably old and static substances to the current view of a dynamic assemblage of organic molecules that interact with each other, trace metals and living organisms over a broad continuum of space and time scales [8].

391

I. V. Perminova et al. (eds.),
Use of Humic Substances to Remediate Polluted Environments: From Theory to Practice, 391–435.
© 2005 *Springer. Printed in the Netherlands.*

NOM plays immensely important roles in the natural world, and it is a key refractory constituent of the global carbon (cf. Figure 1) and other element cycles [4, 5, 9]. To a significant extent, NOM determines the binding and bioavailability of toxic and nutrient metal ions and organic xenobiotics and it is involved in such key processes as modulating temperatures at the globe's surface, weathering rocks to soils, stabilization of atmospheric levels of oxygen over geological time scales [10], and composing precursors for eventual formation of coal and petroleum. Organic compounds embedded in marine sediments and paleosols also provide exquisitely detailed records of natural history, even where macroscopic physical fossils are rare or absent (as in petroleum) [8].

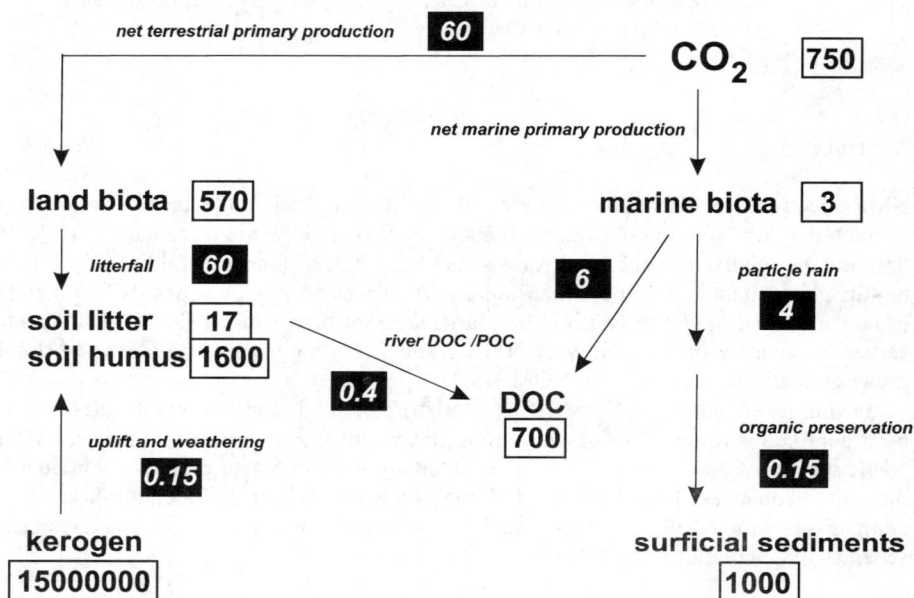

Figure 1. Selected global reservoirs (white boxes) and fluxes (*italics* in black box) of various NOM reservoirs in the geosphere [Gt or Gt/a] [4].

1.1. CHARACTERISTICS AND CONSTRAINTS OF THE SPECTROSCOPIC CHARACTERIZATION OF NOM AND HS

From the perspective of analytical chemistry, both NOM and HS represent the structurally most diverse class of natural organic polymers, and they are characterized by an extensive polydispersity and a pronounced irregularity of their molecular level structure [11-13]. The stochastic synthesis of NOM/HS out of biotic and abiotic precursors is governed by fundamental restraints of thermodynamics and kinetics and follows no genetic code.

The level of intricacy in the analysis of large molecules and of mixtures can be classified according to their polydispersity and heterogeneity (cf. Figure 2). The structures of complicated, but monodisperse, natural products and biopolymers are readily accessible (provided that sufficient amounts of materials are available) by a combination of analytical methods, which primarily rely on NMR spectroscopy and mass spectrometry [14-17]. Supramolecular structures [18], composed of (modified) biopolymers aligned in aggregates, which are supported and defined by weak interactions, require a more elaborate characterization, which requires an adequate definition of covalently bonded molecules and of their non covalent interactions [14]. Consequently, the characterization and structural analysis [19, 20] of geopolymers, which feature a substantial extent of both polydispersity and molecular heterogeneity [11], is most demanding with respect to methodology and concepts.

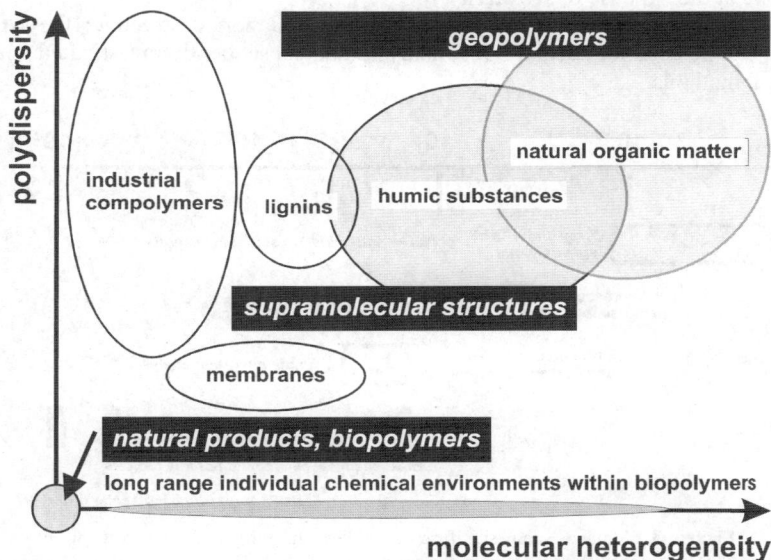

Figure 2. Hierarchical order of complexity in the structural analysis of materials in terms of polydispersity and molecular heterogeneity (cf. text).

A highly resolved three–dimensional structure of a monodisperse biopolymer is based on a precise description of the chemical environment of any single atom [21]. As a consequence, e.g. in a NMR data set of a protein, any atom will show unique values of (multinuclear) chemical shifts δ and (homo- and heteronuclear) coupling constants (D, J), which are defined by the respective short and long range chemical environment. Currently, the molecular level structural analysis of NOM/HS is primarily focused on the definition of the covalent bonds. In an ongoing evolution, future high quality structural analyses of NOM/HS will have to provide a characterization of individual molecules and a description of the extent and mechanisms of their interactions.

Any adequate understanding of geophysical and biological processes requires a sufficient characterization of the participating and constituent materials. Since diffraction methods do not provide usable structural data of mixtures, structural analysis of NOM/HS has to rely on data from complementary spectroscopic methods [17, 22-24], which – in combination – can lead to the deduction of precisely defined chemical (sub)structures.

Many methods of structural spectroscopy rely upon transitions between energy levels in atoms and molecules, and the respective spectra are defined by a substance specific set of energy and intensity values. The interaction of atoms and molecules with light across the electromagnetic spectrum induces an array of physical processes, which can be utilized to reveal structural information. High energy spectroscopic techniques are capable to sensitively determine the content of elements in NOM/HS (AAS, AES) and, with some uncertainty, also their oxidation state [XANES]. When proceeding to ever diminutive splittings of energy levels (XANES [25-28] →UV [5,29] → IR [29,30] → NMR [31-34]), a continually enhanced resolution of organic structures will become accessible at the expense of sensitivity.

NMR spectroscopy is a low energy method and non destructive; the NOM/HS samples can be used for several NMR analyses and subsequently investigated with other analytical methods.

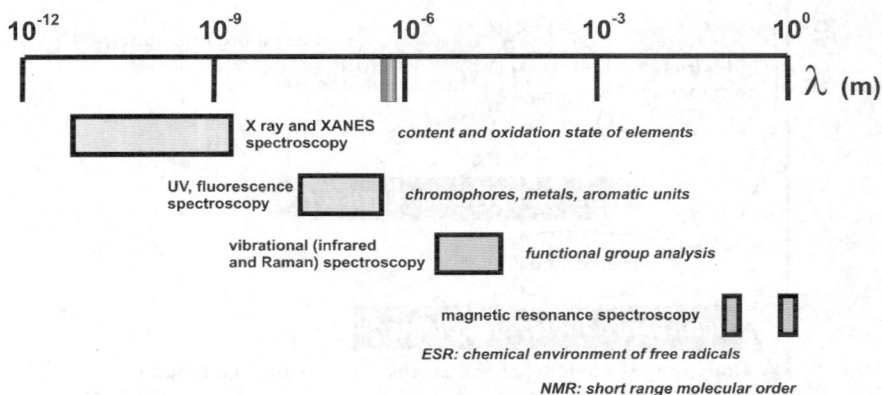

Figure 3. Methods and significance (italics) of structural spectroscopy of NOM/HS across the electromagnetic spectrum.

Analytical methods have recently become capable to separate nearly identical large molecules [35]; however, the analytical chemistry of inseparable mixtures has been neglected and remains significantly underdeveloped in terms of practical methods and concepts even today. Attempts to develop a precise molecular level structural analysis of NOM/HS will therefore address several independent and important topics:

1) An accurate description of the molecular level structure of NOM/HS constitutes the fundamental requirement to understand the mechanisms and significance of its interaction with minerals, organic and inorganic compounds and living organisms in the eco- and geosphere.

2) An exhaustive analysis of data of NOM/HS characterization derived from complementary analytical techniques, aided by mathematical analysis, will contribute significantly to an evaluation of the scope and accuracy of each individual method.

3) A hierarchical order of carefully tested analytical methods and data treatment schemes will contribute to develop and evaluate concepts of the analytical chemistry of inseparable mixtures.

NOM/HS represents an enormously complex mixture of organic and inorganic materials and any spectra of these substances will feature a very significant overlap from individual species present and, consequently, a highly degraded resolution. A significant degree of the uncertainty in the spectral assignment of NOM/HS originates from the fact that both positions of spectral lines (energy levels) and intensities (transition integral values) of a highly complex array of structures and interactions cannot be calculated with an accuracy anywhere near realistic values directly from quantum mechanical *ab-initio* or semi empirical methods [36] (even more so, as precise structural models of NOM/HS are not yet available).

Quantification remains one of the key unsatisfactorily resolved research issues in the analysis of NOM/HS. In any method, complicated non linear and unknown transfer functions are caused by a near continuum of differential responses arising from small and large molecules and strong and/or weak interactions within the NOM/HS system (cf. Figure 4). Even otherwise well established methods, such as elemental analysis, or the determination of the content of heavy metals by atomic absorption spectroscopy, often produce a very wide range of data values [37], when applied in a routine fashion to the analysis of NOM/HS.

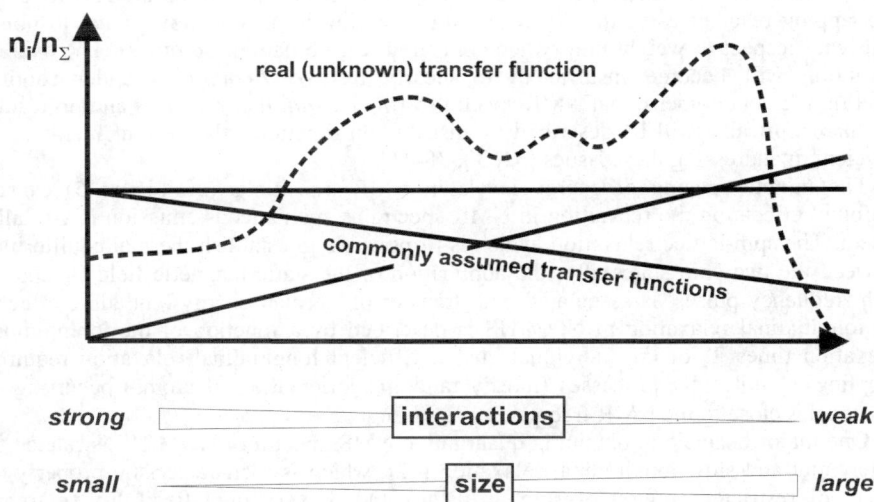

Figure 4. Commonly assumed (linear) and unknown real (non linear) transfer functions in the analytical characterization of NOM/HS.

Therefore, every detail of any spectral analysis of NOM/HS has to be very carefully evaluated and calibrated, both intrinsically and by comparison with realistic reference materials, which are not routinely available on a commercial basis. The International Humic Substances Society IHSS (www.ihss.gatech.edu) provides an ever increasing selection of humic substances reference and standard materials, which are valuable for referencing and testing in method development.

1.2. THE SPECIAL ROLE OF NMR SPECTROSCOPY IN THE MOLECULAR LEVEL CHARACTERIZATION OF NOM/HS

The two most significant and most highly resolving methods of organic structural spectroscopy are magnetic resonance spectroscopy and mass spectrometry. Mass spectrometry involves measurement of mass numbers rather than energy absorption and is not (directly) associated with a region of the electromagnetic spectrum [17]. Magnetic resonance spectroscopy is concerned with the splitting of magnetic spins of electrons (electron spin resonance, ESR) and that of atomic nuclei (nuclear magnetic resonance, NMR) in an external magnetic field B_0. The splitting of the NMR transition is not solely an intrinsic molecular property, but also depends on the magnitude of an external magnetic field B_0. The special role of NMR spectroscopy in the molecular level characterization of NOM/HS resides in its ability to provide unsurpassed in-depth, isotope specific information about short range molecular order.

Owing to the nature of NOM/HS as a complex mixture of organic and inorganic compounds, any NMR parameter will be characterized by a weighed average and by a distribution of values. *NMR shieldings* and *resonance integrals* of individual chemical environments in NOM/HS will be superimposed to produce a broad envelope of overlapping resonances in the NMR spectra, resulting in a rather low signal to noise ratio with respect to weight unit (when compared, e.g. to natural products). The overall resolution will become insufficient to clearly resolve *J-couplings* under routine conditions in one dimensional NMR spectra. *NMR relaxation parameters* and, in result, the *linewidths* also will be described by distribution functions; the current literature is deficient in addressing these issues [31, 33, 38-41].

The minute spacing of energy levels in NMR transitions (cf. Figure 5) causes profound effects on the relaxation in NMR spectra as spontaneous emission is virtually absent. The spin-lattice relaxation, which is responsible to establish thermal equilibrium between the two spin states after the application of the static magnetic field B_0 and of high frequency pulses, is associated with transfer of Zeeman energy (enthalpic effect); the longitudinal relaxation in NOM/HS is described by a function of the longitudinal relaxation times T_1 of the individual nuclei. Efficient longitudinal relaxation requires coupling to molecular processes (mostly random reorientation of atoms) occurring at frequencies close to the NMR frequencies ω_0 or $2\omega_0$.

One major obstacle in obtaining quantitative NMR spectra of NOM/HS is related to differential and slow longitudinal relaxation [42], which is a characteristic property of nuclei of restricted (region of spin diffusion) and of very high flexibility (extreme narrowing limit). Diamonds are rather extreme examples of a rigid lattice and their purity (i. e. absence of lattice defects) can be determined from T_1 values in their ^{13}C NMR spectra [43, 44]. In a crude approximation, the NMR longitudinal relaxation of heteronuclei within NOM/HS is inherently related to their NMR frequencies and to their distance to adjacent protons and free electrons. It is a peculiar feature of NOM/HS that both heteronuclei behaving according to the extreme narrowing limit and to spin diffusion (or in between) may be present in the same material.

Transverse (or spin-spin) relaxation involves mutual exchange of spin energy without altering the total Zeeman energy of the nuclear spins. The net effect of this energy transfer is to cause a loss of phase coherence (entropic process), equivalent to a decay of bulk magnetization in the plane of the detection (xy-plane). The linewidth $\Delta\nu$ in NMR spectra is governed by the rate of transverse relaxation (under the assumption of single exponential decay): $\Delta v = 1/2\pi T_2$, with T_2 representing a function of the transverse relaxation times of individual nuclei. Very fast transverse relaxation, as induced by paramagnetic centres (metal ions [21, 45-47]: especially Fe^{2+} and Fe^{3+}, and organic radicals) and by the limited flexibility of large size molecules and aggregates, may broaden NMR resonances of NOM/HS beyond recognition.

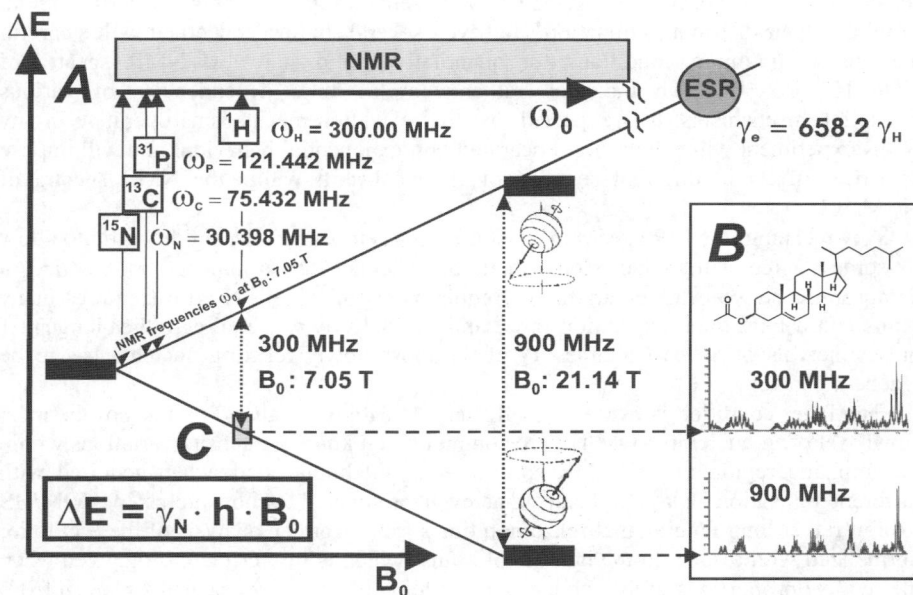

Figure 5. Dependence of some magnetic resonance spectral characteristics on the magnetic field B_0; the electron-proton magnetic moment ratio is 658.2 (**A**) At constant magnetic field B_0, the resonance frequency depends on the relative magnetogyric ratios (or magnetic moments) of the nuclei γ_N (NMR) and the electron γ_e (ESR); (**B**) sections of proton NMR spectra of cholesterol acetate at B_0 = 7.05 T (300 MHz ^{1}H frequency) and at B_0 = 21.14 T (900 MHz ^{1}H frequency); note the *qualitative* difference of high resolution proton NMR spectra acquired at various B_0 values. This variation remains the most significant obstacle for automated NMR resonance assignment in natural products. As coupling constants have less significance in one dimensional NMR spectra of NOM/HS, the variance in the NMR spectra of NOM/HS is much less pronounced, and common assignment procedures, based on integration (cf. section 4) are valid. Pattern recognition in 2D NMR spectra can be automated; NMR spectra of NOM/HS acquired at increased B_0 show enhanced resolution and improved assignment options. (**C**) The relative energy, represented by the chemical shift range, covers a miniscule ["ppm"] range of the (already tiny) NMR energy transition energy; the ratio of total chemical shift range to total NMR transition energy ranges from 20 ppm (^{1}H, diamagnetic molecules) up to 20000 ppm (^{59}Co, ^{195}Pt NMR). Owing to the near equality of the Boltzmann factors for the NMR energy levels, out of 2000000 proton nuclei, only 81 participate in the ^{1}H NMR experiment at B_0 = 11.7 T and 283 in the NMR experiment at B_0 = 21.14 T (at room temperature: 300 K). All other proton nuclei remain silent throughout the NMR experiment. This ratio is even worse for other nuclei, explaining the relative insensitivity of NMR spectroscopy when compared to higher energy spectroscopic methods.

NOM/HS as a complex assemblage of small and large molecules, related by a broad range of various interactions, enacts strong effects on the local mobility of its individual atoms in their respective chemical environments; therefore, the above mentioned fundamental relaxation processes are composed of a broad range of mechanisms [48-51]. In the available literature, the NMR relaxation times determined within high resolution NMR spectra of NOM/HS represent weighed average values without providing their distribution function. However, several studies, concerned with a careful analysis of optimum conditions for quantitative solid state ^{13}C NMR spectra of NOM/HS [52-60] have indicated rather complex time dependencies of various relaxation mechanisms. It is expected that the relaxation mechanisms operative in any NMR experiment will exhibit non-linear and non-exponential behaviour and will impose selective effects in different sections of chemical shift within the NMR spectra of NOM/HS.

Any meaningful NMR spectrum of NOM/HS has to show a decent S/N ratio to allow to some degree valid conclusions about the chemical environments; in practice, a compromise between the contradicting requirements of large S/N ratio (requires many scans and a long total acquisition time) and non-selectivity (requires rather long inter pulse intervals of at least 5 times T_1 of the most slowly relaxing nucleus) has to be reached.

The latter condition is exceedingly difficult to satisfy because T_1 values of the most slowly relaxing nuclei in NOM/HS are commonly not known. Intensity variations within the aliphatic region of ^{13}C NMR spectra of a soil humic acid, when acquired with different relaxation delays indicate, that even protonated carbon atoms in NOM/HS exhibit rather long longitudinal relaxation times [52]. From an analysis of the S/N ratio, normalized with respect to the number of scans (which is further increasing, even when the relative proportions of the fundamental building blocks have reached a plateau [61]) it can be deduced that (^{13}C) T_1 values of individual nuclei in the order of dozens of seconds are rather common in NOM/HS. The parameters used in our laboratory for the acquisition of routine quantitative NMR spectra of NOM/HS with a modern 500 MHz ($B_0 = 11.7$ T) spectrometer, representing the practical compromise between a meaningful S/N ratio and total acquisition time, are given in Table 6 (cf. Section 9).

2. NMR Sensitivity of Nuclei of Interest within NOM/HS Research

The NMR receptivity D is proportional to the signal to noise ratio (S/N) and it is defined as the product of the NMR sensitivity and the number of nuclei [62], and depends on the following variables:

$$D \sim S/N \sim N\, \gamma_{exc}\, \gamma_{det}^{3/2}\, B_0^{3/2}\, (NS)^{1/2}\, T_2^*/T\, [I\,(I+1)] \tag{1}$$

Here, N is the number of nuclei (corresponding either to natural or to a (de)enriched isotopic abundance), γ_{exc} is the magnetogyric ratio of the excited spin [rad s^{-1}T^{-1}]; γ_{det} is the magnetogyric ratio of the detected spin [rad s^{-1}T^{-1}]; B_0 is the static external magnetic field [T]; NS is the number of experiments; $1/T_2^*$ is the homogeneous line width [Hz]; T is the absolute temperature [K]; I is the isotope spin quantum number. Other equations, relating D to spectrometer-linked parameters [63] are useful to further optimise acquisition and sample conditions.

The NMR nuclei of primary interest in NOM/HS research are depicted in Figure 6 according to their significance: *constituent nuclei* form the backbone or the primary functional groups of NOM/HS; *tag nuclei* serve to characterize functional groups in NOM/HS after chemical derivatization (cf. Fig. 18); *metal nuclei* are those with favourable NMR characteristics due to high NMR receptivity with a modest (^{27}Al, ^{51}V) or absent (I = ½ : ^{113}Cd, ^{195}Pt, ^{199}Hg, ^{205}Tl, ^{207}Pb) quadrupole moment, and the *"pore"* *nuclei* are the noble gases ^{3}He and ^{129}Xe, which can be spin polarized [64] to sensitively detect pore size distributions by means of solid – state NMR spectroscopy of NOM/HS [65], but which also show potential to efficiently transfer magnetization to dissolved NOM/HS [66].

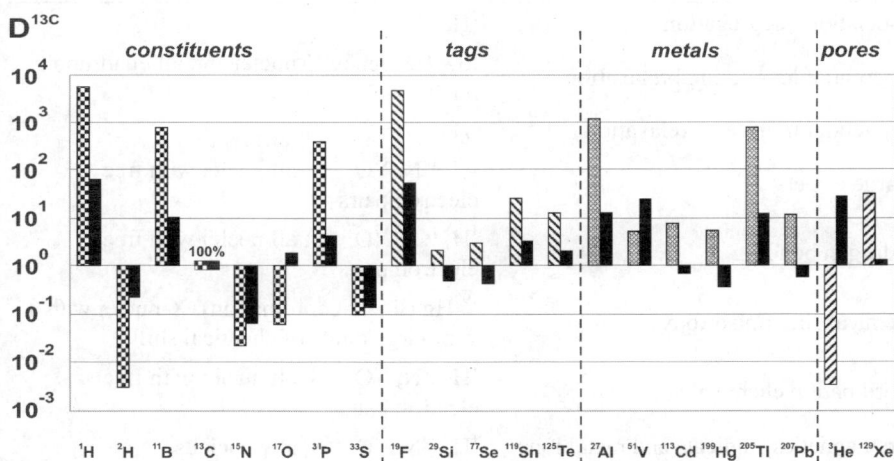

Figure 6. NMR receptivity of nuclei of primary interest in NOM/HS research at natural abundance, normalized to ^{13}C$_{nat}$ = 1 (left columns) and at 100% isotopic abundance, normalized to ^{13}C$_{100\%}$ = 1 [right (black) columns] (cf. text); ^{31}P is a constituent (functional group) [67-69] and a tag [70, 71] nucleus.

3. Relationship of NMR Chemical Shift Values and Resonance Integrals to Structural Parameters in NOM/HS

In the NMR based structural analysis of NOM/HS, the most widely used approach is to relate chemical shifts and NMR resonance integrals as observed in a NMR spectrum to structural features in NOM/HS. This direct approach, however, when applied without appropriate care, can lead to serious misinterpretations, because the observed resonance positions in NMR spectra of NOM/HS are a result of various "direct' and "indirect" effects and interactions. The so-called "direct" effects directly, understandably and quantitatively relate chemical shifts and resonance integrals in experimental NMR spectra with chemical environments in NOM/HS. The so-called "indirect" effects are caused by additional interactions occurring in NOM/HS, which affect chemical shifts and resonance intensities in experimental NMR spectra and therefore obscure the facile conclusions drawn by the sole analysis of "direct" effects. A list of "indirect" effects is provided in Table 1 together with the NMR nuclei most likely to become affected; in sections 4.1 through 4.5 a more in-depth outline will be provided. The peculiar nature of NOM/HS as a complex mixture of small and large molecules in association with minerals and related by a diversity of interactions requires that a dependable NMR analysis has to properly address both "direct" and "indirect" effects on the chemical shift.

Table 1. "indirect' effects on the chemical shift in NOM/HS (cf. text).

name of effect	NMR nuclei most likely to become affected
association / aggregation	^{1}H, ^{19}F
differential longitudinal relaxation	low frequency X-nuclei and all quadrupole nuclei
differential transverse relaxation	^{1}H
charge effects	^{1}H, ^{15}N, ^{17}O, ^{33}S, all nuclei with free electron pairs
hydrogen bonding	^{1}H, ^{15}N, ^{17}O, ^{33}S, all nuclei with free electron pairs
chemical shift anisotropy	^{199}Hg (linear coordination), X-nuclei with a very large range of chemical shift
coordination chemical shifts	^{1}H, ^{15}N, ^{17}O, ^{33}S, all nuclei with free electron pairs
paramagnetics (radicals and metal ions)	^{1}H, ^{15}N, ^{17}O, ^{33}S, any nucleus
chemical exchange	^{1}H, ^{17}O, metals
severe background resonance from probe	^{11}B, ^{17}O, ^{23}Na, ^{27}Al, ^{29}Si

3.1. PROTON NMR SPECTROSCOPY OF NOM/HS

Hydrogen in NOM/HS is bonded to carbon atoms, forming stable covalent bonds (NOM/HS backbone), and to heteroatoms (O > N > S), forming labile bonds (NOM/HS functional groups), which are susceptible to chemical exchange on the NMR timescale (cf. section. 3.4). Proton NMR spectra of NOM/HS in NaOD, acquired under routine conditions, combine all of the labile proton species into a single NMR resonance, which represents the weighed average chemical shift of all chemically exchanging protons in NOM/HS. A highly efficient suppression of this very strong NMR resonance is required and various experimental protocols have been applied [72] to obtain the relative quantities of non-exchangeable hydrogen in various substructures of NOM/HS.

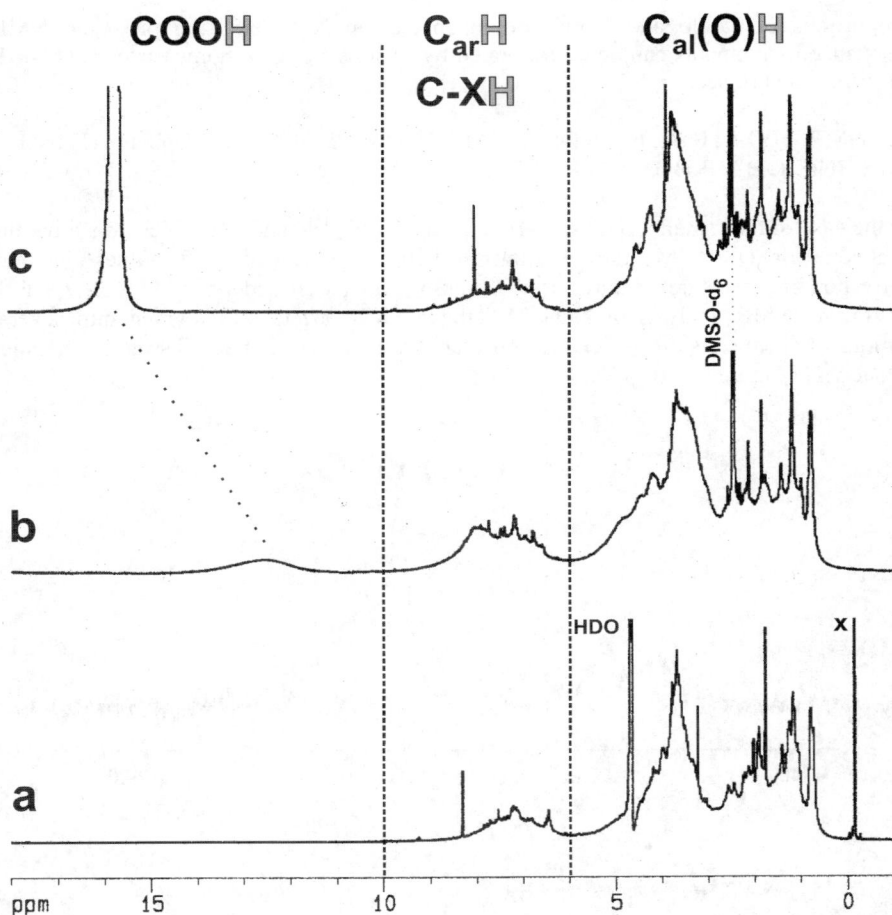

Figure 7. ^{1}H NMR spectra of a soil humic substance Bonzule FA 3; (a) in 0.1N NaOD (x: reference $(H_3C)_3Si-CD_2-CD_2-COONa$: -0.14 ppm) with solvent suppression; (b) in DMSO-d_6 under total exclusion of moisture; single pulse NMR spectrum; (c) in DMSO-d_6 with some CF_3COOD added; single pulse NMR spectrum.

In DMSO-d_6 under total exclusion of moisture, the chemical exchange of labile protons is slowed down sufficiently to allow the observation of different forms of these proton species within NOM/HS, which appear in distinctly separate regions of chemical shift (cf. figure 14) [73]. Consecutive addition of trifluoroacetic acid (preferably CF_3COOD) combines all labile proton NMR resonances into a single far downfield signal ($\delta(^1H)$ = 13-17 ppm). Taking into account subtle variations of chemical shifts caused by effects of pH variation on the charge and the aggregation of NOM/HS, a comparison of the three proton NMR spectra of NOM/HS acquired in NaOD, DMSO-d_6 and DMSO-d_6 / CF_3COOD produces a very detailed assessment of the relative proportions of stable and labile proton species in NOM/HS; the proposed NMR resonance assignments can be corroborated by various forms of homonuclear 2D NMR spectroscopy (cf. section 5.1).

3.2. NMR INVESTIGATIONS OF "SMALL" SPIN-1/2 NUCLEI, CONSTITUTING THE HUMIC BACKBONE STRUCTURE

In the case of the "small sized" spin-1/2 nuclei ^1H, ^{13}C, ^{15}N and ^{31}P, which constitute the backbone of HS/NOM, the relationship between chemical shift values and the corresponding structural features of NOM/HS is rather well understood [31, 32, 68, 69]. In routine NMR analysis of NOM/HS, these NMR spectra are divided into several regions of chemical shift, which are considered to represent coarse substructure regimes as shown in Figures 8-10.

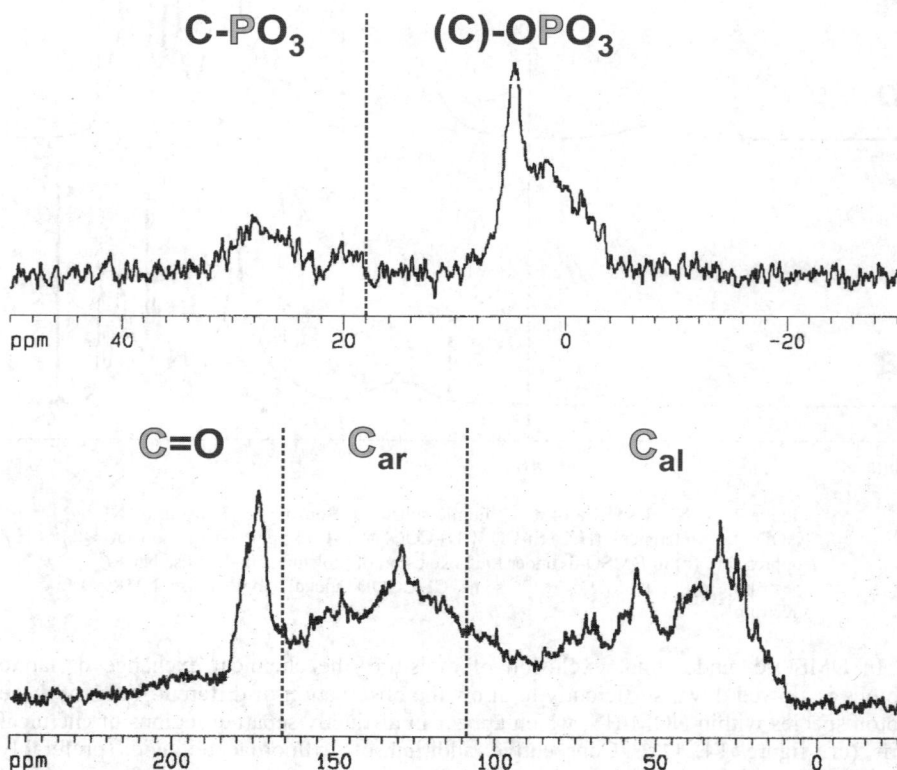

Figure 8. ^{13}C NMR (bottom) of a municipal solid waste humic acid, and ^{31}P NMR (top) spectrum of a marine humic substance, with the chemical shift ranges of the fundamental building blocks indicated.

These quantitative 1D NMR spectra define the relative proportions of the fundamental building blocks of NOM/HS and therefore provide a key and indispensable information and reference against which any structural model of NOM/HS has to be judged [32]. Within these coarse sutructure regimes, a detailed analysis of chemical shift distribution can reveal considerable detail, especially when used in combination with spectral editing (^{13}C and ^{15}N NMR) and higher dimensional NMR spectroscopy (cf. sections 3.3. and 5).

3.3. SPECTRAL EDITING OF CARBON NMR SPECTRA

Spectral editing and, in particular, sorting carbon atoms according to their number of directly bound protons [74-76], is a very powerful assignment tool, primarily within the aliphatic and aromatic sections of the ^{13}C NMR spectra of NOM/HS. The subspectra of methyl, methylene and methine carbon atoms are calculated from linear combinations of three individual DEPT spectra [77].

The two main applications of spectral editing for the structural analysis of NOM/HS are the evaluation of branching in aliphatic units and the determination of the average degree of carbon substitution in aromatic rings. In aliphatic compounds C_nH_{2n+2}, proton chemical shifts are only marginally affected by branching; extensive steric hindrance causes a minor degree of deshielding, which is not clearly related to simple structural details. However, in the carbon frequency, both α- and β-aliphatic substitution cause a deshielding of ~10 ppm, while γ-aliphatic substitution induces an upfield shift of ~4 ppm [78]. Multiplicity information in conjunction with chemical shift data allows to draw detailed conclusions about the remote aliphatic substitution in NOM/HS, which can be further refined by the analysis of HSQC NMR spectra (cf. sections 5.2 and 7.2).

In absence of very fast transverse relaxation [76], the average number of protons per aromatic ring in NOM/HS is obtained by comparing the ratios of the (normalized) areas of a ^{13}C NMR DEPT spectrum (showing protonated carbon only) and the single pulse ^{13}C NMR spectrum (showing all carbon atoms). For the analysis of NOM/HS, it is recommended to use short DEPT pulse sequences, since they are the least susceptible to differential relaxation. However, the subspectral editing according to quaternary carbon atoms via routine QUAT-sequences visibly suffers from the variance of the $^1J(CH)$ values in NOM/HS; within the aliphatic section of the ^{13}C NMR spectrum, crosstalk from protonated carbon atoms obscures the unambiguous definition of genuine quaternary carbon atoms in NOM/HS.

3.4. NMR OF METAL NUCLEI WITH SPIN ½, COORDINATED TO NOM/HS

With a very few exceptions, all metals feature at least one NMR active isotope and thereby provide the opportunity to directly observe metal binding to NOM/HS by metal NMR spectroscopy [79-82]. However, the understanding of the NMR chemical shift - structure relationship of coordination and organometallic compounds, where alterations in coordination number, symmetry and oxidation state can readily occur [83, 84], remains rather restricted for all NMR spectra of metal nuclei until today. Any of these above mentioned effects and, in addition, pronounced concentration [82, 85], pH [82,

86] and temperature dependencies [85, 87, 88] very strongly influence the metal NMR chemical shift – and all of these effects are highly specific for any metal NMR nucleus. In the case of paramagnetic complexes, metal NMR resonances show large shift displacements from those of diamagnetic counterparts [89, 90] and are commonly broadened beyond detection due to very efficient electron-nuclear dipole-dipole interactions. The current lack of understanding how all these processes will translate into chemical shifts of metal nuclei impedes current research: so far many of the metal nuclei with high NMR receptivity have never been used in the investigation of metal binding to NOM/HS.

Chemical processes, like internal rotations in molecules, valence isomerization, chemical exchange and chemical reactions can lead to exchange of nuclei between non-equivalent electronic surroundings. Kinetic processes on time scales ranging from μs to s are effective in causing chemical exchange effects in NMR spectra [91, 92]. Naturally, the near continuous distribution of binding constants and rates of chemical exchange of humic and NOM ligands to metals will impose a strong predominance of effects of chemical exchange on the metal NMR spectra. For that very reason, the chemical shifts and resonance integrals in NMR spectra of metal nuclei as obtained by direct analysis of experimental NMR spectra of metal ions coordinated to NOM/HS do not directly indicate the content and the nature of the chemical environment of the respective nuclei. Instead, these NMR spectra most likely will represent a weighed average of superimposed slow, intermediate and fast chemical exchange characteristics, produced by groups of coordination environments, which are related by chemical shift differences and exchange rates. In metal NMR spectra, which will invariably show broadened resonances and a limited signal-to-noise ratio when coordination of metals to NOM/HS comes into effect, it is not feasible to define an unambiguously defined status of slow chemical exchange, which is an initial requirement to allow any exhaustive mathematical analysis of chemically exchanging systems [93]. Since metal ions are partners in the chemical exchange, the regime of slow chemical exchange is reached only at extremely low metal ion concentrations, when the S/N ratio of the metal NMR spectra becomes prohibitively low. Even conditions of pure fast chemical exchange in metal-NOM/HS systems are very difficult to prove experimentally, as conditions of a dynamic equilibrium, defined by the superposition of the temperature dependent binding constants and exchange rates of each individual metal coordination environment, are reached [93]. It is interesting to note that at a very low metal ion concentration, only the strongest binding sites within NOM/HS will be occupied. This may induce a selectivity of binding vastly different from that observed at elevated metal to NOM/HS ratio [94, 95].

But with these particulars under consideration, metal NMR spectroscopy offers the most direct and most detailed information concerning structure and dynamics of coordination compounds in NOM/HS. Any reliable analysis of metal binding of NOM/HS by metal NMR spectroscopy requires a systematic mapping of the exchanging system by variation of the metal to NOM/HS ratio, pH, temperature and – possibly – concentration in conjunction with an independent determination of the total metal content and the fraction of the (free) metal ions [93]. The acquisition of the metal NMR spectra at different field strengths B_0 is advantageous to further characterize the NMR exchange characteristics [93, 96].

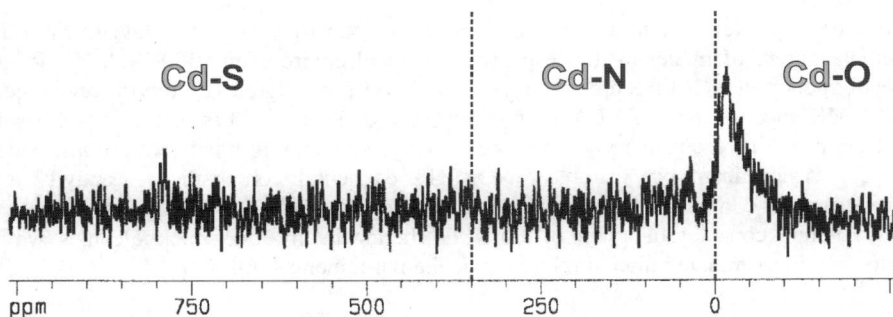

Figure 9. ^{113}Cd NMR spectrum of Suwannee River NOM with Cd(ClO$_4$)$_2$ at a Cd/C ratio of $4.16.10^{-3}$, pH = 10.2, B$_0$ = 11.7 T [93].

Several of the environmentally significant and toxic metal isotopes are spin ½ nuclei (^{113}Cd, ^{195}Pt, ^{199}Hg, $^{203/205}$Tl, ^{207}Pb) with a reasonable NMR receptivity (cf. Fig. 7) and therefore are suitable targets for direct observation of metal coordination to NOM/HS.

3.5. NMR SPECTRA OF QUADRUPOLE NUCLEI WITH SPIN QUANTUM NUMBERS I > ½.

All nuclei with a spin quantum number I>1/2 exhibit an ellipsoidal charge distribution and an electric quadrupole moment eQ and a quadrupole coupling constant ($e^2q_{zz}Q/h$) [62]. Q is positive, if the nucleus is prolate (lengthened) in the direction of its spin angular momentum, negative if it is oblate (flattened). In a molecule, an electric field gradient (efg) at the nucleus is caused because of asymmetry in the local charge distribution, which is imposed by electrons and other nuclei. The energy of a nuclear quadrupole is quantized according to its orientation in the efg even in the absence of an external magnetic field B$_0$ [62]. Like the chemical shift δ, the scalar coupling J and the dipolar coupling D, efg is a tensor and like D it is traceless: the isotropic average of energy terms involving the efg is zero. Therefore, in the liquid phase, the resonance positions in the NMR spectrum are not affected by the nuclear quadrupole coupling and only the averaged value of the shielding tensor δ and the (indirect) spin-spin coupling constants J contribute to the positions of the NMR spectral lines [97]. In the liquid phase, the changes in the local efg with molecular motions induce transitions between different states of the magnetic quantum number, and this is the quadrupole relaxation, which is a very efficient intramolecular NMR relaxation mechanism, depending on the asymmetry parameter η [98] and the quadrupole coupling constant e^2Qq_{zz}/h according to Eq. (2)

$$1/T_1 = 1/T_2 = 3/40 \ \{(2I+3)/[I^2(2I-1)]\} \ [1+\eta^2/3 \ [(e^2Qq_{zz}/h]^2 \ \tau_c \qquad (2)$$

I: nuclear spin quantum number; Q: quadrupole moment [fm^2]; η: asymmetry parameter ($0 < \eta < 1$; $\eta = (q_{xx}-q_{yy})/q_{zz}$); q_{zz}: maximum component of the efg tensor with principal components $|q_{zz}| > |q_{yy}| > |q_{xx}|$.

Owing to pronounced differential line broadening and quadrupole relaxation [87], the observed resonance integrals at sections of chemical shift in NMR spectra of quadrupole nuclei, which are attributed to defined substructures, may deviate from the actual content of nuclei in the respective chemical environments [80, 81]. While ^{11}B NMR spectra of NOM/HS are readily accessible [99] (cf. Figure 10), reports on oxygen-17 NMR spectroscopy of NOM/HS remain scarce. This is at first glance a surprising fact, considering the primary importance of oxygen containing functional groups, which define a significant proportion of NOM/HS elemental composition, reactivity and properties, in combination with a reasonable receptivity of ^{17}O NMR spectroscopy (cf. Fig. 6) and a considerable range of ^{17}O NMR chemical shift of organic molecules, which allows to discriminate rather similar chemical environments [100, 101].

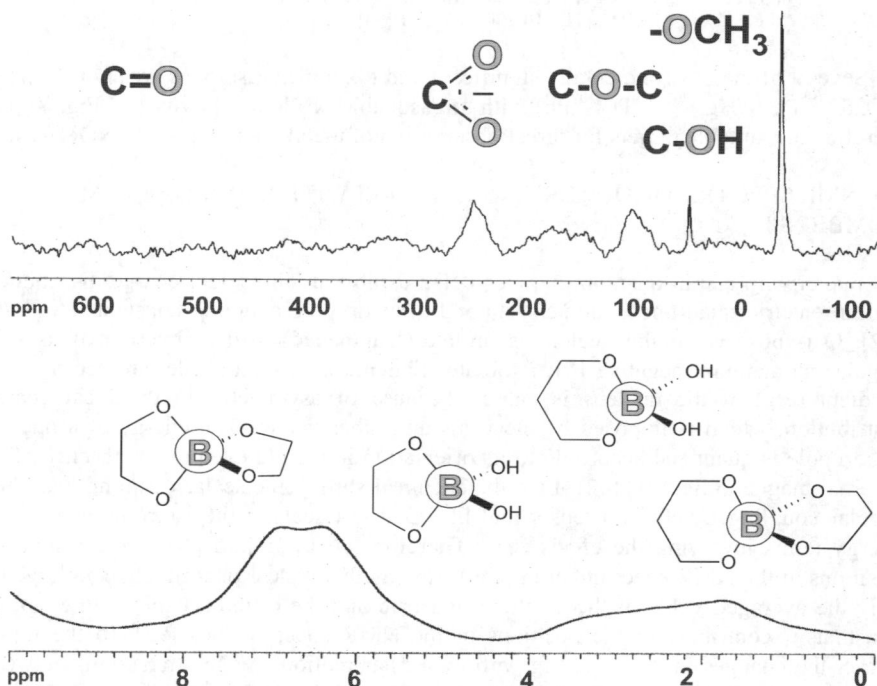

Figure 10. ^{11}B NMR spectrum of a fresh water humic acid (bottom), and a ^{17}O NMR spectrum (top) of a bog lake humic substance (sealed in dry pyridine-d$_5$), together with main substructure regimes.

The content of oxygen derived functional groups in NOM/HS (e. g. aliphatic, aromatic and carboxylic hydroxy, keto, aldehyde, methoxy) can be readily estimated from ^{1}H and ^{13}C NMR spectra in combination with infrared and titration studies. Based upon this information, reasonable estimates of the appearance of the ^{17}O NMR spectra of NOM/HS can be derived [102]. In practice, the combined effects of variable quadrupole coupling constants of individual coordination sites [100, 103], nuclear

Overhauser effects [104], differential local symmetry [100], and, most prominently, the rate of chemical exchange in relationship to the spread in ^{17}O NMR frequency of the respective chemical environments in NOM/HS will lead to severe differential line broadening and intensity distortions [102]. Lack of systematic studies aimed at a better understanding of these effects so far have impeded the use of ^{17}O NMR spectroscopy for the meaningful structural analysis of NOM/HS.

4. NMR Investigations of Interactions of Organic and Inorganic Compounds with NOM/HS

Accurate molecular level information about mechanisms and dynamics of the interactions of NOM/HS with organic molecules can be obtained largely from the NMR spectra of the main constituent nuclei by methods of chemical shift mapping, nuclear-spin relaxation, nuclear Overhauser effect, and study of exchange and diffusion phenomena [105-108]. When labels of ^2H, ^{13}C, ^{15}N in small molecules are used, the low natural abundance of ^2H (0.00015%), ^{13}C (1.1%) and ^{15}N (0.37%) in NOM/HS is of considerable advantage, since the behaviour of specific chemical environments within these small molecules can be probed by an array of one, two and three dimensional NMR spectra at rather low levels of concentrations. The relaxation in ^2H [109], ^{13}C [110] and ^{19}F [111] NMR spectra of small aromatic molecules has been used to analyse the dynamics of their interaction with NOM/HS.

Another promising technique to probe the interaction of NOM/HS with fluorinated organic compounds is ^{19}F NMR spectroscopy [112]. Owing to a very strong back bonding contribution in the carbon-fluorine bond, the effects of fluorination upon representative bulk properties of the organic molecule and its reactivity to NOM/HS are rather small [113]. The ^{19}F NMR label has exceptional sensitivity [114] (cf. Fig. 6); this is even more remarkable as organically bound fluorine is absent in natural NOM/HS: environmentally significant trace levels of fluorinated organic molecules can be readily investigated by ^{19}F NMR spectroscopy [114, 115].

Interaction of NOM/HS with non metal (e. g. ^{11}B, ^{77}Se) and metal ions (most appropriately: ^{27}Al, ^{51}V, ^{113}Cd, ^{199}Hg, $^{203/205}Tl$, ^{207}Pb) can be followed by direct observation of these NMR active nuclei [79, 80, 81, 93]. In NMR spectra of quadrupole nuclei, both chemical exchange and quadrupole relaxation obscure direct relationships between the appearance of the metal NMR spectra and the coordination geometry and the nature of the NOM/HS ligands (cf. section 3.5).

5. General Characteristics of Two Dimensional NMR Spectra of NOM/HS

The invaluable contribution of quantitative one dimensional NMR spectra to NOM/HS structural analysis resides in the ability to define the *relative amounts of fundamental building blocks*. The main significance of the uniquely versatile 2D NMR spectroscopy of NOM/HS is to *enhance the reliability of NMR resonance assignments*, as cross peaks in 2D NMR spectra indicate molecule fragments rather than individual atoms. A combined analysis of several 2D NMR spectra ultimately leads to the assemblage of extended substructures of NOM/HS (cf. Fig. 11, Table 2).

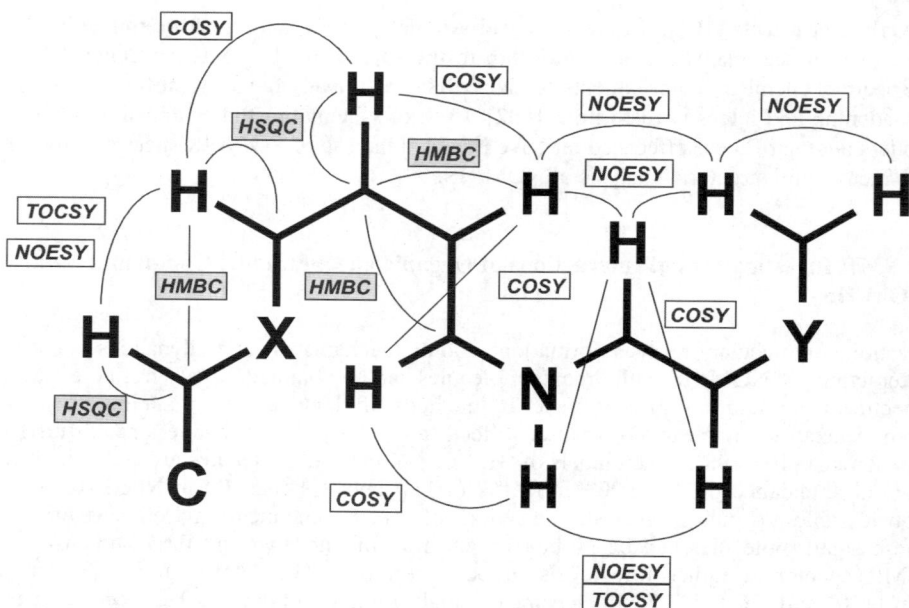

Figure 11. The contributions of different 2D NMR experiments in the assembly of extended substructures of NOM/HS.

Table 2. A selection of 2D NMR experiments, useful for the structural analysis of NOM/HS.

NMR acronym	selected interaction	remarks
JRES	$^nJ(H,H)$	sensitive, but long duration of F1 increments may interfere with fast transverse relaxation at higher resolution (F1)
COSY	$^{2,3}J(H,H)$	antiphase magnetization, susceptible to self-cancellation of cross peaks; optimum transfer: 1/J
TOCSY	$^nJ(H,H)$	extended spin systems; optimum transfer: 1/2J (n = 2-6)
HSQC, HMQC, HETCOR	$^1J(X,H)$	highly informative for X: ^{13}C
HMBC	$^nJ(X,H)$	most useful for X: ^{13}C and ^{15}N
NOESY, HOESY	spatially proximate H...H and H...X interactions	for NOM/HS often accompanied or dominated by chemical exchange

An increased signal dispersion of NMR cross peaks into two frequency dimensions greatly reduces resonance overlap, especially in the case of heteronuclear correlated NMR spectra, where the resonance frequencies of heteronuclei cover a substantial range [116, 117]. In addition, 2D NMR experiments act as filters, which emphasize specific forms of binding (cf. Table 2) and strongly discriminate against other resonances, thereby allowing to selectively probe *bonding interactions, spatial relationships* and *chemical exchange*, respectively [116-119]. This simplification of spectra enables a detailed analysis of heavily crowded regions of chemical shift. The sensitivity of proton detected homo- and heteronuclear 2D NMR spectra is commonly much higher than that of heteronuclear 1D NMR spectra (cf. eq. [1]).

In comparison with classic inverse geometry probes, cryogenic NMR probes provide a distinct increase in signal to noise ratio of proton detected NMR spectra [120-122]. The relative S/N enhancement depends on sample composition, reaching a factor in excess of four in case of organic solvents, which is declining to near unity for solutions with very high ionic strength [39, 123].

Therefore NOM/HS, dissolved in DMSO-d_6 under complete exclusion of air and moisture are promising samples for NMR spectra of very high resolution (cf. Figure 12). The exclusion of oxygen obviates longitudinal relaxation from paramagnetic oxygen. The very large intrinsic sensitivity of the cryogenic probes allows the acquisition of meaningful NMR spectra from samples of low concentration, when the tendency of NOM/HS to align and aggregate in solution is minimized. In addition, hydrogen bonding is severely reduced in dipolar aprotic solvents. Both effects enhance the molecular mobility, increase the longitudinal [38] and spin-spin relaxation times and consequently the attainable resolution in 2D NMR spectra [124].

In extension of the various aspects of high-resolution 2D and 3D [125] NMR spectroscopy of NOM/HS, which have been highlighted in several recent publications [45, 76, 118, 119, 126-135], the emphasis of the following section is to provide representative examples how different two dimensional NMR spectra contribute to the structural analysis of NOM/HS. In practice, 2D NMR spectroscopy of NOM/HS will frequently become an exercise in optimisation of acquisition and calculation parameters for individual samples to enhance NMR sensitivity and resolution, because interaction of NOM/HS constituent molecules – and in result: all relaxation dependent NMR parameters strongly depend on sample conditions, like concentration, solvent, pH and temperature.

5.1. HOMONUCLEAR 2D NMR SPECTRA OF NOM/HS

5.1.1. Introduction; differences between COSY and TOCSY NMR spectra of NOM/HS
Proton homonuclear shift correlated 2D NMR spectra offer high sensitivity and long range connectivity information. The latter is instrumental to assemble NOM/HS substructures (cf. Figure 11). One method is to relate heteronuclear single bond [^{1}J(C,H)] cross peak (which represent isolated pairs of directly connected carbon and proton atoms; cf. section 5.2) via a combined analysis with homonuclear (COSY and TOCSY) cross peaks. The main potential of COSY spectra resides in the structural analysis of vicinal and geminal proton-proton connectivities, which represent substructures of organic molecules with a minimum size of five and six atoms (cf. Table 3).

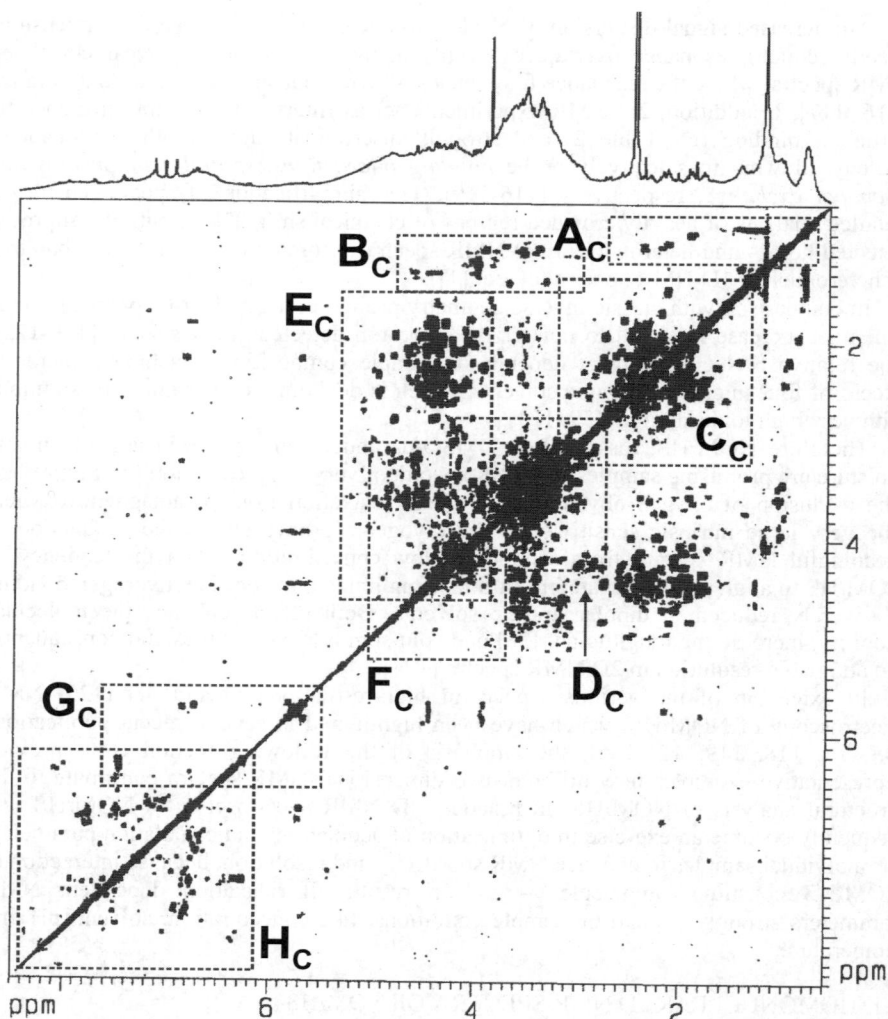

Figure 12. 2Q-COSY spectrum of a peat fulvic acid in dry DMSO-d$_6$ (for assignment of cross peak substructure regimes: cf. Table 3).

The rather long transfer times in COSY and TOCSY NMR spectra [116, 117] imply loss of magnetization caused by fast transverse relaxation. Owing to different mechanisms of magnetization transfer [136], both common and different resonances will appear in COSY and TOCSY NMR spectra of NOM/HS [118]. TOCSY cross peaks represent net transfer of in-phase magnetization with near absorption line shapes for diagonal and cross peaks and they indicate extended spin systems rather than vicinal and geminal connectivities alone as provided in standard COSY NMR spectra. COSY cross peaks represent coherence transfer by scalar couplings (J): antiphase magnetization is created without transfer of net magnetization. Therefore, COSY cross peaks are susceptible to self-cancellation, when the linewidth Δv exceeds the coupling constant J. However, COSY spectra of dilute solutions of soil NOM/HS in dry DMSO-d_6 under total exclusion of moisture exhibit exceptionally well resolved (cf. Fig. 12) cross peaks.

Table 3. Regions of chemical shift in the ^{1}H,^{1}H COSY NMR spectrum (Fig. 12).

region	assignment	humic constituent
A_C	-C-CH-CH$_3$	methyl groups, bound to purely aliphatic carbon
B_C	-C-CH-CH-O-	desoxy sugars, ethers, esters
C_C	-C$_f$-CH-C$_f$H-C-	intra functionalised aliphatic, ß to heteroatoms
D_C	-CH(O)-CH(O)-	intra carbohydrate, without anomerics
E_C	-CH(X)-CH-C$_f$	functionalised aliphatic unit with one heteroatom (X = O, N)
F_C	-O-CH(O)H-CH	intra carbohydrate, with anomerics
G_C	-CH=CH-	double bonds, five-membered heterocycles
H_C	-C$_{ar}$H-C$_{ar}$H-	ortho protons in aromatic rings

Fast and differential transverse relaxation within the polydisperse and molecularly inhomogeneous NOM/HS attenuates COSY cross peaks (optimum transfer time at 1/J) more severely than TOCSY (optimum transfer time at 1/2J) cross peaks [118]. Therefore cross peaks visible solely in TOCSY spectra represent long range couplings in extended spin systems, but also may indicate short range couplings in spin systems of restricted flexibility, where self-cancellation of COSY cross peaks has occurred (at linewidths > 15 Hz). Cross peaks showing up solely in COSY spectra indicate either small spin systems of very high flexibility (with long transverse relaxation times, sharp lines and large S/N ratio) or parts of extended spin systems; in this case the proton resonance integrals can be distributed across many TOCSY cross peaks and then, the S/N ratio of individual TOCSY cross peaks may become attenuated below recognition.

Cross peaks in highly resolved, phase-sensitive COSY NMR spectra of NOM/HS show a fine structure, which allows to determine (under consideration of the effects of the line width on the anti-phase lineshape) the values of $^nJ(H,H)$ coupling constants. In phase sensitive COSY NMR spectra, the determination of active (spins, whose frequencies are connected by the cross peaks, produce anti-phase doublets) and passive spins (spins, which are not directly involved in the magnetization transfer, split the cross peaks into in-phase doublets) allows to specify the coupling network of any given cross peak, a major leap to assemble and confirm extended substructures in NOM/HS (cf. Figure 13).

412

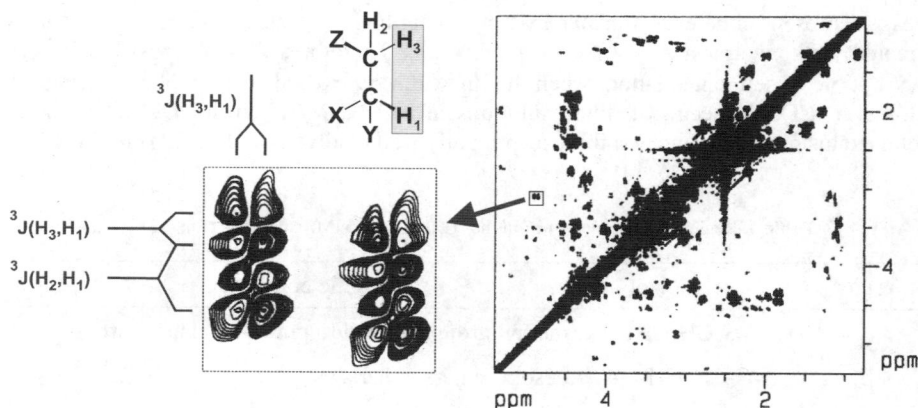

Figure 13. Evaluation of spin systems by analysis of the fine structure of COSY cross peaks in a soil humic substance; the selected cross peaks show splitting by active coupling [$^3J(H_3\text{-}H_1)$] and by passive coupling [$^3J(H_2\text{-}H_1)$]; the nature of this AMX-type spin system indicates non carbohydrate origin (anomeric protons commonly show AB-spin systems).

5.1.2. NOESY/EXSY NMR spectra: identification of exchangeable protons in NOM/HS

NOESY/EXSY NMR spectra rely on the transfer of longitudinal magnetization to establish connectivities between spins either by the nuclear Overhauser effect (nOe) or by chemical exchange. NOESY NMR spectra of NOM/HS exhibit relatively intense cross peaks resulting from chemical exchange and relatively weak cross peaks resulting from spatial proximity (nOe effects) [119, 126]; their relative phase with respect to diagonal peaks depends on the molecular size and correlation time τ_c. In NOM/HS the magnetization transfer caused by cross relaxation and by chemical exchange often results in NOESY cross peaks of alike signs as diagonal peaks, a behaviour typical of large molecules (spin diffusion limit).

Table 4. Regions of chemical shift in the $^1H,^1H$ NOESY/EXSY NMR spectrum shown in Fig. 14.

region	assignment	origin of cross peak
A_N	-C_{al}H...HO-C-, -X-CH and -C_f-CH...HC_f-C, -X-CH...HC-X	chemical exchange, nOe
B_N	-C_{al}-OH...H_2O/HO-C_{al} , C_{al}-H...H-C-X-	chemical exchange, nOe
C_N	-(C=O)NH...H-C_{al}-	nOe
D_N	-C_{ar}-OH...HO-C_{ar} , -C_{ar}H-C_{ar}H-	chemical exchange, nOe
E_N	-COOH...HO-C_{al}-	chemical exchange
F_N	-COOH...HO-C_{ar}-	chemical exchange

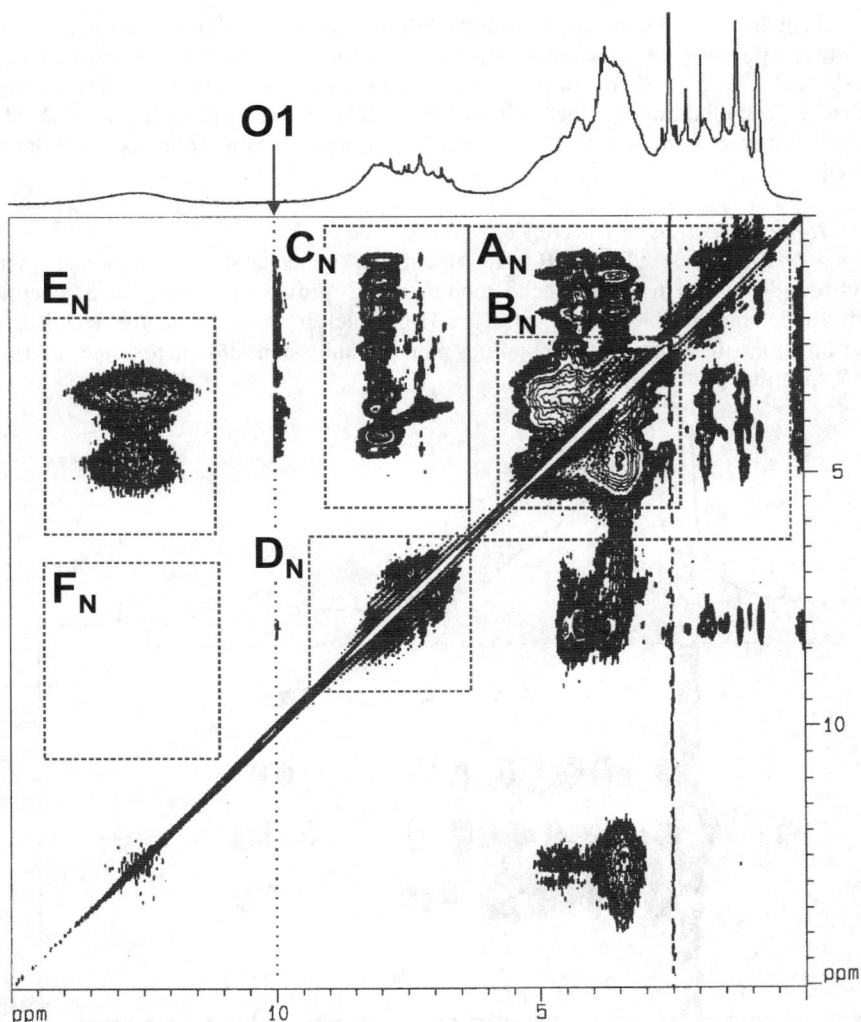

Figure 14. EXSY/NOESY spectrum of a soil humic acid in dry DMSO-d$_6$ (τ_m = 250 ms); O1: excitation offset frequency.

The significance of the NOESY/EXSY spectra for the characterization of NOM/HS resides in the ability to identify exchangeable protons and to determine their exchange rates, which can be calculated from the build-up of the cross peak integral at different mixing times τ_m [137, 138]. The location and intensity distribution of the cross peaks (cf. Fig. 14; Tab. 4) indicates that selective chemical exchange takes place between different labile proton species in NOM/HS [119].

At very short mixing times (τ_m < 10 ms) only cross peaks originating from chemical exchange are observed in NOESY/EXSY spectra of NOM/HS. Under conditions of

414

slow chemical exchange (such as in dry DMSO-d_6 solution under careful exclusion of moisture), spin systems containing HX-C-H units (X: N, S, O) produce cross peaks in COSY and TOCSY NMR spectra, which vanish under conditions of fast chemical exchange. Therefore, a combined analysis of NOESY/EXSY, and COSY/TOCSY cross peaks is required to provide an in depth analysis of non-exchangeable and labile protons in NOM/HS [139].

5.1.3. Homonuclear J-spectroscopy of NOM/HS

Among the simplest 2D NMR experiments to understand and implement is J-spectroscopy, which uses spin echo modulation to provide a separation of chemical shifts and spin-spin coupling effects. Homonuclear J-spectroscopy enables the determination of proton-proton coupling patterns and –constants across the entire ^{1}H NMR spectrum [140, 141].

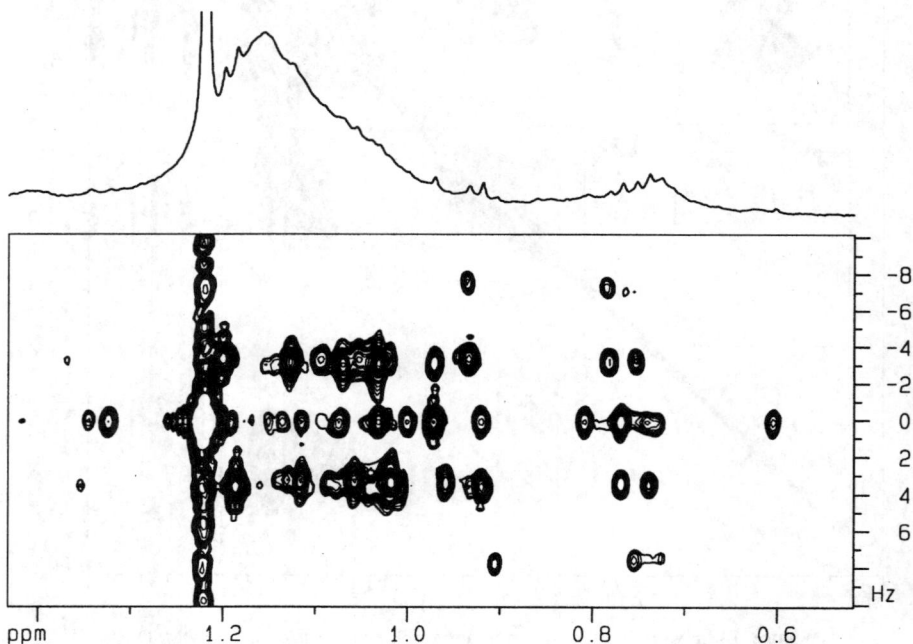

Figure 15. ^{1}H,^{1}H-J-resolved spectrum of a soil leachate fraction, obtained by free flow electrophoresis.

The most significant and useful assignment options in the analysis of NOM/HS are (a) the determination of aromatic substitution patterns, (b) the discrimination of (typically more complex) carbohydrate and (commonly less complex) other chemical environments in the middle region of the spectrum, (c) the assessment of branching in the upfield region of functionalised aliphatics and, perhaps most informative, (d) the determination of the number of protons vicinal to methyl protons (cf. Fig. 15). Heteronuclear variants of J-spectroscopy are considered less informative, because carbon multiplicity information is readily available from ^{13}C DEPT NMR spectra (cf. section 3.3). The sensitivity of homonuclear J-spectroscopy is excellent; however, the attainable resolution of coupling constants strongly depends on the relaxation characteristics of the respective NOM/HS (cf. Table 2).

5.2. HETERONUCLEAR SHIFT CORRELATED 2D NMR SPECTRA

Multidimensional heteronuclear NMR experiments correlate heteronuclear with proton resonances by transfer of coherence (or polarization) between the heteronuclear (S) and the proton (I) spins, with initial excitation of either I or S spin polarization and detection of either I or S spin magnetization. The overall sensitivity of the heteronuclear correlation NMR experiments is proportional [117] to

$$S/N \sim \gamma_{ex} \, \gamma_{det}^{3/2} \, [1 - \exp(-R_{1,ex}T_C)] \qquad (3)$$

γ_{ex} and γ_{det} are the magnetogyric ratios of each nucleus excited at the begin and detected at the end of the pulse sequence; T_C is the recycle time of the experiment; $R_{1,ex}$ is the spin-lattice relaxation rate constant of the excited nucleus [cf. equation (1)].

NMR pulse sequences, which correlate directly bonded proton and carbon atoms with respect to their chemical shift, show optimum transfer amplitudes in the 5-20 ms range (HSQC at ^{1}J(CH) = 150 Hz: 10.9 ms). Their relatively short duration with respect to average T_1 and T_2 relaxation times in humic substances [40, 41, 118] is more favourable to prevent signal loss in case of fast transverse relaxation, when compared to homonuclear 2D NMR sequences. The large spread of frequency in F1 (^{13}C) dimension and the exclusive discrimination in favour of proton-carbon single bonds greatly reduces resonance overlap. These effects combined make HSQC NMR spectra very auspicious candidates to reveal structural information of NOM/HS in great detail.

Table 5. Regions of chemical shift in the ^{1}H,^{13}C HSQC NMR spectrum Fig. 16.

region	assignment	partial structure
A_H	-C-CH$_3$	methyl groups, terminating aliphatic chains
B_H	-C$_f$-CH	aliphatic CH, β to heteroatoms
C_H	-X-CH-	single heteroatom substituted aliphatics
D_H	C-N(H)-CH-	amines, bound to aliphatic carbon
E_H	-(C=O)NH-CH-	amides, e. g. in proteins
F_H	-OCH$_3$	methoxyl (aliphatic and aromatic esters and ethers)
G_H	-O-CH-	single heteroatom (oxygen) substituted aliphatics
H_H	O-CH-X	double heteroatom substituted aliphatics (e.g., acetals, anomeric CH)
I_H	-C$_{ar}$H	aromatic and heterocyclic systems

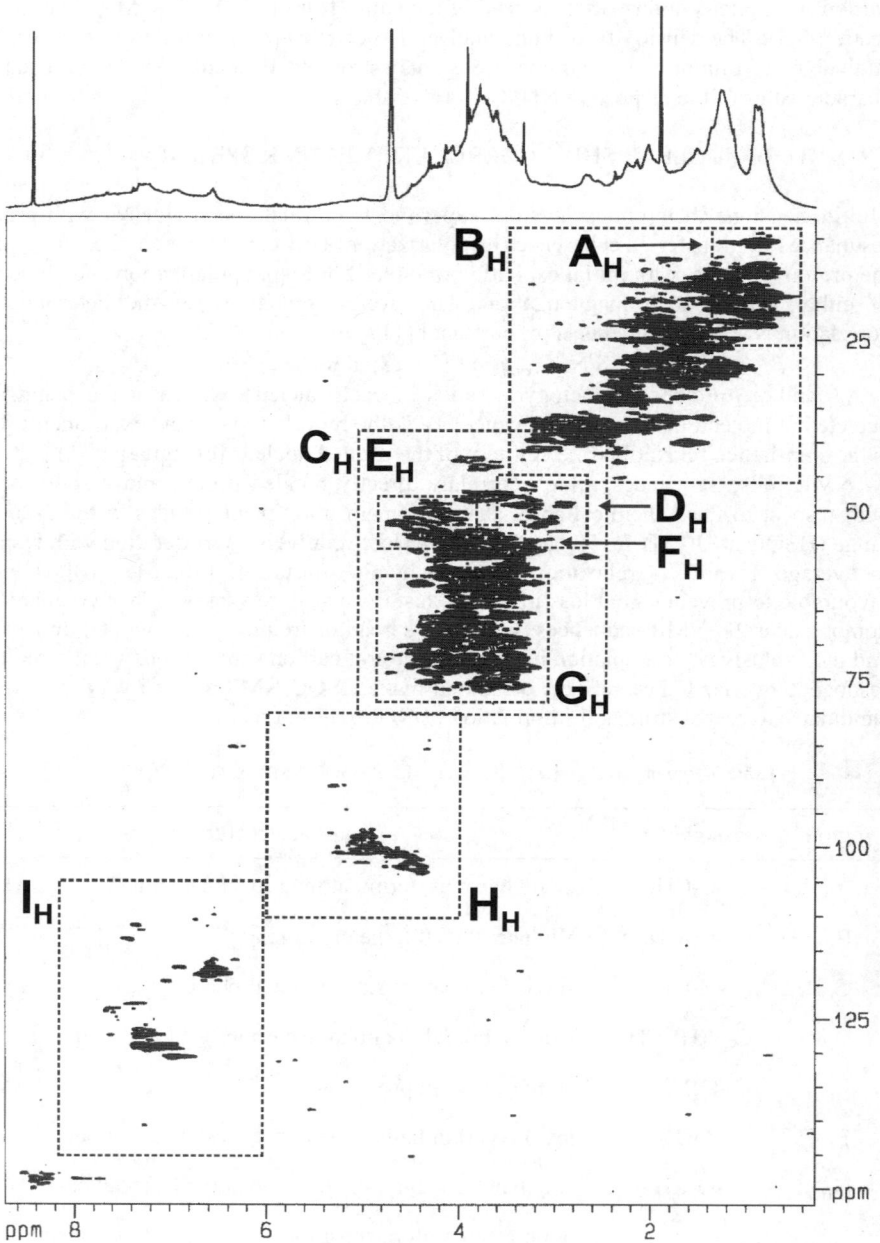

Figure 16. ^{1}H, ^{13}C -HSQC NMR spectrum of a soil humic acid in 0.1 N NaOD with key structures as given in Table 5.

5.2.1. Cross peak fine structure of one bond $^{13}C,^{1}H$ correlation spectra of NOM/HS

In heteronuclear NMR correlation experiments, high resolution in F2 as defined by the digitisation can be easily achieved in the detection period t_2. During the evolution time t_1, the resolution in F1 is limited by the number of experiments in the total acquisition time.

The resolution in ^{13}C NMR spectra of samples with natural isotopic abundance is rather high under conditions of proton decoupling due to rather long transverse relaxation times and the absence of homonuclear $^{1}J(C,C)$ couplings [116,117]. The requirements for the digitisation of the proton NMR spectra are less stringent as the resonances are broadened by fast transverse relaxation and homonuclear $^{n}J(H,H)$ couplings, and because the overall bandwidth of proton NMR spectra (in [Hz]) is smaller. However, carbon detection schemes have considerable sensitivity disadvantages (cf. equations [1] and [3]). Owing to different physical mechanisms operating within the pulse sequences, the four fundamental single bond carbon proton [$^{1}J(C,H)$] correlation NMR spectra of NOM/HS show a variation in relative cross peak amplitudes and resolution [142-144]; an in-depth analysis of these (relaxation) effects provides substantial information about structure and dynamics of NOM/HS [145].

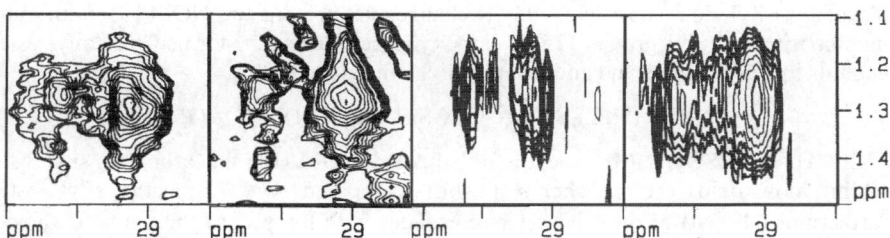

Figure 17. Cross peaks of polymethylene section in IHSS reference material 1R106H, acquired with four different $^{1}J(CH)$ correlation NMR pulse sequences (all four phase sensitive NMR spectra represent identical total acquisition times, cf. Table 4): $^{1}H,^{13}C$-HMQC (left), $^{1}H,^{13}C$-HSQC, (second from left), $^{13}C,^{1}H$-DEPT-HETCOR (second from right), $^{13}C,^{1}H$-INEPT-HETCOR (right).

6. NMR Analysis of Functional Groups in NOM/HS

With a weight percentage ranging from 35 to 50%, the functional groups determine the physicochemical properties and, to a large extent, the ecological behaviour of NOM/HS [146, 147]. The combination of chemical derivatization with one and two dimensional NMR spectroscopy is a powerful new method to determine the composition of functional groups in humic substances [148-150]. Acidic protons of functional groups are reacted with reagents composed of a suitable NMR-active label and a leaving group (cf. Fig. 18); ideally, the tag would substitute any labile proton of a given (class of) functional groups without side reaction – a condition not necessarily met, when NOM/HS is derivatized.

418

A very wide range of reagents can be reacted with NOM/HS for a NMR-based analysis of functional groups (cf. Fig. 6). This method complements existing techniques of functional group analysis in NOM/HS, but provides unprecedented detail with respect to the structural details of their environment. Caused by the very high sensitivity of proton detected NMR experiments, these NMR analyses of functional groups in NOM/HS can be realized with sub-mg amounts of humic materials.

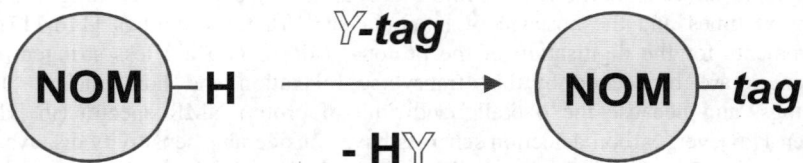

Figure 18. Substitution of an acidic proton in NOM/HS with a NMR active functional group ("tag").

6.1. THE SILYLATION OF NOM/HS

(Trimethyl)silylation substitutes the exchangeable protons in NOM/HS with rather unpolar trimethylsilyl groups [151]; carboxylic acids are transformed into silyl esters, alcohols into silyl ethers and amides into silyl amides.

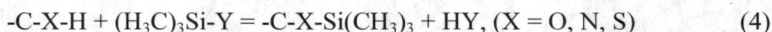

$$-C-X-H + (H_3C)_3Si-Y = -C-X-Si(CH_3)_3 + HY, (X = O, N, S) \tag{4}$$

In ^{1}H,^{29}Si HSQC NMR spectra of silylated NOM/HS, the aromatic silyl esters exhibit downfield proton chemical shift compared with aliphatic silyl esters. Furthermore, $\delta(^{29}$Si$)$ of the silyl esters correlates with the pK_a of the parent carboxylic acid [151, 152]. The aliphatic and aromatic silyl esters, the primary, secondary and tertiary silyl ethers as well as phenol-derived silyl ethers and silyl amides [151, 153] cover distinct areas in the $^2J(^1$H,^{29}Si$)$- and $^1J(^1$H,^{13}C$)$-HSQC NMR spectra and show within their range of chemical shift in the 2D matrix very much detail and resolution. Structural information of this kind is currently not accessible by any other analytical method.

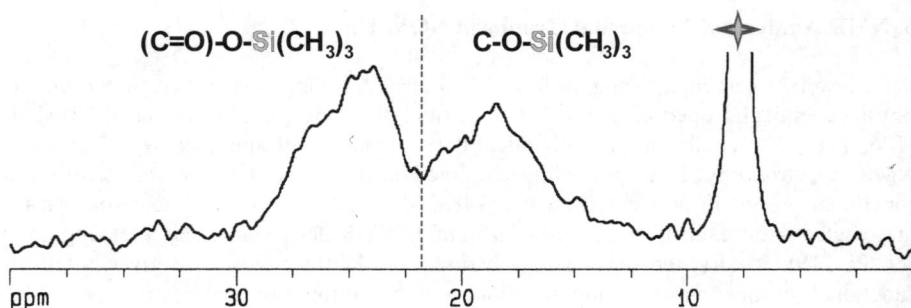

Figure 19. ^2J (^{29}Si, ^{1}H) INEPT NMR spectrum of a silylated Suwannee river organic matter (upfield resonance: excess silylating agent BSTFA).

With a mathematical procedure of 2D NMR subtraction, the *minimum* and *difference* NMR spectra can be calculated from experimental NMR spectra (cf. Figure 20). Minimum spectra (HA = FA) show the common structural elements within two and more humic materials; difference NMR spectra show their specific characteristics, respectively [148].

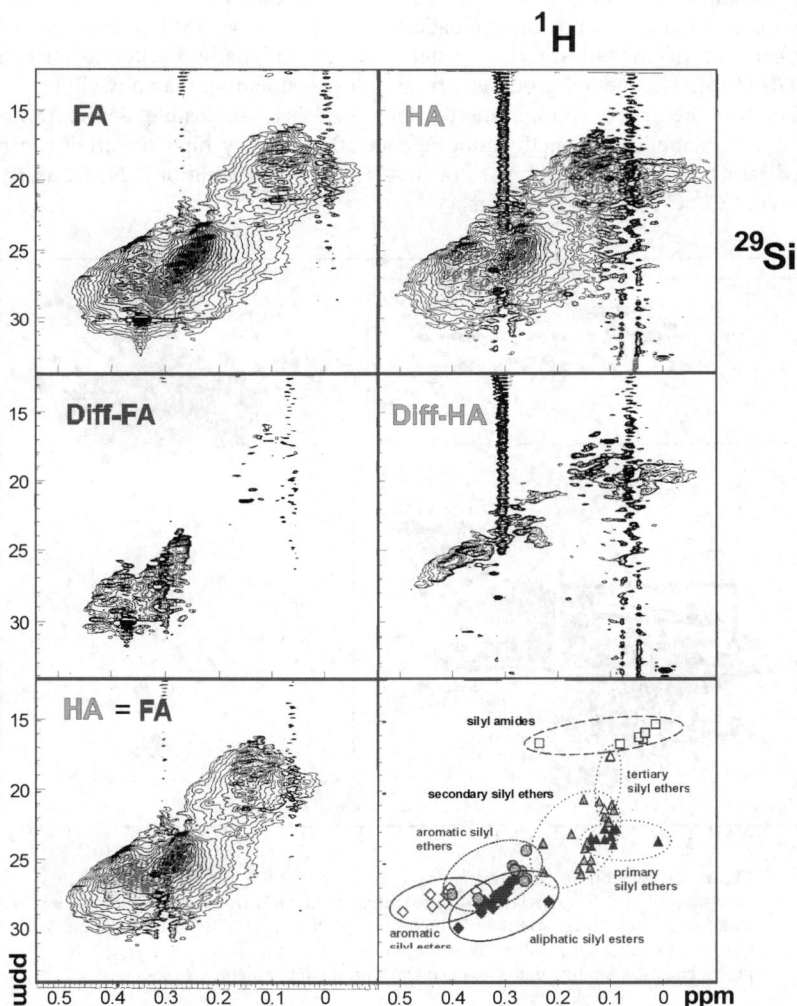

Figure 20. ^2J (^{29}Si,^1H) HSQC NMR spectra of a silylated fulvic (FA, left) and a humic acid (HA, right) from a bog lake (Holohsee HO10FA, HO10HA, Black forest, Germany) ; *upper trace:* full spectra, *middle trace:* difference NMR spectra of HA and FA showing preferential substructures in FA (left) and HA (right); *lower left:* conforming structures in FA and HA; *lower right:* ^1H,^{29}Si-HSQC-NMR chemical shift data of model compounds.

6.2. THE METHYLATION OF NOM/HS

Methylation of NOM/HS in conjunction with one- and two dimensional NMR spectroscopy is another independent method for the characterization of its functional groups according to structural details. The methylation of NOM/HS effects a functionalisation of acidic protons, but also side reactions, like selective cleavages, C-methylation and the 1,3-dipolar cycloaddition, occur [154, 155] that allow a better comprehension of the substitution patterns of the aliphatic backbone structure of NOM/HS [156]. The methyl group is sterically less demanding than any silyl group and may therefore be more appropriate to derivatize less accessible acidic protons in NOM/HS. ^{13}C labelling of methylating agents offers a very high sensitivity in proton detected (and) carbon nD NMR spectra (n = 1, 2, 3), and additional NMR assignment options via $^{13}C,^{13}C$ coupling constants.

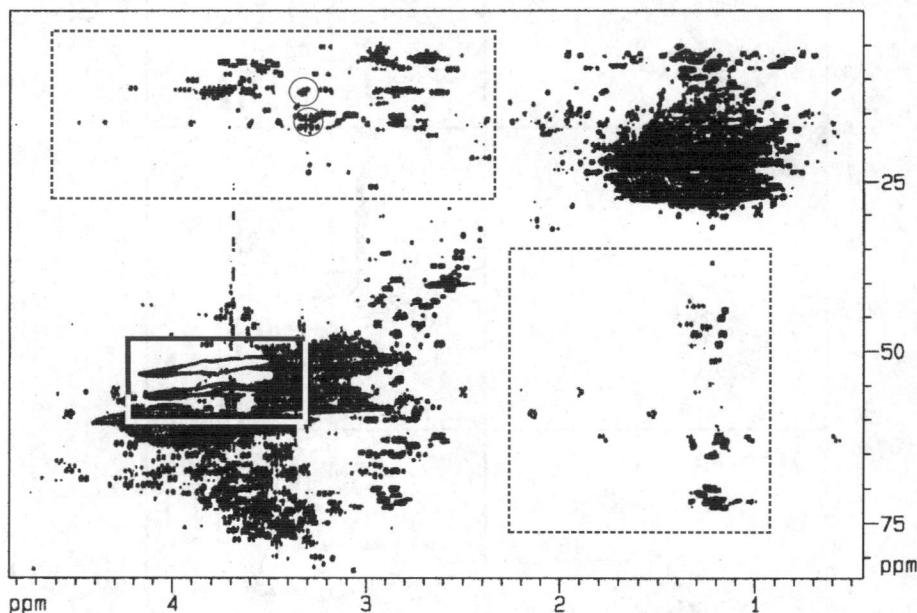

Figure 21. Section of the aliphatic cross peaks in a $^1H,^{13}C$ HSQC-TOCSY NMR spectrum of a ^{13}C-methylated aquatic fulvic acid HO13FA from a bog lake Holohsee, black forest, Germany), representing $^1J(CH)$ and $^nJ(CH)$ correlations; $^nJ(CH)$ correlations are indicated with a dotted box; circles represent HSQC cross peaks, and the solid box indicates chemical shift region of methoxyl cross peaks.

Compared to the NMR spectra of the unmethylated native NOM/HS materials, the NMR spectra of the methylated NOM/HS show an improved resolution as well as a significant increase of the number of resonances and cross peaks with a simultaneous change of the average signal position. The $^1H,^{13}C$ HSQC-TOCSY NMR spectra of a methylated NOM/HS shows one bond carbon-proton couplings and long range carbon-proton correlations that allow the assemblage of extended partial structures (cf.

Figure 21). In the methoxyl region of the HSQC NMR spectrum, aromatic and aliphatic methyl esters and methyl ethers are represented in four distinctly separated regions of chemical shift.

7. NMR Data Processing and Evaluation Schemes

7.1. PROCESSING PARAMETERS FOR THE CALCULATION OF NMR SPECTRA FROM FIDS

The NMR signal typically is acquired in the form of a discrete time domain sampling sequence in digital form (interferogram; free induction decay: FID), and several schemes of numerical digital signal processing are available to extract the information content of the signal [157, 158]. The most commonly used approach to convert the time-domain signal into a frequency-domain NMR spectrum is by applying a Fourier transformation. Alternative processing techniques, like (forward and backward) linear prediction, maximum entropy reconstruction and Bayesian analysis, are not treated in this concise introduction to signal processing [157, 158]. Experimental NMR spectra commonly exhibit a range of shortcomings, like truncation artefacts, low S/N ratios, limited resolution, or undesirable line shapes. Multiplication of the interferogram with an adjusted time-domain filter function prior to Fourier transformation is equivalent to a convolution of the spectrum with a lineshape function. This process is called apodization, windowing, or filtering in the time domain and can be used for the improvement of the S/N ratio, resolution enhancement, lineshape conversion, or a combination of those. Typically, resolution enhancement causes a degradation of the S/N ratio, and vice versa. One of the most widely known apodization function, considered to constitute the optimum balance between noise reduction and the effects of line broadening, is the matched filter: here the window function doubles the linewidth in the frequency domain.

NMR spectra of NOM/HS are composed of superimposed resonances, showing a considerable range of linewidths; therefore, no single window function will be capable to display optimum results across the entire spectral range. However, some general rules apply (cf. Figure 22):

(1) in one dimensional NMR spectra of heteronuclei, an optimum S/N ratio is of primary interest and the corresponding noisy FIDs will be multiplied by rather strongly decaying exponential functions (lb = 4-150 Hz).

(2) the analysis of sections within proton NMR spectra, acquired in dilute solution at optimum B_0 homogeneity, may in special cases benefit from a modest resolution enhancement by a Lorentz-to-Gaussian transformation. More common are modest exponential broadening functions, e.g. in the range of lb = 0.2-1.0 Hz, to reduce baseline noise near the high and low field limits. This improves the reliability and robustness, when the similarity of NMR spectra will be assessed by mathematical methods, like e.g. bucket analysis [28, 159].

(3) Heteroatom NMR spectra of NOM/HS showing groups of resonances, differing widely in their linewidth – a common feature e.g. in ^{13}C NMR spectra of the lower mass fractions of DOC – may require several different apodization functions to separately analyse slow and fast relaxing components in this mixture.

422

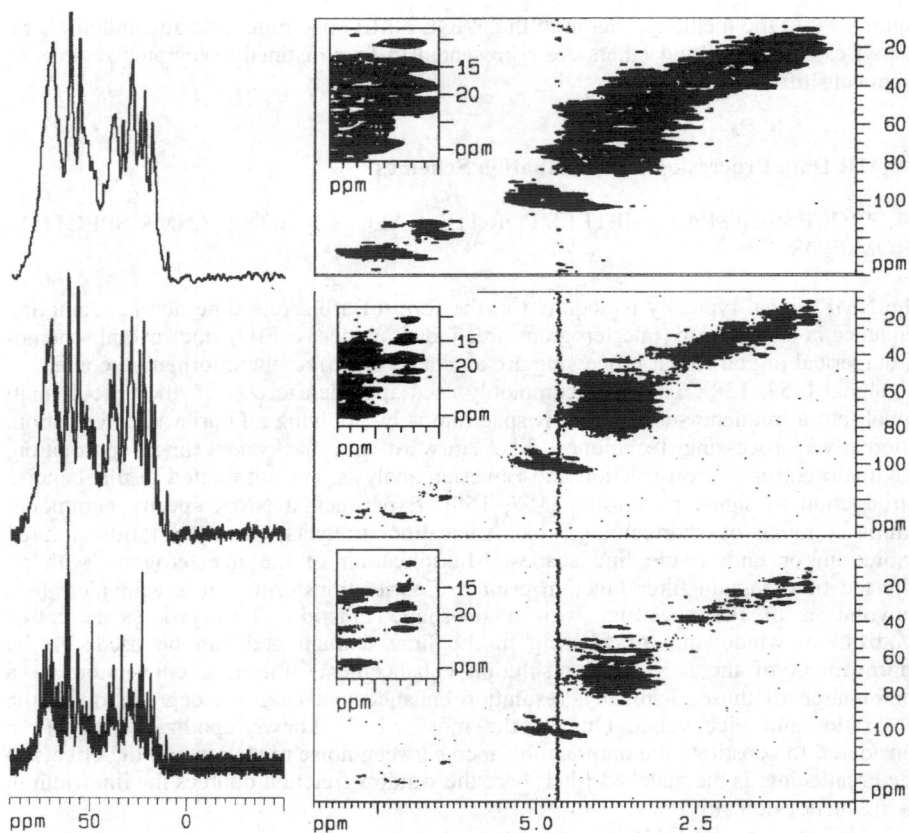

Figure 22. Left: DEPT-90 ^{13}C NMR spectrum (upfield section) of the IHSS reference soil humic acid 1R106H, calculated with different exponential line broadening factors: bottom/middle/top: 3/25/100 Hz; right: gs-^{1}H,^{13}C-HSQC NMR spectrum of the aromatic section of an Italian soil humic acid (Sobretta), calculated with different window functions: bottom/middle/top (F2; F1) EM 3; SSB 6 / EM 10; GM +0.5, -0.5 / EM 50; SSB 2.

(4) The information content of two dimensional spectra of NOM/HS even more depends on judiciously applied window functions [148], which have to be applied separately in F2 and F1 dimension. Two dimensional NMR spectra of NOM/HS with a substantial S/N ratio and a very high digital resolution in F2 and F1 can be calculated with some emphasis on resolution enhancement; this is not feasible for standard resolution 2D NMR spectra of NOM/HS.

(a) High quality absolute value COSY NMR spectra of NOM/HS can withstand unshifted sine bell functions to produce highly resolved cross peaks (cf. Figure 12).

(b) Any phase sensitive heteronuclear 2D NMR spectrum of NOM/HS is most advantageously calculated with adjusted exponential broadening in F2, and with modest exponential broadening or shifted sine bell functions in F1.

(c) Phase sensitive homonuclear 2D NMR spectra of NMO/HS are best calculated with adjusted exponential broadening or shifted sine bells in F2, and – again – with modest exponential broadening or shifted sine bell functions in F1.

(d) The analysis of cross peaks in phase sensitive COSY NMR spectra of NOM/HS may benefit from Lorentz-to-Gaussian resolution enhancement in F2 and F1, but a very considerable digital resolution is required in both F2 and F1.

7.2. THE USE OF SIMULATION OF NMR PARAMETERS IN THE ASSIGNMENT OF NMR RESONANCES AND CROSS PEAKS

The combined analysis of several one and two dimensional NMR spectra of NOM/HS offers a broad range of options for resonance assignment and mutual evaluation. Nowadays, proton and carbon NMR spectra of extended spin systems can be calculated, based on empirical correlations, with reasonable accuracy on desktop computers. An iterative adjustment of model and experimental NMR spectra, corroborated by the current understanding of humic bio- and geosynthetic pathways [22, 160-167], allows to propose model structures [34] conforming to spectral data. In this respect, measurement and data base search of NMR parameters of model compounds [15, 168, 169] and calculation of parameters of NMR spectra of extended organic structures [118] are powerful tools to discriminate among several proposed humic substructures. Even when used only as initial step, which requires verification by independent analytical data, this approach provides very significant resolution in structural detail at atomic resolution, which is not available by other analytical techniques.

7.2.1. Analysis of aromatic substitution patterns

For specific regions of chemical shift, adapted assignment schemes can be developed [170]; an example is the *increment analysis* of 2D NMR spectra to elucidate the aromatic substitution patterns in NOM/HS [171]. The substitution of aromatic rings in HS/NOM, which carry from one to five hydrogen substituents, can be deduced from the increment analysis of homo- and heteronuclear 2D NMR spectra. The $\delta(^1H)$ and $\delta(^{13}C)$ values of any C_{ar}-H bond are most strongly affected by the nature of ortho and para substituents. Oxygen substitution in ortho and para positions causes proton upfield shift, while carbonyl derivatives induce downfield shifts of ortho protons caused by the chemical shift anisotropy of the carbonyl group. However, any cross peak within the aromatic section of a COSY NMR spectrum indicates (at least) two ortho protons; the chemical shift of these two protons is affected by all other substituents of the aromatic ring. Probable aromatic substitution patterns can be deduced from increment analysis of the chemical shift values on a probabilistic and individual structure level (cf. Figure 23). The distribution of cross peaks in homonuclear 2D NMR spectra indicates a broad range of oxygen, carbon and carbonyl derivative substitution in aromatic substructures of NOM/HS; there is no preference for (only) a few specific substitution patterns.

424

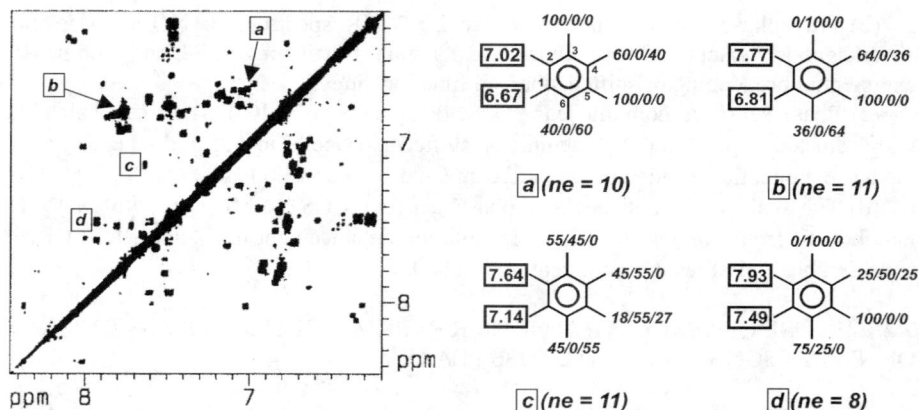

Figure 23. Section of aromatic cross peaks in a 2Q COSY NMR spectrum of a soil humic acid, with a probabilistic level ($\Sigma_R + \Sigma_{COR} + \Sigma_{OR} = 100$ %) of substituent distribution for classes of substituents R / COR / OR indicated for each position [ne = number of hits (out of 2^{16} = 32768 combinations) within a chemical shift window of 0.1 ppm in both F1 and F2 around the center of the cross peak indicated].

8. The Role of NMR-based Structural Analysis of NOM/HS in Remediation

NMR spectroscopy and other molecular level precision analytical methods, like Fourier transform ion cyclotron mass spectrometry FTICR-MS, are indispensable for the structural analysis of NOM/HS and for the elucidation of reaction mechanisms in which these materials participate. However, remediation technologies are rather concerned with an expedient, affordable and swift solution of a practical problem, often within the framework of non – scientific legal and political regulations. In this respect the scope of fundamental research regarding NOM/HS structural analysis is to advance and to distribute knowledge in science and in the public and, on the practical level, to provide guidance for optimising experiments and results. For that reason, a firm relationship needs to be established between the information rich and often highly correlated analytical data and the reactivity of the respective NOM/HS materials towards pollutants. The NMR-based protocols developed for this purpose typically require an initial data reduction step, like bucket analysis or segmental integration, followed by pattern recognition or chemometric methods for classification and maximum information recovery. Various methods of multivariate analysis, like principal component analysis PCA [159, 172], non-linear mapping analysis and probabilistic neural networks PNN [173] are used to interpret NMR data. Very frequently a set of structure-based descriptors is obtained which allows to predict the behaviour and the reactivity of NOM/HS in a specific environment. Along these lines, NMR data of NOM/HS have been employed to classify NOM/HS according to origin [23], performance in analytical methods [174, 175], and reactivity towards organic [176, 177] and inorganic [178] materials.

425

These studies furnish compelling evidence that carefully developed protocols of NMR characterization of NOM/HS in conjunction with mathematic analysis provide a robust and reliable guidance to optimise the experimental design in various stages of remediation.

9. Experimental

All NMR spectra shown have been acquired by the first author with a Bruker DMX 500 NMR spectrometer (Rheinstetten, Germany) operating at 500.13 MHz proton frequency from various NOM/HS materials, according to NMR acquisition and processing parameters as given in Table 7; referencing is performed according to Ref. 97.

Table 6. substance amounts and key acquisition parameters for routine quantitative NMR spectra of NOM/HS (90-degree excitation pulses $B_0 = 11.7$ T; 5 mm NMR sample tubes; cf. section 2).

nucleus	minimum S/N	fair S/N	good S/N	aq [sec]	d1 [sec]
1D ^{1}H NMR	25 µg	500 µg	5 mg*	4.5	16
1D ^{13}C NMR	1 mg	7 mg	40 mg	0.25	8
1D ^{15}N NMR	5 mg	50 mg	135 mg	0.25	45
1D ^{31}P NMR	1 mg	7 mg	40 mg	0.3	4
^{1}H,^{1}H-COSY	150 µg	3 mg	10 mg	0.4	2
^{1}H,^{1}H-TOCSY	75 µg	3 mg	10 mg	0.2	2
^{1}H,^{13}C-HSQC	1 mg	7 mg	35 mg	0.2	2

* higher amounts of NOM / HS (750 µl solvent) would lead to intensity distortions and line broadening, caused by extensive interaction.

10. Glossary

C_f, functionalised aliphatic carbon: functional group, but no heteroatom is directly attached to carbon; gs, gradient enhanced; COSY, Correlated Spectroscopy; F1, vertical frequency axis in 2D NMR spectra (X nucleus calculated data); F2, horizontal frequency axis in 2D NMR spectra (proton detected data); GARP, Globally Optimized Alternating-Phase Rectangular Pulses; HMBC, Heteronuclear Multiple Bond Correlation; HMQC, Heteronuclear Multiple Quantum Correlation; HS: Humic Substances; HSQC, Heteronuclear Single Quantum Coherence; IHSS, International Humic Substances Society; MAXY, Maximum Quantum Spectroscopy; MLEV, Composite Pulse Decoupling Sequence, named after Levitt, M.H.; NOM: Natural Organic Matter; TOCSY, Total Correlation Spectroscopy; WALTZ, Wideband Alternating-Phase Low-Power Technique for Zero Residual Splitting. X – nucleus: any NMR active nucleus except protons;

Table 7. Key acquisition and processing parameters given for NMR spectra displayed in Figures 7-23.

Figure	PK [a]	NS	AQ	D1	NE	WDW1	WDW2	PR1	PR2
7a-c	CR	256	4360	15400		EM		1	
8 (^{13}C)	B5	17684	263	8000		EM		50	
8 (^{31}P)	B5	57241	400	3600		EM		35	
9	B10	54155	233	3000		EM		100	
10 (^{11}B)	B10	26594	341	1000		EM		50	
10 (^{17}O)	B5	2734808	38	40		EM		150	
12	CR	160	341	1329	1779	QS	SI	0	0
13	CR	80	1507	100	4096	EM	EM	4	4
14	CR	48	137	1800	400	QS	SI	3	3
15	CR	640	585	1215	164	QS	SI	0	0
16	I5	416	250	2750	706	EM	SI	35	2.5
17 (^{1}H$_{det}$)	I5	416	250	2750	706	EM	SI	35	2.5
17 (^{13}C$_{det}$)	B5	4096	250	2840	67	EM	GM	5	-1/0.3
19 (^{29}Si)	B5	22849	1101	1500		EM		35	3
20 (^{29}Si)	I5	256	250	1750	512	QS	SI	3	3
20 (^{13}C)	I5	256	250	1750	512	QS	SI	3	3
21	I5	32	187	1500	848	QS	SI	6	6
22a	B5	79552	250	2750		EM		*	
22b	I5	512	171	1500	512	EM	SI	given	given
23	CR	160	341	1329	1779	QS	SI	0	0

* cf. Figure 22; [a] probeheads used for acquisition of NMR spectra, B5: 5 mm broadband observe; CR: cryogenic 5 mm ^{1}H, ^{13}C, ^{15}N inverse geometry TXI/z-gradient; B10: 10 mm broad band observe; I5: 5 mm inverse geometry broad band/z-gradient; NS: number of scans (for 2D NMR: F2); AQ: acquisition time [ms]; D1: relaxation delay [ms]; NE: number of F1 increments in 2D NMR spectra; WDW1, WDW2 windowing function in F1/F2: EM: exponential multiplication [Hz]; GM: Lorentz-to-Gaussian transformation (EM/GM: line broadening factor [Hz] Gaussian multiplication factor GM [Hz]); QS: shifted square sine bell; SI: sine bell; PR1, PR2 coefficients used for windowing functions WDW1, WDW2, EM/GM are given in [Hz], SI/QS derived functions indicate shift by π/n (n is given).

11. Acknowledgements

The authors gratefully acknowledge financial support by the GSF (FE 75184, FE 76186, FE 79836), the German ministry of research (DLR: RUS-143/97), the NSF/DAAD (PPP) and additional funding by the EU INTAS grant No. 1129/97. We thank Eva Holzmann, Heidi Neumeir and Silvia Thaller for skilful technical assistance. Samples of NOM/HS from different ecosystems have been generously provided by the groups of I. V. Perminova (Lomonosov Moscow State University, Russian Federation), J. I. Hedges (University of Washington, Seattle, WA), Ph. Schmitt-Kopplin, J. Junkers (GSF), with sample inputs from A. Günzl (silylation) and F. Jäkle (methylation), M. Wolf (GSF), R. Benner (University of South Carolina, Columbia, SC), R. Artinger (KfK, Karlsruhe, Germany), E. M. Perdue (GeorgiaTech, Atlanta, GA), M. de Nobili (University of Udine, Italy), Y. Chen (The Hebrew University of Jerusalem, Israel), G. R. Aiken (USGS, Boulder, USA), T. Dittmar (NMHFL Tallahasse and University of Florida) and experience gained during work with these samples has indispensably contributed to the development of the ideas given in this manuscript.

12. References

1. Derenne, S., and Largeau, C. (2001) A Review of Some Important Families of Refractory Macromolecules: Composition, Origin, and Fate in Soils and Sediments, *Soil Sci.* **166**, 833-847.
2. Gaffney, J.S., Marley, N.A., Clark, S.B., (1996) *Humic and Fulvic Acids*, American Chemical Society Symposium Series 651, Washington.
3. De Leeuw, J.W. and Largeau, C. (1993) A review of macromolecular organic compounds that comprise living organisms and their role in kerogen, coal, and petroleum formation, in M.H. Engle, S. Macko (eds.), *Organic Geochemistry*, Plenum, New York, pp. 23-72.
4. Hedges, I.J., and Oades, J.M. (1997) Comparative organic geochemistries of soils and marine sediments, *Org. Geochem.* **27**, 319-361.
5. Stevenson, F.J. (1994) *Humus Chemistry: Genesis, Composition, Reactions*, John Wiley, New York.
6. Hansell, D.A., and Carlson, C.A. (2002) *Biogeochemistry of Marine Dissolved Organic Matter*, Academic Press, Elsevier Science.
7 Perdue, E.M., and Ritchie, J.D. (2003) Dissolved Organic Matter in Fresh Waters, in H.D. Holland and K.K. Turekian (eds.), *Treatise on Geochemistry*, Vol. 5, Surface and Ground Water, Weathering, Erosion and Soils, Chapter 11. pp. 273-318.
8. Hedges, J.I. (2002) Why Dissolved Organics Matter, in D.A. Hansell and C.A. Carlson (eds.), *Biogeochemistry of Matine Dissolved Organic Matter*, Academic Press, San Diego, pp. 1-33.
9. Perakis, S.S., and Hedin, L.O. (2002) Nitrogen loss from unpolluted South American forest mainly via dissolved organic compounds, *Nature*, **24**, 415-419.
10. Hedges, J.I. (2002) Sedimentary Organic Matter Preservation and Atmospheric O2 Regulation, *unpublished manuscript.*
11. MacCarthy, P., and Rice, J.A. (1988) An Ecological Rationale for the Heterogeneity of Humic Substances, in S.H. Schneider, P.J. Boston (eds.), *Proceedings of Chapman Conference on the Gaia Hypothesis*, San Diego, CA, March 7-11. MIT Press, Cambridge, pp. 339-345.
12. Hedges, J.I., Eglinton, G., Hatcher, P.G., Kirchman, D.L., Arnosti, C., Derenne, S., Evershed, R.P., Kögel-Knabner, I., deLeeuw, J.W., Littke, R., Michaelis, W., and Rullkötter, J. (2000) The molecularly-uncharacterized component of nonliving organic matter in natural environments, *Org. Geochem.* **31**, 945-958.
13. Perminova, I.V., Frimmel, F.H., Kudryavtsev, A.V., Kulikova, N.A., Abbt-Braun, G., Hesse, S., Petrosyan, V.S. (2003) Molecular weight characteristics of humic substances from different

428

environments as determined by size exclusion chromatography and their statistical evaluation, *Environ. Sci. Technol.* **37**, 2477-2485.

14. Barrientos, L.G., Louis, J.M., Ratner, D.M., Seeberger, P.H., and Gronenborn, A.M. (2003) Solution structure of a circular-permuted variant of the potent HIV-inactivating protein cyanovirin-N: Structural basis for protein stability and oligosaccharide interaction, *J. Mol. Biol.* **32**, 211-223.

15. Clore, G.M., and Gronenborn, A.M. (1998) New methods of structure refinement for macromolecular structure determination by NMR, *Proc. Natl. Acad. Sci. USA* **95**, 5891-5898.

16. Clore, G.M., and Gronenborn, A. M. (1998) Determining the structures of large proteins and protein complexes by NMR, *Tibtech* **16**, 22-34.

17. Kujawinski, E.B. (2002) Electrospray Ionization Fourier Transform Ion Cyclotron Resonance Mass Spectrometry (ESI FT-ICR MS): Characterization of Complex Environmental Mixtures, *Environ. Forensics* **3**, 207-216.

18. Takahashi, H., Nakanishi, T., Kami, K., Arata, Y., and Shimada, I. (2000) A novel NMR method for determining the interfaces of large protein-protein complexes, *Nature Structural Biology* **7**, 220-223.

19. Hertkorn, N., Schmitt-Kopplin, Ph., Junkers, J., and Kettrup, A. (2004) Die strukturchemische Characterisierung von natürlicher organischer Materie, *GIT Laborfachzeitschrift,* **5**, 487-491.

20. Burdon, J. (2001) Are the Traditional Concepts of the Structures of Humic Substances Realistic? *Soil Sci.* **166**, 752-769.

21. Gratias, R., Konat, R., Kessler, H., Crisma, M., Valle, G., Polese, A., Formaggio, F., Toniolo, C., Broxterman, Q.B., Kamphuis, J. (1998) First step toward the quantitative identification of peptide 3(10)-helix conformation with NMR spectroscopy: NMR and X-ray diffraction structural analysis of a fully-developed 3(10)-helical peptide standard, *J. Am. Chem. Soc.* **120**, 4763-4770.

22. Kögel-Knabner, I. (2002) The macromolecular organic composition of plant and microbial residues as inputs to soil organic Matter, *Soil Biol. Biochem.* **34**, 139-162.

23. Thomsen, M., Lassen, P., Dobel, S., Hansen, P.E., Carlsen, L., and Bugel-Mogensen, B. (2002) Characterisation of humic materials of different origin: A multivariate approach for quantifying the latent properties of dissolved organic matter, *Chemosphere* **49**, 1327-1337.

24. Diallo, M. S. (2003) 3-D structural modeling of humic acids through experimental characterization, computer assisted structure elucidation and atomistic simulations. 1. Chelsea soil humic acid, *Environ. Sci. Technol.* **37**, 1783-1793.

25. Lowe, L.E. (1992) Studies on the nature of sulfur in peat humic acids from the Fraser river delta, British Columbia, *Sci. Total Environ.* **113**, 133-145.

26. Morra, M.J., Fendorf, S.E., and Brown, P.D. (1997) Speciation of sulfur in humic and fulvic acids using X-ray absorption near-edge structure (XANES) spectroscopy, *Geochim. Cosmochim. Acta* **61**, 683-688.

27. Xia, K., Skyllberg, U.L., Bleam, W.F., Bloom, P.R., Nater, E.A., and Helmke, P.A. (1999) X-ray absorption spectroscopic evidence for the complexation of Hg(II) by reduced sulfur in soil humic substances, *Environ. Sci. Technol.* **33**, 257-261.

28. Hertkorn, N., Schmitt-Kopplin, Ph., Artinger, R., Buckau, G., Schäfer, T., Geyer, S., Bleam, W.F., and Jacobson, Ch. (2004) Combined ^{1}H/^{13}C NMR, carbon/sulfur K-edge XANES and capillary electrophoresis study on the characterization and fate of aquatic fulvic acids, *Geochim. Cosmochim. Acta,* in preparation.

29. Bloom, P.R., and Leenheer, J.A. (1989) Vibrational, Electronic, High-energy Spectroscopic Methods for Characterizing Humic Substances, in: M.H.B., Hayes, P., Mac Carthy, R.L.,Malcolm, R.S., Swift, (eds.), *Humic Substances II, In Search of Structure*, Wiley, Chichester, pp. 411-446.

30. Hayes, M.H.B., MacCarthy, P., Malcolm, R.L., and Swift, R.S. (1989) *Humic Substances II, In Search of Structure*, John Wiley, Chichester.

31. Preston, C.M. (1996) Applications of NMR to Soil Organic Matter Analysis: History and Prospects, *Soil Sci.* **161**, 144-166.

32. Mahieu, N., Powlson, D.S., and Randall, E.W. (1999) Statistical analysis of published carbon-13 CPMAS NMR spectra of soil organic matter, *Soil Sci. Soc. Am. J.* **63**, 307-319.

33. Cook, R.L. (2004) Coupling NMR to NOM, *Anal. Bioanal. Chem.* **378**, 1484-1503.

34. Simpson, A.J., Kingery, W.L., Hayes, M.B., Spraul, M., Humpfer, E., Dvortsak, P., Kerssebaum, R., and Godejohann, M. (2002) Molecular structures and associations of humic substances in the terrestrial environment, *Naturwiss.* **89**, 84-88.

35. Hong, M.F., Sudor, J., Stefansson, M., and Novotny, M.V. (1998) High resolution studies of hyaluronic acid mixtures through capillary gel electrophoresis, *Anal. Chem.* **7**, 568-573.

36. Helgaker, T., Jaszunski, M., Ruud, K. (1999) Ab initio methods for the calculation of NMR shielding and indirect spin-spin coupling constants, *Chem. Rev.* **99**, 293-352.

37. Heumann, K.G., Abbt-Braun, G., Behrens, K., Burba, P., Frimmel, F.H., Jakubowski, B., Knöchel, A., Mielcke, J., Rädlinger, G., Marx, G., and Vogl, J. (2002) Element Determination and its Quality Control in Fractions of Refractory Organic Substances and the Corresponding Original Water Samples, in F.H. Frimmel, G. Abbt-Braun, K.G. Heumann, B. Hock, H.-D. Lüdemann, and M. Spiteller (eds.), *Refractory Organic Substances in the Environment*, Wiley-VCH, Weinheim, pp. 39-53.

38. Wang, K., Dickinson, L.C., Ghabbour, E.A., Davies G., and Xing, B. (2003) Proton Spin-Lattice Relaxation Times of Humic Acids as Determined by Solution NMR, *Soil Sci.* **168**, 128-136.

39. Flynn, P.F., Mattielo, D.L., Hill, H.D.W. and Wand, A.J. (2000) Optimal Use of Cryogenic Probe Technology in NMR Studies of Proteins, *J. Am. Chem. Soc.* **122**, 4823-4824.

40. Fründ, R., and Lüdeman, H.D. (1989) The quantitative analysis of solution- and CPMAS-C-13 NMR spectra of humic material, *Sci. Total Environ.* **81/82**, 157-168.

41. Preston, C.M., and Blackwell, B.A. (1985) Carbon-13 nuclear magnetic resonance for a humic and a fulvic acid: signal-to-noise optimisation, quantitation, and spin-echo techniques, *Soil Sci.* **139**, 88-96.

42. Tang, H.R., Wang, Y.L., Nicholson, J.K., Lindon, J.C. (2004) Use of relaxation-edited one-dimensional and two dimensional nuclear magnetic resonance spectroscopy to improve detection of small metabolites in blood plasma, *Anal. Biochem.*, **325**, 260-272.

43. Terblanche, C.J., Reynhardt, E.C. and van Wyk, J.A. (2001) C-13 spin-lattice relaxation in natural diamond: Zeeman relaxation at 4.7 T and 300 K due to fixed paramagnetic nitrogen defects, *Solid State Nucl. Mag.* **20**, 1-22.

44. Reynhardt, E.C. (2003) Spin-lattice relaxation of spin-1/2 nuclei in solids containing diluted paramagnetic impurity centers. I. Zeeman polarization of nuclear spin system, *Concept. Magn. Res. A* **19A**, 20-35.

45. Fan, T.W.-M., Higashi, R.M. and Lane, A.N. (2000) Chemical Characterization of a Chelator-Treated Soil Humate by Solution-State Multinuclear Two-Dimensional NMR with FTIR and Pyrolysis-GCMS, *Environ. Sci. Technol.* **34**, 1636-1646.

46. Gèlinas, Y., Baldock, J.A. and Hedges, J.I. (2001) Demineralization of marine and freshwater sediments for CP/MAS ^{13}C NMR analysis, *Org. Geochem.* **32**, 677-693.

47. Smernik, R.J., Oades, J.M. (1999) Effects of added paramagnetic ions on the C-13 CP MAS NMR spectrum of a de-ashed soil, *Geoderma*, **89**, 219-248.

48. Case, D.A. (2002) Molecular Dynamics and NMR Spin Relaxation in Proteins, *Acc. Chem. Res.* **35**, 325-331.

49. Becker, E.D. (2000) *High Resolution NMR. Theory and Chemical Applications*. Academic Press, San Diego, pp. 205-225.

50. Murali, N., Krishnan, V. (2003) A primer for nuclear magnetic relaxation in liquids, *Concepts Magn. Reson. Part A*, **17A**, 86-116.

51. Vugmeyster, L., Pelupessy, P., Vugmeister, B.E., Abergel, D. and Bodenhausen, G. (2004) Cross-correlated relaxation in NMR of macromolecules in the presence of fast and slow internal dynamics, *C.R. Physique* **5**, 377-386.

52. Peuravuori, J., Ingman, P., and Pihlaja, K. (2003) Critical comments on accuracy of quantitative determination of natural humic matter by solid state ^{13}C NMR spectroscopy, *Talanta.* **59**, 177-189.

53. Keeler, C. and Maciel, G. E. (2003) Quantitation in the solid-state C-13 NMR analysis of soil and organic soil fractions, *Anal. Chem.* **75**, 2421-2432.

430

54. Mao, J.D. and Schmidt-Rohr, K. (2004) Accurate quantification of aromaticity and nonprotonated aromatic carbon fraction in natural organic matter by C-13 solid-state nuclear magnetic resonance, *Environ. Sci. Technol.* **38**, 2680-2684.

55. Smernik, R.J. and Oades, J.M. (2003) Spin accounting and RESTORE – two new methods to improve quantitation in solid-state C-13 NMR analysis of soil organic matter, *Eur. J. Soil Sci.* **54**, 103-116.

56. Mao, J.D., Hub, W.G., Ding, G.W., Schmidt-Rohr, K., Davies, G., Ghabbour, E.A., Xing, B. (2002) Suitability of different C-13 solid-state NMR techniques in the characterization of humic acids, *Int. J. Environ. An. Ch.* **82**, 183-196.

57. Preston, C.M. (2001) Carbon-13 solid-state NMR of soil organic matter - using the technique effectively, *Can. J. Soil Sci.* **81**, 255-270.

58. Smernik, R.J. and Oades, J.M. (2001) Solid-state C-13-NMR dipolar dephasing experiments for quantifying protonated and non-protonated carbon in soil organic matter and model systems, *Eur. J. Soil Sci.* **52**, 103-120.

59. Mao, J.D., Hu, W.G., Schmidt-Rohr, K., Davies, G., Ghabbour, E.A., Xing, B., (2000) Quantitative characterization of humic substances by solid-state carbon-13 nuclear magnetic resonance, *Soil Sci. Soc. Am. J.* **64**, 873-884.

60. Cook, R.L. and Langford, C.H. (1998) Structural characterization of a fulvic acid and a humic acid using solid state ramp-CP-MAS C-13 nuclear magnetic resonance, *Environ. Sci. Technol.* **32**, 719-725.

61. Mason, J. (1987), *Multinuclear NMR*. Plenum Press, New York.

62. Kovalevskii, D.V., Permin, A.B., Perminova, I.V. and Petrosyan, V.S. (2000) Choice of the time of pulse delay for quantitative ^{13}C NMR spectroscopy of humic substances, *Bulletin of Moscow University* (*Vestnik MGU*), *Ser. 2 (chem)* **41**, 39-42.

63. Webb, A. G. (1997) Radiofrequency microcoils in magnetic resonance, *Prog. Nucl. Magn. Reson. Spec.* **31**, 1-42.

64. Brunner, E. (1998) Enhancement of Surface and Biological Magnetic Resonance Using Laser-Polarized Noble Gases, *Concept. Magnetic Res.* **11**, 313-335.

65. Magusin, P.C.M.M., Bolz, A., Sperling, K., and Veeman, W.S. (1997) The use of ^{129}Xe NMR spectroscopy for studying soils. A pilot study, *Geoderma.* **80**, 449-462.

66. Navon, G., Song, G.Y., Ròòm, T., Appelt, S., Taylor, A.E., and Pines A. (1996) Enhancement of Solution NMR and MRI with Laser-Polarized Xenon, *Science*, **271**, 1848-1851.

67. Paytan, A., Cade-Menun, B.J., McLaughlin, K., Faul, K.L. (2003) Selective phosphorus regeneration of sinking marine particles: evidence from P-31-NMR, *Mar. Chem.* **82**, 55-70.

68. Makarov, M.I., Malysheva, T.I., Haumaier, L., Alt, H.G., Zech, W. (1997) The forms of phosphorus in humic and fulvic acids of a toposequence of alpine soils in the northern Caucasus, *Geoderma* **80**, 61-73.

69. Cade-Menun, B.J., Liu, C.W., Nunlist, R., McColl, J.G. (2002) Soil and litter phosphorus-31 nuclear magnetic resonance spectroscopy: Extractants, metals, and phosphorus relaxation times, *J. Environ. Qual.* **31**, 457-465.

70. Crestini, C., Argyropoulos, D.S. (1997) Structural analysis of wheat straw lignin by quantitative P-31 and 2D NMR spectroscopy. The occurrence of ester bonds and alpha-O-4 substructures, *J. Agric. Food Chem.* **45**, 1212-1219.

71. Tohmura, S., Argyropoulos, D.S. (2001) Determination of arylglycerol-beta-aryl ethers and other linkages in lignins using DFRC/P-31 NMR, *J. Agric. Food Chem.* **49**, 536-542.

72. Lee, G.S.H., Wilson, M.A., and Young, B.R. (1998) The application of the WATERGATE suppression technique for analyzing humic substances by nuclear magnetic resonance, *Org. Geochem.* **28**, 549-559.

73. Beck, K.C. and Reuter, J.H. (1974) Organic and inorganic geochemistry of some coastal plain rivers of the southeastern United States, *Geochim. Cosmochim. Acta* **38**, 341-364.

74. Shin, H.S., Rhee, S.W., Lee, B. and Moon, H. (1996) Metal binding sites and partial structures of soil fulvic and humic acids compared: aided by Eu(III) luminescence spectroscopy and DEPT/QUAT ^{13}C NMR pulse techniques, *Org. Geochem.* **24**, 523-529.

75. Shin, H. S. and Moon, H. (1996) An "average" structure proposed for soil fulvic acid aided by DEPT/QUAT C-13 NMR pulse techniques, *Soil Sci.* **161**, 250-256.

76. Buddrus, J., Burba, P., Herzog, P.H. and Lambert, J. (1989) Quantification of Partial Structures of Aquatic Humic Substances by One- and Two-Dimensional Solution ^{13}C Nuclear Magnetic Resonance Spectroscopy, *Anal. Chem.* **89**, 628-631.

77. Sorensen, O.W., Jakobsen, H.J. (1988) Polarization Transfer and Editing Techniques, in. W.S. Brey (ed.), *Pulse Methods in 1D and 2D Liquid-Phase NMR*, Academic Press, San Diego, pp. 149-258.

78. Breitmaier, E. and Voelter, W. (1990) *Carbon-13 NMR Spectroscopy, High-Resolution Methods and Applications in Organic Chemistry and Biochemistry*, VCH-Verlagsgesellschaft, Weinheim.

79. Li, J., Perdue, E.M. and Gelbaum, L.T. (1998) Using Cadmium-113 NMR Spectrometry to Study Metal Complexation by Natural Organic Matter, *Environ. Sci. Technol.* **32**, 483-487.

80. Lambert, J., Buddrus, J. and Burba P. (1995) Evaluation of conditional stability constants of dissolved aluminium/humic substance complexes by means of ^{27}Al nuclear magnetic resonance, *Fresenius J. Anal. Chem.* **351**, 83-87.

81. Lu, X., Johnson, W.D. and Hook, J. (1998) Reaction of Vanadate with Aquatic Humic Substances: An ESR and ^{51}V NMR Study, *Environ. Sci. Technol.* **32**, 2257-2263.

82. Howe, R.F., Lu, X.Q., Hook, J. and Johnson, W.D. (1997) Reaction of aquatic humic substances with aluminium: a Al-27 NMR study, *Mar. Freshwater Res.* **48**, 377-383.

83. Öz, G., Pountney, D.L. and Armitage, I.M. (1998) NMR Spectroscopic studies of I = 1/2 metal ions in biological systems, *Biochem. Cell Biol.* **76**, 223-234.

84. Benn, R. and Rufinska, A. (1986) Hochauflösende Metallkern-NMR-Spektroskopie von Organometallverbindungen, *Angew. Chem.* **98**, 851-871.

85. Harrison, P.G., Healy, M.A. and Steel, A.T. (1983) Lead-207 Chemical Shift Data for Bivalent Lead Compounds: Thermodynamics of the Equilibrium Pb(O$_2$CCH$_3$)$_2$; [Pb(O$_2$CCH$_3$)]$^+$ + O$_2$CCH$_3$ in Aqueous Solution in the Temperature Range 303-323 K, *J. Chem. Soc. Dalton Trans*, 1845-1848

86. Nakashima, T.T. and Rabenstein, D.L. (1983) A Lead-207 Nuclear Magnetic Resonance Study of the Complexation of Lead by Carboxylic Acids and Aminocarboxylic Acids, *J. Magn. Reson.* **51**, 223-232.

87. Medek, A., Frydman, V. and Frydman, L. (1997) Solid and liquid phase ^{59}Co NMR studies of cobalamins and their derivatives, *Proc. Natl. Acad. Sci. USA* **94**, 14237-14242.

88. Claudio, E.S., Horst, M.A., Forde, C.E., Stern, Ch.L., Zart, M.K. and Godwin, H.A. (2000) ^{207}Pb-^{1}H Two-Dimensional NMR Spectroscopy: A Useful New Tool for Probing Lead (II) Coordination Chemistry, *Inorg. Chem.* **39**, 1391-1397.

89. Walker, F.A. (2003) Pulsed EPR and NMR Spectroscopy of Paramagnetic Iron Porphyrinates and Related Iron Macrocycles: How to Understand Patterns of Spin Delocalization and Recognize Macrocycle Radicals, *Inorg. Chem.* **42**, 4526-4544.

90. Walker, F.A. (1999) Magnetic spectroscopic (EPR, ESEEM, Mossbauer, MCD and NMR) studies of low-spin ferriheme centers and their corresponding heme proteins, *Coordin. Chem. Rev.* **186**, 471-534.

91. Palmer, A.G. (2002) Chemical exchange effects in biological macromolecules, in D.M. Grant and R.K. Harris (eds.), *Encyclopedia of Nuclear Magnetic Resonance* **9,** Wiley-VCH, pp. 344-353.

92. Palmer, A.G., Kroenke, C.D. and Loria, J.P. (2001) Nuclear magnetic resonance methods for quantifying microsecond-to-millisecond motions in biological macromolecules, *Method. Enzymol.* **339**, 204-238.

93. Hertkorn, N., Perdue, E.M. and Kettrup, A. (2004) The binding of Suwannee River organic matter to cadmium. A potentiometric and ^{113}Cd NMR study at two different magnetic field strengths, *Anal. Chem.*, 6327-6341.

94. Drexel, R.T., Haitzer, M., Ryan, J.N., Aiken, G.R. and Nagy, K.L. (2002) Mercury(II) sorption to two Florida Everglades peats: Evidence for strong and weak binding and competition by dissolved organic matter released from the peat, *Environ. Sci. Technol.* **36**, 4058-4064.

95. Haitzer, M., Aiken, G.R. and Ryan, J.N. (2002) Binding of mercury(II) to dissolved organic matter: The role of the mercury-to-DOM concentration ratio, *Environ. Sci. Technol.* **36**, 3564-3570.

96. Millet, O., Loria, J.P., Kroenke, C.D., Pons, M. and Palmer, A.G. (2000) The static magnetic field dependence of chemical exchange linebroadening defines the NMR chemical shift time scale, *J. Am. Chem. Soc.* **122**(12), 2867-2877.

432

97. Harris, R.K., Becker, E.D., Cabral de Menzes, S.M., Goodfellow, R. and Granger, P. (2002) Nuclear Spin Properties and Conventions for Chemical Shifts, in D.M. Grant and R.K. Harris (eds.), *Encyclopedia of Nuclear Magnetic Resonance* **9**, Wiley-VCH, pp. 5-18.

98. Pregosin, P.S. (1991) *Transition Metal Nuclear Magnetic Resonance*, Elsevier, Amsterdam.

99. Schmitt-Kopplin, Ph., Hertkorn, N., Garrison, A.W., Freitag, D. and Kettrup, A. (1998) Influence of Borate Buffers in the Electrophoretical Behaviour of Humic Substances in Capillary Zone Electrophoresis, *Anal. Chem.* **70**, 3798-3808.

100. Kählig, H. and Robien, W. (1994) ^{17}O NMR Spectroscopic Investigation of Steroids at Natural Abundance, *Magn. Reson. Chem.* **32**, 608-613.

101. Grandy, D.W., Petrakis, L., Young, D.C. and Gates, B.C. (1984) Determination of oxygen functionalities in synthetic fuels by NMR of naturally abundant ^{17}O, *Nature* **308**, 175-177.

102. Hertkorn, N. (2004) ^{17}O NMR spectroscopy of natural organic matter, *Anal. Chem.* (in preparation).

103. Schumacher, M. and Lauterwein, J. (1989) The INEPT Experiment for Nonselective Polarization Transfer in ^{17}O NMR, *J. Magn. Reson.* **83**, 97-110.

104. Sergeyev, N.M., Sergeyeva, N.D. and Raynes, W.T. (1999) Isotope effects on the O-17, H-1 coupling constant and the O-17-{H-1} nuclear Overhauser effect in water, *J. Magn. Reson.* **137**, 311-315.

105. Price, W.S. (2002) Diffusion-based Studies of Aggregation, Binding and Conformation of Biomolecules: Theory and Practice, in D.M. Grant and R.K. Harris (eds.), *Encyclopedia of Nuclear Magnetic Resonance* **9**, Wiley-VCH, pp. 364-374.

106. Morris, K.F., Cutak, B.J., Dixon, A.M. and Larive, C.K. (1999) Analysis of diffusion coefficient distributions in humic and fulvic acids by means of diffusion ordered NMR spectroscopy, *Anal. Chem.* **71**, 5315-5321.

107. Pellecchia, M., Sem, D.S. and Wüthrich, K. (2002) NMR in Drug Discovery, *Nature Reviews.* **1**, 211-219.

108. Zuiderweg, E.R.P. (2002) Mapping protein-protein interactions in solution by NMR Spectroscopy, *Biochemistry* **41**, 1-7.

109. Nanny, M.A. and Maza, J.P. (2001) Noncovalent interactions between monoaromatic compounds and dissolved humic acids: A deuterium NMR T-1 relaxation study, *Environ. Sci. Technol.* **35**, 379-384.

110. Nanny, M.A., Bortiatynski, J.M. and Hatcher, P.G. (1997) Noncovalent interactions between acenaphthenone and dissolved fulvic acid as determined by C-13 NMR T-1 relaxation measurements, *Environ. Sci. Technol.* **31**, 530-534.

111. Dixon, A.M., Mai, M.A. and Larive, C.K. (1999) NMR Investigation of the Interactions between 4-Fluoro-1-acetonaphthone and the Suwannee River Fulvic Acid, *Environ. Sci. Technol.* **33**, 958-964.

112. Anderson, S.J. (1997) Proton and ^{19}F NMR Spectroscopy of Pesticide Intermolecular Interactions, in M. Nanny, R. Minear, J. Leenheer (eds), *Nuclear Magnetic Resonance Spectroscopy in Environmental Chemistry,* pp. 51-71.

113. Campos-Olivas, R., Aziz, R., Helms, G.L., Evans, J.N.S. and Gronenborn, A.M. (2002) Placement of ^{19}F into the centre of GB1: effects on structure and stability, *FEBS Letters* **517**, 55-60.

114. Mortimer, R.D. and Dawson, A. (1991) Using ^{19}F NMR for Trace Analysis of Fluorinated Pesticides in Food Products, *J. Agric. Food Chem.* **39**, 1781-1785.

115. Ellis, D.A., Martin, J.W., Muir, D.C.G. and Mabury, S.A. (2000) Development of an ^{19}F NMR Method for the Analysis of Fluorinated Acids in Environmental Water Samples, *Anal. Chem.* **72**, 726-731.

116. Cavanagh, J., Fairbrother, W.J., Palmer, A.G. and Skelton, N.J. (1996) *Protein NMR Spectroscopy, Principles and Practice*, Academic Press, London.

117. Croasmun, W.R. and Carlson, R.M.K. (1996) *Two-Dimensional NMR Spectroscopy*, Wiley-VCH, Weinheim.

118. Hertkorn, N., Permin, A., Perminova, I., Kovalevskii, D., Yudov, M., Petrosyan, V. and Kettrup, A. (2002) Comparative Analysis of Partial Structures of a Peat Humic and Fulvic Acid Using One- and Two-Dimensional Nuclear Magnetic Resonance Spectroscopy, *J. Environ. Qual.* **31**, 375-387.

119. Hertkorn, N., Schmitt-Kopplin, Ph., Perminova, I.V., Kovalevskii, D. and Kettrup, A. (2001) Two dimensional NMR spectroscopy of humic substances, in R.S. Swift, K.M. Spark (eds.), *Humic Substances Downunder, Understanding and Managing Organic Matter in Soils, Sediments and*

Waters, Sept. 21-25, 1998, International Humic Substances Society, University of Adelaide, Australia, pp. 149-158.

120. Russel, D.J., Hadden, C.E., Martin, G.E., Gibson, A.A., Zens, A.P. and Carolan, J.L. (2000) A Comparison of Inverse-Detected Heteronuclear NMR Performance: Conventional vs Cryogenic Microprobe Performance, *J. Nat. Prod.* **63**, 1047-1049.

121. Crouch, R.E., Llanos, W., Mehr, K.G., Hadden, C.E., Russell, D.J. and Martin, G.E. (2001) Applications of cryogenic NMR probe technology to long-range ^{1}H-^{15}N 2D NMR studies at natural abundance, *Magn. Reson. Chem.* **39**, 555-558.

122. Martin, G.E. (2002) Cryogenic NMR Probes: Applications, in D.M. Grant and R.K. Harris (eds.), *Enzyclopedia of Nuclear Magnetic Resonance* **9**, Wiley-VCH, pp. 33-35.

123. Kelly, A.E., Ou, H.D., Withers, R. and Dotsch, V. (2002) Low-conductivity buffers for high-sensitivity NMR measurements, *J. Am. Chem. Soc.* **124**, 12013-12019.

124. Bovey, F.A. and Mirau, P.A. (1996) *NMR of Polymers*, Academic Press, San Diego, pp. 353-378.

125. Simpson, A.J., Salloum, M.J., Kingery, W.L. and Hatcher, P.G. (2003) The Identification of Plant Derived Structures in Humic Materials Using Three-Dimensional NMR Spectroscopy, *Environ. Sci. Technol.* **37**, 337-342.

126. Chien, Y.Y. and Bleam, W.F. (1998) Two-Dimensional NOESY Nuclear Magnetic Resonance Study of pH-Dependent Changes in Humic Acid Conformation in Aqueous Solution, *Environ. Sci. Technol.*, **32**, 3653-3658.

127. Haiber, S., Burba, P., Herzog, H. and Lambert, J. (1999) Elucidation of aquatic humic partial structures by multistage ultrafiltration and two-dimensional nuclear magnetic resonance spectrometry, *Fresen. J. Anal. Chem.* **364**, 215-218.

128. Kingery, W.L., Simpson, A.J., Hayes, M.H.B., Locke, M.A. and Hicks. R.P. (2000) The application of multidimensional NMR to the study of soil humic substances, *Soil Sci.*, **165**, 483-494.

129. Schmitt-Kopplin, Ph., Hertkorn, N., Schulten, H.R. and Kettrup, A. (1998) Structural Changes in a Dissolved Soil Humic Acid during Photochemical Degradation Processes under O_2 and N_2 Atmosphere, *Environ. Sci. Technol.*, **32**, 2531-2541.

130. Simpson, A.J., Boersma, R.E., Kingery, W.L., Hicks, R.P. and Hayes, M.H.B. (1997) Humic Substances, Peats and Sludges, in M.H.B. Hayes, W.S. Wilson (eds.), *Applications of NMR Spectroscopy for Studies of the Molecular Compositions of Humic Substances*, Royal Society of Chemistry, Cambridge, pp. 46-62.

131. Wang, L., Mao, X. and Yang, Y. (1998) Application of newly-developed ^{1}H NMR techniques to the study of humic acids, *Bopuxue Zazhi (Chinese Journal of Magnetic Resonance)* **15**, 411-420.

132. Simpson, A. (2001) Multidimensional Solution State NMR of Humic Substances: a Practical Guide and Review, *Soil Sci.* **166**, 795-809.

133. Simpson, A.J., Burdon, J., Graham, C.L., Hayes, M.H.B., Spencer, N. and Kingery, W.L. (2001) Inerpretation of heteronuclear and multidimensional NMR spectroscopy of humic substances, *Eur. J. Soil Sci.* **52**, 495-509.

134. Kingery, W.L., Simpson, A.J., Hayes, M.H.B. and Hicks, R.P. (2000) The Application of Multidimensional NMR to the Study of Soil Humic Substances, *Soil Sci.* **165**, 483-494.

135. Simpson, A.J., Salloum, M.J., Kingery, W.L. and Hatcher, P.G. (2002) Improvements in the Two-Dimensional Nuclear Magnetic Resonance Spectroscopy of Humic Substances, *J. Environ. Qual.* **31**, 388-392.

136. Ernst, R. R., Bodenhausen, G. and Wokaun, A. (1987) *Principles of Nuclear Magnetic Resonance in One and Two Dimensions*, Oxford University Press, Oxford.

137. Kumar, A., Wagner, G., Ernst, R.R. and Wüthrich, K. (1981) Buildup Rates of the Nuclear Overhauser Effect Measured by Two-Dimensional Proton Magnetic Resonance Spectroscopy, *J. Am. Chem. Soc.* **103**, 3654-3658.

138. Xiang, B.S. and Markham, G.D. (1996) The conformation of inosine 5'-monophosphate (IMP) bound to IMP dehydrogenase determined by transferred nuclear Overhauser effect spectroscopy, *J. Biol. Chem.* **271**, 27531-27535.

434

139. Hertkorn, N., Kovalevskii, D., Perminova, I.V., Schmitt-Kopplin, Ph., Permin, A., Petrosyan, V., and Kettrup, A. (2004) One and two-dimensional proton NMR methods for the identification of labile and non exchangeable protons in humic substances, *Anal. Chem.*, (in preparation).

140. Tomlins, A.M., Foxall, P.J.D., Lynch, M.J., Parkinson, J., Everett, J.R. and Nicholson, J.K. (1998) High resolution H-1 NMR spectroscopic studies on dynamic biochemical processes in incubated human seminal fluid samples, *Biochim. Biophys. Acta-General Subjects* **1379**, 367-380.

141. Viant, M.R. (2003) Improved methods for the acquisition and interpretation of NMR metabolomic data, *Biochim. Biophys. Res. Commun.* **310**, 943-948.

142. Reynolds, W.F., Lean, S.M., Tay, L., Yu, M., Enriquez, R.G., Estwick, D.M. and Pascoe, K.O. (1997) Comparison of ^{13}C Resolution and Sensitivity of HSQC and HMQC Sequences and Application of HSQC-Based Sequences to the Total ^{1}H and ^{13}C Spectral Assignment of Clionasterol, *Magn. Reson. Chem.* **35**, 455-462.

143. Bax, A., Ikura, M., Kay, L.E., Torchia, D.A. and Tschudin, R. (1990) Comparison of Different Modes of Two-Dimensional Reverse-Correlation NMR for the Study of Proteins, *J. Magn. Reson.* **86**, 304-318.

144. Norwood, T.J., Boyd, J., Heritage, J.E., Soffe, N. and Campbell, I.D. (1990) Comparison of Techniques for ^{1}H-Detected Heteronuclear ^{1}H-^{15}N Spectroscopy, *J. Magn. Reson.* **87**, 488-501.

145. Hertkorn, N. (2004) Differential responses of substructures in the ^{1}J (C,H) correlation NMR spectra of a terrestrial humic substance, *Anal. Chem.* (in preparation)

146. Perdue, E.M. (1985) Acidic Functional Groups of Humic Substances, in G.R. Aiken, D.M. McKnight, R.L. Wershaw, P. MacCarthy (eds.), *Humic Substances in Soil, Sediment, and Water; Geochemistry, Isolation and Characterization*, Wiley, Chichester, pp. 493-526.

147. Ritchie, J.D., Perdue, E.M. (2003) Proton-binding study of standard and reference fulvic acids, humic acids, and natural organic matter, *Geochim. Cosmochim. Acta* **67**, 85-96.

148. Hertkorn, N., Günzl, A., Freitag, D. and Kettrup, A. (2002) Nuclear Magnetic Resonance Spectroscopy Investigations of Silylated Refractory Organic Substances, in F.H. Frimmel, G. Abbt-Braun, K.G. Heumann, B. Hock, H.-D. Lüdemann and M. Spiteller (eds.), *Refractory Organic Substances in the Environment*, Wiley-VCH, Weinheim, pp. 129-145.

149. Hertkorn, N., Günzl, A., Wang, C., Freitag, D. and Kettrup, A. (1996) The Role of Humic Substances in the Ecosystems and in Environmental Protection, NMR Investigations of Silylated Humic Substances, in J. Drozd, S.S. Gonet, N.Senesi, J. Weber (eds.), *Proceedings of the 8th Meeting of the International Humic Substances Society*, Wroclaw, Poland, September 9-14, pp. 139-146.

150. Thorn, K.A., Folan, D.W., Arterburn, J.B., Mikita, M.A. and MacCarthy, P. (1989) Application of INEPT Nitrogen-15 and Silicon-29 Nuclear Magnetic Resonance Spectrometry to Derivatized Fulvic Acids, *Sci. Total Environ.* **81/82**, 209-218.

151. Schraml, J. (1990) ^{29}Si NMR Spectroscopy of Trimethylsilyl Tags, *Prog. NMR Spectrosc.* **22**, 289-348.

152. Schraml, J., Blechta, V., Kvicalová, M., Nondek, L. and Chvalovsky, V. (1986) Polar Functional Group Analysis of Mixtures by Silicon-29 Nuclear Magnetic Resonance, *Anal. Chem.* **58**, 1892-1894.

153. Herzog, H., Burba, P. and Buddrus., J. (1996) Quantification of hydroxylic groups in a river humic substance by ^{29}Si-NMR, *Fresen. J. Anal. Chem.* **354**, 375-377.

154. Thorn, K.A., Steelink, C. and Wershaw, R.L. (1987) Methylation patterns of aquatic humic substances determined by ^{13}C NMR spectroscopy, *Org. Geochem.* **11**, 123-137.

155. Mikita, M.A., Steelink, C. and Wershaw, R.L. (1981) Carbon-13 Enriched Nuclear Magnetic Resonance Method for the Determination of Hydroxyl Functionality In Humic Substances, *Anal. Chem.* **53**, 1715-1717.

156. Lambert, J. and Buddrus, J. (1996) Quantification of Isolated Methyl Groups in Aquatic Humic Substances by Means of ^{1}H and ^{13}C NMR Spectroscopy, *Magn. Reson. Chem.* **34**, 276-282.

157. Rutledge, D.N. (1996) *Signal treatment and signal analysis in NMR*, Elsevier, Amsterdam.

158. Hoch, J.C. and Stern, A.S. (1996) *NMR Data Processing*, Wiley-Liss, New York.

159. Brindle, J.T. (2002) Rapid and noninvasive diagnosis of the presence and severity of coronary heart disease using ^{1}H-NMR-based metabonomics, *Nat. Med.* **9**, 1439-1444.

160. Hedges, I.J. and Keil, G.G. (1999) Organic geochemical perspectives on estuarine processes: sorption reactions and consequences, *Mar. Chem.* **65**, 55-65.

435

161. Hatcher, P.G. and Spiker, E.C. (1988) Selective Degradation of Plant Biomolecules, in F.H. Frimmel, R.F. Christman (eds.), *Humic Substances and Their Role in the Environment*, Wiley, pp. 58-74.
162. Hedges, J.I. (1988) Polymerization of Humic Substances in Natural Environments, in F.H. Frimmel, R.F. Christman (eds.), *Humic Substances and Their Role in the Environment*, Wiley, pp. 45-58.
163. Kögel-Knabner, I., de Leeuw, J.W. and Hatcher, P.G. (1992) Nature and distribution of alkyl carbon in forest soil profiles: implications for the origin and humification of aliphatic biomacromolecules, *Sci. Total Environ.* **117/118**, 175-185.
164. Lu, X.Q., Hanna, J.V. and Johnson, W.D. (2000) Source indicators of humic substances: an elemental composition, solid state ^{13}C CP/MAS NMR and Py-GC/MS study, *Appl. Geochem.* **15**, 1019-1033.
165. Hedges, J.I., Baldock, J.A., Gelinas, Y., Lee, C., Peterson, M.L. and Wakeham, S.G. (2002) The biochemical and elemental compositions of marine plankton: A NMR perspective, *Mar. Chem.* **78**, 47-63.
166. Hertkorn, N., Claus, H., Schmitt-Kopplin, Ph., Perdue, E.M. and Filip, Z. (2002) Utilization and Transformations of Aquatic Substances by Autochthonous Microorganisms, *Environ. Sci. Technol.* **36**, 4334-4345.
167. Hatcher, P.G., Clifford, D.J. (1997) The organic geochemistry of coal: from plant materials to coal, *Org. Geochem.* **27**, 251-274.
168. Wishart, D.S., Bigman, C.G., Holm, A., Hodges, R.S. and Sykes, B.D. (1995) ^{1}H, ^{13}C, and ^{15}N random coil NMR chemical shifts of the common amino acids. I. Investigations of nearest-neighbor effects, *J. Biomol. NMR.* **5**, 67-81.
169. Wishart, D.S., Sykes, B.D. and Richards, F.M. (1991) Relationship between nuclear magnetic resonance chemical shift and protein secondary structure, *J. Mol. Biol.* **222**, 311-333.
170. Simpson, A.J., Lefebvre, B., Moser, A., Williams, A., Larin, N., Kvasha, M., Kingery, W.L. and Kelleher, B. (2004) Identifying residues in natural organic matter through spectral prediction and pattern matching of 2D NMR datasets, *Magn. Reson. Chem.* **42**, 14-22.
171. Hertkorn, N., Perdue, E. M., Schmitt-Kopplin, Ph. and Kettrup, A. (2004) A detailed analysis of aromatic substitution in humic substances by 2D MNR spectroscopy and increment analysis, *Anal. Chem.* (in preparation).
172. Keun, H.C., Ebbels, T.M.D., Antti, H., Bollard, M.E., Beckonert, O., Schlotterbeck, G., Senn, H., Niederhauser, U., Holmes, E., Lindon, J.C. and Nicholson, J.K. (2002) Analytical reproducibility in H-1 NMR-based metabonomic urinalysis, *Chem. Res. Toxicol.* **15**, 1380-1386.
173. Holmes, E., Nicholson, J.K. and Tranter, G. (2001) Metabonomic characterization of genetic variations in toxicological and metabolic responses using probabilistic neural networks, *Chem. Res. Toxicol.* **14**, 182-191.
174. Ussiri, D.A.N. and Johnson, C.E. (2003) Characterization of organic matter in a northern hardwood forest soil by C-13 NMR spectroscopy and chemical methods, *Geoderma* **111**, 123-149.
175. Francioso, O., Ciavatta, C., Montecchio, D., Tugnoli, V., Sanchez-Cortes, S. and Gessa, C. (2003) Quantitative estimation of peat, brown coal and lignite humic acids using chemical parameters, H-1-NMR and DTA analyses, *Biores. Technol.* **88**, 189-195.
176. Kulikova, N.A. and Perminova, I.V. (2002) Binding of atrazine to humic substances from soil, peat, and coal related to their structure, *Environ. Sci. Technol.* **36**, 3720-3724.
177. Perminova, I.V., Grechishcheva, N.Y., Kovalevskii, D.V., Kudryavtsev, A.V., Petrosyan, V.S. and Matorin, D.N. (2001) Quantification and prediction of the detoxifying properties of humic substances related to their chemical binding to polycyclic aromatic hydrocarbons, *Environ. Sci. Technol.* **35**, 3841-3848.
178. Balcke, G.U., Kulikova, N.A., Hesse, S., Kopinke, F.D., Perminova, I.V. and Frimmel, F.H. (2002) Adsorption of humic substances onto kaolin clay related to their structural features, *Soil Sci. Soc. Am. J.* **66**, 1805-1812.

Chapter 22

UNDERSTANDING CAPILLARY ELECTROPHORETIC SEPARATION PROCESSES TO CHARACTERIZE HUMIC SUBSTANCES AND THEIR INTERACTIONS WITH POLLUTANTS

A State of the Art and Perspectives

PH. SCHMITT-KOPPLIN[1], A. KETTRUP[2]
[1] GSF Institute for Ecological Chemistry, Ingolstädter Landstraße 1,
D-85764 Neuherberg, Germany <schmitt-kopplin@gsf.de>
[2] Technische Universität München,Lehrstuhl für Ökologische Chemie und
Umweltanalytik, D-85350 Freising/Weihenstephan, Germany

Abstract

Many structural concepts on natural organic matter and humic substances have been discussed over the last decades within the scientific community, ranging from macromolecular to supramolecular models. The capillary zone electrophoretic (CZE) approach, combines studies with model compounds, simulations of the electrophoretic behaviour of polydisperse system and separations of colloidal mixtures in aqueous buffers. CZE contributes to the understanding of the structure and behaviour of these mixtures in aqueous solutions and offers additional information on the intermolecular processes of NOM (molecular rearrangements, microgel and polymeric behaviour). CE gives mainly information on the charge, conformation and charge density, important parameters when investigating their interactions with pollutants. This presentation aims to present a detailed description of the CE separation processes, in order to interpret the behaviour of humic substances in aqueous media within this analytical technique. Complementary to this article the NMR-spectroscopy approaches, presented by Hertkorn and Kettrup in the same issue gives detailed chemical structural information on the present components rather than on their behaviour in solution.

1. Introduction

High molecular weight polyelectrolytes are of high interest in modern chemistry, biochemistry and geochemistry. Functionalised colloids are not only small entities presenting a huge exchange surface but they also have pH-dependent aggregation potentials, creating "compartments" with specific architectures and functions. Synthetic poly(carboxylates) such as polyacrylic and polymaleic acid are examples of

437

I. V. Perminova et al. (eds.),
Use of Humic Substances to Remediate Polluted Environments: From Theory to Practice, 437–472.
© 2005 *Springer. Printed in the Netherlands.*

438

polyelectrolytes that are used in a technical scale to remove heavy metals from contaminated waters or to avoid calcareous deposits in water-based coolers [1]. The best examples in living organisms are macromolecular systems like collagen, hyaluronic acid or lignin. Within non living systems in the environment, poly(hydroxycarboxylates) like natural dissolved humic substances (HS) are main constituents of the dissolved organic carbon (DOC) pool in surface, ground and soil pore water. Although dissolved organic carbon compose about 25% of the total organic carbon (TOC) on earth and 50% of the organic carbon in marine and fresh waters [2], their structural chemistry is less known than that of any other biopolymer of living origin [3]. HS are formed by (bio)chemical decomposition and reaction of plant and animal residues, as well as by microbial utilization of natural compounds and biopolymers (Figure 1).

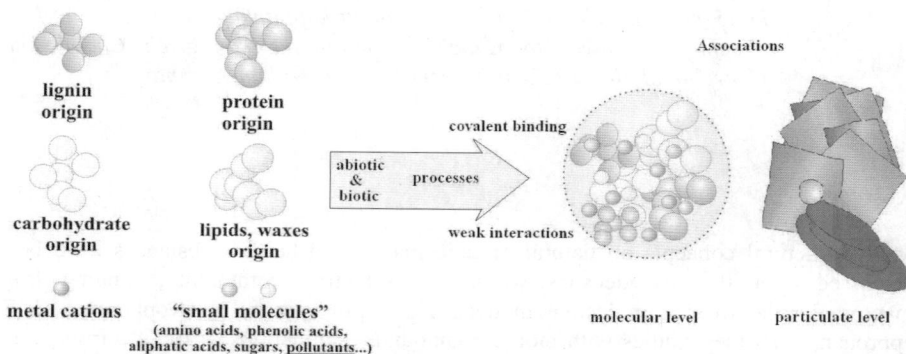

Figure 1. Possible origin of present structural units of humic substances and their association into humic materials on a molecular to particulate level involving weak interactions and covalent bonding.

Both the diagenesis and function of HS in ecologically relevant processes are closely intermingled. HS form an important link between the natural and anthropogenic cycling of elements and molecules. HS closely interact with microbial biomass and influence the microbial diversity in soils, not only affecting soil fertility, but also the ways and rates of transformations of xenobiotic compounds. HS control the migration and bioavailability of trace nutrients and toxic metal ions as well as those of organic pollutants. In particular, the structural characteristics of HS such as the charge and size distribution as well as physico-chemical properties derived from their *secondary/tertiary structures*, affect mobilization (binding to dissolved organic carbon) and immobilization (formation of bound residues) of pollutants, thereby affecting their mobility, bioavailability and toxicity. Assigning to this material a general structure was often proposed [3-6] but should rather be hold as a tool to think in structural concepts and in terms of their functions in the environment rather than in exact chemical structure formula.

The highly complex, polydisperse and irregular structure of HS poses specific and yet unresolved challenges to analytical chemistry in the field of structural and quantitative analysis and data interpretation.

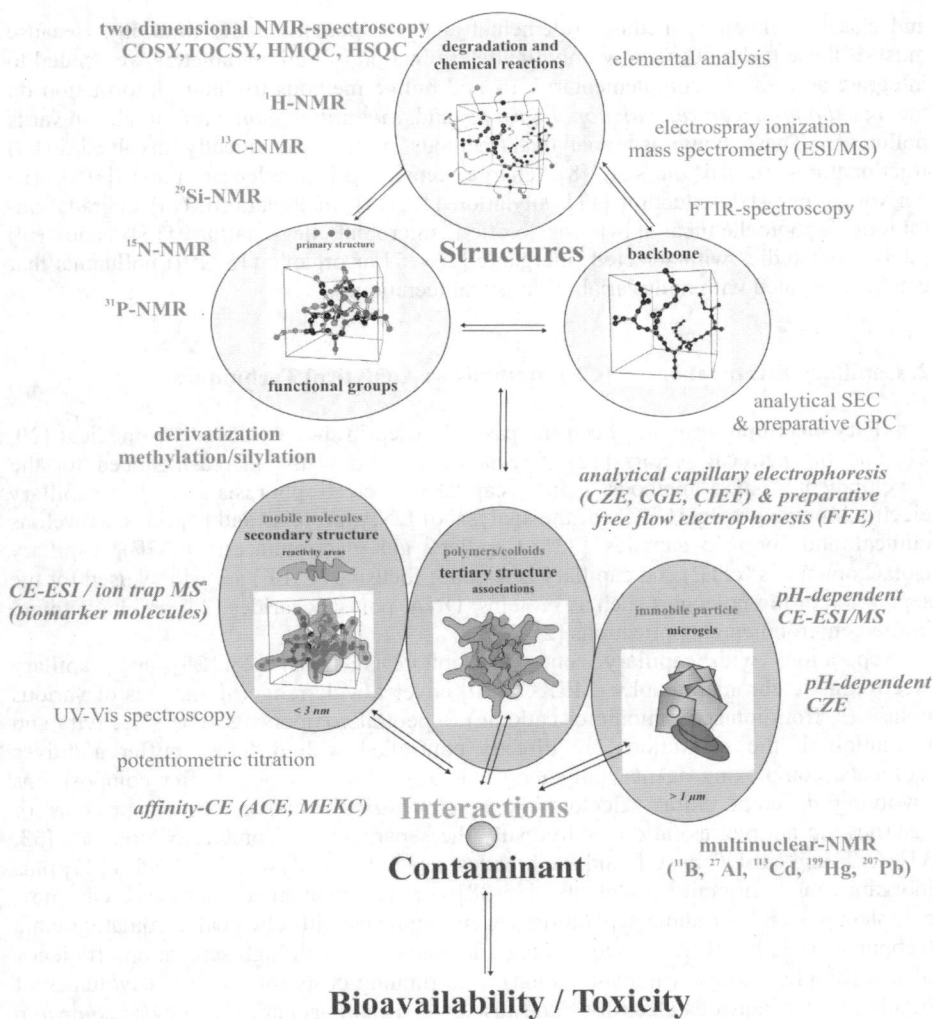

Figure 2. Array of analytical techniques integrated for the characterization of natural organic matter and its influence on the bioavailability/toxicity of pollutants. The integration of the data permits the description of backbone and functional group structures, both responsible for their chemical reactivity and interactions with contaminants.

For the characterization of these substances a systematic, interdisciplinary and multi-analytical approach is necessary [7]. Figure 2 presents some of the approaches we followed over the last years within our group (see also the article of Hertkorn and Kettrup in this issue). It combined the use of separation techniques (chromatographic and electrophoretic), spectrometric techniques (UV-Vis, fluorescence, FTIR, NMR), and classical chemical methods (elemental analysis, potentiometric titration). Because most of these techniques allow only the determination of bulk parameters we needed to integrate an array of complementary off- and online methods to obtain information on the *backbone structure, functional groups* and *interaction potential* of HS towards pollutants. These combined analytical methods were systematically involved in (i) fractionation- (based on size [8], charge density [9], isoelectric point [10]), (ii) derivatization- (HI reduction [11], silylation [12, 13], methylation), (iii) degradation- (abiotic –photochemical [14] -or biotic– microbial degradation [15]) and (iv) interaction-studies with selected (inorganic [16, 17] or organic [18, 19]) pollutants that can be integrated with multivariable statistical techniques.

2. Capillary Electrophoresis (CE) Methods as Analytical Techniques

Capillary electrophoresis has been increasingly used in the last decade in medical [20, 21] and biochemical sectors [22]. Recent reviews show the increasing need for the development of methods in capillary electrophoresis and capillary electrochromatography (CEC) for the analysis of DNA, proteins and peptides, as well as clinical and forensic samples [23]. Capillary gel electrophoresis (CGE), capillary isotachophoresis (cITP) and capillary isoelectric focusing (CIEF) are widely used for the separation of biomolecules such as proteins, DNA, polysaccharides [24] and have found limited environmental applications [25-27].

Separations with capillary zone electrophoresis (CZE) [28, 29] and capillary electrokinetic chromatography (MEKC) [30] cover a wide range of analytes of various polarities, from ionised (anionic or cationic) to neutral components. The selectivity and resolution of the separations are directly controlled with different buffer additives (solvents, complexing agents, polymers) [31, 32]. More complex buffer compositions involving different chiral selectors like cyclodextrins, polysaccharides, proteins or macrocyclic antibiotics, allow additionally the separation of isomeric compounds [33, 34]. CZE and MEKC have found the broadest use in the analysis of pesticides [25] plus inorganic and organic pollutants [35-38] in environmental matrices. CE now complements and for some separations, even competes with classical chromatographic techniques (GC or HPLC) by combining automatisation with high separation efficiency of low sample amounts (miniaturization). The running costs for CE (small volumes of running buffer, aqueous electrolyte and fused silica capillary) are very low as compared to chromatographic packing materials for which high amounts of organic solvents have to be used [35]. Furthermore, rapid conditioning of the CE separation system allows a high flexibility of this analysis method within the same day. All these factors together make CE a rather *"green"* separation technique worthy of being developed for many specific routine applications.

This presentation will focus on the hidden processes and pitfalls that may affect the separation of mixtures using capillary zone electrophoresis – CZE (also called free solution capillary electrophoresis – FSCE) within the frame of the characterization of polydisperse and heterogeneous natural organic matter (NOM). For more information on specific CE applications or CE techniques, the reader is referred to the many books that are presently available [39-45].

2.1. PRINCIPLES OF CAPILLARY ELECTROPHORESIS

2.1.1. Instrumental set-up
The main advantage of CE is the simple instrumentation consisting of a high-voltage power supply, two buffer reservoirs, a capillary and a detector. The basic set-up is usually completed with enhanced features such as multiple injection devices, autosamplers, sample and capillary temperature controls, programmable power supplies, multiple detectors fraction collection and computer interfacing.

CE separation is performed in a flexible fused silica capillary tube (20 to 100 cm; i.D. 20 to 100 µm) that is filled with an appropriate buffer solution (aqueous / non aqueous). A small volume of sample (< 3-4% of column volume) is introduced hydrodynamically into the capillary to which an electrical potential is applied (5 to 30 kV). Charged species of the sample exhibit different electrophoretic mobilities (mobility = field strength reduced velocities) and are thereby separated. Detection is possible with different modii such as UV-Vis, laser induced fluorescence, electrochemistry, mass spectrometry. Different techniques are possible as a function of the type of capillary column relative to the used buffer allowing separation of charged, neutral, polar or hydrophobic analytes. Fraction collection is possible but allows only limited amounts of sample; up scaling to free flow electrophoresis (FFE) is preferred and was described elsewhere [9]. Online detection can be performed by spectroscopic methods at an optical window (local burning of the polyimide coating). Interfacing to mass spectrometry for the analysis of natural products and natural organic matter was described [46, 47].

2.1.2. CE techniques in brief
CZE (Figure 3), *capillary zone electrophoresis* allows the separation of anions and cations as a function of their charge density (effective charge to size ratio). In case of nonaqueous buffers allowing specific selectivity, one speaks of non-aqueous CE (NACE). Selectivity is given by their effective charge governed by the separation pH and the active buffer flow: the electroosmotic flow (EOF).

MEKC (Figure 5), *micellar electrokinetic chromatography (micellar capillary electrophoresis);* the addition of charged surfactants (i.e. SDS, CTAB...) to the separation buffer allows the separation of neutral analytes as a function of their affinity to the micelles (hydrophobicity). The addition of a chiral ligand to the buffer additionally permits the analysis of neutral isomers (enantiomers, stereoisomers) [50]
Chiral-CZE (Figure 4), the addition of a chiral ligand (for example different substituted cyclodextrins, sugar or peptide based ligands) to the CZE-separation buffer allows the separation of ionic isomers (D/L, +/– enantiomers, stereomers) [49].

CIEF (Figure 6), *capillary isoelectric focusing* separates analytes based on differences in their isoelectric points (pI). After a focusing step building up a linear pH gradient in the capillary (controlled with standard zwitterions) [10, 48]

The analytes move as a function of their respective charge until they reach a position of zero charge. The solution is then mobilized to the detector. Possible analytes can be small zwitterions up to proteins [48].

ACE (Figure 7), *affinity capillary electrophoresis* is conducted in different experimental set-ups: generally small variations in ion mobility are monitored as a function of the buffer composition. The obtained binding isotherms allow the calculation of binding (1:1) or partitioning constants of analytes (e.g. pharmaceuticals, pesticides, peptides or metals) with selected ligands (e.g. proteins, humic substances, dissolved organic matter...) [18]. ACE allows the evaluation of binding constants to different analytes simultaneously. This ability, combined with the use of humic substances of different structures, allows the determination of structure – activity relationships (QSAR). Here are illustrated binding isotherms of different antibiotics and metabolites (fluoroquinolones) to a soil humic acid [19].

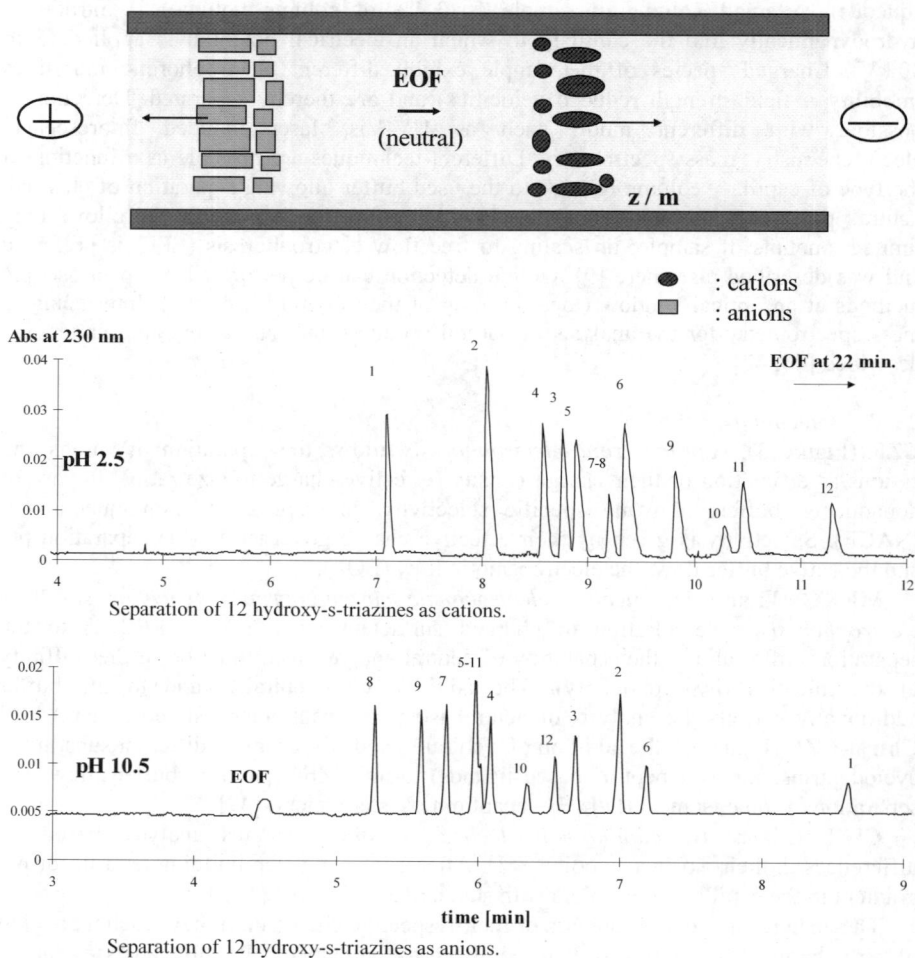

Separation of 12 hydroxy-s-triazines as cations.

Separation of 12 hydroxy-s-triazines as anions.

Figure 3. Separation principles of CZE with separation examples of hydroxytriazines as cations and anions [48].

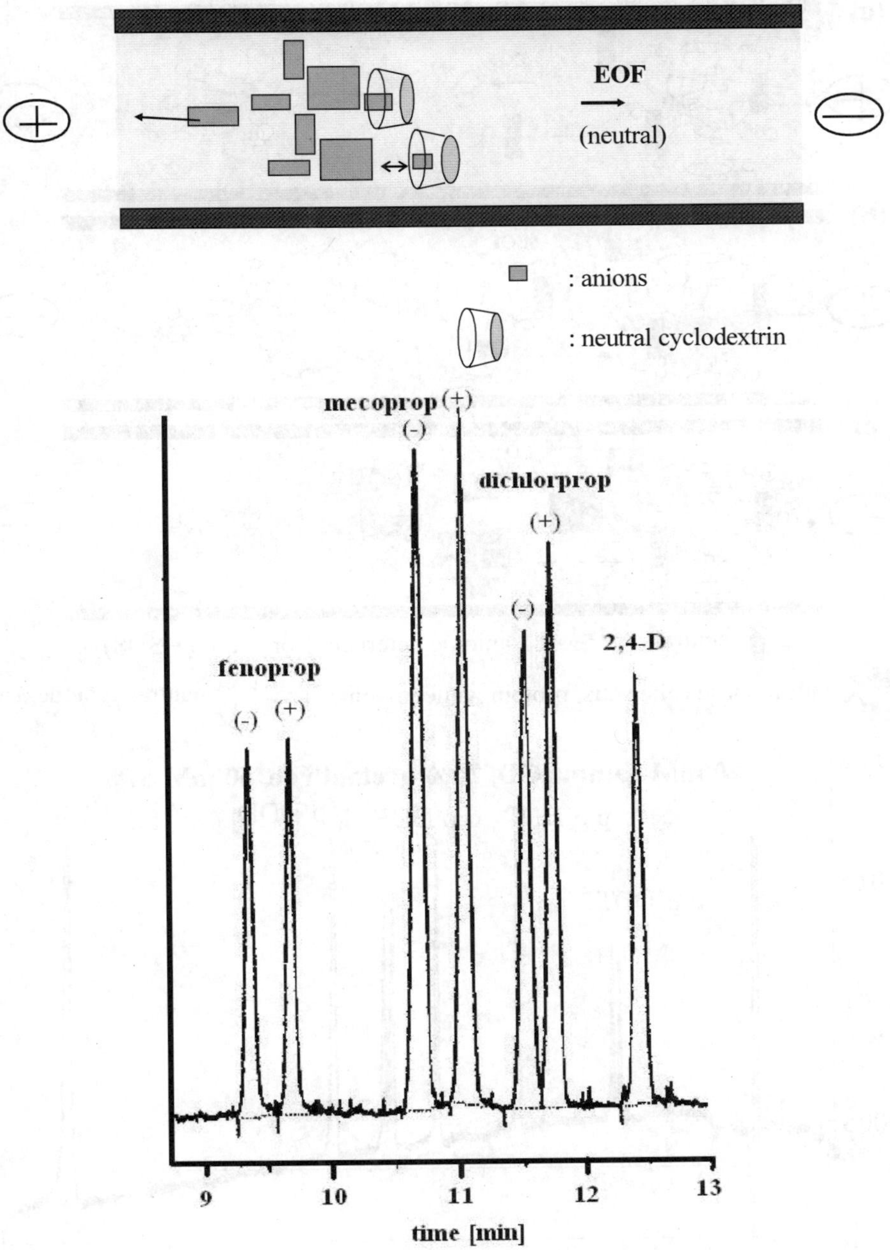

Figure 4. Separation principles of chiral-CZE with a separation example of phenoxyacid herbicides as anions with addition of trimethylated-ß-cyclodextrines in the background buffer (5 mM acetate buffer pH 4.45, 2.5 mM TM-β-CD) [49].

(a)

EOF

(b)

EOF

(c)

EOF

■ neutral ∿∿—○ anionic detergent (for example SDS)

⬛ chiral phase (bile salts, protein, vancomycin ...) ⬭ neutral cyclodextrin

20 mM gamma CD, 20% acetonitrile, 50 mM SDS

p,p´DDT o,p´DDE p,p´DDD p,p´DDE

o,p´DDD

o,p´DDT

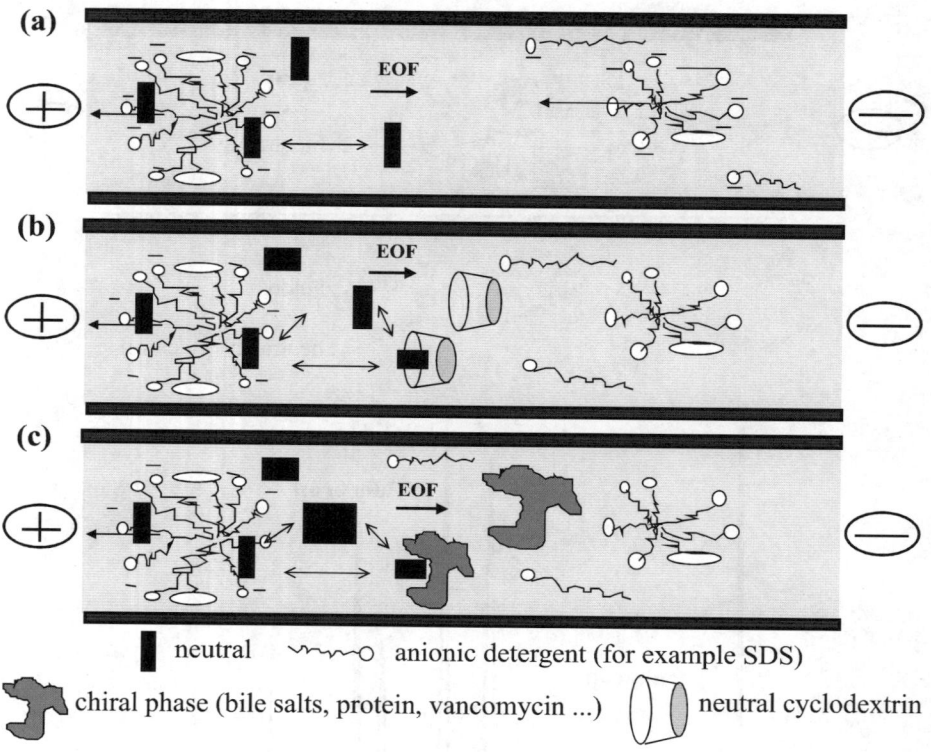

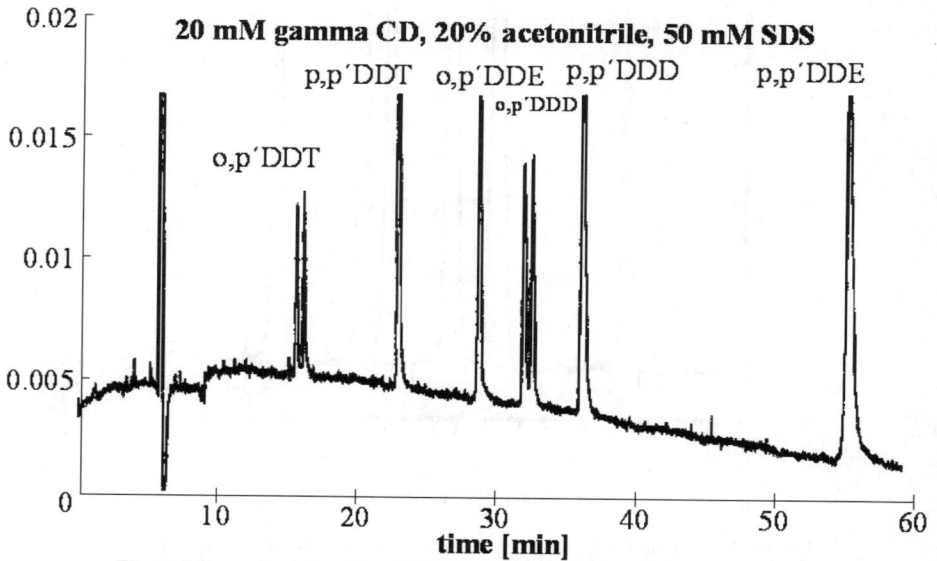

Figure 5. Separation principles of MEKC and CD-MEKC with separation example (DDT, metabolites) [50].

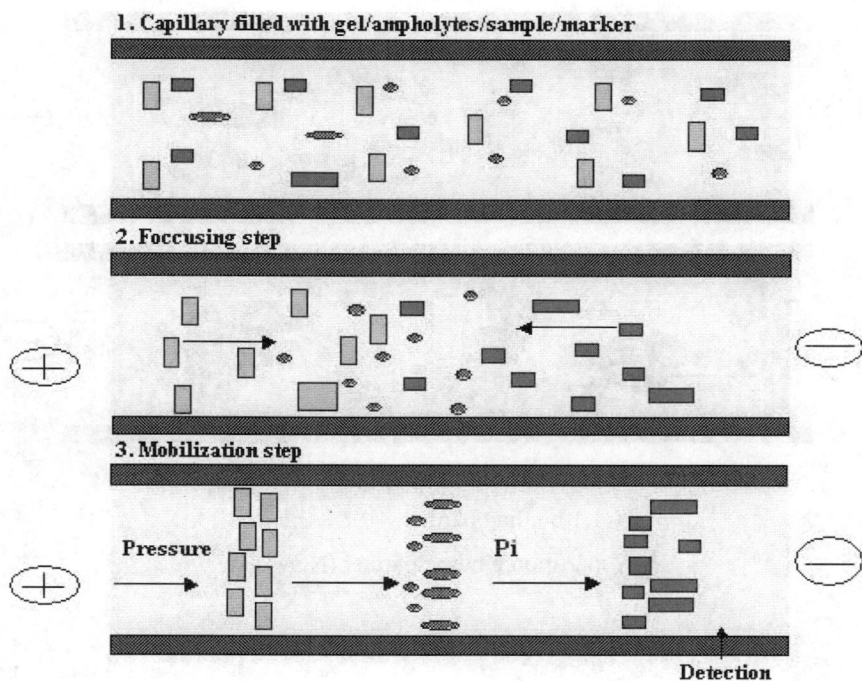

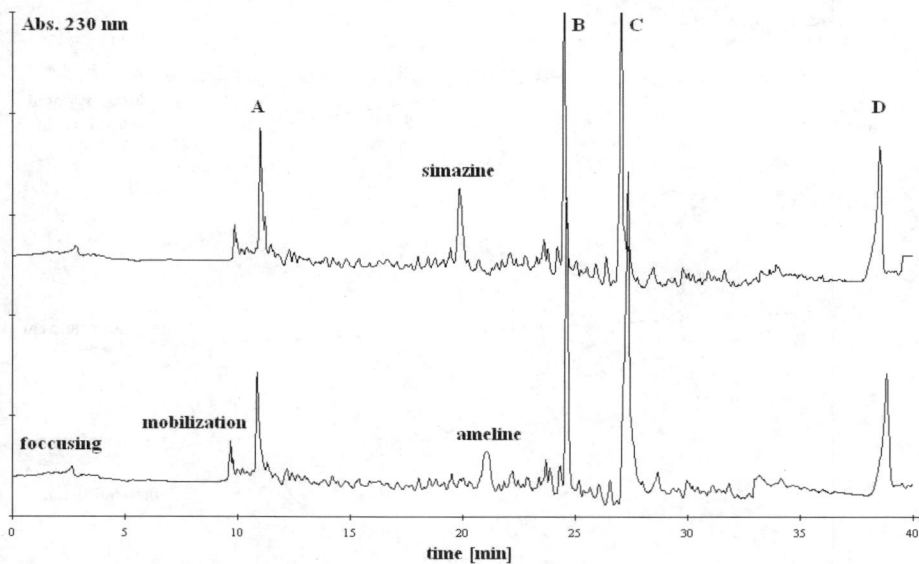

Figure 6. Separation principles of CIEF and separation example of a zwitterionic hydroxytriazine [48].

(a)

EOF

(b)

K   EOF

● anionic probe

■ 1:1 binding partner

⬛ partition binding partner (NOM)

Concentration in HA in the separation buffer [mg/l]

decarboxylated
enrofloxacin

enrofloxacin

danofloxacin

metabolite no. 5

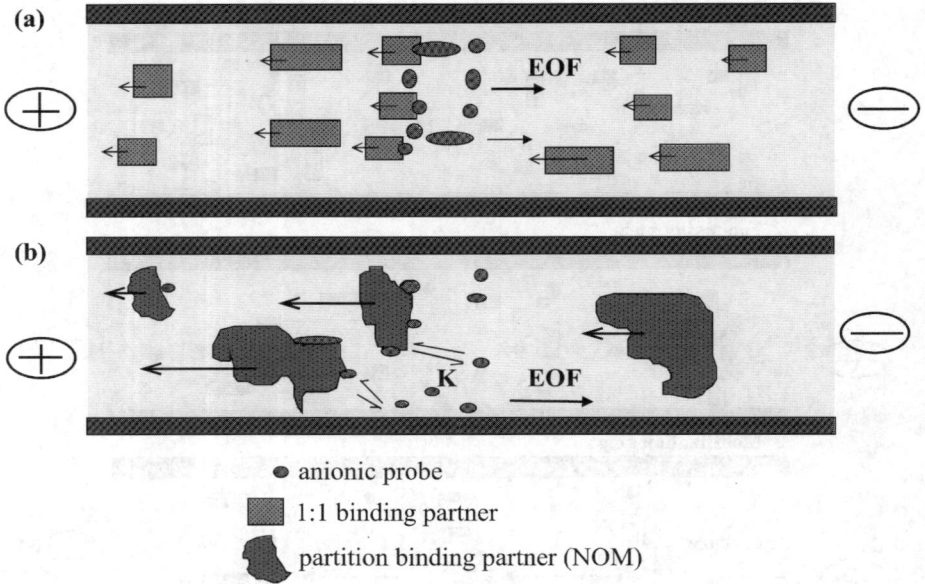

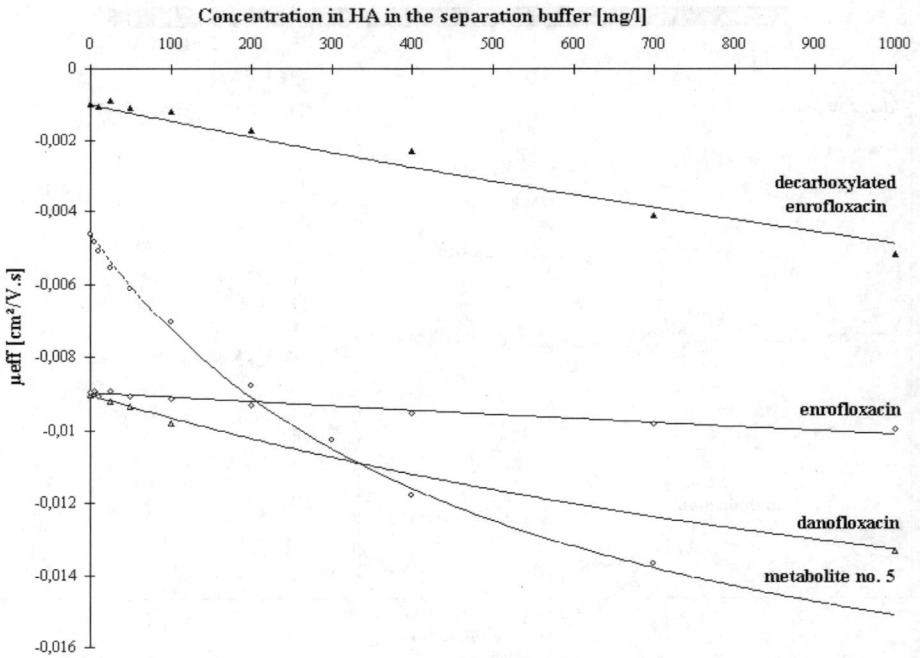

Figure 7. Separation principles of ACE and binding study example with fluoroquinolones; adapted from [19].

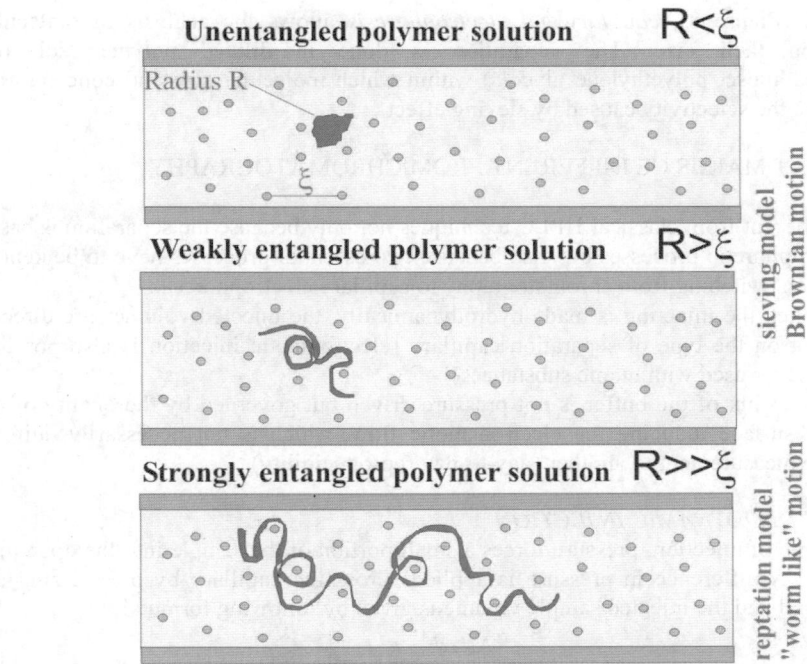

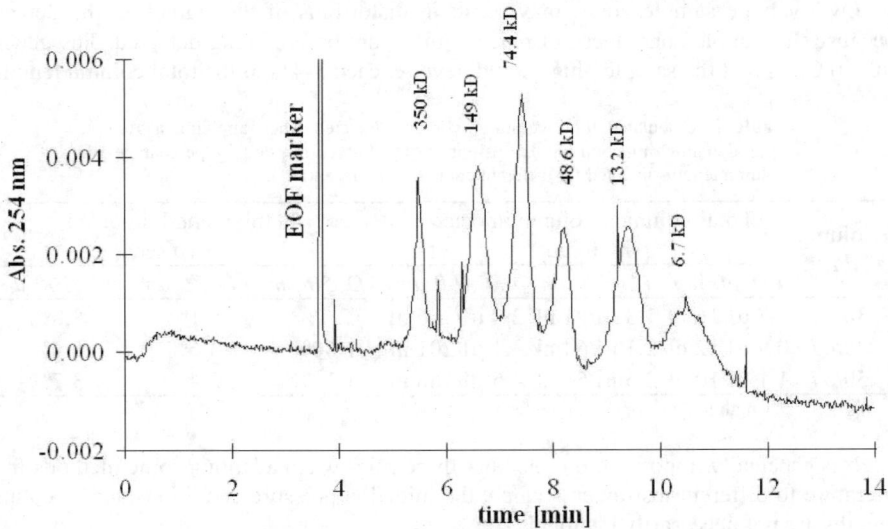

Figure 8. Separation principles of CGE and separation example of polystyrene sulfonates (R: Radius, ξ: mesh size).

CGE (Figure 8), *capillary gel electrophoresis* allows the analysis of molecules based on their size. The separation is done in diluted polymer solutions (methylcellulose, polyethylene glycols) within which molecular size and concentration determine the selectivity caused by sieving effect.

2.2. WHAT MAKES CE DIFFERENT FROM CHROMATOGRAPHY?

CE is different from classical HPLC techniques not only because the separation is based on electrophoretic processes but also because two essential processes have to be kept in mind when switching from chromatography to capillary electrophoresis:

(i) When the injection is made hydrodynamically, the injected volumes are directly dependent on the type of separation capillary (electrokinetic injection is also possible but was never used with humic substances)

(ii) The flux of the buffer is not pressure driven but governed by the quality of the capillary surface inducing the electroosmotic flow, which is not necessarily uniform from one measurement to another (day-to-day reproducibility).

2.2.1. HYDRODYNAMIC INJECTION
In this type of injection, pressure forces a small portion of the sample into the open tube capillary. A difference in pressure is applied across the capillary by pressurizing the sample vial and the injected sample volume is given by following formula:

$$V_{inj} = \frac{\Delta P \cdot d^4 \cdot \pi \cdot t}{128 \cdot \eta \cdot L} \cdot 10^3 \tag{1}$$

with ΔP the difference in pressure across the capillary, d is the capillary inner diameter, t the time of pressure application, η the viscosity and L the capillary length.

Overly large sample zones may result in distortions of the signals in the detector because the sample zone does not reach equilibrium before being detected. The general rule in CE is that the sample plug should never exceed 3-4% of the total column length.

Table 1. Calculated total volumes, volumes injected when applying a 10 sec hydrodynamic injection at 0.5 psi at 25°C, and corresponding percentage of column lengths injected for different columns lengths and i.D.

Column Ld/Lt*	Total volume / volume injected in 10 sec			% total length injected in 10 sec		
	i.D. 50 µm	i.D. 75 µm	i.D. 100 µm	i.D. 50 µm	i.D. 75 µm	i.D. 100 µm
30/37	0.7 µl/16 nl	1.3 µl/81 nl	2.9 µl/256 nl	2.7%	6.1%	8.8%
40/47	0.9 µl/12 nl	2.1 µl/64 nl	3.1 µl/201 nl	1.60%	3.6%	6.4%
50/57	1.1 µl/10 nl	2.5 µl/53 nl	4.5 µl/166 nl	1.10%	2.4%	3.7%

* length to detector (Ld), total length (Lt).

It is especially important to remember these rules when adapting some methods from literature to different instruments where the injection pressure conditions and/or column lengths are not necessarily identical.

Additionally, note that identical injection times with different column i.d.s or length, not only lead to different column volumes but also to different local sample concentrations when passing the detector. This is in particularly important when analysing analytes with concentration dependent aggregation properties such as polymeric materials and natural organic matter (NOM).

2.2.2. THE DRIVING FORCE: THE ELECTROOSMOTIC FLOW

Origin of the electroosmotic flow. Electroosmosis is the fundamental process in CE. The electroosmotic flow (EOF) is a direct consequence of the surface charge on the wall of the uncoated fused silica capillary.

The walls of the fused silica capillary contain silanol groups (pK$_a$ between 3 and 5 depending on the quality of the charge production), which ionise as a function of the pH of the electrolyte solution [48]. This dissociation of silanol groups to SiO$^-$ produces a negatively charged wall. An electrical double layer is established at the solid/liquid interface to preserve electroneutrality (Figure 9).

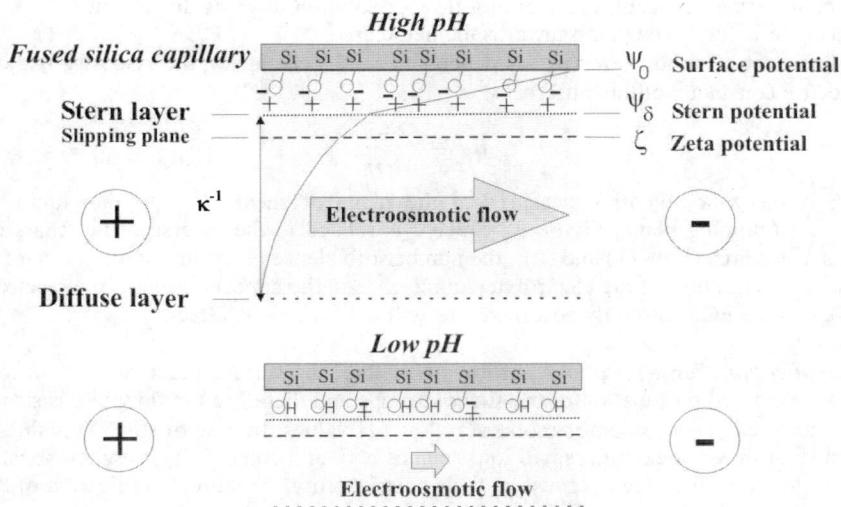

Figure 9. Schematic representation of the fused silica capillary surface at high and low pH values and consequences for the electroosmotic flow.

It can be described by the potential distribution by the Gouy-Chapman model (analogous to the Debye-Hückel model), combining the Boltzmann distribution of energy and the Poison equation, which gives the relation between potential and the charge density [51] . For a flat surface the potential ψ at a distance x from the interface, where ψ_0 is the potential at x = 0 is given by:

$$\psi = \psi_0 e^{-\kappa x} \tag{2}$$

where κ is defined as:

$$\kappa = \sqrt{\frac{8\pi n z^2 e^2}{\varepsilon k T}}, \quad (3)$$

where n – number of ions, e – electron charge, z – valency, ε – dieletric constant, k – Boltzmann constant, T – temperature.

The potential ψ drops to ψ_0/e at a distance of $x = \kappa^{-1}$, which is called the thickness of the diffuse double layer or Debye length.

From (3) it is shown that κ^{-1} is dependent on the ionic strength J which is defined as:

$$J = 0.5 \cdot \Sigma z_i^2 \cdot m_i, \quad (4)$$

where m_i – the molarity of the ions.

For dilute electrolyte systems (10^{-6}, 10^{-5} M) the thickness of the double layer can range from 1000 to 100 nm and for more concentrated electrolyte solutions (1 M to 0.1 M) from 1 to 0.1 nm.

An externally imposed tangential flow of the medium over the surface leads to a distortion of the ions creating a "streaming potential". This process is reversible and when a voltage is applied, the counterions and their associated solvating water molecules migrate towards the cathode. The produced movement of ions and the associated water molecules results in a flow of solution towards the detector. This flow is an electrically driven pump towards the detector.

The electroosmotic flow (μ_{eo}) is dependant on the chemistry, the viscosity η and the dielectric constant ε of the buffer:

$$\mu_{eo} = \frac{\varepsilon \zeta}{4 \pi \eta r}. \quad (5)$$

ξ is the zeta potential measured at the plane of shear close to the liquid-solid interface (slipping plane, Figure 9). Since ξ is related to the inverse of the charge per unit surface area (Equ. (1) and (2)), the number of valence electrons, and the square root of the concentration of the electrolyte, an increase in the concentration of the electrolyte decreases the EOF. Strongly adsorbed ions will have the same effect.

Implications for separations. At low pH values, the surface potential ψ_0 is lowered due to increased protonation of the silanol groups and therefore the flow rate is strongly reduced or can even become zero at very low pH values. In case of small ψ_0 values the Stern layer may not exist for small ions such as Na^+, and complexes or organic ions may adsorb to the wall surface because of their polarizability. The direct implication of these effects is that EOF depends both on pH and capillary size. Some flows are illustrated in theoretical (Table 2) and real values (Figure 10).

The EOF is generated by the entire surface and therefore produces a constant flow rate all along the capillary. As a consequence, the electrophoretic flow profile is plug-like in nature. Because analytes are swept at the same rate in the capillary sample, dispersion is minimized. This is an advantage compared to the flow encountered in pressure-driven systems such as liquid chromatography (LC) where frictional forces at the liquid-solid interface, such as the packing and the walls of the tubing, result in substantial pressure drops. Even in an open tube, frictional forces are severe enough at low flow rates to result in parabolic flow profiles (Figure 11). For laminar flows the solution is pushed from one end of the column and the solution at the column wall is moving slower than in the middle, which results in different solute speeds across the column. Therefore, laminar flow broadens peaks as they travel along the column.

Table 2. Theoretical buffer flow (nl/min) in 50 cm long capillaries of different internal diameters (i.D in μm) as a function of the observed time of the EOF (a t_{eof} of 2.0 minutes corresponds to a buffer velocity of 25 cm/min).

t_{eof}, min.	Theoretical buffer flow, nl/min			
	i.D. 100 μm	i.D. 75 μm	i.D. 50 μm	i.D. 20 μm
2.0	1964	1105	491	79
2.5	1571	884	393	63
3.0	1309	736	327	52
3.5	1122	631	280	45
4.0	982	552	245	39
4.5	873	491	218	35
5.0	785	442	196	31
5.5	714	402	178	29
6.0	655	368	164	26
6.5	604	340	151	24

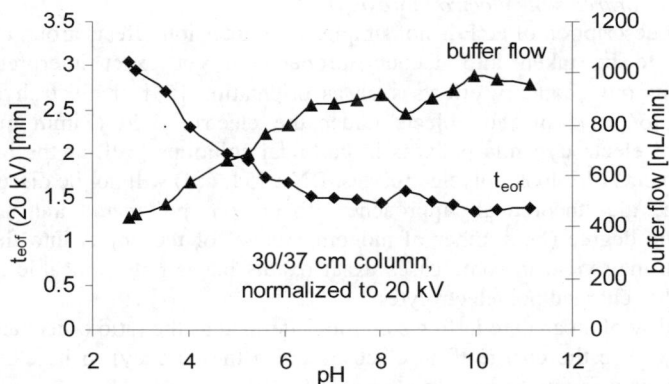

Figure 10. Real t_{eof} and corresponding buffer flow from the CE-titration study on Suwannee River FA.

Cross-Sectional Flow Profile
Due to Electroosmotic Flow

Cross-Sectional Flow Profile
Due to Hydrodynamic Flow

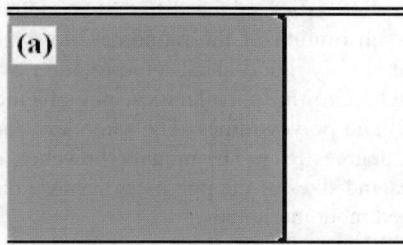
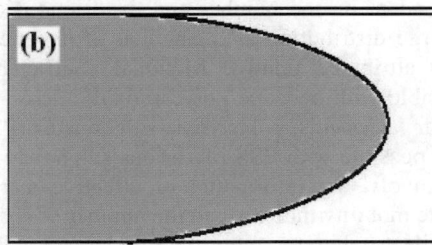

Figure 11. Flow profiles in electrophoretic (a) and pressure driven (b) separation columns.

2.3. CAPILLARY ZONE ELECTROPHORESIS OF POLYELECTROLYTES

CZE separations are routinely used in diverse application fields for the separation of different kinds of analytes ranging from neutral or ionic small molecules up to solid charged particles. This chapter will present the general assumptions that govern the electrophoretic processes of polyelectrolytes; they can be described in a practical way in semi-empirical and phenomenological models that relate the structures of small molecules and polyelectrolytes (as well as mixtures) to their measured electrophoretic mobility. Especially within the fields of *Genomics, Proteomics* and *Peptidomics,* models for a better understanding of the free zone electrophoresis of DNA fragments (few bp up to several thousands of bp), proteins or peptides were developed. These models intended an optimisation of the separation conditions, a prognosis of electrophoretic separations of these mixtures and an identification of structures based on standardized experimental separation conditions (i.e. small peptide structures obtained after tryptic digestion) [52, 53].

2.3.1. Theories of free zone electrophoresis (FZE)

A theoretical description of FZE is not simple, as counter-ion effects around the charged object need to be taken into account properly to get exact interpretation [54]. Considerations on secondary effects such as orientation [55] or electrohydrodynamic deformation [56, 57] of the objects under the electric field (Zimm model) [58], appearance of electrodynamic patterns in colloidal solutions [59], or the behaviour of composite objects (di-block polyelectrolytes, DNA [54, 60]) will not be discussed here.

Many of the theoretical approaches focus on polymers and involve the polymerisation degree (N, number of monomer units) of the object into their models; this approach may yield in some cases exact results but is not applicable in a general manner to other charged polyelectrolytes.

The mobility of an analyte in free solution is defined as the ratio of its electric charge Z (Z = q.e, with e the charge if an electron and q the valency) to its electrophoretic friction coefficient f (equ. 6).

$$\mu = \frac{q \cdot e}{f} \qquad (6)$$

Cottet and Gareil [61] developed a model for different polymeric materials dependent on the number and the frictional characteristics of the different monomers (charge distribution as a function of chemical substitution of the monomers to which they attributed relative frictional coefficients). Fairly good data interpretation was possible with benzene polycarboxylic acids, dichondroitin, polyalanines, polyglycines, linear fatty acids, polystyrene sulphonates (PSS) and polycytidines. The same approach was possible with PSS of various sulphonation degrees [61]. The models, however, do not involve any information on effective charges and sizes of the polymers because they relate mobility indirectly to the number of charged monomeric units.

Electrophoretic mobility of charged colloids in solution when submitted to an external electrical field is governed by the net charge located within the shear surface (Figure 12).

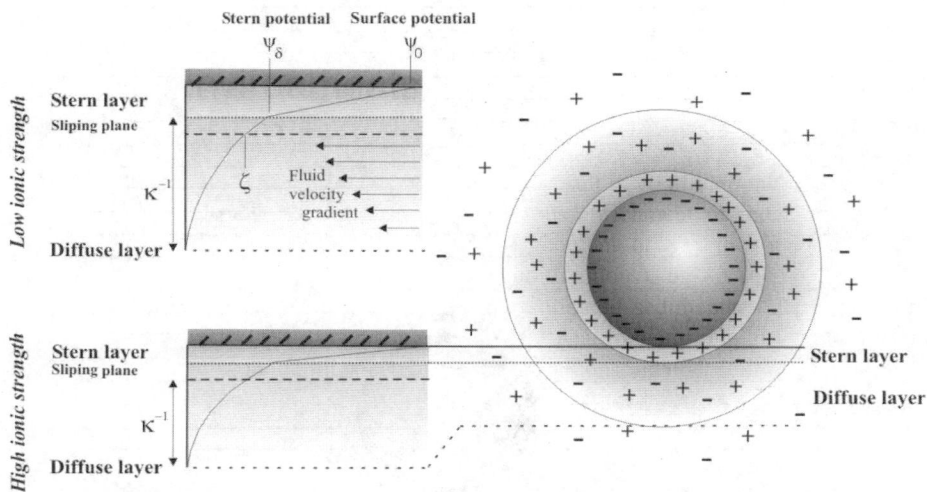

Figure 12. Schematic representation of a charged particle and its different charged layers.

The electrostatic potential at the shear surface (ζ potential) controls to a large extent the mobility of the particle. An estimation of electrophoretic mobility is given by equation (7) [62] in the case of small potentials, assuming a rigid and non conducting spherical particle of Radius R, moving in a medium of viscosity η and permitivity ε, such as:

$$\mu = \frac{2 \cdot \varepsilon \cdot \zeta}{3 \cdot \eta} \cdot f(\kappa \cdot R) \tag{7}$$

As already defined in section 1.2.2, κ is the reciprocal of the Debye length (electric double layer thickness), which is a function of the ionic strength of the solution (see equ. 3). The function $f(\kappa R)$, called *Henry function,* ranges from 1 at $\kappa R \ll 1$ (*Hückel limit*) to 1.5 at $\kappa R \gg 1$ (*Smoluchowski limit*) as a function of the particle shape. As the particle migrates accompanied by a given amount of the surrounding liquid, the counterions and the charged species it contains may create *retardation* effects [63]. Furthermore the distortion of the diffuse layer in the applied field leads to a polarization so that the particle and the counterions tend to be drawn back together. This is especially true for high ζ potentials and is known as a *relaxation* effect [63]. The last effect is due to the distortion of the local electrical field by the particle. When the particle size is small in relation to the double layer ($\kappa R \ll 1$), this effect is negligible (Hückel's theory) (Figure 13).

When $\kappa R \gg 1$ the effect needs to be taken into consideration (Smoluchowski's theory) – Figure 13. As seen in Section 2, κ^{-1} is a function of ionic strength and can correspond to lengths from 1 nm at 0.1 M to 30 nm at 0.0001 M (usual electrophoretic conditions). For particles possessing very high ζ potentials (> 75 mV), the relaxation effect is no longer negligible and the dependence of the mobility on (κR) can be altered. In the case of spherical non-conducting particles the *Overbeek-Booth* theory can be used and has a minimum as illustrated schematically in Figure 14 [62].

454

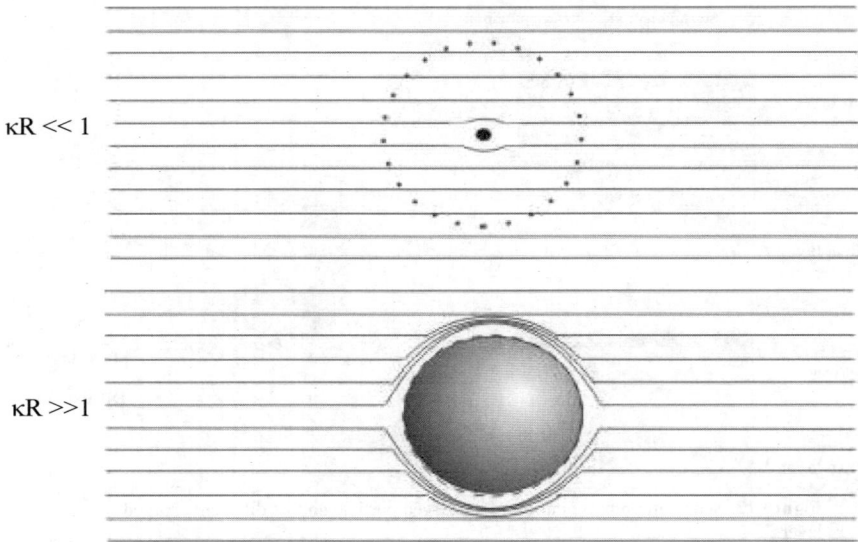

Figure 13. Relative distortions of the local field by particle and double layer.

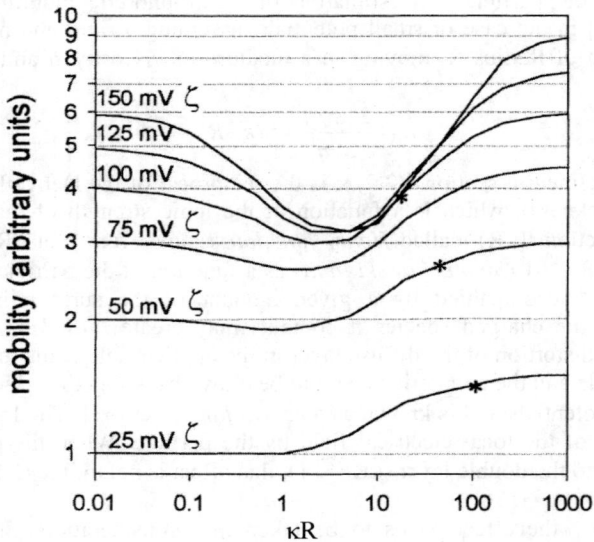

Figure 14. A shematic representation of the dependence of electrophoretic mobility on κR for different ζ potential values. The stars represent the decrease in mobility of a particle of a given size due to a decrease of the electric double layer by changes in ionic strength. Humic substances were reported to have ζ potentials ranging from -39 mV to -69 mV for Suwannee River fulvic acid at pH 4.5 and pH 11 respectivelly [64]. These values are not extreme and relaxation effects thus can be neglected for approximations. (adapted from [62]).

Table 3. Different theoretical models for zone electrophoresis of polymers.

Models (cited ref. in [65])	Authors
Porous sphere model	Hermans and Fujita, Overbeek and Stigter
Poisson-Boltzmann rod model without relaxation effect	Takahashi *et al.*
Debye-Hückel rod model with incorrect orientational averaging	Abramson *et al.*
Poisson-Boltzmann rod model with incorrect orientational and no averaging effects	Mills
Free draining Poisson-Boltzmann coil model with unspecified averaging chain friction constant	Imai and Iwasa
Phenomenological Debye-Hückel theory invoking empirical binary friction coefficient	Schmitt *et al.*
Poisson-Boltzmann rod model with semiempirical relaxation correction	van der Drift *et al*
Zimm model with localized forces and no relaxation correction	Long *et al.*
Boundary element methodology to describe the hydrodynamics and Poisson-Boltzmann electrostatics of arbitrarily shaped and charged poly-ions	Allison *et al.*

Many theories have been developed over the years for polyelectrolyte electrophoresis with different geometrical models (Table 3) and assumptions concerning the characteristics of the studied analytes ([65] and ref. herein). All of them fit well for non-heterogeneous monodisperse model polymers.

2.4. CZE OF HUMIC SUBSTANCES

Over hundred articles cover the separation of humic substances with capillary electrophoretic techniques in the literature (see review [66]). From all investigated humic substances and natural organic matter, the electrophoretic signals (in UV-Vis, Fluorescence, ESI/MS or TOC) were always found in the anionic range with very similar mobility patterns: a (Gaussian like) distribution of charge density (*"hump"*) corresponding to the distribution on charge and size of the present moving entities and, in addition to that (in some cases), single resolved peaks corresponding to single molecules moving freely in the system (polar and charged small molecules).

Briefly the degree of ionisation of their phenolic and carboxylic groups is governed by the capillary electrophoresis (CE) buffer pH. In CE, fulvic acids exhibit a consistent and characteristic set of sharp peaks (phenolic acids), extending from a humic "hump" [67]. The average electrophoretic mobility (AEM) of these humps depends on humic structure, experimental conditions and buffer composition [68, 69]. Humic acids give only the "hump" (sometimes multiple humps). Examples of electropherograms and their interpretation will be shown further down (see significance ahead).

2.4.1. A question of data representation: the mobility scale
The plot of UV-absorbance versus effective mobility (μ_{eff}) shows the Gaussian-like distribution around an average electrophoretic mobility (AEM). This representation of the primary electrophoretic data in the μ_{eff} domain is a useful visualization of effective mobility because it takes into account the changes in electroosmotic flow that can occur from one measurement to the other (dependant on buffer chemistry – pH, ionic strength, type of buffer) [8]. An electropherogram in this new scale can be considered as a frequency distribution of individual molecules (or "*molecular associates*" in these experimental conditions) having a given effective electrophoretic mobility. From the relation between mobility with charge and size using model compounds, structural information can be derived directly from the electropherograms in mobility-scale [47].

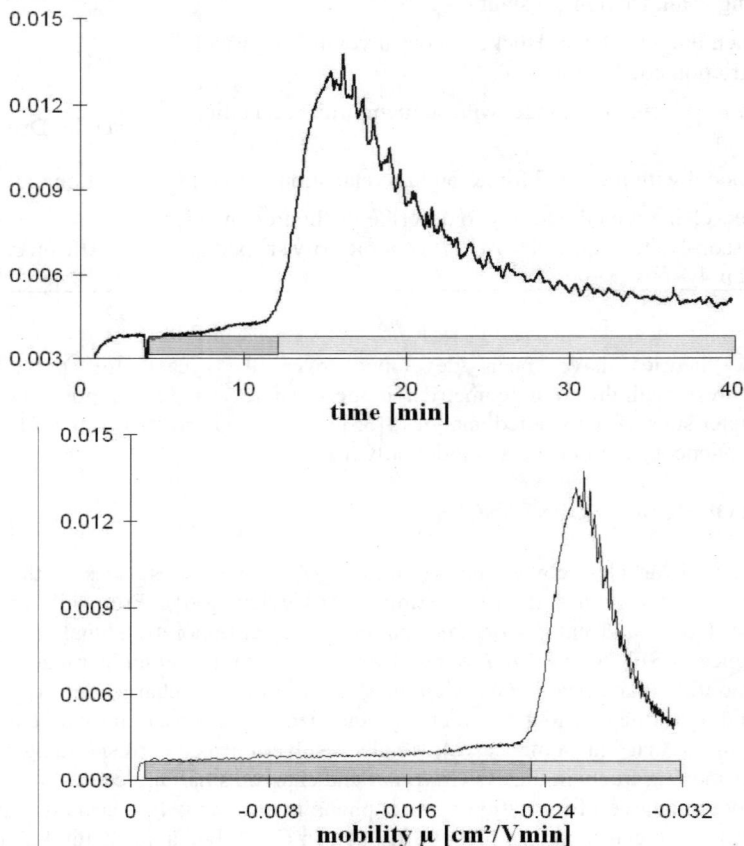

Figure 15. Catechol polymer separated with CZE in a carbonate buffer at pH 11.4 in time and corresponding effective mobility scale; the gray bares illustrates the distorsions from the time to the mobility domain.

Mobility scaling includes (i) the use of an internal standard (charged *p*-hydroxybenzoic acid – phb, polystyrene sulphonate or neutral EOF marker – mesithyl oxide), (ii) a baseline correction (iii) a scale transformation from migration time to effective mobility and (iv) the deletion of the internal standard peak. The sign of the mobility scale is negative for anions and positive for cations; to avoid confusion speaking of high mobility always implies the absolute value of the mobility. Figure 15 illustrates a catechol polymer [68] in time and mobility scale separated in a 25 mM carbonate buffer pH 11.4; at this pH different oligomers can be separated within the humic hump.

2.4.2. Consequences of the mobility scale
Utilizing the mobility scale reveals tremendous effects on the qualitative and quantitative interpretation possibilities of the data when analysing single components [70, 71]. Changing the time scale to mobility can have tremendous effects on the electropherogram shape and the resulting interpretations when the samples are mixtures. The data acquisition is in the time scale (between 5 to 10 data points per sec). When converting to mobility scale (mobility = 1/ time), the electropherogram´s shape is *"compressed"* and is significantly changed. More data points describe peaks in the high mobility region than near the EOF (as a consequence, the good description of analytes migrating near the EOF should be done with a high data acquisition rate). The mobility scale is a representation of the mobility distribution in the sample and the time scale is overestimating the contribution of high mobility components (see example in Figure 15).

Fulvic acids generally show higher polydispersity (wider peaks) than humic acids. Several separated sharp peaks, corresponding to lower molecular mass compounds, were often found in the fulvic acid fractions and rarely in the humic acid mixtures. Some of the sharp peaks rising out of the humps were identified as phenolic acids such as syringic (a), vanillic (b) and *p*-hydroxybenzoic (c) acids by spiking the fulvic samples and comparing the UV-spectra [72]. The presence of such low molecular weight acids in the humic mixture was previously demonstrated with ^{1}H-NMR spectroscopy [73] and recently with capillary electrochromatography combined with 2D-NMR techniques [74]. These phenolic acids could have been released in solution by partial hydrolysis of the fulvic acid core or/and coextracted from the natural soil matrix; they ultimately result from the oxidation of lignin structures (soft or hard wood origins) and are found in different amounts characteristic of the vegetation of the studied soils [75]. This low molecular mass fraction, which can account for up to 30% of the dissolved organic carbon of the FA mixture (less than 5 to 10% in humic acids), can be nicely separated with CZE into single molecular peaks (molecular behaviour) as compared to the humic hump (colloidal behaviour) [67, 76]. Studies are still in progress for the structural identification of these ionised hydroxycarboxylates with CE-ESI/MS [70].

In polymer chemistry, a number of average statistical parameters describing molecular mass distribution of a sample are used [77]. In general they are calculated as

$$M_i = \frac{\int_0^\infty M^i p(M)dM}{\int_0^\infty M^{i-1} p(M)dM}, \tag{8}$$

where $p(M)$ is a signal (physical property) proportional to a weight fraction of a polymer with molecular mass M. Most commonly used are M_0, M_1 and M_2, called number average, weight-average and z-average molecular mass, respectively.

By analogy, we propose average parameters of effective electrophoretic mobility defined as follows [8]:

number average effective mobility

$$M_0 = \mu_n = \frac{\int_0^\infty A(\mu).d\mu}{\int_0^\infty A(\mu).\mu^{-1}.d\mu} \qquad (9)$$

weight-average effective mobility

$$M_1 = \mu_w = \frac{\int_0^\infty A(\mu).\mu.d\mu}{\int_0^\infty A(\mu).d\mu} \qquad (10)$$

z-average effective mobility

$$M_2 = \mu_z = \frac{\int_0^\infty A(\mu).\mu^2 d\mu}{\int_0^\infty A(\mu).\mu.d\mu} \qquad (11)$$

Here $A(\mu)$ corresponds to the UV-absorbance signal at 254 nm at a given effective mobility μ. In addition, the peak-average effective mobility μ_p may be defined analogue to the peak average molecular weight M_p, as the mobility calculated from the migration time of the peak maximum.

For any distribution $|\mu_n| \leq |\mu_w| \leq |\mu_z|$ (the mobility of anionic compounds is negative in sign); more generally, $M_i \leq M_{i+1}$. Strict equality takes place for infinitely narrow distribution (monodisperse system) only. Thus, the ratios μ_w/μ_n and μ_z/μ_w (≥ 1) can be used to describe the degree of polydispersity in mobility. The ratio μ_p/μ_w can serve as an indicator of the shape of distribution. For symmetrical (e.g., Gaussian) distributions $\mu_p = \mu_w$ and $\mu_p/\mu_w = 1$. For non-symmetrical distributions this ratio differs from 1, being greater than unit for left-sided and smaller than unit for right-sided asymmetry. Therefore, the value of μ_p/μ_w reflects the relative contribution of different fractions to the entire electrophoretic peak: $\mu_p/\mu_w > 1$ means the greater contribution of less mobile and $\mu_p/\mu_w < 1$ that of more mobile fractions. Figure 16 illustrates usage of the defined parameters for description of mobility distributions. In the case of a wide distribution of mobilities (higher polydispersity in mobility), like with the Suwannee River fulvic acids at pH 5.04, the calculated average mobilities are noticeably different from each other. When the humic hump present a more homogeneous distribution like the Suwannee River NOM at pH 9.03, the calculated average mobilities are closer in value.

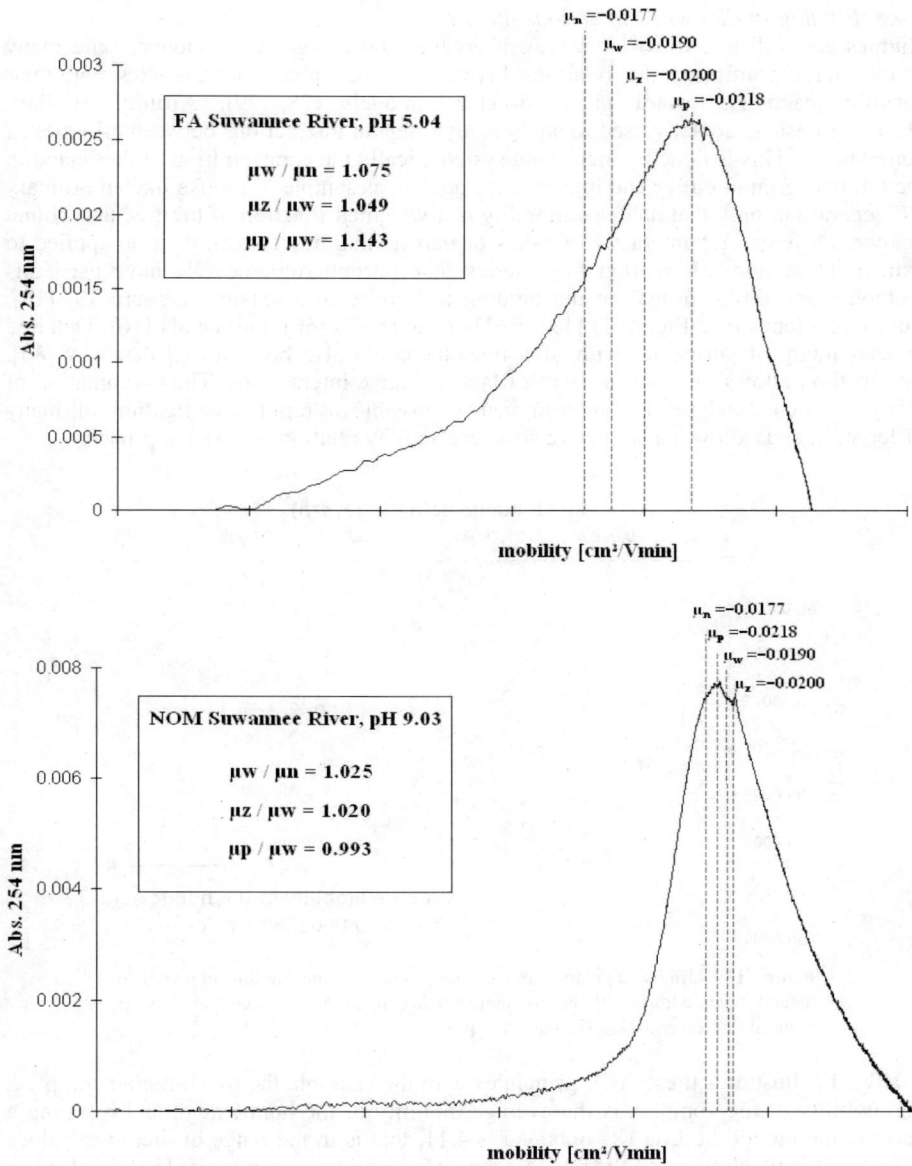

Figure 16. Illustration of the significance of the defined number-average effective mobility (μ_n), the weight-average effective mobility (μ_w), z-average effective mobility (μ_z), the peak-average effective mobility (μ_p) and the corresponding ratios μ_w/μ_n, μ_z/μ_w and μ_p/μ_w (($\mu_p/\mu_w > 1$ means left-sided asymmetry and $\mu_p/\mu_w < 1$ means right-sided asymmetry) from [8].

2.4.3. Binding studies with buffer constituents
Humics are well known to interact with organic and inorganic components and many studies have confirmed the relations between active humic binding sites and their specific reactivity toward given structural models [78, 79]. Affinity capillary electrophoresis is actually used to study analyte-ligand interactions between all types of components. This is done by increasing systematically the concentration of the ligand in the CE background buffer and injecting the analyte as sample – it looks like an ordinary CE separation, only that analytes mobility is a weighted function of the free and bound analyte. Different 1:1 interaction models or partitioning models can then be applied to extract interaction information from series of electropherograms. We have used this method many times to analyse the binding to humics of s-triazines (Figure 18) [18], fluoroquinolones (see Figure 7) [19], PAHs (Figure 17) [66] and metals [16]. Detailed understanding of s-triazines with SDS micelles could also be followed that way [80]. The method allows the measurement of fast exchange interactions. The combination of many different analytes in one run makes possible a rapid investigation of many different ligands allow a quantitative structure activity relation (QSAR) approach.

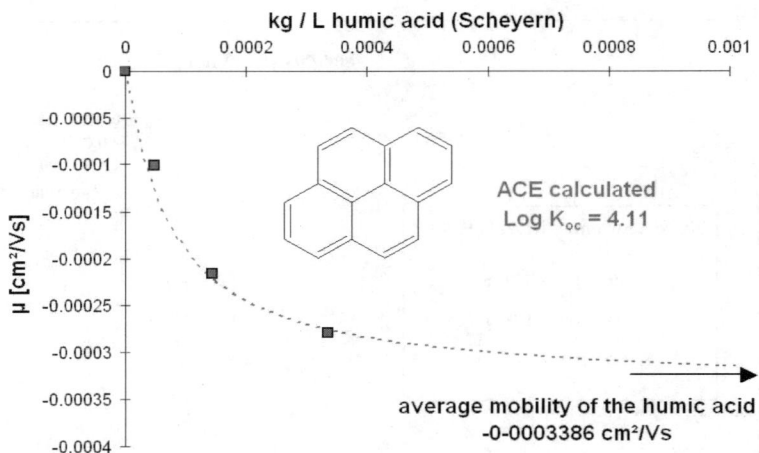

Figure 17. Affinity capillary electrophoresis study of the binding of pyren to natural organic matter (Scheyern humic acid); pyren was detected with laser induced fluorescence (λ_{ex} 325 nm) from [66].

Fig. 17 illustrates these ACE principles with the example the pyren-humic complex: the mobility of the complex is the average mobility of the humic acid, and by using a partitioning model the Log K_{oc} obtained is 4.11, that is in the range of literature values [81,81]. This illustrates also the phenomenon of solubility enhancement [83] that lead to the membrane or micellar concepts of humic substances presented by Wershaw [84, 85].

Further example is shown in Figure 18 with the simultaneously measured partition coefficients K_d of ameline, hydroxyatrazine, atraton and ametryn with different dissolved humic substances at pH 4.6 [18]. Combining in a principal component analysis the affinity data with selected structural data of the humic substances showed high correlations of the partition coefficients with the aromaticity of the humics.

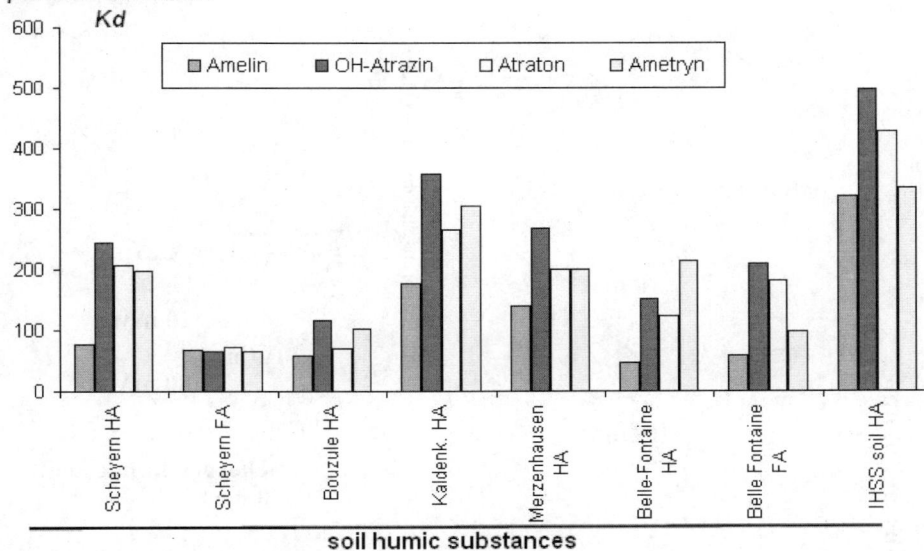

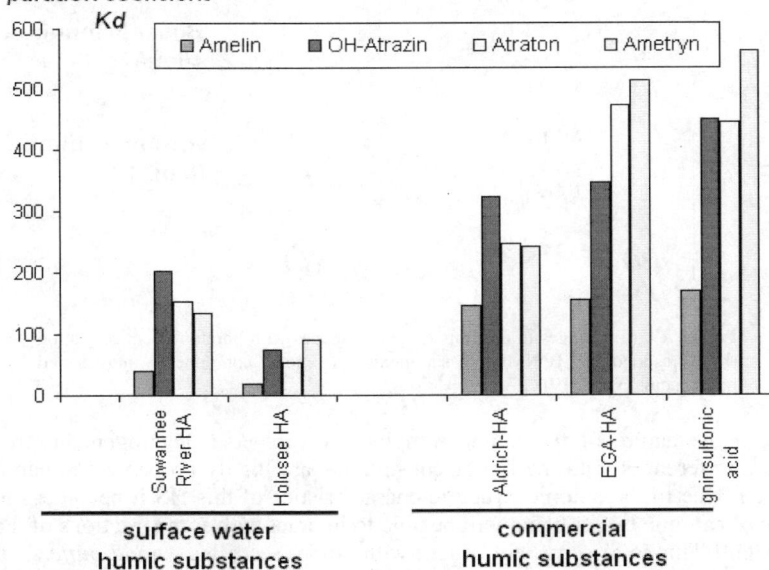

Figure 18. Partition coefficients of s-triazines with selected dissolved humic materials as measured with affinity capillary electrophoresis (ACE).

462

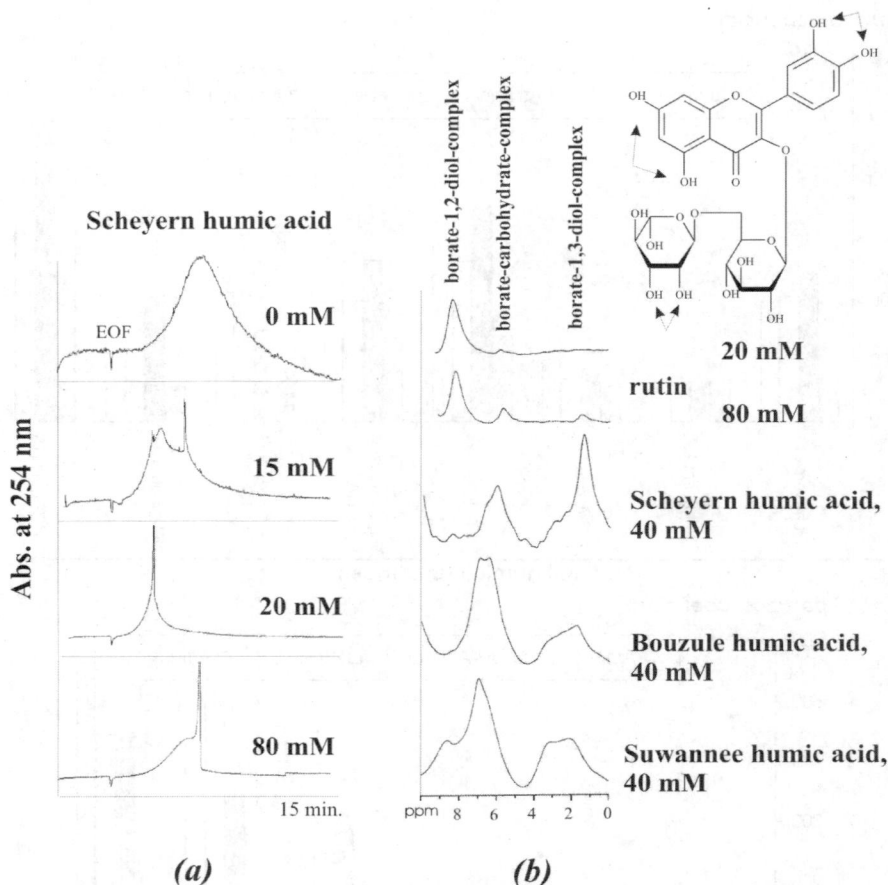

Figure 19. (a) changes in electropherograms shapes with borate ion concentration (b) corresponding ^{11}B NMR measurements with model components and selected humic acids (from [17]).

Due to the nature of the humic samples (polydisperse, heterogeneous, reactive), possible interferences with the buffer constituents are highly probable and can only be minimized. Due to the anionic structure characteristics of this HS it becomes clear that any type of cationic buffer being will be able to interact with some fractions of the HS to some extent! This is also true for other zwitterionic, so called *"good buffers"* that are chemically similar to the amphoteric *ampholytes* used in isoelectric focusing (IEF). The latter were shown to interact strongly with humics [69, 76]. These complexing buffers may be used to fingerprint NOM, but no structural interpretation should be made. The influence of tetrahydroxyborate ions on the electrophoretic mobility of humic acids has also been proposed in 1995 [67] and systematically evaluated by capillary electrophoresis (CE) in following studies [17, 72]. Depending on the molarity of borate ions in the separation buffer, the humic acids exhibit electropherograms with sharp peaks consistently extending from a "humic hump". Variations in the migration times of these peaks depend on the concentration of borate ions in the separation buffer (Figure 19(a)). The complexation of borate ions and humic acid fractions was also analysed with ^{11}B and ^{1}H NMR spectroscopy as well as UV-spectrophotometry in solutions of the same composition as the CE separation buffers (Figure 19(b)). Supplementary studies with model compounds (flavonoids, phenolic and sugar acids) [17] indicate reaction mechanisms that include the formation of bidentate esters (monocomplexes) as well as spiranes (tetradentate esters or dicomplexes) within the humic substructure.

It was thus shown that special attention must be given to the interpretation of CE electropherograms while fingerprinting humic substances with borate buffers since observed peaks do not necessarily indicate distinct humic components but may be artefacts (that can also be used for fingerprinting purposes) caused by the interaction of the buffer ions with the humic substances.

We systematically used acetate and carbonate buffers over a wide pH range because these buffer were found to be non-complexing towards humics. In contrast to many studies stating that small acids (i.e. acetic acid) are able to disrupt supramolecular complexes of humic substances [86, 87], we were never able to observe such a phenomenon in our concentration ranges.

2.5. CZE OF HUMIC SUBSTANCES: POSSIBLE APPLICATIONS

2.5.1. Ideal cases and extravagant mixtures

Within the electropherogram collection some samples showed very ideal, symmetrical or continuous mobility distributions, as the humic acid obtained from mature cattle compost CSW (Figure 20) that can be considered as a typical example of a behaviour of these humic acids in solution as molecular associations over a wide pH range.

On the other hand, the most intriguing sample was extracted from a deposits of a blue lake from the ice-cap in Greenland just above the arctic circle (6.6% OC) [88]. These lakes are isolated from terrestrial influence and the bottom deposits mainly result from meteoric dust. The organic matter originates exclusively from blue algae (cyanobacteria) and thus has a particular character containing essentially polysaccharides and peptidic fragments. They also show a double hump with essentially carboxylic acidity. Because of their extreme character, these substances are still under investigation, especially with 2D- NMR and CE-ESI/MS.

2.5.2. Aging of humic substances

Aging is an important unknown factor in evolution of natural organic matter under environmental conditions. Aging concerns the evolution of intramolecular interactions and the formation of stable associates or aggregates over time. Figure 21 illustrates such a system time evolution with 2 peat humic acids (T4, T5) kept in solution for over one year compared to the same material freshly dissolved in 0.05 N NaOH.

Decreases in mobility can clearly be observed and indicate changes in both surface charge and size of the aged associates. Reorganisation of the molecules in solution happened during more than one year; a stable system could emerge from the mixture and be analysed with CZE. Aggregation and precipitation of humic acids was already reported to occur over several hours, days or weeks [89]. Identical double peak signals

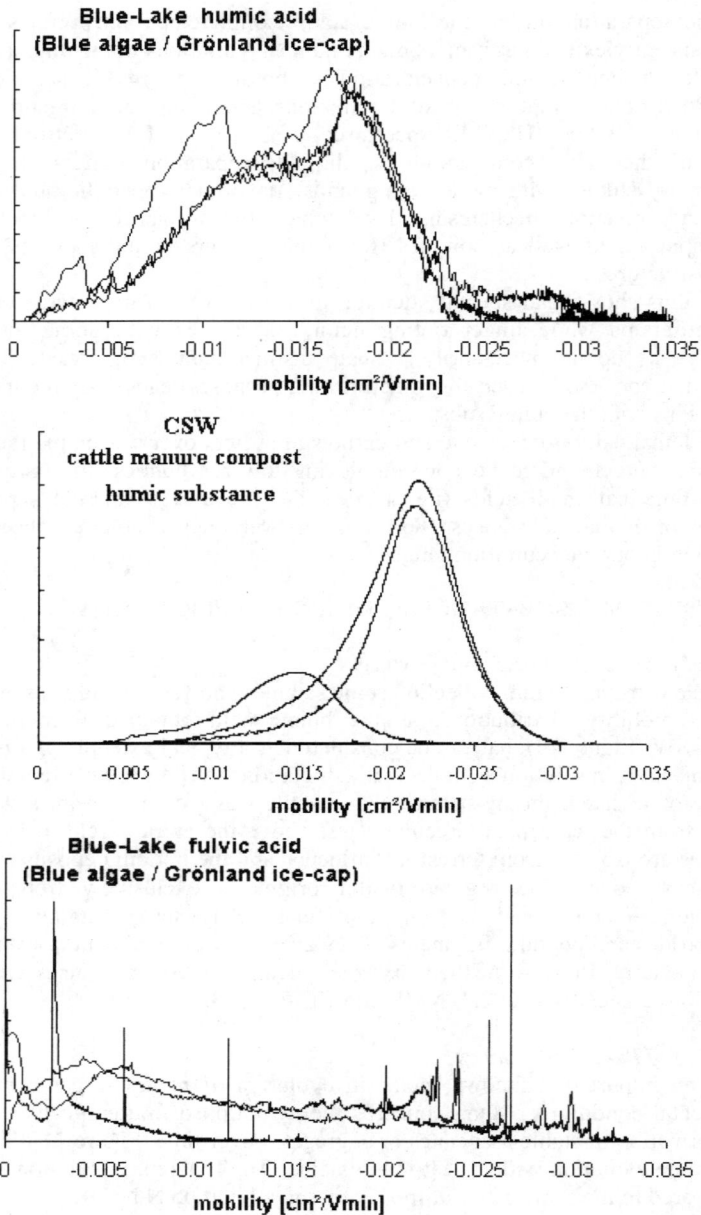

Figure 20. Two extremes: the Blue lake HA and FA (Dr. F. Gadel (✝) Université de Perpignan, France) and the mature cattle compost humic acid (Prof. Y. Chen, University of Jerusalem, Israel) [electropherograms from left to right: pH 5, pH 9.3, pH 11.4].

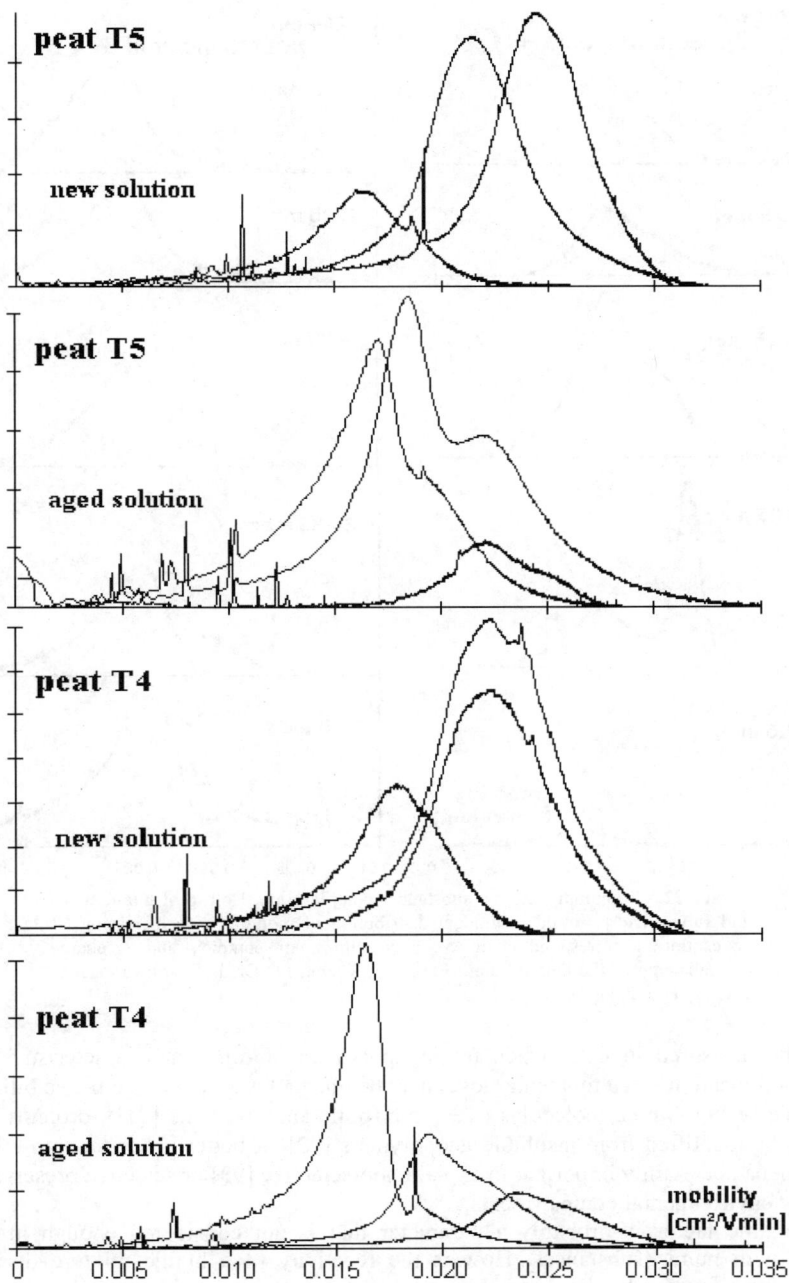

Figure 21. New and aged solutions of peat humic acids (obtained from Dr. I. Perminova) [electropherograms from left to right: pH 5, pH 9.3, pH 11.4].

466

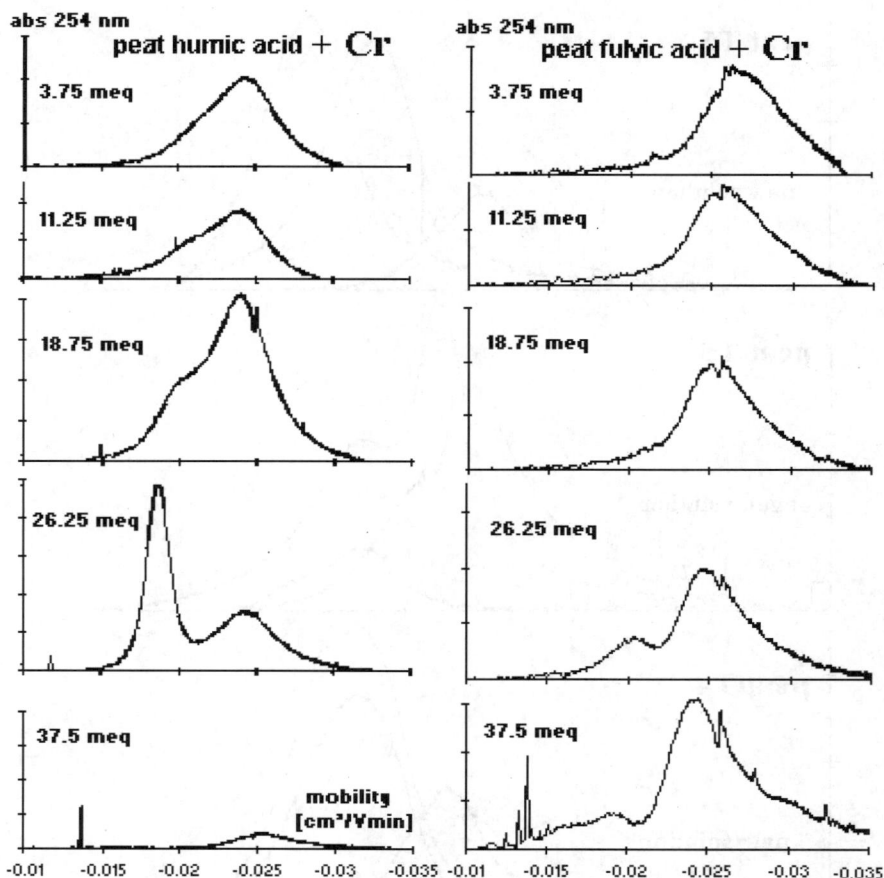

Figure 22. Chromium – humic substances complexation. Both peat humic acids and fulvic acids showed a bimodal distribution of mobility with increasing Cr concentration corresponding to a stable chromium complex and a charge neutralisation of the humic hump (PD Dr. J. Kyziol, PD Dr. I. Twardowska, IPIS PAN, Zabrze Poland).

could be measured in CZE when mixing polymers of different characteristics [90]. These stabilisations on a molecular level may be one pathway leading to the stabilization of humic or non-humic molecules over time (pollutants, proteins [91]); proteins could already be identified from insoluble geopolymers [92]. A better understanding of these phenomena is certainly important to explain non-selective [93] or selective preservations in many environmental compartments.

The time factor is probably a parameter that is not considered enough in many studies with humic substances. How is the reactivity of a humic mixture dissolved already for a few weeks (kept in solution in the fridge) compared to a freshly dissolved one? Analysed bulk parameters may stay constant but changes in secondary/tertiary structures would affect significantly its reactivity potential.

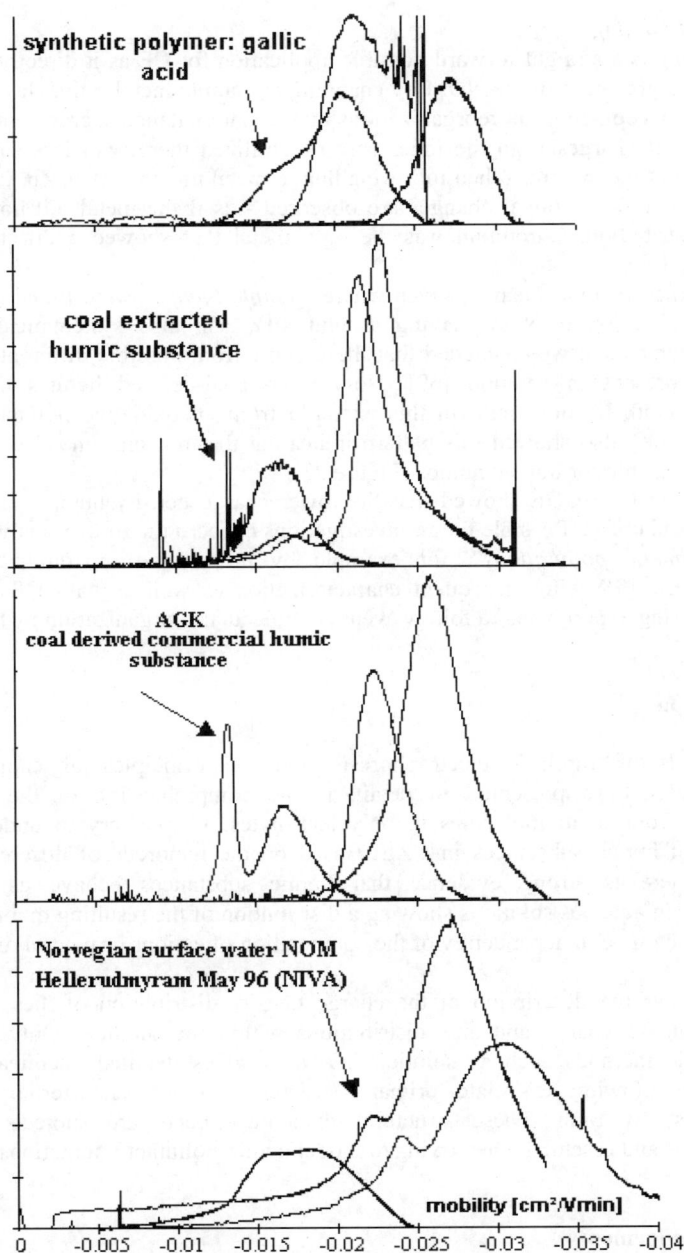

Figure 23. Double-humped mobility patterns of humic substances of different origins; the coal humics were obtained from Dr. Perminova, the synthetic humic acid was prepared for medical use by Prof. Kloecking [electropherograms from left to right: pH 5, pH 9.3, pH 11.4. Arrow shows the additional hump.

468

2.5.3. Metal binding

Metal binding is a straight forward possible application for CE as it directly affects the charge and aggregation of the NOM by changing the humic/metal ratio. Having in mind Manning´s concept, constant reorganizations may counterbalance charge neutralisations, given the pool charges from the inner core (neutralized there with less strong bound counterions) of the system. When following humic-metal mixtures (Cu, Zn, Cd) with CE before flocculation, minimal changes are observed - as if the metal did not affect the mobility distribution. Chromium was the only metal that showed a different pattern (Figure 22).

Many other examples can be given where "*double humps*" were found from native humics. Surface water NOM having around 40% ash also systematically showed additional humps and we concluded that these come from possible differentiated metal complexes present in solution [69]. Very often coal-derived humics display this behaviour. Synthetic polymers (in this example from an oxidative polymerisation of gallic acid) [68] also showed this pattern indicating that not only metal complexation may be responsible for double humps (Figure 23).

Analysis in CZE/CGE showed that the lower charge density hump was always of higher molecular size. Possible future investigations concern the further analysis of these different "*humic populations*", for example by preparative separation (free flow electrophoresis [9]) with subsequent characterization as well as with CE analysis in different mixing experiments to follow eventual molecular reorganization by time.

3. Conclusion

The methods of capillary electrophoresis and the principles of capillary zone electrophoresis were presented in detail as the comprehension of the separation phenomena from small molecules to polyelectrolytes is necessary to understand the behaviour of humic substances in CZE. Based on the hundreds of different samples analysed, there is strong evidence that humic substances behave as *molecular associations* in aqueous solutions showing a distribution of the resulting mobility strictly in the anionic range, independently of the composition of cations or neutral molecules in the mixture.

CZE allows the description of the charge density distribution of these associates corresponding to charge and size distributions within the mixture. Only combining spectroscopic methods such as solution 2D-NMR gives detailed chemical structure information, allowing to relate primary structures to secondary/tertiary structure (associations) by time. These structural dynamic aspects are subjects of actual investigations and essential when trying to model humic-pollutant interactions.

4. Acknowledgements

H. Neumeier, B. Look and A. Wust are thanked for their skilful technical assistance over the last decade.

5. References

1. n.n. (1998) Neue Horizonte in der Polymerforschung, *Nachrichten Chemische Technische Labor* **46**, 1169-1172.
2. Aiken, G.R. (1985) *Humic Substances in Soil, Sediment, and Water.*, John Wiley & Sons.
3. Shevchenko, S.M. and Bailey, G.W. (1996) Life after death: lignin-humic relationships reexamined, *Critical Reviews in Environmental Science and Technology* **26**, 95-153.
4. Kubicki, J.D. and Apitz, S.E. (1999) Models of natural organic matter and interactions with organic contaminants, *Organic Geochemistry* **30**, 911-927.
5. Shin, H.S. and Moon, H. (1996) An "average" structure proposed for soil fulvic acid aided by dept/quat ^{13}C NMR pulse techniques, *Soil Science* **161**, 250-258.
6. Schulten, H.-R. (1998) Interactions of dissolved organic matter with xenobiotic compounds: Molecular modeling in water, *Environmental Toxicology and Chemistry* **18**, 1643-1655.
7. Kögel-Knabner, I. (2000) Analytical approaches for characterizing soil organic matter, *Organic Geochemistry* **31**, 609-625.
8. Schmitt-Kopplin, P., Hertkorn, N., Freitag, D., Kettrup, A., Garmash, A.V., Kudryavtsev, A.V., Perminova, I. and Pettrosyan, V.S. (1999) Mobility distribution of synthetic and natural polyelectrolytes with capillary zone electrophoresis, *Journal of AOAC International* **82**, 1594-1603.
9. Junkers, J., Schmitt-Kopplin, P., Munch, J.C. and Kettrup, A. (2002) Up-scaling capillary zone electrophoresis (CZE) separations of polydisperse anionic polyelectrolytes with preparative free flow electrophoresis (FFE) exemplified with a soil fulvic acid., *Electrophoresis* **23**, 2872-2879.
10. Schmitt, P., Garrison, A.W., Freitag, D. and Kettrup, A. (1997) Capillary isoelectric focusing (CIEF) for the characterization of humic substances, *Water Research* **31**, 2037-2049.
11. Akim, L.G., Schmitt-Kopplin, P. and Bailey, G.W. (1998) Reductive splitting of humic substances with dry hydrogen iodide, *Organic Geochemistry* **28**, 325-336.
12. Hertkorn, N., Günzl, A., Freitag, D. and Kettrup, A. (2002) Nuclear magnetic resonance spectroscopy investigations of silylated refractory organic substances, in F.H. Frimmel, G. Abbt-Braun, H.K.G. Heumann, B. Hock, H.D. Lüdemann and M. Spiteller (eds.), *Refractory organic substances in the environment*, Wiley-VCH, pp. 129-145.
13. Hertkorn, N., Schmitt-Kopplin, P., Perminova, I.V., Kovalevskii, D.V. and Kettrup, A. (2001) Two dimensional NMR spectroscopy of humic substances, in R.S. Swift and K.M. Spark (eds.), *Understanding and managing organic matter in soils, sediments and waters*, IHSS, pp. 149-159.
14. Schmitt-Kopplin, P., Hertkorn, N., Schulten, H.-R. and Kettrup, A. (1998) Structural changes in a dissolved soil humic acid during photochemical degradation processes under O_2 and N_2 atmosphere, *Environmental Science and Technology* **32**, 2531-2541.
15. Hertkorn, N., Claus, H., Schmitt-Kopplin, P., Perdue, M.E., and Filip, Z. (2002) Utilization and transformation of aquatic humic substances by autochthonous microorganisms, *Environmental Science & Technology* **36**, 4334-4345.
16. Schmitt, P., Garrison, A.W., Freitag, D. and Kettrup, A. (1996) Flocculation of humic substances with metal ions as followed by capillary electrophoresis (CZE), *Fresenius Journal of Analytical Chemistry* **354**, 915-920.
17. Schmitt-Kopplin, P., Hertkorn, N., Garrison, A. W., Freitag, D. and Kettrup, A. (1998) Influence of borate buffers on the electrophoretic behavior of humic substances in capillary zone electrophoresis, *Analytical Chemistry* **70**, 3798-3808.
18. Schmitt, P., Trapp, I., Garrison, A. W., Freitag, D. and Kettrup, A. (1997) Binding of s-triazines to dissolved humic substances: Electrophoretic approaches using affinity capillary electrophoresis (ACE) and micellar electrokinetic chromatography (MEKC), *Chemosphere* **35**, 55-75.
19. Schmitt-Kopplin, P., Burhenne, J., Freitag, D., Spiteller, M. and Kettrup, A. (1999) Development of capillary electrophoresis methods for the analysis of fluoroquinolones and applications to the study of the influence of humic substances on their photodegradation in aqueous phase, *Journal of Chromatography A* **837**, 253-265.
20. Wang, C.Y., Huang, K.L., Kuo, Y.Z. and Hsieh, J. (1998) Analysis of Puerariae radix and its medicinal preparations by capillary electrophoresis, *Journal of Chromatography A* **802**, 225-231.
21. Yang, J., Long, H., Liu, H., Huang, A. and Sun, Y. (1998) Analysis of tetrandine and fangchinoline in traditional Chinese medicines by capillary electrophoresis, *Journal of Chromatography A* **811**, 274-279.

470

22.   Kataoka, H. (1998) Chromatographic analysis of lipoic acid and related compounds, *Journal of Chromatography B* **717**, 247-262.
23.   Altria, K.D. (1999) Overview of capillary electrophoresis and capillary electrochromatography, *Journal of Chromatography A* **856**, 443-463.
24.   Deyl, Z., Miksik, I., Tagliaro, F. and Tesarova, E. (1998) Advanced Chromatographic and Electromigration Methods in BioSciences, *Journal of Chromatography A* **60**.
25.   Menzinger, F., Schmitt-Kopplin, P., Freitag, D. and Kettrup, A. (2000) Analysis of agrochemicals by capillary electrophoresis. Tabulated review, *Journal of Chromatography A* **891**, 45-67.
26.   Padarauskas, A., Olsauskaité, V. and Paliulionyté, V. (1998) Simultaneous determination of inorganic anions and cations in waters by capillary electrophoresis, *Journal of Chromatography A* **829**, 359-365.
27.   Michalke, B. and Schramel, P. (1999) Antimony species in environmental samples by interfacing capillary electrophoresis on-line to an inductively coupled plasma mass spectrometer, *Journal of Chromatography A* **834**, 341-348.
28.   Jorgenson, J.W. and Lukacs, K.D. (1981) Zone electrophoresis in open-tubular glass capillaries, *Analytical Chemistry* **53**, 1298-1302.
29.   Jorgenson, J.W. and Lukacs, K.D. (1983) Capillary zone electrophoresis, *Science* **222**, 266-272.
30.   Terabe, S., Otsuka, K., Ichikawa, K., Tsuchija, A. and Ando, T. (1984) Electrokinetic separations with micellar solutions and open-tubular capillaries, *Analytical Chemistry* **56**, 111-113.
31.   Sarmini, K. and Kenndler, E. (1997) Influence of organic solvents on the separation selectivity in capillary electrophoresis, *Journal of Chromatography A* **792**, 3-11.
32.   Fanali, S.J. (1997) Controlling enantioselectivity in chiral capillary electrophoresis with inclusion-complexation, *Journal of Chromatography A* **792**, 227-267.
33.   Gübitz, G. and Schmid, M.G. (1997) Chiral separation principles in capillary electrophoresis, *Journal of Chromatography A* **792**, 179-225.
34.   Vespalec, R. and Bocek, P. (1997) Chiral separations by capillary zone electrophoresis: present state of the art, *Electrophoresis* **18**, 843-852.
35.   Brumley, W.C. (1995) Environmental applications of capillary electrophoresis for organic pollutant determination, *LC/GC Europe* **13**, 556-568.
36.   El Rassi, Z. (1997) Capillary electrophoresis of pesticides, *Electrophoresis* **18**, 2465-2481.
37.   Karcher, A. and El Rassi, Z. (1999) Capillary electrophoresis and electrochromatography of pesticides and metabolites, *Electrophoresis* **20**, 3280-3296.
38.   Sovocool, G.W., Brumley, W.C. and Donnelly, J.R. (1999) Capillary electrophoresis and capillary electrochromatography of organic pollutants, *Electrophoresis* **20**, 3297-3310.
39.   Baker, D.R. (1995) *Capillary Electrophoresis*, John Wiley & Sons.
40.   Chankvetadze, B. (1997) *Capillary Electrophoresis in Chiral Analysis*, John Wiley & Sons.
41.   Guzman, N.A. (1993) *Capillary Electrophoresis Technology*, Marcel Decker Inc.
42.   Khaledi, M.G. (1998) *High-performance Capillary Electrophoresis: Theory, Techniques, and Applications*, John Wiley & Sons.
43.   Rhighetti, P.G. (1996) *Capillary Electrophoresis in Analytical Biotechnology*, CRC Press.
44.   Shintani, H. and Polonski, J. (1997) *Handbook of Capillary Electrophoresis Applications*, Blackie Academic & Professional.
45.   Li, S.F.Y. (1993) *Capillary Electrophoresis. Principles, Practice and Applications*, Elsevier.
46.   Bianco, G., Schmitt-Kopplin, P., De Benedetto, G., Cataldi, T.R.I. and Kettrup, A. (2002) Determination of glycoalkaloids and relative a glyconesby non aqueous capillary electrophoresis (NACE) coupled with electrospray ion-trap mass spectrometry (ESI-ion trap-MS), *Electrophoresis* **23**, 2904-2912.
47.   Schmitt-Kopplin, P. (2003) *Interpreting capillary electrophoresis - electrospray / mass spectroscopy (CZE-ESI/MS) of Suwannee River natural organic matter (NOM)*, Taylor and Francis, Inc.
48.   Schmitt, P., Poiger, T., Simon, R., Garrison, A.W., Freitag, D. and Kettrup, A. (1997) Simultaneous ionization constants and isoelectric points determination of 12 hydroxy-s-triazines by capillary zone electrophoresis (CZE) and capillary electrophoresis isoelectric focusing (CIEF), *Analytical Chemistry* **69**, 2559-2566.
49.   Garrison, A.W., Schmitt, P. and Kettrup, A. (1994) Analysis of phenoxy acid herbicides and their enantiomers by high performance capillary electrophoresis, *Journal of Chromatography A* **688**, 317-327.

471

50. Schmitt, P., Garrison, A.W., Freitag, D. and Kettrup, A. (1997) Application of Cyclodextrine-modified micellar electrokinetic capillary chromatography (CD-MEKC) to the separations of selected neutral pesticides and their enantiomers, *Journal of Chromatography A* **792**, 419-429.

51. Fischer, C.-H. and Kenndler, E. (1997) Analysis of colloids IX. Investigation of the electrical double layer of colloidal inorganic nanometer-particles by size-exclusion chromatography, *Journal of Chromatography A* **773**, 179-187.

52. Cifuentes, A. and Poppe, H. (1995) Effect of pH and ionic strength of running buffer on peptide behavior in capillary electrophoresis: Theoretical calculation and experimental evaluation, *Electrophoresis* **16**, 516-524.

53. Cifuentes, A. and Poppe, H. (1997) Behavior of peptides in capillary electrophoresis: effect of peptide charge, mas and structure, *Electrophoresis* **18**, 2362-2376.

54. Long, D. and Ajdari, A. (1996) Electrophoretic mobility of composite objects in free solution: application to DNA separation, *Electrophoresis* **17**, 1161-1166.

55. Grossman, P.D. and Soane, D.S. (1990) Orientation effects on the electrophoretic mobility of rod-shaped molecules in free solution, *Analytical Chemistry* **62**, 1592-1596.

56. Isambert, H., Ajdari, A., Viovy, J.-L. and Prost, J. (1997) Electrohydrodynamic patterns in charged colloidal solutions, *Physical Review Letters* **78**, 971-974.

57. Long, D., Viovy, J.-L. and Ajdari, A. (1996) Simultaneous action of electric fields and nonelectric forces on an polyelectrolyte: motion and deformation, *Physical Review Letters* **76**, 3858-3861.

58. Long, D., Viovy, J.-L. and Ajdari, A. (1996) A Zimm model for polyelectrolytes in an electric field, *Journal of Physics* **8**, 9471-9475.

59. Isambert, H., Ajdari, A., Viovy, J.-L. and Prost, J. (1997) Electrohydrodynamic patterns in macroion dispersion under a strong electric field, *Physical Review Letters* **56**, 5688-5704.

60. Desruisseaux, C., Long, D., Drouin, G. and Slater, G.W. (2001) Electrophoresis of composite molecular objects. 1. Relation between friction, charge and ionic strength in free solution, *Macromolecules* **34**, 44-52.

61. Cottet, H. and Gareil, P. (2000) From small charged molecules to oligomers: A semiempirical approach to the modeling of actual mobility in free solution, *Electrophoresis* **21**, 1493-1504.

62. Radko, S.P., Stastna, M. and Chrambach, A. (2000) Capillary zone electrophoresis of sub-μm-sized particles in electrolyte solutions of various ionic strengths: Size-dependent electrophoretic migration and separation efficiency, *Electrophoresis* **21**, 3583-3592.

63. Fitch, R.M. (1997) *Pollymer colloids: a comprehensive introduction*, Academic Press.

64. Hosse, M. and Wilkinson, K.J. (2001) Determination of electrophoretic mobilities and hydrodynamic radii of three humic substances as a function of pH and ionic strenght, *Environmental Science and Technology* **35**, 4301-4306.

65. Hoagland, D.A., Arvanitidou, E. and Welch, C. (1999) Capillary electrophoresis measurements of the free solution mobility for several model polyelectrolyte systems, *Macromolecules* **32**, 6180-6190.

66. Schmitt-Kopplin, P. and Junkers, J. (2003) Capillary electrophoresis of natural organic matter, *Journal of Chromatography A*, review article, **998**, 1-20.

67. Garrison, A.W., Schmitt, P. and Kettrup, A. (1995) Capillary electrophoresis for the characterization of humic substances, *Water Research* **29**, 2149-2159.

68. Schmitt-Kopplin, P., Freitag, D., Kettrup, A., Hertkorn, N., Schoen, U., Klöcking, R., Helbig, B., Andreux, F. and Garrison, A.W. (1999) Analysis of synthetic humic substances for medical and environmental applications by capillary zone electrophoresis, *Analusis* **27**, 6-11.

69. Schmitt-Kopplin, P., Freitag, D., Kettrup, A., Schoen, U. and Egeberg, P. (1999) Capillary zone electrophoretic studies on Norwegian surface water natural organic matter, *Environment International* **25**, 259-274.

70. Schmitt-Kopplin, P., Menzinger, F., Freitag, D. and Kettrup, A. (2001) Improving the use of CE in a chromatographer's world, *LC-GC Europe* **14**, 284-388.

71. Schmitt-Kopplin, P., Garmash, A.V., Kudryavtsev, A.V., Menzinger, F., Perminova, I.V., Hertkorn, N., Freitag, D., Petrosyan, V.S. and Kettrup, A. (2001) Quantitative and qualitative precision improvements by effective mobility-.scale data transformation in capillary electrophoresis analysis, *Electrophoresis* **22**, 77-87.

72. Schmitt-Kopplin, P., Garrison, A.W., Perdue, E. M., Freitag, D. and Kettrup, A. (1998) Capillary electrophoresis in humic substances analysis, facts and artifacts, *Journal of Chromatography A* **807**, 101-109.

472

73. Wilson, M.A., Collin, P.J., Malcom, R.L., Perdue, E.M. and Cresswell, P. (1988) Low molecular weight species in humic and fulvic fractions, *Organic Geochemistry* **12**, 7-12.
74. Ping, G., Schmitt-Kopplin, P., Zhang, W., Whang, Y. and Kettrup, A. (2003) Separation of selected humic degradation compounds by capillary electrochromatography with monolithic and packed columns, *Electrophoresis* **24**, 958-969.
75. Manan, O., Marseille, F., Guillet, B., Disnar, J.R. and Morin, P. (1996) Separation of aldehydes, ketones and acids from lignin degradation by capillary zone electrophoresis, *Journal of Chromatography A* **755**, 89-97.
76. Schoen, U. (1999) Erklärungsansatz zum Phenomen der isoelektrischen Fokussierung (IEF) von Huminstoffen auf Polyacrylamid Gelen in der Flachbett-Elektrophorese, PhD-Thesis in Geosciences, Katolische Universität Eischstät.
77. Belenkii, B.G. and Vilenchik, L.Z. (1983) *Modern Liquid Chromatography of Macromolecules*, Elsevier.
78. Peuravuori, J. (2001) Binding of pyrene on lake aquatic humic matter: the role of structural descriptors, *Analytica Chimica Acta* **429**, 75-89.
79. Steinberg, C.E.W., Haitzer, M., Brüggemann, R., Perminova, I.V., Yashchenko, N.Y. and Petrosyan, V.S. (2000) Towards a quantitative structure activity relationship (QSAR) of dissolved humic substances as detoxifying agents in freshwaters, *International Review in Hydrobiological* **85**, 253-266.
80. Freitag, D., Schmitt-Kopplin, P., Simon, R., Kaune, A. and Kettrup, A. (1999) Interactions of hydroxy-s-triazines with SDS-micelles by micellar electrokinetic capillary chromatography (MEKC). *Electrophoresis* **20**, 1568-1577.
81. Chin, Y.-P., Aiken, G.R. and Danielsen, K. M. (1997) Binding of pyrene to aquatic and commercial humic substances: The role of molecular weight and aromatic, *Environmental Science Technology* **31**, 1630-1635.
82. Chiou, C.T., Kile, D.E., Brinton, T.I., Malcom, M., Leenheer, J.A. and MacCarthy, P. (1987) A comparison of water solubility enhancements of organic solutes by aquatic humic materials and commercial humic acids, *Environmental Science & Technology* **21**, 1231-1234.
83. Chiou, C.T., McGroddy, S.E. and Kile, D.E. (1998) Partition characteristics of polycyclic aromatic hydrocarbons on soils and sediments, *Environmental Science Technology* **32**, 264-269.
84. Wershaw, R.L. (1986) A new Model for humic materials and their interations with hydrophobic chemicals in soil-water or sediment-water systems, *Journal of Contam. Hydrol.* **1**.
85. Wershaw, R.L. (1999) Molecular aggregation of humic substances, *Soil Science* **164**, 803-813.
86. Conte, P. and Piccolo, A. (1999) High pressure size exclusion chromatography (HPSEC) of humic substances: molecular sizes, analytical parameters, and column performance, *Chemosphere* **38**, 517-528.
87. Piccolo, A., Nardi, S. and Concheri, G. (1996) Macromolecular changes of humic substances induced by interaction with organic acids, *European Journal of Soil Sience* **47**, 319-328.
88. Gadel, F., Torri, G. and Bruchet, A. (1987) Humic substances from deposits of a natural laboratory: a blue lake on the ice-cap (Greenland), *The Science of the Total Environment* **62**, 107-109.
89. Manning, T.J., Bennett, T. and Milton, D. (2000) Aggregation studies of humic acid using multiangle laser light scattering, *The Science of the Total Environment* **257**, 171-176.
90. Bohrisch, J., Grosche, O., Wendler, U., Jaeger, W. and Engelhardt, H. (2000) Electroosmotic mobility and aggregation phenomena of model polymers with permanent cationic groups, *Macromolecular Chemistry and Physics* **201**, 447-452.
91. Hedges, J.I., Eglinton, G., Hatcher, P.G., Kirchman, D.L., Arnosti, C., Derenne, S., Evershed, R.P., Kögel-Knabner, I., de Leeuw, J.W., Littke, R., Michaelis, W. and Rullkötter, J. (2000) The molecularly-uncharacterized component of nonliving organic matter in natural environments, *Organic Geochemistry* **31**, 945-958.
92. Knicker, H., del Rio, J.C., Hatcher, P.G. and Minard, R.D. (2001) Identification of protein remnants in insoluble geopolymers using TMAH thermochemolysis / GC-MS, *Organic Geochemistry* **32**, 397-409.
93. Hedges, J.I., Baldock, J.A., Gelinas, Y., Lee, C., Peterson, M. and Wakeham, S. G. (2001) Evidence for non-selective preservation of organic matter in sinking marine particles, *Nature* **409**, 801-804.

Chapter 23

OZONE APPLICATION FOR MODIFICATION OF HUMATES AND LIGNINS

M.M. KSENOFONTOVA, A.V. KUDRYAVTSEV,
A.N. MITROFANOVA, I.V. PERMINOVA, A.N. PRYAKHIN,
V.V. LUNIN
*Department of Chemistry, Lomonosov Moscow State University,
Leninskie Gory, 119992 Moscow, Russia <mgusenk@kge.msu.ru>*

Abstract

The modifications of lignins and humic substances under ozone action were investigated in this paper. The rate constants and ozone consumption in these reactions were estimated. UV/Vis, IR-, and NMR-spectroscopy methods were used to follow functional groups changes in the course of ozonation. The pathways of lignins and humic substances ozonation were suggested.

1. Introduction

The synthesis of new and environmentally benign complexing agents to heavy metals on a base of humic substances (HS) and lignins attracts great interest. Lignin is irregular aromatic polymer of phenolic nature consisting of phenyl-propane units substituted predominanlty with OCH_3 and OH groups, and linked together via carbon-carbon and ether bonds. HS are also natural irregular polymers of acidic nature consisting of aromatic core highly substituted with carboxylic and hydroxylic groups and bound to it carbohydratic-peptidic peripheral fragments [1]. Both lignin and humic compounds are rich in oxygen-containing functional groups and can be considered as promising resources for production of complexing agents. The latter can be used as detoxifying agents for metal polluted environments or as chelates for production of micronutrients (Fe, Mn, Zn -complexes). To make such complexing agents competitive with the highly specific low molecular weight chelates, the modification of lignin and humic materials should be conducted aimed at an increase of the content of oxygen-containing functions. A use of ozonation technique for this purpose deserves particular consideration while a use of ozone as an oxidant has a whole number of advantages. Among those are high reactivity and selectivity which allow to conduct oxidation under relatively mild conditions facilitating a control over oxidation degree of reactive products. The latter is of particular importance to avoid formation of toxic by-products.

473

I. V. Perminova et al. (eds.),
Use of Humic Substances to Remediate Polluted Environments: From Theory to Practice, 473–484.
© 2005 *Springer. Printed in the Netherlands.*

The literature data concerning the kinetics and mechanism of ozonation of such complicated macromolecular compounds are scarce. Mostly qualitative observations on the kinetics have been published [2-4] based on chromatographic data and numerical simulations. The problem is that the ozonation is multistage process, and a wide variety of intermediates formed is difficult to identify. An additional complication is a strong dependence of the ozonation products on the experimental conditions (pH, temperature, etc.). For example, the ozonation pathways of phenolic compounds depend greatly on pH [5].

It is known that molecular ozone preferably oxidizes electron-rich sites (aromatic carbon-carbon double bond, aromatic alcohols, and non-saturated structures) [6]. As a result, a dramatic decrease in ultraviolet (UV) absorbance at 254/280 nm is observed during ozonation of the aromatic compounds [7]. The application of small doses of ozone was shown to result in polymerisation processes [8].

The major end products of lignin ozonation are reported to be carbonyl compounds, low molecular weight (MW) carboxylic acids, mainly acetic, formic, and oxalic acids, which are stable to ozonation [9, 10]. In spite of many studies on the structure of HS and of their oxidation products, the reaction pathways that HS undergo during ozonation process remain poorly understood. The production of lower MW fulvic acid (FA) using ozone oxidation of higher MW humic acid (HA) has been discussed by several researchers [11-13]. The obtained FA showed a high content of oxygen and carboxylic groups.

It is known that during ozonation in water solutions, hydroxyl radicals (OH•) reactions proceed along with direct ozonation [14, 15]. OH• radicals react relatively unselectively. The radical generation is usually ascribed either to ozone self-decomposition in neutral and basic media, or to substrate-ozone reactions. In its turn, OH• radicals produce organic radicals as a result of their reactions with organic compounds. The reactions with OH• radicals may bring about substantial structural changes in macromolecules including hydroxylation, decarboxylation, and depolymerisation of the initial materials, producing oxidized structures that are less hydrophobic and less aromatic (16).

At present, the advanced oxidation processes (AOPs) are being developed to improve the kinetics of ozone reactions with natural organic matter. A strong and selective oxidation could be brought about by addition of the effective catalysts, e.g. transition metal ions. Some heterogeneous catalysts have also been found to enhance the ozonation efficiency of aquatic HS [17] and lignin-containing materials [18]. The catalysts facilitate the oxidation of less reactive organic compounds in aqueous solutions and diminish the ozone consumption [19]. Some studies illustrate that a use of metal ions in the ozone reactions with aromatic compounds could lead to reactions of hydroxylation, alkyl radicals oxidation, oxidative condensation, etc. [20] without aromatic ring destruction.

To deepen the understanding of the kinetics and mechanism of ozone reactions with lignins and HS, the objectives of this paper were: to estimate the kinetic parameters of ozonation; to study structural changes of lignin and HS brought about by ozonation; to suggest the ozone oxidation pathways for lignin and HS.

2. Materials and methods

Materials. Sulphate lignin (LG) (Germany), sodium ligninsulphonate (LS) (Sloka paper mill, Latvia) and potassium humate (CHP-K) produced from leonardite (Humintech, Germany) were used in this study. Elemental composition and molecular weight of the samples used are given in Table 1.

Table 1. Elemental composition and molecular weight (M_w) of humic and lignin samples used in this study.

Sample	Content of elements, %				M_w, kDa
	C	H	N	Ash	
LG	64.4	5.81	0.26	n.d.	3.0
CHP-K	42.2	3.04	1.02	26.0	3.2

Preparation of solutions. The solutions of LS and CHP-K were prepared in the range of concentrations of 0.5-5.0 g/L. To obtain solutions at pH 8.6, 0.13 M phosphate buffer was used. To attain strongly basic or acidic pH, concentrated NaOH or H_2SO_4, respectively, was used.

Ozonation procedure. Ozonation was carried out in a bubble reactor at 20°C. The latter was a thermostated glass cylinder with a porous filter bottom equipped with an outlet tube and a special valve, to facilitate solution sampling during the reaction. Ozone generated in the air-fed ozonator was bubbled through the liquid or solid sample placed in the reactor. The volume of the reaction mixture was 20 mL. Ozonation was performed at an initial ozone concentration of $1.2 \cdot 10^{-3}$ M, the volume flow velocity of the gas mixture was 10.0 L/h. The reaction was stopped when inlet and outlet ozone concentrations became equal. After the ozone treatment, the contactor was purged with the air to remove residual ozone. The inlet and outlet ozone concentrations in the ozone-air gas mixture were determined using absorbance measurements at 260 nm. The ozone concentration was calculated from the known molar extinction value of 3300 mole/(L·cm). After ozonation, the reaction mixture was rotor-evaporated to obtain the solid preparations for further structural investigations.

Structural characterization of the samples before and after ozonation. To follow transformations of LS and CHP-K structures, the *UV/Vis absorption spectra* were recorded with Cary 3E UV/Vis spectrometer (Varian, USA). The optical length was 1 cm. Prior to the analysis, the samples collected in the course of the reaction were diluted up to the concentration of 0.1 g/L.

Elemental Analyses (C, H, N) were performed on a Carlo Erba Strumentazione elemental analyser. Ash content was determined manually by dry combustion.

Infrared (IR) spectra were recorded with a Specord 75 IR spectrophotometer in KBr pellets. The optical density was determined using $K_4Fe(CN)_6$ as an external standard. The relative optical density (a_v) for a frequency v was determined as a ratio of the optical density for a given band to that of $K_4Fe(CN)_6$ at 2015 cm^{-1}.

^{13}C *solution-state NMR* spectra of LG and CHP-K samples were measured on solutions of these materials in 0.1M NaOD/D$_2$O at a concentration of 10 g/L. Measurements were made on a Varian VXR-400 spectrometer operating at 100 MHz ^{13}C observation frequency. Each spectrum was a result of 12000-14000 scans. Sodium trimethylsilylpronanesulphonate was used as an internal standard. The spectra were recorded at a 4-s delay time. These conditions were shown to provide quantitative determination of carbon distribution among the main structural fragments of HS [21].

To quantify the obtained spectra, for HS samples the assignments were made according to [21] (in ppm): 5-50 – aliphatic H and C-substituted C atoms (C$_{Alk}$), 50-95 – alcoholic and ether groups (C-OH, C-OR), 95-105 – C atoms of -O-C-O- groups; 105-145 – aromatic H and C-substituted atoms (C$_{Ar}$-H,C), 145-160 – aromatic O-substituted C-atoms (C$_{Ar}$-O), 160-188 – C atoms of carboxylic and esteric groups (COO), 188-220 – C atoms of quinonic and ketonic groups (C=O). For LG sample, the assignments were made according to [^{22}] and were as follows: 6-54 – aliphatic H and C-substituted C atoms (C$_{Alk}$), 54-62 – aliphatic O-substituted C atoms (C$_{Alk}$-O), 62-100 – alcoholic and ether groups (C-OH, C-OR); 100-140 – aromatic H and C-substituted atoms (C$_{Ar}$-H,C), 140-169 – aromatic O-substituted C-atoms (C$_{Ar}$-O), 169-190 – C atoms of carboxylic and esteric groups (COO), 190-220 – C atoms of quinonic and ketonic groups (C=O).

3. Results and Discussion

3.1. KINETIC STUDIES ON LIGNIN AND HS OZONATION

The ozonation reactions in solution were studied using lignosulphonate (LS) as a water-soluble representative of lignin compounds, and potassium humate (CHP-K) as a representative of humic compounds. Lignin (LG) was use as a suspension. The solubility of LS is provided by -SO$_3$H groups introduced into initial lignin backbone during sulphite treatment, while the solubility of CHP-K is provided by its presence in the form of the salt of alkali metal – by the high ionisation degree of acidic groups of HS.

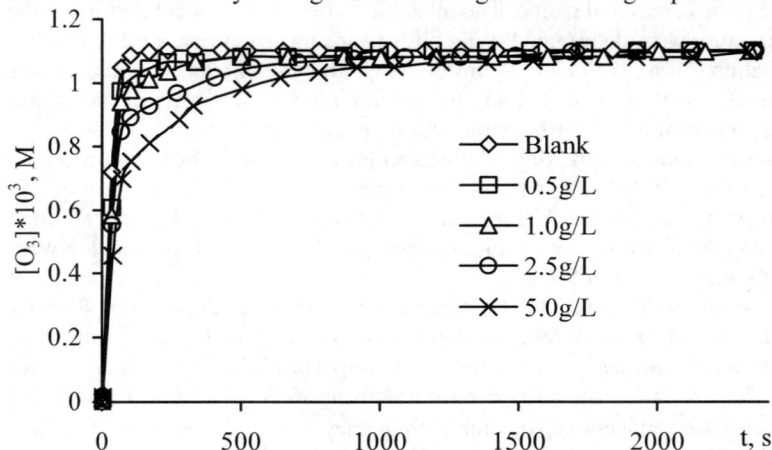

Figure 1. Kinetic curves for LS ozonation at pH = 5,5 for various initial concentrations.

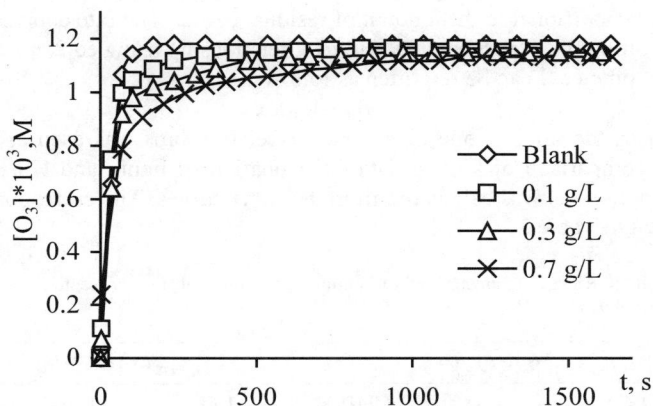

Figure 2. Kinetic curves for CHP-K ozonation at pH=5,5 for various initial concentrations.

The typical ozonation curves obtained for the LS and CHP-K solutions at neutral pH are shown in Figs. 1 and 2, respectively. The data on outlet ozone concentration ([O_3], 10^{-3} M) were further plotted against treatment duration (t, sec). The similar curves were obtained for solutions of these preparations at pH 1.0 (only for LS), 8.6, and 12.0.

To treat the obtained results numerically, the kinetic model for ozone reactions with chemical compounds in a bubble reactor was used [23]. The model assumes that (1) the rate of reaching the equilibrium ozone distribution between the gas and solution phases substantially exceeds the reaction rate, and (2) the dissolved compounds interact with dissolved ozone only.

To calculate the quantitative kinetic parameters – rate constants and ozone consumption – out from the obtained kinetic curves using the above model, the approach developed in our previous papers [24, 25] was used. For the reaction:

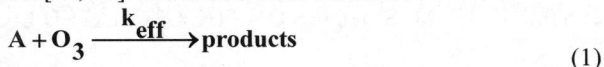

$$A + O_3 \xrightarrow{\ k_{eff}\ } products \tag{1}$$

Solution of the corresponding system of differential equations can be written as follows:

$$y/[A]_0 = a \cdot (1 - \exp(-k_{eff} K \cdot x)) \tag{2}$$

where $[A]_0$ is the concentration of the initial compound with dimensional units (g/L), **a** is the total ozone consumption, k_{eff} is the effective rate constant ($M^{-1}sec^{-1}$), **K** is the equilibrium constant of ozone distribution between gas phase and solution, **y** and **x** are parameters calculated from the integration of the kinetic curves according to the following expressions:

$$y = W \int_0^t (c_0 - c) dt - Kc(t) \tag{3}$$

$$x = \int_0^t c \, dt \tag{4}$$

where **W** is the specific gas flow rate (sec^{-1}), c_0 and **c** are inlet and outlet ozone concentrations, respectively (M), **t** – ozonation time (sec).

As follows from eqs 3 and 4, **y** is the amount of ozone consumed up to the time **t**; **x** is a parameter proportional to the amount of residual ozone. The rate constant k_{eff} can be calculated by fitting the experimental **y** *versus* **x** relationship using eq 2.

At the start time eq 2 can be rewritten as follows:

$$y/[A]_0 \sim k_{eff}Kx \qquad (5)$$

In this paper, the initial slope of **y** *versus* **x** relationships was used as a numerical value for the comparison of kinetic data on ozonation of lignin and HS preparations. Ozone consumption (**a**) was calculated from the same curves. The corresponding results are summarized in Table 2.

Table 2. Kinetic parameters of the lignin and humic substances reactions with ozone at various pH.

pH	a, g O_3/g			k_{eff} ($\times 10^2$), L·g^{-1}·sec^{-1}		
	LG*	LS	CHP-K	LG	LS	CHP-K
1.0	0.48±0.07	0.34±0.05	–	2.8±0.8	4.7±1.4	–
5.5	3.60±0.54	0.91±0.14	1.82±0.27	9.0±2.7	7.9±2.4	16.2±4.9
8.6	3.00±0.45	1.92±0.29	1.73±0.26	8.0±2.4	19.5±5.9	27.9±8.4
12.0	–	2.26±0.34	1.87±0.28	–	71.0±21.3	71.9±21.6

* LG was used as a suspension – heterogeneous reaction.

Table 2 indicates that the rate constants k_{eff} for LS and CHP-K oxidation by ozone are on the same order of magnitude, and increase substantially with an increase in pH. Concerning the ozone demand (**a**), it strongly depends on pH in the case of LS ozonation, and, in contrast, it does not change for the CHP-K reaction with ozone. The similar dependence – increase of oxidation rate and ozone demand along with an increase in pH – was observed for the LG suspension.

3.2. STRUCTURAL STUDIES ON THE OZONATED LIGNIN AND HS

To follow the pathways of ozonation reactions, the structural changes in the ozonated lignin samples – LS and LG, and HS were characterized using UV/Vis-, IR- and ^{13}C NMR-spectroscopy.

Figs. 3 and 4 display UV/Vis spectra of LS and CHP-K samples, respectively, in the course of ozonation at pH 5.5. Very similar spectra were observed at pH 1.0 (for LS only), 8.6, and 12.0.

For CHP-K, the UV absorbance at 280 nm started abruptly decreasing even at initial moments of ozonation. For LS, the relationship of the absorbance at 280 nm versus time of ozone consumption, has more complicated character (Fig. 5). Ozone consumption was calculated per mole of LS structural units (M_r = 240).

IR- spectra of ozonated LS samples were treated using the semi-quantitative data treatment technique for determination of the content of functional groups [26]. The results obtained show that ozonation caused oxidation of LS sulpho-groups to sulphate ions. The reduction in intensity of spectral bands corresponding to aromatic carbon along with an increase in the intensity of carbonyl groups suppose opening the rings due to oxidation with follow up formation of additional carboxyls or ketone. The formation of quinones cannot be excluded as well.

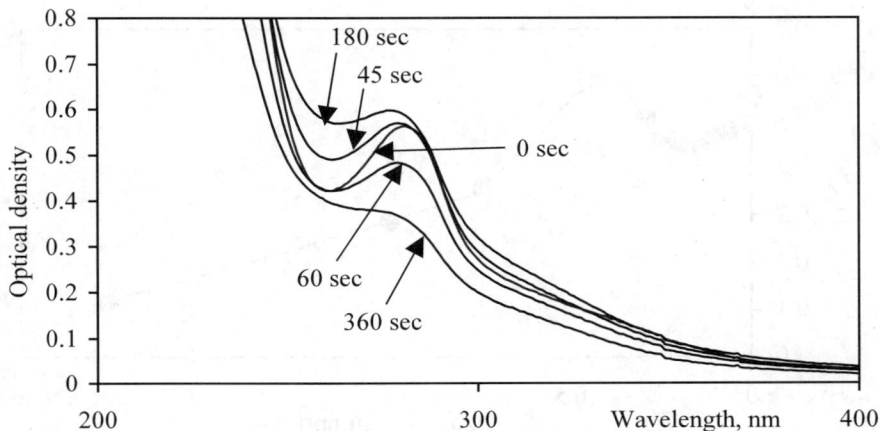

Figure 3. UV/Vis spectra of LS in the course of ozonation (pH = 5.5).

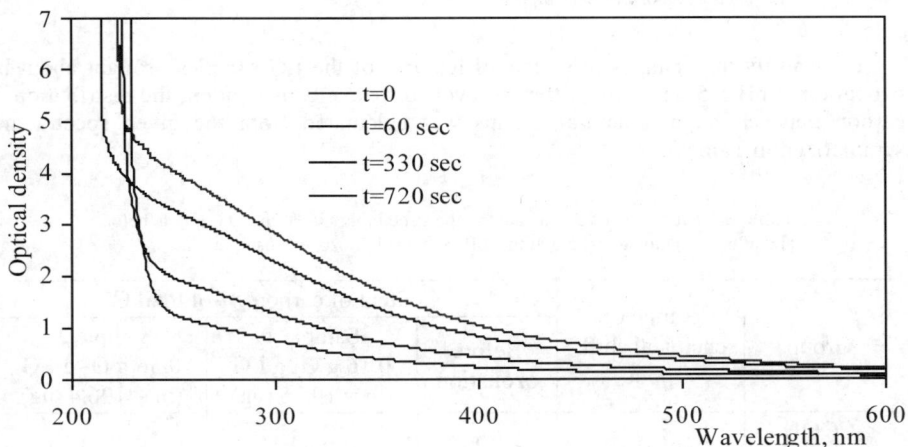

Figure 4. UV/Vis-spectra of CHP-K in the course of ozonation (pH=5.5).

^{13}C NMR spectroscopy requires high sample concentrations and solubility in NaOH. So, LG rather than LS was taken for this analysis. The ^{13}C NMR spectra of initial LG sample and of the samples ozonated in neutral and alkali solution (pH 12.0) are given in Figure 6. As it can be seen, the spectra of ozonated LG samples are characterized with a substantial decrease in spectral intensity in the range of non-substituted aromatic carbons (100-140 ppm) along with an increase in the spectral intensity of O-substituted aromatics (140-169 ppm) and carboxylic carbon (169-190 ppm). The most significant

changes in the peak shapes – broadening and substantial overlapping – are observed for the sample ozonated in the alkali solution indicating its higher degradation during the ozone treatment compared to that in neutral solution.

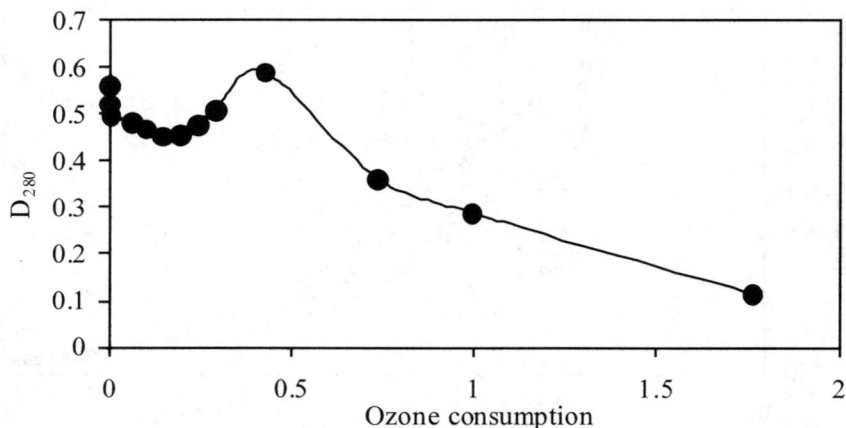

Figure 5. Dependence of UV absorption at 280 nm of LS solution (pH 5.5, c_{LS}=1.0 g/L) on ozone consumption.

To quantify the changes in structural features of the LG samples brought about by ozonation at pH 5.5 under the different levels of ozone consumption, the distribution of carbon between main structural groups was calculated from the given spectra and summarized in Table 3.

Table 3. Distribution of carbon among the structural units in the LG sample before and after ozonation at pH 5.5 at the different level of ozone consumption.

Group	Range of chemical shift, ppm	Content of carbon, % of total C		
		Before ozonation	Sample 1 0.16 g O_3/g LG (m = 195.5 mg)	Sample 2. 0.36 g O_3/g LG (m = 196.4 mg)
-COOH, -COOR	169-190	2.1	9.34	13.3
$-C_{Ar}$-OH, $-C_{Ar}$-OR	140-169	20.1	16.21	13.60
C_{Ar}-H,C	100-140	42.2	36.06	25.80
-C-OH, -C-OR	62-100	9.9	9.72	18.80
$-OCH_3$	54-62	12.8	10.50	8.70
C_{Alk}	6-54	13.0	18.17	19.60

481

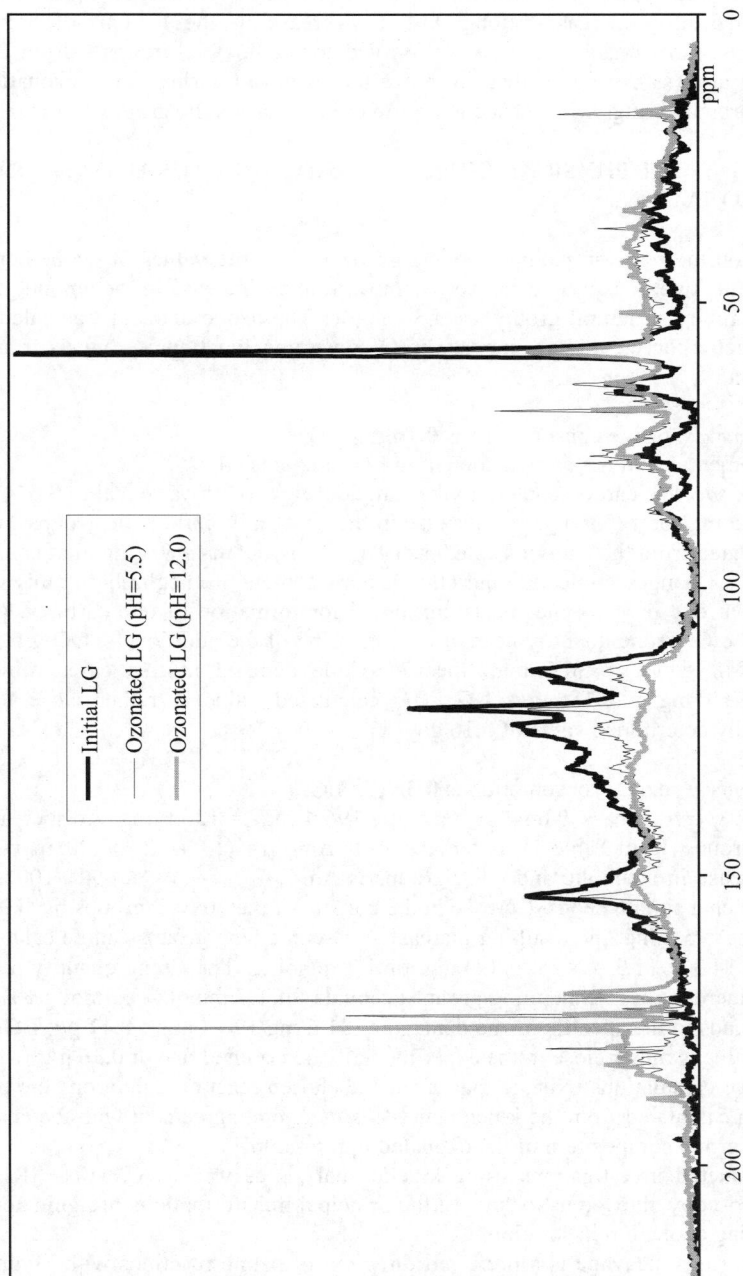

Figure 6. ^{13}C NMR-spectra of LG samples in 0.1 M NaOD/D_2O ozonated at neutral (5.5) and alkali (12.0) pH.

As follows from Table 3, the content of carboxylic, alcoholic, and ether groups in the course of ozonation increases along with a decrease in the content of aromatic structures. The most drastic increase is observed for carboxylic groups – from 2.1 to 13.3%. This allows to suppose that the prevailing pathway during lignin ozonation is splitting of aromatic rings with subsequent formation of carboxylic groups.

3.3. COUPLING THE PHYSICAL-CHEMICAL AND STRUCTURAL DATA ON LIGNIN OZONATION

To surmise on the reaction pathways of lignin ozonation, the values of ozone demand obtained in the kinetic experiments, were compared to an increase in the amount of the oxygen-containing structural groups for LS sample. The ozone demand was calculated from the kinetic curves, and an increase in O-containing functional groups - from the NMR-spectra.

Sample 1:

Kinetic curves: ozone consumption is **0.16 g/g LG**.

From elemental analysis carbon content in LG sample is 64.4%.

^{13}*C-NMR spectra*: carbon content in the sample weight (m_C) was equal to $0.644*m_{LG}$ (m_{LG} = 195.5 mg) or 125.9 mg. An increase in the content of carboxylic groups Δm_{COO} can be estimated from the data in Table 3 as $(9.3 - 2.1)/100*m_C$ = 9.1 mg, that equals to 0.76 mmol C. Changes in alcohol and ether groups content are negligible at this stage. Assuming that one ozone molecule is consumed for formation of one carboxyl group from aromatic C, ozone quantity needed for formation of the additional COO groups is 0.76 mmol*48 g/mol = 36.5 mg. Specific ozone demand in this case would be 36.5 mg/195.50 mg = **0.186 g/g LG**. The calculated value is rather close to the experimentally determined value of 0.16 g/g LG.

Sample 2:

Kinetic curves: ozone consumption is **0.36 g/g LG**.

^{13}*C-NMR spectra*: m_C = $0.644*m_{LG}$ (m_{LG} = 196.4 mg) = 126.5 mg. An increase in carboxylic groups (from Table 3) content equals to Δm_{COO} = $(13.4 - 2.1)/100*m_C$ = 14.3 mg. An increase in alcoholic and ether groups is $\Delta m_{COH, COR}$ = $(18.8 - 9.9)/100*m_C$ = 11.3 mg. Given a simultaneous decrease in the content of methoxylic groups by $(12.76 - 8.70)/100*m_C$ = 5.2 mg, the resulting increase in O-containing groups can be calculated as follows: $(14.3 + 11.3 - 5.2)$ = 20.4 mg, or 1.7 mmol C. The ozone quantity needed for such an increase in O-containing groups accounts for 1.7 mmol*48 g/mol = 81.6 mg that corresponds to the specific ozone demand of 81.6 mg/196.4 mg = **0.42 g/g LG**. The calculated value is rather close to the experimentally determined one of 0.36 g/g LG.

The above calculations indicate that under the chosen reaction conditions the ozone consumption calculated from the kinetic curves is in a good agreement with the changes in structural group composition of the ozonated lignin samples.

The conducted investigations using kinetic analysis as well as UV/Vis-, IR-, and NMR-spectroscopy allowed us to suggest the principal transformations of lignin and HS samples during ozonation in solution.

Aromatic rings cleavage is a preferred direction of ozone reactions with lignin and HS in solutions. As a result, unsaturated oxidized compounds are formed. They further transform into aliphatic structures. The literature data show that small ozone doses leads to condensation processes. The decrease of LS absorbance at 280 nm at initial time moments could be due to condensed structures formation or various modifications of LS functional groups (C=O, C–OH).

The experimental data presented allow us to conclude that the variation of operating conditions (pH, ozone dose, etc.) enables regulation of the ozonation process, producing oxidized molecules with desirable contents of O-containing functional groups.

4. Conclusion

The rate constants for LS and CHP-K ozonation in solutions increase substantially with an increase in pH. Total ozone consumption strongly depends on pH for LS ozonation, and, does not change depending on pH media for the CHP-K reaction with ozone. The ozonation in solutions is much deeper than in solid state. The content of O-containing groups (carboxylic, alcoholic, and ether groups) increases in the course of LG, PHF, and CHP-K ozonation along with aromatic ring destruction.

5. Acknowledgements

This research was supported by the grants of the Lomonosov Moscow State University in the framework of "Interdisciplinary scientific projects" program in 2002-2003.

6. References

1. Clapp, C.E., Hayes, M.H.B. and Swift, R.S. (1993) Isolation, fractionation, functionalities, and concepts of structure of soil organic macromolecules, in A J. Beck, K.C. Jones, M.B.H. Hayes, and U. Mingelgrin (eds.), *Organic Substances in Soil and Water*, Royal Society of Chemistry, Cambridge.
2. Sweeney, M. (1981) Comparative studies in the ozonolysis of lignin and coal, *Termochimica Acta* **48** (3), 263-275.
3. Naoto, H. and Matsumoto J. (1983) Studies on chemical structure of lignin by ozonation, *Int. Symp. on Wood and Pulp. Chem. (Jap. TAPPI)* **4**, 68-71.
4. Melo, R., Mariani, S. and Arias, A. (1997) Kinetic study of the reaction of Klason lignin with ozone, *Invest. Tec. Pap.* **34** (134), 791-800.
5. Bailey P.S. (1982) *Ozonation in Organic Chemistry*, Vd. 2 Academic Press, New York.
6. Hoighe, J. and Bader, H. (1983) Rate constants of reactions of ozone with organic and inorganic compounds in water. – I: non-dissociating organic compounds, *Wat. Res.* **17**, 173-183.
7. Ohlenbusch, G., Hesse, S. and Frimmel, F.H. (1998) Effects of ozone treatment on the soil organic matter on contaminated sites, *Chemosphere* **37** (8), 1557-1569.
8. Amy, G.L., Kuo, C.J. and Sjerca, R.A. (1987) Ozonation of humic substances: effects on molecular weight distributions of organic carbon and trihalomethane formation potential, *Ozone Sci. & Eng.* **10**, 39-54.
9. Kusakabe, K., Aso, S., Hayashi, J.I., Isomura, K. and Morooka, S. (1990) Decomposition of humic acid and reduction of trihalomethane formation potential in water by ozone with u.v. irradiation, *Wat. Res.* **24** (6), 781-785.
10. Takahashi, N., Nakai, T., Satoh, Y. and Katoh, Y. (1995) Ozonolysis of humic acid and its effect on decoloration and biodegradability, *Ozone Sci. & Eng.* **17**, 511-525.

484

11. Yamada, P., Akiba, T., Yazawa, Y. and Yamaguchi, T. (2002) Characteristics of humic and fulvic acids extracted from weathered coal of China, *Nippon Kagaku Kaishi* **3**, 351-358.

12. Kobayashi, S., Yuan, H., Kuragane, K., Gomi, K., Takiguchi, Y., Onoe, K. and Yamaguchi, T. (2000) Preparation of active carbon from brown coal and weathered coal with alkali, *Nippon Kagaku Kaishi* **5**, 329-335.

13. Shinozuka, T., Ito, A., Sasaki, O., Yazawa, Y. and Yamaguchi, T. (2002) Preparation of fulvic acid and low-molecular organic acids by oxidation of weathered coal humic acid, *Nippon Kagaku Kaishi* **3**, 345-350.

14. Staehelin, J. and Hoighe, J. (1985) Decomposition of ozone in water in the presence of organic solutes acting as promoters and inhibitors of radical chain reactions, *Environ. Sci. Technol.* **9** (12), 1206-1213.

15. Hoighe, J. (1982) Mechanisms, rates and selectivities of oxidations of organic compounds initiated by ozonation of water, in R.G. Rice and A. Netzer (eds.), *Handbook of Ozone Technology and Applications*, Vol. 1. Ann Arbor Science Publishers, Ann Arbor, MI, 341-379.

16. Kerc, A., Bekbolet, M. and Saatci, A.M. (2002) Sequential oxidation of humic acids by ozonation and photocatalysis, *Proceedings of the International Conference on Ozone in Global Water Sanitation*, 1-3 October, Amsterdam, the Netherlands.

17. Allemane, H., Delouane, B., Paillard, H. and Legube, B. (1993) Comparative efficiency of three systems (O_3, O_3/H_2O_2, O_3/TiO_2) for the oxidation of natural organic matter in water, *Ozone Sci. & Eng.* **15**, 419-432.

18. Letumier F., Barbe J.-M., Trichet A. and Guilard R. (1999) Ozonation Reactions of Monomer and Dimer Lignin Models: Influence of a Catalytic Amount of a Manganese Cyclam Derivative on the Ozonation Reaction, *Ozone Sci. Eng.* **21**, 53-67.

19. Andreozzi, R., Insola, A., Caprio, V. and D'Amore, M.G. (1992) The kinetics of Mn(II)-catalyzed ozonation of oxalic acid in aqueous solution, *Water Res.* **26** (7), 917-921.

20. Galstyan, G.A., Yakobi, V.A., Dvortsevoi, M.M. and Galstyan, T.M. (1978) Liquid-phase catalytic oxidation of *p*-nitrotoluene by ozone-oxygen mixture, *Zhurnal prikladnoi khimii* **51** (1), 133-136.

21. Kovalevskii, D.V., Permin, A.B., Perminova, I.V. and Petrosyan V.S. (2000) Conditions for acquiring quantitative ^{13}C NMR spectra of humic substances, *Moscow State University Bulletin (Vestnik Moskovskogo Universiteta), Series 2 (Chemistry)* **41**, 39-42.

22. Kalabin, G.A., Kanitskaya, L.V. and Kushnarev D.F. (2000) *Quantitative NMR Spectroscopy of Natural Organic Matter and Its Transformation Products*, Khimiya, Moscow (in Russian).

23. Razumovskii, S.D. and Zaikov, G.E. (1974) *Ozone and Its Reactions with Organic Compounds*, Nauka, Moscow (in Russian).

24. Pryakhin, A.N., Gusenkova, M.M., Kasterin, K.V., Mitrofanova, A.N., Benko, E.M. and Lunin, V.V. (1998) The new method for the description of consequent ozone reactions with chemical compounds solutions in a bubble reactor, *Russ. J. Phys. Chem.* **72** (10), 1841-1843.

25. Kovaleva, V.V., Mitrofanova, A.N., Benko, E.M., Mamleeva, N.A., Pryakhin, A.N. and Lunin, V.V. (1999) The new method for the theoretical analysis of ozonation kinetics in solutions in a bubble reactor, *Russ. J. Phys. Chem.* **73** (1), 35-38.

26. Ksenofontova, M.M., Mitrofanova, A.N., Mamleeva, N.A., Pryakhin, A.N. and Lunin, V.V. (2001) Reactions of Ozone with Lignins, *Proceedings of the 15th World Ozone Congress*, IOA, Imperial College, London, England, September 11-15, Vol. 3.

Chapter 24

SYNTHESIS, METAL-BINDING PROPERTIES AND DETOXIFYING ABILITY OF SULPHONATED HUMIC ACIDS

M.V. YUDOV[1], D.M. ZHILIN[1], A.P. PANKOVA[1], A.G. RUSANOV[2], I.V. PERMINOVA[1], V.S. PETROSYAN[1], D.N. MATORIN[2]

[1]*Department of Chemistry, Lomonosov Moscow State University, Moscow 119992, Russia <mat@org.chem.msu.ru>*
[2]*Department of Biology, Lomonosov Moscow State University, Moscow 119992, Russia*

Abstract

The complexing properties of humic substances are of primary importance for their application as detoxifying agents and microfertilizers. The promising approach for enhancing solubility of metal-humics complexes is incorporation of SO_3H-groups into the structure of humics. The water soluble sulphonated humic materials can be used as flushing agents for heavy metal polluted sites and as chelating agents for production of microfertilizers. The objectives of this study were: 1) to synthesize sulphonated humic materials; 2) to evaluate solubility and stability of their complexes with Fe(III) and Hg(II); 3) to assess potential toxicity and detoxifying properties of the sulphoderivatives with respect to Hg(II) and Cu(II). Humic acids (HA) from leonardite and peat were used for modification. Concentrated sulphuric acid and chlorosulphonic acid were used as sulphonating agents. Chlorosulphonic acid was found to be much more efficient sulphonating agent causing introduction of 1.6 and 0.4 mmol SO_3H per g of humic material from leonardite and peat, respectively. The binding properties of the sulphonated HA were studied in relation to Hg(II) and Fe(III). The solubility of metal-humic complexes was estimated as an amount of strong metal-binding sites (SBS) in the humic material. The sulphonated materials were characterized by twofold increase in the amount of SBS: 2-4 mmol/g versus 1-2 mmol/g in the parent material. Stability constants were determined using the ligand exchange technique. For Hg(II) complexes, the log K values accounted for 13.1 and 12.4 L/mol SBS for leonardite HA and their sulphoderivative, respectively. For Fe(III) complexes, the log K values accounted for 12.0 and 11.5 for the parent and chlorsulphonated coal HA, respectively, and 12.4 and 11.4 – for the parent and chlorsulphonated peat HA. Assessment of the toxicity of the obtained derivatives and of their detoxifying properties was performed using algological tests. The sulphoderivatives did not exhibit adverse effects onto algae. Both parent and sulphonated humic materials displayed high detoxifying properties with respect to Hg(II) and Cu(II), leonardite HA were more effective than peat HA. The obtained results allow to consider sulphonation as a promising tool for obtaining soluble metal-HA complexes.

I. V. Perminova et al. (eds.),
Use of Humic Substances to Remediate Polluted Environments: From Theory to Practice, 485–498.
© 2005 *Springer. Printed in the Netherlands.*

1. Introduction

The presence of a wide range of functional groups (carboxylic, hydroxylic, carbonylic) in the structure of humic substances (HS) provides their ability to form the complexes with metals [1]. As a result, HS play a key role in both migration of metals and their uptake by higher plants. Of particular importance is, that the properties of metal complexes of HS depend greatly on the content of functional groups and on the molecular weight of humics. So, the lower molecular weight (MW) fraction of HS – fulvic acids (FA) enriched with oxygen containing functional groups produce soluble metal complexes of much higher mobility and bioavailability than those of humic acids (HA). The latter is higher MW fraction of HS enriched with the condensed aromatic structures. Given the above properties of metal complexes of HA and FA, G.M. Varshal and co-workers distinguished greatly the geochemical roles of HA and FA in metal migration characterizing the former as "accumulators" or "immobilisers" of metals in the environment, whereas the latter as "scatters or dissipaters" of metals [2].

It has been numerously discussed in the literature, that HS application for the pollution control has a significant potential [3]. They can be effectively used as natural, cheap and regenerative ion-exchangers for cleaning-up waste waters and gases [4]. The complexing properties of HS determine their potential to be used as detoxifying agents and microfertilizers. Even though the available supply of raw humic materials is enormous, two fundamental reasons can be cited as to why HS have not been widely used for the above applications. First, few natural HS possess the specific reactive properties required to form highly mobile and soluble metal complexes. Second, HS are by definition polydispersive and heterogeneous; consequently, the reactive properties vary between natural sources and between industrial suppliers. The presented research has addressed both problems using a novel approach of using chemical modification for designing humic materials with specific properties. Previous investigations involving HS modifications (hydrolytical cleavage, oxidation, methylation, acetylation, silylation, and others) were focused on an understanding of HS structure [5, 6, 7, 8], or bioavailability [9], not to enhance or change reactive properties of HS. Sulfonation was used as a promising way of producing highly soluble HS enriched with functional groups.

Sulphonation (introduction of -SO_3H group) is a well-known technique of chemical modification of organic materials [10]. Sulphonated organic materials are applied as drilling fluids [11]. The sulphonating agents that are widely used are chlorosulphonic acid, concentrated sulphuric acid and sulphur trioxide [12, 13, 14]. Previously, coals have been directly sulphonated using sulphuric acid, to introduce sulphonic groups, but the sulphonated products were not water-soluble and could not be dissolved, purified and otherwise treated as an organic chemical [15]. They have usually found application as ion exchange materials. Water-soluble humic acids were previously obtained by various oxidation treatments of humic materials [16] and some coals [17, 18] but far-reaching molecular breakdown has always accompanied the oxidation, and the water-soluble products were, therefore, extensively degraded materials [19]. A number of agents such as chlorsulphonic acids or sulphur trioxide solution in pyridine allow to introduce sulphuric groups without significant oxidation or degradation of treated humic acids [20]. Another technique of synthesis of water-soluble sulphoderivatives of HS is based on reaction with a sulphonating agents like sulphite or bisulphite precursors which form a sulphite or bisulphite *in situ*, in alkaline medium and recovering a water-soluble product from the reaction medium. Sulphomethylation was also reported as the method of solubilization of water-insoluble humic acids [10]. The synthesis of sulphoderivatives of HA is of particular interest because they can be used as drilling mud thinners, soil conditioners, and tanning agents [12] and also in various flushing technologies [21].

The objectives of this study were: 1) to synthesize sulphonated humic materials; 2) to evaluate solubility and stability of their complexes with Fe(III) and Hg(II); 3) to assess potential toxicity and detoxifying properties of the sulphoderivatives with respect to Hg(II) and Cu(II).

2. Experimental Section

2.1. MATERIALS AND METHODS

Humic materials used were isolated from highland peat (Sk3-00, Sakhtysh Lake, Ivanovo Region, Russia) and leonardite – oxidized coal, kindly provided by Humintech Ltd. (Duesseldorf, Germany). Isolation procedure included a preliminary treatment of a peat sample with ethanol-benzene (1:1) mixture followed up by an alkaline (0.1 M NaOH) extraction as described elsewhere [22], leonardite was treated with 0.1 M NaOH. The obtained alkaline extract was acidified to pH 1 using concentrated HCl. The precipitated HA were centrifuged, washed out with distilled water and dialysed. A set of the parent and modified samples is described in Table 1.

Table 1. Set of parent and modified humic materials under study.

Sample	Description
CHA-Leo	Leonardite HA
CHA-S	Leonardite HA, sulphonated with H_2SO_4
CHA-ClS	Leonardite HA, sulphonated with $ClHSO_3$
PHA-Sk300	Peat HA
PHA-S	Peat HA, sulphonated with H_2SO_4
PHA-ClS	Peat HA, sulphonated with $ClHSO_3$

Sulphonation of humic acids with concentrated H_2SO_4. A weight of 100 g of humic acids was wetted with water until paste-like state and then it was added dropwise with 400 ml concentrated H_2SO_4 The obtained mixture was stirred during 2 hours using mechanical stirrer. When reaction was completed, the mixture was poured into 4 litres of distilled water, stirred, and left for 10 hours. The supernatant was discarded, and the precipitate was purified by dialysis and dried out for the further analytical studies.

Sulphonation of humic acids with chlorosulphonic acid. A weight of 100 g of humic acids was dried over P_2O_5, placed into 500 ml beaker, cooled in the ice bath and then was added with 200 ml of cooled $ClSO_3H$ by small portions with permanent stirring during 6 hours. When the reaction was completed, the mixture was poured into 4 L of iced distilled water to hydrolyse the rest of chlorosulphonic acid, and the obtained solution was left for 10 hours. Then, the supernatant was discarded, and the precipitate was washed out with water 4 times until the supernatant solution gave negative reactions to $BaCl_2$ and $AgNO_3$. The obtained precipitate of the sulphonated humic acids was further purified using dialysis and then dried.

Determination of the content of strong acidic groups (COOH and SO_3H). Ca acetate method was used to determine the content of strong acid groups in the humic samples as described in [23]. For this purpose, an aliquot of HS solution (5-10 mL) containing 5-20 mg HS was transferred into a vial (~ 22 mL), and 10 mL of 0.6 M $Ca(CH_3COO)_2$ was added. The vial was tightly sealed, shaken well and left for equilibration for 24 hours at room temperature. An aliquot of supernatant above the precipitate of Ca humates was transferred into titration cell and titrated with NaOH standard solution (~0.05 M) using autotitrator (Metrohm 716 DMS Titrino). Acidity of the samples (A, mmol/g) was calculated as follows:

$$A = \frac{(V_{HS} - V_0) \cdot c_{NaOH}}{m} \tag{1}$$

where V_{HS} and V_0 are the volumes of NaOH used for titration of the HS solution and the blank solution containing same amount of $Ca(CH_3COO)_2$, L, C_{NaOH} – concentration of NaOH, mM, m – a weight of HS sample, g. *A* has dimensional units of mmol/g.

Determination of the stability constants of iron humates. Stability constants of the iron(III) complexes with parent and sulphonated HA were determined using ligand exchange technique with photometric detection. Tartrate was used as a competitive ligand. The method is based on detecting the difference in the optical density of HA and of their Fe(III) complexes. Two types of titrations of HS (0.5 g/L) by $Fe_2(SO_4)_3$ (10 mM Fe) solution were conducted: (1) in the presence of 0.024 M tartrate buffer (pH 4.37); and (2) in the absence of the buffer, but in the presence of 0.048M KNO_3 used to maintain the same ionic strength as in the buffer containing solution. During titration of the latter solution, pH was kept at 4.37±0.1 by adding 0.05 M NaOH. After each addition of Fe^{3+} and NaOH, optical density was measured at 650 nm using Metrohm 662 Photometer (Switzerland) equipped with a submersible optical cell (1 cm optical path). Solution of $Fe_2(SO_4)_3$ containing 10 mM Fe was prepared in 0.01M H_2SO_4. 0.4 M tartrate buffer was prepared by dissolving 6 g of $C_4H_6O_6$ in 50 ml of water, adding it with 1 M NaOH to adjust pH to 4.37, and then making it up to 100 mL with distilled water. At pH 4.37, tartrate exits as a 1:1 mixture of divalent and monovalent anions.

Determination of the content of strong metal-binding sites. To determine the content of strong iron-binding sites in parent and sulphonated humic materials, 0.8 M $Fe_2(SO_4)_3$ was added to 0.25 g/l solution of HA until the beginning of precipitation (pH was maintained at 4.37±0.1 using NaOH). The concentration of iron causing the precipitate formation was assumed to be equal to the concentration of strong iron-binding sites in the HA solution. The content of strong mercury-binding sites was determined similarly, using 0.1 M $Hg(NO_3)_2$ and maintaining pH at 6.68±0.1. To obtain a numerical value of the content of strong metal-binding sites in the humic materials, the metal concentration causing HA precipitation (mmol/L) was normalized to the mass concentration of HS in the solution (g/L) yielding dimensional units of mmol Me/g HA.

Determination of the stability constants of mercury humates. Stability constants of Hg(II) complexes of the parent and modified HA were determined using ligand exchange technique with photometric detection similar to the described above for Fe(III). Chloride was used as a competitive ligand. The HA solutions (0.3 g/L) were titrated photometrically at 650 nm by 0.01M $Hg(NO_3)_2$ (1) in the presence of 0.01 M NaCl; and (2) in the absence of NaCl, but in the presence of 0.01 M KNO_3 used to maintain the same ionic strength as in case of chloride. In both experiments, pH was adjusted to 6.2 using NaOH or HNO_3. During titration, pH was maintained at 6.2±0.1 by adding 0.05 M NaOH. To maintain the constant concentration of Cl^- in the reacting systems during titrations in the presence of NaCl, solutions of both NaOH and $HgCl_2$ contained 0.01M NaCl. $Hg(NO_3)_2$ solution was prepared by dissolving a weight of HgO in minimal amount of concentrated HNO_3.

Acute toxicity tests. Humic materials tested were CHA-Leo, CHA-ClS, PHA-Sk300 and PHA-ClS. HA solutions had concentration in the range of 0.5-20 mg/L. Hg(II) and Cu(II) were used as model heavy metals. Acute toxicity tests were performed according to [24]. Green unicellular algae *Chlorella pyrenoidosa* was used as a biotarget, photosynthetic activity – as a response. *Chlorella pyrenoidosa* (thermophilic strain CALU-175 from the collection of the Biology Institute, Saint-Petersburg State University, Russia) was grown in 1/5 Tamiya medium (pH 6.8) [25] at 32°C, under 60 μmol photons $m^{-2}s^{-1}$ continuous irradiance and bubbling with moisturized air. The culture was maintained in log growth phase by daily dilution with fresh medium to maintain a cell density of about $1-10 \cdot 10^5$ cells·ml^{-1}. For the bioassays, the algae was concentrated by centrifugation and resuspended to a final concentration of 5-$7 \cdot 10^5$ cells·mL^{-1} in the 1/10 Tamiya medium without phosphates and EDTA.

After three hours of growth, photosynthetic activity of algae was measured. For this purpose, the algae was adapted to darkness for 30 s. Chlorophyll fluorescence induction was recorded for 300 s using a lab-made fluorometer. F_0 was estimated from the value of fluorescence detected immediately after opening of the shutter. F_m was calculated by averaging the values of the last 50 points of the fluorescence induction curve. Photosynthetic activity (R) was calculated as a relative yield of the variable fluorescence using following expression:

$$R = \frac{F_v}{F_m} = \frac{F_m - F_0}{F_m} \qquad (2)$$

where F_v, F_m and F_0 are intensity of variable, maximum and background chlorophyll fluorescence, respectively. Three replicates were made for each experimental point.

The toxicity of model heavy metals was tested in the range of concentrations of (0.4-3)·10^{-6} and (0.3-2.6)·10^{-6} M for $HgCl_2$ and $CuSO_4$, respectively. The corresponding dose-response relationship fits satisfactorily with the linear model. The slope *k* (mean±SD, *n* = 3) was (0.35±0.03)·10^{-6} and (0,25±0.04)·10^{-6} M^{-1} for Hg (II) and Cu (II), respectively. The corresponding r^2 were 0.86 and 0.78. The effective concentrations causing 50% reduction in the photosynthetic activity of algae (EC_{50}) were 1.2 and 1.5·10^{-6} M for Hg (II) and Cu (II), respectively.

The above concentrations were used in the experiments on studies of detoxifying properties of the humic materials. Concentration of humic materials in the solution was in the range of 0.5-4 mg/L and 0.5-20 mg/L for Hg(II) and Cu(II), respectively. Each experimental series included four types of test solutions: control, HA, heavy metal, and heavy metal + HA. The corresponding responses were designated as R_0, R_{Me}, R_{HA} and R_{Me+HA}. They were used to calculate detoxification coefficient (relative decrease in heavy metal toxicity in the presence of HA) as described in [26]:

$$D = \frac{T_{Me} - T_{Me+HA}}{T_{Me}} \tag{3}$$

where T_{Me} and T_{Me+HA} are toxicities of the heavy metal solution in the absence and presence of HA, respectively.

The toxicity values can be calculated using the following expressions:

$$T_{Me} = \frac{R_0 - R_{Me}}{R_0} \tag{4}$$

$$T_{Me+HA} = \frac{R_{HA} - R_{Me+HA}}{R_{HA}} \tag{5}$$

where R_0, R_{HA}, R_{Me}, and R_{Me+HA} is photosynthetic activity of algae in the control, in the humic material solution, and in heavy metal solution in the absence and presence of HA, respectively.

3. Results and discussion

3.1. SYNTHESIS AND STRUCTURAL ANALYSIS OF THE SULPHONATED HUMIC MATERIALS

HA samples from leonardite – highly oxidized coal, and peat were selected for modification. They were chosen as representatives of the humics isolated from the raw materials of the greatest industrial value. Leonardite HA are enriched with aromatic fragments, whereas peat HA are enriched with carbohydratic units. It was expected that introduction of sulphonic groups will increase water-solubility of both HA and their complexes with metals [27].

The sulphonated humic materials obtained with a use of both sulphuric and chlorosulphonic acid were amorphous powders of the dark brown colour. The samples treated with chlorosulphonic acid (CHA-ClS and PHA-ClS) were much more soluble in water than those treated with sulphuric acid. Their solubility laid in the range of 100-130 g/L. It could be indicative of a higher sulphonation efficiency of chlorosulphonic acid compared to that of concentrated sulphuric acid.

Elemental composition of the parent and sulphonated humic materials is given in Table 2.

The parameters of major interest were S content and C/S ratio which can serve as indicators of the sulphonation efficiency. As it can be seen, for the initial samples and for the derivatives treated with concentrated sulphuric acid, C/S ratio is very similar (CHA-Leo – 256, CHA-S – 196 and PHA – 265, PHA-S – 210). This indicates that the sulphuric-acid treatment did not lead to an increase in the content of sulphur, hence, sulphonation had not occurred. At the same time, for the derivatives treated with chlorosulphonic acid, the C/S ratio was substantially less compared to the parent materials, and accounted for 30 and 43 for CHA-ClS and PHA-ClS, respectively. This shows much higher efficiency of chlorsulphonic acid as sulphonating agent of the humic materials compared to sulphuric acid.

Table 2. Elemental composition of the parent and sulphonated humic materials on ash-free basis.

Humic material	Content of elements, % mass.					Ash	Atomic ratio		
	C	H	N	S	O	%	C/S	H/C	O/C
CHA-Leo	56.4	4.10	1.23	0.59	37.6	7.78	256	0.87	0.50
CHA-S	62.5	4.12	1.07	0.56	31.7	4.28	296	0.79	0.38
CHA-ClS	58.5	3.58	1.31	5.12	31.5	5.85	30	0.74	0.40
PHA-Sk300	48.7	4.65	3.27	0.49	42.9	3.28	265	1.15	0.66
PHA-S	62.2	5.23	2.72	0.79	29.1	2.06	210	1.01	0.35
PHA-ClS	46.9	3.03	3.66	2.93	43.5	2.14	43	0.78	0.70

Efficiency of sulphonation was also estimated via determination of the content of strong acidic groups using the Ca-acetate technique. The latter determines all acidic groups present in humic material that have dissociation constant < 4.76 (pK_a of acetic acid). In case of the sulphonated humic materials, the pool of strong acidic groups is composed of carboxylic groups of the parent material and of the introduced sulphonic groups. Hence, an increase in the content of the strong acidic group could serve as another parameter of sulphonation efficiency. The corresponding data of the Ca-acetate determinations are given in Table 3.

Table 3. The content of strong acidic groups in the humic materials used in this study.

Sample	COOH + SO$_3$H, mmol/g	ΔSO$_3$H, mmol/g
CHA-Leo	2.9±0.5*	–
CHA-S	3.3±0.2	0.4
CHA-ClS	4.5±0.1	1.6
PHA-Sk300	3.3±0.1	–
PHA-S	2.1±0.1	–
PHA-ClS	3.7±0.2	0.4

* ± stands for a one standard deviation (n = 3).

The titration data show that similar to C/S ratio, the amount of strong acid groups has substantially increased after sulphonation with chlorsulphonic acid (1.6 and 0.4 mmol/g for CHA-ClS and PHA-ClS, respectively), and to much lesser extent – after sulphonation with sulphuric acid. Of particular importance is that the derivatives obtained by the treatment with chlorosulphonic acid were completely soluble in water (up to 100-130 g/L), whereas those treated with sulphuric acid were not soluble.

3.2. METAL-BINDING PROPERTIES OF THE SULPHONATED HUMIC MATERIALS

The metal-binding properties of the sulphonated peat and leonardite HA were studied using Hg(II) and Fe(III) as model metals. It was of particular importance to answer the following questions: 1) how does sulphonation influence solubility and stability of the metal-complexes; and 2) how do these changes influence detoxifying properties of HA. The enhanced solubility of HA complexes with nutritional metals would contribute greatly in their quality as microfertilizers. At the same time, sulphonation could cause a decrease in stability constants of those complexes. Therefore, interaction of parent and sulphonated humic materials should be investigated to estimate an influence of sulphonation on both complexation parameters of interest: on the content of strong binding sites determining the solubility of HS and on the stability constants. Iron was chosen as a nutrition metal, and mercury – as a model toxic metal.

To quantitate solubility of the metal complexes of the sulphonated humic materials, the HA solutions were titrated with the corresponding metal until the beginning of the precipitate formation. It was assumed that the metal-HA complex remains in solution until strong binding sites are not saturated with metal; once these sites are saturated, the complex precipitates. This allowed us to consider the metal:HS ratio (mol/g) of the beginning of precipitation as a quantitative estimate of the content of strong binding sites (SBS). The obtained results are presented in Table 4.

Table 4. Content of strong metal-binding sites in the parent and sulphonated humic materials.

Sample	Amount of Fe(III)-SBS, mmol/g	Amount of Hg(II)-SBS, mmol/g
CHA-Leo	3.2	2.2
CHA-S	2.8	n.d.
CHA-ClS	3.4	4.5
PHA-Sk300	0.9	n.d.
PHA-S	1.5	n.d.
PHA-ClS	1.9	n.d.

The obtained contents of SBS for Fe(III) and Hg(II) are quite similar and lay in the same range of values as total amount of strong acidic groups determined by Ca-acetate technique (Table 3). It should be noted that much higher content of SBS is observed in the coal humic material compared to peat. In case of peat HA, sulphonation with chlorosulphonic acid leads to twofold increase in the amount of SBS. Hence, introduction of the sulphonic groups into humic materials increases the solubility of complexes as a result of an increase in the amount of SBS.

Stability constants of Fe(III) and Hg(II) complexes with the parent and sulphonated materials were determined using the ligand exchange technique with spectrophotometric detection. Tartrate-ion was used as a competitive ligand for Fe(III) and chloride-ion - for Hg(II). The determination is based on different absorptive properties of the humic material and of its metal-complexes: the complexes absorb light stronger. Hence, adding Me to HA solution increases its optical density. If competitive ligand is added to HA solution that forms non-absorbing complexes with metal, an increase in optical density of HA solution upon addition of metal will be much less. The results of Fe(III)-titration of HA solution with and without competitive ligand (tartrate) are shown in Fig. 1. Concentration of the competitive ligand was selected in such a manner to assure existence of only one predominant metal species in the solution. For iron, it was tartrate complex 1:2, and for mercury – $HgCl_2$. These species do not adsorb light at 650 nm. As it can be seen, the shown titration curve in the presence of tartrate (Fig. 1) has much lesser slope than in the absence of tartrate.

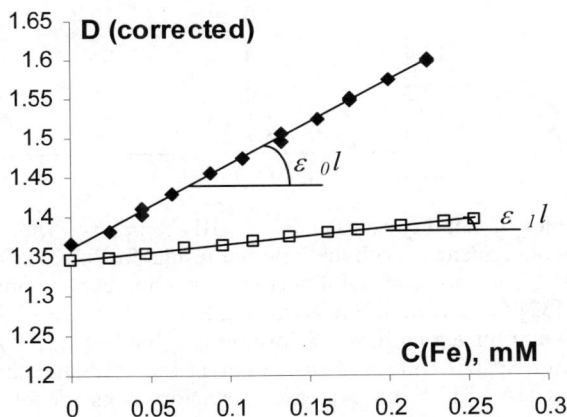

Figure 1. Photometric titration of PHA-Sk300 (0.548 g/L) by 10 mM Fe^{3+} in the presence (open) and absence (closed) of tartrate buffer (0.024 M, pH 4.37) at 650 nm.

The given plots can be used to calculate the stability constant of metal(M)-HA complex K(MHA) by the following expression:

$$\log K(MHA) = \left(\frac{\varepsilon_0 l}{\varepsilon l} - 1 \right) \cdot \frac{C^n(L)}{C(SBS)} \cdot \beta(ML_n) \tag{6}$$

where the terms εl and $\varepsilon_0 l$ are proportional to molar absorptivity of the MHA and ML complexes, respectively, and can be determined from the slope of titration curve as it is shown in Fig. 1, C(L) and C(SBS) are the total concentrations of competitive ligand (L) and of metal-binding sites (SBS) in HA solution, β (ML_n) is the stability constant of the corresponding predominant complex.

For calculating a value of stability constant of MHA complexes using equation (6), the value of $\log\beta(HgCl_2)$ was put equal to 13.4. It was obtained by averaging the reported data from [28] with follow up correction to the ionic strength of 0.048M using Davies equation [29]. The value of $\log\beta(Fe(tartrate)_2^-)$ was put equal to 10.5 as given in [30] without corrections. The stability constant values were also recalculated into dimensional units of L per g of HS using mass concentration of HS instead of molar concentration of metal binding sites. The obtained results are given in Table 5.

Table 5. Stability constants of the Fe(III)- and Hg(II)-complexes with the parent and sulphonated humic materials.

Sample	Log K, Fe(III)		Log K, Hg(II)	
	L/mol SBS	L/g HA	L/mol SBS	L/g HA
CHA-Leo	12.0	9.5	13.1	10.4
CHL-S	11.3	8.8	–	–
CHL-ClS	11.5	9.0	12.4	10.1
PHA-Sk300	12.4	9.4	–	–
PHS-S	12.2	9.4	–	–
PHS-ClS	11.4	8.7	–	–

The found values of stability constants of Fe(III)-complexes with the parent humic materials are in good agreement with the reported results for fulvic acids: log K = 11.5-14.0 L/mol BS [31]. The values of stability constants of mercury complexes are close to those reported in [32] for aquatic fulvic acids (log K = 11 L/mol BS). As it can be seen, the sulphonated humic materials have slightly weaker binding affinity for both metals studied (Fe(III) and Hg(II)) compared to the parent humic materials. This is true for both coal and peat HA. The given decrease in stability constant for coal HA did not exceed half an order of magnitude, and for peat HA – an order of magnitude. Hence, sulphonated humic derivatives still possess high binding affinity for metals.

3.3. TOXICITY ASSESSMENT OF THE SULPHONATED HUMIC MATERIALS AND THEIR DETOXIFYING PROPERTIES WITH RESPECT TO HG(II) AND CU(II)

To assess toxicity of the sulphonated humic materials from coal and peat (CHA-ClS and PHA-ClS, respectively), acute toxicity tests were used. Microalga *Chlorella pyrenoidosa* was used as a test organism, its photosynthetic activity (R) – as a response. A range of the tested concentrations of humic materials was 0.5-20 mg/L. The dose-response relationship for the parent and sulphonated humic materials are shown in Fig. 2.

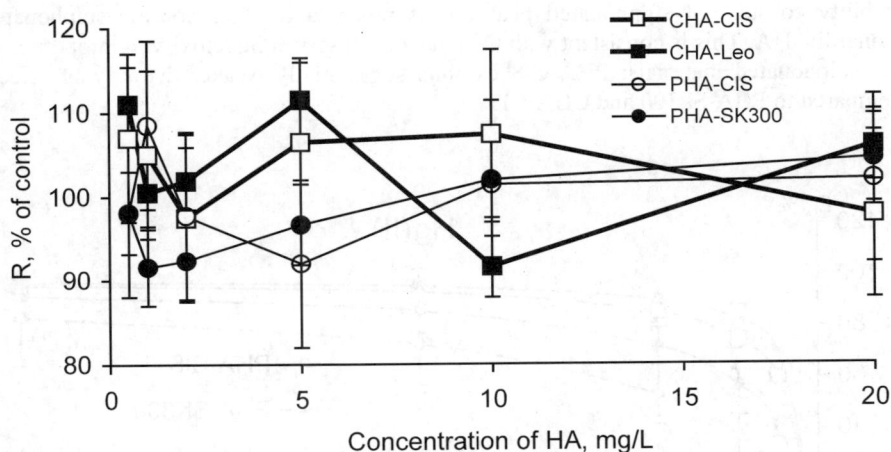

Figure 2. The relationships of the photosynthetic activity of *Clorella pyrenoidosa* versus concentration of humic acids (HA). Bars represent ± 1 SD (n = 3).

As it can be seen, a ratio of the fluorescence intensity of the HA solutions to control (R) varied in the range of 0.91–1.12 indicating a lack of either stimulation or inhibition effect of HS on the photosynthetic activity of *Chlorella*. The obtained results – a lack of the significant direct effects of HS onto *Chlorella* are in line with our previous investigations [33]. Of particular importance is that sulphonation did not lead to an increase in toxicity of the humic materials.

Acute toxicity tests conducted with the cultivation media containing model toxic metals – Hg(II) and Cu(II) in the absence and presence of the humic materials allowed us to estimate changes in detoxifying properties of humic materials as a result of their sulphonation. The corresponding data are shown in Figs. 3a, b as the relationships of detoxification coefficient D *versus* HA concentration.

It can be seen that the detoxification effects of the sulphonated humic materials are slightly weaker compared to those of the parent materials. A decrease in detoxifying effect is much bigger for Cu(II) than for Hg(II), and among the humic preparations it is more distinct for peat HA compared to coal HA. These observations are confirmed by the statistical analysis. So, covariance analysis shows that in case of Hg (II), detoxification curves are significantly different for PHA and PHA-ClS: F = 10.30 at p = 0.003, where F – Fisher criterion, p – confidence level. At the same time, there is no significant difference between detoxifying properties of the parent and sulphonated coal HA (F = 3.21, p = 0.08). In case of Cu (II), detoxification properties of the sulphonated HA were significantly weaker both for peat (F = 16.55, p < 0.001) and coal HA (F = 13.30, p = 0.001).

The lower detoxifying efficiency of the sulphonated HA compared to the parent materials corroborates well the data on metal-binding properties of the sulphonated HA with respect to Fe(III) and Hg(II). So, as it follows from Table 5, sulphonation causes a decrease in binding affinity of HA for both metals. At the same time, a decrease in the stability constant of sulphonated peat HA is much larger than for the sulphonated leonardite HA. This is consistent with the tendency observed in detoxifying properties of the sulphonated materials: PHA-ClS exhibits substantially weaker detoxifying effects compared to PHA-Sk300 and CHA-ClS.

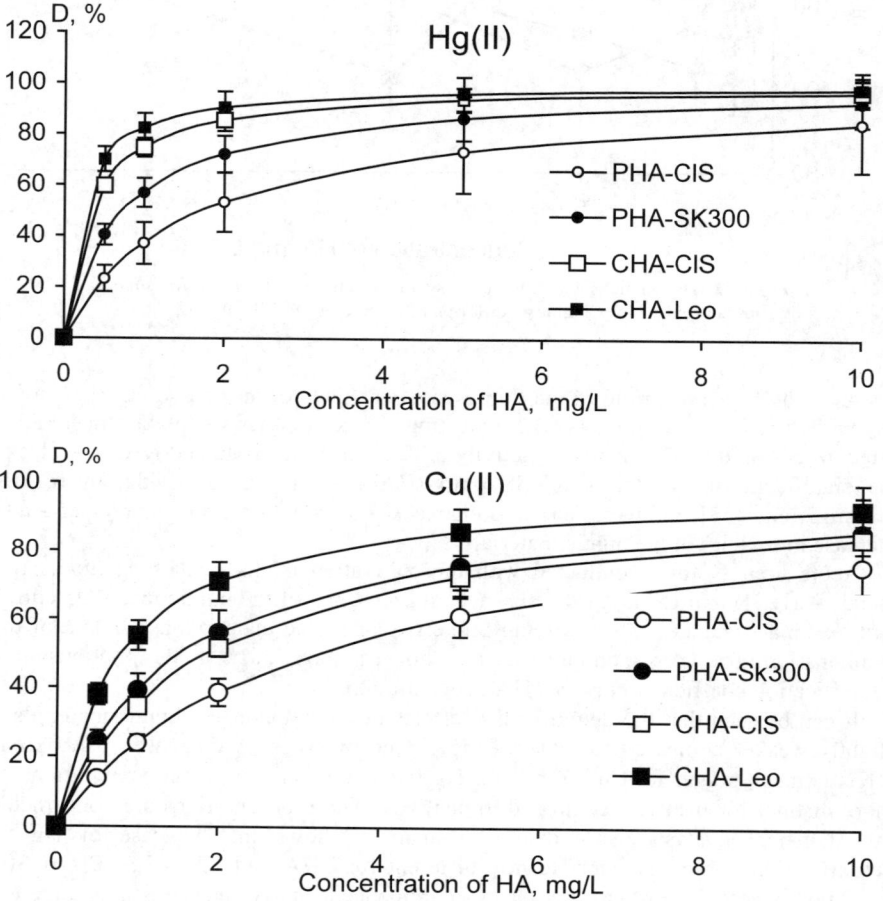

Figure 3. The relationships of the detoxification coefficient D versus concentration of HS in the presence of heavy metals at the EC_{50} concentration: a) Hg(II); $1.2 \cdot 10^{-6}$ M; and b) Cu(II), $1.5 \cdot 10^{-6}$ M. Bars represent $\pm$ SD (n = 3).

Given that the sulphonated coal humic material retains its complexing properties, the combinatory action of these HS as detoxifying agents to heavy metals in the contaminated environments and as chelating agents improving the bioavailability of nutritional metals can be expected.

4. Conclusions

The obtained results allow to consider sulphonation as a promising tool for obtaining soluble, mobile complexes of HA with metals. Chlorosulphonic acid is more efficient sulphonating agent than concentrated H_2SO_4, and leonardite humic acids yield higher modification degree than peat HA. Sulphonated humic materials are highly soluble and form soluble metal complexes with high "metal to humic acid" ratio. The metal binding affinity of the sulphonated humic materials do not drop substantially compared to the parent materials. Sulphonation does not induce toxicity of the corresponding derivatives. Sulphonation causes only a slight decrease in the detoxifying properties of humic material with respect to Hg(II) and a more substantial decrease – with respect to Cu(II). Given that the sulphonated HA from leonardite retain complexing and detoxifying properties almost on the level of the parent material, the combinatory action of these humic materials as detoxifying agents to heavy metals in the contaminated environments and as chelating agents improving the bioavailability of nutritional metals can be expected. They can be also used as environmentally sound flushing agents for heavy metal polluted sites. For the purpose of industrial production of the sulphonated humic materials, the preference should be given to a use of leonardite humic material.

5. Acknowledgements

This research was supported by the grant of International Science and Technology Centre (ISTC) KR-964.

6. References

1. Stevenson, F.I. (1982) Reactive Functional Groups of Humic. Substances, in *Humic Substances Genesis, Isolation, Composition, Reactions*, Wiley, New York, pp. 221-267.
2. Varshal, G.M., Bugaevsky, A.A., Holin, Yu.V., Merny S.A., Veliukhanova, T.K., Kosheeva, I. and Krasovitsky, A.V. (1990) Modeling the equilibrium in solutions of fulvic acids in natural waters, *Chem. Technol. Water* **12**, 979-986 (in Russian).
3. Gerse, R., Csicsor, J. and Pinter, L. (1994) Application of humic acids and their derivatives in environmental pollution control, in *Humic Substances in the Global Environment and Implication for Human Health*, Elsevier Science B.V., pp. 1297-1303.
4. Scherer, M.M., Richer, S., Valentine, R.L. and Alvarez, P.J.J. (2000) Chemistry and microbiology of permeable reactive barriers for in situ groundwater clean up, *Critical Rev. Environ. Sci. Technol.* **30**, 363-411.
5. Gonzalez-Vila, F.J., Luedemann, H.-D. and Martin F. (1983) ^{13}C-NMR structural features of soil humic acids and their methylated, hydrolized and extracted derivatives, *Geoderma* **31**, 3-15.
6. Leenheer J.A. and Noyes T.I. (1989) Derivatisation of humic substances for structural studies, in M.H.B. Hayes, P. MacCarthy, R.L. Malcolm, R.S. Swift (eds.), *Humic substances: in search of structure*, Wiley, New York, pp. 257-281.
7. Simpson, M. J., Chefetz, B. and Hatcher, P.G. (2003). Phenanthrene sorption to structurally modified humic acids, *J. Environ. Qual.* **32**, 1750-1758.
8. Chefetz, B., Salloum, M.J., Deshmukh, A.P. and G. Hatcher, P.G. (2002). Structural components of humic acids as determined by chemical modifications and carbon-13 NMR, pyrolysis-, and thermochemolysis- gas chromatography/mass spectrometry, *Soil Sci. Soc. Am. J.* **66**, 1159-1171.

498

9.   Almendos G. and Dorado, J. (1999) Molecular characteristics related to the biodegradability of humic acid preparation, *Eur. J. Soil Sci.* **50**, 227-236.
10. U.S. Patent No. 3,028,333 issued April 3,1962 to Charles A. Stratton et al.
11. U.S. Patent No. 3,309,958 issued June 19, 1962 to Kenneth P. Monroe.
12. U.S. Patent No. 3,190,837 issued June 22, 1965 to Joseph U. Messenger.
13. Canadian Pat. No. 722,720 issued November 30, 1965 to Speros E. Moschopedis.
14. U.S. Patent No 5,663,425 issued September 2, 1997 to Detroit, W.J., Lebo, Jr, Stuart, E., Bushar, L.L.
15. Zhambal, D. (1991) Composition and structural features of sulphonated humic acids, *Khimiya tverdogo topliva* (*Chemistry of Solid Fuel*) **2**, 70-72. (in Russian).
16. Rhee, D.S. and Jung, Y.-R. (2000) Characterization of humic acid in the chemical oxidation technology (II) Characterization by ozonation, *Analyt. Sci. Technol.* **13**, 241-249.
17. Sasina, V.N., Rumyanceva, Z.A. and Pevzner, Z.I. (1985) Water-soluble products of oxidation of brown coals and humic acids, *Chemistry of Solid Fuel* **3**, 30-36 (in Russian)
18. Shishkov, V.F., Verkhodanova, N.N., Egor'kov, A.N. and Tuturina, V.V. (1984) Ozonation of brown coals and humic acids, *Chemistry of Solid Fuel* **5**, 35-39 (in Russian).
19. Wang, G.S., Hsieh, S.T. and Hong, C.S. (2000) Destruction of humic acid in water by UV light-catalyzed oxidation with hydrogen peroxide, *Wat. Res.* **34(15)**, 3882-3887.
20. Aronov, S.G., Sklyar, M.G. and Tiutiunnikov, Yu.B. (1968) *Complex chemical technological treatment of coal*, Kiev, Tekhnika (in Russian).
21. Scherer, M.M., Richer, S., Valentine, R.L. and Alvarez, P.J.J. (2000) Chemistry and microbiology of permeable reactive barriers for in situ groundwater clean up, *Critical Rev. Environ. Sci. Technol.* **30**, 363-411.
22. Lowe, L.E. (1992) Studies on the nature of sulfur in peat humic acids from Froser river delta, British Columbia, *Sci. Total Environ.* **113**, 133–145.
23. Perdue E.M. (1985) Acidic functional groups of humic substances, in Aiken, G.R., McKnight, D.M., Wershaw, R.L., MacCarthy, P. (eds.) *Humic substances in soil, sediment and water*, Wiley Interscience, New York, pp. 493-525.
24. Vavilin, D.V., Polynov, V.A., Matorin, D.N. and Venediktov, P.S. (1995) Sublethal concentrations of copper stimulate photosystem II photoinhibition in *Chlorella pyrenoidosa*, *J. Plant Physiol.* **146**, 609-614.
25. Tamiya, H.K., Morimura, K., Yokota, M. and Kunieda, R. (1961) The mode of nuclear division in synchronous culture of Chlorella: comparison of various methods of synchronization, *Plant Cell Physiol.* **2**, 383-403.
26. Perminova, I.V., Grechishcheva, N.Yu., Kovalevskii, D.V., Kudryavtsev, A.V., Matorin, D.N. and Petrosyan, V.S. (2001) Quantification and prediction of detoxifying properties of humic substances to polycyclic aromatic hydrocarbons related to chemical binding, *Environ. Sci. Technol.* **35**, 3841-3848.
27. Kumok, V.N. (1977) *Regularities in stability of complex compounds in solution*, Tomsk Univ. Publ., Tomsk (in Russian).
28. IUPAC Stability constants data base (computer edition) (1997-2000) IUPAC and Academic Software.
29. Davies, C.W. (1962) *Ion Association*, Butterworths, London.
30. Ramamoorthy, S. and Manning, P.G. (1973) Equilibrium studies of metal-ion complexes of interest to natural waters - VI. Simple and mixed complexes of Fe(III) involving NTA as primary ligand and a series of oxygen-bonding organic anions as secondary ligands, *Inorg. Nucl. Chem.* **35**, 1571-1575.
31. Pandeya, S.B. (1993) Ligand competition method for determining stability constants of fulvic acid iron complexes, *Geoderma* **58**, 219-231.
32. Varshal, G.M. and Buachidze, N.S. (1983). Investigation of coexisting forms of mercury(II) in surface waters, *J. Anal. Chem.* **38**, 2155-2167.
33. Perminova, I.V., Kulikova, N.A., Zhilin, D.M., Gretschishcheva, N.Y., Holodov, V.A., Lebedeva, G.F., Matorin, D.N., Venediktov, P.S. and Petrosyan, V.S. (2004) Mediating effects of humic substances in aquatic and soil environments, in Ph. Baveye (ed.), *Environmentally Acceptable Pollution and Reclamation Endpoints*, Kluwer Academic Publisher, NATO ASI - Series, Dordrecht (in Press).

AUTHOR INDEX

SUBJECT INDEX

actinides
  binding to HS, 86, 168
  migration, 175, 180
  reduction by HS, 18, 177
  retention onto mineral surfaces, 166
  solubility in the presence of HA, 178
  speciation in the presence of HS, 163, 165, 166
  stability constants of humates, 176
anodic stripping voltammetry (ASV) 93
aquifer
  immobilization of HS, 222, 236
  remediation, 203, 233
atrazine
  complexing capacity of FA, 61
  detoxification by humates, 383
  hydrolysis in the presence of FA, 56, 57, 58
  toxicity, 317

bioavailability
  of ecotoxicants, 13, 14, 439
  of metals, 13, 103, 147
bioremediation, 9, 11
  limitations of, 11, 260
  principles of, 260
  use of HS, 250, 353
brown coal, 5
  adsorption of metals, 188, 193
  detoxification by, 188, 382
  influence on humus state, 195, 198
  influence on microorganisms, 197
  transformation in soil, 196

calculations, predictive, 56
capillary electrophoresis (CE), 440

principles, 421
techniques, 421
capillary zone electrophoresis (CZE), 452

case study
  actinide migration in the presence of HA, 180
  biodegradation of PAH in soil, 259
  bioremediation of soil polluted with TNT, 356
  clean-up of water polluted with TNT, 363
  detoxification of soil in Dnepropetrovsk region, 323
  heavy metals speciation in the Dnieper reservoirs, 137
  influence of HS on Fe uptake by plants, 101
  influence of humates on radioluclides uptake, 386
  pilot-scale laboratory experiment with diesel, 245
  remediation of soil using brown coal, 192
  water treatment using ZEOPAR-technology, 272
chelating units (sites). See metal binding sites
coal
  brown. See brown coal
  rank, 5
  reserves, 5
coalification, 5
complexation with HS
  americium (Am), 163, 176
  copper (Cu), 59, 67, 87
  curium (Cm), 163
  dysprosium (Dy), 163

501

502